Engineering Materials

This series provides topical information on innovative, structural and functional materials and composites with applications in optical, electrical, mechanical, civil, aeronautical, medical, bio- and nano-engineering. The individual volumes are complete, comprehensive monographs covering the structure, properties, manufacturing process and applications of these materials. This multidisciplinary series is devoted to professionals, students and all those interested in the latest developments in the Materials Science field, that look for a carefully selected collection of high quality review articles on their respective field of expertise.

Indexed at Compendex (2021) and Scopus (2022)

Shahid ul Islam · Mohammad Shahid
Editors

Green Materials for Active Food Packaging

Springer

Editors
Shahid ul Islam
Department of Applied Science and Humanities
Jamia Millia Islamia
New Delhi, Delhi, India

Mohammad Shahid
Department of Chemistry
Faculty of Science and Technology
Mewar University
Chittorgarh, Rajasthan, India

ISSN 1612-1317 ISSN 1868-1212 (electronic)
Engineering Materials
ISBN 978-981-96-0371-8 ISBN 978-981-96-0369-5 (eBook)
https://doi.org/10.1007/978-981-96-0369-5

This Springer imprint is published by the registered company Springer Nature Singapore Pte Ltd.
The registered company address is: 152 Beach Road, #21-01/04 Gateway East, Singapore 189721, Singapore

Preface

In an era marked by an ever-growing awareness of environmental sustainability and the critical need for innovation in the food industry, the quest for greener alternatives to conventional packaging has become more urgent. This book, *Green Materials for Active Food Packaging*, represents a comprehensive exploration of the forefront of research and development in this dynamic field. It aims to provide an in-depth understanding of how advanced materials and technologies can be harnessed to address the pressing environmental challenges posed by traditional packaging solutions. A global demand for more sustainable practices has driven the journey from traditional, petroleum-based packaging to innovative, eco-friendly alternatives. This transition responds to environmental concerns and is necessary to ensure our food systems remain viable and efficient amid mounting ecological pressures. The chapters within this volume collectively address the multifaceted aspects of this transition, offering insights into the latest advancements and emerging trends in green materials for active food packaging. Chapter 1 introduces the concept of green materials and underscores the significance of transitioning from conventional, non-biobased packaging to more sustainable alternatives, setting the stage for understanding the broader implications of these materials. Chapter 2 discusses biopolymers, detailing their sources, properties, and applications in food packaging, and highlights their role in meeting sustainability goals while offering practical benefits in food preservation. Chapter 3 investigates biodegradable polymers as viable replacements for petroleum-based packaging, emphasizing their potential to offer environmentally friendly solutions through improved mechanical and barrier properties. Chapter 4 examines the integration of antimicrobial agents from essential oils and nanomaterials into biopolymer films, focusing on their effectiveness in enhancing food safety and extending shelf life. Chapter 5 discusses the development of edible packaging made from natural polymers, evaluating its potential to replace traditional plastics and addressing challenges related to its mechanical properties and functionality. Chapter 6 reviews replacing conventional packaging materials with green polymers derived from natural sources, exploring their environmental benefits and applications in active food packaging. Chapter 7 provides an overview of recent advances in polymer nanocomposites, highlighting their versatile properties and applications in active food packaging to enhance food quality and safety. Chapter 8 focuses on sustainable coatings and films, detailing the use of bio-based compounds and polymers to create eco-friendly packaging solutions that replace conventional materials. Chapter 9 offers insights into bio-based barrier materials, discussing their effectiveness in controlling moisture, gas, and microbial migration while addressing their suitability for active packaging. Chapter 10 reviews the role of bioactive compounds in active food packaging, examining their functions and safety considerations to ensure they enhance food quality while meeting health standards. Chapter 11 explores the development and application of smart sensors in green packaging solutions, emphasizing their potential to monitor food quality and safety throughout the supply chain. Chapter 12 investigates edible packaging as a sustainable alternative to traditional materials, highlighting its composition, technological advancements, and regulatory considerations. Chapter 13 addresses eco-friendly active packaging solutions for preserving fresh and minimally processed fruits and vegetables, focusing on their impact on quality, safety, and sustainability. Chapter 14 examines active packaging for convenience foods, exploring recent innovations that enhance food quality and align with sustainability goals while navigating regulatory challenges. Chapter 15 discusses the use of natural dyes in food packaging, providing an overview of plant-based colorants and their benefits for both visual appeal and environmental impact. This book aims to provide a thorough understanding of these advancements and inspire further research and development in the field of green food packaging. The book will serve as a crucial resource in the ongoing quest for more sustainable food packaging solutions, offering an in-depth exploration of innovative materials and technologies designed to tackle pressing environmental challenges. This book underscores the importance of transitioning from traditional petroleum-based packaging to greener alternatives by addressing the latest advancements in biopolymers, biodegradable materials, and smart technologies. Its comprehensive analysis of antimicrobial agents, edible packaging, and bioactive compounds will give researchers, practitioners, and policymakers valuable insights to drive future innovations and make informed decisions in the food packaging industry. The contributions from experts in the field provide a thorough analysis of both the potential and the challenges associated with these materials and technologies. This volume will be an essential guide for those committed to advancing sustainability, fostering a deeper understanding of how green materials can effectively enhance environmental and food safety outcomes.

Mohammad Shahid
Mewar University
Chittorgarh, India

Shahid ul Islam
Jamia Millia Islamia
New Delhi, India

Contents

About the Editors

Dr. Shahid ul Islam Ph.D., is an Assistant Professor in the Department of Applied Sciences and Humanities at Jamia Millia Islamia (A Central University), New Delhi, India. Before joining Jamia Millia Islamia, he worked as a Fulbright Postdoctoral Fellow at the University of California, Davis, United States, and a National Postdoctoral Fellow at IIT Delhi. He is on the editorial board of several journals. He is a member of many groups, including the American Chemical Society (USA), and is a life member of the Asian Polymer Association. He authored several scientific articles, including peer-reviewed journal papers, reviews, and chapters published by the American Chemical Society, the Royal Society of Chemistry, Elsevier, Wiley, and Springer. He has edited and authored numerous books with reputed international publishers.

Dr. Mohammad Shahid Ph.D., is an Assistant Professor at the Department of Chemistry at Mewar University, Chittorgarh, India. He earned his Master's degree in Chemistry from Shibli National College, Azamgarh, in 2006 and his Ph.D. in Chemistry from Jamia Millia Islamia, New Delhi, in 2014. He served as a Postdoctoral Fellow at Soochow University, China, a Marie Curie Postdoctoral Fellow at the University of Glasgow, UK, a Kothari Postdoctoral Fellow at the Institute of Chemical Technology, Mumbai, India, and Assistant Professor & Head, Department of Applied Science, Dr. K. N. Modi University, Rajasthan, India. His research was supported by numerous esteemed fellowships and grants, including the UGC Fellowship, China Postdoctoral Fellowship, EU Marie Curie Fellowship, and Dr. D. S. Kothari Fellowship. His prolific academic career includes over 50 scientific publications in peer-reviewed journals, reviews, reports, patents, and book chapters. He has also edited and authored multiple books with renowned international publishers. His research interests are diverse, spanning textile chemistry, green chemistry, and the development of sustainable materials.

Chapter 1
Introduction: Green Materials and Sustainability in Active Food Packaging

Mohammad Raza Miah, Jinggang Wang, and Jin Zhu

Abstract Active, smart, and intelligent packaging focuses on innovative strategies to enable food preservation, freshness, easy detection, quality control, and health standards in the global packaging market. Moreover, various types of components applied in food wrapping are non-biobased and cannot satisfy society's expanding demands for sustainability and environmental protection. Sustainability is a worldwide aspiration that addresses societal challenges such as environment, raw materials, resource efficiency, climate protection, etc. The application of non-biobased materials in led packaging creates environmental issues. Furthermore, rising environmental problems have put packaging under intense scrutiny owing to the ongoing source of huge quantities of plastic waste. Bio-based polymer nanocomposites were recently developed to address these issues using nanoparticles with excellent performance. Finally, this chapter briefly discusses the possibilities for sustainable materials in the 21st century in diverse packaging applications.

Keywords Active packaging · Bio-polyester · Materials optimization · Green technology · Biodegradable and sustainable packaging

---

M. R. Miah · J. Wang (✉) · J. Zhu (✉)
Key Laboratory of Bio-Based Polymeric Materials Technology and Application of Zhejiang Province, Ningbo Institute of Materials Technology and Engineering, Chinese Academy of Sciences, Ningbo 315201, People's Republic of China
e-mail: wangjg@nimte.ac.cn

J. Zhu
e-mail: jzhu@nimte.ac.cn

M. R. Miah
University of Chinese Academy of Sciences, Beijing 100049, People's Republic of China

S. ul Islam and M. Shahid (eds.), *Green Materials for Active Food Packaging*, Engineering Materials, https://doi.org/10.1007/978-981-96-0369-5_1

1.1 Introduction

In recent years, the food packaging industry has joined the rest of the world's industries in adopting a more sustainable model [1, 2]. Active food packaging with green materials has emerged as a promising option due to escalating environmental consciousness and consumer desire for eco-friendly substitutes [3, 4]. This change is a big step toward minimizing the negative environmental effects of packaging without sacrificing the quality and safety of food goods [5]. Food packaging is essentially among the most important processes in the food sector since it retains the items at the greatest possible level of quality and nutrition while also attracting the attention of customers [6–8]. Primary, secondary, and tertiary wrappings are the three tiers of wrapping. Primary packaging is the layering or film that immediately encloses and interacts with it [9]. Because primary packaging comes into close contact with the materials, it is the primary layer that affects food quality [10]. Products wrapped in the main packaging are covered with secondary packaging [11]. The outer packing used for bulk handling, distribution, and storage is tertiary packaging. Food safety may not be ensured even with these three layers of packaging [12]. Moreover, it is believed that plastics and bioplastics are widely utilized coating components, especially in packaging uses [13]. Plastics are produced in around 320 million tons annually worldwide, with demand rising because of their many uses across numerous industries [14]. Because the plastics sector depends so heavily on rising prices for natural gas and fossil fuels, it could have affected the plastics market's profitability [15]. Ethylene vinyl alcohol (EVOH), Polyvinyl chloride (PVC), polyamide (PA), polyethylene (PE), polypropylene (PP), polystyrene (PS), and polyethylene terephthalate (PET) are frequently employed polymers for this reason [16–18]. Their superior mechanical and barrier qualities, as well as their affordability, make them the material of choice. However, when synthetic plastics are disposed of, emissions of greenhouse gases, such as methane and carbon dioxide, accumulate in the atmosphere and damage the ecosystem [19]. Because fossil-based polymers are not recyclable or biodegradable, their use is restricted. As a result, switching to alternative raw resources for the production of plastic is crucial [20]. To provide environmentally beneficial alternatives to conventional resources, developing and using appropriate, abundant, and sustainable bio-resources is becoming important [21]. Compostability is important for polymer resources because recycling needs energy. Compostability, in its broadest sense, refers to a series of processes that use the ability of organic matter to biodegrade to produce a specific product known as "compost." As a result, compostability promotes the breakdown of soil elements by releasing water, carbon dioxide, and other inorganic chemicals [22, 23]. Renewable resources are not the only source for making compostable or biodegradable plastics [24]. Biodegradation is positively connected with the substance's chemical composition instead of its source; therefore, biodegradable polymers are not always based on renewable materials [25]. In particular, the chemical bonds inside molecules determine how long a biodegradation process takes. Whether derived synthetically or biologically, carbon-containing polymers such as cellulose, starch, lignin, gelatin, and keratin are all biodegradable or compostable [26].

This chapter briefly introduces bio-based and sustainable materials for active food packaging. Along with these topics, we also covered a variety of bio-based polymers, nano-fillers, active and smart packaging terminology, processing bio-based packaging materials, features, and uses for bio-plastic packaging.

1.2 Green Components for Active and Smart Food Wrapping

Plastic has been utilized extensively in the manufacturing of wrapping components over the past 50 years because it is simple to produce, store, and transport [27]. Petroleum-based packaging materials, including polyethylene terephthalate (PET), polyvinyl chloride (PVC), polyamide (PA), polypropylene (PP), and polystyrene (PS), are in high demand due to the growth and expansion of the food business [28, 29]. These plastic items offer many benefits but have created major environmental issues because they don't break down in the environment [30]. Developing biopolymers, which mimic the characteristics of plastics but are biodegradable and only temporarily present in the environment, is an effective way to address the ecological issues that plastics cause [31]. Based on their origin, biopolymers are classified into three groups: polymers made from plant material, those manufactured from monomers originating from plant material, and those generated from microbes [32]. Concerning the food industry's uses, food coating, packaging, and edible films are the three primary areas in the food packaging system (FPS) where these bio-based materials are very useful [33]. Furthermore, the fruits and vegetables gas permeability coefficients may change due to the coating, which could impact their respiration rate and shelf life [34]. Most of the time, these coatings are edible, biodegradable, and environmentally benign. Three different coating techniques- spread, spray, and dip-have been investigated. The environmentally safe materials for creative and innovative food packaging are shown in Fig. 1.1.

1.2.1 Polymers Originating from Microbes

Among the three primary biopolymers derived from fermentation are exopolysaccharides (EPS), polyhydroxyalkanoates (PHA), and polylactic acid (PLA) [35]. Renewable biomaterials can be used to generate the monomers required to synthesize PHA, PLA, and bio-polyethylene (Bio-PE). Wastes from maize and sugarcane are beneficial since separating the monomers from the plant source is simple due to their low degree of polymerization [36, 37].

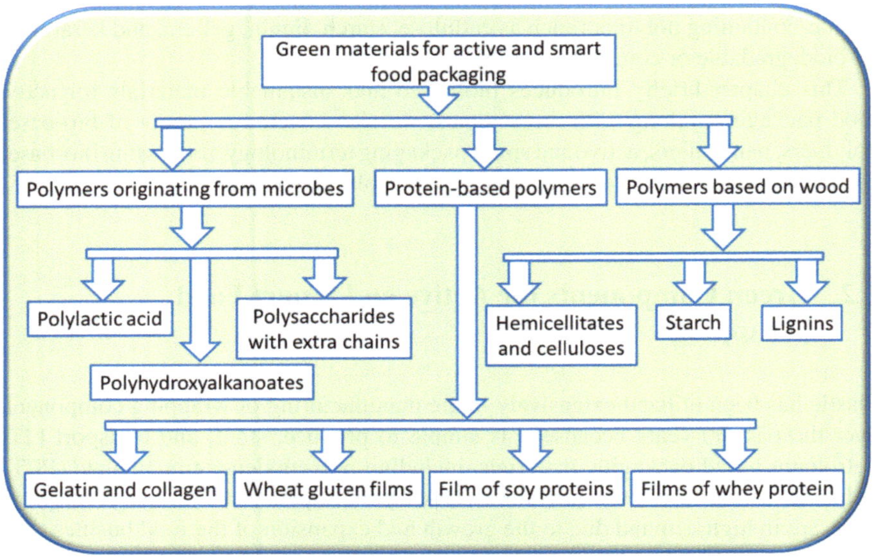

Fig. 1.1 The green materials for active and smart food packaging

1.2.1.1 Polylactic Acid

Lactic acid and biologically derived lactide are produced by fermenting carbohydrates, most commonly corn starch [38]. Following hydrolysis, lingo-cellulosic biomass appears to be a fantastic route for generating lactic acid and lactides. Compared with lactic acid (PLA), conventional thermoplastics require a greater processing temperature [39, 40]. Reverse osmosis, adsorption, solvent extraction, distillation, and electrodialysis are needed to purify PLA. Increasing input material will lower conversion process costs and increase fermentation and enzyme hydrolysis costs [41]. Filtration procedures are used to remove any remaining proteins following fermentation. Lactate is bio-converted to lactic acid and sodium hydroxide by bipolar electro-dialysis. Vacuum distillation is used to purify lactide. To produce PLA, lactic acid condensation and polymerization are done [42]. Several PLA-based technologies have recently arisen to achieve the necessary chemical, biological, and mechanical qualities comparable to fossil-based polyesters [43]. PLA has garnered significant scientific interest in recent years since it requires physical and/or chemical changes to achieve the necessary qualities for its intended consumer uses [44]. A growing field of study is the creation of "green" or ecologically friendly components through combining with various fibers and polyesters, like plant fiber reinforced polyester matrix [45]. Furthermore, creating nanocomposites based on PLA using a range of distinct nanostructures in industrial and biomedical biotechnological uses is a useful strategy that lowers the cost of experiments and facilitates the induction of required functionality [46].

1.2.1.2 Polyhydroxyalkanoates

Polyhydroxyalkanoates, or PHAs, are a class of bio-based polymers. PHAs are also thermally convertible to PLA and are produced through biotechnological conversion using various microorganisms and sustainable raw ingredients (e.g., maltose, glucose, and fatty acids) [47]. The selection of microorganisms, carbon sources, additives, and environmental factors all influence the formation of polymers with varying building blocks and attributes [48]. PHAs of different kinds, i.e., it has proven possible to produce polyhydroxybutyrate and its copolymer, 3-hydroxybutyrate co-3-hydroxyvalerate. Their uses include the packaging industry, medical, and textile implant sectors. The high expense of PHA synthesis and recovery may be why PHAs have not been used extensively as bio-plastics. Researchers are looking for more affordable feedstocks to replace PHA production [49]. For instance, synthesizing bacterial PHAs can be facilitated by utilizing wood-derived raw materials, including hemicelluloses [50].

1.2.1.3 Polysaccharides with Extra Chains

Exopolysaccharides (EPSs) are complex biopolymers released outside cell walls and mostly composed of carbohydrates. They are manufactured by a range of microbial species [51]. These organisms include blue-green algae, fungi, gram-positive and -negative bacteria. EPS is available in multiple configurations involving glucans, dextrin, alginate, xanthan, levan, etc. Kefiran has benefits over other forms of EPS since it is biodegradable and water-soluble [52]. It is generated within the kefir synthesis process when milk ferments. Studies on Kefiran's potential have been conducted to improve human health in several ways, including its appealing healing and antibacterial properties, anti-inflammatory properties, etc. [53]. Furthermore, its uses as a substance for gelling, stabilizer, and emulsifier in the food business are extensively researched. Kefiran-based films are receiving much attention from scientists because of their unique properties, which include strong mechanical and excellent water vapor permeability, safety, ability to decompose and be biocompatible, and functions of emulsification and stabilization [54]. Furthermore, bioplastics made of kefiran-based films offer outstanding visual qualities and may be made efficiently with edible plasticizers such as glycerol. Thus, developing films based on kefiran could lead to more appropriate packing materials and coatings with better qualities that are also environmentally benign [55].

1.2.2 Polymers Based on Proteins

1.2.2.1 Gelatin and Collagen

Gelatin and collagen are derived from animal products. The most prevalent protein found in nature is collagen [56]. It makes up roughly 20–25% of the overall mass of an animal. Three cross-linked α-chains make up its structure, whereas gelatin, a denatured derivative of collagen composed of many proteins and polypeptides, is constituted to collagen [57]. Methionine, hydroxyproline/proline, and glycine amino acids are abundant in collagen. While gelatin-based films have to be made using a wet method that involves the development of a film-making agent, collagen-based bio-plastics are produced through the extrusion procedure and have numerous applications [58]. The mechanical features of bio-plastic films based on collagen are beneficial; for example, it has been discovered that hydrolyzed collagen films offer remarkable tensile strength. However, the hydrophilic character of gelatin films is demonstrated by their poor mechanical and barrier properties [59].

1.2.2.2 Films Made of Wheat Gluten

Plasticized wheat gluten can create bio-plastic films through intense extrusion, compression, and molding [60]. The preparation of hydrogen, disulfide, as well as hydrophobic interactions takes place across the creation of films. Sulfhydryl groups stabilize disulfide bonds. Heating leads to the polymer becoming denaturized and breaks the natural disulfide and hydrophobic units in it [61]. As the gluten oxidizes once more, additional disulfide linkages are formed throughout the drying process. The gluten mass quality and the casting medium, for example, acidic or alkaline, determine the clarity of the film. The resulting films possess strong mechanical abilities, outstanding gas barrier performance, as well as uniformity [62].

1.2.2.3 Film of Soy Proteins

Additionally, soy proteins are used to create bio-plastic films that are used in packaging [63]. Compared to other protein-based bio-plastics, soy-based films are more affordable, transparent, flexible, and smoother. Additionally, soy-based films have strong oxygen barrier qualities when the moisture content is low [64]. We have also comparing soy-based films to polyethylene with a low density. The main drawbacks are their decreased mechanical performance, less thermal stability, and allergic nature. The aqueous solution of soy protein isolates can also form films at high temperatures when placed on stainless steel plates [65]. Several scientists have also reported on the creation of films from different proteins as raw ingredients, examples include globulin protein from pistachios, pumpkin oil cake, canola protein, and pea protein [66].

1.2.2.4 Films of Whey Protein

Whey protein possesses remarkable operational properties and the capacity to produce films [67]. Whey-based films were investigated for their exceptional plasticity, the barrier between oil and gas, and transparency at comparatively minimal humidity levels [26]. Nevertheless, the water barrier qualities of whey protein-based films are subpar. It has been demonstrated that adding essential oils or lipids to whey-based films enhances their water barrier and other active characteristics [68]. However, to produce wrapping materials alongside the appropriate characteristics as well as functionality, understanding the interactions between various biopolymers is essential during the entire process of developing coatings [69]. The active elements in bio-plastic films were lemon and bergamot essential oils. The outcomes demonstrated the outstanding antibacterial potential of adding essential oils. The results showed that the addition of essential oils has outstanding antibacterial potential [70].

1.2.3 Polymers Based on Wood

1.2.3.1 Hemicellitates and Celluloses

By weight, cellulose makes up about 40–50% of lingo-cellulosic wood fibers, while hemicelluloses comprise 25–30% [71, 72]. Cellulose films have outstanding transparency, a high surface gloss, exceptional tensile strength, and good toughness. Cellulose is chemically changed throughout the dissolution process to help break the polymer chains [73]. Cellulose derivatives are produced when chains break down and can be recycled into bio-plastic films for barrier applications such as packing. Hemicelluloses are hydrophilic polysaccharides derived from wood with the least thermal resistance. They are amorphous, multidimensional, heterogeneous polysaccharides with a less organized structure [74]. Hemicellulosic-based films are brittle, yet it has been shown that adding plasticizer increases the films' toughness, flexibility, and low oxygen permeability [26]. These films' low oxygen permeability makes them interesting for packaging applications. It has been reported that the glucuronoxylan (hardwood-hemicelluloses) and the softwood hemicelluloses can be combined with a bio-plasticizer to create packaging materials with increased flexibility and oxygen permeability. However, the films were susceptible to absorption by water. By decreasing film flexibility, enhanced moisture uptake resistance and increased mechanical performance were facilitated by the increased content of polymer (CMC or alginate) [75].

1.2.3.2 Starch

Starch is the commonly used plant-derived polymer for creating bio-plastic films owing to its affordability, price, and important film-making abilities [76]. It has been suggested that various starches from undiscovered sources can be used alone or alongside more natural polymers to create packing films or coatings that decompose naturally, prolonging product duration on shelf. Red Crimson grape post-harvest preservation was enhanced with a composite covering according to sorbitol-plasticized maize starch and gelatin, according to Susmitha et al. (2021) [77]. The resulting biocomposite, a starch matrix combined with gelatin, had better mechanical and solubility properties. Furthermore, a considerable reduction in the permeability of the water vapor in the composite film was seen when relative to altered blends and waxy corn starch. Mango kernel starch was utilized as a substance for packaging by Nawab et al. [78] to increase the longevity of tomatoes [78]. By examining many physical and chemical characteristics, such as the fruit's firmness, reduction in weight, tartaric acid (TA) amount, and soluble solids concentration (SSC), the packing material effectively postponed the ripening process of tomatoes [79]. The modification of this remarkable biopolymer has proven useful not only in the bio-plastic and packaging industries but also in the porous media that can be created in the biomedical, pharmacological, tissue engineering, drug delivery, and even metallurgical fields [80].

1.2.3.3 Lignins

Lignin is the component 10–25% of the mass of lingo-cellulosic wood. An estimated lignin in the amount of fifty to sixty million tons is synthesized annually in the pulp and paper industries; of this, only 2% is used economically, and the remaining 98% is squandered as inexpensive fuel for chemical recovery boilers [81]. To swap out polymers derived from fossil fuels in polyurethane preparations and so enhance the material's thermal and mechanical characteristics of polyurethane, biomass-based lignin may be employed as an affordable reinforcement [82]. Lignin has the ability to substitute phenolic chemicals in the manufacture of phenolformaldehyde (PF) resins. Lignin can be used in PF resins based on petrochemicals as a biodegradable phenol solution [83]. Compared to pristine lignins, phenolated lignin-PF resins revealed superior mechanical and physical properties. Aldehydes derived from lignin biomass have recently been synthesized and used as a great formaldehyde substitute [84].

1.3 Nanoparticles

Nanoparticles increase food's storage life by acting on the headspace or directly on food when incorporated into the active packaging [85]. On the other hand, nanoparticles in intelligent packaging do not interact with food; instead, they share information and keep an eye on the packaged product's condition [86]. Smart packaging is made up of components that are both cognitive and active.

1.3.1 Metal and Metal Oxide Nanoparticles

When it comes to bacteria that are both Gram-positive and -negative, metal and metal-oxide nanoparticles (NPs) show varying antibacterial efficiency [87]. They may perhaps have distinct impacts on bacteria strains belonging to identical groups. This impact is most likely caused by the existence of distinct components of membranes in various bacteria, which serve as the initial obstacle and NP target [88]. For example, the surface of bacteria that are Gram-negative includes lipoproteins and phospholipids. In contrast, the surface of bacteria that are Gram-positive has many holes and a far increased surface charge in negative terms [89]. Oxidative stress caused by reactive oxygen species (ROS) is a notable system linked to the antibacterial activity of nanoparticles (NPs) made of metal and metal oxide [90]. Singlet oxygen, hydrogen peroxide, superoxide radical, and hydroxide radical are the four forms of ROS that are most commonly formed. Different NPs can cause the hyper-production of one or many ROS [91]. For example, superoxide radicals are produced by CaO and MgO NPs, hydrogen peroxide is mostly produced by ZnO NPs, and all of those ROS can be produced by CuO NPs. Ag NPs cause the production of hydroxyl radicals and hydrogen peroxide. Oxidative stress, disruption suppression of membrane integrity and biomolecules of enzymes, and damage to nucleic acids are all caused by increased production of ROS [92]. Additionally, ROS has the ability to puncture the bacterial barrier, allowing the contents to seep out. Figure 1.2 provides an overview of several metal and metal oxide nanoparticles.

1.3.1.1 Silver Nanoparticles

Silver (Ag) nanoparticles (NPs) represent the most often employed nano-filler to give the antibacterial characteristics among the composite metrics [93]. To disinfect bottled water, the US Food and Drug Administration (FDA) has approved the direct addition of up to 17 μg/kg of Ag salts [91]. Ag NPs attach themselves to enter bacterial cells or pass through bacterial membranes to provide their antibacterial action. Ag NPs may interact with internal cell proteins, DNA, and membrane lipids. Additionally, it has been demonstrated that Ag NPs suppress the development of enzymes that are antioxidants, including catalase, glutathione, and superoxide. Dismutase inhibits the bacterial cell's ability to its defense mechanism in opposition to oxidative stress [94]. Another action method is releasing extremely poisonous silver ions, which intercalate inside the DNA chain or bind protein sulfhydryl groups. Furthermore, (Ag^+)-ions emitted by Ag NPs have the ability to engage in interactions with respiratory chain bacterial proteins to cause cell demise and produce a high concentration of reactive oxygen species [95].

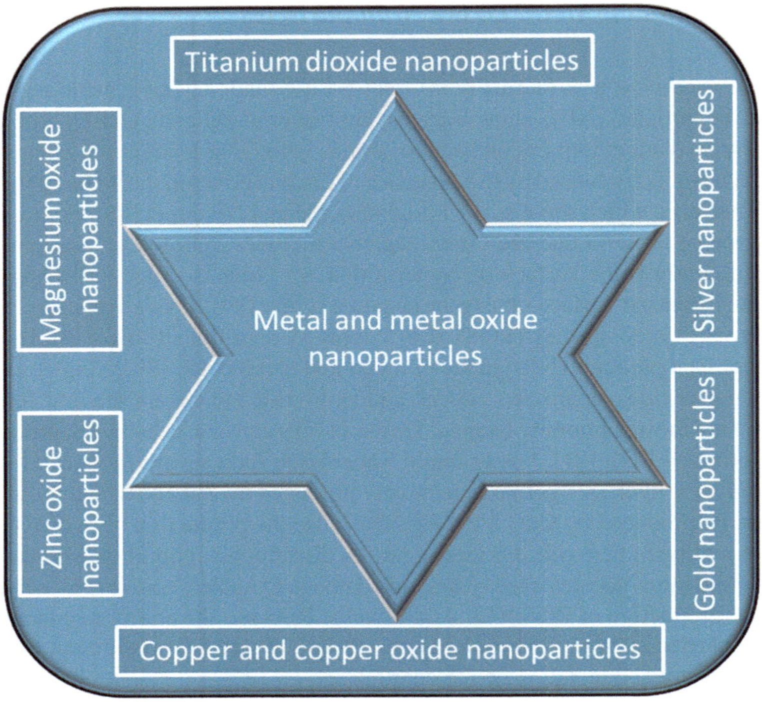

Fig. 1.2 An overview of various metal and metal oxide nanoparticles

1.3.1.2 Gold Nanoparticles

Additionally, a gold nanoparticle-infused antimicrobial package was developed. The gold (Au) nanoparticles kill the bacterial cells without needing reactive oxygen species (ROS), in contrast to most antibacterial nano-materials that produce cytotoxic ROS [96]. Moreover, it was demonstrated that the bacterial strain that Au nanoparticles (NPs) are applied to, their size, and their functionalization significantly impact the antibacterial activity of the particles [97]. When nano-gold particles integrate into cells, they either lower adenosine triphosphate levels or prevent transfer RNA (tRNA) from attaching to the ribosome [98]. Alternatively, they attach to the bacterial membrane and alter its function. It was observed that Au NPs harmed bacterial membranes when they came into contact with *S. aureus* and *E. coli*. Glyoxal cross-linking of polyvinyl acetate (PVA) films and reinforcement with Au NPs as nano-fillers resulted in composites with enhanced hydrophobicity, mechanical strength, and barrier qualities [99]. The creation of a packed pore structure, which reduced the water vapor transmission rate and enhanced the film's water and thermal stability, was facilitated by intermolecular cross-linking [100]. Because Au NPs blocked the thiol groups of the bacterial enzymes and caused holes in the cell walls, the nano-composite exhibited strong antibacterial action against E. coli [101]. The substance proved to be more effective at preserving wrapped organic bananas than PVA film using graphene oxide (GO) to create crosslinks and glutaraldehyde [102].

1.3.1.3 Copper and Copper Oxide Nanoparticles

It has been shown that carbohydrate polymers' mechanical (e.g., their elastic modulus and tensile strength) and bactericidal capabilities are strengthened through copper oxide nanoparticles, or CuO NPs [103]. Employing chitosan, alginate, carrageenan, agar, and carboxymethyl cellulose, the impact of adding CuO NPs to biopolymers of carbohydrates was investigated [104]. Except for the agar/CuO NP nanocomposite film, which displayed CuO NP agglomerates and clusters, all of the generated films of CuO NP composites displayed homogenous, even, condensed, and silky surface structures. The fact that only CuO/agar nano-composite films lacked antibacterial action suggests that the NPs' accessible surface area was small [105]. Interestingly, neither the mechanical characteristics nor the thickness of the films was considerably enhanced by including Cu NPs in the agar polymer [106].

1.3.1.4 Titanium Dioxide Nanoparticles

Titanium dioxide (TiO_2) particles are frequently added to meals as an ingredient in food, producing a whitened result that brightens and enhances the dish's appearance [107]. TiO_2 NPs are utilized in food packaging due to their broad spectrum of antibacterial characteristics. Earlier TiO_2 NPs were thought to be harmless. However, TiO_2 NPs' capacity to modify the intestinal barrier raised certain worries, as the European Food Safety Agency (EFSA) reported, leading some EU nations to outlaw the substance in food [108]. The detrimental effects of TiO_2 NPs on human cells have been verified by several investigations. According to in vivo experiments, TiO_2 NPs can be collected in the kidneys, spleen, liver, heart, lungs, and digestive system after being inhaled or consumed orally [109]. Still, the food business uses a limited number of authorized nanomaterials, including nanosized TiO_2.

1.3.1.5 Zinc Oxide Nanoparticles

FDA stands for the US Food and Drug Administration has categorized zinc oxide (ZnO) as a micronutrient necessary for both animal and human health and has designated it as broadly acknowledged as secure [110]. Because of its white color, ZnO is utilized as a food ingredient and in the food industry as a nutritional supplement. ZnO is utilized in food packaging in nanoscale form because its significant antibacterial and antifungal activity, enhanced mechanical strength, barrier qualities, and material stability come from its incorporation into the polymeric matrix [111]. When exposed to light, ZnO NPs play a dynamic role in food preservation by producing reactive oxygen species (ROS), particularly hydrogen peroxide [112].

1.3.1.6 Magnesium Oxide Nanoparticles

Large-scale production of magnesium oxide (MgO) is a natural mineral that the US Food and Drug Administration (FDA) has classified as generally recognized as safe (GRAS) to be utilized in food [113]. A highly transparent film may often be fabricated when MgO NPs are included in a biocomposite. Even while MgO NPs produce reactive oxygen species (ROS), their less potent antibacterial activity makes them less popular in uses involving food packaging than zinc oxide (ZnO) or titanium dioxide (TiO_2) NPs [114]. As a result of the demonstration that MgO NPs harmed cell membranes without experiencing oxidative stress, a ROS-independent mechanism for MgO NPs' antibacterial activity on E. coli was also established. Numerous in vitro and in vivo investigations have proven the great biocompatibility of MgO NPs [115].

1.4 Bio-Based Packaging Material Processing

A multi-step process is employed in creating bio-based materials for packaging, including intermolecular bonds disintegrating, synthesizing novel configurations of molecules, and utilizing the recently established links to develop a three-dimensional polymeric network [26]. The polymer's shape (i.e., length, width, and ratio) and the processing conditions determine the formation of new molecular bonds. The connections resulting from interactions include covalent, electrostatic, hydrophobic, and H-bonding to maintain the freshly formed films [116]. For the production of bio-based, there are two primary methods for processing plastic materials: wet processing and dry processing regularly. The foundation of wet processing is dissolution and the kind and pH of the solvent used, which might alter the polymer verification. Dry processing is based on the polymer's thermoplastic characteristics, where thermo-mechanical processing yields the commencement of the reaction of sulfhydryl or disulfide conversion [117]. Figure 1.3 illustrates the process of manufacturing packing materials made of biomass.

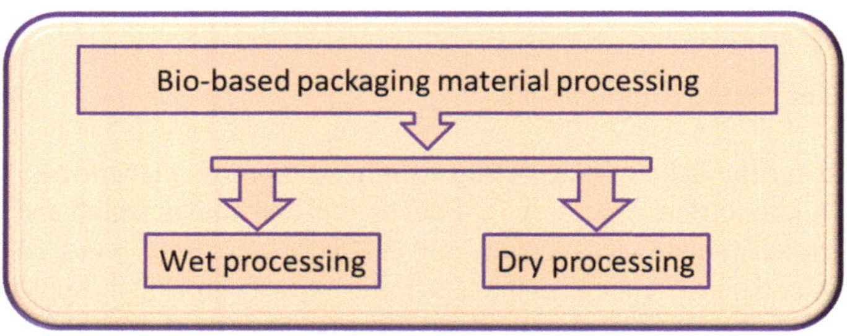

Fig. 1.3 The processing of bio-based packaging materials

1.4.1 Wet Processing

Casting, spreading continuously, or wet processing have all been widely employed in creating bio-based plastics utilizing renewable resources such as proteins, carbohydrates, and lipids [118]. The biopolymer must dissolve in the proper solvent to create a film-forming solution during wet processing. Bio-plastic films can be supplemented with various resources, including fillers, micro/nanostructures, antioxidants, cross-linking, antimicrobials, and plasticizers. This process can be divided into two categories: solvent evaporation and film-forming solution casting [119]. A plasticizer is added, which works breaks down intermolecular bonds and hardness, resulting in a more flexible and smooth material. The packaging industry benefits from wet processing by enhancing the bio-plastic films' mechanical qualities [120].

1.4.2 Dry Processing

The dry processing depends on a polymer's thermoplastic qualities and is important in packaging materials [121]. This process concerns the glass transition theory, which states that at a specific temperature, a glassy material changes into a semi-solid shape [122]. This transition state, also known as semi-solid, essentially causes molecular mobility and disorderliness, changing polymers' physicochemical and mechanical properties [26]. Protein molecules denatured due to broken intermolecular connections, and the material's characteristics change as new bonds and linkages form Schmid et al. [61]. Packaging made of polymers can be produced using extrusion and heat processing [123]. Heat processing is used to make the finished product, while extrusion is used for minor modifications and mixing [124]. It is possible to carry out both processes simultaneously or independently.

1.5 Features of the Bio-Plastic Films' Reinforcing for Active and Smart Food Wrapping

Bio-plastic films are growing in popularity as eco-friendly substitutes for conventional plastics. To maximize their impact and usefulness, one must comprehend their reinforcing characteristics. The visual, barrier, mechanical, and other qualities of typical packaging materials are not the only functional features incorporated into active food packaging materials [125]. To make food last longer on the shelf, they could be made to, for example, prevent microbiological development, delay unfavorable chemical reactions, or regulate enzyme activity [126]. This is usually accomplished by adding antioxidants and/or antimicrobials to the packaging materials, focusing on using naturally occurring botanically derived functional compounds for clean labeling [127]. Active packaging materials have the potential to lower the

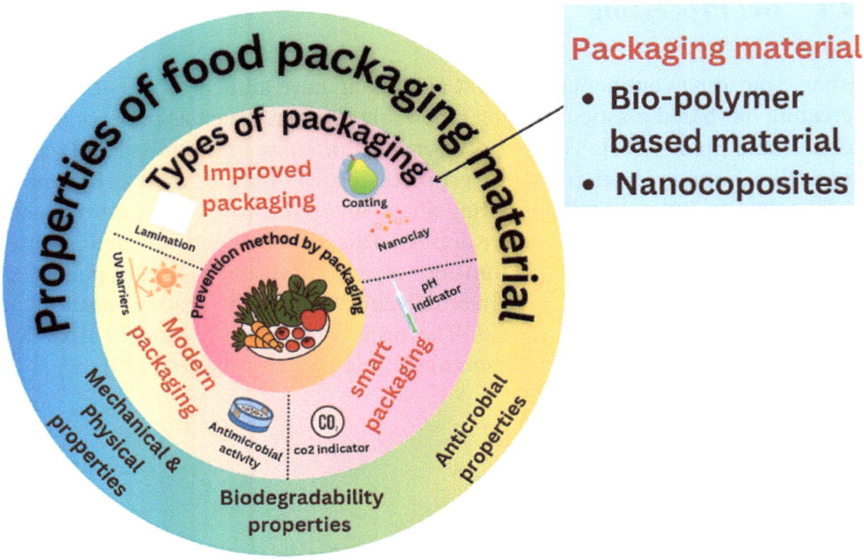

Fig. 1.4 Food packaging materials with remarkable qualities can be made using biocompatible green packaging, Reproduced with permission from Ref. [128], Copyright © 2023 Elsevier Ltd.

amount of antioxidants and antimicrobials needed to protect food during storage and reduce the amount consumed because these compounds are found in the film rather than the food itself. Biocompatible green packaging can be used to create food packaging materials with exceptional properties, as seen in Fig. 1.4.

1.5.1 Food Packaging's Biodegradability

Scientists have been interested in natural polymers because of their higher-degradability properties and natural sources [129]. These non-toxic, environmentally benign biomaterials include chitosan, polysaccharides, starches, alginate, and gelatins. Biomaterials produced through the actions of microbes, like xanthan, polyhydroxybutyrate (PHB), and bacterial cellulose, have a range of industrial and medical uses [130, 131]. Natural sources of oil and biomass (bio-polyester or lactic acid) or renewable monomers like polyethylene terephthalate (PET) through polymerization are the sources of synthetic goods. Packaging is still essentially produced only by employing petroleum-derived polymers like polyamide and PET in food packaging systems (FPS) [33]. Due to its inability to degrade naturally, plastic trash is a global issue. If plastic output doubles globally over the next ten years, this is anticipated to have a substantial negative impact on the environment [132].

1.5.2 Food Packaging's Antimicrobial Properties

A major global worry about food safety is foodborne illness, which can be attributed to various pathogens, including bacteria, fungi, and viruses [133]. Foodborne viruses contaminate certain items, like unprocessed agricultural goods, which comprise many food items (such as fruits), because there is no method to raise their safety [134]. One more crucial use for novel food packaging (NFP) is its antibacterial properties. Keeping food from spoiling due to microbial growth is one of NFP's primary objectives. Introducing new technology for storing and packing food results from expanding customer demand for organic and healthful foods [135]. Several researches have been conducted on the effectiveness of numerous powerful substances against a wide range of species of bacteria, both Gram-positive and -negative, involving *Vibrio cholera, Enterococcus faecalis, Pseudomonas aeruginosa, Staphylococcus aureus,* and *Staphylococcus epidermidis* [136].

1.5.3 Food Freshness and Spoilage Indicators

To let customers assess the quality and spoilage of their products, novel food packaging, is investigating active packaging techniques [137]. Much research has been conducted to develop packaging methods that can detect corruption. Introducing oxygen scavengers and antimicrobial chemicals into food packaging in the 20th century has set new standards for extending food shelf life by shielding it from the elements [138]. However, it would be intriguing if shoppers became aware of food quality, nutrition, or spoilage while shopping without experts' assistance. Furthermore, the quality of the food baskets would improve, and smart food packaging (SFP) would demonstrate its ability to implement this strategy if the vendors could quickly evaluate the caliber of their goods and remove any bad food that is displayed. Temperature is one of the most important factors in determining a food product's shelf life. Temperature indicators with time (TTIs) are one of the food packaging systems (FPSs) tactics for updating the client on the product's state [33]. These devices take the shape of color dots that change color in response to variations in temperature. Most products that need particular temperature settings for storage or consumption employ these systems.

1.5.4 Food Packaging's Scavengers

Food packing can be made to prevent oxygen buildup by using vacuum packaging. Certain novel techniques are frequently utilized to stop packaged foods' lipids from oxidizing [139]. This method limits the amount of oxygen in packaging by directly adding antioxidants or using modified atmospheres. Active food packaging's (AFPs) are designed to keep food from spoiling by encasing the relevant agents in the packaging (film/coating). Controlling carbon dioxide is crucial in prolonging food's shelf life [140]. Carbon dioxide scavenging (CS) agents, like zeolites and calcium hydroxide, primarily aim to eliminate generated CO_2 gas by respiration, microbial development, and degradation processes [141]. Vegetables' respiration rate and microbial growth are reduced by CO_2 emitters found in dietary acids and sodium bicarbonate [142]. To mitigate the harmful effects of oxidation and heat degradation, various natural and synthetic antioxidants, such as spices, herbs, organophosphates, and polyphenols, are incorporated into the packaging [143]. In short, scavenging-based AFP, like novel food packaging (NFP), is a strong contender to lessen food spoilage and lipid oxidation in packaged foods.

1.5.5 The Film's Optical Characteristics

The films' optical characteristics were assessed by calculating their transparency, which was stated as the transmittance of light at 600 nm, an intermediate visible wavelength and the one that is easier to compare to results from the literature [144, 145]. Transparent packaging materials are undoubtedly advantageous from a marketing perspective because consumers are more likely to buy products readily visible inside the package [146]. However, because ultraviolet (UV) light can have a negative impact on foods that deteriorate quickly, it is preferable to use materials that can act as a light barrier for foods that are susceptible to rapid deterioration. With optical qualities halfway between zein and chitosan, the combination appears to meet both requirements in this specific situation [147].

1.5.6 Mechanical Characteristics

After determining the thickness of the films, six specimens for each sample of zein, chitosan, and polymeric blend were examined using a tensile test to ascertain the mechanical characteristics [148, 149]. The study assessed the impact of blend formulation on the performance of pure zein and pure chitosan films by evaluating Young's modulus, elongation at break, and tensile strength [150].

1.5.7 Features of the Oxygen Barrier

These findings demonstrated how the polymer blend could, in fact, enhance several qualities where the single polymer falls short [151]. Additionally, the solubility and diffusivity drop. Specifically, the solubility of oxygen in the blend is significantly reduced when compared to pure zein [152]. This is most likely because of the blend's weaker chemical compatibility, which results in lower oxygen condensability and permeability [152].

1.5.8 Antioxidants

Addition of antioxidants to food packaging supplies to prevent oxidation reactions in food, especially those that involve lipids or proteins and degrade [153]. Antioxidants usually work by chelating transition metal ions, decreasing hydrogen peroxide, neutralizing singlet oxygen, or squelching free radicals to prevent oxidation [154]. Botanical extracts have been utilized as organic antioxidant additions in materials for active packaging, derived from various plants, such as potatoes, tomatoes, strawberries, cabbage, garlic, and saffron [155]. Numerous compounds with potential antioxidant action can be found in these extracts. For example, it has been demonstrated that the maximum antioxidant content in extracts from green tea is found in epigallocatechin and epigallocatechin gallate [156]. Using a radical scavenging experiment using 2,2-diphenyl-1-picrylhydrazyl (DPPH), an additional study investigated the anti-oxidative capacity of several plant extracts found in gelatin films, such as extracts from green tea, ginkgo leaves, ginger, and grape seed. According to this study, the most powerful antioxidant activity was found in ginkgo leaf extracts [157].

1.5.9 Antimicrobials

Antimicrobial compounds are useful because they can be added to active packing materials to prevent the growth of pathogenic or rotting bacteria [158]. Biodegradable packaging materials can incorporate natural antimicrobial chemicals, like vital oils (like thyme, oregano, cloves, and oils of rosemary) as well as botanical extracts (like extracts of barberries, saffron, potatoes, strawberries, garlic, tomatoes, and lettuce and cabbage) [155]. During storage, these antimicrobial substances could seep within the meal or stay in the packaging materials. To extend their activity, the antimicrobials' rate of dietary penetration should ideally be regulated [159]. Antimicrobial compounds can either be integrated into or immobilized on the surfaces of packaging materials. Thermal processing may cause some naturally occurring antimicrobial compounds to lose their antibacterial action because they are heat-sensitive. In this instance, this kind of packaging material should be prepared using non-thermal fabrication techniques like electro-spinning, solvent evaporation, or casting techniques [160]. Because they may break down the cytoplasmic membranes of pathogens and enter cell membranes, anthocyanin's have demonstrated strong antimicrobial effects in materials for active packing [161]. They also inhibit extracellular enzymes.

1.5.10 Gas Regulators

Food packaging materials can allow oxygen molecules to seep through and quicken the oxidation and discoloration processes in packaged foods [162]. As a result, it's critical to have techniques for managing food oxygen levels. Oxygen-depleting agents can be made from the scavenging of oxygen substances, including enzymes, tocopherols, ascorbic acid, ferrous acids, sulfites, catechols, and unsaturated hydrocarbons like oxidase of glucose [163]. Additional strategies to prevent harmful oxygen impacts include applying plant-based compounds such as Gallic acid, salicylic acid, phenolics, and flavonoids [164]. The food supply's economics and sustainability can be enhanced by obtaining these compounds from food processing enterprises' waste streams [165]. A material must possess several advantageous structural characteristics, such as hydroxyl groups and double bonds between carbon and oxygen, to block oxygen [166]. Therefore, it's critical to be able to regulate the amount of produce packed with ethylene gas while it's being stored. Adding materials that take in, oxidize, or break down the gasses generated via berries and vegetables will lower the amount of ethylene present. Activated carbon, clay, palladium (Pd), titanium dioxide (TiO_2), potassium permanganate ($KMnO_4$), and palladium are compounds that may extract and adsorb ethylene gas [167]. These additives can be employed singly or in combination to achieve synergistic effects. For example, TiO_2 nanoparticles, chitosan, and extract from black plum peel have been combined to create nano-composite films, demonstrating strong ethylene scavenging, antioxidant, and antibacterial capabilities.

1.5.11 pH Measurements

The pH of a packaged product is significantly altered, this type of indicator offers a detectable shift [168]. A food's pH might fluctuate due to enzyme activity, chemical interactions, or microbial development; hence pH sensors can indicate changes in food safety or quality [169]. Natural pigments are typically favored over synthetic dyes because of the growing customer desire for clean-label products [170]. Many natural pigments undergo noticeable color changes in response to variations in the pH of the surrounding medium. Known alternatively as a halo-chromic sensor, this kind of colorimetric pH sensor usually relies on the protons (H+) exchanged between the pigments and their surroundings. Certain colorimetric indicators have the ability to change color in reaction to particular volatile molecules, which may also indicate changes in the food's quality [171].

1.5.12 Temperature–Time Indicators

Indicators of temperature and time, or ITTs, are generally employed to track and monitor the effectiveness and security of the packaging goods while Gao et al. [172] are being stored and distributed [172]. They do this by identifying if the foods have been in contact with high temperatures for prolonged periods. Natural food pigments have been used as indicators in developing smart packaging materials, including ITTs [173]. The curve of temperature with the time to which the packed food has come into contact determines how many indicator colors have changed. Therefore, an oblique indicator regarding whether a food item has most likely declined while being stored can be obtained using these ITTs. Because ITTs are easy for consumers to understand and can be designed reasonably straightforward and affordable, Zhang et al. [174] have found widespread use in the food packaging business [174]. The fundamentals of the temperature-detection technique-polymerization and diffusion, microbiological development, enzymatic process, thermo-chromic process, surface Plasmon resonance, electrical, and photochromic, classify ITTs into several groups.

1.6 Application of Bio-Plastics Active and Smart Food Packaging

Developing packaging materials that biodegrade has improved the safety, quality and longevity of a wide range of goods [175]. Figure 1.5 illustrates several applications for packaging materials for different food goods. We have using bio-plastics to create thoughtful and functional food packaging.

1.6.1 Seafood and Meat

Meat products' shelf life has been increased and their quality has been enhanced by utilizing packaging materials that are biodegradable [176]. These packaging materials are commonly employed to control the environment surrounding red meat to put an end to the unintended color changes that occur when myoglobin is affected [177]. As such, they should have the ability to control the flow of gases, including oxygen, into and out of the package. Moreover, antimicrobial or antioxidant components may be used in active packaging materials to reduce microbiological contamination or oxidation reactions, while sensors may be included in smart packaging materials to offer information on the product's quality or safety [178]. For instance, red cabbage extract is used in biodegradable meat packaging materials made of starch and whey protein, which acts as a naturally occurring antioxidant to prevent oxidation of lipids and enhance the quality of the meat [179]. Similarly, to improve lamb meat's quality and prolong its shelf life through refrigeration by preventing microbial development,

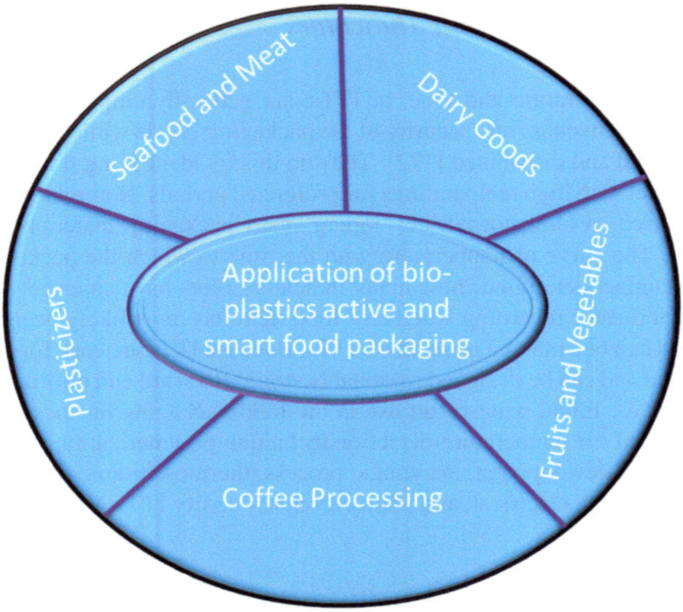

Fig. 1.5 Using bio-plastics for active and smart food packaging

whey protein films, including nanoparticles of titanium dioxide and antimicrobial vital oils (rosemary), have been produced. According to reports, these films extend the beef products' shelf life from roughly 6–13 days at 4 °C [180]. In a different investigation, films containing λ-carrageenan with an indication of color using a plant extract (Lycium ruthenicum Murr) were developed as smart packaging materials [181]. The color of the films changed based on how much the shrimp inside the films had spoiled. To indicate the state of fresh beef quality as it is being stored, a substance that is sensitive to temperature has been created [172]. Foods that are particularly perishable due to oxidative reactions and microbial spoiling include seafood, fish, and other goods [182]. It has also been demonstrated that intelligent and active packaging materials work effectively to safeguard these goods and keep an eye on their condition while being stored [141].

1.6.2 Dairy Goods

Nutrient-dense foods like dairy items are extremely vulnerable with relation to chemicals and microbiological deterioration while being stored, which compromises their quality and safety [8]. Several studies have demonstrated that adding antimicrobial or antioxidant ingredients to active packaging components help increase shelf life and improve the standard of dairy products [158]. For example, it has been demonstrated that the development of spoilage bacteria inside mozzarella cheese is inhibited by packing materials made of films made of sodium alginate and lemon essence, thus prolonging the cheese's shelf life [183]. It has also been demonstrated that Saccharomyces cerevisiae growth on Cheddar cheese is inhibited by packing materials made of films made of starch infused with essential oils that are antibacterial (thymol, linalool, and carvacrol) [184]. Anthocyanins and limonene-containing starch/polyvinyl alcohol sheets that turned red instead of purple as soon as pasteurized milk's pH fluctuated when being stored were an additional type of clever packing material produced. Using chitosan/PVA films infused with red cabbage anthocyanins, smart packaging materials were created in a different study [185]. These materials allowed for color changes to indicate the temperature history of milk products during storage [186].

1.6.3 Fruits and Vegetables

Highly perishable goods, like raw or chopped vegetables and fruits, can lose quality and safety while stored due to biochemical processes, including respiration, microbial contamination, or insect infestation [187]. As a result, innovative and intelligent packaging materials are being created to both report on the quality status of these items and safeguard them throughout storage and transportation [188]. For example, it has been demonstrated that adding titanium dioxide nanoparticles to chitosan films extends the shelf life of tomatoes by postponing their ripening [34, 189]. It has also been observed that cherry tomato shelf life can be increased by adding blackberry extracts high in anthocyanins to carboxymethylcellulose (CMC) films [190]. Analogously, it has been demonstrated that adding essential oils contained in β-cyclodextrins to packaging materials improves cherry tomato quality and shelf life [191]. It has been demonstrated that using antimicrobial packaging materials with star anise or palmarosa essential oils extends the shelf life of apples and inhibits the formation of penicillium expansum [192].

1.6.4 Coffee Processing

Renewable-sourced biological materials are the source of bio-based polymers used in coffee processing and packaging [26]. As customer demand for sustainable products and behaviors grows, they provide an environmentally favorable substitute for conventional plastics made from petroleum. Ultimately, the coffee industry has made great strides toward sustainability using bio-based polymers for processing and packaging [193]. In supporting the expanding trend toward eco-friendly and sustainable consumer goods, Nguyen et al. [194] solve the environmental issues raised by conventional plastic packaging [194].

1.7 Plasticizers

Bio-based polymer plasticizers are additives made from sustainable biological sources that enhance polymers' workability, flexibility, and durability [195]. Because they are poisonous and not biodegradable, traditional petroleum-based plasticizers have caused health and environmental issues. As an eco-friendly and sustainable substitute, bio-based plasticizers are available [196]. All things considered, bio-based polymer plasticizers, which solve the health and environmental issues related to traditional plasticizers, provide a promising and quickly developing field within sustainable materials [197].

1.8 Prospective Obstacles and Future Challenges

According to World meter, the World's population will increase and reach 10 billion people by 2057 [198]. The demand for food would rise by more than 50% by the middle of the century toward food in general and through about 70% for food items derived from animals [199]. Therefore, more food will be required to meet this need. However, constant pressure has been placed on agriculture and marine resources to ensure global food security. The results have been disastrous, with concerns about biodiversity preservation, land degradation, the alarming depletion of natural resources, climate change, water scarcity, and rising temperatures [200]. The continued search for biodegradable packaging components, especially active and edible ones, has increased due to recent disturbing revelations concerning the harmful ecological implications of routine usage of plastic packaging materials [201]. The formulation of "green" packaging includes food-grade chemicals derived from biomass. However, food processing residue biomass could be employed to not compete with food production. Industrial bio-wastes will inevitably be produced, and they are a valuable source of biopolymers and other bioactive substances [202]. Because of this, using them in packaging technology to create new materials is a fantastic solution to deal with the problem of disposing of these leftovers inconveniently while also recovering high-value chemicals with functional capabilities. The drawbacks have included cost, production scaling, allergic responses, barrier problems, and religious or dietary restrictions that prevent the utilization of animal-based coatings and films. For example, cross-linking, which can be physical, enzymatic, or chemical, nano-composites, and mixing hydrophobic bio-based polymers have been employed to overcome barrier issues, and surplus materials from food processing have been utilized to lower manufacturing costs [203–205]. Better labeling has been employed to prevent unintentional consumption, and more suitable technology and equipment have been tested to accomplish industrial output. Improvements have been made to some degree, enabling the effective commercial application of certain edible packaging designs and materials. However, there hasn't been much replacement for petroleum-based materials, and more study is required to guarantee a sustainable packaging sector free of plastic.

1.9 Conclusions

In the food business, natural materials can be used to create intelligent and active packaging materials that not only prolong food shelf life and save wastage but also enhance food quality and safety. By changing color in response to particular changes in pH, gas levels, or temperature, natural pigments can be added to these materials to serve as indications of changes in freshness, quality, or safety. Using natural antimicrobials or antioxidants can help prolong the shelf life of packaged foods by preventing unwanted chemical reactions or microbial growth. In some circumstances, a single addition may serve as a sensor, antioxidant, and antibacterial. The primary benefit of smart packing supplies is the ability to track a product's safety and freshness in real time without opening the box. Furthermore, information about the product's prior history, including its exposure to light, oxygen, pH, and temperature fluctuations, can be determined. This chapter concludes with a brief discussion of the 21st-century packaging applications that could use sustainable materials.

Acknowledgements This work is supported by the National Key Research and Development Program of China (Grant No. 2021YFB3700300), 2018 CAS-TWAS President's Fellowship Program for International Doctoral Candidate (Grant No. 2018CTF2969), the National Natural Science Foundation of China (Grant No. U21B2093, 52103015, 51773218 and 31970041), Youth Innovation Promotion Association of CAS (Grant No. 2018338), Programs Supported by Ningbo Natural Science Foundation (Grant No. 2021J207) and Public Welfare Science and Technology Projects of Ningbo (Grant No. 2021S071).

Declaration The authors declare that they have no known competing financial interests or personal relationships that could have appeared to influence the work reported in this chapter. Generative AI and AI-assisted technologies have been used in the writing process to improve the readability and language of the work only. After using this tool, the authors reviewed and edited the content as needed and take full responsibility for the content of the publication.

References

1. Hu, Y., Tu, T., Lei, Z.: Impact of corporate environmental goals on compliance behaviors in European Union food industry: a cross-country analysis. J. Clean. Prod. 141212 (2024)
2. Rahman, M., Emon, M.E.H., Antor, M.H., Haque, S.A., Talapatra, S.: Identification and prioritization of barriers to Industry 4.0 adoption in the context of food and beverage industries of Bangladesh. Benchmarking: Int. J. (2024)
3. Al Mahmud, M.Z., Mobarak, M.H., Hossain, N.: Emerging Trends in Biomaterials for Sustainable Food Packaging: A Comprehensive Review. Heliyon (2024)
4. Verma, S.K., Prasad, A., Katiyar, V.: State of art review on sustainable biodegradable polymers with a market overview for sustainability packaging. Mater. Today Sustain. 100776 (2024)

5. Barik, M., BhagyaRaj, G., Dash, K.K., Shams, R.: A thorough evaluation of chitosan-based packaging film and coating for food product shelf-life extension. J. Agric. Food Res. 101164 (2024)
6. Courbet, D., Jacquemier, L., Hercberg, S., Touvier, M., Sarda, B., Kesse-Guyot, E., Galan, P., Buttafoghi, N., Julia, C.: A randomized controlled trial to test the effects of displaying the Nutri-Score in food advertising on consumer perceptions and intentions to purchase and consume. Int. J. Behav. Nutr. Phys. Act. **21**(1), 38 (2024)
7. Kaur, S., Kumar, Y., Singh, V., Kaur, J., Panesar, P.S.: Cold plasma technology: reshaping food preservation and safety. Food Control 110537 (2024)
8. Augustin, M.A., Riley, M., Stockmann, R., Bennett, L., Kahl, A., Lockett, T., Osmond, M., Sanguansri, P., Stonehouse, W., Zajac, I.: Role of food processing in food and nutrition security. Trends Food Sci. Technol. **56**, 115–125 (2016)
9. Majid, I., Nayik, G.A., Dar, S.M., Nanda, V.: Novel food packaging technologies: innovations and future prospective. J. Saudi Soc. Agric. Sci. **17**(4), 454–462 (2018)
10. Thomassen, G., Peeters, E., Van Hee, N., Noëth, E., Du Bois, E., Boone, L., Compernolle, T.: The environmental impacts of reusable rice packaging: an extended comparative life cycle assessment. Sustain. Prod. Consum. **45**, 333–347 (2024)
11. Dani, R., Tiwari, K., Prabhakar, A.P.B.: A review of food packaging materials and its impact on environment. In: AIP Conference Proceedings. AIP Publishing (2024)
12. Wibowo, C., Salsabila, S., Muna, A., Rusliman, D., Wasisto, H.S.: Advanced biopolymer-based edible coating technologies for food preservation and packaging. Compr. Rev. Food Sci. Food Saf. **23**(1), 1–31 (2024)
13. Youssef, A.M., El-Sayed, S.M.: Bionanocomposites materials for food packaging applications: concepts and future outlook. Carbohyd. Polym. **193**, 19–27 (2018)
14. Chen, Y., Awasthi, A.K., Wei, F., Tan, Q., Li, J.: Single-use plastics: production, usage, disposal, and adverse impacts. Sci. Total Environ. **752**, 141772 (2021)
15. Tilsted, J.P., Bauer, F., Birkbeck, C.D., Skovgaard, J., Rootzén, J.: Ending fossil-based growth: confronting the political economy of petrochemical plastics. One Earth **6**(6), 607–619 (2023)
16. Piergiovanni, L., Limbo, S., Piergiovanni, L., Limbo, S.: Plastic packaging materials. Food Packag. Mater. 33–49 (2016)
17. Lagaron, J., Catalá, R., Gavara, R.: Structural characteristics defining high barrier properties in polymeric materials. Mater. Sci. Technol. **20**(1), 1–7 (2004)
18. Meng, W., Sun, H., Su, G.: Plastic packaging-associated chemicals and their hazards–an overview of reviews. Chemosphere 138795 (2023)
19. Royer, S.-J., Ferrón, S., Wilson, S.T., Karl, D.M.: Production of methane and ethylene from plastic in the environment. PLoS ONE **13**(8), e0200574 (2018)
20. Pellis, A., Malinconico, M., Guarneri, A., Gardossi, L.: Renewable polymers and plastics: performance beyond the green. New Biotechnol. **60**, 146–158 (2021)
21. Yara-Varón, E., Li, Y., Balcells, M., Canela-Garayoa, R., Fabiano-Tixier, A.-S., Chemat, F.: Vegetable oils as alternative solvents for green oleo-extraction, purification and formulation of food and natural products. Molecules **22**(9), 1474 (2017)
22. Wang, Q., Awasthi, M.K., Zhang, Z., Wong, J.W.: Sustainable composting and its environmental implications, pp. 115–132. Elsevier, Sustainable Resource Recovery and Zero Waste Approaches (2019)
23. Goldan, E., Nedeff, V., Barsan, N., Culea, M., Panainte-Lehadus, M., Mosnegutu, E., Tomozei, C., Chitimus, D., Irimia, O.: Assessment of manure compost used as soil amendment—a review. Processes **11**(4), 1167 (2023)
24. Kržan, A.: Biodegradable polymers and plastics. Innov. Value Chain. Dev. Sustain. Plast. Cent. Eur. (Plast.) **12**, 12–29 (2012)
25. Haider, T.P., Völker, C., Kramm, J., Landfester, K., Wurm, F.R.: Plastics of the future? Impact Biodegrad. Polym. Environ. Soc. Angew. Chem. Int. Ed. **58**(1), 50–62 (2019)
26. Asgher, M., Qamar, S.A., Bilal, M., Iqbal, H.M.: Bio-based active food packaging materials: sustainable alternative to conventional petrochemical-based packaging materials. Food Res. Int. **137**, 109625 (2020)

27. Verma, M., Shakya, S., Kumar, P., Madhavi, J., Murugaiyan, J., Rao, M.: Trends in packaging material for food products: historical background, current scenario, and future prospects. J. Food Sci. Technol. 1–14 (2021)
28. Tajeddin, B., Arabkhedri, M.: Polymers and food packaging, pp. 525–543. Elsevier, Polymer Science and Innovative Applications (2020)
29. Kim, Y.T., Min, B., Kim, K.W.: General characteristics of packaging materials for food system. In: Innovations in Food Packaging, pp. 13–35. Elsevier (2014)
30. Rhodes, C.J.: Solving the plastic problem: from cradle to grave, to reincarnation. Sci. Prog. **102**(3), 218–248 (2019)
31. Kabir, E., Kaur, R., Lee, J., Kim, K.-H., Kwon, E.E.: Prospects of biopolymer technology as an alternative option for non-degradable plastics and sustainable management of plastic wastes. J. Clean. Prod. **258**, 120536 (2020)
32. Kawaguchi, H., Ogino, C., Kondo, A.: Microbial conversion of biomass into bio-based polymers. Biores. Technol. **245**, 1664–1673 (2017)
33. Ahari, H., Soufiani, S.P.: Smart and active food packaging: insights in novel food packaging. Front. Microbiol. **12**, 657233 (2021)
34. Jafarzadeh, S., Nafchi, A.M., Salehabadi, A., Oladzad-Abbasabadi, N., Jafari, S.M.: Application of bio-nanocomposite films and edible coatings for extending the shelf life of fresh fruits and vegetables. Adv. Coll. Interface. Sci. **291**, 102405 (2021)
35. Nicolescu, C.M., Bumbac, M., Buruleanu, C.L., Popescu, E.C., Stanescu, S.G., Georgescu, A.A., Toma, S.M.: Biopolymers produced by lactic acid bacteria: characterization and food application. Polymers **15**(6), 1539 (2023)
36. Chen, G.-Q., Patel, M.K.: Plastics derived from biological sources: present and future: a technical and environmental review. Chem. Rev. **112**(4), 2082–2099 (2012)
37. Rebouillat, S., Pla, F.: Recent strategies for the development of biosourced-monomers, oligomers and polymers-based materials: a review with an innovation and a bigger data focus. J. Biomater. Nanobiotechnology **7**(4), 167–213 (2016)
38. Marques, S., Moreno, A.D., Ballesteros, M., Gírio, F.: Starch biomass for biofuels, biomaterials, and chemicals. Biomass Green Chem.: Build. Renew. Pathw. 69–94 (2018)
39. Lee, C., Sapuan, S., Ilyas, R., Lee, S., Khalina, A.: Development and processing of PLA, PHA, and other biopolymers. In: Advanced Processing, Properties, and Applications of Starch and Other Bio-Based Polymers, pp. 47–63. Elsevier (2020)
40. Mokhena, T.C., Sefadi, J.S., Sadiku, E.R., John, M.J., Mochane, M.J., Mtibe, A.: Thermoplastic processing of PLA/cellulose nanomaterials composites. Polymers **10**(12), 1363 (2018)
41. Brethauer, S., Wyman, C.E.: Continuous hydrolysis and fermentation for cellulosic ethanol production. Biores. Technol. **101**(13), 4862–4874 (2010)
42. Karande, R.D., Abitha, V., Rane, A.V., Mishra, R.K.: Preparation of polylactide from synthesized lactic acid and effect of reaction parameters on conversion. J. Mater. Sci. Eng. Adv. Technol **12**(1–2), 1–37 (2015)
43. Naser, A.Z., Deiab, I., Darras, B.M.: Poly (lactic acid) (PLA) and polyhydroxyalkanoates (PHAs), green alternatives to petroleum-based plastics: a review. RSC Adv. **11**(28), 17151–17196 (2021)
44. Farah, S., Anderson, D.G., Langer, R.: Physical and mechanical properties of PLA, and their functions in widespread applications—a comprehensive review. Adv. Drug Deliv. Rev. **107**, 367–392 (2016)
45. Khalid, M.Y., Al Rashid, A., Arif, Z.U., Ahmed, W., Arshad, H., Zaidi, A.A.: Natural fiber reinforced composites: sustainable materials for emerging applications. Results Eng. **11**, 100263 (2021)
46. Raquez, J.-M., Habibi, Y., Murariu, M., Dubois, P.: Polylactide (PLA)-based nanocomposites. Prog. Polym. Sci. **38**(10–11), 1504–1542 (2013)
47. Albuquerque, P.B., Malafaia, C.B.: Perspectives on the production, structural characteristics and potential applications of bioplastics derived from polyhydroxyalkanoates. Int. J. Biol. Macromol. **107**, 615–625 (2018)

48. Saravanan, K., Umesh, M., Kathirvel, P.: Microbial polyhydroxyalkanoates (PHAs): a review on biosynthesis, properties, fermentation strategies and its prospective applications for sustainable future. J. Polym. Environ. **30**(12), 4903–4935 (2022)
49. Tebaldi, M.L., Maia, A.L.C., Poletto, F., de Andrade, F.V., Soares, D.C.F.: Poly (-3-hydroxybutyrate-co-3-hydroxyvalerate) (PHBV): current advances in synthesis methodologies, antitumor applications and biocompatibility. J. Drug Deliv. Sci. Technol. **51**, 115–126 (2019)
50. Adeleye, A.T., Odoh, C.K., Enudi, O.C., Banjoko, O.O., Osiboye, O.O., Odediran, E.T., Louis, H.: Sustainable synthesis and applications of polyhydroxyalkanoates (PHAs) from biomass. Process Biochem. **96**, 174–193 (2020)
51. Rana, S., Upadhyay, L.S.B.: Microbial exopolysaccharides: synthesis pathways, types and their commercial applications. Int. J. Biol. Macromol. **157**, 577–583 (2020)
52. Arserim Ucar, D.K., Konuk Takma, D., Korel, F.: Exopolysaccharides in food processing industrials. Microb. Exopolysaccharides Nov. Signif. Biomater. 201–234 (2021)
53. Angelin, J., Kavitha, M.: Exopolysaccharides from probiotic bacteria and their health potential. Int. J. Biol. Macromol. **162**, 853–865 (2020)
54. Júnior, L.M., Vieira, R.P., Anjos, C.A.R.: Kefiran-based films: fundamental concepts, formulation strategies and properties. Carbohyd. Polym. **246**, 116609 (2020)
55. de Carvalho, A.P.A., Conte-Junior, C.A.: Food-derived biopolymer kefiran composites, nanocomposites and nanofibers: emerging alternatives to food packaging and potentials in nanomedicine. Trends Food Sci. Technol. **116**, 370–386 (2021)
56. Liu, D., Nikoo, M., Boran, G., Zhou, P., Regenstein, J.M.: Collagen and gelatin. Annu. Rev. Food Sci. Technol. **6**, 527–557 (2015)
57. Gómez-Guillén, M., Giménez, B., López-Caballero, M.A., Montero, M.: Functional and bioactive properties of collagen and gelatin from alternative sources: a review. Food Hydrocoll. **25**(8), 1813–1827 (2011)
58. Álvarez-Castillo, E., Felix, M., Bengoechea, C., Guerrero, A.: Proteins from agri-food industrial biowastes or co-products and their applications as green materials. Foods 2021, 10 981. (s Note: MDPI stays neu-tral with regard to jurisdictional claims in …, 2021)
59. Xia, C., Wang, W., Wang, L., Liu, H., Xiao, J.: Multilayer zein/gelatin films with tunable water barrier property and prolonged antioxidant activity. Food Packag. Shelf Life **19**, 76–85 (2019)
60. Sajjad, A., Rasheed, F., Xiao, X., Olsson, R.T., Capezza, A.J., Zia, M.: Integration of zinc oxide nanoparticles in wheat gluten hydrolysates-development of multifunctional films with pliable properties. J. Inorg. Organomet. Polym Mater. **33**(4), 914–929 (2023)
61. Schmid, M., Proels, S., Kainz, D.M., Hammann, F., Grupa, U.: Effect of thermally induced denaturation on molecular interaction-response relationships of whey protein isolate based films and coatings. Prog. Org. Coat. **104**, 161–172 (2017)
62. Zhang, J., Lei, W., Schutz, J., Liu, D., Tang, B., Wang, C.H., Wang, X.: Improving the gas barrier, mechanical and thermal properties of poly (vinyl alcohol) with molybdenum disulfide nanosheets. J. Polym. Sci., Part B Polym. Phys. **57**(7), 406–414 (2019)
63. Senthilkumaran, A., Babaei-Ghazvini, A., Nickerson, M.T., Acharya, B.: Comparison of protein content, availability, and different properties of plant protein sources with their application in packaging. Polymers **14**(5), 1065 (2022)
64. Zheng, M., Tajvidi, M., Tayeb, A.H., Stark, N.M.: Effects of bentonite on physical, mechanical and barrier properties of cellulose nanofibril hybrid films for packaging applications. Cellulose **26**, 5363–5379 (2019)
65. Ciannamea, E.M., Stefani, P.M., Ruseckaite, R.A.: Physical and mechanical properties of compression molded and solution casting soybean protein concentrate based films. Food Hydrocoll. **38**, 193–204 (2014)
66. Akhtar, S., Ismail, T., Layla, A., Hussain, M., Qamar, M.: An overview of plant-based protein rich products. Plant Protein Foods 27–60 (2022)
67. Wang, C., Killpatrick, A., Humphrey, A., Guo, M.: Whey protein functional properties and applications in food formulation. Whey Protein Prod. Chem. Funct. Appl. 157–204 (2019)

68. Rosseto, M., Rigueto, C.V.T., Alessandretti, I., de Oliveira, R., Raber Wohlmuth, D.A., Loss, R.A., Dettmer, A., Richards, N.S.P.D.S.: Whey-based polymeric films for food packaging applications: a review of recent trends. J. Sci. Food Agric. **103**(7), 3217–3229 (2023)
69. Cazón, P., Velazquez, G., Ramírez, J.A., Vázquez, M.: Polysaccharide-based films and coatings for food packaging: a review. Food Hydrocoll **68**, 136–148 (2017)
70. Ju, J., Chen, X., Xie, Y., Yu, H., Guo, Y., Cheng, Y., Qian, H., Yao, W.: Application of essential oil as a sustained release preparation in food packaging. Trends Food Sci. Technol. **92**, 22–32 (2019)
71. Immonen, K.: Ligno-cellulose fibre poly (lactic acid) interfaces in biocomposites (2018)
72. Das, A., Mondal, C., Roy, S.: Pretreatment methods of Ligno-cellulosic biomass: a review. J. Eng. Sci. & Technol. Rev. **8**(5), (2015)
73. Zhang, J., Wu, J., Yu, J., Zhang, X., He, J., Zhang, J.: Application of ionic liquids for dissolving cellulose and fabricating cellulose-based materials: state of the art and future trends. Mater. Chem. Front. **1**(7), 1273–1290 (2017)
74. Ibn Yaich, A., Edlund, U., Albertsson, A.-C.: Transfer of biomatrix/wood cell interactions to hemicellulose-based materials to control water interaction. Chem. Rev. 117(12) (2017) 8177–8207.
75. Han, Y., Wang, L.: Sodium alginate/carboxymethyl cellulose films containing pyrogallic acid: physical and antibacterial properties. J. Sci. Food Agric. **97**(4), 1295–1301 (2017)
76. Kumar, R., Lalnundiki, V., Shelare, S.D., Abhishek, G.J., Sharma, S., Sharma, D., Kumar, A., Abbas, M.: An investigation of the environmental implications of bioplastics: recent advancements on the development of environmentally friendly bioplastics solutions. Environ. Res. 117707 (2023)
77. Susmitha, A., Sasikumar, K., Rajan, D., Nampoothiri, K.M.: Development and characterization of corn starch-gelatin based edible films incorporated with mango and pineapple for active packaging. Food Biosci. **41**, 100977 (2021)
78. Nawab, A., Alam, F., Haq, M.A., Haider, M.S., Lutfi, Z., Kamaluddin, S., Hasnain, A.: Innovative edible packaging from mango kernel starch for the shelf life extension of red chili powder. Int. J. Biol. Macromol. **114**, 626–631 (2018)
79. Prasad, K., Jacob, S., Siddiqui, M.W.: Fruit maturity, harvesting, and quality standards, pp. 41–69. Elsevier, Preharvest modulation of postharvest fruit and vegetable quality (2018)
80. Sharma, S., Sharma, P., Sharma, V., Bajaj, B.K.: Polyhydroxybutyrate as an eco-friendly alternative of synthetic plastics. Environ. Agric. Microbiol. Appl. Sustain. 101–149 (2021)
81. Umakanth, A., Datta, A., Reddy, B.S., Bardhan, S.: Biomass feedstocks for advanced biofuels: sustainability and supply chain management. Adv. Biofuel Technol. 39–72 (2022)
82. Ma, Y., Xiao, Y., Zhao, Y., Bei, Y., Hu, L., Zhou, Y., Jia, P.: Biomass based polyols and biomass based polyurethane materials as a route towards sustainability. React. Funct. Polym. **175**, 105285 (2022)
83. Peng, Z., Jiang, X., Si, C., Joao Cárdenas-Oscanoa, A., Huang, C.: Advances of modified Lignin as substitute to develop Lignin-based phenol-formaldehyde resin adhesives. ChemSusChem **16**(15), e202300174 (2023)
84. Lawoko, M., Gioia, C.: Recent strategies for Lignin-based thermosets, Lignin Utilization Strategies: From Processing to Applications, pp. 175–206. ACS Publications (2021)
85. Rezić, I., Haramina, T., Rezić, T.: Metal nanoparticles and carbon nanotubes—perfect antimicrobial nano-fillers in polymer-based food packaging materials, pp. 497–532. Elsevier, Food Packaging (2017)
86. Amin, U., Khan, M.K.I., Maan, A.A., Nazir, A., Riaz, S., Khan, M.U., Sultan, M., Munekata, P.E., Lorenzo, J.M.: Biodegradable active, intelligent, and smart packaging materials for food applications. Food Packag. Shelf Life **33**, 100903 (2022)
87. Kong, J., Zhang, S., Shen, M., Zhang, J., Yoganathan, S.: Evaluation of copper (I)-doped zinc oxide composite nanoparticles on both gram-negative and gram-positive bacteria. Colloids Surf. A **643**, 128742 (2022)
88. Epand, R.M., Walker, C., Epand, R.F., Magarvey, N.A.: Molecular mechanisms of membrane targeting antibiotics. Biochim. Biophys. Acta (BBA)-Biomembr. **1858**(5), 980–987 (2016)

89. Wang, J., Li, J., Shen, Z., Wang, D., Tang, B.Z.: Phospholipid-mimetic aggregation-induced emission luminogens for specific elimination of gram-positive and gram-negative bacteria. ACS Nano **17**(5), 4239–4249 (2023)
90. Abdal Dayem, A., Hossain, M.K., Lee, S.B., Kim, K., Saha, S.K., Yang, G.-M., Choi, H.Y., Cho, S.-G.: The role of reactive oxygen species (ROS) in the biological activities of metallic nanoparticles. Int. J. Mol. Sci. **18**(1), 120 (2017)
91. Omerović, N., Djisalov, M., Živojević, K., Mladenović, M., Vunduk, J., Milenković, I., Knežević, N.Ž, Gadjanski, I., Vidić, J.: Antimicrobial nanoparticles and biodegradable polymer composites for active food packaging applications. Compr. Rev. Food Sci. Food Saf. **20**(3), 2428–2454 (2021)
92. Yu, Z., Li, Q., Wang, J., Yu, Y., Wang, Y., Zhou, Q., Li, P.: Reactive oxygen species-related nanoparticle toxicity in the biomedical field. Nanoscale Res. Lett. **15**(1), 115 (2020)
93. Sharma, R., Jafari, S.M., Sharma, S.: Antimicrobial bio-nanocomposites and their potential applications in food packaging. Food Control **112**, 107086 (2020)
94. Younis, N.S., Mohamed, M.E., El Semary, N.A.: Green synthesis of silver nanoparticles by the Cyanobacteria synechocystis sp.: characterization, antimicrobial and diabetic wound-healing actions. Mar. Drugs **20**(1), 56 (2022)
95. Xu, Z., Zhang, C., Wang, X., Liu, D.: Release strategies of silver ions from materials for bacterial killing. ACS Appl. Bio Mater. **4**(5), 3985–3999 (2021)
96. Alexander, J.W.: History of the medical use of silver. Surg. Infect. **10**(3), 289–292 (2009)
97. Ranjan Sarker, S., Polash, S.A., Boath, J., Kandjani, A.E., Poddar, A., Dekiwadia, C., Shukla, R., Sabri, Y., Bhargava, S.K.: Functionalization of elongated tetrahexahedral Au nanoparticles and their antimicrobial activity assay. ACS Appl. Mater. & Interfaces **11**(14), 13450–13459 (2019)
98. Ahmad, B., Hafeez, N., Bashir, S., Rauf, A.: Phytofabricated gold nanoparticles and their biomedical applications. Biomed. Pharmacother. **89**, 414–425 (2017)
99. Dutta, H.: Potential of polysaccharide nanoparticles in foods. In: Nanotechnology Horizons in Food Process Engineering, pp. 85–155. Apple Academic Press (2023)
100. Zhang, W., Jiang, Q., Shen, J., Gao, P., Yu, D., Xu, Y., Xia, W.: The role of organic acid structures in changes of physicochemical and antioxidant properties of crosslinked chitosan films. Food Packag. Shelf Life **31**, 100792 (2022)
101. Tian, Y., Cai, R., Yue, T., Gao, Z., Yuan, Y., Wang, Z.: Application of nanostructures as antimicrobials in the control of foodborne pathogen. Crit. Rev. Food Sci. Nutr. **62**(14), 3951–3968 (2022)
102. Dat, N.M., Nam, N.T.H., Cong, C.Q., Hai, N.D., An, H., Duy, B.T., Dat, N.T., Dai Viet, V.N., Duong, H.T., Phong, M.T.: Chitosan membrane drafting silver-immobilized graphene oxide nanocomposite for banana preservation: Fabrication, physicochemical properties, bioactivities, and application. Int. J. Biol. Macromol. **242**, 124607 (2023)
103. Shankar, S., Wang, L.-F., Rhim, J.-W.: Preparation and properties of carbohydrate-based composite films incorporated with CuO nanoparticles. Carbohyd. Polym. **169**, 264–271 (2017)
104. Shankar, S., Wang, L.-F., Rhim, J.-W.: Preparation and properties of carbohydrate-based. Polymer **62**, 35–41 (2017)
105. Duta, L., Grumezescu, V.: The effect of doping on the electrical and dielectric properties of hydroxyapatite for medical applications: from powders to thin films. Materials **17**(3), 640 (2024)
106. Arfat, Y.A., Ahmed, J., Jacob, H.: Preparation and characterization of agar-based nanocomposite films reinforced with bimetallic (Ag-Cu) alloy nanoparticles. Carbohyd. Polym. **155**, 382–390 (2017)
107. Wasilewska, A., Bielicka, M., Klekotka, U., Kalska-Szostko, B.: Nanoparticle applications in food–a review. Food Funct. **14**(6), 2544–2567 (2023)
108. Fiordaliso, F., Bigini, P., Salmona, M., Diomede, L.: Toxicological impact of titanium dioxide nanoparticles and food-grade titanium dioxide (E171) on human and environmental health. Environ. Sci. Nano **9**(4), 1199–1211 (2022)

109. Baranowska-Wójcik, E., Szwajgier, D., Oleszczuk, P., Winiarska-Mieczan, A.: Effects of titanium dioxide nanoparticles exposure on human health—a review. Biol. Trace Elem. Res. **193**, 118–129 (2020)
110. Madkour, L.H.: A review: metal nanoparticles and their safety processing in functional foods. J. Chem. Sci. Chem. Eng **1**, 18–36 (2021)
111. Sharma, N., Gupta, P.C., Upadhyay, S., Rai, S., Mishra, P.: Antimicrobial applications of zinc oxide nanoparticles in food packaging industry. In: Metal and Metal-Oxide Based Nanomaterials: Synthesis, Agricultural, Biomedical and Environmental Interventions, pp. 67–93. Springer (2024)
112. Du, M., Zhao, W., Ma, R., Xu, H., Shan, C., Liu, K., Zhuang, J., Jiao, Z.: Visible-light-driven photocatalytic inactivation of S. aureus in aqueous environment by hydrophilic zinc oxide (ZnO) nanoparticles based on the interfacial electron transfer in S. aureus/ZnO composites. J. Hazard. Mater. **418**, 126013 (2021)
113. He, X., Deng, H., Hwang, H.-M.: The current application of nanotechnology in food and agriculture. J. Food Drug Anal. **27**(1), 1–21 (2019)
114. Ballesteros, L.F., Lamsaf, H., Sebastian, C.V., Cerqueira, M.A., Pastrana, L., Teixeira, J.A.: Active packaging systems based on metal and metal oxide nanoparticles. Nanotechnol.-Enhanc. Food Packag. 143–181 (2022)
115. Khalid, A., Norello, R., Abraham, A.N., Tetienne, J.-P., Karle, T.J., Lui, E.W.C., Xia, K., Tran, P.A., O'Connor, A.J., Mann, B.G.: Biocompatible and biodegradable magnesium oxide nanoparticles with in vitro photostable near-infrared emission: short-term fluorescent markers. Nanomaterials **9**(10), 1360 (2019)
116. Yang, Y., Ding, X., Urban, M.W.: Chemical and physical aspects of self-healing materials. Prog. Polym. Sci. **49**, 34–59 (2015)
117. Kaur, A., Sharma, S.: A sustainable replacement for conventional petrochemical-based packaging materials as bio-based food packaging. Pharma Innov. J. **12**(5), 3347–3357 (2023)
118. Álvarez-Castillo, E., Bengoechea, C., Felix, M., Guerrero, A.: Protein-based bioplastics from biowastes: sources, processing, properties and applications. Bioplastics Sustain. Dev. 137–176 (2021)
119. Borbolla-Jiménez, F.V., Peña-Corona, S.I., Farah, S.J., Jiménez-Valdés, M.T., Pineda-Pérez, E., Romero-Montero, A., Del Prado-Audelo, M.L., Bernal-Chávez, S.A., Magaña, J.J., Leyva-Gómez, G.: Films for wound healing fabricated using a solvent casting technique. Pharmaceutics **15**(7), 1914 (2023)
120. Kumar, Y., Shukla, P., Singh, P., Prabhakaran, P., Tanwar, V., Kumar, Y.: Bio-plastics. A perfect tool for eco-friendly food packaging: a review. J. Food Prod. Dev. Packag. **1**(1), 01–06 (2014)
121. Banerjee, R., Ray, S.S.: Sustainability and life cycle assessment of thermoplastic polymers for packaging: a review on fundamental principles and applications. Macromol. Mater. Eng. **307**(6), 2100794 (2022)
122. Petters, M., Kasparoglu, S.: Predicting the influence of particle size on the glass transition temperature and viscosity of secondary organic material. Sci. Rep. **10**(1), 15170 (2020)
123. Maguire, A., Pottackal, N., Saadi, M., Rahman, M.M., Ajayan, P.M.: Additive manufacturing of polymer-based structures by extrusion technologies. Oxf. Open Mater. Sci. **1**(1), itaa004 (2021)
124. Alam, M., Kaur, J., Khaira, H., Gupta, K.: Extrusion and extruded products: changes in quality attributes as affected by extrusion process parameters: a review. Crit. Rev. Food Sci. Nutr. **56**(3), 445–473 (2016)
125. Nilsen-Nygaard, J., Fernández, E.N., Radusin, T., Rotabakk, B.T., Sarfraz, J., Sharmin, N., Sivertsvik, M., Sone, I., Pettersen, M.K.: Current status of biobased and biodegradable food packaging materials: impact on food quality and effect of innovative processing technologies. Compr. Rev. Food Sci. Food Saf. **20**(2), 1333–1380 (2021)
126. Behera, S.S., Ray, R.C., Zdolec, N.: Lactobacillus plantarum with functional properties: an approach to increase safety and shelf-life of fermented foods. BioMed Res. Int. 2018 (2018)

127. Alexandri, M., Kachrimanidou, V., Papapostolou, H., Papadaki, A., Kopsahelis, N.: Sustainable food systems: the case of functional compounds towards the development of clean label food products. Foods **11**(18), 2796 (2022)
128. Priyanka, S., Namasivayam, S.K.R., Bharani, R.A., John, A.: Biocompatible green technology principles for the fabrication of food packaging material with noteworthy mechanical and antimicrobial properties a sustainable developmental goal towards the effective, safe food preservation strategy. Chemosphere 139240 (2023)
129. Mishra, T., Mandal, P., Rout, A.K., Sahoo, D.: A state-of-the-art review on potential applications of natural fiber-reinforced polymer composite filled with inorganic nanoparticle. Composites Part C Open Access **9**, 100298 (2022)
130. Arif, Z.U., Khalid, M.Y., Sheikh, M.F., Zolfagharian, A., Bodaghi, M.: Biopolymeric sustainable materials and their emerging applications. J. Environ. Chem. Eng. **10**(4), 108159 (2022)
131. Insuasti-Cruz, E., Suárez-Jaramillo, V., Mena Urresta, K.A., Pila-Varela, K.O., Fiallos-Ayala, X., Dahoumane, S.A., Alexis, F.: Natural biomaterials from biodiversity for healthcare applications. Adv. Healthc. Mater. **11**(1), 2101389 (2022)
132. Shen, M., Huang, W., Chen, M., Song, B., Zeng, G., Zhang, Y.: (Micro) plastic crisis: unignorable contribution to global greenhouse gas emissions and climate change. J. Clean. Prod. **254**, 120138 (2020)
133. Abebe, E., Gugsa, G., Ahmed, M.: Review on major food-borne zoonotic bacterial pathogens. J. Trop. Med. 2020 (2020)
134. Alum, E.A., Urom, S., Ben, C.M.A.: Microbiological contamination of food: the mechanisms, impacts and prevention. Int. J. Sci. Technol. Res. **5**(3), 65–78 (2016)
135. Wyrwa, J., Barska, A.: Innovations in the food packaging market: active packaging. Eur. Food Res. Technol. **243**, 1681–1692 (2017)
136. Durán, N., Nakazato, G., Seabra, A.B.: Antimicrobial activity of biogenic silver nanoparticles, and silver chloride nanoparticles: an overview and comments. Appl. Microbiol. Biotechnol. **100**, 6555–6570 (2016)
137. Ahmed, M.W., Haque, M.A., Mohibbullah, M., Khan, M.S.I., Islam, M.A., Mondal, M.H.T., Ahmmed, R.: A review on active packaging for quality and safety of foods: current trends, applications, prospects and challenges. Food Packag. Shelf Life **33**, 100913 (2022)
138. Hemavathi, A., Siddaramaiah, H.: Food packaging: polimers as packaging materials in food supply chains. In: Encyclopedia of Polymer Applications, pp. 1374–1397. CRC Press Boca Raton (2018)
139. Yildirim, S., Röcker, B., Pettersen, M.K., Nilsen-Nygaard, J., Ayhan, Z., Rutkaite, R., Radusin, T., Suminska, P., Marcos, B., Coma, V.: Active packaging applications for food. Compr. Rev. Food Sci. Food Saf. **17**(1), 165–199 (2018)
140. Boz, Z., Brecht, J.K., Welt, B.A., Pelletier, W., McLamore, E., Kiker, G., Butler, J.E.: Challenges and advances in development of active components to modify headspace gases in packaging of fresh produce and muscle foods. J. Appl. Packag. Res. **10**(1), 62–97 (2018)
141. Firouz, M.S., Mohi-Alden, K., Omid, M.: A critical review on intelligent and active packaging in the food industry: research and development. Food Res. Int. **141**, 110113 (2021)
142. B. Kuswandi, Active and intelligent packaging, safety, and quality controls, Fresh-cut fruits and vegetables (2020) 243–294.
143. Gómez-Estaca, J., López-de-Dicastillo, C., Hernández-Muñoz, P., Catalá, R., Gavara, R.: Advances in antioxidant active food packaging. Trends Food Sci. Technol. **35**(1), 42–51 (2014)
144. Simona, J., Dani, D., Petr, S., Marcela, N., Jakub, T., Bohuslava, T.: Edible films from carrageenan/orange essential oil/trehalose—structure, optical properties, and antimicrobial activity. Polymers **13**(3), 332 (2021)
145. Cazón, P., Vázquez, M., Velazquez, G.: Cellulose-glycerol-polyvinyl alcohol composite films for food packaging: evaluation of water adsorption, mechanical properties, light-barrier properties and transparency. Carbohyd. Polym. **195**, 432–443 (2018)

146. Ghaani, M., Cozzolino, C.A., Castelli, G., Farris, S.: An overview of the intelligent packaging technologies in the food sector. Trends Food Sci. Technol. **51**, 1–11 (2016)
147. Yi, F., Chen, X., Hou, F., Song, L., Zhan, S., Wang, X., Zhang, R., Yang, Q., Wang, X., Liu, Z.: Chitosan/zein-based sustained-release composite films: fabrication, physicochemical properties and release kinetics of tea polyphenols from polymer matrix. Int. J. Biol. Macromol. 131970 (2024)
148. Giteru, S.G., Ali, M.A., Oey, I.: Solvent strength and biopolymer blending effects on physicochemical properties of zein-chitosan-polyvinyl alcohol composite films. Food Hydrocoll. **87**, 270–286 (2019)
149. Vahedikia, N., Garavand, F., Tajeddin, B., Cacciotti, I., Jafari, S.M., Omidi, T., Zahedi, Z.: Biodegradable zein film composites reinforced with chitosan nanoparticles and cinnamon essential oil: physical, mechanical, structural and antimicrobial attributes. Colloids Surf. B **177**, 25–32 (2019)
150. Sun, Y., Liu, Z., Zhang, L., Wang, X., Li, L.: Effects of plasticizer type and concentration on rheological, physico-mechanical and structural properties of chitosan/zein film. Int. J. Biol. Macromol. **143**, 334–340 (2020)
151. Zhu, J., Birgisson, B., Kringos, N.: Polymer modification of bitumen: advances and challenges. Eur. Polymer J. **54**, 18–38 (2014)
152. Teklehaimanot, W.H., Ray, S.S., Emmambux, M.N.: Characterization of pre-gelatinized maize starch-zein blend films produced at alkaline pH. J. Cereal Sci. **95**, 103083 (2020)
153. Wu, H., Richards, M.P., Undeland, I.: Lipid oxidation and antioxidant delivery systems in muscle food. Compr. Rev. Food Sci. Food Saf. **21**(2), 1275–1299 (2022)
154. Loganathan, V., Akbar, I., Ahmed, M.Z., Kazmi, S., Raman, G.: Spectroscopic studies on the antioxidant and anti-tyrosinase activities of anthraquinone derivatives. J. King Saud Univ.-Sci. **35**(10), 102971 (2023)
155. Sani, M.A., Azizi-Lalabadi, M., Tavassoli, M., Mohammadi, K., McClements, D.J.: Recent advances in the development of smart and active biodegradable packaging materials. Nanomaterials **11**(5), 1331 (2021)
156. Masek, A., Chrzescijanska, E., Latos, M., Zaborski, M., Podsędek, A.: Antioxidant and anti-radical properties of green tea extract compounds. Int. J. Electrochem. Sci. **12**(7), 6600–6610 (2017)
157. Rasid, N., Nazmi, N., Isa, M., Sarbon, N.: Rheological, functional and antioxidant properties of films forming solution and active gelatin films incorporated with Centella asiatica (L.) urban extract. Food Packag. Shelf Life **18**, 115–124 (2018)
158. Deshmukh, R.K., Gaikwad, K.K.: Natural antimicrobial and antioxidant compounds for active food packaging applications. Biomass Convers. Biorefinery **14**(4), 4419–4440 (2024)
159. Bouarab Chibane, L., Degraeve, P., Ferhout, H., Bouajila, J., Oulahal, N.: Plant antimicrobial polyphenols as potential natural food preservatives. J. Sci. Food Agric. **99**(4), 1457–1474 (2019)
160. Kumar, A., Sinha-Ray, S.: A review on biopolymer-based fibers via electrospinning and solution blowing and their applications. Fibers **6**(3), 45 (2018)
161. Deng, H., Zhu, J., Tong, Y., Kong, Y., Tan, C., Wang, M., Wan, M., Meng, X.: Antibacterial characteristics and mechanisms of action of Aronia melanocarpa anthocyanins against Escherichia coli. Lwt **150**, 112018 (2021)
162. Motelica, L., Ficai, D., Oprea, O.C., Ficai, A., Andronescu, E.: Smart food packaging designed by nanotechnological and drug delivery approaches. Coatings **10**(9), 806 (2020)
163. Gupta, P.: Role of oxygen absorbers in food as packaging material, their characterization and applications. J. Food Sci. Technol. **61**(2), 242–252 (2024)
164. Kumar, S., Abedin, M.M., Singh, A.K., Das, S.: Role of phenolic compounds in plant-defensive mechanisms. Plant Phenolics Sustain. Agric. **1**, 517–532 (2020)
165. Sharma, P., Gaur, V.K., Sirohi, R., Varjani, S., Kim, S.H., Wong, J.W.: Sustainable processing of food waste for production of bio-based products for circular bioeconomy. Biores. Technol. **325**, 124684 (2021)

166. Duan, X., Sun, H., Ao, Z., Zhou, L., Wang, G., Wang, S.: Unveiling the active sites of graphene-catalyzed peroxymonosulfate activation. Carbon **107**, 371–378 (2016)
167. Chudasama, M.H.: Ethylene Scavengers in Food Packaging Technology: A Review
168. Kalpana, S., Priyadarshini, S., Leena, M.M., Moses, J., Anandharamakrishnan, C.: Intelligent packaging: trends and applications in food systems. Trends Food Sci. Technol. **93**, 145–157 (2019)
169. Mustafa, F., Andreescu, S.: Chemical and biological sensors for food-quality monitoring and smart packaging. Foods **7**(10), 168 (2018)
170. Roobab, U., Inam-Ur-Raheem, M., Khan, A.W., Arshad, R.N., Zeng, X.-A., Aadil, R.M.: Innovations in high-pressure technologies for the development of clean label dairy products: a review. Food Rev. Intl. **39**(2), 970–991 (2023)
171. Luo, X., Zaitoon, A., Lim, L.T.: A review on colorimetric indicators for monitoring product freshness in intelligent food packaging: indicator dyes, preparation methods, and applications. Compr. Rev. Food Sci. Food Saf. **21**(3), 2489–2519 (2022)
172. Gao, T., Tian, Y., Zhu, Z., Sun, D.-W.: Modelling, responses and applications of time-temperature indicators (TTIs) in monitoring fresh food quality. Trends Food Sci. Technol. **99**, 311–322 (2020)
173. Mohammadian, E., Alizadeh-Sani, M., Jafari, S.M.: Smart monitoring of gas/temperature changes within food packaging based on natural colorants. Compr. Rev. Food Sci. Food Saf. **19**(6), 2885–2931 (2020)
174. Zhang, X., Guo, M., Ismail, B.B., He, Q., Jin, T.Z., Liu, D.: Informative and corrective responsive packaging: advances in farm-to-fork monitoring and remediation of food quality and safety. Compr. Rev. Food Sci. Food Saf. **20**(5), 5258–5282 (2021)
175. Jafarzadeh, S., Salehabadi, A., Nafchi, A.M., Oladzadabbasabadi, N., Jafari, S.M.: Cheese packaging by edible coatings and biodegradable nanocomposites; improvement in shelf life, physicochemical and sensory properties. Trends Food Sci. Technol. **116**, 218–231 (2021)
176. Hawthorne, L.M., Beganović, A., Schwarz, M., Noordanus, A.W., Prem, M., Zapf, L., Scheibel, S., Margreiter, G., Huck, C.W., Bach, K.: Suitability of biodegradable materials in comparison with conventional packaging materials for the storage of fresh pork products over extended shelf-life periods. Foods **9**(12), 1802 (2020)
177. Faustman, C., Suman, S.P., Ramanathan, R.: The eating quality of meat: I color, pp. 363–392. Elsevier, Lawrie's Meat Science (2023)
178. Drago, E., Campardelli, R., Pettinato, M., Perego, P.: Innovations in smart packaging concepts for food: an extensive review. Foods **9**(11), 1628 (2020)
179. Sanches, M.A.R., Camelo-Silva, C., da Silva Carvalho, C., de Mello, J.R., Barroso, N.G., da Silva Barros, E.L., Silva, P.P., Pertuzatti, P.B.: Active packaging with starch, red cabbage extract and sweet whey: characterization and application in meat. LWT **135**, 110275 (2021)
180. Esmaeili, H., Cheraghi, N., Khanjari, A., Rezaeigolestani, M., Basti, A.A., Kamkar, A., Aghaee, E.M.: Incorporation of nanoencapsulated garlic essential oil into edible films: a novel approach for extending shelf life of vacuum-packed sausages. Meat Sci. **166**, 108135 (2020)
181. Liu, J., Wang, H., Guo, M., Li, L., Chen, M., Jiang, S., Li, X., Jiang, S.: Extract from Lycium ruthenicum Murr. Incorporating κ-carrageenan colorimetric film with a wide pH–sensing range for food freshness monitoring. Food Hydrocoll. **94**, 1–10 (2019)
182. Hassoun, A., Çoban, Ö.E.: Essential oils for antimicrobial and antioxidant applications in fish and other seafood products. Trends Food Sci. Technol. **68**, 26–36 (2017)
183. Lima, R.C., Carvalho, A.P.A.D., Vieira, C.P., Moreira, R.V., Conte-Junior, C.A.: Green and healthier alternatives to chemical additives as cheese preservative: natural antimicrobials in active nanopackaging/coatings. Polymers 13(16), 2675 (2021)
184. Irkin, R., Esmer, O.K.: Novel food packaging systems with natural antimicrobial agents. J. Food Sci. Technol. **52**, 6095–6111 (2015)
185. Koosha, M., Hamedi, S.: Intelligent Chitosan/PVA nanocomposite films containing black carrot anthocyanin and bentonite nanoclays with improved mechanical, thermal and antibacterial properties. Prog. Org. Coat. **127**, 338–347 (2019)

186. Chudy, S., Bilska, A., Kowalski, R., Teichert, J.: Colour of Milk and Milk Products in CIE L* a* b* Space (2020)
187. El-Ramady, H.R., Domokos-Szabolcsy, É., Abdalla, N.A., Taha, H.S., Fári, M.: Postharvest management of fruits and vegetables storage. Sustain. Agric. Rev. **15**, 65–152 (2015)
188. Sohail, M., Sun, D.-W., Zhu, Z.: Recent developments in intelligent packaging for enhancing food quality and safety. Crit. Rev. Food Sci. Nutr. **58**(15), 2650–2662 (2018)
189. Kaewklin, P., Siripatrawan, U., Suwanagul, A., Lee, Y.S.: Active packaging from chitosan-titanium dioxide nanocomposite film for prolonging storage life of tomato fruit. Int. J. Biol. Macromol. **112**, 523–529 (2018)
190. Sganzerla, W.G., Ribeiro, C.P.P., Uliana, N.R., Rodrigues, M.B.C., da Rosa, C.G., Ferrareze, J.P., de Lima Veeck, A.P., Nunes, M.R.: Bioactive and pH-sensitive films based on carboxymethyl cellulose and blackberry (Morus nigra L.) anthocyanin-rich extract: a perspective coating material to improve the shelf life of cherry tomato (Solanum lycopersicum L. var. cerasiforme). Biocatal. Agric. Biotechnol. **33**, 101989 (2021)
191. Khalid, S., Hassan, S.A., Javaid, H., Zahid, M., Naeem, M., Bhat, Z.F., Abdi, G., Aadil, R.M.: Factors responsible for spoilage, drawbacks of conventional packaging, and advanced packaging systems for tomatoes. J. Agric. Food Res. (2024) 100962.
192. da Rocha Neto, A.C., Beaudry, R., Maraschin, M., Di Piero, R.M., Almenar, E.: Double-bottom antimicrobial packaging for apple shelf-life extension. Food Chem. 279, 379–388 (2019)
193. Maspul, K., Moga, K.: Biodegradable Coffee Bag Use in Increasing the Coffee Roastery Sustainability in the United Arab Emirates. Academia Letters (2021)
194. Nguyen, A.T., Parker, L., Brennan, L., Lockrey, S.: A consumer definition of eco-friendly packaging. J. Clean. Prod. **252**, 119792 (2020)
195. Hu, Y., Liu, C., Wang, P., Li, G., Wang, A., Cong, Y., Liang, X., Li, W., Zhang, X., Li, N.: Sustainable production of safe plasticizers with bio-based fumarates and 1, 3-dienes. Ind. Eng. Chem. Res. **59**(16), 7367–7374 (2020)
196. Jeong, H., Yuk, J.S., Lee, H., Kang, S., Park, H., Park, S.H., Shin, J.: Lactide-derived ester oligomers for highly compatible poly (lactide) plasticizer produced through an eco-friendly process: renewable resources, biodegradation, enhanced flexibility, and elastomeric performance. Green Chem. **23**(19), 7549–7565 (2021)
197. Bocqué, M., Voirin, C., Lapinte, V., Caillol, S., Robin, J.J.: Petro-based and bio-based plasticizers: chemical structures to plasticizing properties. J. Polym. Sci., Part A: Polym. Chem. **54**(1), 11–33 (2016)
198. Varma, R., Vajaranant, T.S., Burkemper, B., Wu, S., Torres, M., Hsu, C., Choudhury, F., McKean-Cowdin, R.: Visual impairment and blindness in adults in the United States: demographic and geographic variations from 2015 to 2050. JAMA Ophthalmol. **134**(7), 802–809 (2016)
199. Gouel, C., Guimbard, H.: Nutrition transition and the structure of global food demand. Am. J. Agr. Econ. **101**(2), 383–403 (2019)
200. Singh, R.L., Singh, P.K.: Global Environmental Problems, Principles and Applications of Environmental Biotechnology for a Sustainable Future, pp. 13–41 (2017)
201. Evans, D.M., Parsons, R., Jackson, P., Greenwood, S., Ryan, A.: Understanding plastic packaging: the co-evolution of materials and society. Glob. Environ. Chang. **65**, 102166 (2020)
202. Mehmood, T., Nadeem, F., Qamar, S.A., Bilal, M., Iqbal, H.M.: Bioconversion of agro-industrial waste into value-added compounds. Sustain. Bioconversion Waste Value Added Prod. 349–368 (2021)
203. Silva, F.A., Dourado, F., Gama, M., Poças, F.: Nanocellulose bio-based composites for food packaging. Nanomaterials **10**(10), 2041 (2020)
204. Wypij, M., Trzcińska-Wencel, J., Golińska, P., Avila-Quezada, G.D., Ingle, A.P., Rai, M.: The strategic applications of natural polymer nanocomposites in food packaging and agriculture: chances, challenges, and consumers' perception. Front. Chem. **10**, 1106230 (2023)
205. Lizundia, E., Sipponen, M.H., Greca, L.G., Balakshin, M., Tardy, B.L., Rojas, O.J., Puglia, D.: Multifunctional lignin-based nanocomposites and nanohybrids. Green Chem. **23**(18), 6698–6760 (2021)

Chapter 2
Biopolymers in Active Food Packaging: Current Landscape and Future Directions

Anirudh Singh Bhathiwal, Harsh Sharma, Geetanjali Yadav, Neera Raghav, and Anjaneyulu Bendi

Abstract Fourth-generation biopolymers, the most recent advancement in packaging, are becoming increasingly in demand for food packaging because of their effectiveness, biodegradability, and non-contaminating qualities when in contact with food. Biomass, biological monomers, and persistent microbes are direct sources of these biopolymers. The affordability of the ingredients needed as raw materials to produce these biopolymers, some of which come from industrial and agricultural waste, is one of the economic benefits. Plant-based protein biopolymers are notable for their exceptional mechanical properties and gas permeability. The literature on the use of biopolymers made from biomass, such as proteins, lipids, and polysaccharides, in food packaging is covered in this chapter, emphasizing their antibacterial and antioxidant qualities.

Keywords Biopolymers · Biomass · Antibacterial · Antioxidant

2.1 Introduction

Food packaging has become more significant since it prolongs the food's shelf life and preserves its quality when it is moved from one location to another. For the same reason, man-made packaging films are widely employed because they are highly practical, affordable, and exhibit superior oil and liquid barrier qualities. Since these artificial packaging films are mostly generated or synthesized with non-renewable,

A. S. Bhathiwal
Department of Chemistry, MMV, Banaras Hindu University, Varanasi 221005, Uttar Pradesh, India

H. Sharma · G. Yadav · N. Raghav
Department of Chemistry, Kurukshetra University, Kurukshetra 136119, Haryana, India

A. Bendi (✉)
Innovation and Translational Research Hub (iTRH) & Department of Chemistry, Presidency University, Rajanukunte, Itgalpura, Bangalore 560064, Karnataka, India
e-mail: anjaneyulu.bendi@gmail.com

S. ul Islam and M. Shahid (eds.), *Green Materials for Active Food Packaging*, Engineering Materials, https://doi.org/10.1007/978-981-96-0369-5_2

mainly oil ingredients, their high environmental cost is one of the major concerns. Therefore, using biopolymers in packaging is one way to reduce the amount of packaging trash and petroleum derivatives consumed, providing us with an alternative to these harmful materials. Biopolymers, fourth-generation packaging materials, are valued for their zero-food contamination, biodegradability, processing ability, and combinability [1–5]. Biopolymers are mainly derived from biomass (Fig. 2.1), bio-derived monomers, and from microorganisms. The current book chapters focus on using biomass-derived biopolymers for food packaging, including polysaccharides, lipids, and proteins.

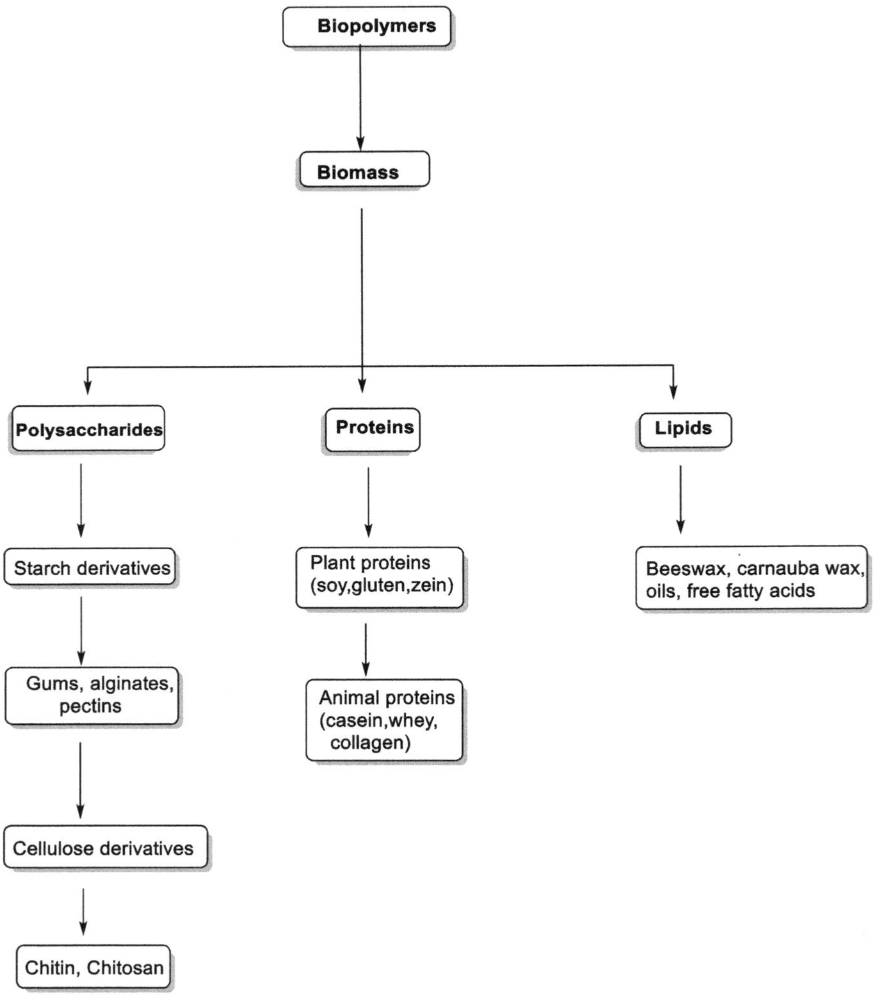

Fig. 2.1 Biopolymers derived from biomass

Numerous polysaccharides, such as kefiran, cellulose, starch, chitosan, pullulan, alginate, pectin, and carrageenan, have shown their ability to create films [6–8]. It is well known that polysaccharide-based biodegradable films effectively block the flow of gases like CO_2 and O_2. Alginate and carrageenan are two polysaccharides that retain moisture despite poor water vapor barrier qualities and can be used as reasonably thick coatings to absorb water from food surfaces and provide momentary protection against more moisture loss. This prevents the product from losing a lot of moisture until the coating dehydrates. Furthermore, the polysaccharide-based films and gas barrier qualities are crucial for preventing solvent leakage or delaying the disappearance of organic vapors, or fragrance compounds, during storage [9–11].

When dissolved in methanol or chloroform, lipids are obtained directly from plants and animals. Lipids are separated by dissolving in non-polar solvents when plant and animal components are pulverized and dissolved. Lipids are the non-polar compounds that are obtained using this approach. They are defined not only by a single structure but also by the method used to isolate them. Nonpolar solvents do not dissolve proteins or hydrocarbons. Lipids are one of the three primary food ingredient types and are the energy source. Due to their concentration in compact areas, carbon atoms on lipid acyl chains were strongly reduced; they store energy even more efficiently than carbohydrates. Energy transfer per mole is enhanced when they get oxidized. The most common lipids found in fats and oils are great energy storers that can be utilized as needed [12, 13].

The best biopolymers are those derived from proteins because of their better gas barrier and mechanical characteristics. Several amino acids, which are connected by amide bonding, make up proteins. Since the α-carbon atom has numerous outside groups linked to it, amino acids can form a wide variety of proteins with varying sizes, shapes, structures, and activities. Proteins are a great choice for creating biopolymer materials because of their unique combinations of chemical, physical, and enzymatic changes. It is easily obtained and possible to employ materials with specified qualities in their place for packaging. Proteins are classified into two sub-categories regarding their origin: animal and plant proteins [2, 14–19].

2.2 Polysaccharide Based Biopolymers

Ghahfarrokhi et al. [20] modified starch solutions produced through photochemical reactions and examined them as biodegradable materials in food packaging. Results showed that when the films were exposed to UV radiation for up to 12 h, starch film's hydrophobicity, moisture sensitivity, and water vapor permeability all decreased. In contrast, the specimen's tensile strength and Young's modulus dropped at the time of exposure to increased UV radiation. Furthermore, the film's tensile energy to break and elongation at break were raised simultaneously. Most of the film's physical characteristics were shown to be influenced by their microstructures in scanning electron microscope images (Fig. 2.2).

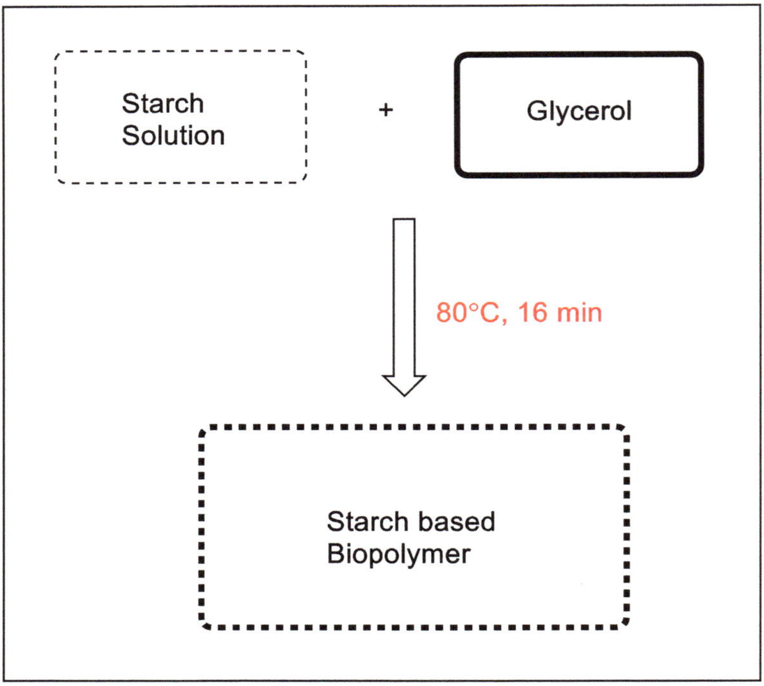

Fig. 2.2 Synthesis of starch-based biopolymers using glycerol

Castro-Yobal et al. [21] analyzed and studied the biofilms produced using alginate, glycerol, oleic acid, and calcium chloride; synthesized films were evaluated as suitable packaging materials. The results showed that the biofilms showed a high modulus of storage (E) and stability at −50 °C. Oleic acid was found as a second plasticizer, and thermal analysis identified several degradation stages due to functional group affinity. FTIR analysis revealed that active sites with carboxylate and carboxy groups for chemical binding, with the alginate polymer's pyranoside ring have the greatest affinity for formulation chemicals interactions (Fig. 2.3).

Dash et al. [22] created edible films with biodegradable starch-pectin (3:1) and titanium dioxide (TiO_2) nanoparticles, which were added at different concentrations. The films were formed by reacting starch from sweet potato and pectin from lemon waste in a 3:1 ratio. Glycerol was used as a plasticizer, and the obtained gelatine solution was later cooled and kept at the same conditions for approximately an hour. The film was then dried in a temperature and humidity control chamber for 24 h. From the results, it could be concluded that the interconnection among TiO_2 nanoparticles and starch-pectin compounds significantly influenced the mechanical and physical properties of the films. The insertion of TiO_2 enhanced the films' mechanical and moisture barrier qualities. The biodegradable films showed the potential to have

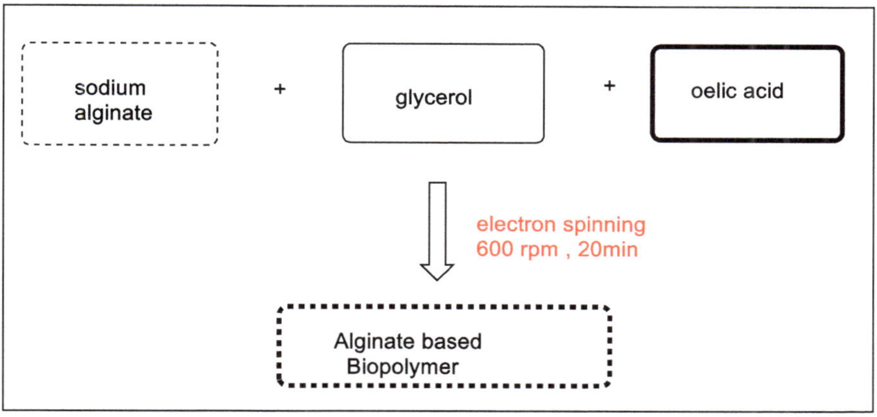

Fig. 2.3 Formation of alginate-based biopolymers

superior properties than traditional artificial plastic packaging and can be used as alternative edible films in the food industry (Fig. 2.4).

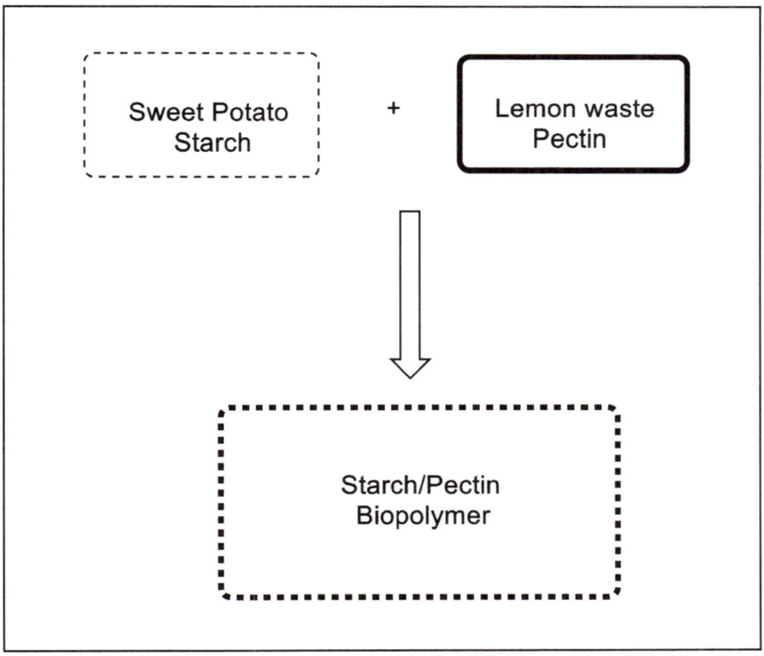

Fig. 2.4 Biopolymer formed by amalgamation of starch from sweet potato with pectin of Lemon waste

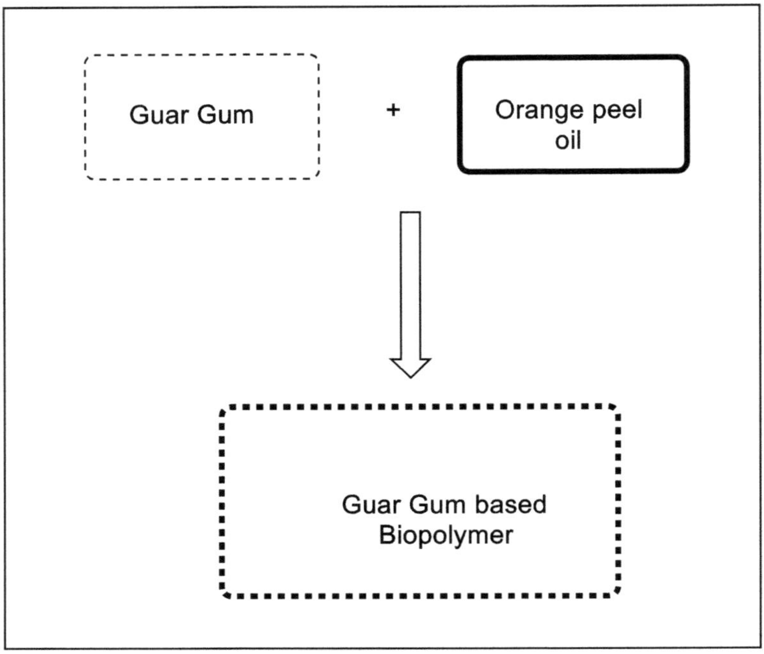

Fig. 2.5 Guar gum-based biopolymer development using orange peel oil

Kirtil et al. [23] noted the effect of Guar gum (GG)–based films for better biopolymers used in food packaging. The results showed that mixing GG and glycerol with orange peel oil, halloysite nanotubes, and sodium tri meta phosphate improved hydrophobicity, water barrier, and mechanical strength while maintaining the film's antibacterial properties. The results also showed that using techniques like crosslinking would help improve the Guar gum (GG) film's physical characteristics (Fig. 2.5).

Kurek et al. [24] investigated the impact of red prickly pears on the antioxidant properties of chitosan and pectin film formulations. The authors studied the four types of red prickly pears, i.e., Peel or fruit pulp, aqueous extracts, and powdered extracts. The results showed that fruit pulp had lower antioxidant qualities than peel extract, whereas the antioxidant power of all films was around 100 mg AAE 100 g^{-1}, with aqueous extracts causing changes in oxygen permeability (OP), solubility in water (WS), amount of water present, thickness, and also in water vapor permeability (WVP). Further, results showed that the reason behind mechanical problems were due to powdered extracts, with rose violet and yellow-orange taints on pectin and chitosan films. The study also noted the potential of cactus agriculture food waste as a functional antioxidant-rich additive. The pomace and peels of red cacti could be used to enhance bio-based polymers for food packaging. Both natural sources

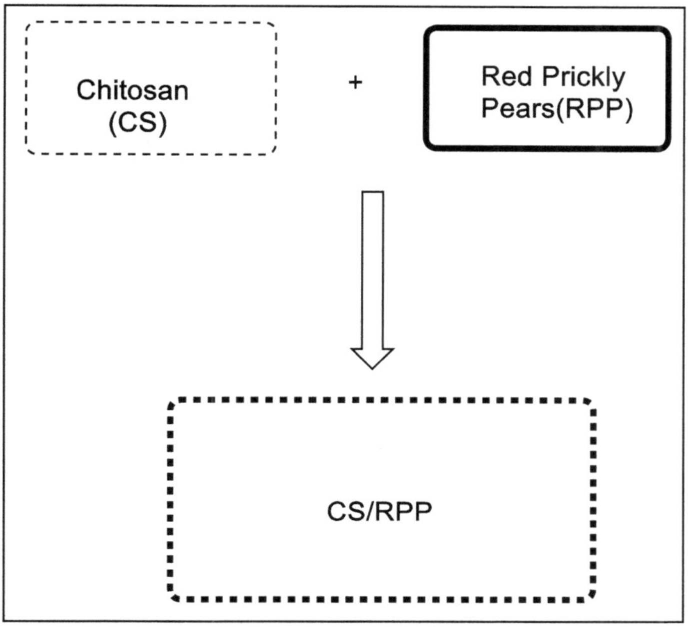

Fig. 2.6 Synthesis of chitosan/red prickly pear biopolymer

could add antioxidants to films, providing a positive outlook for their potential use as antioxidant-active food coatings (Figs. 2.6 and 2.7).

Ma et al. [25] created tara gum (TG)-based films for food packaging by treating them with oleic acid (OA) to increase resistance against the water (WR) of the films. The result depicted the impact of concentration of OA (0, 5, 10, 15, and 20%, w/w, Tara Gum foundation) on TG-based film's color, opacity as well as water vapor permeability (WVP), moisture content along with mechanical characteristics as well as thermal properties. Moreover, Fourier transform infrared (FTIR) spectroscopy and scanning electron microscopy (SEM) were used to study the surface morphologies and chemical structure of the TG/OA films, respectively. The findings indicate no such interaction among the TG and OA in the films and that films with a greater OA concentration showed more pores. When OA was added, the films developed a yellow tint with decreased moisture content and water vapor permeability. It was also shown that their Tensile strength dropped and reached up to 26.8 MPa, whereas the contact angle improved from 76.75° to 107.28° and increased thermal stability (Fig. 2.8).

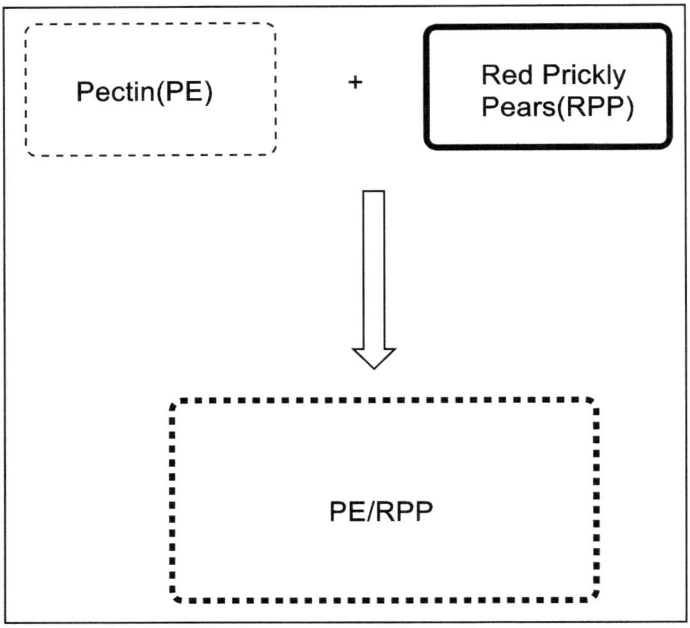

Fig. 2.7 Synthesis of pectin/red prickly pear biopolymer

2.3 Lipid-Based Biopolymers

Haruna et al. [26] studied konjac glucomannan (KGM) based biopolymers amalgamated with carnauba wax. The solvent casting method was used to form composite films of KGM incorporated with different Carnauba Wax (CW) concentration levels, such as 0%, 4%, 8%, 12%, 16%, and 20%. The synthesized films showed improved hydrophobicity, moisture barrier, tensile strength, and mechanical properties as the CW was increased emulsified composite films (KW). The changes observed in the KGM films recorded were almost five times the reduction in solubility, twice the contact angle, and increased strength. Also, there was a significant increase in stretch up to 61% and a drop in the transmission rate of water vapors, i.e., 48%. From the scanning electron microscopy results it demonstrated that when CW concentrations are higher in KW aggregates, cohesiveness, smoothness, and homogeneity all improved (Fig. 2.9).

Filho et al. [27] investigated the use of carnauba wax in emulsion technologies such as microemulsion (ME) and nanoemulsion (NE) (0–15 wt %) in arrowroot starch biopolymers. The results showed that Carnauba wax increased water vapor permeability, moisture, solubility, thermal stability, transparency, and light barrier/absorbing properties. Adding any microemulsion or nanoemulsion affects the mechanical properties along with film elongation. Moreover, nanoemulsion was found to be a suitable procedure for the incorporation of hydrophilic compounds in

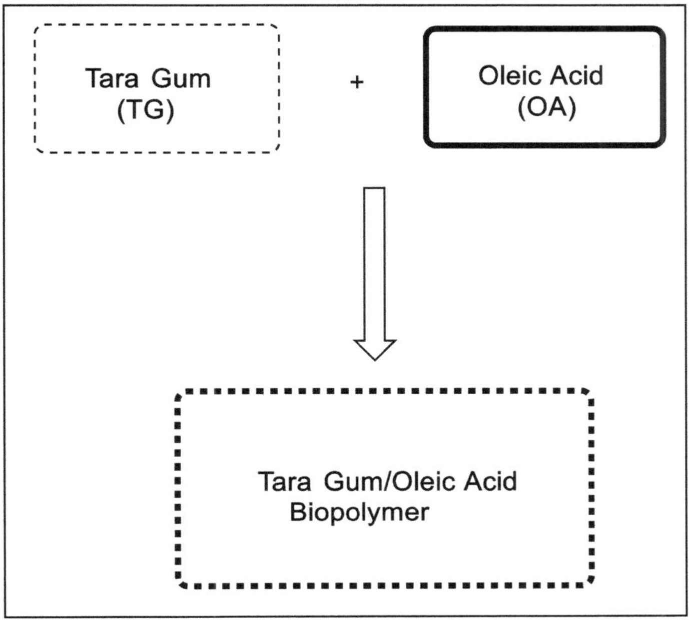

Fig. 2.8 Synthesis of tara gum (TG)-based films for food packaging by treating them with oleic acid (OA)

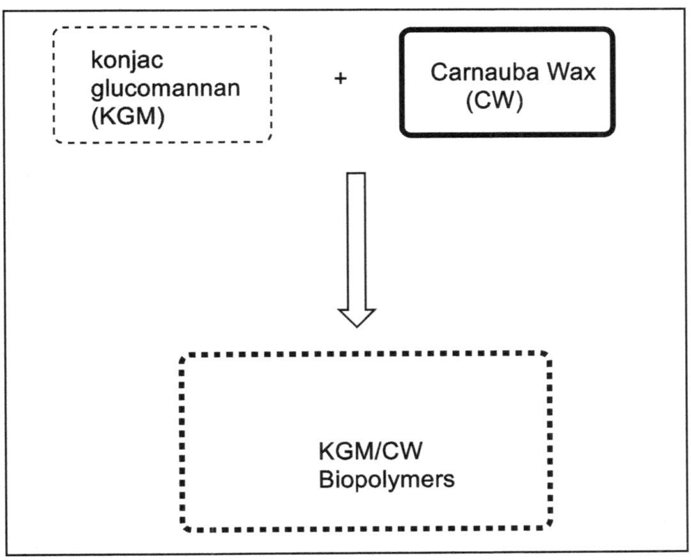

Fig. 2.9 Formation of konjacglucomannan (KGM) based biopolymers amalgamated with carnauba wax

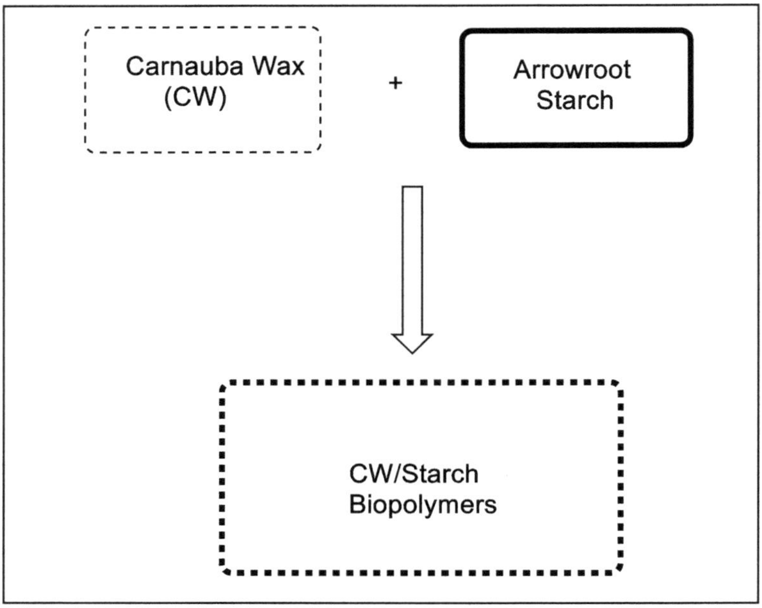

Fig. 2.10 Use of microemulsion (ME) and nanoemulsion (NE) for the addition of the carnauba wax

starch-based films. It was also shown that the transition of the carnauba phase from microemulsion to nanoemulsion proved more effective in reducing the water vapor permeability and opacity (Fig. 2.10).

Bahrami et al. [28] noted the synthesis of trinary film-based bio-composite, i.e., beeswax/Hydroxypropyl methylcellulose/Tragacanth, along with s nanoparticles of silver (AgNPs). The results showed that the tensile strength of the composite ranges from 33.64 to 16.12 MPa due to the presence of silver nanoparticles. In contrast, the inclusion of nanoparticles led to an improvement in water vapor permeability. The films were characterized using SEM, which showed the film's topography had no chemical interaction between the biopolymer and AgNPs. Furthermore, the incorporation of AgNPs led to a reduction in the thermal stability of the film. Also, it was concluded from the results that nano-composite films strongly acted against both gram-positive and gram-negative bacteria, which suggests nano-composites can be actively used as packaging materials for foods. The results of Gram-negative microorganisms show stronger antimicrobial activity than Gram-positive ones (Fig. 2.11).

Cruces et al. [29] prepared biopolymers for food packaging consisting of beeswax/colophony mixtures. Several methods, including XRD, FTIR, DSC, PLM, and water vapor penetration tests, were used to characterize the synthesized films, which showed that interactions between beeswax and colophony were polar, which occurred in the amorphous phase. There was a significant change in the crystalline phase of

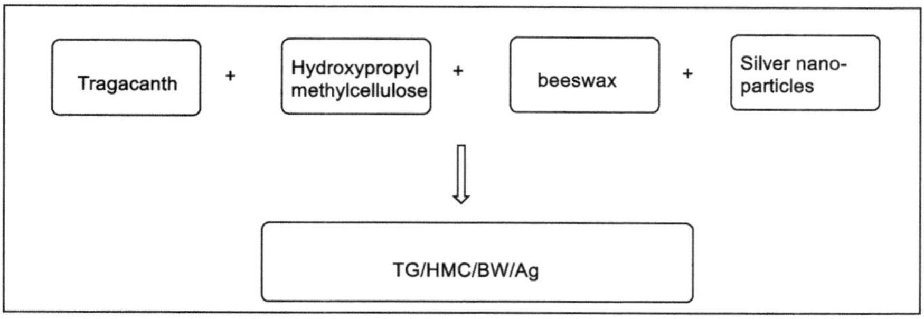

Fig. 2.11 Creating a trinary bio-composite film using silver nanoparticles and tragacanth, hydroxypropyl methylcellulose, and beeswax

beeswax, with structural changes from sponge to needle-like. Lastly, eutectic was formed at BC3 in a 70:30 ratio. The results showed that the pectin layers provided mechanical support. Polar heads of beeswax consisting of ester within the amorphous phase were blocked by polar carboxylic acid groups on colophony, and beeswax hydrophobic tails were crystallized in a structure similar to a needle when colophony was present, which behaved like the eutectic element. The water vapor permeability of the formed film was 56×10^{-13} gm m^{-2} s^{-1} Pa^{-1} (Fig. 2.12).

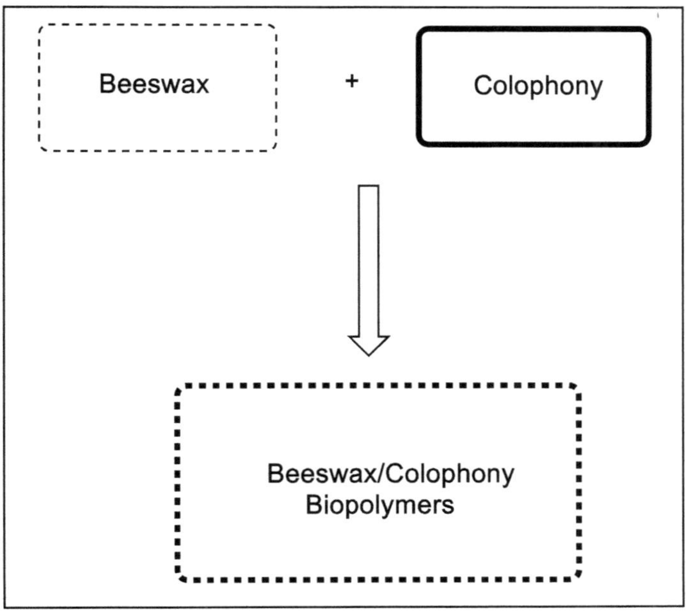

Fig. 2.12 Developing biopolymers for packaging of food consisting of beeswax/colophony

Amin et al. [30] synthesized biopolymers for food packaging using different amounts of chitosan and Aloe Vera. The results showed that Aloe Vera was responsible for the antimicrobial and antioxidant effect in biodegradable films. It also showed that the films with the presence of 20% Aloe Vera showed low thickness, and the dispersion of beeswax was in Chitosan Aloe Vera solution, reaching the concentration of 2.0%, leading to the formation of film with the help of the casting method. The results concluded that with the increase in the concentration of beeswax, the thickness and water vapor permeability also improved. When the concentration of beeswax was kept between 0.5% and 2.0%, the tensile strength of the packaging film also improved with no considerable change in the particle size of beeswax. Also, emulsion was stable at 20–25 °C (Fig. 2.13).

Hendrawati et al. [31] used beeswax to reduce water resistance in the starch polymers, as beeswax is hydrophobic. The films were synthesized at different concentrations, and the beeswax coating duration contained mechanical properties that differed at 0,2,4,6,8 and 10% wt concentrations with time 30, 60, 90, 120, and 150 s. A biodegradable foam was created after baking at 80 °C for one hour and twenty minutes. The results showed that one of the finest biodegradable foams was produced with a 4% (beeswax) coating and 2.5 min. The percentage of water absorbed was close to 2%. The greater result for hardness was 0.09 MPa, and the biodegradability was 73–87% in 28 days (Fig. 2.14).

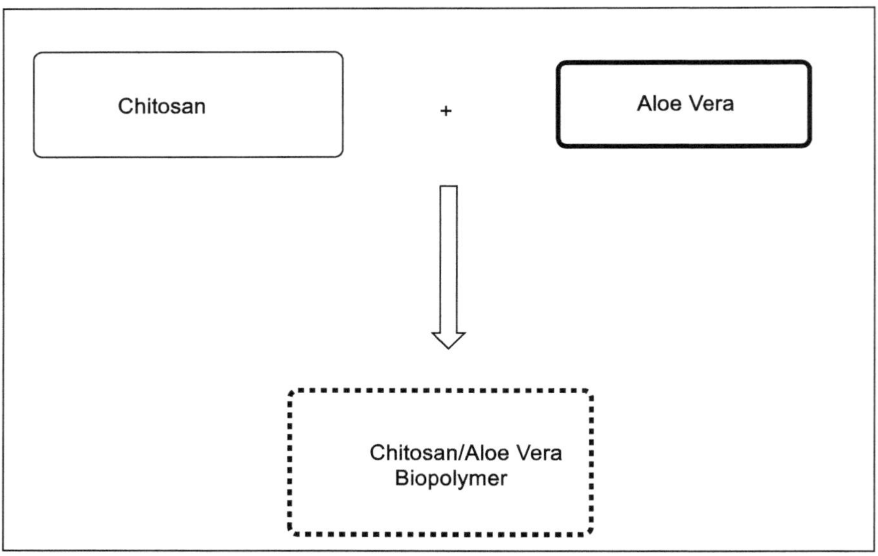

Fig. 2.13 Synthesis of biopolymer using Chitosan and Aloe Vera

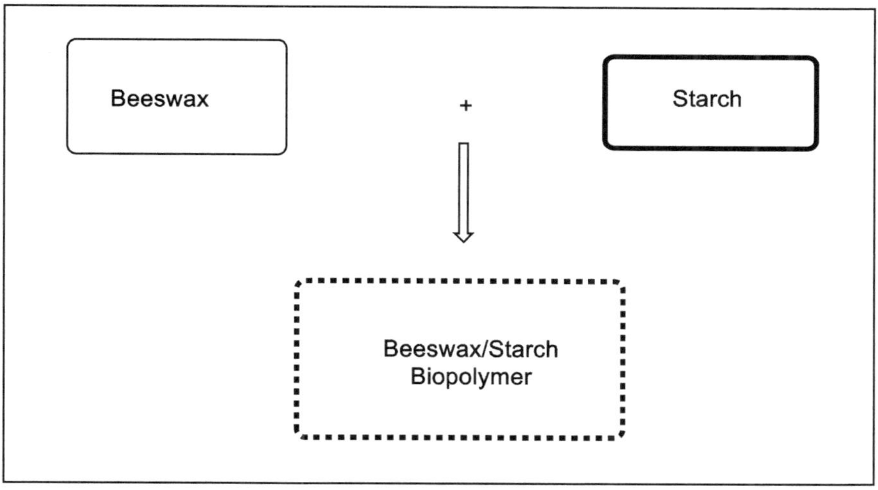

Fig. 2.14 Use of beeswax as a coating to reduce the water resistance properties of starch polymers

2.4 Protein-Based Biopolymers

Su et al. [32] studied the characteristics by combining a man-made polymer, i.e., polyvinyl alcohol, with soy protein isolates (SPI), creating mixed films, either with or without glycerol compatibilization. The blend film was prepared using an aqueous solution of soy protein isolate, with a pH controlled between 6.8 and 10, using NaOH solution. Similar to this, a polyvinyl alcohol-water solution was created. Next, the glycerol, SPI, and PVA solutions were combined to create a fill fabrication resin with a pH of 10.0. The target mix films were created using this resin on a Teflon-coated sheet, drying it for many hours at high temperatures in the oven. PVA weight ratio in blend films led to the formation of a series of blend films named P-0, P-10, P-15, P-20, P-25, and P-30. Upon assessing the parameters of the produced films, it was found that (in the absence of glycerol) the polymer melting temperature (Tm) had increased to a range of 150–175 °C. Compared to the original Soy Protein Isolate (SPI) films, the Tm of those films that were combined with glycerol reduced. The films were also subjected to circular notch tensile tests to determine their tensile strength. The findings demonstrated that, in both cases, those containing and those lacking glycerol, the tensile strength rose as the temperature rose, and it was observed that the sample temperature reached 240 °C represents the maximum tensile strength (Fig. 2.15).

Galus et al. [33] provided a range of soy protein-based films (SPI) combined with starch and maltodextrin to examine how these substances affected the SPI films and what kinds of modifications to the film's characteristics were made possible by their inclusion. After creating the mix films using a water solution of soy protein isolate (SPI), maltodextrin (DE 10.2)/DE 15.6, starch acetate, and acetylated glucose

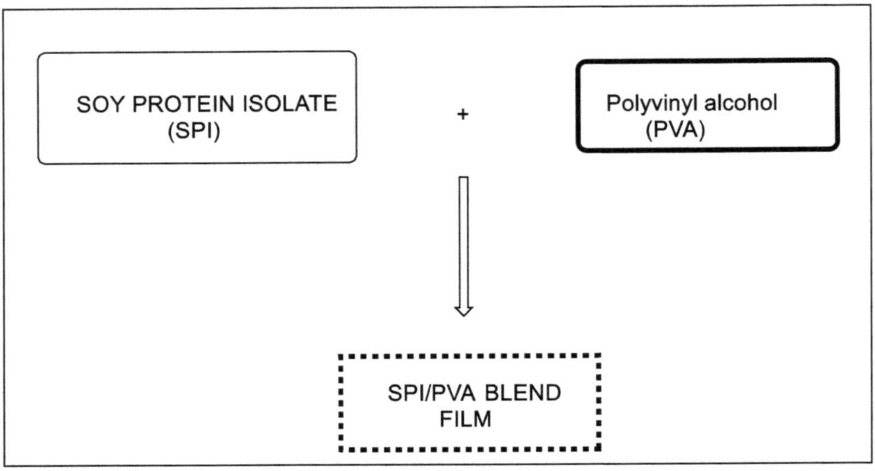

Fig. 2.15 Combining soy protein-containing films with polyvinyl alcohol

phosphate were added. SPI.S1, SPI.S2, SPI.M1, and SPI.M2 were the respective markings on the obtained series of films. The plasticizer was glycerol, and sodium hydroxide solutions were used to keep the film's pH at 10.0. The resulting film temperature was increased to 70 °C, placed on a plate, and dried for a few hours before peeling off. The obtained series of packing films were now subjected to property testing. Comparing modified films to SPI films, the results concluded that adding starch and maltodextrin did not significantly alter the modified films' thickness or surface density. With the Dapp value rising, the data demonstrate a change in the water vapor diffusion coefficient. Additionally, film visibility was improved compared to SPI films by including starch and maltodextrin (Fig. 2.16).

Ciannamea et al. [34] combined soy protein film with red grape extract to influence different functional aspects. Solution casting and compression molding were the two distinct techniques to synthesize the integrated films. When their qualities were examined, it was discovered that red grape extract improved the films' strength, moisture content, and water vapor permeability. Making the films anti-oxidant was one of the primary goals of adding RGE. Screening the films for this characteristic revealed that the anti-oxidant properties differed depending on the processing technique. The antioxidant activity of the films produced by the solution-casting method increased from 24.5 to 43.2%. In contrast, the activity increased far more under the compression molding approach, with values as high as 81.4% (Fig. 2.17).

Zuo et al. [35] created blend films utilizing wheat gluten (WG) and methylcellulose (MC) for food packaging biopolymers with improved mechanical, tensile, and water vapor permeability properties. Casting and compression molding were the techniques used to create the target biopolymer. Wheat gluten and methyl cellulose granules were dissolved throughout the casting process using an aqueous ammonia solution. The

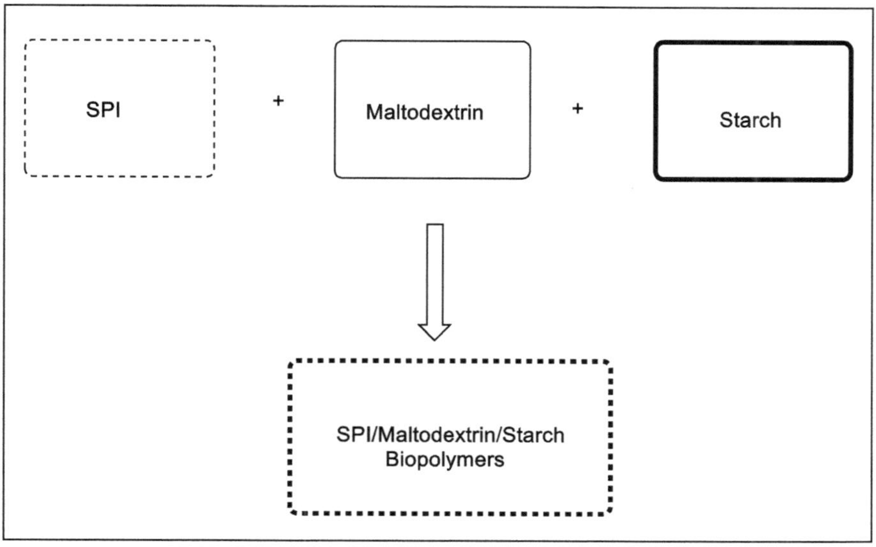

Fig. 2.16 Synthesis of SPI incorporated with starch and maltodextrin films

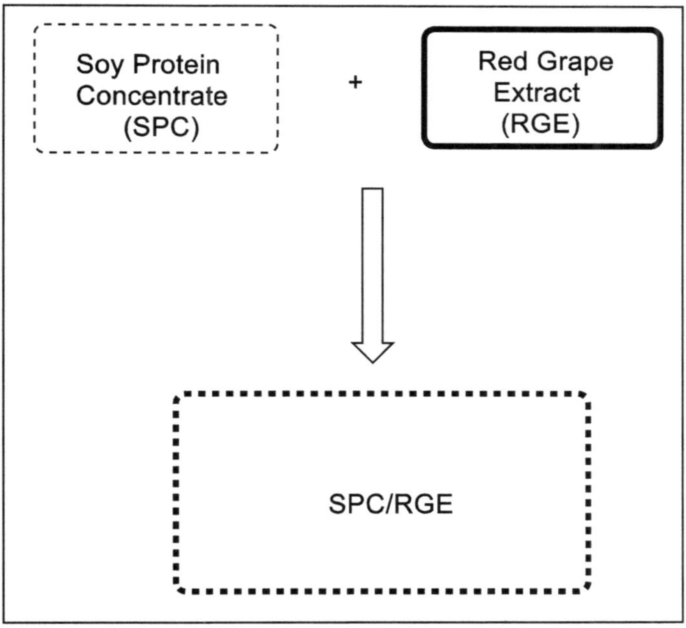

Fig. 2.17 Integration of red grape extract with soy protein film to affect various functional properties

final mixture was swirled with a magnetic stirrer and progressively mixed with glycerol solutions. After pouring the solution onto Petri plates, the mixture was allowed to cool and dry at room temperature before being removed to create clear films. Then, to initiate the process of glycerol evaporating from the mix films, these films were treated at higher temperatures of 100–125 °C. In the compression molding process, glycerol, methylcellulose, and wheat gluten were manually combined using a mortar and placed into a three-rolling mixer set at 25 °C. The obtained product underwent a brief treatment at 125 °C and 15 MPa. A variety of characteristics of the synthetic blend films were screened. The outcome showed that, in terms of mechanical qualities, the films made using the casting approach were significantly better than those made using the compression molded technique. As mc (methylcellulose to total polymer ratio) grew, it was shown that the young modulus, elongation at break, and tensile strength-like characteristics also increased (Fig. 2.18).

El-Wakil et al. [36] created a bio nanocomposite coated on unbleached kraft paper sheets using nanocrystals of cellulose along with wheat gluten (WG) and titanium dioxide nanoparticles. First, sulfuric acids and chloroform were used to create the cellulose nanocrystals through acid hydrolysis. To create the wheat gluten films, ethanol, distilled water, and ammonium hydroxide were combined with wheat gluten at a pH of 10. Then, in different ratios ranging from 2.5 to 12.5%, CNCs and glycerol were added to this wheat gluten solution. Finally, the titanium dioxide nanoparticles were added to this WG-CNC solution to obtain the target films. The resulting bio

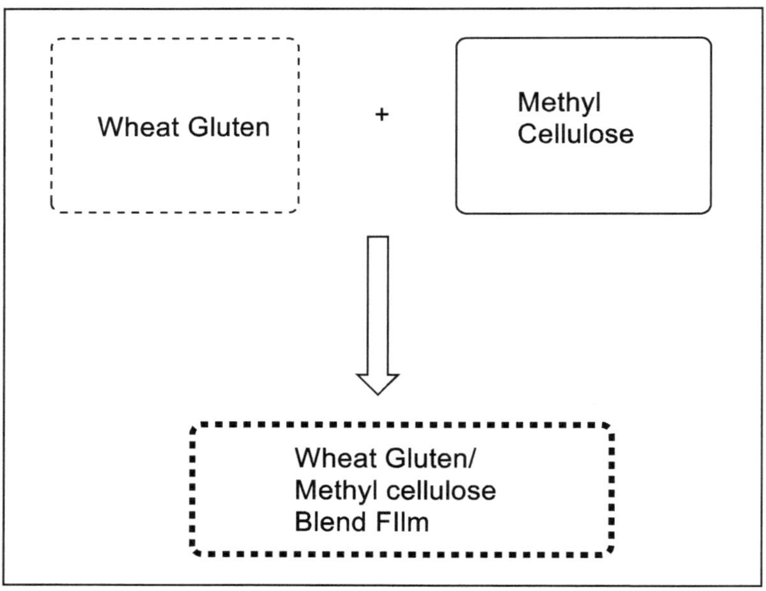

Fig. 2.18 Synthesis of blend films using wheat gluten (WG) and methylcellulose (MC)

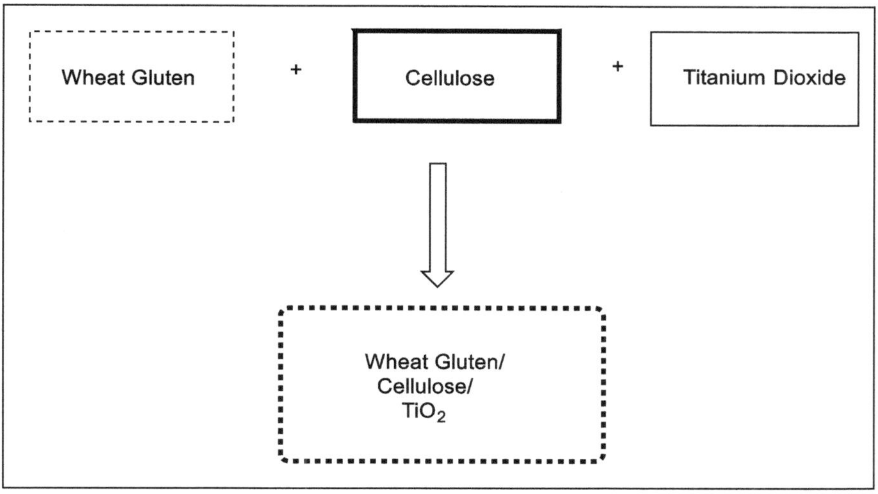

Fig. 2.19 Formation of wheat gluten based films having cellulose and titanium dioxide

nanocomposites were examined for their anti-microbial qualities against *S. cerevisiae, E. coli*, and *S. aureus*, as well as their tensile strength, young modulus, and water sensitivity. The results showed that the bio nanocomposite, which contained 7.5% CNCs and 0.6% TiO_2, demonstrated superior water sensitivity and tensile strength. Additionally, they demonstrated encouraging anti-microbial action against various bacterial strains with values of 100, 100, and 98.5% (Fig. 2.19).

Sanchez-Garcia et al. [37] created a range of blend films by combining mica i.e., clay, carrageenan (CG), and zein (ZN) prolamine. Zein prolamine was dissolved in either ethanol or water to create the film. In a similar manner, a mixer solution was made with 10% glycerol, 15% mica, 10% cellulose, and 1, 5, 10, 20% mica. Both solutions were added and mixed thoroughly for a few minutes and then poured onto Petri dishes. After the synthesized films were examined for various functional characteristics, it was discovered that they were transparent and colored. Additionally, it was reported that those films, which included more carrageenan and zein, had a higher tensile strength. Samples with low levels of nano clays showed a significant decrease in water permeability (Fig. 2.20).

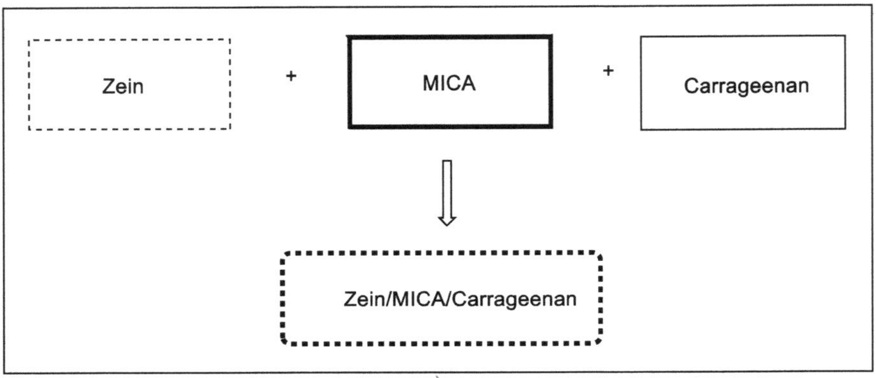

Fig. 2.20 Synthesis of blend films by mixing zein prolamine with carrageenan and mica (clay)

2.5 Conclusion

In food applications, especially packaging, biopolymers are increasingly considered preferable substitutes for synthetic materials. An overview of several biopolymers made from biomass, such as those made of proteins, lipids, and polysaccharides, is given in this chapter. These biopolymers have shown remarkable efficacy in producing packaging films for food items. The research in this area has a wide scope, aiming to develop nanocomposites with appropriate cross-linkages for improving the functional properties of the films, like mechanical strength and water resistance. Both individuals and industries dependent on food are impacted by its packaging. It is essential that the food sector switch from synthetic packaging materials to biopolymer-based alternatives to reduce pollution, as human activities continue to harm our environment. Biopolymers from biomass have proven highly effective in various food applications, particularly in food packaging, providing economical and environmentally beneficial solutions.

Acknowledgements The authors want to convey their appreciation to the management of SGT University, Gurugram, Haryana, India, Presidency University, Bangalore, Karnataka, India, and Kurukshetra University, Kurukshetra, India for providing the resources to finish this chapter on time.

Conflict of Interest No conflict of interest was declared by the authors.

References

1. Zhang, H., Mittal, G.: Biodegradable protein-based films from plant resources: a review. Environ. Prog. Sustain. Energy **29**, 203–220 (2010). https://doi.org/10.1002/ep.10463
2. Siracusa, V., Lotti, N.: Biobased Plastics for Food Packaging. Reference Module in Food Science. Elsevier (2018). https://doi.org/10.1016/B978-0-08-100596-5.22413-X
3. Popović, S.Z., Lazić, V.L., Hromiš, N.M., Šuput, D.Z., Bulut, S.N.: Biopolymer packaging materials for food shelf-life prolongation. In: Biopolymers for Food Design, pp. 223–277. Elsevier Inc. (2018). https://doi.org/10.1016/B978-0-12-811449-0.00008-6
4. Haosagul, S., Boonyawanich, S., Pisutpaisal, N.: Biomethane production from co-fermentation of agricultural wastes. Int. J. Hydrog. Energy **44**, 5355–5364 (2019). https://doi.org/10.1016/j.ijhydene.2018.09.080
5. Tharanathan, R.N.: Biodegradable films and composite coatings: past, present and future. Trends Food Sci. Technol. **14**, 71–78 (2003). https://doi.org/10.1016/S0924-2244(02)00280-7
6. Elsabee, M.Z., Abdou, E.S.: Chitosan based edible films and coatings: a review. Mater. Sci. Eng. C Mater. Biol. Appl. **33**, 1819–1841 (2013). https://doi.org/10.1016/j.msec.2013.01.010
7. Espitia, P.J.P., Du, W.X., de Jesús Avena-Bustillos, R., Soares, N.D.F.F., McHugh, T.H.: Edible films from pectin: physical-mechanical and antimicrobial properties-a review. Food Hydrocoll. **35**, 287–296 (2014). https://doi.org/10.1016/j.foodhyd.2013.06.005
8. Zolfi, M., Khodaiyan, F., Mousavi, M., Hashemi, M.: The improvement of characteristics of biodegradable films made from kefiran-whey protein by nanoparticle incorporation. Carbohydr. Polym. **109**, 118–125 (2014). https://doi.org/10.1016/j.carbpol.2014.03.018
9. Bertuzzi, M.A., Castro Vidaurre, E.F., Armada, M., Gottifredi, J.C.: Water vapor permeability of edible starch based films. J. Food Eng. **80**, 972–978 (2007). https://doi.org/10.1016/j.jfoodeng.2006.07.016
10. Fabra, M.J., Talens, P., Chiralt, A.: Effect of alginate and λ-carrageenan on tensile properties and water vapour permeability of sodium caseinate-lipid based films. Carbohydr. Polym. **74**, 419–426 (2008). https://doi.org/10.1016/j.carbpol.2008.03.010
11. Cazón, P., Velazquez, G., Ramírez, J.A., Vázquez, M.: Polysaccharide-based films and coatings for food packaging: a review. Food Hydrocoll. **68**, 136–148 (2017). https://doi.org/10.1016/j.foodhyd.2016.09.009
12. Burdge, G.C., Calder, P.C.: Introduction to fatty acids and lipids. World Rev. Nutr. Diet. **112**, 1–16 (2015). https://doi.org/10.1159/000365423
13. Ul-Ain, Q., Zia, K.M., Zia, F., Ali, M., Rehman, S., Zuber, M.: Lipid functionalized biopolymers: a review. Int. J. Biol. Macromol. **93**, 1057–1068 (2016). https://doi.org/10.1016/j.ijbiomac.2016.09.071
14. Nagarajan, S., Radhakrishnan, S., Kalkura, S.N., Balme, S., Miele, P., Bechelany, M.: Overview of protein-based biopolymers for biomedical application. Macromol. Chem. Phys. **220**, 1900126 (2019). https://doi.org/10.1002/macp.201900126
15. Bhaskar, R., Zo, S.M., Narayanan, K.B., Purohit, S.D., Gupta, M.K., Han, S.S.: Recent development of protein-based biopolymers in food packaging applications: a review. Polym. Testing **124**, 108097 (2023). https://doi.org/10.1016/j.polymertesting.2023.108097
16. Tirrell, J.G., Tirrell, D.A.: Synthesis of biopolymers: proteins, polyesters, polysaccharides and polynucleotides. Curr. Opin. Solid State Mater. Sci. **1**, 407–411 (1996). https://doi.org/10.1016/S1359-0286(96)80033-7
17. Rebelo, R., Fernandes, M., Fangueiro, R.: Biopolymers in medical implants: a brief review. Procedia Eng. **200**, 236–243 (2017). https://doi.org/10.1016/j.proeng.2017.07.034
18. Du, Y., Li, S., Zhang, Y., Rempel, C., Liu, Q.: Treatments of protein for biopolymer production in view of processability and physical properties: a review. J. Appl. Polym. Sci. **133**, (2016). https://doi.org/10.1002/app.43351
19. Gupta, P., Nayak, K.K.: Characteristics of protein-based biopolymer and its application. Polym. Eng. Sci. **55**, 485–498 (2015). https://doi.org/10.1002/pen.23928

20. Shahabi-Ghahfarrokhi, I., Goudarzi, V., Babaei-Ghazvini, A.: Production of starch based biopolymer by green photochemical reaction at different UV region as a food packaging material: physicochemical characterization. Int. J. Biol. Macromol. **122**, 201–209 (2019). https://doi.org/10.1016/j.ijbiomac.2018.10.154
21. Azucena Castro-Yobal, M., Contreras-Oliva, A., Saucedo-Rivalcoba, V., Rivera-Armenta, J.L., Hernández-Ramírez, G., Salinas-Ruiz, J., Herrera-Corredor, A.: Evaluation of physicochemical properties of film-based alginate for food packing applications. e-Polymers **21**, 82–95 (2021). https://doi.org/10.1515/epoly-2021-0011
22. Dash, K.K., Ali, N.A., Das, D., Mohanta, D.: Thorough evaluation of sweet potato starch and lemon-waste pectin based-edible films with nano-titania inclusions for food packaging applications. Int. J. Biol. Macromol. **139**, 449–458 (2019). https://doi.org/10.1016/j.ijbiomac.2019.07.193
23. Kirtil, E., Aydogdu, A., Svitova, T., Radke, C.J.: Assessment of the performance of several novel approaches to improve physical properties of guar gum-based biopolymer films. Food Packag. Shelf Life **29**, 100687 (2021). https://doi.org/10.1016/j.fpsl.2021.100687
24. Kurek, M., Benbettaieb, N., Ščetar, M., Chaudy, E., Elez-Garofulić, I., Repajić, M., Klepac, D., Valić, S., Debeaufort, F., Galić, K.: Novel functional chitosan and pectin bio-based packaging films with encapsulated opuntia-ficus indica waste. Food Biosci. **41**, 100980 (2021). https://doi.org/10.1016/j.fbio.2021.100980
25. Ma, Q., Hu, D., Wang, H., Wang, L.: Tara gum edible film incorporated with oleic acid. Food Hydrocoll. **56**, 127–133 (2016). https://doi.org/10.1016/j.foodhyd.2015.11.033
26. Haruna, M.H., Wang, Y., Pang, J.: Konjac glucomannan-based composite films fabricated in the presence of carnauba wax emulsion: hydrophobicity, mechanical and microstructural properties evaluation. J. Food Sci. Technol. **56**, 5138–5145 (2019). https://doi.org/10.1007/s13197-019-03932-1
27. de Oliveira Filho, J.G., Albiero, B.R., Cipriano, L., de Oliveira Nobre Bezerra, C.C., Oldoni, F.C.A., Egea, M.B., de Azeredo, H.M.C., Ferreira, M.D.: Arrowroot starch-based films incorporated with a carnauba wax nanoemulsion, cellulose nanocrystals, and essential oils: a new functional material for food packaging applications. Cellulose **28**, 6499–6511 (2021). https://doi.org/10.1007/s10570-021-03945-0
28. Bahrami, A., Rezaei Mokarram, R., Sowti Khiabani, M., Ghanbarzadeh, B., Salehi, R.: Physico-mechanical and antimicrobial properties of tragacanth/hydroxypropyl methylcellulose/beeswax edible films reinforced with silver nanoparticles. Int. J. Biol. Macromol. **129**, 1103–1112 (2019). https://doi.org/10.1016/j.ijbiomac.2018.09.045
29. Cruces, F., García, M.G., Ochoa, N.A.: Reduction of water vapor permeability in food multilayer biopackaging by epitaxial crystallization of beeswax. Food Bioprocess Technol. **14**, 1244–1255 (2021). https://doi.org/10.1007/s11947-021-02628-9
30. Amin, U., Khan, M.A., Akram, M.E., Al-Tawaha, A.R.M.S., Laishevtcev, A., Shariati, M.A.: Characterization of compisote edible films from aloe vera gel, beeswax and chitosan. Potravinarstvo Slovak J. Food Sci. **13**, 854–862 (2019). https://doi.org/10.5219/1177
31. Hendrawati, N., Wibowo, A.A., Chrisnandari, R.D., Adawiyah, R.: Biodegradable foam tray based on sago starch with beeswax as coating agent. IOP Conf. Ser. Mater. Sci. Eng. **1073**, 012006 (2021). https://doi.org/10.1088/1757-899x/1073/1/012006
32. Su, J.F., Wang, X.Y., Huang, Z., Yuan, X.Y., Li, M., Xia, W.L., Feng, C.L.: Heat-sealing properties of soy protein isolate/poly(vinyl alcohol) blend films: Effect of the heat-sealing temperature. J. Appl. Polym. Sci. **115**, 1901–1911 (2010). https://doi.org/10.1002/app.31150
33. Galus, S., Mathieu, H., Lenart, A., Debeaufort, F.: Effect of modified starch or maltodextrin incorporation on the barrier and mechanical properties, moisture sensitivity and appearance of soy protein isolate-based edible films. Innov. Food Sci. Emerg. Technol. **16**, 148–154 (2012). https://doi.org/10.1016/j.ifset.2012.05.012
34. Ciannamea, E.M., Stefani, P.M., Ruseckaite, R.A.: Properties and antioxidant activity of soy protein concentrate films incorporated with red grape extract processed by casting and compression molding. LWT **74**, 353–362 (2016). https://doi.org/10.1016/j.lwt.2016.07.073

35. Zuo, M., Song, Y., Zheng, Q.: Preparation and properties of wheat gluten/methylcellulose binary blend film casting from aqueous ammonia: a comparison with compression molded composites. J. Food Eng. **91**, 415–422 (2009). https://doi.org/10.1016/j.jfoodeng.2008.09.019
36. El-Wakil, N.A., Hassan, E.A., Abou-Zeid, R.E., Dufresne, A.: Development of wheat gluten/nanocellulose/titanium dioxide nanocomposites for active food packaging. Carbohydr. Polym. **124**, 337–346 (2015). https://doi.org/10.1016/j.carbpol.2015.01.076
37. Sanchez-Garcia, M.D., Hilliou, L., Lagaron, J.M.: Nanobiocomposites of carrageenan, zein, and mica of interest in food packaging and coating applications. J. Agric. Food Chem. **58**, 6884–6894 (2010). https://doi.org/10.1021/jf1007659

Chapter 3
Biodegradable Polymers for the Preparation of Active Food Packaging

Nadim Ibn Sayed, Md Imran Howlader, A Z M Mofasser, Md Mahbubul Alam, and Rony Mia

Abstract Food industries have a substantial demand for non-biodegradable petroleum-derived packages due to their easy processability, higher strength, and durability. However, the environment faces severe ecological challenges triggered by polyolefin packages as they do not undergo natural decomposition. Using biopolymers in food packages is a promising way to solve these environmental issues. Recently, investigations have been conducted to impart mechanical, physical, thermal, and barrier properties to natural and synthetic biopolymeric films to be a perfect biodegradable replacement for polyolefin films. This chapter explores the natural and synthetic polymers that can be utilized in biodegradable active food packaging, emphasizing their features and mechanisms of natural decomposition. Here, the biopolymers current and potential applications are also mentioned. It also comprises detailed explanations regarding the methods to incorporate those biodegradable polymers into the food packages.

Keywords Biopolymer · Active packaging · Food industry · Film production · Sustainability · Biodegradability

N. Ibn Sayed · M. I. Howlader · A. Z. M. Mofasser · M. M. Alam
Department of Textile Engineering, European University of Bangladesh, 2/4 Gabtoli, Mirpur 1216, Dhaka, Bangladesh

R. Mia (✉)
Wilson College of Textiles, North Carolina State University, 1020 Main Campus Dr, Raleigh 27606, NC, USA
e-mail: mroni_mia@yahoo.com

Department of Textile Engineering, National Institute of Textile Engineering and Research, University of Dhaka, Dhaka 1000, Bangladesh

S. ul Islam and M. Shahid (eds.), *Green Materials for Active Food Packaging*, Engineering Materials, https://doi.org/10.1007/978-981-96-0369-5_3

3.1 Introduction

Polymers are essential to today's lifestyle because of their flexibility, durability, and ease of manufacturing [1]. In 2019, non-biodegradable plastic manufacturing reached nearly 368 million metric tons worldwide, with a significant portion subsequently ending up in landfills and the ecosystem [2, 3]. Global plastic production projections, including diverse categories such as adhesives, thermosets, thermoplastics, elastomers, PP fibers, coatings, and sealants, anticipate an increase to approximately 445 and 590 million metric tons in 2025 and 2050, respectively [4]. A major portion of non-degradable plastics are used in the food packaging industries, and traditional food packaging materials cause long-lasting pollution, which is hazardous to the ecosystem. This issue underscores the importance of biopolymers in food packaging. Biopolymers are biodegradable alternatives to conventional packaging materials as their ability to be naturally decomposed significantly reduces their environmental impacts [2, 5].

Biopolymers are classified based on their sources. One originates from natural resources, including microorganisms, vegetables, animals, and proteins. This category includes polysaccharides (cellulose, starch, chitosan, and pectin), proteins (Whey protein, and soy protein), lipids, and adhesives. The other category comprises synthetic biopolymers like PHA (polyhydroxyalkanoates), PLA (polylactic acid), polycaprolactone, PBS (polybutylene succinate), PVA (polyvinyl alcohol), PGA (polyglycolic acid), PBT (polybutylene terephthalate) and Polybutylene adipate-co-terephthalate [6]. These biopolymers are employed to fabricate food packaging films that exhibit color, flavor, moisture barrier, antimicrobial, and antioxidant properties [5, 7, 8].

The need for active food packaging arises from the limitations of conventional packaging in addressing crucial challenges related to food quality, safety, and sustainability. While traditional food packaging primarily focuses on physical protection, barrier properties, and containment during transportation and storage, active food packaging transcends these basic functions [9]. By incorporating various active components, active food packaging interacts with the material inside the package to enhance shelf life, safety, and sensory qualities [10].

This chapter discusses the characteristics, chemistry of production, and applications of biopolymers catering to active food packaging. This also includes an extensive discussion on the merits and demerits of packages made of biodegradable polymers. The factors that affect biodegradation and the methods that are applied to examine the biodegradability of a polymeric material have been included. The anticipated trends of and innovations regarding the execution of biopolymers in active food packaging have been demonstrated. In the concluding section, some exclusive ideas have been mentioned that can accelerate the commercialization of many biodegradable polymer-based food packages.

3.2 Origin of Active Food Packaging

Active food packaging has ancient origins, tracing back to early civilizations where people employed rudimentary methods for preserving food. One such method was the wineskin, which, despite its imperfections, aimed to minimize oxygen exposure by collapsing as the wine was consumed. However, its effectiveness as an oxygen barrier remains uncertain. A significant advancement occurred with the use of tin plate for can construction, where tin acted as a sacrificial corrosion barrier, protecting the iron base and preventing food contamination. Tin also served as an autoxidation catalyst in the presence of residual oxygen and contributed to flavor development in canned orange juice. This importance of tin was highlighted with the introduction of aluminum cans, where its absence led to a loss of the traditional flavor [11, 12].

3.3 Types of Active Food Packaging

Active packaging technologies are classified into two primary categories: scavenging systems, often called 'absorbers', and releasing systems, also known as 'emitters'. Active packaging prioritizes preserving volatile constituents and moisture of the food products, hindering microbial growth, oxidation, and controlling respiration rate. Diverse compounds are incorporated into these packaging systems to achieve these functionalities [13]. The types of active food packaging systems, active compounds, and their functions in these packaging systems are demonstrated in Table 3.1 [10, 13–15].

3.4 Biodegradable Polymers

A compelling solution to reduce environmental impact and conserving natural resources is choosing biodegradable polymers over their non-biodegradable counterparts. Biodegradable polymers are derived from natural sources, including those produced by microorganisms and synthetic sources [16]. Biodegradable polymers can be decomposed into substances like water, methane, CO_2, and biomass through the activity of microorganisms (mainly bacteria and fungi) [17]. This decomposition, achieved through anaerobic and aerobic degradation, returns non-toxic elements to the ecosystem. Aerobic and anaerobic degradation of biopolymers refers to the degradation with the presence and absence of oxygen. Enzymes secreted by microorganisms dismantle the polymers into smaller units, and photo-degradation, oxidation, and hydrolysis can also play a vital role during this biodegradation process. Ideally, biodegradable polymers should be durable, non-toxic, and easily decomposable after disposal. The degradation process often begins at the sides of the polymer chain due to the difficulty of breaking down their strong carbon backbones [18].

Table 3.1 Type of active food packaging systems

Active food packaging systems		Active compounds	Functions
Absorber (Active scavenging systems)	Oxygen scavengers	Iron, ascorbic acid, catechole, iron-sulfur, Pd (Palladium), sulfite salt-copper sulfate and Photosensitive dyes	Reduces the impact of oxygen permeation through the packaging material while the product is stored, hinders unsuitable oxidation and aerobic growth microorganisms, prevent oxidative rancidity, controls discoloration occurred by enzymes, inhibits the lipid oxidation and growth of molds
	Moisture scavenger	$CaCl_2$, Bentonite, sorbitol, Polyacrylic acid, sodium salt, $NaCl_2$ etc.	Controls humidity and excess moisture in food packages
	Ethylene absorber	Zeolite-based minerals, Activated charcoal, SiO_2 powder, Nano-ZnO, Nano-TiO_2, Nano-Ag, Potassium Permanganate impregnated Zeolite, bentonite etc.	Reduces ripening and deterioration of foods specially fruits by controlling the concentration of ethylene leading to extending their shelf life
Emitter-active releasing systems	Antioxidant releaser	Citric acid, ascorbic acid, Butylated Hydroxytoluene, Catechin, Butylated Hydroxyanisole, Tertiary Butylhydroquinone, Tocopherol, Quercetin, etc.	Enhances oxidative stability
	Carbon dioxide emitter	CO_2 concentrated with N_2 or O_2 in different MAP (Modified Atmosphere Packaging)	Ensures the maintenance of sensory shelf-life, decreases the loss of drip of meat and fish items in the package
	Antimicrobial packaging systems	Quaternary ammonium salts, sorbates, benzoates, sulfur dioxide, propionates, Ag-zeolite, Chitosan, ozone, ethanol etc.	Hinders the growth of microorganisms, decreases the after harvest decay and storage quality thus enhances the shelf-life of packaged food

3.5 Factors Affecting Biodegradation

The enhanced biodegradability of polymers containing both hydrophobic and hydrophilic regions compared to purely hydrophobic or hydrophilic structures is affected by several key factors, including macromolecular weight, hydrophobicity, crystalline domain size, and enzymatic and chemical actions of microorganisms [18].

For the biomaterials to be considered truly biodegradable, they should fulfill the following criteria [19, 20]:

- The degradation rate should align with the materials' intended function, ensuring complete breakdown within the desired timeframe.
- The materials should not induce persistent responses to inflammation.
- The processability and permeability must be suitable for their designated application.
- Degradation should only generate non-toxic byproducts.

3.6 Biodegradation Testing and Analysis

Various methods are employed to assess the biodegradability of biopolymers. These techniques evaluate mass loss, oxygen consumption, CO_2 formation, microbial proliferation, and structural alterations within the material. While field testing offers a realistic assessment of degradation in natural environments, its high cost and inherent variability lead to limitations. Conversely, laboratory testing provides reproducible results but may not fully capture the complexities of real-world conditions. Therefore, a multi-layered system, including lab screening, faster testing, and virtual field experiments, is considered optimal for comprehensive biodegradation evaluation. Several techniques are used for biodegradation assessment, such as respirometry, spectrometry analysis, physical methods, and thermogravimetric analysis [21].

Respirometry is an approach to quantify the metabolic respiration rates of diverse biological entities. This technique can be strategically applied to investigate biodegradation mechanisms, specifically focusing on the breakdown of polymeric materials by microbial activity [22]. For instance, evolved carbon dioxide to quantify the rate and kinetics of biopolymer mineralization over time can be measured by respirometry technique [21]. Spectroscopy techniques, for example, NMR and FTIR spectroscopy, can exhibit information regarding chemical structure changes of biopolymers such as chain scission that occurs during biodegradation [22]. Thermogravimetric analysis (TGA) is a potent method for evaluating the biodegradability of substances. It quantifies alterations in the mass of a specimen in relation to either temperature or duration within a regulated environment. This data facilitates comprehension of the material's thermal endurance and enables the delineation of its degradation phases [23].

3.7 Standards for Biodegradability and Compostability

The evaluation of biodegradation and compostability in packaging materials made of plastic bio-polymers is guided by established international standards, such as EN 13432 (EU) and ASTM D6400 (USA). These universal standards employ a 4-pronged approach [21]:

- Requires at least 90% of the material's carbon to be converted into CO_2 within 6 months, reflecting complete breakdown by microorganisms.
- Demands over 90% fragmentation into pieces smaller than 2 mm within 3 months, ensuring physical decomposition alongside biodegradation.
- Limits the content of heavy metals to predetermined levels, safeguarding against environmental harm.
- Ensures the product's positive impact, maturity, and grade of the final compost product.

Determining the true compostability of materials requires a comprehensive evaluation across diverse environmental conditions. This necessitates employing a combination of standardized testing protocols. Within this framework, ISO 14855-1 plays a vital role by measuring evolved CO_2 to assess oxygen-based biodegradability under organized composting conditions. Additionally, ISO 14851 offers established methods for evaluating biodegradability in aqueous environments, while ISO 17556 provides a dedicated testing system for assessing biodegradability in land. By incorporating these diverse standards, a more comprehensive picture of a material's biodegradability can be established, solidifying compostability claims [21].

3.8 Natural Biopolymers for Active Food Packaging Films

Natural biopolymers possess numerous advantages. For example, they are biodegradable, contain nutritional food value, are abundant in nature, have minimal cost, and impart antimicrobial and antioxidant properties also, they have no harmful impact on the environment [24].

3.8.1 Polysaccharide Films

Polysaccharide films are mainly carbohydrate-based compounds that are popular in the packaging sectors due to their abundance in nature and slim manufacturing cost. Although polysaccharides exhibit low moisture barrier and brittle properties, their biodegradability and compatibility with other polymers to enhance their properties make them ideal candidates for active food packaging [25]. When polysaccharide films are formed, firstly, polymer segments are broken, and then by creating hydrogen bonding, solvents are evaporated. Several types of polysaccharides (chitosan, carrageenan, pectin, cellulose, and starch) are employed in the food wrapping industry [26].

3.8.1.1 Starch

Starch, an annual renewable crop, is extensively employed as an active food packaging material owing to its easy obtainability and cost-effectiveness. However, hydrophilicity, insufficient tensile strength, and poor moisture barrier limit its application [27]. To overcome these limitations, modification was required, so some researchers suggested plasticization or producing biocomposites where starch is used as a matrix [28]. For several reasons, absorbing moisture and oxidation is mainly responsible for the quality deterioration of food inside a package [29]. To enhance the mechanical and antioxidant properties of starch, TPS (Thermoplastic Starch) or starch matrix-based biocomposites can be used in active food packaging applications where starch is incorporated with other polymers [27].

To bring a thermoplastic character to starch, several types of plasticizers, such as sucrose, maltodextrin, glucose, glycerol, urea, amino acids ethane-1,2-diol, sorbitol, and fructose are used. However, the enhanced properties of starch by applying plasticizers are insufficient to apply as a food packaging material. So, further improvement in mechanical strength and hydrophobicity is required to get satisfactory results, which can be done physically, chemically, or in a dual method. However, physical improvements are preferable to chemical ones because of environmental issues. The researchers suggest various natural and synthetic reinforcing agents to improve the hydrophobicity of thermoplastic starch [27]. To impart hydrophobicity in starch, the hydrophilic group must be replaced by a bulky hydrophobic group, which can be done by esterification, etherification, or oxidation. Table 3.2 shows the suggested reactions or processes for the properties enhancement of starch.

3.8.1.2 Cellulose

Cellulose biopolymer, formed by 1,4-β-D glycosidic linkage of D-Glucose abundant in nature, can be derived from biomass, which costs high energy and chemicals for extraction and purification [46]. Cellulose exhibits some wonderful characteristics, like biodegradability, biocompatibility, and renewability, making it an ideal food packaging material candidate. Due to high crystallinity and linearity, cellulose cannot be fused and is transformed into derivatives using chemicals. Figure 3.1 shows the common derivatives of cellulose [47].

Among numerous derivatives of cellulose, it is reported by several researchers that hydroxypropyl cellulose is a thermoplastic resin that has film-forming capability and can be applied as an edible food packaging material [48–50]. In addition, cellulose acetate can produce fiber or film, which can be combined with other polymers and applied as active food packaging applications. The main reason behind the interest in

Table 3.2 Reaction or processes for the properties enhancement of starch

Source of starch	Process applied	Required chemicals/ conditions	Remarks	References
Barley (*Hordeum sativum*)	Acetylation	Acetic anhydride, NaOH	Starch chain is altered to acetyl groups	[30]
Sugar Palm (*Arenga pinnata*)	Acetylation	Glycerol and sorbitol	Replacement of hydroxyl group of starch to acetyl groups	[31]
Cassava (*Manihot esculenta*)	Acetylation	Starch used in three different forms native cassava starch, octenyl-succinated and acetylated starch	Starch combined with (LDPE) low-density polyethylene to enhance wetting and mechanical properties	[32]
Achira (*Canna indica*)	Acetylation	Poly (lactic acid) polymers	Acetylated starch combined with polylatic acid	[33]
Tapioca (*Manihot esculenta*)	Octenyl succinic anhydride modification	1000–1500 MPa, 135–140 °C, for 9 min	25% Improvement in mechanical properties	[34]
Taro starch (*Colocasia esculenta*)	Octenyl succinic anhydride modification	pH 4.4 and 60 °C	Self-assembly of short chain amylose unbranched	[35]
Wheat starch. (*Amylum Tritici*)	Cross-linking	Sodium tripolyphosphate and sodium trimetaphosphate	Showed better properties than native starch	[36]
Maize starch (*Zea mays*)	Hydroxypropylation and debranching	ZnO-chitosan nanoparticles	Reduction of water vapor permeability and improved strength	[37]
Sweet potato starch (*Ipomoea batatas*)	Oxidation	Sodium hypochlorite	Hydroxyl groups transformed to the carbonyl and carboxyl groups Gel forming capacity improved	[38]
Corn starch (*Zea mays*)	Hydroxypropylation	Propylene oxide	Hydrogen bond breaks from starch and form hydroxy-propylated starch film	[39]

(continued)

Table 3.2 (continued)

Source of starch	Process applied	Required chemicals/ conditions	Remarks	References
Tapioca (*Manihot esculenta*)	Reinforcement	Organic fillers (lignocellulosic fibers, cellulose nanofibers, Polysaccharide-based crystal, Chitosan)	Increase hydophobicity and mechanical strength of thermoplastic starch	[40–43]
Corn Starch (*Zea mays*)	Reinforcement	In organic fillers (chalcogenides, metal oxides, nano-clay, graphene quantum dots and nano-silica)	Enhanced properties found than the original polymer	[44, 45]

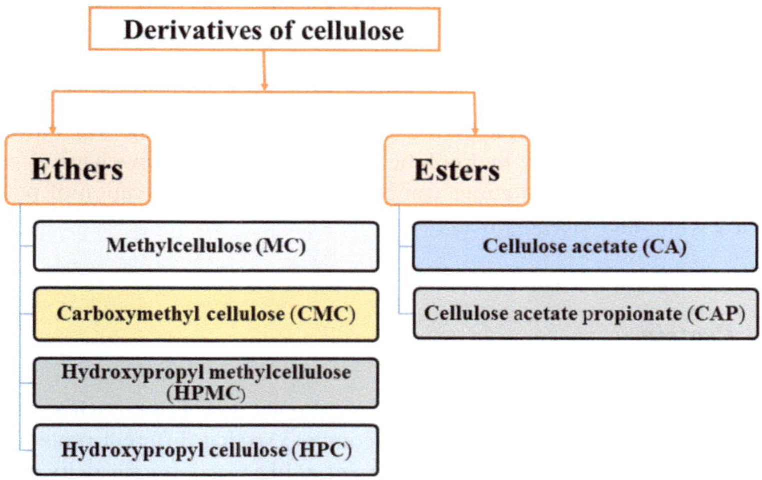

Fig. 3.1 Common derivatives of cellulose

cellulose acetate is its enhanced mechanical and thermal properties. Erceg et al. [51] showed that cellulose acetate/polycaprolactone diol film can be applied to ready-to-eat food packaging.

3.8.1.3 Chitosan

Chitosan can be considered the second most accessible biopolymer following cellulose, which is a derivative of chitin and formed by β-1,4 glycosidic bonds of glucosamine and N-acetylglucosamine units. Chitosan is used in active food packaging and other industrial purpose for its biodegradability, biocompatibility and nontoxicity, film-forming capability, and antibacterial activity [52]. It is produced by utilizing waste material; however, because of high processing costs in comparison to typical plastic limit its application. Researchers reported that chitosan-based packaging material can increase the shelf life of food [53]. As chitosan exhibits film-forming capability, however, its poor mechanical strength and water sensitivity is reported by the researchers [54]. The mechanical and barrier properties of chitosan can be ameliorated by combining it with phenolic compounds, and the antioxidant and antimicrobial properties can be enhanced by mixing with oils, fruit extracts, and phenolic compounds [55].

3.8.1.4 Pectin

Pectin, a plentiful polysaccharide derived from various plants, is largely used in the food packaging industry as a thickener, gelling agent, and colloid stabilizer due to biodegradability and edibility. As protein displays outstanding physical and chemical characteristics (good polymerization, non-toxic), it adapts to fit numerous applications, including active food packaging [56]. It has been focused on making pectin-based films/coating and applying it in active food packaging [57]. Pure pectin is thermally unstable, highly hydrophilic, and has insufficient mechanical properties. Nevertheless, these limitations can be overcome by modification of pectin or integrating with other bio-polymers [56].

3.8.2 Protein

The formation of proteins involves the process of polycondensation, where different arrangements of amino acids repeat units combine so that different proteins show distinct properties. Proteins exhibit some properties like plasticity, elasticity, and film-forming ability, which are important for becoming food packaging materials. In terms of film forming, some plant proteins (cottonseed protein, wheat gluten, soy protein, protein produced from corn zein, sorghum kafrin, pea, rice bran, and peanuts) have taken the attention of the researchers [58–60]. Researchers also depicted that some proteins such as whey protein, collagen, fish myofibrillar protein, gelatin, keratin, casein, and egg white etc. can be extracted from animal sources and they can produce films [25]. In comparison to polysaccharide films, protein films exhibit better O_2 and CO_2 barrier properties [61].

3.8.2.1 Whey Protein

Whey protein, a water-soluble yellowish-green liquid, collected from the dairy industry which is obtained as a by-product during casein extraction in cheese production [62]. It consists of five primary components which include, β-Lg(β-lactoglobulin), α-La(α-lactalbumin), GMP (glycomacropeptide), immunoglobulin (Ig), and BSA (bovine serum albumin) [63, 64]. It can be employed as an edible film which can be applied on food [65]. Whey protein exhibits some characteristics like film-forming efficiency, acid stability, gelation and aeration which lead it for coating material [65].

Combining with other materials researchers have used whey protein in active food packaging material. Andrade et al. [66] have applied seaweed extract with whey protein film for the preservation of chicken breast. The result has shown improved strength, thickness and elastic modulus as well as the active film was able to work as an antioxidant of lipid for 25 days. Incorporating whey protein with chitosan an edible film has been developed by Brink et al. [67] and employed on fresh cut turkey pieces, then tested for antibacterial activity. They found that the film can inhibit bacterial growth for six days on turkey pieces.

3.8.2.2 Soy Protein

Soy proteins, derived from soybean seeds, can be used as food packaging. They consist of twenty different amino acids such as tyrosine, glutamic acid, phenylalanine, leucine, lysine, aspartic, and so on [68]. Soy proteins have grabbed the attention of the researchers due to their low cost, sustainability, functionality and abundance in the environment as well as their degradation easily and compatibility with other polymers [61]. Sufficient mechanical properties, water vapor barrier, low oxygen permeability are the primary factors for food packaging. However, soy protein cannot meet the barrier and mechanical properties without using plasticizers or other polymers [61].

The film produced by soy proteins exhibits excellent gas barrier properties compared to polysaccharides and lipids [69]. Igarzabal et al. [70] have produced a film with soy protein/polylactic acid for active packaging application of food items which showed antibacterial and antifungal effects when treated with natamycin and thymol agents. Yu et al. [71] also developed a soy protein based film incorporated with cellulose nanocrystals which depicted enhanced mechanical properties and antibacterial activity, the authors suggested the film in an active food packaging application.

Besides, whey protein and soy protein some researchers also used gluten and zein protein for protein film formation and finally applied it in active food packaging material [72–75].

3.8.3 Lipid Films and Additives

The source of lipids are diversified, they can be extracted from plants, animals, and insects [76]. Lipids were used as a coating material on fruits but they exhibit poor mechanical properties so composite edible films are prepared by them. Lipids are used in composite film production to reduce the production and processing cost [77].

Incorporating with other film-forming agents, lipids can produce moisture barriers due to low polarity which increases the shelf life of food.

Plasticizersizers and some other ingredients like colorants, antimicrobials, antioxidants, flavors, and nutraceuticals are added with film-forming material for the enhancement of properties [26, 78].

3.9 Synthetic Biopolymers for Active Food Packaging Films

Synthetic polymers such as polyethylene, polypropylene, polyethylene terephthalate, and polysulfides become flexible or shapeable into packaging like plastic containers, bottles, and films after a certain temperature and harden as they cool [79–81]. Nevertheless, the majority of traditional thermoplastics do not break down naturally. Manipulating biosynthetic enzymes through genetic engineering presents distinct prospects for generating synthetic variations of biopolymers and enables additional polymer customization [82]. Therefore, modifications such as plasticization, blending with nanomaterials, or employing multilayer construction might be required to achieve the required functional characteristics for different food packaging uses in the context of biopolymers [21]. The variety of composition units is influenced by factors such as the carbon source, microorganism type, carbon-to-nitrogen ratio, dilution rate, duration, and temperature [83, 84].

3.9.1 Polylactic Acid (PLA)

Carothers initiated the production of PLA through the ring-opening polymerization of lactic acid back in 1932, with subsequent advancements by DuPont and Ethicon [85]. The synthesis of PLA is being practiced in two approaches such as condensation and ring opening polymerization of lactic acids (Fig. 3.2) among which the latter is more preferable due to better efficiency [86]. Biodegradable PLA can also be obtained from sugarcane, corn, and various other agricultural sources. Findings from multiple experiments indicated that PLA when employed for gas barrier films intended for food or medical packaging, demonstrates comparatively lower resistance to the permeation of oxygen and water vapor unlike traditional non-biodegradable polymer resins [87]. PLA has become increasingly recognized as a highly promising and extensively utilized biopolymer within the realm of food packaging, attributed to its notable features such as superior strength, thermoplastic moldability, transparent nature, and ability to undergo composting [88]. The stereocomplex structure of PLA can be either highly crystalline or amorphous, contingent upon the proportions of L-lactide and D-lactide enantiomers employed [89]. This permits adjustment of its mechanical traits, propensity for interacting with water, and rates of decomposition [90]. Despite its effectiveness in creating a barrier against grease and retaining

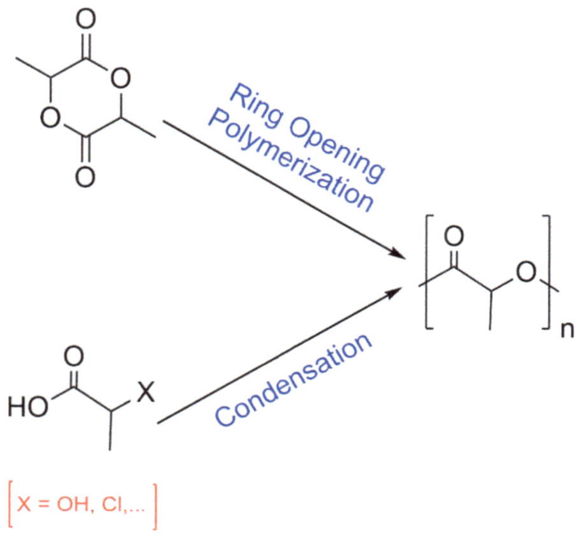

Fig. 3.2 General scheme of PLA synthesis through ring opening and condensation polymerization

aromas, PLA food packaging demonstrates inferior performance in blocking moisture and gas when contrasted with PET and oriented PS. Leading contributors to PLA packaging production include Taghleef Industries [91], NatureWorks LLC [92], and Total Corbion [93].

3.9.2 Polyhydroxyalkanoates (PHAs)

PHAs represent linear polyesters derived from hydroxyalkanoic acids through bacteria-induced fermentation occurring amidst imbalanced bacterial growth environments [94]. Featuring hydrophobicity and microbial resilience, they can match Low-density polyethylene (LDPE) in providing gas barrier functionality [95]. Polyhydroxyalkanoates stand out as the most renowned biopolymer types, particularly characterized by polyhydroxybutyrates (PHBs) and combining hydroxybutyrate (HB) and hydroxyvalerate (HV) copolymers. Modifying the proportion of HB to HV allows for tailoring the film's properties, a significant concentration of polyhydroxybutyrate (PHB) results in the formation of a robust and rigid substance, while polyhydroxyvalerate (PHV) enhances suppleness and durability. Furthermore, the characteristics of PHBV may change plasticization processes, such as the introduction of citrate ester [96]. The application of PHAs in food packaging encompasses the preservation of oils, cheese, and meat through diverse techniques such as injection molding, film extrusion, and coatings [94]. The utilization of PHAs is hindered by their elevated production expenses, proneness to brittleness, and deficiencies in gas barrier attributes [85]. Danimer Scientific [97], TianAn Biopolymers [98], and

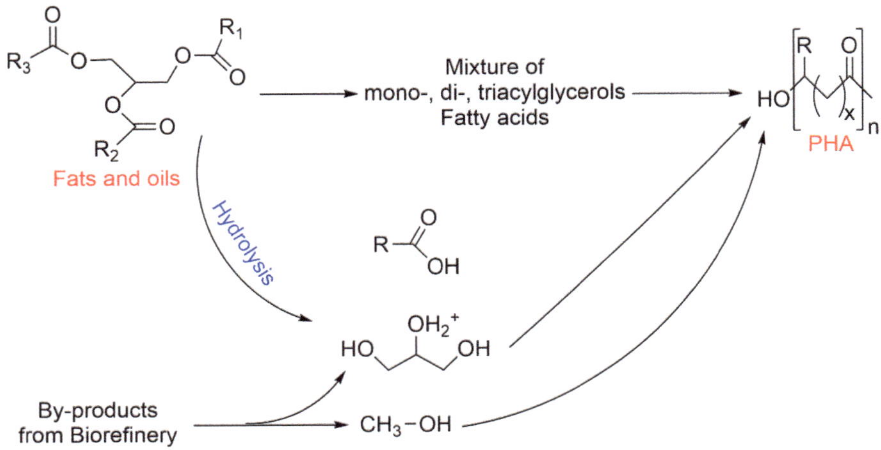

Fig. 3.3 Chemistry of PHA production through fermentation

Newlight Technologies [99] are among the entities involved in the manufacturing of commercial PHAs (Fig. 3.3) [100].

3.9.3 Polybutylene Succinate (PBS)

Polybutylene succinate is a biodegradable aliphatic polyesters derived from petroleum, appropriate for the use in food packaging under high-temperature conditions. Its degradation rate per month was quantified at weight loss of 13 mg per cm^2, which makes it a promising candidate for such applications [101]. Despite its appeal, PBS carries a higher production cost in comparison to conventional petroleum-based polymers like PA (polyamide), PE (polyethylene), polystyrene and PET (polyethylene terephthalate) [102]. Remarkably, the oxygen and water vapor permeability of polybutylene succinate (PBS) exhibit values comparable to those of polylactic acid (PLA). Notably, for semi-crystalline PBS with a crystallinity level of approximately 35%, the oxygen barrier property falls within the range of 200–300 cc mil/ m^2-day-atm [103]. Polybutylene succinate can be manufactured through a two-step polycondensation involving succinic acid and butanediol (Fig. 3.4) [104]. Initially, esterification occurs between succinic acid and the 1,4-butanediol, resulting in the formation of oligomers with water as a byproduct. Subsequently, these oligomers undergo polycondensation at elevated temperatures, leading to the formation of high-molecular-weight PBS [105].

Fig. 3.4 Chemistry of PBS synthesis

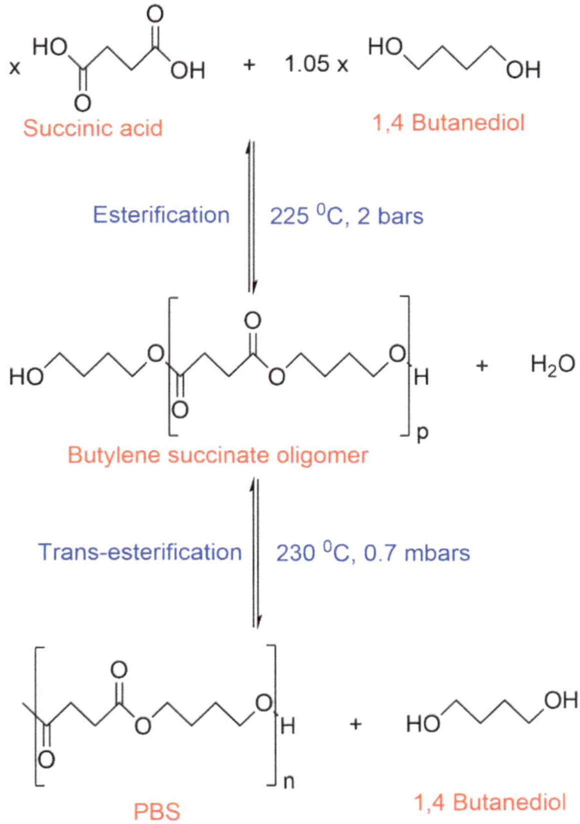

3.9.4 *Polyvinyl Alcohol*

In 1924, polyvinyl alcohol (PVA), a hydrophilic polymer, was first synthesized through the hydrolysis of polyvinyl acetate in ethanol, using potassium hydroxide as a catalyst [106]. The hydrolysis of acetate groups involves an interchange of the ester with methanol under the influence of aqueous sodium hydroxide or anhydrous sodium methylate (Fig. 3.5) [107]. PVA films exhibit enhanced adhesive strength and mechanical and barrier properties when they have high molecular weight and degree of hydrolysis [108]. Adding nanofillers like chitin nanowhiskers, graphene oxide, halloysite, nanolignin, nanocellulose, TEMPO-oxidized cellulose, and zinc sulfide improves the thermal and barrier properties of PVA film. To reduce production expenses and enhance biodegradability, PVA can be blended with several biopolymers like carboxymethyl cellulose, chitosan, gelatin, starch, xylan, etc. Antioxidant, antimicrobial, UV resistance and transparency properties can be attributed to PVA-based functional nanocomposite films by adding various functional materials [109].

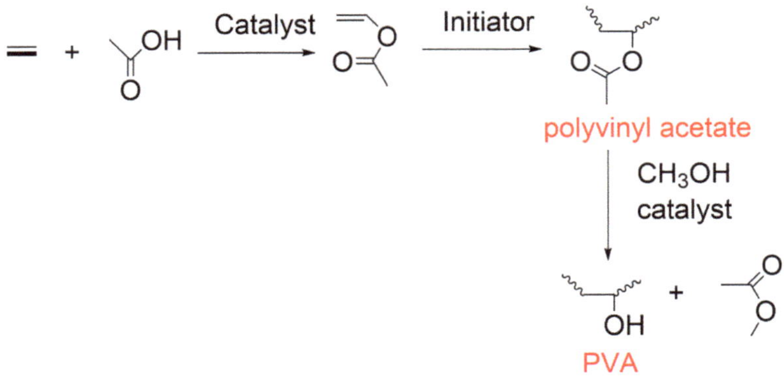

Fig. 3.5 Chemistry of the industrial PVA production

Silver, zinc oxide, and copper oxide nanoparticles are employed to impart antimicrobial properties to PVA or PVA/biopolymer composite films [110]. Associated with natural products (especially essential oils), PVA films exhibit activity against numerous microorganisms and hence lengthen the storing time of foodstuffs [111].

3.9.5 *Polyglycolic Acid (PGA)*

Among the linear aliphatic polyesters, polyglycolide (PGA) is the simplest one that is produced either by ring-opening polymerization of glycolide or by catalytic polycondensation of linear glycolic acid, as shown in Fig. 3.6 [112].

Owing to the high degree of crystallinity (45–55%) of PGA, it possesses exceptional moisture, carbon dioxide, and oxygen barrier properties. The permeability of PGA for carbon dioxide is roughly 1000 times lower than that of PLA. Consequently, in the multilayer bottles containing carbonated drinks, polyglycolide is used as the protective layer [113, 114]. Polyglycolide possesses superior mechanical properties due to the planar zig-zag conformation of its crystalline structure [115].

The biodegradation of PGA occurs in two stages such as hydrolysis and breaking down of ester linkages. Initially, the hydrolyzable backbone of PGA reacted with water and hydrolyzed, and the ester linkages in the polymer break down. Due to the random breakdown of ester linkages, new carboxylic acid groups appear, and thus, the pH value decreases, which further triggers the degradation process [116]. The biodegradation rate of PGA (roughly 1.5–3 months) is higher compared to that of PLA [117].

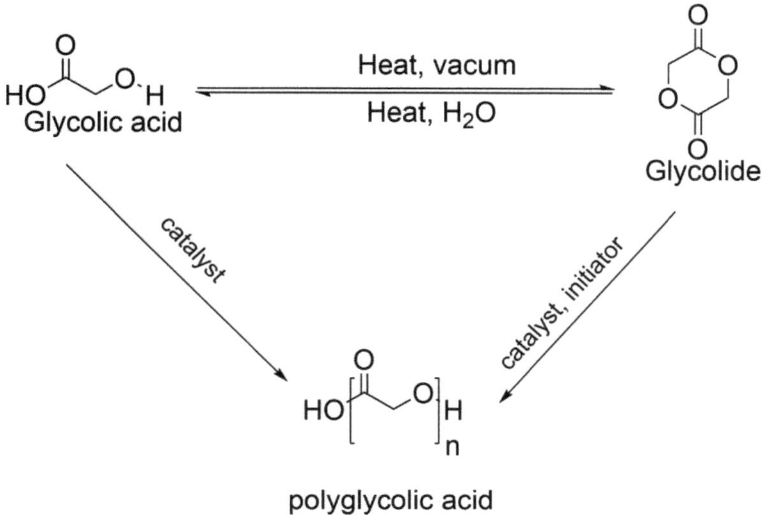

Fig. 3.6 PGA synthesis from glycolic acid and glycolide

3.9.6 *Poly (ε-Caprolactone)*

Poly (ε-caprolactone) or PCL is an aliphatic polyester having superior toughness because of its semi-crystalline structure characterized by amorphous domains in its rubbery state [118]. It is produced by the polymerization of ε-caprolactone (Fig. 3.7). It enhances the dyeability, adhesion property and resistance to crack under stress, being blended with other polymers like PLA and cellulose derivatives [119]. PCL/PLA blend is less brittle and more thermally stable than PLA film [120].

PCL decomposes by the influence of aerobic and/or anaerobic microbes and produces carbon dioxide, water, methane, inorganic salts and biomasses [121]. The hydrolysis of PCL occurs in the same way as that of PGA, however, the difference lies in their rate of biodegradability. As PCL exhibits hydrophobic methylene groups, the rate of degradation of PCL is substantially lower than that of PGA [122]. PCL sees implementation in the production of bottles and bags [123]. It is a good carrier of active materials that enhances the functional properties of active food containers [123]. In the food packaging industry, PLC is blended with a less pricey material like starch to minimize the cost of producing biodegradable composite films [124].

3.9.7 *Polytrimethylene Terephthalate (PTT)*

Polytrimethylene terephthalate (PTT), a linear thermoplastic aromatic polyester, is derived from condensation polymerization of terephthalic acid or its dimethyl ester

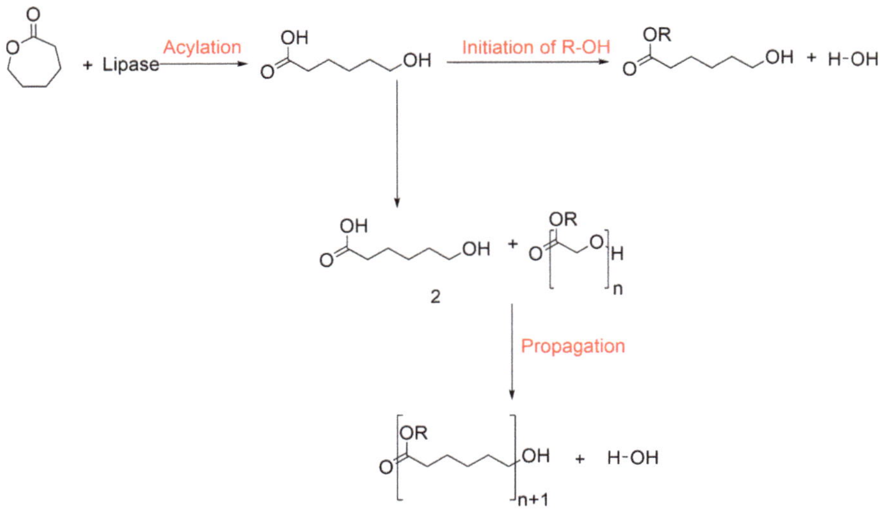

Fig. 3.7 PCL synthesis by ring-opening polymerization

with 1,3-propanediol. The 1,3-propanediol employed in the reaction is biobased material extracted from the wastes of plants [114]. Figure 3.8 shows the reaction pathways of PTT. It has become popular due to its biodegradability and comparable mechanical strength with polybutylene terephthalate (PBT) and PET [125].

Several researchers reported numerous implementations of PTT in food packaging [126, 127] because PTT exhibits outstanding chemical and physical properties, moisture barrier properties, renewability, and processability [125].

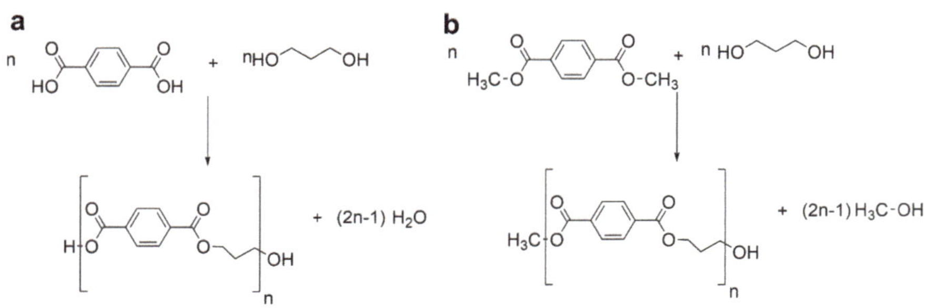

Fig. 3.8 Schematic presentation of synthesis of PPT through esterification (**a**) and transesterification (**b**)

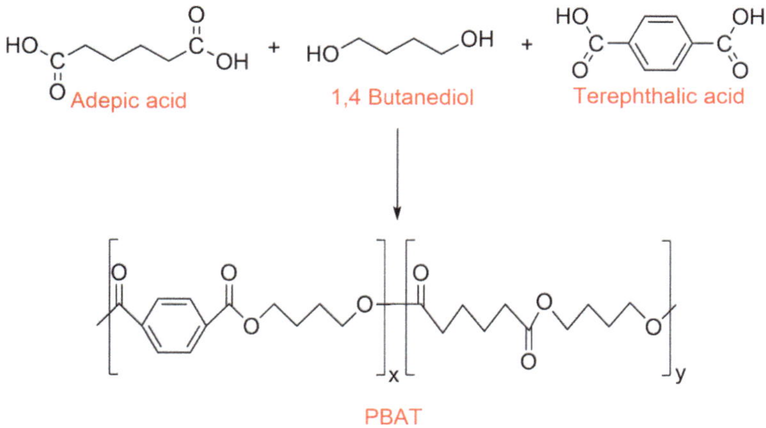

Fig. 3.9 Schematic presentation of PBAT synthesis

3.9.8 Polybutylene Adipate-Co-Terephthalate

There are two basic units of Polybutylene adipate-co-terephthalate (PBAT), flexible and rigid units, where 1,4-butanediol and adipic acid monomers and 1,4-butanediol and terephthalic acid monomers work as the flexible and rigid parts respectively [128]. PBAT is combined with ferulic acid (FA) and used as a food packaging material to increase its mechanical and antioxidant properties. It is reported that PBAT incorporated with poly (lactide) (PLA) has a better UV barrier, mechanical properties and antibacterial activity [129]. Figure 3.9 depicts the PBAT synthesis reaction [130].

Several companies (KINGFA, BIOTECH, Novamont, and BASF) have been producing PBAT-based packaging material. KINGF, a giant company that produces biodegradable materials, has developed compostable products with a combination of PBAT, starch PLA, etc. [130]. Starch-PBAT-based shopping bags made by KINGFA are very much popular in China.

3.9.9 Bacterial Cellulose

Bacterial cellulose, a linear polysaccharide, is composed of β-D-glucopyranose monomers joined by β-1,4-glycosidic linkages to form cellobiose. Bacterial cellulose (BC) and vegetable cellulose are comparable but the production of BC is a pure synthesis method, so no additional plant materials like lignin, hemicellulose, or pectin are present. No environmentally hazardous chemicals are used in the production process as well as the purity is also better than other cellulose [131].

Zhu et al. [132] reported a packaging material for sausage composed of BC/ε-polylysine and found the antibacterial activity which increases the storage time as well as concluded that it would extend the shelf life of meat. BC/nisin and BC/postbiotic lactic acid also inhibit bacterial growth in meat and fruits [133, 134].

3.10 Incorporation of Biodegradable Polymers in Active Food Packaging

Biodegradable polymeric packages can be manufactured in diverse approaches. The selection of the production methods depends on the type, shape and expected features of the final product. Film extrusion [135], thermoforming [21], lamination [136], coating [137], biofoam formation [138], Compression molding [139] and Injection molding [140] are the mentionable methods for producing packages using biopolymers.

3.10.1 Film Extrusion

Film extrusion, the most prevalent method in the production of films from biopolymers, refers to the continual formation of films by the extrusion of molten polymers through dies and immediately solidifying the film by cooling [135]. This process involves extruder(s) (Fig. 3.10a) [141], die and some other tools in order to solidify, cut and wind the films immediately after extrusion. The initial stage of the procedure is feeding the granular polymeric materials through a hopper to the revolving screw at the throat. The screw pushes the materials into the extruder barrel that comprises a number of regulated heating zones that provide gradually intensified temperature and pressure during melting. Precisely controlled temperature and pressure is noteworthy as it diminishes the degradation of polymers. Afterwards, the polymer melt is converted into a film of higher thickness compared to the targeted thickness by pushing through the die.

Extrusion process can be applied in manufacturing biodegradable films tailored by blending of two or more polymers such as polyethylene/collagen [142], polylactic acid/polybutylene adipate-co-terephthalate [143], Polylactic acid/butylene succinate [144], thermoplastic starch blends [145] and hydroxypropyl distarch phosphate/polyhydroxyalkanoate [146].

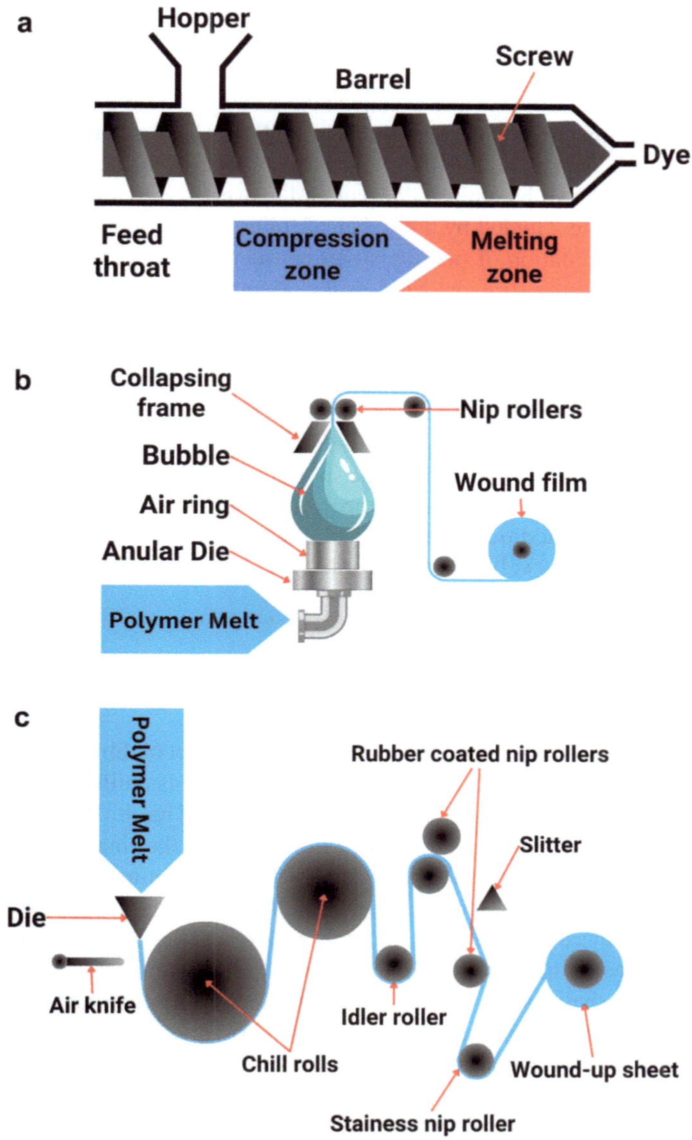

Fig. 3.10 Schematic presentation of extruder (**a**) and material path diagram (from extrusion to final product winding) of Blown film (**b**) and cast film (**c**) extrusion methods

3.10.2 Blown Film Extrusion

Blown film extrusion employs a ring-shaped die through which the polymer melt is vertically (in the upward direction) extruded in a tubular shape as shown in Fig. 3.10b [147]. The tube is then air-inflated to make a bubble which is cooled by an air ring and subsequently drawn by a nip roller pair and converted into a flat film by means of a collapsing frame. Guided by a number of rollers, the film is finally wound up. The lengthwise orientation of polymers can be maintained by tuning the velocity ratio of nip roller pair to extrusion, whereas the orientation around the circumference can be controlled by tuning the ratio of the diameter of the bubble to that of the die. Furthermore, the mechanical and physical features can be customized by setting the speed of the screw and nip rollers, the diameter of the die and pressure of blown air. By concurrent cooling of the exterior and interior of the bubble, optical characteristics can be attributed to the film. The synergistic effect of internal and external cooling can improve productivity as a result of the shortening of time for crystallization. To ensure the uniformity of thickness and temperature, the design of the die has garnered significant interest. In addition, precise regulation of film thickness can be achieved by incorporating airflow valves to the periphery of air rings [147].

3.10.3 Cast Film/Sheet Extrusion

In the cast film (Fig. 3.10c) extrusion process, the thick sheet of thin rectangle-shaped cross-section obtained from the extrusion through slit die is chilled through immediate contact with the smooth surfaces of chill rolls [147]. The chill rolls offer a rapid cooling rate that facilitates stabilization of dimension of the final film. A way for better transfer of heat from the film to the chill rolls, an air knife is incorporated that blows chilled air jets in between Huang et al. [148]. An electrostatic pinning device can also be used there to enhance the contact between the chill roll and the film by generating an electric field in the film [149]. Another way to meet this purpose is to use a vacuum box that ensures the film to chill roll contact by creating vacuum [150]. While increasing the stretching length, a natural cooling action by atmospheric air can be obtained and this action can even be improved by either involving the supply of inert gas/convection air or running the film through a fluid medium [151]. To manipulate polymer orientation, the degree of stretching can be controlled by adjusting the draw ratio, which is the ratio of the surface speed of drawing rollers to the mean speed of the molten sheet as it leaves the die opening. On one hand, stress-induced crystallization occurs with the increase of draw ratio and the length of the stretching zone; on the other hand, thermally induced crystallization can be imparted by escalating the rate of cooling. With the intention of manufacturing packages with substantial functionalities like strength; stiffness; heat sealability, resistance to moisture, resistance to oxygen and resistance to chemicals, multilayer film are produced coextruding and stretching two or more distinct polymer melts. Each layer of the coextruded film exhibits its designated functions [152]. Moreover, the manufacturers can improvise the film by plasma treatment to impart desired features of the end product [153].

Unlike conventional plastics, biopolymers are vulnerable to degradation at high temperature and pressure. Consequently, with a view to establishing proper melting and homogenization, precise adjustments of temperature, screw configuration and dual time is required during the extrusion of biyopolymeric films. An approach to process the biopolymers in an optimized manner is the incorporation of twin screws. The counter-rotating twin screws offer lateral and longitudinal shearing and plasticizing effects whereas the co-rotating twin screws only exert lateral force [154]. The self-wiping mechanism of twin screws ensures proper flow of materials through the extruder. In addition, twin screw facilitates the extrusion process with the discharge gasses from the melt through vent ports [155]. Moreover, application of plasticizers can ensure the feasibility of the biopolymer extrusion process [156].

The films of natural polymers derived from polysaccharides such as cellulose, chitosan, polylactic acid, and starch can be produced by film extrusion process [94]. The best way to produce cellulose film is to follow the twin-screw extrusion method [157]. Cost effective production of thermoplastic starch sheets can be carried out in the film extrusion process involving two stages. In the first stage, an extruder is used to transform the polymer melts into spaghetti form which is subsequently palletized and fed to the second extruder. The polymer pallets are then passed through the second screw and extruded as a film [158]. Pranata et al. [159] developed a food grade thin film from egg white protein by twin-screw extrusion method where the temperature from feed throat to die was raised from 40 to 75 °C. The extrudate was then calendared at 115–120 °C by means of double drum dryers. The film produced by this process resembles the color, clarity, lightness, responsiveness to relative humidity and ethanol resistance of polylactic acid film which is the most popular material for manufacturing biodegradable food packages in the industrial scale.

3.10.4 Reactive Extrusion

Reactive extrusion is an effectual and continuous process that rolls forward by blend melting, homogenization, and some chemical reactions such as polymerization, functionalization, branching and grafting, interchain copolymer formation [155, 160–163]. In this process, multiple reactants are injected in either liquid or gaseous form into the extruder barrel and the by-products are discharged before the polymer is extruded through the die in solidified, quenched or palletized form Raquez et al. [154].

Reactive extrusion is impactful in the economical production of polylactide films. $Sn(Oct)_2$ is used to accelerate the polymerization of lactic acids [164]. Furthermore, addition of triphenylphosphine with $Sn(Oct)_2$ substantially improves polymerization rate of lactic acids and resistance to the transesterification-induced degradation of polymer [165].

3.10.5 Thermoforming

Thermoforming refers to softening of the extrudate by heating, drawing onto a mold by pressurization or vacuuming, and achieving aspired form by cooling sequentially. To avoid rupture, precision in controlling temperature and thickness of the film is of great significance [21]. Thermoforming can be carried out to produce packages from polylactides, starch and aliphatic polyesters [166].

3.10.6 Lamination

Apart from coextrusion, an approach to fabricate multilayer film is to overlay synthetic polymers on a film of natural biopolymer. The multilayer films manufactured in this way possess superior functional properties including strength, appearance, stability and the like [136]. Laminates can be prepared in two methods such as extrusion lamination and adhesive lamination.

3.10.6.1 Extrusion Lamination

After the extrusion of a polymeric material, the extrudate is fed in such a way that it lies in between a primary and a secondary web constructed by polymeric or non-polymeric materials. Subsequently, the layers are passed through chill rolls where pressurization and cooling is done. As a consequence of the solidification of the polymer positioned within the interior of the film, the primary and secondary layers are attached to it. Finally the multilayer films are wound posterior to edge neatening [136].

3.10.6.2 Adhesive Lamination

Compared to extrusion lamination, adhesive lamination is more feasible and a time-saving process. To produce aluminum-infused multilayer films a suitable approach is adhesive lamination which can be carried out in three distinct techniques. 'Dry bond lamination', the first one, allows the incorporation of solvent-based adhesives which are applied on a layer of the film. After the completion of this process, the film undergoes drying so that the solvent can evaporate leaving the adhesive as a dry and gluey layer. The second way for adhesive lamination is the 'solventless lamination technique' where solid adhesives are used and hence the drying process is eliminated. Nevertheless, slight heat is required in order to lower the viscosity of the adhesive for the ease of bonding [167]. 'Wet bond lamination technique', the third approach, involves water-based adhesives (dispersion of protein or starch) for making a single web. Immediately, two single webs are affixed to one another and then they are dewatered by an oven [168].

3.10.7 Coating

In the coating method, a solution capable of forming food-grade and edible or non-edible films is brushed or sprayed right on the food. In some cases, coating is carried out by dipping the food into the solution. Furthermore, coating can also be done by the deposition of a source material on a web [137]. Afterwards, the coating is dried following a suitable approach [169]. Vacuum coating, extrusion coating, solvent-based coating and aqueous dispersion coating are the processes that can be applied in food packaging.

Vacuum coating is a technique principally applied in the manufacturing of metallized films that functions as protective barriers against odor, light, moisture and gas. Here, silicon oxide or aluminum is evaporated by means of strong vacuum and the vapors produced in such a way are placed on the moving web of biopolymers [137]. Extrusion coating refers to the casting of the freshly extruded biopolymer film to a primary web to produce a film consisting of two layers. Paper/polyethylene films are produced in this method which find implementation in making food take-out packages and coffee cups [152]. As a biodegradable replacement of paper/polyethylene, Cheng et al. [170] developed a paper/polylactic acid film using plant or animal-derived gelatin in the extrusion coating method. Solvent-based coating technique is applied with a view to coating a web with a solvent-based material and subsequently oven drying to remove the solvent. In the context of aqueous dispersions, the drying process is carried out in different ways like blowing hot air, ultrasonic or infrared heating [152].

3.10.8 Biofoam Formation

Current world finds an extensive implementation of polymer foams in the production of food packages. Particularly, the biodegradable synthetic polymers like starch, polylactic acid, polyglycolic acid and aliphatic polyester are widely used since their edible or non-edible foams offer better load bearing capacity; digestibility; moisture absorption, organoleptic, sensory or mechanical properties [171, 172]. The process of foam production commences with the generation of wet foam and ends up with the conversion of solid foam through freeze-drying or baking process [138, 173]. To impart desirable properties of the polymer foam, selection of foaming agent, modifying foam structure and tuning the configuration of pores are of importance [174]. Light and stable foams can be generated by using blowing agents such as supercritical carbon dioxide or nitrogen that is responsible for the formation of bubbles in the biopolymer extrudate [175]. Solid foams are used as edible constituents of various foodstuffs like meringue, bread, ice cream and so on [174]. In addition, Foams of polylactic acid, starch and aliphatic polyester are applied in the production of biodegradable cups and trays. Polymer blends are also used in the foam formation process with the intent of lowering the cost of production [21].

3.10.9 Compression and Injection Molding

Compression molding involves an upper mold and a lower mold between which the biopolymer is placed in order to produce the end product of a certain shape by means of pressure and/or temperature. This method is applied in the manufacturing of food- and liquid containers and bottle caps [139]. In the injection molding technique, biopolymers in granular form may be fed to the extruder. Immediately after the extrusion the molten polymers reach a mold cavity and eventually solidify [176]. This technique is a conventional one that is followed to produce complex-shaped products having smooth surfaces [140].

3.11 Environmental Impact and Sustainability of Biopolymer-Based Active Food Packaging

Unlike non-biodegradable petroleum-based polymers, biopolymers undergo chemical or enzymatic degradation and form biomasses, inorganic compounds, water, carbon dioxide and eventually are converted into humus [114]. Polyolefins remain accumulated in the environment for decades and they break down at a lethargic pace to form microplastics which form garbage patches moving through rivers to the oceans [177]. Utilization of biodegradable polymers hold significance in solving the worldwide oil-based plastic waste accumulation problem [178]. The active food packages made of biodegradable polymers are the sustainable replacements of the packages produced from petrochemical products. The kinetics of biodegradability of the packages are dependent on the amount of microorganisms, temperature and humidity of the environment [114].

3.12 Future Trends and Innovations

In the modern age, implementation of biodegradable materials in producing packages has appeared indispensable. However, biodegradable polymer-based packages are still unable to resemble the barrier and mechanical characteristics of the packages made of petroleum resources. Several efforts have been made to upgrade the features of biopolymer-based food packages by blending multiple types of biopolymers so that each individual polymer type can impart a specific property to the packages. Effects of adding various additives in the film extrusion process has been explored with a view to improving some functional properties of the food packages. To elevate the mechanical properties and biodegradation kinetics of biopolymer-based food packages, the grafting technique has been invented. Moreover, the use of biodegradable fiber or plasticizer reinforcement adds a new dimension in inducing extra strength of the final product. Besides, the remarkable rise of oil price motivates researchers to focus on developing alternatives for oil-based polymers for reducing manufacturing cost. Therefore, there is a scope of research on developing low-cost biopolymer-based products that can completely replace the petrochemical polymers [179]. Polylactic acid (PLA), starch, cellulose and aliphatic polyesters are notable references of biodegradable replacement of polyolefins that is being produced currently in large scale at low production cost [180]. Currently, nanoclay/polymer composite films seize the attention of researchers as these have the potential to be applied in the production of biodegradable packages with superior mechanical properties and functionalities [181].

Bio-based polymers are sustainable and they have extraordinary potential in active food packaging but they have some drawbacks regarding mechanical and thermal properties, sensitivity of moisture and the incorporation of active compounds [182]. Further studies can be done to resolve these issues.

To design more novel macromolecules susceptible to microbial breakdown, an enhanced understanding of microbial communities existing in diversified environments and their particular mechanisms of biodegradation is needed. Furthermore, the selection of biodegradable polymers for specific applications demands careful consideration of the key factors that affect their biodegradation rates [178].

3.13 Conclusion

As the utilization of biopolymers in food packaging continues to expand, it is imperative to advance biopolymer-based package production to a larger scale. Biodegradable polymers have only succeeded in replacing a small fraction of conventional plastics. Directing fundamental and applied research to ameliorate physicochemical and thermal attributes of biodegradable polymers, cost-effectiveness and simplicity of their manufacturing process is essential to make them competent for implementation in food packaging. Proper investigation of the interaction of the biopolymers with the food products before incorporating bio-derived packaging materials is of great importance. Moreover, efforts should be made to utilize smart sensors and nanotechnology to preserve the authenticity and to provide pertinent details of the food products. To invent new decomposable biopolymers, obtaining clear concepts of microorganisms inhabiting diverse environments is required.

Declaration The authors declare that they have no known competing financial interests or personal relationships that could have appeared to influence the work reported in this paper. Generative AI and AI-assisted technologies have been used in the writing process to improve the readability and language of the work only. After using this tool, the author(s) reviewed and edited the content as needed and take(s) full responsibility for the content of the publication.

References

1. Mangaraj, S., Yadav, A., Bal, L.M., Dash, S., Mahanti, N.K.: Application of biodegradable polymers in food packaging industry: a comprehensive review. J. Packag. Technol. Res. **3**, 77–96 (2019)
2. Ahari, H., Soufiani, S.P.: Smart and active food packaging: insights in novel food packaging. Front. Microbiol. **12**, 657233 (2021)
3. Zhara, H., Adeel, S., Özomay, Z., Mia, R.: Properties and Performance Relationship of Biopolymers in Textile Industry. In: Biopolymers in the Textile Industry: Opportunities and Limitations, pp. 87–121. Springer (2024)
4. Tiseo, I.: Plastic Production Forecast Worldwide 2025–2050. Statista (2021)
5. Atta, O.M., Manan, S., Shahzad, A., Ul-Islam, M., Ullah, M.W., Yang, G.: Biobased materials for active food packaging: a review. Food Hydrocoll. **125**, 107419 (2022)
6. Salerno, A., Cesarelli, G., Pedram, P., Netti, P.A.: Modular strategies to build cell-free and cell-laden scaffolds towards bioengineered tissues and organs. J. Clin. Med. **8**(11), 1816 (2019)
7. Aider, M.: Chitosan application for active bio-based films production and potential in the food industry. LWT-Food Sci. Technol. **43**(6), 837–842 (2010)
8. Mia, R., Mofasser, A., Bhat, M.A., Howlader, M.I., Sayed, N.I., Bakar, M.A.: Utilizing neem wood waste as a natural dye in the development of bio-based functionalized colored jute fabric. Sustain. Chem. Pharm. **38**, 101489 (2024). https://doi.org/10.1016/j.scp.2024.101489
9. Vermeiren, L., Devlieghere, F., van Beest, M., de Kruijf, N., Debevere, J.: Developments in the active packaging of foods. Trends Food Sci. Technol. **10**(3), 77–86 (1999)
10. Wyrwa, J., Barska, A.: Innovations in the food packaging market: active packaging. Eur. Food Res. Technol. **243**, 1681–1692 (2017)
11. Rooney, M.: Overview of active food packaging. In: Active Food Packaging, pp. 1–37. Springer (1995)
12. Ahmed, T., et al.: New trends in printing applications of natural dyes and pigments. Renew. Dyes Pigment., pp. 139–163, 2024, https://doi.org/10.1016/B978-0-443-15213-9.00003-X.
13. Ozdemir, M., Floros, J.D.: Active food packaging technologies. Crit. Rev. Food Sci. Nutr. **44**(3), 185–193 (2004)
14. Yildirim, S., et al.: Active packaging applications for food. Compr. Rev. Food Sci. Food Saf. **17**(1), 165–199 (2018)
15. Mahapatra, D., Goswami, S., Das, M.: New-age packaging for foods and food products. In: Food Process Engineering and Technology: Safety, Packaging, Nanotechnologies and Human Health, pp. 165–187. Springer (2024)
16. Yadav, R. B.: Biodegradable packaging: recent advances and applications in food industry. In: Food Process Engineering and Technology: Safety, Packaging, Nanotechnologies and Human Health, pp. 189–213 (2024)
17. Ndahebwa Muhonja, C., Magoma, G., Imbuga, M., Makonde, H.M.: Molecular characterization of low-density polyethene (LDPE) degrading bacteria and fungi from Dandora dumpsite, Nairobi, Kenya. Int. J. Microbiol. **2018** (2018)
18. Patwary, M.S., Surid, S., Gafur, M.: Properties and applications of biodegradable polymers. J. Res. **9**, 33 (2020)
19. Ulery, B.D., Nair, L.S., Laurencin, C.T.: Biomedical applications of biodegradable polymers. J. Polym. Sci., Part B Polym. Phys. **49**(12), 832–864 (2011)

20. Karlsson, S., Albertsson, A.C.: Biodegradable polymers and environmental interaction. Polym. Eng. & Sci. **38**(8), 1251–1253 (1998)
21. Cheng, J., Gao, R., Zhu, Y., Lin, Q.: Applications of biodegradable materials in food packaging: a review. Alex. Eng. J. **91**, 70–83 (2024)
22. Baidurah, S.: Methods of analyses for biodegradable polymers: a review. Polymers **14**(22), 4928 (2022)
23. Nurazzi, N., et al.: Thermogravimetric analysis properties of cellulosic natural fiber polymer composites: a review on influence of chemical treatments. Polymers **13**(16), 2710 (2021)
24. Sam, S., Nuradibah, M., Chin, K., Hani, N.: Current application and challenges on packaging industry based on natural polymer blending. Nat. Polym. Ind. Tech. Appl. 163–184 (2016)
25. Krochta, J.M., Mulder-Johnston, C.D.: Edible and Biodegradable Polymer Films: Challenges and Opportunities (1997)
26. Rhim, J.-W., Ng, P.K.: Natural biopolymer-based nanocomposite films for packaging applications. Crit. Rev. Food Sci. Nutr. **47**(4), 411–433 (2007)
27. Bangar, S.P., Whiteside, W.S., Ashogbon, A.O., Kumar, M.: Recent advances in thermoplastic starches for food packaging: a review. Food Packag. Shelf Life **30**, 100743 (2021)
28. Lopez-Gil, A., Rodriguez-Perez, M., De Saja, J.A.: Strategies to improve the mechanical properties of starch-based materials: plasticization and natural fibers reinforcement. Polímeros **24**, 36–42 (2014)
29. Jacob, J., Thomas, S., Loganathan, S., Valapa, R.B.: Antioxidant incorporated biopolymer composites for active packaging. In: Processing and Development of Polysaccharide-Based Biopolymers for Packaging Applications, pp. 239–260. Elsevier (2020)
30. El Halal, S.L.M., et al.: Structure, morphology and functionality of acetylated and oxidised barley starches. Food Chem. **168**, 247–256 (2015)
31. Sanyang, M.L., Sapuan, S.M., Jawaid, M., Ishak, M.R., Sahari, J.: Effect of plasticizer type and concentration on tensile, thermal and barrier properties of biodegradable films based on sugar palm (Arenga pinnata) starch. Polymers **7**(6), 1106–1124 (2015)
32. Panrong, T., Karbowiak, T., Harnkarnsujarit, N.: Effects of acetylated and octenyl-succinated starch on properties and release of green tea compounded starch/LLDPE blend films. J. Food Eng. **284**, 110057 (2020)
33. Fonseca-García, A., Osorio, B.H., Aguirre-Loredo, R.Y., Calambas, H.L., Caicedo, C.: Miscibility study of thermoplastic starch/polylactic acid blends: thermal and superficial properties. Carbohyd. Polym. **293**, 119744 (2022)
34. Phiriyawirut, M., Duangsuwan, T., Uenghuab, N., Meena, C.: Effect of octenyl succinate starch on properties of tapioca thermoplastic starch blends. Key Eng. Mater. **751**, 290–295 (2017)
35. Jiang, S., Dai, L., Qin, Y., Xiong, L., Sun, Q.: Preparation and characterization of octenyl succinic anhydride modified taro starch nanoparticles. PLoS ONE **11**(2), e0150043 (2016)
36. Yousefi, A., Savadkoohi, B., Zahedi, Y., Hatami, M., Ako, K.: Fabrication and characterization of hybrid sodium montmorillonite/TiO_2 reinforced cross-linked wheat starch-based nanocomposites. Int. J. Biol. Macromol. **131**, 253–263 (2019)
37. Hu, X., Jia, X., Zhi, C., Jin, Z., Miao, M.: Improving the properties of starch-based antimicrobial composite films using ZnO-chitosan nanoparticles. Carbohyd. Polym. **210**, 204–209 (2019)
38. Chibuzo, I.M.: Physicochemical and retrogradation characteristics of modified sweet potato (Ipomoea batatas L. (Lam)) starch. J. Agric. Food Technol. **2**(3), 49–55 (2012)
39. Kim, H.Y., Lamsal, B., Jane, J.L., Grewell, D.: Sheet-extruded films from blends of hydroxypropylated and native corn starches, and their characterization. J. Food Process. Eng. **43**(3), e13216 (2020)
40. Mussatto, S.I., Machado, E.M., Martins, S., Teixeira, J.A.: Production, composition, and application of coffee and its industrial residues. Food Bioprocess Technol. **4**, 661–672 (2011)
41. Matet, M., Heuzey, M.-C., Ajji, A.: Morphology and antibacterial properties of plasticized chitosan/metallocene polyethylene blends. J. Mater. Sci. **49**, 5427–5440 (2014)

42. Jumaidin, R.: Development and Characterization of Thermoplastic Sugar Palm Starch/Agar Polymer Blend, Reinforced Seaweed Waste and Sugar Palm Fiber Hybrid Composite. Universiti Putra Malaysia Selangor, Malaysia (2017)
43. Wattanakornsiri, A., Pachana, K., Kaewpirom, S., Traina, M., Migliaresi, C.: Preparation and properties of green composites based on tapioca starch and differently recycled paper cellulose fibers. J. Polym. Environ. **20**, 801–809 (2012)
44. Hammache, Y., Serier, A., Chaoui, S.: The effect of thermoplastic starch on the properties of polypropylene/high density polyethylene blend reinforced by nano-clay. Mater. Res. Express **7**(2), 025308 (2020)
45. Javanbakht, S., Namazi, H.: Solid state photoluminescence thermoplastic starch film containing graphene quantum dots. Carbohyd. Polym. **176**, 220–226 (2017)
46. Westlake, J.R., Tran, M.W., Jiang, Y., Zhang, X., Burrows, A.D., Xie, M.: Biodegradable biopolymers for active packaging: demand, development and directions. Sustain. Food Technol. **1**(1), 50–72 (2023)
47. Doelker, E.: Cellulose derivatives. In: Biopolymers I, pp. 199–265. Springer (2005)
48. Bigi, F., Haghighi, H., Siesler, H.W., Licciardello, F., Pulvirenti, A.: Characterization of chitosan-hydroxypropyl methylcellulose blend films enriched with nettle or sage leaf extract for active food packaging applications. Food Hydrocoll. **120**, 106979 (2021)
49. Imran, M., El-Fahmy, S., Revol-Junelles, A.-M., Desobry, S.: Cellulose derivative based active coatings: effects of nisin and plasticizer on physico-chemical and antimicrobial properties of hydroxypropyl methylcellulose films. Carbohyd. Polym. **81**(2), 219–225 (2010)
50. Moghimi, R., Aliahmadi, A., Rafati, H.: Antibacterial hydroxypropyl methyl cellulose edible films containing nanoemulsions of Thymus daenensis essential oil for food packaging. Carbohyd. Polym. **175**, 241–248 (2017)
51. Erceg, T., et al.: Preparation and characterization of biodegradable cellulose acetate-based films with novel plasticizer obtained by polyethylene terephthalate glycolysis intended for active packaging. Cellulose 1–20 (2023)
52. Wang, H., Qian, J., Ding, F.: Emerging chitosan-based films for food packaging applications. J. Agric. Food Chem. **66**(2), 395–413 (2018)
53. No, H., Meyers, S.P., Prinyawiwatkul, W., Xu, Z.: Applications of chitosan for improvement of quality and shelf life of foods: a review. J. Food Sci. **72**(5), R87–R100 (2007)
54. Biji, K., Ravishankar, C., Mohan, C., Srinivasa Gopal, T.: Smart packaging systems for food applications: a review. J. Food Sci. Technol. **52**, 6125–6135 (2015)
55. Flórez, M., Guerra-Rodríguez, E., Cazón, P., Vázquez, M.: Chitosan for food packaging: recent advances in active and intelligent films. Food Hydrocoll. **124**, 107328 (2022)
56. Huang, J., Hu, Z., Hu, L., Li, G., Yao, Q., Hu, Y.: Pectin-based active packaging: a critical review on preparation, physical properties and novel application in food preservation. Trends Food Sci. Technol. **118**, 167–178 (2021)
57. Roy, S., Priyadarshi, R., Łopusiewicz, Ł., Biswas, D., Chandel, V., Rhim, J.-W.: Recent progress in pectin extraction, characterization, and pectin-based films for active food packaging applications: a review. Int. J. Biol. Macromol. 124248 (2023)
58. Hernández-Muñoz, P., Kanavouras, A., Ng, P.K., Gavara, R.: Development and characterization of biodegradable films made from wheat gluten protein fractions. J. Agric. Food Chem. **51**(26), 7647–7654 (2003)
59. Gennadios, A., Brandenburg, A.H., Weller, C.L., Testin, R.F.: Effect of pH on properties of wheat gluten and soy protein isolate films. J. Agric. Food Chem. **41**(11), 1835–1839 (1993)
60. Marquis, R.: Oxygen metabolism, oxidative stress and acid-base physiology of dental plaque biofilms. J. Ind. Microbiol. Biotechnol. **15**(3), 198–207 (1995)
61. Koshy, R.R., Mary, S.K., Thomas, S., Pothan, L.A.: Environment friendly green composites based on soy protein isolate–a review. Food Hydrocoll. **50**, 174–192 (2015)
62. Yadav, J.S.S., Yan, S., Pilli, S., Kumar, L., Tyagi, R.D., Surampalli, R.Y.: Cheese whey: a potential resource to transform into bioprotein, functional/nutritional proteins and bioactive peptides. Biotechnol. Adv. **33**(6), 756–774 (2015)

63. Schmid, M., Müller, K.: Whey protein-based packaging films and coatings. In: Whey Proteins, pp. 407–437. Elsevier (2019)
64. de Castro, R.J.S., et al.: Whey protein as a key component in food systems: physicochemical properties, production technologies and applications. Food Struct. **14**, 17–29 (2017)
65. Kandasamy, S., et al.: Application of whey protein-based edible films and coatings in food industries: an updated overview. Coatings **11**(9), 1056 (2021)
66. Andrade, M.A., et al.: Novel active food packaging films based on whey protein incorporated with seaweed extract: development, characterization, and application in fresh poultry meat. Coatings **11**(2), 229 (2021)
67. Brink, I., Šipailienė, A., Leskauskaitė, D.: Antimicrobial properties of chitosan and whey protein films applied on fresh cut turkey pieces. Int. J. Biol. Macromol. **130**, 810–817 (2019)
68. Félix, M., Martín-Alfonso, J., Romero, A., Guerrero, A.: Development of albumen/soy biobased plastic materials processed by injection molding. J. Food Eng. **125**, 7–16 (2014)
69. Song, F., Tang, D.-L., Wang, X.-L., Wang, Y.-Z.: Biodegradable soy protein isolate-based materials: a review. Biomacromol **12**(10), 3369–3380 (2011)
70. González, A., Igarzabal, C.I.A.: Soy protein–poly (lactic acid) bilayer films as biodegradable material for active food packaging. Food Hydrocoll. **33**(2), 289–296 (2013)
71. Yu, Z., Sun, L., Wang, W., Zeng, W., Mustapha, A., Lin, M.: Soy protein-based films incorporated with cellulose nanocrystals and pine needle extract for active packaging. Ind. Crops Prod. **112**, 412–419 (2018)
72. Maroufi, L.Y., Ghorbani, M., Tabibiazar, M., Mohammadi, M., Pezeshki, A.: Advanced properties of gelatin film by incorporating modified kappa-carrageenan and zein nanoparticles for active food packaging. Int. J. Biol. Macromol. **183**, 753–759 (2021)
73. Sedlarikova, J., Janalikova, M., Peer, P., Pavlatkova, L., Minarik, A., Pleva, P.: Zein-based films containing monolaurin/eugenol or essential oils with potential for bioactive packaging application. Int. J. Mol. Sci. **23**(1), 384 (2021)
74. Gómez-Heincke, D., Martínez, I., Partal, P., Guerrero, A., Gallegos, C.: Development of antimicrobial active packaging materials based on gluten proteins. J. Sci. Food Agric. **96**(10), 3432–3438 (2016)
75. El-Wakil, N.A., Hassan, E.A., Abou-Zeid, R.E., Dufresne, A.: Development of wheat gluten/nanocellulose/titanium dioxide nanocomposites for active food packaging. Carbohyd. Polym. **124**, 337–346 (2015)
76. Mohamed, S.A., El-Sakhawy, M., El-Sakhawy, M.A.-M.: Polysaccharides, protein and lipid-based natural edible films in food packaging: a review. Carbohyd. Polym. **238**, 116178 (2020)
77. Amin, U., et al.: Potentials of polysaccharides, lipids and proteins in biodegradable food packaging applications. Int. J. Biol. Macromol. **183**, 2184–2198 (2021)
78. Suppakul, P., Miltz, J., Sonneveld, K., Bigger, S.W.: Active packaging technologies with an emphasis on antimicrobial packaging and its applications. J. Food Sci. **68**(2), 408–420 (2003)
79. Rostamian, R., Khalilzadeh, M.A., Zareyee, D.: Wood ash biocatalyst as a novel green catalyst and its application for the synthesis of benzochromene derivatives. Sci. Rep. **12**(1), 1145 (2022)
80. Khalilzadeh, M.A., Hosseini, S., Rad, A.S., Venditti, R.A.: Synthesis of grafted nanofibrillated cellulose-based hydrogel and study of its thermodynamic, kinetic, and electronic properties. J. Agric. Food Chem. **68**(32), 8710–8719 (2020)
81. Amirsoleimani, M., Khalilzadeh, M.A., Sadeghifar, F., Sadeghifar, H.: Surface modification of nanosatrch using nano silver: a potential antibacterial for food package coating. J. Food Sci. Technol. **55**, 899–904 (2018)
82. Kim, E., Moore, B.S., Yoon, Y.J.: Reinvigorating natural product combinatorial biosynthesis with synthetic biology. Nat. Chem. Biol. **11**(9), 649–659 (2015)
83. Hartmann, R., Hany, R., Geiger, T., Egli, T., Witholt, B., Zinn, M.: Tailored biosynthesis of olefinic medium-chain-length poly [(R)-3-hydroxyalkanoates] in pseudomonas p utida GPo1 with improved thermal properties. Macromolecules **37**(18), 6780–6785 (2004)
84. Kim, D.-Y., Kim, H.-W., Chung, M.-G., Rhee, Y.-H.: Biosynthesis, modification, and biodegradation of bacterial medium-chain-length polyhydroxyalkanoates. J. Microbiol. **45**(2), 87–97 (2007)

85. Petersen, K., et al.: Potential of biobased materials for food packaging. Trends Food Sci. Technol. **10**(2), 52–68 (1999)
86. Dubey, S.P., Thakur, V.K., Krishnaswamy, S., Abhyankar, H.A., Marchante, V., Brighton, J.L.: Progress in environmental-friendly polymer nanocomposite material from PLA: synthesis, processing and applications. Vacuum **146**, 655–663 (2017)
87. Iotti, M., Fabbri, P., Messori, M., Pilati, F., Fava, P.: Organic–inorganic hybrid coatings for the modification of barrier properties of poly (lactic acid) films for food packaging applications. J. Polym. Environ. **17**, 10–19 (2009)
88. Ncube, L.K., Ude, A.U., Ogunmuyiwa, E.N., Zulkifli, R., Beas, I.N.: Environmental impact of food packaging materials: a review of contemporary development from conventional plastics to polylactic acid based materials. Materials **13**(21), 4994 (2020)
89. Cheng, J., Lin, X., Wu, X., Liu, Q., Wan, S., Zhang, Y.: Preparation of a multifunctional silver nanoparticles polylactic acid food packaging film using mango peel extract. Int. J. Biol. Macromol. **188**, 678–688 (2021)
90. Mirosanloo, A., Zareyee, D., Khalilzadeh, M.A.: Recyclable cellulose nanocrystal supported Palladium nanoparticles as an efficient heterogeneous catalyst for the solvent-free synthesis of coumarin derivatives via von Pechmann condensation. Appl. Organomet. Chem. **32**(12), e4546 (2018)
91. Debeaufort, F., Riondet, J., Brachais, C.-H., Benbettaieb, N.: Influence of gelatin-based coatings crosslinked with phenolic acids on PLA film barrier properties. Coatings **12**(7), 993 (2022)
92. Rigotti, D., et al.: Novel biobased polylactic acid/poly (pentamethylene 2, 5-furanoate) blends for sustainable food packaging. ACS Sustain. Chem. & Eng. **9**(41), 13742–13750 (2021)
93. Morão, A., De Bie, F.: Life cycle impact assessment of polylactic acid (PLA) produced from sugarcane in Thailand. J. Polym. Environ. **27**(11), 2523–2539 (2019)
94. Kumari, S.V.G., Pakshirajan, K., Pugazhenthi, G.: Recent advances and future prospects of cellulose, starch, chitosan, polylactic acid and polyhydroxyalkanoates for sustainable food packaging applications. Int. J. Biol. Macromol. (2022)
95. Tripathi, A.D., Mishra, P.K., Darani, K.K., Agarwal, A., Paul, V.: Hydrothermal treatment of lignocellulose waste for the production of polyhydroxyalkanoates copolymer with potential application in food packaging. Trends Food Sci. Technol. **123**, 233–250 (2022)
96. Tang, X., Kumar, P., Alavi, S., Sandeep, K.: Recent advances in biopolymers and biopolymer-based nanocomposites for food packaging materials. Crit. Rev. Food Sci. Nutr. **52**(5), 426–442 (2012)
97. Amelia, T.S.M., Govindasamy, S., Tamothran, A.M., Vigneswari, S., Bhubalan, K.: Applications of PHA in agriculture. Biotechnol. Appl. Polyhydroxyalkanoates 347–361 (2019)
98. Koller, M., Mukherjee, A.: A new wave of industrialization of PHA biopolyesters. Bioengineering **9**(2), 74 (2022)
99. Kunamaneni, S.: Bioplastics innovation: commercialization strategies for polyethylene furanoate (PEF) and polyhydroxy alkanoates (PHA). Biofuels Bioprod. Biorefin. **17**(3), 421–436 (2023)
100. Winnacker, M.: Polyhydroxyalkanoates: recent advances in their synthesis and applications. Eur. J. Lipid Sci. Technol. **121**(11), 1900101 (2019)
101. Thurber, H., Curtzwiler, G.W.: Suitability of poly (butylene succinate) as a coating for paperboard convenience food packaging. Int. J. Biobased Plast. **2**(1), 1–12 (2020)
102. Rafiqah, S.A., et al.: A review on properties and application of bio-based poly (butylene succinate). Polymers **13**(9), 1436 (2021)
103. Wu, F., Misra, M., Mohanty, A.K.: Challenges and new opportunities on barrier performance of biodegradable polymers for sustainable packaging. Prog. Polym. Sci. **117**, 101395 (2021)
104. Jacquel, N., et al.: Synthesis and properties of poly (butylene succinate): efficiency of different transesterification catalysts. J. Polym. Sci., Part A: Polym. Chem. **49**(24), 5301–5312 (2011)
105. Jiang, L., Sabzi, M., Zhang, J.: Biodegradable and Biobased Polymers, pp. 133–165. Applied Plastics Engineering Handbook (2024)
106. Mousa, M., Dong, Y.:Multiscaled PVA Bionanocomposite Films. Springer (2021)

107. Nagarkar R, Patel J.: Polyvinyl alcohol: A comprehensive study. Acta Sci. Pharm. Sci. **3**(4), 34–44 (2019)
108. Tang, X., Alavi, S.: Recent advances in starch, polyvinyl alcohol based polymer blends, nanocomposites and their biodegradability. Carbohyd. Polym. **85**(1), 7–16 (2011)
109. Oun, A.A., Shin, G.H., Rhim, J.-W., Kim, J.T.: Recent advances in polyvinyl alcohol-based composite films and their applications in food packaging. Food Packag. Shelf Life **34**, 100991 (2022)
110. Xu, X., Yang, Y.-Q., Xing, Y.-Y., Yang, J.-F., Wang, S.-F.: Properties of novel polyvinyl alcohol/cellulose nanocrystals/silver nanoparticles blend membranes. Carbohyd. Polym. **98**(2), 1573–1577 (2013)
111. Kwon, S.-J., Chang, Y., Han, J.: Oregano essential oil-based natural antimicrobial packaging film to inactivate Salmonella enterica and yeasts/molds in the atmosphere surrounding cherry tomatoes. Food Microbiol. **65**, 114–121 (2017)
112. Budak, K., Sogut, O., Aydemir Sezer, U.: A review on synthesis and biomedical applications of polyglycolic acid. J. Polym. Res. **27**, 1–19 (2020)
113. Coles, R., Kirwan, M.J.: Food and Beverage Packaging Technology. Wiley & Sons (2011)
114. Rydz, J., Musioł, M., Zawidlak-Węgrzyńska, B., Sikorska, W.: Present and future of biodegradable polymers for food packaging applications. Biopolym. Food Des. 431–467 (2018)
115. Lee, S., Hongo, C., Nishino, T.: Crystal modulus of poly (glycolic acid) and its temperature dependence. Macromolecules **50**(13), 5074–5079 (2017)
116. Zhang, Z., Ortiz, O., Goyal, R., Kohn, J.: Biodegradable polymers. In: Handbook of Polymer Applications in Medicine and Medical Devices, pp. 303–335 (2014)
117. Gentile, P., Chiono, V., Carmagnola, I., Hatton, P.V.: An overview of poly (lactic-co-glycolic) acid (PLGA)-based biomaterials for bone tissue engineering. Int. J. Mol. Sci. **15**(3), 3640–3659 (2014)
118. Mishra, N., et al.: Biodegradable polymer based particulate carrier (s) for the delivery of proteins and peptides. Anti-Inflamm. Anti-Allergy Agents Med. Chem. (Former. Curr. Med. Chem.-Anti-Inflamm. Anti-Allergy Agents) **7**(4), 240–251 (2008)
119. Zhong, Y., Godwin, P., Jin, Y., Xiao, H.: Biodegradable polymers and green-based antimicrobial packaging materials: a mini-review. Adv. Ind. Eng. Polym. Res. **3**(1), 27–35 (2020)
120. Bikiaris, D.N., Papageorgiou, G.Z., Achilias, D.S., Pavlidou, E., Stergiou, A.: Miscibility and enzymatic degradation studies of poly (ε-caprolactone)/poly (propylene succinate) blends. Eur. Polymer J. **43**(6), 2491–2503 (2007)
121. Vilaplana, F., Strömberg, E., Karlsson, S.: Environmental and resource aspects of sustainable biocomposites. Polym. Degrad. Stab. **95**(11), 2147–2161 (2010)
122. Bhattarai, S.R., Bhattarai, N., Viswanathamurthi, P., Yi, H.K., Hwang, P.H., Kim, H.Y.: Hydrophilic nanofibrous structure of polylactide; fabrication and cell affinity. J. Biomed. Mater. Res. Part A Off. J. Soc. Biomater. Jpn. Soc. Biomater. Aust. Soc. Biomater. Korean Soc. Biomater. **78**(2), 247–257 (2006)
123. Khan, R.A., Beck, S., Dussault, D., Salmieri, S., Bouchard, J., Lacroix, M.: Mechanical and barrier properties of nanocrystalline cellulose reinforced poly (caprolactone) composites: Effect of gamma radiation. J. Appl. Polym. Sci. **129**(5), 3038–3046 (2013)
124. Deshmukh, K., Ahamed, M.B., Deshmukh, R., Pasha, S.K., Bhagat, P., Chidambaram, Biopolymer composites with high dielectric performance: interface engineering. In Biopolymer Composites in Electronics, pp. 27–128. Elsevier (2017)
125. Kuddus, M.: Bioplastics for Sustainable Development. Springer (2021)
126. Pisula, W., Pigłowski, J., Kummerlöwe, C.: Preparation and characterization of copolyesters of poly (tetramethylene succinate) and poly (butylene terephthalate). Polimery **51**(5), 341–350 (2006)
127. Roohi, Srivastava, P., Bano, K., Zaheer, M.R., Kuddus, M.: Biodegradable smart biopolymers for food packaging: sustainable approach toward green environment. Bio-Based Mater. Food Packag. Green Sustain. Adv. Packag. Mater. 197–216 (2018)

128. Bohlmann, G.M.: General characteristics, processability, industrial applications and market evolution of biodegradable polymers. Handb. Biodegrad. Polym. 183–218 (2005)
129. Wang, L.-F., Rhim, J.-W., Hong, S.-I.: Preparation of poly (lactide)/poly (butylene adipate-co-terephthalate) blend films using a solvent casting method and their food packaging application. LWT-Food Sci. Technol. **68**, 454–461 (2016)
130. Jian, J., Xiangbin, Z., Xianbo, H.: An overview on synthesis, properties and applications of poly (butylene-adipate-co-terephthalate)–PBAT. Adv. Ind. Eng. Polym. Res. **3**(1), 19–26 (2020)
131. Cazón, P., Vázquez, M.: Bacterial cellulose as a biodegradable food packaging material: a review. Food Hydrocoll. **113**, 106530 (2021)
132. Zhu, H., Jia, S., Yang, H., Tang, W., Jia, Y., Tan, Z.: Characterization of bacteriostatic sausage casing: a composite of bacterial cellulose embedded with ε-polylysine. Food Sci. Biotechnol. **19**, 1479–1484 (2010)
133. Nguyen, V., Gidley, M., Dykes, G.: Potential of a nisin-containing bacterial cellulose film to inhibit Listeria monocytogenes on processed meats. Food Microbiol. **25**(3), 471–478 (2008)
134. Yordshahi, A.S., Moradi, M., Tajik, H., Molaei, R.: Design and preparation of antimicrobial meat wrapping nanopaper with bacterial cellulose and postbiotics of lactic acid bacteria. Int. J. Food Microbiol. **321**, 108561 (2020)
135. Gutiérrez, T.J., Valencia, G.A.: Reactive extrusion-processed native and phosphated starch-based food packaging films governed by the hierarchical structure. Int. J. Biol. Macromol. **172**, 439–451 (2021)
136. Anukiruthika, T., Sethupathy, P., Wilson, A., Kashampur, K., Moses, J.A., Anandharamakrishnan, C.: Multilayer packaging: advances in preparation techniques and emerging food applications. Compr. Rev. Food Sci. Food Saf. **19**(3), 1156–1186 (2020)
137. Vartiainen, J., et al.: Bio-based multilayer barrier films by extrusion, dispersion coating and atomic layer deposition. J. Appl. Polym. Sci. **133**(2), (2016)
138. Araque, L.M., Alvarez, V.A., Gutiérrez, T.J.: Composite foams made from biodegradable polymers for food packaging applications. Polym. Food Appl. 347–355 (2018)
139. Ghaani, M., Cozzolino, C.A., Castelli, G., Farris, S.: An overview of the intelligent packaging technologies in the food sector. Trends Food Sci. Technol. **51**, 1–11 (2016)
140. Stanley, J., John, A., Pušnik Črešnar, K., Fras Zemljič, L., Lambropoulou, D.A., Bikiaris, D.N.: Active agents incorporated in polymeric substrates to enhance antibacterial and antioxidant properties in food packaging applications. Macromol **3**(1), 1–27 (2022)
141. Azevedo, A.G., et al.: Active flexible films for food packaging: a review. Polymers **14**(12), 2442 (2022)
142. Puccini, M., Seggiani, M., Vitolo, S.: Polyethylene and hydrolyzed collagen blend films produced by blown extrusion. Chem. Eng. Trans. **43**, 1705–1710 (2015)
143. Li, X., et al.: Improvement of compatibility and mechanical properties of the poly (lactic acid)/poly (butylene adipate-co-terephthalate) blends and films by reactive extrusion with chain extender. Polym. Eng. Sci. **58**(10), 1868–1878 (2018)
144. Mallegni, N., Phuong, T.V., Coltelli, M.-B., Cinelli, P., Lazzeri, A.: Poly (lactic acid)(PLA) based tear resistant and biodegradable flexible films by blown film extrusion. Materials **11**(1), 148 (2018)
145. Thunwall, M., Kuthanova, V., Boldizar, A., Rigdahl, M.: Film blowing of thermoplastic starch. Carbohyd. Polym. **71**(4), 583–590 (2008)
146. Sun, S., Liu, P., Ji, N., Hou, H., Dong, H.: Effects of low polyhydroxyalkanoate content on the properties of films based on modified starch acquired by extrusion blowing. Food Hydrocoll. **72**, 81–89 (2017)
147. Hilliou, L., Covas, J.A.: Production and processing of polymer-based nanocomposites. Nanomater. Food Packag. 111–146 (2018)
148. Huang, H., et al.: Room temperature brittle-to-ductile transition in polystyrene induced by extrusion casting melt stretching: contribution of free volume and chain entanglement. Polymer 126745 (2024)

149. Christodoulou, K., Hatzikiriakos, S.G., Mitsoulis, E.: Numerical simulation of the wire-pinning process in PET film casting: steady-state results. AIChE J. **58**(7), 1979–1986 (2012)
150. Resch, K., Wallner, G.M., Teichert, C., Gahleitner, M.: Highly transparent polypropylene cast films: relationships between optical properties, additives, and surface structure. Polym. Eng. Sci. **47**(7), 1021–1032 (2007)
151. Tadmor, Z., Gogos, C.G.: Principles of Polymer Processing. Wiley & Sons (2013)
152. Ajitha, A., Aswathi, M., Maria, H.J., Izdebska, J., Thomas, S.: Multilayer polymer films. Multicomponent Polym. Mater. 229–258 (2016)
153. Fleming, R., Pardini, L., Brito, C., Oliveira, M., Alves, N., Massi, M.: Plasma treatment of polyacrylonitrile/vinyl acetate films obtained by the extrusion process. Polym. Bull. **66**, 277–288 (2011)
154. Raquez, J.M., Narayan, R., Dubois, P.: Recent advances in reactive extrusion processing of biodegradable polymer-based compositions. Macromol. Mater. Eng. **293**(6), 447–470 (2008)
155. Huda, M., Mohanty, A., Drzal, L., Schut, E., Misra, M.: "Green" composites from recycled cellulose and poly (lactic acid): physico-mechanical and morphological properties evaluation. J. Mater. Sci. **40**, 4221–4229 (2005)
156. Montilla-Buitrago, C.E., Gómez-López, R.A., Solanilla-Duque, J.F., Serna-Cock, L., Villada-Castillo, H.S.: Effect of plasticizers on properties, retrogradation, and processing of extrusion-obtained thermoplastic starch: a review. Starch-Stärke **73**(9–10), 2100060 (2021)
157. Gutiérrez, T.J., Alvarez, V.A.: Cellulosic materials as natural fillers in starch-containing matrix-based films: a review. Polym. Bull. **74**, 2401–2430 (2017)
158. Vedove, T.M., Maniglia, B.C., Tadini, C.C.: Production of sustainable smart packaging based on cassava starch and anthocyanin by an extrusion process. J. Food Eng. **289**, 110274 (2021)
159. Pranata, M.P., et al.: Egg white protein film production through extrusion and calendering processes and its suitability for food packaging applications. Food Bioprocess Technol. **12**, 714–727 (2019)
160. Tzoganakis, C.: Reactive extrusion of polymers: a review. Adv. Polym. Technol. J. Polym. Process. Inst. **9**(4), 321–330 (1989)
161. Nield, S.A., Budman, H.M., Tzoganakis, C.: Control of a LDPE reactive extrusion process. Control. Eng. Pract. **8**(8), 911–920 (2000)
162. Xie, F., Yu, L., Liu, H., Chen, L.: Starch modification using reactive extrusion. Starch-Stärke **58**(3–4), 131–139 (2006)
163. Wu, L., Jia, Y., Sun, S., Zhang, G., Zhao, G., An, L.: Study on reactive extrusion processes of block copolymer. Mater. Sci. Eng. A **454**, 221–226 (2007)
164. Kowalski, A., Duda, A., Penczek, S.: Kinetics and mechanism of cyclic esters polymerization initiated with tin (II) octoate, 1. Polymerization of ε-caprolactone. Macromol. Rapid Commun. **19**(11), 567–572 (1998)
165. Jacobsen, S., Fritz, H., Degée, P., Dubois, P., Jérôme, R.: Single-step reactive extrusion of PLLA in a corotating twin-screw extruder promoted by 2-ethylhexanoic acid tin (II) salt and triphenylphosphine. Polymer **41**(9), 3395–3403 (2000)
166. Jamshidian, M., Tehrany, E.A., Imran, M., Jacquot, M., Desobry, S.: Poly-lactic acid: production, applications, nanocomposites, and release studies. Compr. Rev. Food Sci. Food Saf. **9**(5), 552–571 (2010)
167. Li, X., et al.: Active packaging for the extended shelf-life of meat: perspectives from consumption habits, market requirements and packaging practices in China and New Zealand. Foods **11**(18), 2903 (2022)
168. Butler, T.I., Morris, B.A.: PE-based multilayer film structures. In: Multilayer Flexible Packaging, pp. 281–310. Elsevier (2016)
169. Silva, E.G., Cardoso, S., Bettencourt, A.F., Ribeiro, I.A.: Latest trends in sustainable polymeric food packaging films. Foods **12**(1), 168 (2022)
170. Cheng, H.Y., Yang, Y.J., Li, S.C., Hong, J.Y., Jang, G.W.: Modification and extrusion coating of polylactic acid films. J. Appl. Polym. Sci. 132(35), (2015)

171. Wu, Q., Lindh, V.H., Johansson, E., Olsson, R.T., Hedenqvist, M.S.: Freeze-dried wheat gluten biofoams; scaling up with water welding. Ind. Crops Prod. **97**, 184–190 (2017)
172. Cvrček, L., Horáková, M.: Plasma modified polymeric materials for implant applications. In: Non-Thermal Plasma Technology for Polymeric Materials, pp. 367–407. Elsevier (2019)
173. Saiz-Arroyo, C., Rodríguez-Pérez, M.Á., Velasco, J.I., de Saja, J.A.: Influence of foaming process on the structure–properties relationship of foamed LDPE/silica nanocomposites. Compos. B Eng. **48**, 40–50 (2013)
174. Jarpa-Parra, M., Chen, L.: Applications of plant polymer-based solid foams: current trends in the food industry. Appl. Sci. **11**(20), 9605 (2021)
175. Jin, F.-L., Zhao, M., Park, M., Park, S.-J.: Recent trends of foaming in polymer processing: a review. Polymers **11**(6), 953 (2019)
176. Zheng, R., Tanner, R.I., Fan, X.-J.: Injection Molding: Integration of Theory and Modeling Methods. Springer Science & Business Media (2011)
177. Eriksen, M., et al.: Plastic pollution in the South Pacific subtropical gyre. Mar. Pollut. Bull. **68**(1–2), 71–76 (2013)
178. Haider, T.P., Völker, C., Kramm, J., Landfester, K., Wurm, F.R.: Plastics of the future? The impact of biodegradable polymers on the environment and on society. Angew. Chem. Int. Ed. **58**(1), 50–62 (2019)
179. Asgher, M., Qamar, S.A., Bilal, M., Iqbal, H.M.: Bio-based active food packaging materials: sustainable alternative to conventional petrochemical-based packaging materials. Food Res. Int. **137**, 109625 (2020)
180. Schneiderman, D.K., Hillmyer, M.A.: 50th anniversary perspective: there is a great future in sustainable polymers. Macromolecules **50**(10), 3733–3749 (2017)
181. Cesur, S., Köroğlu, C., Yalçın, H.T.: Antimicrobial and biodegradable food packaging applications of polycaprolactone/organo nanoclay/chitosan polymeric composite films. J. Vinyl Add. Tech. **24**(4), 376–387 (2018)
182. Basavegowda, N., Baek, K.-H.: Advances in functional biopolymer-based nanocomposites for active food packaging applications. Polymers **13**(23), 4198 (2021)

Chapter 4
Antimicrobial Biopolymers for Food Preservation: Recent Developments and Applications Based on Essential Oils and Nanomaterials

Anik Das, Gazi Farhan Ishraque Toki, and Rony Mia

Abstract In the 21st century, sustainability is a global buzzword for the food industry to ensure a secure and healthy world. This book chapter explores the potential of essential oils (EOs) and nanomaterials for sustainable food packaging applications. EOs, particularly oregano and thyme, exhibit promising antimicrobial and antioxidant properties due to their phenolic components. Their incorporation into biopolymer films can enhance food safety and shelf life. However, challenges remain, including encapsulation to overcome their hydrophobic nature and ensuring safe interaction with food contents. Nanomaterials offer another promising avenue for sustainable food packaging. They can enhance the mechanical, barrier, and functional characteristics of biopolymers. Nevertheless, the possibility of nanoparticles migrating from packaging to food gives rise to worries. Further research is needed to understand the safety and migration behavior of nanomaterials. Both EOs and nanomaterials offer exciting possibilities for the future of food packaging. However, addressing research gaps related to encapsulation, safety, and scalability is crucial for their widespread adoption. By addressing their classification, mechanisms, advantages, disadvantages, recent developments, and prospects, the chapter contributes to the ongoing narrative of sustainable and effective food preservation methodologies.

A. Das
School of Textile Science and Engineering, Tiangong University, Tianjin 300387, People's Republic of China

G. F. I. Toki
Nanotechnology Center, School of Fashion and Textiles, The Hong Kong Polytechnic University, Kowloon 999077, Hong Kong

R. Mia (✉)
Wilson College of Textiles, North Carolina State University, 1020 Main Campus Dr, Raleigh 27606, NC, United States
e-mail: mroni_mia@yahoo.com

Department of Textile Engineering, National Institute of Textile Engineering and Research, University of Dhaka, Dhaka 1000, Bangladesh

S. ul Islam and M. Shahid (eds.), *Green Materials for Active Food Packaging*, Engineering Materials, https://doi.org/10.1007/978-981-96-0369-5_4

Keywords Food preservation · Antimicrobial biopolymers · Recent developments · Essential oils · Nanomaterials

4.1 Introduction

In the ever-evolving landscape of food preservation, the demand for safer and more sustainable methods has led to exploring innovative solutions. Antimicrobial biopolymers, harnessing the power of nature's defense mechanisms, have emerged as a promising avenue in this quest. Biopolymers, obtained from renewable sources like plants, animals, or microbes, have natural antibacterial qualities that help prevent the formation of hazardous infections in food items [1]. Furthermore, integrating essential oils (EOs) into packaging materials has revolutionized how we approach food preservation. EOs, extracted from various plants, exhibit potent antimicrobial activity due to their complex chemical compositions. When integrated into packing materials, they provide a defensive barrier against microbial contamination, prolonging the storage duration of perishable products and reducing the need for synthetic preservatives [2]. Moreover, advancements in nanotechnology have paved the way for nanomaterial-based packaging solutions with antimicrobial properties. Nanomaterials, engineered at the nanoscale, offer unique advantages such as high surface area-to-volume ratios and enhanced reactivity. By incorporating antimicrobial agents into nanocomposites, these packaging materials can effectively combat spoilage microorganisms, safeguarding the quality and safety of food products throughout the supply chain [3–6].

EOs, extracted from aromatic herbs, are increasingly considered a possible addition to food packaging. There are several benefits associated with utilizing these natural constituents. The inherent antibacterial properties of these substances can combat microbes responsible for food degradation, resulting in reduced food waste and extended shelf life [7]. EOs serve as a more environmentally friendly alternative to synthetic preservatives, as they are typically considered safe. Food packaging composed of protein mostly focuses on natural components [8]. Researchers are now examining the potential use of EOs as a component in protein films derived from whey or soy protein. These bio-based products are already quite biodegradable, and it can make them even more antimicrobial. They provide an eco-friendly and doable strategy for food storage when used together. Innovations in food packaging are being propelled by the quest for environmentally friendly substitutes for traditional plastic packaging [9]. Utilizing starch and EOs is a promising technique. Biodegradable packaging films are produced using starch, a plentiful substance that can be replenished. Additionally, these films may be enhanced by using EOs, which increase their antibacterial properties and extend the shelf life of packed food [10]. This blend offers an environmentally beneficial and natural substitute when contrasted with traditional packaging materials. Chitosan, a naturally occurring biopolymer derived from chitin, a protein found in crab shells, is a noteworthy recent finding. Chitosan films have excellent biodegradability and inherent antibacterial characteristics. EOs enhance the antibacterial action, creating a strong barrier that prevents the growth of germs responsible for food spoilage. These sheets, composed of chitosan and fragrant oils, offer an environmentally benign and potentially superior alternative to conventional packaging [11, 12].

The study of nanotechnology is evolving in food preservation technologies, causing a revolution [13, 14]. Researchers are now devising innovative methods to inhibit food spoilage and extend its longevity using nanomaterials, materials engineered at the atomic and molecular scale, into food packaging [15]. Nanomaterials possess the capacity to exhibit unique attributes, such as enhanced barrier properties, enabling them to build a more efficient barrier against gases and moisture that can lead to food degradation [16]. In addition, some nanomaterials have potent antibacterial properties, enabling them to specifically target and eradicate harmful bacteria and fungi that contribute to spoilage [17–19]. These technical breakthroughs enable the development of innovative packaging methods. Consider, for instance, food packaging that incorporates nanoparticles capable of actively sensing and responding to the presence of decay. These nanoparticles can release natural antimicrobials upon detecting the development of dangerous pathogens [20]. Implementing this would extend the durability of the food and greatly decrease food waste. Although research in this area is ongoing, there is a significant potential for nanomaterial-based packaging to improve the preservation and safety of food.

This chapter entitled "Antimicrobial Biopolymers for Food Preservation: Recent Developments and Applications based on Essential oils and nanomaterials" explores the synergistic potential of antimicrobial biopolymers, essential oil-based packaging, and nanomaterial-based packaging in revolutionizing food preservation practices. By harnessing the power of nature and cutting-edge technology, these innovations not only address the pressing challenges of food safety and shelf-life extension but also contribute to a more sustainable and environmentally conscious approach to food packaging and preservation.

4.2 Application of Essential Oil in Food Packaging

EOs are emerging as a promising natural alternative to synthetic preservatives for various food products, including cereals, grains, pulses, fruits, and vegetables. This chapter delves into the existing research to explore the potential of EOs as food preservatives. We will specifically focus on their diverse antimicrobial properties, highlighting their effectiveness against bacteria (bactericidal and antibacterial) and fungi (fungicidal and antifungal) [21]. Figure 4.1 demonstrates all the active EOs constituents utilized over the years.

The intricate makeup of EOs is responsible for these extraordinary qualities. Terpenes, terpenoids, carotenoids, coumarins, and curcumins are just a few of the active ingredients in this diverse mixture. These ingredients work together to prevent food spoiling by drastically reducing the number of dangerous microbes that can

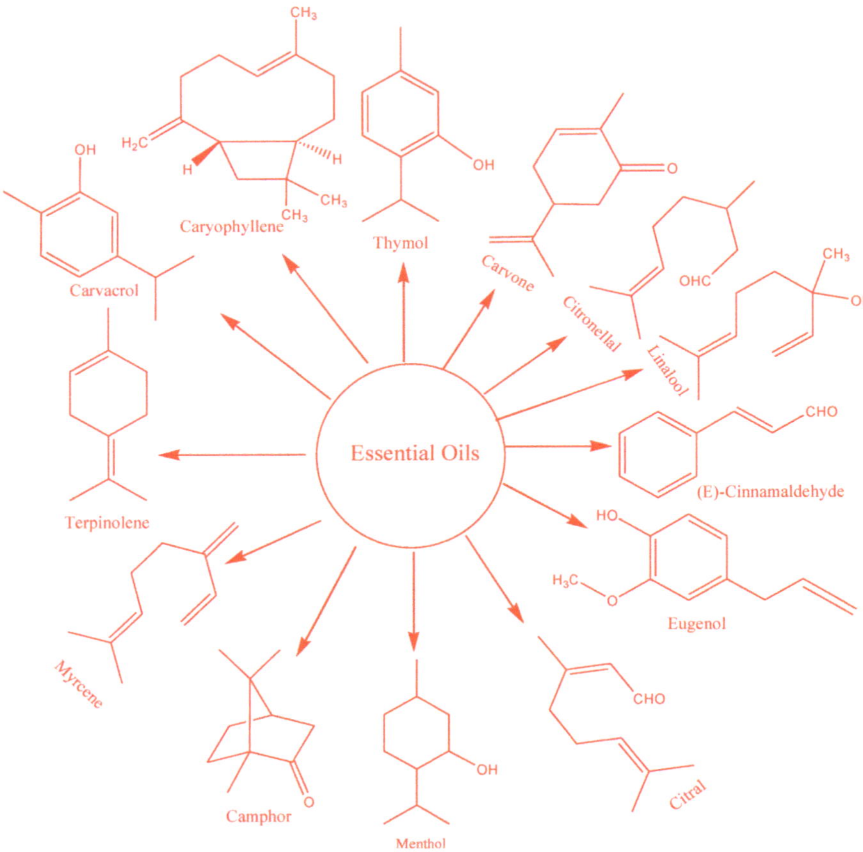

Fig. 4.1 Several active components of essential oils, Reproduced from Ref. [22], Copyright CC-BY terms, 2022

develop in food. The natural approach to food preservation offers a multitude of benefits, for instance:

- **Safety**: EOs are generally considered safe (GRAS) for consumption, making them a consumer-friendly alternative.
- **Eco-Friendliness**: Derived from plants, EOs are a sustainable and renewable resource with minimal environmental impact.
- **Cost-Effectiveness**: Compared to synthetic preservatives, EOs can offer a cost-competitive solution.
- **Biodegradability**: Once their role is complete, EOs break down naturally, further minimizing their environmental footprint [23, 24].

By leveraging the power of EOs, we can potentially develop a new generation of food preservation strategies that are effective and align with the growing consumer demand for natural and sustainable food solutions.

4.2.1 *Protein-Based Food Packaging Improvement Adding Essential Oil*

Protein-based food packaging can benefit greatly from EOs (Fig. 4.2). It improves protein-based food packaging by boosting its barrier against moisture, microbes, and oxidation. EOs are hydrophobic (water-repellent), so incorporating them into protein-based films reduces water vapor permeability, keeping food from drying out [25, 26]. Some studies show that EOs strengthen the film structure by influencing protein networks and forming bonds with polymer chains, improving tensile strength [25, 26]. Gelatin, soy protein isolates, and whey protein isolates are three different types of protein that may be used to make films. Several EOs, such as lemongrass, citronella, bergamot, and basil leaf, can be added to gelatin films to make them more appealing [27–29]. Mint essential oil, known for its volatile components, including flavonoids, organic acids, and quinones, is another option for these films [30]. Mint essential oil enhances films' antifungal properties and reduces water vapor permeability [31, 32]. Citrus EOs (like lemon, lime, etc.) improve flexibility and water vapor barrier properties while maintaining good antioxidant activity [33]. EOs are volatile, meaning they can evaporate. To address this, synthetic surfactants or biopolymer blends are used to disperse and hold them within the film [34, 35]. EOs can negatively impact the mechanical and thermal properties of protein films. To overcome this, researchers are using nanofillers like MMT (montmorillonite) and ZnO, which improve strength and control the release of antimicrobial agents [36–38].

4.2.1.1 Recent Progress of Protein-Based Food Packaging

Gelatin/MMT films with ginger EOs showed improved rheological properties and excellent antioxidant activity [37]. Films made of gelatin and chitosan that included oregano essential oil displayed superior water and light barrier qualities, making them extremely efficient against various bacteria [40, 41]. Antibacterial and antioxidant capabilities were improved by incorporating clove essential oil into soy protein films, but the film ability to maintain strong water vapor barrier qualities was preserved [42, 43]. Despite having lower mechanical qualities, hake protein films containing essential oils (citronella, coriander, and others) demonstrated a high level of antioxidant activity [44]. Although there was a modest decrease in water vapor permeability, there was no increase in lipid oxidation protection demonstrated by sodium caseinate films that contained essential oils (ginger and cinnamon) [45]. While maintaining the integrity of sardine patties, sunflower protein films containing clove essential oil

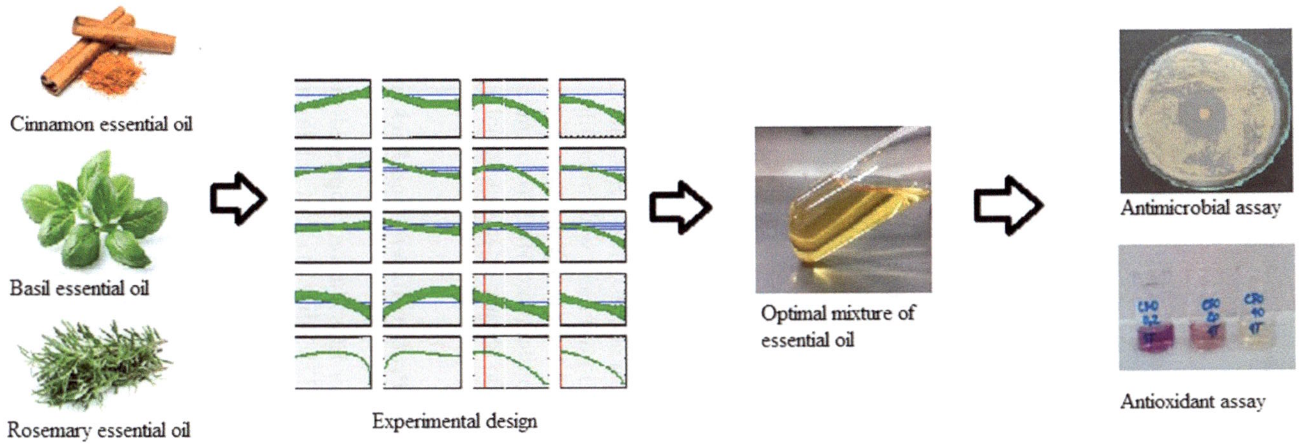

Fig. 4.2 Protein-based food packaging improvement by adding essential oil [39]. Republished with permission from Elsevier. Copyright Elsevier, 2017

demonstrated enhanced antioxidant and antibacterial effects [46]. Table 4.1 highlights the functions and applications of Protein and essential oil composites utilized in recent years.

4.2.2 Essential Oil/Cellulose Composite for Diverse Food Preservation Application

Abundant, cheap, and naturally occurring cellulose derivatives (like chitosan, starch, pectin, and alginates) are widely used in bio-based films due to their biocompatibility, safety, and stability [56, 57]. Cellulose sources like hemp, cotton, and wood are readily available from plants with chemical structures, as shown in Fig. 4.3 [58]. The limited solubility of cellulose in some solvents can hinder processing [59, 60]. Compatibility between EOs and the chosen polymer needs evaluation before application [61]. Several cellulose-based essential oil-added composite features are in Table 4.2.

Enhanced properties: EOs can improve water vapor barrier properties and mechanical strength, which act as plasticizers [63–65].
Antimicrobial activity: EOs offer a natural solution to fight food spoilage caused by various microorganisms (bacteria, fungi) [57, 63, 66–77]
Antioxidant activity: EOs can help extend shelf life by reducing oxidation in food items [78, 79].

The type and amount of EO significantly impact the film's properties [68, 70, 71]. Nanomaterials like cellulose nanocrystals and montmorillonite can improve film properties and control EO release [36–38, 57, 80]. Different food systems may require specific EO combinations based on their composition [63].

4.2.3 Starch and Essential Oils: A Sustainable Alternative for Food Packaging

Starch is a natural biopolymer commonly used to create transparent, thermoplastic films for food packaging (Fig. 4.4). However, these films are typically brittle and have poor water resistance [83]. There are several strategies to improve starch-based films. Water or glycerol can be added as plasticizers to improve flexibility [83]. Chemical crosslinkers may enhance film structure and water resistance [84, 85]; for example: Ferulic acid used to crosslink starch-chitosan films [109] and citric acid is another example [84–86]. Cellulose fibers increase tensile strength and reduce water vapor permeability [87, 88]. Starch films can incorporate and release antioxidants and antimicrobials into food [89–91]. The successful addition of antioxidants into starch-based films has been documented in extensive research [90–92].

Table 4.1 Functions and applications of protein/essential oil composites

Protein matrix with additives	Application	Significant functions	References
Garlic/whey protein	Baked sausages	Liposomal EOs effectively preserved refrigerated vacuum-packed sausages. Common bacteria such as lactic acid bacteria, psychrotrophic bacteria, Staphylococcus aureus, and coliforms were able to decrease the rate at which the food went bad as a result of their presence. Interestingly, even at low concentrations, the liposomal encapsulation helped maintain and even improve the antioxidant properties of the EOs	[47]
Functional coatings made with gluten and zein, incorporating additives for specific properties like potassium caseinate (adhesive), rennet casein (water resistance), xanthan gum (thickening), and locust bean gum (gelling)	Fresh trout fillets, boneless trout fillets, or skin-on trout fillets	Flavor, aroma, texture, and the underlying chemical compounds	[48]
Gelatin containing 0.15 g of thymol nanoemulsions	GRAS-certified biodegradable packaging designed to extend food product shelf life	The therapy demonstrates powerful inhibitory effects on a wide variety of bacterial strains, including Gram-positive (for example, Bacillus subtilis) and Gram-negative (for example, E. coli O157:H7) strains	[35]
Whey protein-based coatings formulated with additives like glycerol (plasticizer) and NaOH (for pH adjustment)	Atlantic salmon	Freshness, texture, fat content, or flavor profile	[49]
Whey protein isolate bio-composite films incorporating Thymbra leaves sourced specifically from the Ablus and Qabatiya regions	Food packaging	Effective against a broad spectrum of bacteria	[50]

(continued)

Table 4.1 (continued)

Protein matrix with additives	Application	Significant functions	References
Functionally-enhanced whey protein coatings featuring oregano and clove EOs for antimicrobial activity, and glycerol for improved film properties	Breast fillets-chicken	Enhance the quality and extend the shelf life of chicken breast while minimizing the reliance on harsh antimicrobial techniques	[51]
Thyme/clove/WPI	Kashar cheese	The formulation showed sustained release of the essential oil (EO), leading to long-lasting antibacterial activity against E. coli O157:H7, Staphylococcus aureus, and L. monocytogenes	[52]
Functionally-tailored calcium caseinate coatings incorporating acetylated monoglycerides for enhanced emulsification, fatty acid esters for improved film flexibility, and sodium carboxymethylcellulose for thickening properties	Zucchini	Optimization of edible coating formulations to minimize water loss in zucchini	[53]
Bio-based materials: European eel gelatin and protein isolate for film formation	Reduced food waste through improved shelf life with bio-packaging	Promoted their ability to preserve freshness	[54]
Caseinate-based coatings formulated with ascorbic acid (vitamin C) for enhanced properties, including potential antioxidant activity	Beef	Investigation of gamma irradiation's influence on the microbiological profile of ground beef	[55]

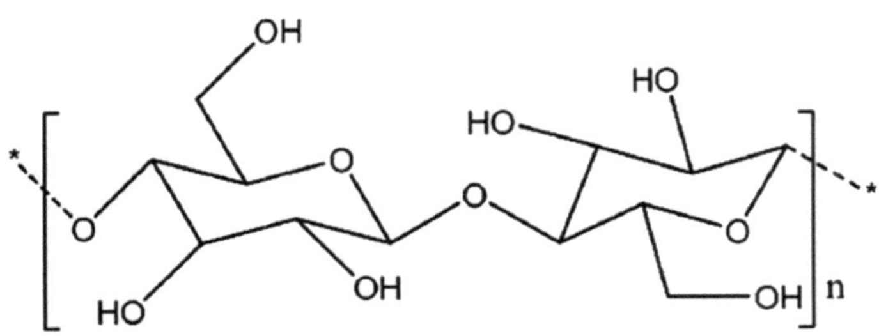

Fig. 4.3 Nanocellulose chemical structure, Reproduced from Ref. [62], Copyright CC-BY terms, 2023

Table 4.2 Properties of cellulose-based films with EOs

Property	Improvement	Examples	References
Transparency	Decrease	Rosemary, pepper, basil, lemongrass EOs	[81]
Mechanical strength	Increase	Lemongrass, rosemary, basil EOs	[81]
Antimicrobial activity	Increase	Various EOs (cinnamon, oregano, etc.)	[57, 63, 66–78, 82]
Antioxidant activity	Increase	Cinnamon EO	[78]
Shelf life of food	Increase	Various EOs	[68, 72, 78]
Water vapour permeability	Decrease	Rosemary, pepper, basil, lemongrass EOs	[81]

Nanoparticles of zirconium oxide and microencapsulated *Zataria multiflora* essential oil were added to the films made from potato starch and pectin derived from apple peel. These coatings' antioxidant, antibacterial, mechanical, and thermal characteristics were significantly enhanced. Consequently, these films are suitable for packing quail meat [94]. An inquiry was carried out to assess the capability of cinnamon essential oil, clay nanoparticles, glycerol, cassava starch, and additional substances to activate packaging sheets. The films showed promising antifungal efficacy against the bacteria responsible for bread spoilage [95]. *Aspergillus niger* and *Penicillium digitatum* are fungal species that are responsible for the deterioration of several food items. This study examined the feasibility of using starch, chitosan, and amaranth films containing EOs such as Mexican oregano, cinnamon, or lemongrass to inhibit the growth of fungi [96].

Sago starch films were incorporated with cinnamon essential oil and titanium dioxide nanoparticles [97], showed improved barrier and mechanical properties, and inhibited *E. coli, S. Typhimurium*, and *S. aureus* growth. Thermoplastic starch films contained nanocellulose-stabilized Pickering emulsions of various EOs [98]. Cinnamon and cardamom essential oils were found to have a substantial impact on the material's tensile strength. On the other hand, the essential oils of ho wood were particularly remarkable for their ability to improve the material's overall mechanical qualities. Through the incorporation of corn starch films, the essential oil of Zanthoxylum bungeanum was used [99], and showed improved antibacterial properties and elongation at break, but decreased water resistance and tensile strength.

Starch-chitosan composite films contained various EOs including lemongrass, kaffir lime peel, plai, fingerroot, and guava leaf [100]. Lemongrass oil was most effective against *E. coli, S. Typhimurium, S. aureus*, and *A. niger*. Corn starch films were incorporated with bamboo leaf essential oil [101], and showed enhanced antibacterial properties, surface roughness, opacity, thickness, and elongation at break. Sugar palm starch/sugar palm nanocrystalline cellulose films were incorporated with cinnamon essential oil [102], increasing antibacterial and mechanical assets, while decreasing

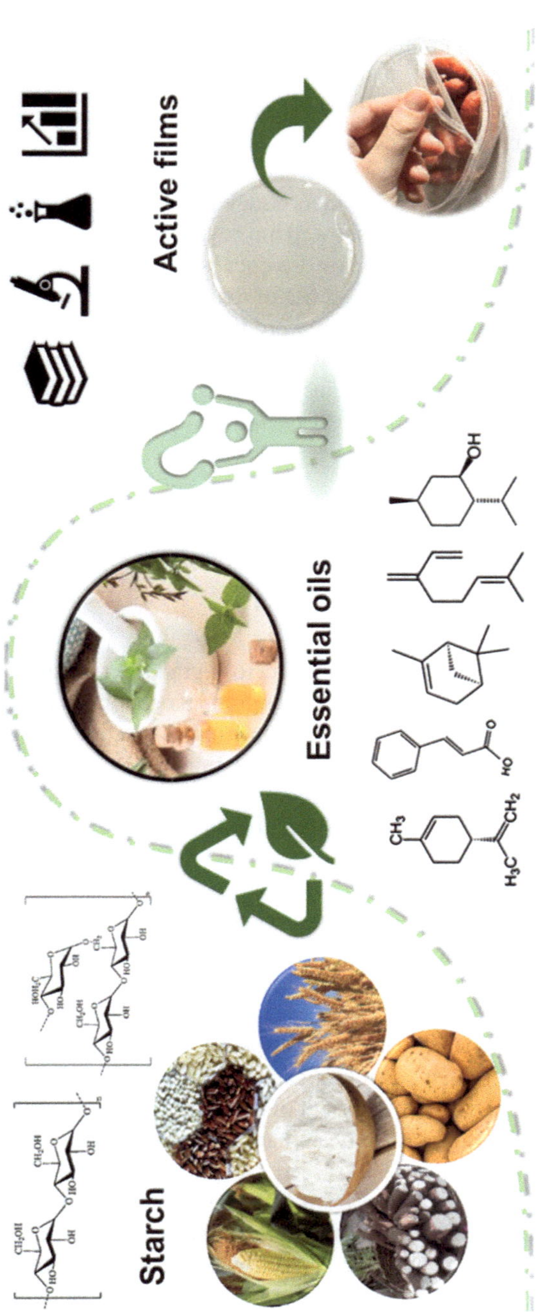

Fig. 4.4 Essential oils as additives in active starch-based food packaging films [93]. Republished with permission from Elsevier. Copyright Elsevier, 2021

density and moisture content. Modified starch/agave fructans microcapsules were encapsulated with thyme oil for controlling mango decay [103], and showed 100% growth inhibition of various fungal pathogens. Cassava starch films, both native and acetylated, were combined with whey protein isolate, rambutan peel extract, and cinnamon essential oil to create composite films [86].

These films, containing both whey protein isolate (WPI) and the blended components, demonstrated excellent antibacterial and antioxidant properties [104]. Starch-montmorillonite composite films were incorporated with a hybrid compound encapsulating carvacrol essential oil [105], showed improved thermal properties, and inhibited *E. coli* growth. Starch-based films were encapsulated with thyme essential oil microcapsules [106]; extending the shelf life of mangoes by 10 days at 25 °C. Sugar palm starch and chitosan composite films were incorporated with extra-virgin olive oil [107]; which enhanced antioxidant activity but slightly decreased antibacterial properties. Cassava starch films were incorporated with lemongrass essential oil [108], showed good antibacterial properties against *S. aureus* and *E. coli*, and did not affect biodegradability.

4.2.4 The Potential of Chitosan and Essential Oils in Sustainable Food Packaging

EOs are known for their volatility, susceptibility to degradation in light or oxygen, and high affinity for fats, making their use challenging, especially in humid conditions [109–113]. Thus, preserving their active components and controlling their release in aqueous environments is essential. Encapsulation using natural polymers like chitosan has emerged as a promising solution, protecting the active compounds and aiding their dispersion, enhancing their functionality shown in Fig. 4.5 [114–116]. Chitosan nanoparticles can effectively encapsulate EOs. Some indispensable oils include peppermint, clove, and green tea. Moreover, this not only aids in the conservation of the oils but also amplifies their antioxidant properties, perhaps leading to an augmentation of their usefulness in many contexts [117, 118]. Interestingly, nanoparticles derived from green tea oil exhibited superior antibacterial activity compared to those from peppermint oil [118]. Similarly, nano-encapsulated clove essential oil demonstrated enhanced performance against specific fungi [117].

In addition, current research has investigated the possibility of synthesizing films based on chitosan that incorporate a variety of essential oils to enhance features such as thermal stability, antioxidant activity, and antibacterial activity (Table 4.3) [119]. Incorporating oils like olive, false flax seed, or lemon essential oil improved the mechanical and antimicrobial properties of the films [120]. Additionally, combining chitosan with EOs such as fennel or cinnamon leaf showed potential for food packaging applications, reducing moisture content and enhancing antioxidant properties [121]. Furthermore, edible films incorporating EOs like *Eucalyptus globulus* or

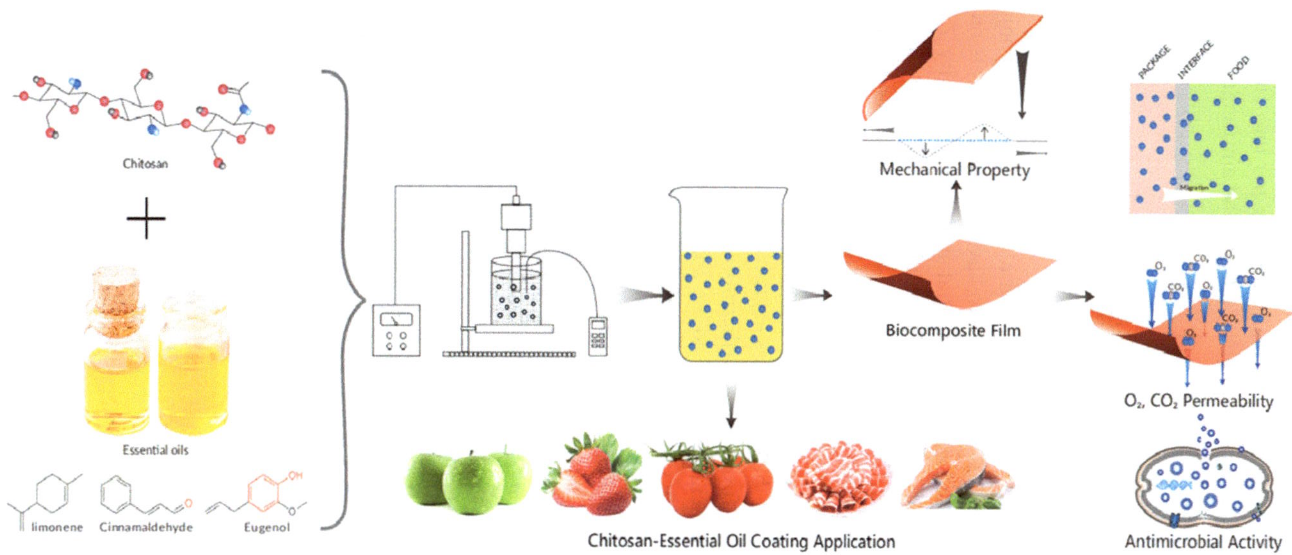

Fig. 4.5 Emerging chitosan-essential oil films and coatings for food preservation [123]. Republished with permission from Elsevier. Copyright Elsevier, 2021

Melissa officinalis exhibited improved antibacterial activity and antioxidant properties, offering new avenues for enhancing food safety and shelf life. Similarly, the incorporation of EOs like ginger or *Thymus vulgaris* into chitosan-based films demonstrated enhanced radical scavenging activity and antifungal properties [122].

4.2.5 Recent Progress on Polysaccharide/Essential Oil Composites

The use of EOs, which are predominantly antimicrobial agents, in conjunction with various fillers or matrices has been shown to have beneficial effects on mechanical and chemical properties and exhibit substantial antibacterial activity. An illustration of this would be the formation of a Pickering emulsion by the inclusion of oregano EO into cellulose nanocrystals (CNC) [138, 139]. The chemical and thermal durability of this emulsion has been demonstrated to be exceptional, and it is effective against bacteria such as *Staphylococcus aureus*, *Staphylococcus cerevisiae*, *Escherichia coli*, and *Bacillus subtilis*. Furthermore, the combination of cinnamon EO and cellulose nanofibers (CNF) proved the emulsion's capacity to maintain its stability over time and selectively targeted *B. subtilis*. Furthermore, the use of ginger essential oil, in conjunction with citric acid, in a CNF edible coating not only improved the flavor, fragrance, and consistency of the meat but also extended the amount of time that it could be stored [140].

In addition, it has been demonstrated that the utilization of various combinations of EOs like oregano, thyme, and cinnamon with various substances like cellulose acetate or methylcellulose can improve the flexibility of meat and poultry products, reduce the transmission rates, and extend the shelf life of these products [142, 143]. This contributes to the effective combat against a wide range of bacterial species, which is a significant benefit. These cutting-edge combinations have the potential to alleviate worries regarding the safety of food, improve the quality of the product, and further extend the product's lifetime on the market. An investigation has been conducted to determine how the utilization of a number of different antimicrobial agents, specifically EOs, in conjunction with a number of different fillers or matrices, such as hydroxyl-propyl methyl cellulose, chitosan, and chitosan nanoparticles, can influence the mechanical and chemical properties of the material, in addition to the antimicrobial activity of the material [144–147]. The introduction of the essential oil of *Thyme daenensis* into hydroxyl-propylmethyl cellulose edible films resulted in a reduction in the tensile strength and Young's modulus of the films. This was in addition to the fact that the oil shown antibacterial activities against both Gram-positive and Gram-negative bacteria [145]. Mentha spicata EO was added to chitosan as an edible coating, which resulted in improved sensory features and increased the amount of time that fresh strawberries could be stored [144].

Additionally, it was able to successfully inhibit the development of *L. monocytogenes* and other bacteria. The incorporation of red thyme and oregano extracts

Table 4.3 Chitosan/essential oil-based films: casting method and antimicrobial effects

Essential oil	Incorporation method	Effect on films	Application	References
Peppermint	Emulsification/ ionic gelation	Maintained phenolic content, improved antioxidant activity		[118]
Green tea	Emulsification/ ionic gelation	Enhanced antibacterial activity (esp. against *S. aureus*)		[118]
Clove	Emulsification/ ionic gelation	Superior performance against *A. niger*		[118]
Lippia	ρ-coumaric acid modification	Improved essential oil's antioxidant ability		[117]
Cyperus articulatus	Ionic gelation	Extended-period antioxidant activity	Food industry (natural fungicides, antioxidants)	[124]
Edible nanochitosan, *Byrsonima crassifolia*	Spray coating	Improved bell pepper quality, antimicrobial effects	Food preservation (coating)	[125]
Various animal fats/plant oils	Solvent casting	Improved surface morphology, thermal stability, mechanical properties, and antibacterial activity	Environmentally friendly chitosan packaging films	[119]
Perilla frutescens (L.) Britt. essential oil	Solvent casting	Enhanced light barrier and antimicrobial properties, decreased mechanical properties		[126]
False flax seed oil		The material exhibited enhanced thermal stability, improved antioxidant and antimicrobial properties, and increased elongation at break		[120]
Berberis crataegina DC. seed oil, fruit extract		Improved thermal stability, antioxidant, antimicrobial properties, and anti-quorum sensing activity		[127]
Anise essential oil		The material became stronger, stretchier, held moisture better, and resisted water vapor and dissolving	Chicken burger storage (improved quality)	[128]

(continued)

Table 4.3 (continued)

Essential oil	Incorporation method	Effect on films	Application	References
Artemisia annua oil	Liposome encapsulation	Controlled *E. coli* O157:H7, excellent potential for food preservation	Food preservation (coating)	[129]
Carum copticum essential oil		Enhanced antioxidant, antibacterial activities, mechanical and water vapor barrier properties	Food packaging (antioxidant, antimicrobial)	[130]
Lemon essential oil		Fungal prevention, negative impact on fruit aroma		[131]
Eucalyptus globulus essential oil	Casting, solvent evaporation	Decreased moisture content, water solubility, improved antioxidant properties	Improved food safety, shelf life	[132]
Zinc oxide (ZnO) particles, Melissa essential oil	Central composite design (CCD)	Increased tensile strength, reduced water solubility/vapor permeability, improved antimicrobial properties	Food packaging (antimicrobial, biodegradable films)	[133]
Cinnamon essential oil, chitosan nanoparticles		Considerable inhibition of *E. coli* and *S. aureus* growth	Food packaging (antimicrobial)	[134]
Fennel, peppermint essential oil		Reduced moisture content, solubility, water swelling, improved thermal stability, antioxidant activity	Packaging that allows vegetables to breathe while minimizing moisture loss	[135]
Cinnamon leaf, clove oil	Phase inversion, solvent casting	Improved antimicrobial activity against *S. aureus*	Chronic wound therapy	[136]
Pomegranate peel extract, Melissa officinalis essence		Cream cheese packaging with boosted antioxidants, stronger Melissa oil against *B. cereus*, and a smart quality indicator	Food packaging (antimicrobial, smart packaging)	[121]
Ceylon cinnamon, cloves (essential oil components)	Coacervation	Grafting onto cotton fabric yielded excellent antibacterial effects, stronger and more flexible material, and better wettability	Cotton fabrics (antibacterial)	[137]

into a chitosan matrix resulted in an increase in the storage life of strawberries and showed antifungal properties [148]. A decrease in solubility, water vapor permeability (WVP), and ultraviolet light transmission was seen as a consequence of the incorporation of chitosan into rosemary EO [149]. *L. monocytogenes*, *Streptococcus agalactiae*, and *E. coli* were the bacterial strains that were especially targeted by this combination. When incorporated into a chitosan edible film, the EO of *Ziziphora clinopodioides* showed decreased solubility, water vapor permeability (WVP), UV-light transmission, swelling index, and tensile strength [150]. Additionally, it was able to successfully combat *Listeria monocytogenes*, *Staphylococcus aureus*, and *Bacillus cereus*. These combinations indicate the capability of extending the period of time that particular food goods may be stored without going bad, while also retaining or increasing the quality features of the products (Fig. 4.6).

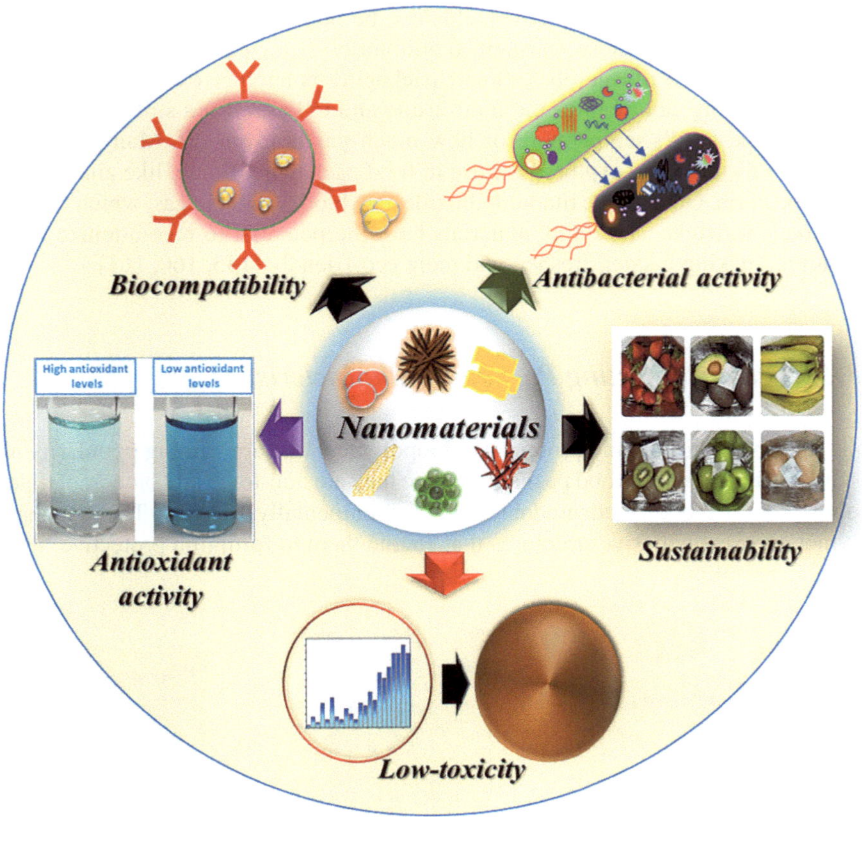

Fig. 4.6 Roles of polysaccharides-based nanomaterials in food preservation and extension of shelf-life of food products [141]. Republished with permission from Elsevier. Copyright Elsevier, 2023

4.3 Nanomaterials Inclusion for Preservation

Tiny particles, naturally occurring or man-made, called nanomaterials, have been around for a long time [151]. Milk and fruits are full of them. Scientists are now using these super-small materials to improve food packaging. For instance, they've created tiny salt crystals that deliver full flavor with less salt. Nanomaterials can also carry vitamins and minerals or improve food's color, taste, and shelf life (Fig. 4.7) [152–154]. They can even help make packaging stronger and more heat-resistant. Consumers are demanding better packaging that's good for the environment. In response, scientists are developing new types of packaging from natural materials like starches and cellulose. These new packages can also contain natural antimicrobials like EOs to fight food spoilage [155].

Researchers are even creating super tiny antimicrobial shields made of metals or metal oxides [157–159]. These shields can be built into packaging to fight off harmful bacteria, even those resistant to antibiotics. They can also be adjusted to target specific types of microbes. Smart packaging is another exciting area. This type of packaging can monitor the food inside and tell you if it's still fresh. It can even track the food's temperature or how much gas is given off. Many different nanomaterials are being explored for use in food packaging. Some, like zinc oxide, can kill bacteria. Others, like titanium dioxide, can block ethylene gas, which speeds up spoilage in fruits. These tiny materials have the potential to revolutionize food packaging, making it safer, fresher, and more eco-friendly [155, 160, 161].

4.3.1 *Factors Affecting Packaging Characteristics*

Polysaccharides, lipids, and proteins are biopolymers that are gaining popularity as sustainable materials for food packaging since they are non-toxic and biocompatible. Biopolymers are both biodegradable and environmentally friendly. These natural polymers possess cohesive structures that enable them to form thin protective films

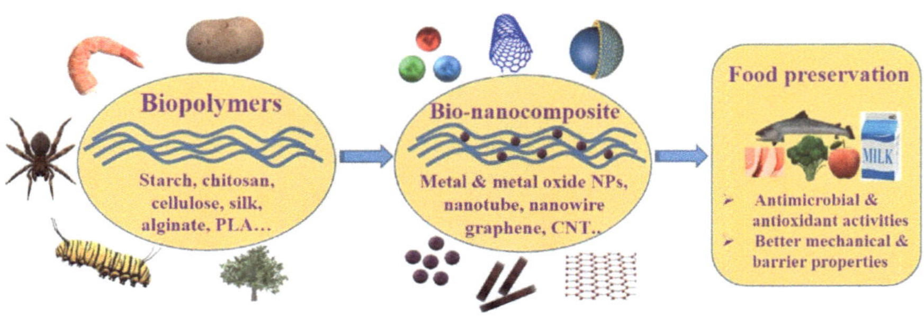

Fig. 4.7 Functional biopolymer-based nanocomposites for active food packaging application, Reproduced from Ref. [156], Copyright CC-BY terms, 2021

on food surfaces. The protective layers formed by these biopolymer-based materials play a vital role in maintaining food stability and extending postharvest shelf-life through several mechanisms:

- Regulation of gas, moisture, and volatile component exchange between the food and surroundings.
- Protection against microbial contamination.
- These innovative coatings prevent the loss of desirable flavors and aromas while acting as carriers for beneficial ingredients like antioxidants, antimicrobials, vitamins, and coloring/flavoring agents. This helps extend shelf life and enhance food products overall nutritional and sensory qualities.

Despite the dominance of petroleum-based synthetic polymers in the packaging industry, the increasing strain on finite resources and environmental concerns have spurred the adoption of naturally occurring biopolymers and bio-based polymers. A significant number of studies have been conducted on the incorporation of nanoparticles into both types of polymers to improve their mechanical, thermal, barrier, and optical characteristics. Nanocomposites based on biopolymers, also known as BNCs, are multiphase materials that contain nanoscale objects scattered inside a biopolymer matrix. As a result of the wide interface that exists between the nanoscale components and the matrix, these materials have superior mechanical, thermal strength, and barrier characteristics when compared to pure biopolymers and traditional composite films [154].

Nanoscale objects, referred to as "nanofillers," can be organic or inorganic and vary in geometric shape:

- One-dimensional nanofillers have two dimensions less than 100 nm.
- Two-dimensional nanofillers have one dimension less than 100 nm.
- Zero-dimensional nanofillers have all three dimensions less than 100 nm [162, 163].

The optical properties of biopolymer-based nanocomposites, including UV-light barrier properties, transparency, brightness, haze, and refractive index, are crucial for food packaging applications and consumer satisfaction. Incorporating nanomaterials such as metals and their oxides or nanoclays can enhance UV-barrier properties. The optical properties depend on factors like concentration, size, shape, and distribution pattern of nanomaterials [164]. Scientists have developed improved packaging materials with enhanced UV-light barrier properties. These materials utilize nanotechnology, incorporating nanoparticles like ZnO and TiO_2 into films made from natural polymers such as agar, carrageenan, carboxymethyl cellulose (CMC), starch, and polyvinyl alcohol [165]. Additionally, the transmittance of WPI-based nanocomposites can be adjusted by varying the concentration of cellulose nanofibers [166]. Researchers have developed eco-friendly packaging solutions that offer enhanced protection against UV light. This is achieved by incorporating nanoparticles like ZnO and TiO_2 into films made from natural polymers, such as agar and carrageenan. This technology can extend food products' shelf life and improve their overall quality[167, 168].

Various nanofillers have been investigated to improve the properties of biopolymer nanocomposites for food packaging (Table 4.4). Metal oxides like ZnO and TiO_2 offer a one-two punch, providing both antimicrobial activity and UV protection for packaged foods [165, 169, 170]. Other nanofillers like MnO_2 and CuO focus on specific functionalities, such as MnO_2's bacteriostatic effect or CuO's antifungal properties [171, 172]. Nanoparticles like silicon dioxide (nano-SiO_2) enhance both UV-shielding and water barrier properties while also improving the mechanical strength of the biopolymer matrix [173, 174]. Similarly, alumina (Al_2O_3) improves water resistance and mechanical properties [175]. Iron oxide nanoparticles (Fe_3O_4) contribute to tensile strength, thermal stability, and oxygen scavenging, making them valuable for maintaining food quality [176]. Clays like MMT and bentonite offer a combination of moisture barrier properties and enhanced mechanical strength [177]. Nano-kaolinite offers a unique benefit by reducing photo-oxidative degradation in packaged foods, while halloysite nanotubes improve rigidity, strength, and Young's modulus of the biopolymer composite [178]. Natural biopolymers like chitosan nanoparticles (CH NPs) can be incorporated as nanofillers. These offer antibacterial activity, improved water barrier properties, and UV protection [179]. Chitin nanofibers (CNF) contribute antifungal properties, barrier properties, and improved tensile strength to biopolymer nanocomposites [180, 181]. They can even enhance thermal and mechanical properties [182, 183].

4.3.2 Recent Progress on Innovative Packaging to Reduce Food Spoilage and Extending Shelf-Life

An increase in the number of sustainable alternatives is being driven by the worries of consumers around plastic packaging. According to recent evaluations, this has resulted in a significant growth in the number of food packaging systems that are based on biopolymers. Biopolymers, which include carbohydrates, proteins, organic acids, and esters, are often utilized in the research and development of environmentally friendly and biodegradable food packaging [153]. The utilization of bio-nanocomposite materials in food packaging represents a promising and innovative development. Scientists are consistently achieving achievements in all directions as they utilise nanomaterials to enhance biopolymers.

Several advancements have been made, including enhanced barrier properties to restrict the penetration of gases and moisture, improved mechanical properties for increased strength and flexibility, and novel functionalities like resistance to microbes and extension of shelf life through antioxidants. These advanced materials often contain components such as nanoclays, nanoparticles of different metals, and even nanocarbons. It is important to examine the increasing popularity of cellulose nanofibers and chitosan nanoparticles, which are natural alternatives. The strong interaction between biopolymers and nanomaterials with high aspect ratios and huge

Table 4.4 Different kinds of nanofillers and their functionalities

Nanofiller	Functionality	References
Clay minerals		
Montmorillonite (MMT)	High surface area, good dispersion, cation exchange capacity	[184]
Kaolinite, Saponite, Hectorite	Barrier properties	[185]
Processing techniques		
In-situ polymerization, Solution-based processing		[186]
Melt processing	More advantages	[186]
Property impacts		
Mechanical strength	Increases up to an optimal loading point	[154, 187]
Tensile Strength, Young's Modulus	Varies depending on concentration	[188–190]
Elongation-at-Break	Decreases with increased clay	[189]
Thermal Stability (Tg)	Increases	[189]
Water Vapor Permeability	Varies depending on concentration	[189, 191]
Barrier Properties	Improves against water vapor, O_2, CO_2	[189, 192]
Other nanofillers		
Graphene	High thermal, mechanical, and barrier properties	[193]
Graphene Oxide (GO)	Blends well with water-soluble biopolymers	[192, 194]
Carbon Nanotubes	Enhanced thermal and mechanical properties	[195–197]
Cellulose Nanocrystals (CNCs)	Improves mechanical and barrier properties	[198–200]
Chitin Nanofibrils	Enhanced mechanical, barrier, and antimicrobial properties	[201]

surface areas leads to a highly efficient food packaging solution. This is the case due to the distinctiveness of their shape and size [154, 202, 203].

Besides creating films and coatings made from biopolymers that improve the durability of products, bio-nanocomposites also have several other applications. These new materials may be thermoformed into utensils such as bowls, platters, and cups, making them versatile for various applications in the food industry. Research has confirmed that they can prolong the durability of many food items, such as fresh vegetables and lean meats. Research on composites of sodium alginate and gum acacia containing silver nanoparticles found that these materials effectively inhibited microbial growth and reduced weight loss in preserved black grapes. Bio-nanocomposite packaging holds great promise for enhancing food preservation and minimizing food waste [204]. This study examines grape varieties beyond black grapes. Despite being maintained at room temperature, green grapes enclosed in bio-nanocomposite films composed of agar and zinc oxide nanoparticles (ZnO) substantially increased their shelf life. All available research indicates that these cutting-edge packing materials can extend the duration for which fresh fruits and vegetables can remain fresh throughout storage [161].

Mangoes now have a greatly extended shelf life due to the innovative packaging that utilizes titanium dioxide (TiO_2) nanoparticles, bergamot essential oil, and bioplastics (PLA). The inherent sweetness and firmness of the fruit are preserved, and the bags' ability to safeguard against bacteria that lead to fruit spoilage is an added benefit. By employing this innovative packing technique, you can extend the duration during which you can savor the delectable mangoes and potentially minimize food wastage [205]. Moreover, strawberry films composed of carboxymethyl cellulose (CMC), guar gum, and silver nanoparticles show great potential. Apples covered with a nanocomposite consisting of silicon dioxide (SiO_2) and soy protein have the same outcome, namely, an extension of their remaining shelf life. This study demonstrates the versatility of bio-nanocomposite packaging for preserving a wide range of fruits. This packaging offers the advantage of both extending the lifespan of perishable goods and decreasing food waste [206, 207].

Food preservation is undergoing a revolution due to the use of packaging made from bio-nanocomposite materials. This cutting-edge method combines natural materials with nanoparticles to create packaging that effectively avoids the deterioration of different food items. Starch and EO films derived from sweet potatoes were found to effectively hinder the growth of dangerous microorganisms on baby spinach while simultaneously maintaining its freshness and visual attractiveness [208]. In addition, chicken preserved using various bio-nanocomposite films had beneficial outcomes. Studies have demonstrated that probiotic films can extend the shelf life of chicken [209], whilst zinc oxide nanoparticles can reduce the occurrence of spoiled chicken [210]. A particular nanocomposite significantly improves the storage quality of cheese, demonstrating the potential for improvement in this area [211]. Research has shown that applying coatings containing rosemary EO can help preserve fish fillets for a longer period. This indicates that this strategy is not restricted to only fresh food. The findings of this study demonstrate the significant potential of bio-nanocomposite packaging in prolonging the durability of various products, hence reducing food waste and enhancing food safety [212].

Researchers are exploring a variety of nanomaterials to improve food packaging and extend shelf life. Zinc oxide nanoparticles (ZnO NPs) incorporated into chitosan films or coatings have effectively reduced microbial growth and spoilage in guava, okra, and ladyfinger [203, 213]. Similar results were achieved with chitosan nanoparticles (Chitosan NPs) in starch films for cherry tomatoes [214]. TiO_2 NPs were also useful in extending the shelf life of crude fish oil and lamb meat by retarding oxidation and microbial growth [215, 216].

Beyond controlling microbes, other nanomaterials offer additional benefits. Silicon dioxide nanoparticles (SiO_2 NPs) enhanced the mechanical and barrier properties of gelatin films, making them more suitable for active food packaging [217]. Melanin nanoparticles (Melanin NPs) increased the strength and water vapor barrier properties of agar films while also exhibiting high antioxidant activity [218]. In another study, combining zinc oxide nanoparticles with fish skin gelatin and ginger EO in a wrapping film effectively extended the shelf life of fresh meat by delaying lipid oxidation and suppressing bacterial growth [219]. Nanomaterials can even be used in combination. A biocomposite film containing rosemary EO, titanium dioxide nanoparticles, and cellulose nanofibrils (CNF) was successful in reducing microbial spoilage, lipid oxidation, and lipolysis in meat stored under refrigerated conditions [216]. Finally, incorporating titanium dioxide nanoparticles into PLA pouches significantly extended the shelf life of cottage cheese without impacting its taste or texture [220]. These studies demonstrate the promise of nanomaterials for creating more effective food packaging systems.

4.4 Challenges and Future Perspectives

Food manufacturers are increasingly looking for natural alternatives to chemical preservatives. EOs from plants and spices are gaining interest due to their antimicrobial and antioxidant properties, as well as their safety for human consumption (GRAS status). However, their strong flavor can be a problem. One solution is to incorporate them into edible films [221, 222]. These edible films can improve food preservation by reducing water vapor permeability and offering antimicrobial/antioxidant effects. Recent research suggests that encapsulating EOs in tiny spheres (nanomicelles) can further enhance their antimicrobial activity and slow down their release [223]. While EOs can be easily incorporated into film materials, their effectiveness can be affected by taste. Some EO components, like citral and eugenol, have shown promise in controlling food spoilage microorganisms when used in edible coatings [224]. Aloe vera gel is another emerging natural ingredient with potential in the food industry. However, processing methods need to be optimized to preserve its beneficial properties. Interestingly, studies indicate that aloe vera gel, despite being mostly a polysaccharide (sugar), can be surprisingly effective as a moisture barrier on its own, without needing additional lipids [225].

Despite their benefits, edible coatings have some drawbacks. Thick coatings can limit oxygen exchange and lead to off-flavors in food. Additionally, some coatings are moisture-attracting, which can promote microbial growth [226]. Further research is needed to address these limitations and make edible coatings a more practical and versatile solution for food preservation. As the demand for fresh, minimally processed food grows, so does the need for better packaging. Innovative packaging is crucial for keeping food safe during storage and transport. Bioplastics, made from natural materials, are being explored as a more sustainable alternative to traditional packaging. However, they're not quite ready for widespread use because they aren't as strong or effective as conventional options, and they can be expensive to produce. Scientists are working on ways to improve bioplastics for food packaging. This includes mixing them with other materials or using tiny particles (nanoparticles) to make them stronger and more effective at preserving food. While there's been a lot of research on improving bioplastics, more work is needed to make them a practical option for large-scale food production. One promising area of research is using nanotechnology to create even better bioplastics. Nanoparticles could be used to improve the flavor, shelf life, and overall quality of food. They could also help prevent spoilage and the growth of harmful bacteria. However, safety is a major concern, and more research is needed to ensure these new technologies are safe for consumers.

The awareness and expectations of consumers regarding the quality of food have increased, resulting in a greater demand for packaging solutions that are both appropriate and sustainable. Hence, the creation of packaging necessitates careful consideration of elements such as the origins of raw materials, compatibility with living organisms, longevity, methods for preserving food, approaches for discarding, and effects on the environment. Within the context of active packaging, the sources of active components, which may include antibacterial and antioxidant components, are of the utmost necessary importance. When integrating active components into packaging, it is essential to take into consideration a variety of factors, including but not limited to safety issues, regulatory restrictions, economic consequences, manufacturing capacity, commercial viability, sensory characteristics, and consumer acceptability [227]. Active packaging in the United States is governed by the Federal Food, Drug, and Cosmetic Act (FFDCA) and regulated by the US Food and Drug Administration (USFDA) [228]. Active substances deliberately integrated into packaging sheets are categorized as food additives and are listed under the Generally Recognized as Safe (GRAS) category. The regulation of antimicrobial chemicals in packaging is governed by distinct amendments, namely Regulation 1282/2011 and Regulation-1935/590/EC.

Because biopolymers are non-hazardous and favorable to the environment, they are increasingly recognized as potential replacements for conventional materials used in packaging the product. Nevertheless, the general deployment of large-scale commercialization is limited by limits in barrier, thermal, and mechanical qualities, as well as excessive production prices [226]. Research is required to improve the physical and microbiological characteristics of biopolymer-based packaging materials using methods such as nanocomposites and polymeric mixing. Nanotechnology has the potential to enhance the efficiency and longevity of biopackaging materials, particularly in edible films and coatings. Nanoencapsulated bioactive compounds demonstrate promise for precise release and enhanced food quality. Future studies should prioritize the identification of synergistic interactions between various bioactive agents, the optimization of formulations, and the examination of interactions between food components and antimicrobial nanocarriers [227, 228]. The widespread usage of bio-nanocomposites in packaging applications is limited due to the high costs associated with the manufacturing process and the low production rates. Further investigation is necessary to examine the most effective compositions and methods of processing, as well as conducting toxicological studies and implementing safety precautions prior to the widespread use of nanomaterials in food items.

4.5 Conclusion

Possible advancements in food packaging, such as using EOs and nanomaterials, might offer a more environmentally friendly and effective approach to food preservation. EOs align to provide safe and ecologically sustainable choices because of their natural antibacterial and antioxidant properties. An important advantage is that they can enhance the efficacy of biopolymer packaging. However, there are impediments. Encapsulation techniques are necessary to overcome the hydrophobic nature of EOs and ensure their long-term effectiveness. In addition, to ensure widespread application, it is necessary to employ production techniques on a big scale and possess a comprehensive understanding of potential interactions between EOs and food additives. Nanomaterials provide several potential applications, including the augmentation of mechanical and barrier qualities in packaging, as well as the incorporation of antioxidant and antibacterial characteristics. However, further study is necessary to investigate and resolve issues regarding the safety of nanoparticles in packaging and their potential migration into food, both from a consumer and environmental perspective. Both EOs and nanomaterials have great promise for the future of food packaging. However, it is imperative that growth is carried out in a responsible manner, prioritizing safety. In order to address the gaps in our knowledge, establish secure protocols, and fully harness the potential of these advanced technologies for a food system that is both safer and more sustainable, it is imperative for researchers, entrepreneurs, and policymakers to collaborate.

Declaration of Competing Interest The authors declare that they have no known competing financial interests or personal relationships that could have appeared to influence the work reported in this paper. Generative AI and AI-assisted technologies have been used in the writing process to improve the readability and language of the work only. After using this tool, the author(s) reviewed and edited the content as needed and take(s) full responsibility for the publication's content.

References

1. Shankar, S., Rhim, J.W.: Polymer nanocomposites for food packaging applications. In: Functional and Physical Properties of Polymer Nanocomposites, pp. 29–55 (2016)
2. Sharma, S., Barkauskaite, S., Jaiswal, A.K., Jaiswal, S.: Essential oils as additives in active food packaging. Food Chem. **343**, 128403 (2021)
3. Joshi, J.T., Harsha, V., Arakal, J.J., Krishnan, A.S.: Engineered nanomaterials in food packaging: synthesis, safety issues, and assessment. In: Food Process Engineering and Technology: Safety, Packaging, Nanotechnologies and Human Health, pp. 245–266. Springer (2024)
4. Giannakas, A.E.: Bionanocomposites with hybrid nanomaterials for food packaging applications. In: Advances in Biocomposites and Their Applications, pp. 201–225. Elsevier (2024)
5. Nabizadeh, F., Kashtiban, A.E., Amiri, S.: Starch-based nanomaterials for food packaging. In: Starch Based Nanomaterials for Food Packaging, pp. 51–82. Elsevier (2024)
6. El-Gammal, M.A., Mohammed, M.S., Hawary, S.A., Youssef, D.G., Aly, F.M., Fahmy, H.M.: Nanomaterials in food contact materials. In: Handbook of Nanomaterials, vol. 2, pp. 715–744. Elsevier (2024)

7. Shah, Y.A., et al.: Mechanical properties of protein-based food packaging materials. Polymers **15**(7), 1724 (2023)
8. Deshmukh, R.K., Gaikwad, K.K.: Natural antimicrobial and antioxidant compounds for active food packaging applications. Biomass Convers. Biorefinery **14**(4), 4419–4440 (2024)
9. Liu, F., et al.: Future foods: alternative proteins, food architecture, sustainable packaging, and precision nutrition. Crit. Rev. Food Sci. Nutr. **63**(23), 6423–6444 (2023)
10. Bhaskar, R., Zo, S.M., Kanan, B.N., Purohit, S., Gupta, M.K., Han, S.S.: Recent development of protein-based biopolymers in food packaging applications: a review. Polym. Test. 108097 (2023)
11. Sudheer, S., Bandyopadhyay, S., Bhat, R.: Sustainable polysaccharide and protein hydrogel-based packaging materials for food products: a review. Int. J. Biol. Macromol. 125845 (2023)
12. Kumar, N., et al.: Recent trends in edible packaging for food applications—perspective for the future. Food Eng. Rev. **15**(4), 718–747 (2023)
13. Mia, R., Sk, M.S., Oli, Z.B.S., Ahmed, T., Kabir, S., Waqar, M.A.: Functionalizing cotton fabrics through herbally synthesized nanosilver. Clean. Eng. Technol. **4**, 100227 (2021). https://doi.org/10.1016/j.clet.2021.100227
14. Jiang, H., et al.: Eco-friendly dyeing and finishing of organic cotton fabric using natural dye (gardenia yellow) reduced-stabilized nanosilver: full factorial design. Cellulose **29**(4), 2663–2679 (2022). https://doi.org/10.1007/s10570-021-04401-9
15. de Sousa, M.S., Schlogl, A.E., Estanislau, F.R., Souza, V.G.L., dos Reis Coimbra, J.S., Santos, I.J.B.: Nanotechnology in packaging for food industry: past, present, and future. Coatings **13**(8), 1411 (2023)
16. Singh, R., et al.: Future of nanotechnology in food industry: challenges in processing, packaging, and food safety. Global Chall. **7**(4), 2200209 (2023)
17. Jiang, H., et al.: Fabrication and stabilization of green nanosilver using gardenia yellow natural dyes for efficient degradation of bacteria. Environ. Prog. Sustain. Energy **42**(3), e14048 (2023). https://doi.org/10.1002/ep.14048
18. Wan, H., Huang, Q., Mia, R., Tao, X., Mahmud, S., Liu, H.: Bioreduction and stabilization of nanosilver using Chrysanthemum phytochemicals for antibacterial and wastewater treatment. ChemistrySelect **7**(29), e202200649 (2022). https://doi.org/10.1002/slct.202200649
19. Lin, L., Wan, H., Mia, R., Jiang, H., Liu, H., Mahmud, S.: Bioreduction and stabilization of antibacterial nanosilver using Radix Lithospermi phytonutrients for azo-contaminated wastewater treatment: synthesis, optimization and characterization. J. Cluster Sci. **34**(2), 1141–1155 (2023). https://doi.org/10.1007/s10876-022-02280-z
20. Gupta, R.K., Abd El Gawad, F., Ali, E.A., Karunanithi, S., Yugiani, P., Srivastav, P.P.: Nanotechnology: current applications and future scope in food packaging systems. Meas. Food 100131 (2023)
21. Hossen, M.A., et al.: Essential oil–loaded biopolymeric particles on food industry and packaging: a review. Int. J. Biol. Macromol. 130765 (2024)
22. Zubair, M., Shahzad, S., Hussain, A., Pradhan, R.A., Arshad, M., Ullah, A.: Current trends in the utilization of essential oils for polysaccharide- and protein-derived food packaging materials. Polymers **14**(6), 1146. https://www.mdpi.com/2073-4360/14/6/1146 (2022). (Online)
23. Saeed, K., Pasha, I., Jahangir Chughtai, M.F., Ali, Z., Bukhari, H., Zuhair, M.: Application of essential oils in food industry: challenges and innovation. J. Ess. Oil Res. **34**(2), 97–110 (2022)
24. Rout, S., et al.: Recent trends in the application of essential oils: the next generation of food preservation and food packaging. Trends Food Sci. Technol. **129**, 421–439 (2022)
25. Han, Y., Yu, M., Wang, L.: Preparation and characterization of antioxidant soy protein isolate films incorporating licorice residue extract. Food Hydrocoll. **75**(NA), 13–21 (2018). https://doi.org/10.1016/j.foodhyd.2017.09.020
26. Atarés, L., Chiralt, A.: Essential oils as additives in biodegradable films and coatings for active food packaging. Trends Food Sci. & Technol. **48**(NA), 51–62 (2016). https://doi.org/10.1016/j.tifs.2015.12.001

27. Ahmad, M., Benjakul, S., Prodpran, T., Agustini, T.W.: Physico-mechanical and antimicrobial properties of gelatin film from the skin of unicorn leatherjacket incorporated with essential oils. Food Hydrocoll. **28**(1), 189–199 (2012). https://doi.org/10.1016/j.foodhyd.2011.12.003
28. Arfat, Y.A., Benjakul, S., Prodpran, T., Sumpavapol, P., Songtipya, P.: Properties and antimicrobial activity of fish protein isolate/fish skin gelatin film containing basil leaf essential oil and zinc oxide nanoparticles. Food Hydrocoll. **41**(NA), 265–273 (2014). https://doi.org/10.1016/j.foodhyd.2014.04.023
29. Tongnuanchan, P., Benjakul, S., Prodpran, T.: Structural, morphological and thermal behaviour characterisations of fish gelatin film incorporated with basil and citronella essential oils as affected by surfactants. Food Hydrocoll. **41**(NA), 33–43 (2014). https://doi.org/10.1016/j.foodhyd.2014.03.015
30. Hussain, A.I., Anwar, F., Nigam, P.S.N., Ashraf, M., Gilani, A.H.: Seasonal variation in content, chemical composition and antimicrobial and cytotoxic activities of essential oils from four Mentha species. J. Sci. Food Agric. **90**(11), 1827–1836 (2010). https://doi.org/10.1002/jsfa.4021
31. Guerra, I.C.D., et al.: Coatings comprising chitosan and Mentha piperita L. or Mentha × villosa Huds essential oils to prevent common postharvest mold infections and maintain the quality of cherry tomato fruit. Int. J. Food Microbiol. **214**(NA), 168–178 (2015). https://doi.org/10.1016/j.ijfoodmicro.2015.08.009
32. Abbaszadeh, S., Sharifzadeh, A., Shokri, H., Khosravi, A.R., Abbaszadeh, A.: Antifungal efficacy of thymol, carvacrol, eugenol and menthol as alternative agents to control the growth of food-relevant fungi. Journal de Mycologie Medicale **24**(2), e51–e56 (2014). https://doi.org/10.1016/j.mycmed.2014.01.063
33. Tongnuanchan, P., Benjakul, S., Prodpran, T.: Properties and antioxidant activity of fish skin gelatin film incorporated with citrus essential oils. Food Chem. **134**(3), 1571–1579 (2012). https://doi.org/10.1016/j.foodchem.2012.03.094
34. Liu, Q., Wang, W., Qi, J.-R., Huang, Q., Xiao, J.: Oregano essential oil loaded soybean polysaccharide films: effect of pickering type immobilization on physical and antimicrobial properties. Food Hydrocoll. **87**(NA), 165–172 (2019). https://doi.org/10.1016/j.foodhyd.2018.08.011
35. Li, X., Yang, X., Deng, H., Guo, Y., Xue, J.: Gelatin films incorporated with thymol nanoemulsions: physical properties and antimicrobial activities. Int. J. Biol. Macromol. **150**(NA), 161–168 (2020). https://doi.org/10.1016/j.ijbiomac.2020.02.066
36. Ahmed, J., Hiremath, N., Jacob, H.: Antimicrobial efficacies of essential oils/nanoparticles incorporated polylactide films against L. monocytogenes and S. typhimurium on contaminated cheese. Int. J. Food Prop. **20**(1), 53–67 (2016). https://doi.org/10.1080/10942912.2015.1131165
37. Alexandre, E.M.C., Lourenço, R.V., Bittante, A.M.Q.B., Moraes, I.C.F., do Amaral Sobral, P.J.: Gelatin-based films reinforced with montmorillonite and activated with nanoemulsion of ginger essential oil for food packaging applications. Food Packag. Shelf Life **10**(NA), 87–96 (2016). https://doi.org/10.1016/j.fpsl.2016.10.004
38. Ejaz, M., Arfat, Y.A., Mulla, M., Ahmed, J.: Zinc oxide nanorods/clove essential oil incorporated Type B gelatin composite films and its applicability for shrimp packaging. Food Packag. Shelf Life **15**(NA), 113–121 (2018). https://doi.org/10.1016/j.fpsl.2017.12.004
39. Ribeiro-Santos, R., et al.: Combined use of essential oils applied to protein base active food packaging: Study in vitro and in a food simulant. Eur. Polym. J. **93**, 75–86 (2017). https://doi.org/10.1016/j.eurpolymj.2017.03.055
40. Wu, J., et al.: Properties and antimicrobial activity of silver carp (Hypophthalmichthys molitrix) skin gelatin-chitosan films incorporated with oregano essential oil for fish preservation. Food Packag. Shelf Life **2**(1), 7–16 (2014). https://doi.org/10.1016/j.fpsl.2014.04.004
41. Hosseini, S.F., Rezaei, M., Zandi, M., Farahmandghavi, F.: Bio-based composite edible films containing Origanum vulgare L. essential oil. Ind. Crop. Prod. **67**(NA), 403–413 (2015). https://doi.org/10.1016/j.indcrop.2015.01.062

42. Echeverría, I., López-Caballero, M.E., del Carmen Gómez-Guillén, M., Mauri, A.N., Montero, M.P.: Active nanocomposite films based on soy proteins-montmorillonite- clove essential oil for the preservation of refrigerated bluefin tuna (Thunnus thynnus) fillets. Int. J. Food Microbiol. **266**(1), 142–149 (2017). https://doi.org/10.1016/j.ijfoodmicro.2017.10.003
43. Ortiz, C.M., Salgado, P.R., Dufresne, A., Mauri, A.N.: Microfibrillated cellulose addition improved the physicochemical and bioactive properties of biodegradable films based on soy protein and clove essential oil. Food Hydrocoll. **79**(NA), 416–427 (2018). https://doi.org/10.1016/j.foodhyd.2018.01.011
44. Pires, C., Ramos, C., Teixeira, B., Batista, I., Nunes, M.L., Marques, A.: Hake proteins edible films incorporated with essential oils: physical, mechanical, antioxidant and antibacterial properties. Food Hydrocoll. **30**(1), 224–231 (2013). https://doi.org/10.1016/j.foodhyd.2012.05.019
45. Atarés, L., Bonilla, J., Chiralt, A.: Characterization of sodium caseinate-based edible films incorporated with cinnamon or ginger essential oils. J. Food Eng. **100**(4), 678–687 (2010). https://doi.org/10.1016/j.jfoodeng.2010.05.018
46. Salgado, P.R., López-Caballero, M.E., Gómez-Guillén, M.C., Mauri, A.N., Montero, M.P.: Sunflower protein films incorporated with clove essential oil have potential application for the preservation of fish patties. Food Hydrocoll. **33**(1), 74–84 (2013). https://doi.org/10.1016/j.foodhyd.2013.02.008
47. Esmaeili, H., et al.: Incorporation of nanoencapsulated garlic essential oil into edible films: a novel approach for extending shelf life of vacuum-packed sausages. Meat Sci. **166**(NA), 108135–108135 (2020). https://doi.org/10.1016/j.meatsci.2020.108135
48. Kilincceker, O., Dogan, İS., Kucukoner, E.: Effect of edible coatings on the quality of frozen fish fillets. LWT-Food Sci. Technol. **42**(4), 868–873 (2009)
49. Rodriguez-Turienzo, L., Cobos, A., Diaz, O.: Effects of edible coatings based on ultrasound-treated whey proteins in quality attributes of frozen Atlantic salmon (Salmo salar). Innov. Food Sci. Emerg. Technol. **14**, 92–98 (2012)
50. Abdalrazeq, M., et al.: Physicochemical and antimicrobial properties of whey protein-based films functionalized with Palestinian Satureja capitata essential oil. Coatings **11**(11), 1364–NA (2021). https://doi.org/10.3390/coatings11111364
51. Fernández-Pan, I., Carrión-Granda, X., Maté, J.I.: Antimicrobial efficiency of edible coatings on the preservation of chicken breast fillets. Food Control **36**(1), 69–75 (2014). https://doi.org/10.1016/j.foodcont.2013.07.032
52. Kavas, G., Kavas, N., Saygili, D.: The effects of thyme and clove essential oil fortified edible films on the physical, chemical and microbiological characteristics of Kashar cheese. J. Food Qual. **38**(6), 405–412 (2015). https://doi.org/10.1111/jfq.12157
53. de Jesús Avena-Bustillos, R., Krochta, J.M., Saltveit, M.E., de Jesús Rojas-Villegas, R., Sauceda-Pérez, J.: Optimization of edible coating formulations on zucchini to reduce water loss. J. Food Eng. 21(2), 197–214 (1994)
54. Taktak, W., et al.: Design of bioinspired emulsified composite European eel gelatin and protein isolate-based food packaging film: thermal, microstructural, mechanical, and biological features. Coatings **10**(1), 26–NA (2019). https://doi.org/10.3390/coatings10010026
55. Lacroix, M., Ouattara, B., Saucier, L., Giroux, M., Smoragiewicz, W.: Effect of gamma irradiation in presence of ascorbic acid on microbial composition and TBARS concentration of ground beef coated with an edible active coating. Radiat. Phys. Chem. **71**(1–2), 73–77 (2004)
56. Wang, S., Lu, A., Zhang, L.: Recent advances in regenerated cellulose materials. Prog. Polym. Sci. **53**(53), 169–206 (2016). https://doi.org/10.1016/j.progpolymsci.2015.07.003
57. Montero, Y., de Souza, A.G., de Oliveira, É.R., dos Santos Rosa, D.: Nanocellulose functionalized with cinnamon essential oil: a potential application in active biodegradable packaging for strawberry. Sustain. Mater. Technol. **29**(NA), e00289–NA (2021). https://doi.org/10.1016/j.susmat.2021.e00289
58. Kolybaba, M., Tabil, L., Panigrahi, S., Crerar, W., Powell, T., Wang, B.: Biodegradable polymers: past, present, and future. In: ASABE/CSBE North Central Intersectional Meeting: American Society of Agricultural and Biological Engineers, p. 1 (2006)

59. Lindman, B., Karlström, G., Stigsson, L.: On the mechanism of dissolution of cellulose. J. Mol. Liq. **156**(1), 76–81 (2010)
60. Ren, H., Chen, C., Wang, Q., Zhao, D., Guo, S.: The properties of choline chloride-based deep eutectic solvents and their performance in the dissolution of cellulose. BioResources **11**(2), 5435–5451 (2016). https://doi.org/10.15376/biores.11.2.5435-5451
61. Gonçalves, S.M., Motta, J.F.G., Ribeiro-Santos, R., Chávez, D.W.H., de Melo, N.R.: Functional and antimicrobial properties of cellulose acetate films incorporated with sweet fennel essential oil and plasticizers. Curr. Res. Food Sci. **3**(NA), 1–8 (2020). https://doi.org/10.1016/j.crfs.2020.01.001
62. Casalini, S., et al.: Antimicrobial activity of different nanocellulose films embedded with thyme, cinnamon, and oregano essential oils for active packaging application on raspberries. Front. Sustain. Food Syst. Orig. Res. **7**, 2023 (2023). https://doi.org/10.3389/fsufs.2023.1190979. (in English)
63. Requena, R., Vargas, M., Atarés, L., Chiralt, A.: Biopolymers carrying essential oils, or their compounds, for food antimicrobial packaging. Curr. Org. Chem. **22**(12), 1141–1156 (2018). https://doi.org/10.2174/1385272822666171129153011
64. do Socorro Rocha Bastos, M., et al.: Physical and mechanical testing of essential oil-embedded cellulose ester films. Polym. Test. **49**(NA), 156–161 (2016). https://doi.org/10.1016/j.polymertesting.2015.11.006
65. Moghimi, R., Aliahmadi, A., Rafati, H.: Antibacterial hydroxypropyl methyl cellulose edible films containing nanoemulsions of Thymus daenensis essential oil for food packaging. Carbohydr. Polym. **175**(NA), 241–248 (2017). https://doi.org/10.1016/j.carbpol.2017.07.086
66. Honarvar, Z., et al.: Application of cold plasma to develop carboxymethyl cellulose-coated polypropylene films containing essential oil. Carbohydr. Polym. **176**(NA), 1–10 (2017). https://doi.org/10.1016/j.carbpol.2017.08.054
67. Muppalla, S.R., Kanatt, S.R., Chawla, S.P., Sharma, A.: Carboxymethyl cellulose–polyvinyl alcohol films with clove oil for active packaging of ground chicken meat. Food Packag. Shelf Life **2**(2), 51–58 (2014). https://doi.org/10.1016/j.fpsl.2014.07.002
68. Shahbazi, Y., Shavisi, N.: Effects of sodium alginate coating containing Mentha spicata essential oil and cellulose nanoparticles on extending the shelf life of raw silver carp (Hypophthalmichthys molitrix) fillets. Food Sci. Biotechnol. **28**(2), 433–440 (2018). https://doi.org/10.1007/s10068-018-0486-y
69. da Silva Dannenberg, G., Funck, G.D., dos Santos Cruxen, C.E., de Lima Marques, J., da Silva, W.P., Fiorentini, Â.M.: Essential oil from pink pepper as an antimicrobial component in cellulose acetate film: Potential for application as active packaging for sliced cheese. LWT-Food Sci. Technol. **81**(NA), 314–318 (2017). https://doi.org/10.1016/j.lwt.2017.04.002
70. Noshirvani, N., et al.: Cinnamon and ginger essential oils to improve antifungal, physical and mechanical properties of chitosan-carboxymethyl cellulose films. Food Hydrocoll. **70**(NA), 36–45 (2017). https://doi.org/10.1016/j.foodhyd.2017.03.015
71. Simsek, M., Eke, B., Demir, H.: Characterization of carboxymethyl cellulose-based antimicrobial films incorporated with plant essential oils. Int. J. Biol. Macromol. **163**(NA), 2172–2179 (2020). https://doi.org/10.1016/j.ijbiomac.2020.09.075
72. Hossain, F., et al.: Antifungal activity of combined treatments of active methylcellulose-based films containing encapsulated nanoemulsion of essential oils and γ–irradiation: in vitro and in situ evaluations. Cellulose **26**(2), 1335–1354 (2018). https://doi.org/10.1007/s10570-018-2135-2
73. Abbasi, H., Fahim, H., Mahboubi, M.: Fabrication and characterization of composite film based on gelatin and electrospun cellulose acetate fibers incorporating essential oil. J. Food Meas. Charact. **15**(2), 2108–2118 (2021). https://doi.org/10.1007/s11694-020-00799-1
74. Zhou, L., Wang, Y.: Physical and antimicrobial properties of zein and methyl cellulose composite films with plasticizers of oleic acid and polyethylene glycol. LWT **140**(NA), 110811–NA (2021). https://doi.org/10.1016/j.lwt.2020.110811
75. de Oliveira Filho, J.G., et al.: Arrowroot starch-based films incorporated with a carnauba wax nanoemulsion, cellulose nanocrystals, and essential oils: a new functional material for

food packaging applications. Cellulose **28**(10), 6499–6511 (2021). https://doi.org/10.1007/s10570-021-03945-0
76. Bahrami, A., Fattahi, R.: Biodegradable carboxymethyl cellulose–polyvinyl alcohol composite incorporated with Glycyrrhiza Glabra L. essential oil: physicochemical and antibacterial features. Food Sci. Nutr. **9**(9), 4974–4985 (2021). https://doi.org/10.1002/fsn3.2449
77. Mei, L., et al.: Characterization of carboxymethyl cellulose films incorporated with Chinese fir essential oil and their application to quality improvement of shine Muscat grape. Coatings **11**(1), 97–NA (2021). https://doi.org/10.3390/coatings11010097
78. Han, Y., Yu, M., Wang, L.: Physical and antimicrobial properties of sodium alginate/carboxymethyl cellulose films incorporated with cinnamon essential oil. Food Packag. Shelf Life **15**(NA), 35–42 (2018). https://doi.org/10.1016/j.fpsl.2017.11.001
79. Yourghanloo, R.A., Hemmati, H., Hamedani, N.F.: Study investigate physicochemical and antioxidant properties of biodegradable film prepared from hydroxy propyl methyl cellulose containing alginate and essential oil of Acanthophyllum and its effect on oxidative resistance of local butter oil (2020)
80. Pola, C.C., et al.: Cellulose acetate active films incorporated with oregano (Origanum vulgare) essential oil and organophilic montmorillonite clay control the growth of phytopathogenic fungi. Food Packag. Shelf Life **9**(NA), 69–78 (2016). https://doi.org/10.1016/j.fpsl.2016.07.001
81. Bastos, M.D.S.R., et al: Physical and mechanical testing of essential oil-embedded cellulose ester films. Polym. Test. **49**, 156–161 (2016)
82. Wiastuti, T., Khasanah, L., Kawiji, W.A., Manuhara, G., Utami, R.: Characterization of active paper packaging incorporated with ginger pulp oleoresin. IOP Conf. Ser. Mater. Sci. Eng. **107**(1), 012057 (2016). (IOP Publishing)
83. Lumdubwong, N.: Applications of Starch-Based Films in Food Packaging (2019)
84. Olsson, E., Menzel, C., Johansson, C., Andersson, R., Koch, K., Järnström, L.: The effect of pH on hydrolysis, cross-linking and barrier properties of starch barriers containing citric acid. Carbohyd. Polym. **98**(2), 1505–1513 (2013). https://doi.org/10.1016/j.carbpol.2013.07.040
85. Menzel, C., et al.: Molecular structure of citric acid cross-linked starch films. Carbohyd. Polym. **96**(1), 270–276 (2013). https://doi.org/10.1016/j.carbpol.2013.03.044
86. Seligra, P.G., Jaramillo, C.M., Famá, L., Goyanes, S.: Biodegradable and non-retrogradable eco-films based on starch-glycerol with citric acid as crosslinking agent. Carbohydr. Polym. **138**(NA), 66–74 (2015). https://doi.org/10.1016/j.carbpol.2015.11.041
87. Müller, C.M.O., Laurindo, J.B., Yamashita, F.: Effect of cellulose fibers addition on the mechanical properties and water vapor barrier of starch-based films. Food Hydrocoll. **23**(5), 1328–1333 (2009). https://doi.org/10.1016/j.foodhyd.2008.09.002
88. Wilpiszewska, K., Czech, Z.: Citric acid modified potato starch films containing microcrystalline cellulose reinforcement–properties and application. Starch-Stärke **66**(7–8), 660–667 (2014). https://doi.org/10.1002/star.201300093
89. Ali, A., et al.: Starch-based antimicrobial films functionalized by pomegranate peel. Int. J. Biol. Macromol. **129**(NA), 1120–1126 (2018). https://doi.org/10.1016/j.ijbiomac.2018.09.068
90. Luchese, C.L., Uranga, J., Spada, J.C., Tessaro, I.C., de la Caba, K.: Valorisation of blueberry waste and use of compression to manufacture sustainable starch films with enhanced properties. Int. J. Biol. Macromol. **115**(NA), 955–960 (2018). https://doi.org/10.1016/j.ijbiomac.2018.04.162
91. Menzel, C., González-Martínez, C., Chiralt, A., Vilaplana, F.: Antioxidant starch films containing sunflower hull extracts. Carbohydr. Polym. **214**(NA), 142–151 (2019). https://doi.org/10.1016/j.carbpol.2019.03.022
92. Menzel, C., González-Martínez, C., Vilaplana, F., Diretto, G., Chiralt, A.: Incorporation of natural antioxidants from rice straw into renewable starch films. Int. J. Biol. Macromol. **146**(146), 976–986 (2019). https://doi.org/10.1016/j.ijbiomac.2019.09.222
93. Vianna, T.C., Marinho, C.O., Marangoni Júnior, L., Ibrahim, S.A., Vieira, R.P.: Essential oils as additives in active starch-based food packaging films: a review. Int. J. Biol. Macromol. **182**, 1803–1819 (2021). https://doi.org/10.1016/j.ijbiomac.2021.05.170

94. Sani, I.K., Geshlaghi, S.P., Pirsa, S., Asdagh, A.: Composite film based on potato starch/apple peel pectin/ZrO_2 nanoparticles/microencapsulated Zataria multiflora essential oil; investigation of physicochemical properties and use in quail meat packaging. Food Hydrocoll. **117**(NA), 106719–NA (2021). https://doi.org/10.1016/j.foodhyd.2021.106719
95. de Souza, A.C., Goto, G.E.O., Mainardi, J.A., Coelho, A.C.V., Tadini, C.C.: Cassava starch composite films incorporated with cinnamon essential oil: antimicrobial activity, microstructure, mechanical and barrier properties. LWT Food Sci. Technol. **54**(2), 346–352 (2013). https://doi.org/10.1016/j.lwt.2013.06.017
96. Avila-Sosa, R., Palou, E., Munguía, M.T.J., Nevárez-Moorillón, G.V., Cruz, A.R.N., López-Malo, A.: Antifungal activity by vapor contact of essential oils added to amaranth, chitosan, or starch edible films. Int. J. Food Microbiol. **153**(1–2), 66–72 (2012)
97. Arezoo, E., Mohammadreza, E., Maryam, M., Abdorreza, M.N.: The synergistic effects of cinnamon essential oil and nano TiO_2 on antimicrobial and functional properties of sago starch films. Int. J. Biol. Macromol. **157**(NA), 743–751 (2019). https://doi.org/10.1016/j.ijbiomac.2019.11.244
98. de Souza, A.G., Ferreira, R.R., Paula, L.C., Mitra, S.K., dos Santos Rosa, D.: Starch-based films enriched with nanocellulose-stabilized Pickering emulsions containing different essential oils for possible applications in food packaging. Food Packag. Shelf Life **27**(NA), 100615–NA (2021). https://doi.org/10.1016/j.fpsl.2020.100615
99. Wang, B., et al.: Physicochemical properties and antibacterial activity of corn starch-based films incorporated with Zanthoxylum bungeanum essential oil. Carbohydr. Polym. **254**(NA), 117314–117314 (2020). https://doi.org/10.1016/j.carbpol.2020.117314
100. Perdana, M.I., Ruamcharoen, J., Panphon, S., Leelakriangsak, M.: Antimicrobial activity and physical properties of starch/chitosan film incorporated with lemongrass essential oil and its application. LWT **141**(NA), 110934–NA (2021). https://doi.org/10.1016/j.lwt.2021.110934
101. Wang, B., et al.: Antibacterial activity, optical, and functional properties of corn starch-based films impregnated with bamboo leaf volatile oil. Food Chem. **357**(NA), 129743–129743 (2021). https://doi.org/10.1016/j.foodchem.2021.129743
102. Syafiq, R., Sapuan, S.M., Zuhri, M.R.M.: Antimicrobial activity, physical, mechanical and barrier properties of sugar palm based nanocellulose/starch biocomposite films incorporated with cinnamon essential oil. J. Mater. Res. Technol. **11**(NA), 144–157 (2021). https://doi.org/10.1016/j.jmrt.2020.12.091
103. Esquivel-Chávez, F., et al.: Control of mango decay using antifungal sachets containing of thyme oil/modified starch/agave fructans microcapsules. Future Foods **3**(NA), 100008–NA (2021). https://doi.org/10.1016/j.fufo.2020.100008
104. Chollakup, R., et al.: Antioxidant and antibacterial activities of cassava starch and whey protein blend films containing rambutan peel extract and cinnamon oil for active packaging. LWT **130**(NA), 109573–NA (2020). https://doi.org/10.1016/j.lwt.2020.109573
105. de Souza, A.G., dos Santos, N.M.A., da Silva Torin, R.F., dos Santos Rosa, D.: Synergic antimicrobial properties of Carvacrol essential oil and montmorillonite in biodegradable starch films. Int. J. Biol. Macromol. **164**(NA), 1737–1747 (2020). https://doi.org/10.1016/j.ijbiomac.2020.07.226
106. Cai, C., Ma, R., Duan, M., Deng, Y., Liu, T., Lu, D.: Effect of starch film containing thyme essential oil microcapsules on physicochemical activity of mango. LWT **131**(NA), 109700–NA (2020). https://doi.org/10.1016/j.lwt.2020.109700
107. Hasan, M., Rusman, R., Khaldun, I., Ardana, L., Mudatsir, M., Fansuri, H.: Active edible sugar palm starch-chitosan films carrying extra virgin olive oil: Barrier, thermo-mechanical, antioxidant, and antimicrobial properties. Int. J. Biol. Macromol. **163**(NA), 766–775 (2020). https://doi.org/10.1016/j.ijbiomac.2020.07.076.
108. Mendes, J.F., et al.: Correlating emulsion characteristics with the properties of active starch films loaded with lemongrass essential oil. Food Hydrocoll. **100**(NA), 105428–NA (2020). https://doi.org/10.1016/j.foodhyd.2019.105428.
109. Navarro, R., Arancibia, C., Herrera, M.L., Matiacevich, S.: Effect of type of encapsulating agent on physical properties of edible films based on alginate and thyme oil. Food Bioprod. Process. **97**(NA), 63–75 (2016). https://doi.org/10.1016/j.fbp.2015.11.001

110. de Oliveira, D.R., Leitão, G.G., Fernandes, P.D., Leitão, S.G.: Ethnopharmacological studies of Lippia origanoides. Rev. Bras **24**(2), 206–214 (2014). https://doi.org/10.1016/j.bjp.2014.03.001
111. Fernandes, L.P., Candido, R.C., de Oliveira, W.P.: Spray drying microencapsulation of Lippia sidoides extracts in carbohydrate blends. Food Bioprod. Process. **90**(3), 425–432 (2012). https://doi.org/10.1016/j.fbp.2011.12.001
112. Marques, H.C.: A review on cyclodextrin encapsulation of essential oils and volatiles. Flavour Fragr. J. **25**(5), 313–326 (2010). https://doi.org/10.1002/ffj.2019
113. Wadhwa, G., Kumar, S., Chhabra, L., Mahant, S., Rao, R.: Essential oil–cyclodextrin complexes: an updated review. J. Incl. Phenom. Macrocycl. Chem. **89**(1), 39–58 (2017). https://doi.org/10.1007/s10847-017-0744-2
114. Damasceno, E.T.S., et al.: Lippia origanoides Kunth. essential oil loaded in nanogel based on the chitosan and ρ-coumaric acid: Encapsulation efficiency and antioxidant activity. Ind. Crop. Prod. **125**(NA), 85–94 (2018). https://doi.org/10.1016/j.indcrop.2018.08.074
115. Zubair, M., Arshad, M., Pradhan, R.A., Ullah, A.: Chitosan/chitin-based composites for food packaging applications. In: Handbook of Chitin and Chitosan, pp. 641–670. Elsevier (2020)
116. Zubair, M., Arshad, M., Ullah, A.: Chitosan-based materials for water and wastewater treatment. In: Handbook of Chitin and Chitosan, pp. 773–809. Elsevier (2020)
117. Hasheminejad, N., Khodaiyan, F., Safari, M.: Improving the antifungal activity of clove essential oil encapsulated by chitosan nanoparticles. Food Chem. **275**(NA), 113–122 (2018). https://doi.org/10.1016/j.foodchem.2018.09.085
118. Shetta, A., Kegere, J., Mamdouh, W.: Comparative study of encapsulated peppermint and green tea essential oils in chitosan nanoparticles: encapsulation, thermal stability, in-vitro release, antioxidant and antibacterial activities. Int. J. Biol. Macromol. **126**(1), 731–742 (2018). https://doi.org/10.1016/j.ijbiomac.2018.12.161
119. Akyuz, L., et al.: Effect of different animal fat and plant oil additives on physicochemical, mechanical, antimicrobial and antioxidant properties of chitosan films. Int. J. Biol. Macromol. **111**(NA), 475–484 (2018). https://doi.org/10.1016/j.ijbiomac.2018.01.045
120. Gürsoy, M., et al.: False flax (Camelina sativa) seed oil as suitable ingredient for the enhancement of physicochemical and biological properties of chitosan films. Int. J. Biol. Macromol. **114**(NA), 1224–1232 (2018). https://doi.org/10.1016/j.ijbiomac.2018.04.029
121. Pirsa, S., Sani, I.K., Pirouzifard, M.K., Erfani, A.: Smart film based on chitosan/Melissa officinalis essences/ pomegranate peel extract to detect cream cheeses spoilage. Food Addit. & Contam. Part A Chem. Anal. Control. Expo. & Risk Assess. **37**(4), 634–648 (2020). https://doi.org/10.1080/19440049.2020.1716079
122. Dinu, M.V., et al.: Physically cross-linked chitosan/dextrin cryogels entrapping Thymus vulgaris essential oil with enhanced mechanical, antioxidant and antifungal properties. Int. J. Biol. Macromol. **184**(NA), 898–908 (2021). https://doi.org/10.1016/j.ijbiomac.2021.06.068
123. Zhang, X., et al.: Emerging chitosan-essential oil films and coatings for food preservation-a review of advances and applications. Carbohydr. Polym. **273**, 118616 (2021). https://doi.org/10.1016/j.carbpol.2021.118616
124. Kavaz, D., Idris, M.A., Onyebuchi, C.: Physiochemical characterization, antioxidative, anticancer cells proliferation and food pathogens antibacterial activity of chitosan nanoparticles loaded with Cyperus articulatus rhizome essential oils. Int. J. Biol. Macromol. **123**(NA), 837–845 (2018). https://doi.org/10.1016/j.ijbiomac.2018.11.177
125. González-Saucedo, A., Barrera-Necha, L.L., Ventura-Aguilar, R.I., Correa-Pacheco, Z.N., Bautista-Baños, S., Hernández-López, M.: Extension of the postharvest quality of bell pepper by applying nanostructured coatings of chitosan with Byrsonima crassifolia extract (L.) Kunth. Postharvest Biol. Technol. **149**(NA), 74–82 (2019). https://doi.org/10.1016/j.postharvbio.2018.11.019
126. Zhang, Z., et al.: Preparation and characterization of biocomposite chitosan film containing Perilla frutescens (L.) Britt. essential oil. Ind. Crop. Prod. **112**(NA), 660–667 (2018). https://doi.org/10.1016/j.indcrop.2017.12.073

127. Kaya, M., et al.: Production and characterization of chitosan based edible films from Berberis crataegina's fruit extract and seed oil. Innov. Food Sci. & Emerg. Technol. **45**(NA), 287–297 (2018). https://doi.org/10.1016/j.ifset.2017.11.013
128. Mahdavi, V., Hosseini, S.E., Sharifan, A.: Effect of edible chitosan film enriched with anise (Pimpinella anisum L.) essential oil on shelf life and quality of the chicken burger. Food Sci. Nutr. **6**(2), 269–279 (2017). https://doi.org/10.1002/fsn3.544
129. Cui, H., Yuan, L., Li, W., Lin, L.: Edible film incorporated with chitosan and Artemisia annua oil nanoliposomes for inactivation of Escherichia coli O157:H7 on cherry tomato. Int. J. Food Sci. Technol. **52**(3), 687–698 (2016). https://doi.org/10.1111/ijfs.13322
130. Jahed, E., Khaledabad, M.A., Almasi, H., Hasanzadeh, R.: Physicochemical properties of Carum copticum essential oil loaded chitosan films containing organic nanoreinforcements. Carbohydr. Polym. 164(NA), 325–338 (2017). https://doi.org/10.1016/j.carbpol.2017.02.022
131. Perdones, Á., Escriche, I., Chiralt, A., Vargas, M.: Effect of chitosan–lemon essential oil coatings on volatile profile of strawberries during storage. Food Chem. **197**(Pt A), 979–986 (2015). https://doi.org/10.1016/j.foodchem.2015.11.054
132. Hafsa, J., et al.: Physical, antioxidant and antimicrobial properties of chitosan films containing Eucalyptus globulus essential oil. LWT-Food Sci. Technol. **68**(NA), 356–364 (2016). https://doi.org/10.1016/j.lwt.2015.12.050
133. Thaker, M., Hanjabam, M.D., Gudipati, V., Kannuchamy, N.: Protective effect of fish gelatin-based natural antimicrobial coatings on quality of Indian salmon fillets during refrigerated storage. J. Food Process. Eng. 40(1), NA–NA (2015). https://doi.org/10.1111/jfpe.12270
134. Vahedikia, N., et al.: Biodegradable zein film composites reinforced with chitosan nanoparticles and cinnamon essential oil: Physical, mechanical, structural and antimicrobial attributes. Colloids Surf. B Biointerfaces **177**(NA), 25–32 (2019). https://doi.org/10.1016/j.colsurfb.2019.01.045
135. Liu, T., Wang, J., Fumin, C., Tan, Z., Liu, L.: Development and characterization of novel active chitosan films containing fennel and peppermint essential oils. Coatings 10(10), 936–NA (2020). https://doi.org/10.3390/coatings10100936
136. Antunes, J.C., et al.: Eugenol-containing essential oils loaded onto chitosan/polyvinyl alcohol blended films and their ability to eradicate staphylococcus aureus or pseudomonas aeruginosa from infected microenvironments. Pharmaceutics 13(2), 195–NA (2021). https://doi.org/10.3390/pharmaceutics13020195
137. Dridi, D., et al.: Enhanced antibacterial efficiency of cellulosic fibers: microencapsulation and green grafting strategies. Coatings 11(8), 980–NA (2021). https://doi.org/10.3390/coatings11080980
138. Zhou, Y., Sun, S., Bei, W., Zahi, M.R., Yuan, Q., Liang, H.: Preparation and antimicrobial activity of oregano essential oil pickering emulsion stabilized by cellulose nanocrystals. Int. J. Biol. Macromol. **112**, 7–13 (2018)
139. de Souza, A. G., Ferreira, R.R., Aguilar, E.S.F., Zanata, L., de Rosa, D.S.: Cinnamon essential oil nanocellulose-based pickering emulsions: Processing parameters effect on their formation, stabilization, and antimicrobial activity. Polysaccharides 2(3), 608–625 (2021)
140. Khaledian, Y., Pajohi-Alamoti, M., Bazargani-Gilani, B.: Development of cellulose nanofibers coating incorporated with ginger essential oil and citric acid to extend the shelf life of ready-to-cook barbecue chicken. J. Food Process. Preserv. **43**(10), e14114 (2019)
141. Manikandan, V., Min, S.C.: Roles of polysaccharides-based nanomaterials in food preservation and extension of shelf-life of food products: a review. Int. J. Biol. Macromol. **252**, 126381 (2023). https://doi.org/10.1016/j.ijbiomac.2023.126381
142. Agrimonti, C., White, J.C., Tonetti, S., Marmiroli, N.: Antimicrobial activity of cellulosic pads amended with emulsions of essential oils of oregano, thyme and cinnamon against microorganisms in minced beef meat. Int. J. Food Microbiol. **305**, 108246 (2019)
143. Santos, D.C., et al.: Antimicrobial activity studies and characterization of cellulose acetate films containing essential oils. Ital. J. Food Sci. **28**(2), 248 (2016)
144. Shahbazi, Y.: Application of carboxymethyl cellulose and chitosan coatings containing Mentha spicata essential oil in fresh strawberries. Int. J. Biol. Macromol. **112**, 264–272 (2018)

145. Moghimi, R., Aliahmadi, A., Rafati, H.: Antibacterial hydroxypropyl methyl cellulose edible films containing nanoemulsions of Thymus daenensis essential oil for food packaging. Carbohyd. Polym. **175**, 241–248 (2017)
146. Shahbazi, Y.: The properties of chitosan and gelatin films incorporated with ethanolic red grape seed extract and Ziziphora clinopodioides essential oil as biodegradable materials for active food packaging. Int. J. Biol. Macromol. **99**, 746–753 (2017)
147. Mohammadi, A., Hashemi, M., Hosseini, S.M.: Nanoencapsulation of Zataria multiflora essential oil preparation and characterization with enhanced antifungal activity for controlling Botrytis cinerea, the causal agent of gray mould disease. Innov. Food Sci. Emerg. Technol. **28**, 73–80 (2015)
148. Vu, K., Hollingsworth, R., Leroux, E., Salmieri, S., Lacroix, M.: Development of edible bioactive coating based on modified chitosan for increasing the shelf life of strawberries. Food Res. Int. **44**(1), 198–203 (2011)
149. Abdollahi, M., Rezaei, M., Farzi, G.: Improvement of active chitosan film properties with rosemary essential oil for food packaging. Int. J. Food Sci. Technol. **47**(4), 847–853 (2012)
150. Shavisi, N., Khanjari, A., Basti, A.A., Misaghi, A., Shahbazi, Y.: Effect of PLA films containing propolis ethanolic extract, cellulose nanoparticle and Ziziphora clinopodioides essential oil on chemical, microbial and sensory properties of minced beef. Meat Sci. **124**, 95–104 (2017)
151. Ijaz, I., Gilani, E., Nazir, A., Bukhari, A.: Detail review on chemical, physical and green synthesis, classification, characterizations and applications of nanoparticles. Green Chem. Lett. Rev. **13**(3), 223–245 (2020)
152. Swargiary, M., Kumar, S.: One pot phytosynthesis of gold nanoparticles using aqueous extract of elephant apple-an eco-friendly approach. Orient. Pharm. Exp. Med. **17**(3), 285–289 (2017)
153. Kumar, S., Mukherjee, A., Dutta, J.: Chitosan based nanocomposite films and coatings: Emerging antimicrobial food packaging alternatives. Trends Food Sci. Technol. **97**, 196–209 (2020)
154. Basumatary, I.B., Mukherjee, A., Katiyar, V., Kumar, S.: Biopolymer-based nanocomposite films and coatings: recent advances in shelf-life improvement of fruits and vegetables. Crit. Rev. Food Sci. Nutr. **62**(7), 1912–1935 (2022)
155. Ameta, S.K., Rai, A.K., Hiran, D., Ameta, R., Ameta, S.C.: Use of nanomaterials in food science. In: Biogenic Nano-Particles and Their Use in Agro-Ecosystems, pp. 457–488 (2020)
156. Basavegowda, N., Baek, K.-H.: Advances in Functional Biopolymer-Based Nanocomposites for Active Food Packaging Applications. Polymers 13(23), 4198 (2021). https://www.mdpi.com/2073-4360/13/23/4198. (Online)
157. Kumar, S., Ye, F., Mazinani, B., Dobretsov, S., Dutta, J.: Chitosan nanocomposite coatings containing chemically resistant ZnO–SnOx core–shell nanoparticles for photocatalytic antifouling. Int. J. Mol. Sci. **22**(9), 4513 (2021)
158. Kumar, S., Bhattacharya, W., Singh, M., Halder, D., Mitra, A.: Plant latex capped colloidal silver nanoparticles: a potent anti-biofilm and fungicidal formulation. J. Mol. Liq. **230**, 705–713 (2017)
159. Kumar, S., Mudai, A., Roy, B., Basumatary, I., Mukherjee, A., Dutta, J. (eds.): Biodegradable hybrid nanocomposite of chitosan/gelatin and green synthesized zinc oxide nanoparticles for food packaging. Foods **9**, 1143 2020
160. Gaikwad, K.K., Singh, S., Negi, Y.S.: Ethylene scavengers for active packaging of fresh food produce. Environ. Chem. Lett. **18**, 269–284 (2020)
161. Kumar, S., Boro, J.C., Ray, D., Mukherjee, A., Dutta, J.: Bionanocomposite films of agar incorporated with ZnO nanoparticles as an active packaging material for shelf life extension of green grape. Heliyon **5**(6), 2019
162. Hornyak, G.L., Tibbals, H.F., Dutta, J., Moore, J.J.: Introduction to Nanoscience and Nanotechnology. CRC Press (2008)
163. Mosnáčková, K., Kollár, J., Huang, Y.-S., Huang, C.-F., Mosnáček, J.: Synthesis routes of functionalized nanoparticles. Polym. Compos. Funct. Nanoparticles 1–46 (2019)

164. Rhim, J.-W., Kim, Y.-T.: Biopolymer-based composite packaging materials with nanoparticles. In: Innovations in Food Packaging, pp. 413–442. Elsevier (2014)
165. Kanmani, P., Rhim, J.-W.: Properties and characterization of bionanocomposite films prepared with various biopolymers and ZnO nanoparticles. Carbohyd. Polym. **106**, 190–199 (2014)
166. Carvalho, R.A., Santos, T.A., de Azevedo, V.M., Felix, P.H.C., Dias, M.V., Borges, S.V.: Bio-nanocomposites for food packaging applications: Effect of cellulose nanofibers on morphological, mechanical, optical and barrier properties. Polym. Int. **67**(4), 386–392 (2018)
167. Kim, I., Viswanathan, K., Kasi, G., Sadeghi, K., Thanakkasaranee, S., Seo, J.: Poly (lactic acid)/ZnO bionanocomposite films with positively charged ZnO as potential antimicrobial food packaging materials. Polymers **11**(9), 1427 (2019)
168. Sothornvit, R.: Nanostructured materials for food packaging systems: new functional properties. Curr. Opin. Food Sci. **25**, 82–87 (2019)
169. Al-Sherbini, A.-S.A., Ghannam, H.E., El-Ghanam, G.M., El-Ella, A.A., Youssef, A.M.: Utilization of chitosan/Ag bionanocomposites as eco-friendly photocatalytic reactor for Bactericidal effect and heavy metals removal. Heliyon **5**(6), (2019)
170. Azizi-Lalabadi, M., Garavand, F., Jafari, S.M.: Incorporation of silver nanoparticles into active antimicrobial nanocomposites: release behavior, analyzing techniques, applications and safety issues. Adv. Coll. Interface. Sci. **293**, 102440 (2021)
171. Anwar, Y.: Antibacterial and lead ions adsorption characteristics of chitosan-manganese dioxide bionanocomposite. Int. J. Biol. Macromol. **111**, 1140–1145 (2018)
172. Safaei, M., Taran, M., Imani, M.M.: Preparation, structural characterization, thermal properties and antifungal activity of alginate-CuO bionanocomposite. Mater. Sci. Eng., C **101**, 323–329 (2019)
173. Hou, X., et al.: Effect of SiO_2 nanoparticle on the physical and chemical properties of eco-friendly agar/sodium alginate nanocomposite film. Int. J. Biol. Macromol. **125**, 1289–1298 (2019)
174. Tabatabaei, R.H., Jafari, S.M., Mirzaei, H., Nafchi, A.M., Dehnad, D.: Preparation and characterization of nano-SiO_2 reinforced gelatin-k-carrageenan biocomposites. Int. J. Biol. Macromol. **111**, 1091–1099 (2018)
175. Moradi, Z., Esmaiili, M., Almasi, H.: Development and characterization of kefiran-Al_2O_3 nanocomposite films: morphological, physical and mechanical properties. Int. J. Biol. Macromol. **122**, 603–609 (2019)
176. Yadav, M.: Study on thermal and mechanical properties of cellulose/iron oxide bionanocomposites film. Compos. Commun. **10**, 1–5 (2018)
177. Rohini, B., Ishwarya, S.P., Rajasekharan, R., VijayaKumar, A.K.: Ocimum basilicum seed mucilage reinforced with montmorillonite for preparation of bionanocomposite film for food packaging applications. Polym. Test. **87**, 106465 (2020)
178. Neji, A.B., Jridi, M., Nasri, M., Sahnoun, R.D.: Preparation, characterization, mechanical and barrier properties investigation of chitosan-kaolinite nanocomposite. Polym. Test. **84**, 106380 (2020)
179. Ju, S., Zhang, F., Duan, J., Jiang, J.: Characterization of bacterial cellulose composite films incorporated with bulk chitosan and chitosan nanoparticles: a comparative study. Carbohyd. Polym. **237**, 116167 (2020)
180. Ifuku, S., et al.: Preparation of high-strength transparent chitosan film reinforced with surface-deacetylated chitin nanofibers. Carbohyd. Polym. **98**(1), 1198–1202 (2013)
181. Azarifar, M., et al.: The optimization of gelatin-CMC based active films containing chitin nanofiber and Trachyspermum ammi essential oil by response surface methodology. Carbohyd. Polym. **208**, 457–468 (2019)
182. Hossain, R., Tajvidi, M., Bousfield, D., Gardner, D.J.: Multi-layer oil-resistant food serving containers made using cellulose nanofiber coated wood flour composites. Carbohyd. Polym. **267**, 118221 (2021)
183. Tian, X., Yan, D., Lu, Q., Jiang, X.: Cationic surface modification of nanocrystalline cellulose as reinforcements for preparation of the chitosan-based nanocomposite films. Cellulose **24**, 163–174 (2017)

184. Gorrasi, G., et al.: Vapor barrier properties of polycaprolactone montmorillonite nanocomposites: effect of clay dispersion. Polymer **44**(8), 2271–2279 (2003)
185. Garcia, M.S., Lagaron, J.: Nanocomposites for food and beverage packaging materials. In: Nanotechnology in the Food, Beverage and Nutraceutical Industries, pp. 335–361. Elsevier (2012)
186. Ray, S.S.: Recent trends and future outlooks in the field of clay-containing polymer nanocomposites. Macromol. Chem. Phys. **215**(12), 1162–1179 (2014)
187. Chan, M.-L., Lau, K.-T., Wong, T.-T., Ho, M.-P., Hui, D.: Mechanism of reinforcement in a nanoclay/polymer composite. Compos. B Eng. **42**(6), 1708–1712 (2011)
188. Zhou, M., Liu, Q., Wu, S., Gou, Z., Wu, X., Xu, D.: Starch/chitosan films reinforced with polydopamine modified MMT: effects of dopamine concentration. Food Hydrocoll. **61**, 678–684 (2016)
189. Abdurrahim, I.: Water sorption, antimicrobial activity, and thermal and mechanical properties of chitosan/clay/glycerol nanocomposite films. Heliyon **5**(8), 2019
190. Kwaśniewska, A., Chocyk, D., Gładyszewski, G., Borc, J., Świetlicki, M., Gładyszewska, B.: The influence of kaolin clay on the mechanical properties and structure of thermoplastic starch films. Polymers **12**(1), 73 (2020)
191. Sanchez-Garcia, M.D., Hilliou, L., Lagaron, J.M.: Nanobiocomposites of carrageenan, zein, and mica of interest in food packaging and coating applications. J. Agric. Food Chem. **58**(11), 6884–6894 (2010)
192. Goh, K., et al.: Sandwich-architectured poly (lactic acid)–graphene composite food packaging films. ACS Appl. Mater. Interfaces. **8**(15), 9994–10004 (2016)
193. Novoselov, K.S., et al.: Electric field effect in atomically thin carbon films. Science **306**(5696), 666–669 (2004)
194. Yang, Y.-H., Bolling, L., Priolo, M.A., Grunlan, J.C.: Super gas barrier and selectivity of graphene oxide-polymer multilayer thin films. Adv. Mater. (Deerfield Beach, Fla.) **25**(4), 503–508 (2012)
195. Wu, C.-S., Liao, H.-T.: Study on the preparation and characterization of biodegradable polylactide/multi-walled carbon nanotubes nanocomposites. Polymer **48**(15), 4449–4458 (2007)
196. Kuan, C.-F., Kuan, H.-C., Ma, C.-C.M., Chen, C.-H.: Mechanical and electrical properties of multi-wall carbon nanotube/poly (lactic acid) composites. J. Phys. Chem. Solids **69**(5–6), 1395–1398 (2008)
197. Azizi-Lalabadi, M., Jafari, S.M.: Bio-nanocomposites of graphene with biopolymers; fabrication, properties, and applications. Adv. Coll. Interface. Sci. **292**, 102416 (2021)
198. Frone, A., Berlioz, S., Chailan, J.F., Panaitescu, D., Donescu, D.: Cellulose fiber-reinforced polylactic acid. Polym. Compos. **32**(6), 976–985 (2011)
199. de Mesquita, J.P., Donnici, C.L., Pereira, F.V.: Biobased nanocomposites from layer-by-layer assembly of cellulose nanowhiskers with chitosan. Biomacromolecules **11**(2), 473–480 (2010)
200. De Mesquita, J.P., Donnici, C.L., Teixeira, I.F., Pereira, F.V.: Bio-based nanocomposites obtained through covalent linkage between chitosan and cellulose nanocrystals. Carbohyd. Polym. **90**(1), 210–217 (2012)
201. Shankar, S., Reddy, J.P., Rhim, J.-W., Kim, H.-Y.: Preparation, characterization, and antimicrobial activity of chitin nanofibrils reinforced carrageenan nanocomposite films. Carbohyd. Polym. **117**, 468–475 (2015)
202. Kumar, S., et al.: Plant extract mediated silver nanoparticles and their applications as antimicrobials and in sustainable food packaging: a state-of-the-art review. Trends Food Sci. Technol. **112**, 651–666 (2021)
203. Al-Naamani, L., Dutta, J., Dobretsov, S.: Nanocomposite zinc oxide-chitosan coatings on polyethylene films for extending storage life of okra (Abelmoschus esculentus). Nanomaterials **8**(7), 479 (2018)
204. Kanikireddy, V., et al.: Development of alginate-gum acacia-Ag0 nanocomposites via green process for inactivation of foodborne bacteria and impact on shelf life of black grapes (Vitis vinifera). J. Appl. Polym. Sci. **136**(15), 47331 (2019)

205. Chi, H., et al.: Effect of PLA nanocomposite films containing bergamot essential oil, TiO_2 nanoparticles, and Ag nanoparticles on shelf life of mangoes. Sci. Hortic. **249**, 192–198 (2019)
206. Kanikireddy, V., Varaprasad, K., Rani, M.S., Venkataswamy, P., Reddy, B.J.M., Vithal, M.: Biosynthesis of CMC-Guar gum-Ag0 nanocomposites for inactivation of food pathogenic microbes and its effect on the shelf life of strawberries. Carbohyd. Polym. **236**, 116053 (2020)
207. Liu, R., Liu, D., Liu, Y., Song, Y., Wu, T., Zhang, M.: Using soy protein SiOx nanocomposite film coating to extend the shelf life of apple fruit. Int. J. Food Sci. Technol. **52**(9), 2018–2030 (2017)
208. Issa, A., Ibrahim, S.A., Tahergorabi, R.: Impact of sweet potato starch-based nanocomposite films activated with thyme essential oil on the shelf-life of baby spinach leaves. Foods **6**(6), 43 (2017)
209. Zabihollahi, N., Alizadeh, A., Almasi, H., Hanifian, S., Hamishekar, H.: Development and characterization of carboxymethyl cellulose based probiotic nanocomposite film containing cellulose nanofiber and inulin for chicken fillet shelf life extension. Int. J. Biol. Macromol. **160**, 409–417 (2020)
210. Mohammadi, H., Kamkar, A., Misaghi, A., Zunabovic-Pichler, M., Fatehi, S.: Nanocomposite films with CMC, okra mucilage, and ZnO nanoparticles: extending the shelf-life of chicken breast meat. Food Packag. Shelf Life **21**, 100330 (2019)
211. Youssef, A.M., El-Sayed, S.M., El-Sayed, H.S., Salama, H.H., Dufresne, A.: Enhancement of Egyptian soft white cheese shelf life using a novel chitosan/carboxymethyl cellulose/zinc oxide bionanocomposite film. Carbohyd. Polym. **151**, 9–19 (2016)
212. Abdollahi, M., Rezaei, M., Farzi, G.: Influence of chitosan/clay functional bionanocomposite activated with rosemary essential oil on the shelf life of fresh silver carp. Int. J. Food Sci. Technol. **49**(3), 811–818 (2014)
213. Kalia, A., et al.: Nettle-leaf extract derived ZnO/CuO nanoparticle-biopolymer-based antioxidant and antimicrobial nanocomposite packaging films and their impact on extending the post-harvest shelf life of guava fruit. Biomolecules **11**(2), 224 (2021)
214. Shapi'i, R. A., Othman, S.H., Nordin, N., Basha, R.K., Naim, M.N.: Antimicrobial properties of starch films incorporated with chitosan nanoparticles: in vitro and in vivo evaluation. Carbohydr. Polym. **230**, 115602 (2020)
215. Vejdan, A., Ojagh, S.M., Abdollahi, M.: Effect of gelatin/agar bilayer film incorporated with TiO_2 nanoparticles as a UV absorbent on fish oil photooxidation. Int. J. Food Sci. Technol. **52**(8), 1862–1868 (2017)
216. Alizadeh-Sani, M., Mohammadian, E., McClements, D.J.: Eco-friendly active packaging consisting of nanostructured biopolymer matrix reinforced with TiO_2 and essential oil: application for preservation of refrigerated meat. Food Chem. **322**, 126782 (2020)
217. Batra, M., Malik, G.K., Mitra, J.: Enhancing the properties of gelatin–chitosan bionanocomposite films by incorporation of silica nanoparticles. J. Food Process Eng **43**(2), e13329 (2020)
218. Roy, S., Rhim, J.-W.: Agar-based antioxidant composite films incorporated with melanin nanoparticles. Food Hydrocoll. **94**, 391–398 (2019)
219. Zhang, L., et al.: Characterisation of microemulsion nanofilms based on Tilapia fish skin gelatine and ZnO nanoparticles incorporated with ginger essential oil: Meat packaging application. Int. J. Food Sci. Technol. **52**(7), 1670–1679 (2017)
220. Li, W., Li, L., Zhang, H., Yuan, M., Qin, Y.: Evaluation of PLA nanocomposite films on physicochemical and microbiological properties of refrigerated cottage cheese. J. Food Process. Preserv. **42**(1), e13362 (2018)
221. Viuda-Martos, M., et al.: Chemical composition and antioxidant and anti-Listeria activities of essential oils obtained from some Egyptian plants. J. Agric. Food Chem. **58**(16), 9063–9070 (2010)
222. Ruiz-Navajas, Y., Viuda-Martos, M., Sendra, E., Perez-Alvarez, J., Fernández-López, J.: In vitro antibacterial and antioxidant properties of chitosan edible films incorporated with Thymus moroderi or Thymus piperella essential oils. Food Control **30**(2), 386–392 (2013)

223. Wu, J., et al.: The preparation, characterization, antimicrobial stability and in vitro release evaluation of fish gelatin films incorporated with cinnamon essential oil nanoliposomes. Food Hydrocoll. **43**, 427–435 (2015)
224. Guerreiro, A.C., Gago, C.M., Faleiro, M.L., Miguel, M.G., Antunes, M.D.: The use of polysaccharide-based edible coatings enriched with essential oils to improve shelf-life of strawberries. Postharvest Biol. Technol. **110**, 51–60 (2015)
225. He, Q., Changhong, L., Kojo, E., Tian, Z.: Quality and safety assurance in the processing of aloe vera gel juice. Food Control **16**(2), 95–104 (2005)
226. El Ghaouth, A., Arul, J., Ponnampalam, R., Boulet, M.: Use of chitosan coating to reduce water loss and maintain quality of cucumber and bell pepper fruits. J. Food Process. Preserv. **15**(5), 359–368 (1991)
227. Ilyas, R., et al.: Regulations for food packaging materials. Bio-Based Packag. Mater. Environ. Econ. Asp. 467–494 (2021)
228. Marone, P.: Food safety and regulatory concerns. In: Insects as Sustainable Food Ingredients, pp. 203–221. Elsevier (2016)

Chapter 5
Designing Smart and Sustainable Edible Packaging Materials from Biopolymers, Proteins, and Polysaccharides

Shahid Iqbal, Azhar Ayyub, Rizwan Ahmed Bhutto, and Wajid Rehman

Abstract Edible food packaging has emerged as an innovative solution to address the environmental issues of plastic pollution and excessive packaging. Traditional petroleum-based plastics, commonly used for food packaging due to their availability, strength, and cost-effectiveness, contribute significantly to landfill waste, water pollution, and health problems. In contrast, edible packaging made from natural polymers like proteins and polysaccharides offers a sustainable alternative. However, these materials often have weaker mechanical properties, durability, and barrier characteristics than plastics. This chapter reviews the development of edible packaging, such as films, coatings, and fibers, emphasizing their environmental benefits and recent advancements in the field. Despite their potential, edible packaging faces challenges in protecting food products effectively due to their inferior gas barrier and water resistance properties. To overcome these limitations, innovative methods, such as advanced emulsion technology under controlled conditions (pH, temperature, and ionic strength), are being researched to enhance the barrier properties, texture, antimicrobial effects, and rheological behavior of lipid-based films. This chapter highlights the importance of addressing these challenges to create effective, sustainable edible packaging solutions that offer both environmental benefits and improved functionality.

S. Iqbal (✉)
Myddelton College Jinhua, Rongguang Road, Wucheng, Jinhua 321025, Zhejiang, China
e-mail: shahid_chem7@yahoo.com; shahidiqbal@suda.edu.cn

School of Chemical and Environmental Engineering, College of Chemistry, Chemical Engineering and Materials Science, Soochow University, Suzhou 215123, Jiangsu, China

A. Ayyub
State Key Laboratory of Chemical Resource Engineering, Beijing University of Chemical Technology, Beijing 100029, China

R. A. Bhutto (✉)
Department of Food Science and Engineering, College of Chemistry and Environmental Engineering, Shenzhen University, Shenzhen 518060, Guangdong, China
e-mail: rizwan.bhutto@hotmail.com

W. Rehman
Department of Chemistry, Hazara University, Mansehra 21120, Pakistan

S. ul Islam and M. Shahid (eds.), *Green Materials for Active Food Packaging*, Engineering Materials, https://doi.org/10.1007/978-981-96-0369-5_5

Keywords Natural polymers · Environment · Health · Film · Mechanical strength · Barrier properties

5.1 Introduction

In recent decades, there has been an increasing interest in developing food packaging materials developed from proteins, polysaccharides, resins, and lipids because of their echo-friendly substitutes to conventional packaging [193, 226, 235]. Since edible packaging is prepared from natural polymers (biopolymers), these materials are biodegradable—and can be eaten with food [105]. Natural polymer-derived packaging materials are deemed safe and hygienic for consumers [105, 193, 235]. Biopolymers and their mixture-based edible packaging materials are utilized to reduce landfill waste and water pollution caused by conventional petroleum-based packaging [210, 212]. However, designing smart and sustainable edible packaging that includes strong nutraceuticals, antimicrobial and climate change characteristics presents a considerable challenge for food researchers [53, 212, 239].

Edible packaging (such as coating and films) has a long history, dating back to the early applications of packaging using skins of vegetables and fruits [164]. In ancient China, the practice of applying wax coatings on oranges and lemons was employed to reduce or inhibit water and moisture loss in these fruits [164, 154]. The first cellulose-based packaging materials were developed in 1856 [154]. Plastics have been utilized as traditional packaging for several years due to their exceptional functionality, which includes versatility, stability, durability, and affordability [249]. Among the commonly used food packaging are polyethylene (PE), polypropene (PP), polystyrene, etc. [181]. The synthesis of plastic began in 1907, with the introduction of the first plastic material called "Bakelite" by Leo Baekeland in the market [172, 205]. The growth of global plastic production remained slow until the 1950s [249]. This slow production was because of the low availability of raw materials and the high cost of producing plastic materials. The utilization of plastic has considerably increased in the past 73 years, particularly after the 1950s, and doubled in the last 20 years [249]. Today, plastic is synthesized for various uses, including shopping bags, food packaging, construction, and electronic materials. Plastic materials, such as shopping bags and food packaging, have become essential to life. However, the extensive use of plastics led to harmful effects on the environment and human health when not properly managed and recycled [205, 249] (Fig. 5.1). It was noted that 5–13 million tons of plastic (1.5–4%) enter the ocean [210]. Despite the known environmental and health hazards linked with plastics, their widespread worth continues for various purposes, with price being a major factor [193, 226]. In addition, users and industries lack awareness regarding the detrimental effects of food packaging on the ecosystem [53].

In the food industry, alternative food packaging is fabricated exhibiting similar physicochemical and functional characteristics such as rheological, mechanical, toughness, optical, and barrier as their traditional counterparts' plastic materials

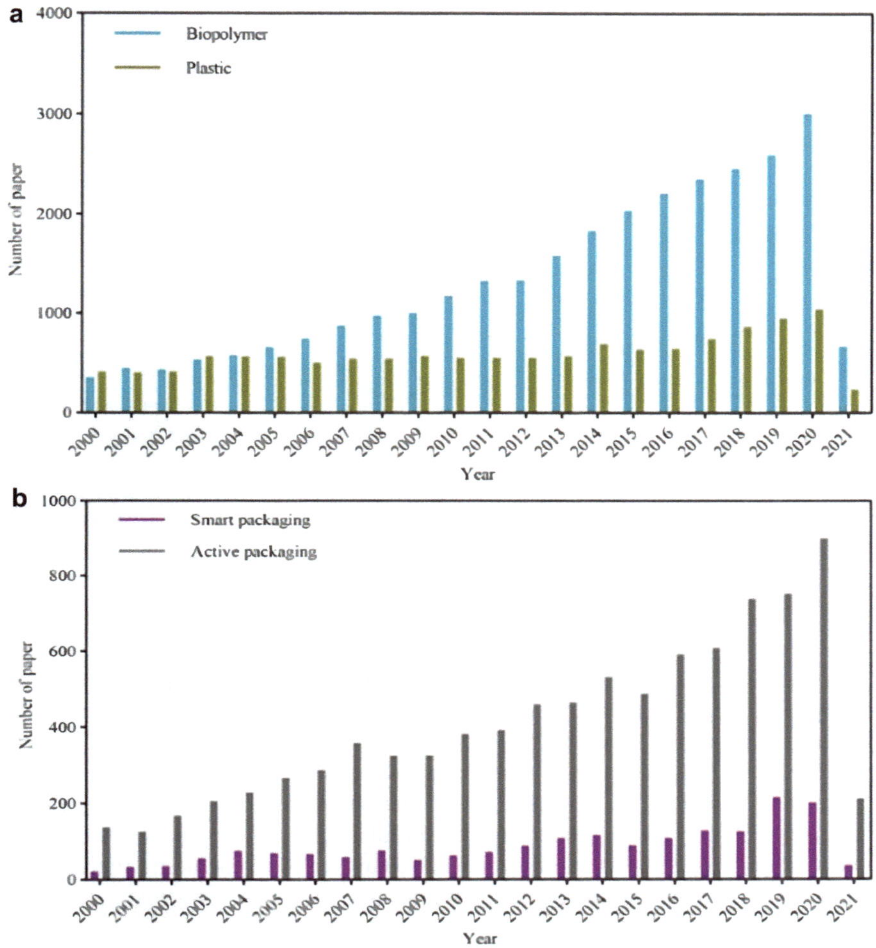

Fig. 5.1 Progress of publications over 21 years comparing biopolymers-based edible packaging materials, including proteins, polysaccharides, and lipids, against plastic-based packaging materials (**a**) and highlights the difference in trends between smart packaging and active packaging materials (**b**) (Sani et al. [212]). Comparison of the health and environmental effects of traditional plastic and edible food packaging materials (**c**). Reproduced with permission from Li et al. [140], Copyright Elsevier

[53, 193, 212]. Edible packaging materials are composed of natural polymers or biopolymers, which include proteins, polysaccharides, and their combinations with other natural compounds, such as lipids or surfactants (Table 5.1) [212]. Biopolymers-based edible packaging offers several advantages over conventional non-biodegradable packaging materials. The Food and Agriculture Organization of the United Nations (FAO) reports that every year, 1.3 billion tons of food, showing 45% of produced food is discarded globally [185]. Therefore, biopolymers-based

c

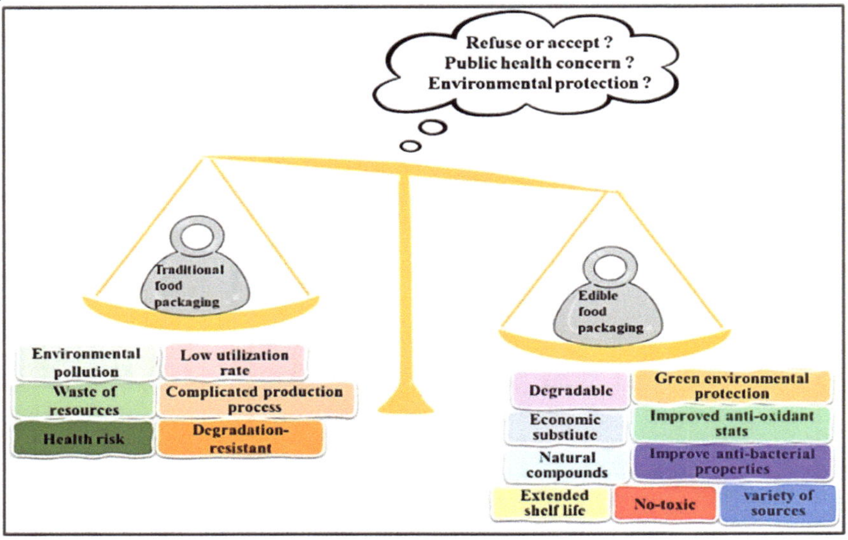

Fig. 5.1 (continued)

edible packaging is consumable with food and enhances the stability of food products by inhibiting the physical, chemical, and environmental factors that influence the detrimental of the food products (Fig. 5.2) [212]. Several studies describe the production of edible packaging materials as biodegradable and prolonged shelf-life materials using protein, polysaccharide, and their mixture [193, 210, 212] (Singh and Packirisamy 2022). Biopolymers can be helpful for the development of advanced food packaging materials that show several characteristics, including mechanical, barrier, antimicrobial, antioxidant flexibility, and sustainability (Fig. 5.2) [116]. Proteins such as gelatin, collagen, keratin, sodium caseinate, zein, whey, and soy proteins have been utilized as packaging materials because of their unique characteristics [193, 210, 212]. Polysaccharides, such as sodium alginate (SA), cellulose, pectin, etc., have been widely employed for the fabrication of food packaging because of their film-making, viscosity modifying, and barrier characteristics (Table 5.1) [212]. These polysaccharides offer a range of benefits that make them suitable for creating strong and sustainable packaging solutions [116, 210, 212].

Edible food packaging offers numerous benefits over their counterparts, plastic materials due to the following functionality: (i) they can be eaten with the foods, and are biodegradable and environmentally friendly materials [210]; (ii) The emulsifying and gelling properties of biopolymers in packaging materials enhance the stability, and shelf life of various products; (iii) They are recognized as a thickening agent, and enhance the density and viscosity of the final product [215]; (iv) The protein and polysaccharide not only improve the texture, consistency, and nutritional value of the product but also induce satiety, which can be useful for managing weight and

Table 5.1 Utilization of biopolymers, composition, fabrication method, and suggested applications in food packaging

Biopolymers	Composition	Edible packaging	Preparation method	Suggested applications	References
Gelidium amansii polysaccharide (GAP)	Carbohydrate (75.37%), 36-AG (17.33%), Uronic acid (5.94%), Sulfate (5.31%), Protein (2.53%), Phenolics (<0.1%)	GAP-based composite films with cyclic adenosine monophosphate (cAMP), cytidine-5′-monophosphate (CMP), and adenosine-5′-monophosphate (AMP)	Dissolving polysaccharides in distilled water, adding plasticizer and nucleotides or their derivatives, homogenizing, and drying	Food packaging, especially for fish, to improve shelf-life and provide antimicrobial properties	Wang et al. [247]
Seaweed-Derived Insoluble Cellulose Fibers and Soluble Polysaccharides	Cellulose (65.86%), Hemicellulose (5.13%), Acid Detergent Lignin (4.82%), Ash (0.29%), Polysaccharides (mainly consisting of total carbohydrates (81.30%), uronic acid (10.75%), protein (5.85%), and sulfate (1.47%))	Edible, hydrostable, and degradable seaweed-based straws	Seaweed is processed to extract soluble polysaccharides and insoluble cellulose fibers, which are then formed into films and rolled into straws with Ca2+ as a crosslinking agent. The structure is reinforced through freeze–thaw cycles	Used as drinking straws that indicate acid–base properties of beverages, decompose quickly in natural environments, especially in marine conditions, and can discriminate the freshness of milk	Ni et al. [171]
SSPS/Gelatin (SG)	Soluble soybean polysaccharide and gelatin, curcumin	SG and SG/Cur films	Soluble casting method: Dissolving SSPS and gelatin in deionized water, adding glycerol as a plasticizer, mixing with curcumin solutions, homogenizing, pouring into dishes, and drying	Active packaging material for soybean oil, quickly dissolving in hot water	Dong et al. [59]
CMC enriched with MLP (CMC-MLP)	Carboxymethyl cellulose (CMC), Mulberry leaves polysaccharides (MLP), Glycerol (plasticizer)	Biodegradable packaging films	Solution casting method: CMC dissolved in water, the addition of MLP at varying concentrations (1, 5, 10% w/w relative to CMC), incorporation of glycerol as a plasticizer, pouring into Petri dishes, and drying at 25 °C	Packaging material for food products, particularly effective in enhancing antioxidant and antimicrobial properties, suitable for sunflower oil packaging due to oil resistance capacity	Akhtar et al. [9]
Chitosan-Citric Acid Based	Chitosan (CS), Citric acid, β-Cyclodextrin, Pomace extract	Edible films	Film-forming solution, casting, drying	Packaging material with antioxidant properties, suitable to improve the shelf-life of foods	Loukri et al. [143]

(continued)

Table 5.1 (continued)

Biopolymers	Composition	Edible packaging	Preparation method	Suggested applications	References
Aloe vera polysaccharide (AVP), β-Hydroxy-β-methyl butyrate calcium (HMBCa), Nanocellulose (CNCs)	Aloe vera polysaccharide, β-Hydroxy-β-methylbutyrate calcium, Nanocellulose	AVP-HMBCa-CNCs films	Casting and solvent evaporation	Freshness preservation of blueberries, green food packaging with antioxidant and antibacterial abilities	Tang et al. [232]
Lily Polysaccharide (LP) and Sodium Alginate (SA)	Lily polysaccharide, Sodium Alginate, Glycerin	LP-based edible films	Solvent evaporation method	Packaging, especially for chicken breast, to enhance shelf-life and provide antioxidants and antibacterial characteristics	Cui et al. [50]
High Methoxyl Pectin and Pea Protein Isolate (PPI) or Sodium Caseinate (SC)	High methoxy pectin, pea protein isolates or Sodium Caseinate, Glycerol (plasticizer)	Edible films	Film-forming solution casting	Suitable for food packaging as an environmentally friendly alternative to conventional packaging, enhancing product life quality and safety	Zioga et al. [264]
Gelatin	Native corn starch, low methoxyl pectin, bovine gelatin (bloom value ~220 g), CMC, Tween 80, glycerol	Single-serve soluble edible pouches	Conventional solvent casting technique using a continuous conveyor dryer; Gelatin (7% w/w) dissolved with processing aids in distilled water, stirred at 60 °C (30 min), cast on the continuous belt, dried at 65 °C	Suitable for packaging individual coffee pouches and other dry food ingredients, ensures product integrity during storage	Gamboni et al. [80]
Soluble Soybean Polysaccharide (SSPS) with Plasticizers	SSPS, Glycerol, Sorbitol, PEG400	Edible films	Solvent-casting method: Dissolving SSPS powder in deionized water at 60 ± 2 °C, adding glycerol, sorbitol, or PEG400 at various concentrations (0%, 15%, 30%, by weight), stirring, degassing, casting, and drying at 60 °C for 7 h	Packaging material for food due to biodegradability, improved ductility, and moisture content; suitable for applications requiring UV barrier and transparent	Dong et al. [60]

(continued)

Table 5.1 (continued)

Biopolymers	Composition	Edible packaging	Preparation method	Suggested applications	References
CS-Cellulose Nanofiber with SPI	CS, Cellulose Nanofibers, Soybean Protein Isolate	CS CNF SPI films	CS dissolved in acetic acid, SPI dissolved in water, mixed with CNF solution, adjusted pH, added glycerin, cast, and dried at 50 °C for 6 h	Food packaging, especially for improving mechanical, hydrophobic, and antioxidants characteristics	Qi et al. [190]
Malva sylvestris Polysaccharide Mucilage and Postbiotic from Saccharomyces cerevisiae var. boulardii	Malva sylvestris seed polysaccharide mucilage (MSM) and cell-free supernatant postbiotic (PSB) from Saccharomyces cerevisiae var. boulardii	Edible coating for lamb meat	Tween 80 combined with MSM, dissolved in distilled water, heated, and stirred. For coating, PSB added at various concentrations (0%, 2%, 4%, 6%, 8%, 10% v/v)	Preservation of lamb meat, enhancing shelf life while maintaining sensory and microbiological quality	Abbasi et al. [1]
Grapefruit Soluble Dietary Fiber (GSDF) and Grapefruit Insoluble Dietary Fiber (GIDF) with Nanocellulose (GNCC)	Grapefruit peel pomace dietary fiber, Nanocellulose (GNCC)	Dietary fiber-based edible packaging film	1. Extraction of GSDF, GIDF, and GNCC from grapefruit peels 2. Preparation of the film solution with GSDF, GIDF, GNCC, and plasticizers (glycerol) 3. Casting the film solution into Petri dishes and drying at 40 °C (12 h) 4. Process optimization was achieved through principal component analysis (PCA), response surface methods, and membership function evaluation. The optimal GSDF/SA/GNCC composition was determined	Edible packaging material for food due to its biodegradability and nutritional enhancement capabilities	Chen et al. [38]
Wheat Gluten Protein (WGP)	Wheat gluten protein	Base film	WGP dissolved in deionized water, adjusted to pH 12, stirred, glycerol added as a plasticizer, cast in dishes, dried at 50 °C for 5 h	Base material for edible packaging	Dong et al. [61]

(continued)

Table 5.1 (continued)

Biopolymers	Composition	Edible packaging	Preparation method	Suggested applications	References
WGP with Tartaric Acid (WGP-TA)	Wheat gluten protein, tartaric acid	Film with improved UV resistance and thickness	WGP dissolved in tartaric acid solution, heated at 90 °C for 2 h, glycerol added, cast in dishes, dried at 50 °C for 5 h	Edible packaging with enhanced UV resistance	Dong et al. [61]
WGP with Carboxymethyl Cellulose (WGP-CMC)	Wheat gluten protein, carboxymethyl cellulose	A film with enhanced mechanical properties	WGP and CMC dissolved in deionized water, adjusted to pH 12, then to pH 7, glycerol added, cast in dishes, dried at 50 °C for 5 h	Edible packaging with improved mechanical strength	Dong et al. [61]
WGP with Egg White Protein (WGP-EWP)	Wheat gluten protein, egg white protein	A film with higher physical strength	WGP and EWP dissolved in deionized water, adjusted to pH 12, then to pH 7, glycerol added, cast in dishes, dried at 50 °C for 5 h	Edible packaging with improved strength and opacity	Dong et al. [61]
WGP with Apple Pectin (WGP-AP)	Wheat gluten protein, apple pectin	Smooth and homogeneous film	WGP and apple pectin dissolved in deionized water, adjusted to pH 12, then to pH 7, glycerol added, cast in dishes, dried at 50 °C for 5 h	Edible packaging with a smooth texture	Dong et al. [61]
Carboxymethyl Cellulose (CMC)	CMC, Water, Glycerol	Edible coating for ALT tomatoes	Dipping ALT ripe fruits in polysaccharide solutions at concentrations of 2% and 4% w/v, adding glycerol as a plasticizer	Extending shelf-life and improving postharvest quality of ALT tomatoes at ambient and refrigerated conditions	Asiamah et al. [15]
Xanthan Gum (XG)	Xanthan Gum, Water, Glycerol	Edible coating for ALT tomatoes	Like the CMC method, using Xanthan Gum at concentrations of 2% and 4% w/v	Similar to CMC, with the potential for enhancing antioxidant stability in ALT tomatoes	Asiamah et al. [15]
Pectin	High Methoxyl Pectin, Water, Glycerol	Edible coating for ALT tomatoes	Similar to CMC and XG methods, using High Methoxyl Pectin at concentrations of 2% and 4% w/v	Similar to CMC and XG, particularly effective under refrigerated storage conditions	Asiamah et al. [15]

(continued)

Table 5.1 (continued)

Biopolymers	Composition	Edible packaging	Preparation method	Suggested applications	References
Whey Protein Isolate (WPI), Jujube Polysaccharide (JPS), Starch Nanocrystals (SNC)	WPI (3.3–11.7%), JPS (1.59–18.41%), SNC (0.32–3.68%)	WPI/JPS composite films reinforced with SNC	Casting; WPI solutions were heated and then mixed with JPS solutions; glycerol was added as a plasticizer; SNCs dispersed in water and added to the mix; films were cast on Teflon plates and dried	Shelf-life extension of fruits, especially bananas, due to the films' low oxygen and water vapor permeability rates	Gharibzahedi et al. [88]
Zein, sodium caseinate (SC), sodium alginate (SA)	Zein, SC, SA, Curcumin (1%, 2%, 3%, 4%)	Zein/SC/SA film loading curcumin	Zein dissolved in ethanol, curcumin added, glycerol as a plasticizer, continuous stirring; addition of SC and SA solutions; poured into the mold, following drying at 40 °C, conditioned at 25 ± 1 °C	Preservation of food due to improved antioxidant and mechanical properties	Tian et al. [238]
CMC-Gel	CMC, Gelatin, Glycerol, PPE (Pomegranate Peel Extract)	Composite films		Casting method: gelatin and CMC were mixed in water in equal ratios, then 40% glycerol (v/w weight of CMC-Gel) was added followed by different concentrations of PPE. The mixture was stirred for 60 min, ultrasonically treated, poured into Petri dishes, and dried	Nabeel Ahmad et al. [163]

(continued)

Table 5.1 (continued)

Biopolymers	Composition	Edible packaging	Preparation method	Suggested applications	References
CS, Cinnamaldehyde	CS with a degree of deacetylation of 75–85%, Trans-cinnamaldehyde, Glycerol	Polysaccharide film containing cinnamaldehyde-chitosan nanoparticles	Low molecular weight chitosan dissolved in acetic acid, mixed with Tween 80 and TPP under stirring. Cinnamaldehyde encapsulated in CS-nanoparticles using an O/W emulsion followed by ionic gelation. Films made by adding NPCH-CIN to chitosan solution with glycerol	Meat preservation, is effective against various bacteria including E. coli and Listeria, reducing environmental impact compared to plastic packaging	Mondéjar-López et al. [159]
SC- CS	SC (2% w/v), CS (2% w/v), Glycerol (1% v/v), Black Pepper Essential Oil (BPO) (various concentrations)	Composite Edible Films loaded with BPO	SC-CS solutions were prepared separately, mixed, added Glycerol, divided into batches for different BPO contents, poured into a petri dish following dried at ambient temperature (48 h)	Active food packaging material, with improved antioxidant activity for better preservation	Bhatia et al. [24]
CS, Pullulan (PL), Taro Starch (TS)	CS (1% solution), PL (1% solution), TS (various weights), Glycerol (2 mL as plasticizer)	CS-PL-TS (CPT) blend films	CS and PL solutions were prepared separately, mixed, and stirred; TS was gelatinized with hot water added to the solution; glycerol was included as a plasticizer; mixture stirred overnight; solutions cast on Petri plates, solvent evaporated, dried films were peeled off	Active packaging materials for antimicrobial and antioxidant characteristics	Pinto et al. [182]
Sodium Alginate, CS	Sodium Alginate, CS, Glycerol, Red Beetroot Anthocyanin Extract	Smart films for chicken fillets	Sodium Alginate dissolved in water, mixed with CS solution, Glycerol added as a plasticizer, red beetroot anthocyanin extract added, homogenized, cast onto plates, and air-dried	Packaging for chicken fillets, smart labeling, monitoring freshness, and extending shelf life	Ranjbar et al. [197]

(continued)

Table 5.1 (continued)

Biopolymers	Composition	Edible packaging	Preparation method	Suggested applications	References
Gelatin, Agar	Gelatin, ZnO nanoparticles, Glycerol; Agar, ZnO nanoparticles, Glycerol	Gelatin/ZnO and Agar/ZnO films	Solution Casting: Gelatin or agar dissolved in water, heated; glycerol added as a plasticizer; ZnO nanoparticles mixed in; solution subjected in petri dish, following dried at room temperature (25 °C, 48 h). For Gelatin/ZnO films: 2.5 g gelatin, 0.5 g ZnO; for Agar/ZnO films: 2.8 g agar, 0.1 g ZnO	Packaging for bread to enhance shelf life by reducing microbial growth and maintaining sensory quality	Yasmeen et al. [255]
MP/ANT/NCC	Myofibrillar Protein, Anthocyanins, Nanocrystalline Cellulose	Myofibrillar protein-based films	Incorporation and blending: 20 g/L MP concentration, modification with ultrasound, glycerol as a plasticizer, dried for 48 h at 25 °C, RH 50 ± 5%	Food packaging with improved barrier and mechanical properties	Chen et al. [38]
MP-Myr/ANT/NCC	Myofibrillar Protein, Myricetin, Anthocyanins, Nanocrystalline Cellulose	Myofibrillar protein-based films with Myricetin	Incorporation and blending with ultrasound treatment: Same as MP/ANT/NCC but modified with Myricetin (56.25 μmol/g protein)	Active packaging with superior mechanical, water vapor barrier, and ammonia sensitivity capabilities	Chen et al. [39]
MP-Lut/ANT/NCC	Myofibrillar Protein, Luteolin, Anthocyanins, Nanocrystalline Cellulose	Myofibrillar protein-based films with Luteolin	Incorporation and blending with ultrasound treatment: Similar to MP-Myr/ANT/NCC, modified with Luteolin instead of Myricetin	Edible packaging with enhanced color and mechanical properties	Chen et al. [38]
MP-Que/ANT/NCC	Myofibrillar Protein, Quercetin, Anthocyanins, Nanocrystalline Cellulose	Myofibrillar protein-based films with Quercetin	Incorporation and blending with ultrasound treatment: Similar to MP-Myr/ANT/NCC, modified with Quercetin instead of Myricetin	Packaging films with enhanced UV-blocking and mechanical properties	Chen et al. [39]

(continued)

Table 5.1 (continued)

Biopolymers	Composition	Edible packaging	Preparation method	Suggested applications	References
Single-layer Film	Furcellaran, Soy Protein Hydrolysates	Edible vegan films	Layer formation and drying: Furcellaran mixed with soy protein hydrolysates; drying at controlled conditions	Packaging for tofu and vegan products	Tkaczewska et al. [239]
Double-layer Film	Furcellaran, Soy Protein Hydrolysates, Soybean Bran Ethanol Extract	Edible vegan double-layered films	The first layer of turbellarian and soy protein hydrolysates; the second layer with soybean bran extract; dried at controlled conditions	Packaging for tofu and vegan products with enhanced antioxidant properties	Tkaczewska et al. [239]
Pullulan, CS, Chitin Nanofibers	Pullulan, CS, Nisin, Anthocyanins	Nanocomposite films	Casting Method: Mix Pullulan (PUL) and Chitosan (CS) solutions, add Chitin Nanofibers (CN) and Nisin, incorporate different percentages of Red Cabbage Anthocyanins (RCA), cast on plates, dry, and store. Films prepared with varying RCA concentrations	Active-intelligent food packaging	Wu et al. [251] Dursun Capar [65]
Sodium Alginate-based Films	Sodium Alginate, Glycerol, Lecithin, Quercetin, Vitis vinifera Leaf Extract	Biodegradable edible films with antioxidative and antimicrobial properties	Ultrasound-Assisted Technology: 2% sodium alginate solution with glycerol and lecithin. Quercetin and VVLE were added separately. Films treated with ultrasound at 40 kHz for 10 min. Dried for 24 h at 30 °C and 58% humidity	Edible active films for food packaging	Dursun Capar [65]
Starch (S), potato peels (P), Quince peels (Q)	Starch, Glycerol, Potato Peels, Quince Peels	Biodegradable Edible Films	Solvent Casting Method: Films were fabricated by S-P-Q-glycerol. Mixtures were vacuum filtered, cast into plastic Petri dishes, dried, and stored at controlled humidity	Cheese preservation, specifically cured cheeses	Coimbra et al. [46]

(continued)

Table 5.1 (continued)

Biopolymers	Composition	Edible packaging	Preparation method	Suggested applications	References
Basil Seed Mucilage and Sodium Alginate	Basil Seed Mucilage, Sodium Alginate, Glycerol, Clove Oil	Antimicrobial Edible Film	Casting Technique: Basil seed mucilage and sodium alginate in a 1:1 ratio, dissolved in water with glycerol and clove oil, dried at room temperature	Packaging material for food with added antimicrobial protection, enhancing shelf stability	Nazir and Wani [168]
Native Starch-Based Oleogels	Native Corn Starch, Soybean Oil, Water, Xanthan Gum	Edible inks for 3D food printing	Capillary Bridges Formation: Starch dispersed in oil; water added as a secondary liquid; variations with xanthan gum in water. The process involves stirring and mixing	3D food printing, personalizable food products, delivery of nutrients, and nutraceuticals	Miao et al. [155]
Balangu Seed Mucilage (SBM)-based Coating	Shirazi Balangu Seed Mucilage, Cell-Free Supernatant (CFS) of Levilactobacillus brevis G145, Tween 80, Distilled Water	Edible coating for lamb meat	Coating and Cold Storage: Lamb meat slices coated with different concentrations of CFS (1%, 2%) combined with SBM, air dried following storage under controlled conditions, such as 4 °C (1, 4, 7, 10 days)	Enhancing the shelf-life of lamb meat slices in cold conditions	Barzegar et al. [19]
Whey Protein- Lepidium perfoliatum Gum	Whey Protein Concentrate (WPC), Lepidium perfoliatum Gum (LPG), Transglutaminase Enzyme, Glycerol, Oregano Essential Oil	Biodegradable and Antimicrobial Films	Optimum film was created with 0.5% w/v LPG concentration, 2.9 WPC/LPG ratio, 10% enzyme content, and 15% v/v glycerol concentration. Distributed in molds, dried at 30 °C for 48 h. Oregano essential oil was added for antimicrobial properties	Biodegradable food packaging	Alipour et al. [11]
Gelatin, Xanthan, Echinacea purpurea	Gelatin, Xanthan, Echinacea purpurea Extract (EPE)	Multilayer and Edible Films	Gelatin and xanthan dissolved in water and different concentrations of EPE were added. For LBL, Gel/EPE solution was cast on a plate, dried, and then the Xan solution layered on top. For multilayer, the process repeated to form [Gel/EPE/Xan] sequence. Dried for 48 h at 25 °C before use	Ricotta cheese preservation	Rofeal and Abdelmalek [207]

(continued)

Table 5.1 (continued)

Biopolymers	Composition	Edible packaging	Preparation method	Suggested applications	References
Polyvinyl Alcohol-Acetylated Pullulan Polysaccharides	Polyvinyl Alcohol (PVA), Acetylated Pullulan Polysaccharides, Alpinia Galanga Essential Oil, Glycerol	Ternary Composite Degradable Plastics	Blending-Scraping Method: PVA mixed with acetylated pullulan solution in different ratios, adding 2% Pickering emulsion and glycerol, spread on glass plates, and dried in oven	Multifunctional active packaging for food preservation	Liang et al. [141]
Soy Protein Isolate, Sodium Caseinate, Polysaccharides (Sodium Alginate, Sodium Carboxymethylcellulose)	Soy Protein Isolate, Sodium Caseinate, Sodium Alginate, Sodium Carboxymethylcellulose	Microencapsulated Green Coffee Oil Powders	Complex coacervation combined with freeze-drying: GCO encapsulated using different combinations of soy protein isolate, sodium caseinate, sodium alginate, and sodium carboxymethylcellulose. The complexation between biopolymers was enhanced by freeze-drying	Useful for the production of powdered foods encapsulated with bioactive compounds	Mu et al. [162]
Polyvinyl Alcohol and Carrageenan	Polyvinyl Alcohol, Carrageenan, Glucose, Sugars, Carminic Acid	Films plasticized with edible gel	Gel Preparation, Film Casting: Gel based on glucose, sugar, and carrageenan as a gelling agent, mixed with PVA and carminic acid for antioxidant properties. Films dried under ambient conditions	Food packaging with improved flexibility and antioxidant properties	Bayer et al. [21]
Chitosan (CS) Film	Chitosan	Active edible film	Casting method with drying: 0.625% CS solution in 1% glacial acetic acid, with variations for NS-Et and NS-Met loaded films. Dried at 37 °C for 24 h	Food packaging, coating	Karkar et al. [124]

(continued)

Table 5.1 (continued)

Biopolymers	Composition	Edible packaging	Preparation method	Suggested applications	References
NS-Et-loaded-CS Film	Chitosan, Nigella sativa (ethanol)	Active edible film	Film casting and drying: Same as CS film, but with 3 mL NS (ethanol extract), and 0.2 mL glycerol added	Food packaging, coating	Karkar et al. [124]
NS-Met-loaded-CS Film	Chitosan, Nigella sativa (methanol)	Active edible film	Same as CS film, but with 3 mL NS (methanol extract), and 0.2 mL glycerol added	Food packaging, coating	Karkar et al. [124]
GGC	Guar gum (0.7% w/v), Coconut Oil (2 mL)	Composite coating	Dipping application method: Guar gum dissolved in water with continuous stirring, addition of coconut oil, and other components	Coating for Khasi mandarin	Goswami et al. [94]
GGC-Cl	Guar gum (0.7% w/v), Coconut Oil (2 mL), Clove Oil (1 mL)	Composite coating	Dipping application method: Similar to GGC, but with added clove oil	Coating for Khasi mandarin	Goswami et al. [94]
GGC-Cl-Cn	Guar gum (0.7% w/v), Coconut Oil (2 mL), Clove Oil (1 mL), Cinnamon Oil (1 mL)	Composite coating	Dipping application method: Preparation similar to GGC, with additional clove and cinnamon oils	Coating for Khasi mandarin	Goswami et al. [94]
GGChC-Cl-Cn	Guar gum (0.7% w/v), CS (0.3% w/v), Coconut Oil (2 mL), Clove Oil (1 mL), Cinnamon Oil (1 mL)	Composite coating	Dipping application method: Guar gum and chitosan mixed with water, glycerol, Tween-80, essential oils, and other components, under continuous stirring. Mandarins were immersed in composite formulations for 5 min, and the surplus solution was allowed to drip off on perforated plates	Coating for Khasi mandarin	Goswami et al. [94]

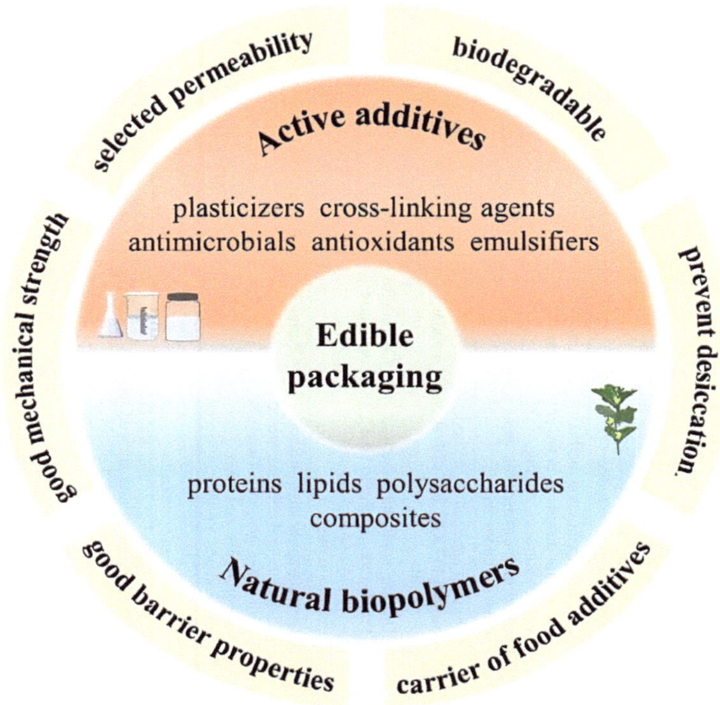

Fig. 5.2 Schematic diagram illustrates the production process of edible food packaging materials using natural ingredients, such as proteins, polysaccharides, and lipids, as an eco-friendly alternative to plastic materials in the food and beverage industries [245]

overall health of consumers [86, 111]. However, formulating smart edible packaging that shows characteristics of traditional plastic packaging materials needs to address several challenges, such as mechanical strength, and elasticity [212]. This chapter provides insights into the harmful impact of plastic packaging, and the development of edible packaging using biopolymers, including proteins, and polysaccharides. It was further highlighted that these natural polymers enhance several properties of materials, including barrier, texture, physicochemical, digestive, and rheological properties. In addition, these polymers impart substantially for the improvement of the material, whether it is packaging, film, or emulsions. By incorporating natural polymers into films and emulsions, researchers can achieve high-quality outcomes with enhanced stability and longer shelf-life. Nevertheless, the chapter delves into the current developments and state-of-the-art insights on the creation and application of edible packaging, challenges faced in straightforward implementation, and future perspectives on smart and sustainable packaging materials.

5.2 Natural Polymers for Smart and Sustainable Edible Packaging Materials

Edible packaging is emerging as a promising solution for green and sustainable packaging, utilizing a combination of natural and smart materials to fulfill consumer preferences for environmentally friendly options [193, 210, 212]. Natural ingredients like chitosan, alginate, starch, and protein-based films provide biodegradability and safety, minimizing ecological impact and ensuring food safety [226, 235] (Fig. 5.3 and Table 5.1). Moreover, the incorporation of smart materials such as pH-responsive polymers, antimicrobial agents, and oxygen scavengers enhances applications and prolongs the stability of edible packaging [235]. These smart components allow for customized responses to environmental factors and food preservation requirements, guaranteeing the packaged foods' quality and safety [212]. By synergistically employing natural and smart materials, edible packaging not only addresses sustainability challenges but also presents innovative strategies to improve food packaging efficiency and reduce reliance on plastic [105, 193, 212, 235].

Edible smart materials are created from nontoxic material derived from lipids proteins (sourced from either animals or plants), polysaccharides, and their combinations. The natural polymers-based food packaging materials trade is predicted to have considerable growth, with a forecasted value of 2.14 billion US dollars by the year 2030 [7]. This shows a considerable increase from its value of 783.32 million US dollars in 2021. The market is estimated to exhibit a compound yearly growth rate, CAGR of 6.79% (2022–2030). North America is expected to dominate this market during the forecast period, with significant growth also expected in the UK, Japan, Indonesia, and Israel [8]. Extensive research has been conducted in developing and characterizing edible films from various plant-based polymers (Table 5.1). For the past three decades, there has been a consistent effort to generate edible material that can rival traditional plastic films for commercial uses. The subsequent section will cover the different natural polymers used to produce edible packaging [7].

5.2.1 Edible Packaging from Proteins

Edible packaging made from plant-based proteins encompasses sesame protein, corn zein, quinoa protein, wheat gluten, and wheat gluten (Fig. 5.2). Animal-derived proteins such as whey proteins (WP), casein, gelatin, collagen, and several others are utilized to protect and preserve foods because of their film-forming characteristics (Table 5.1) [237]. Among the various sources of edible film, protein-based materials stand out due to their nutrient content [79]. Additionally, edible packaging from proteins displays superior mechanical and barrier characteristics compared to lipid and polysaccharide material, albeit with lower moisture barrier characteristics (Fig. 5.2). The tightly arranged and organized network of hydrogen bonds provides

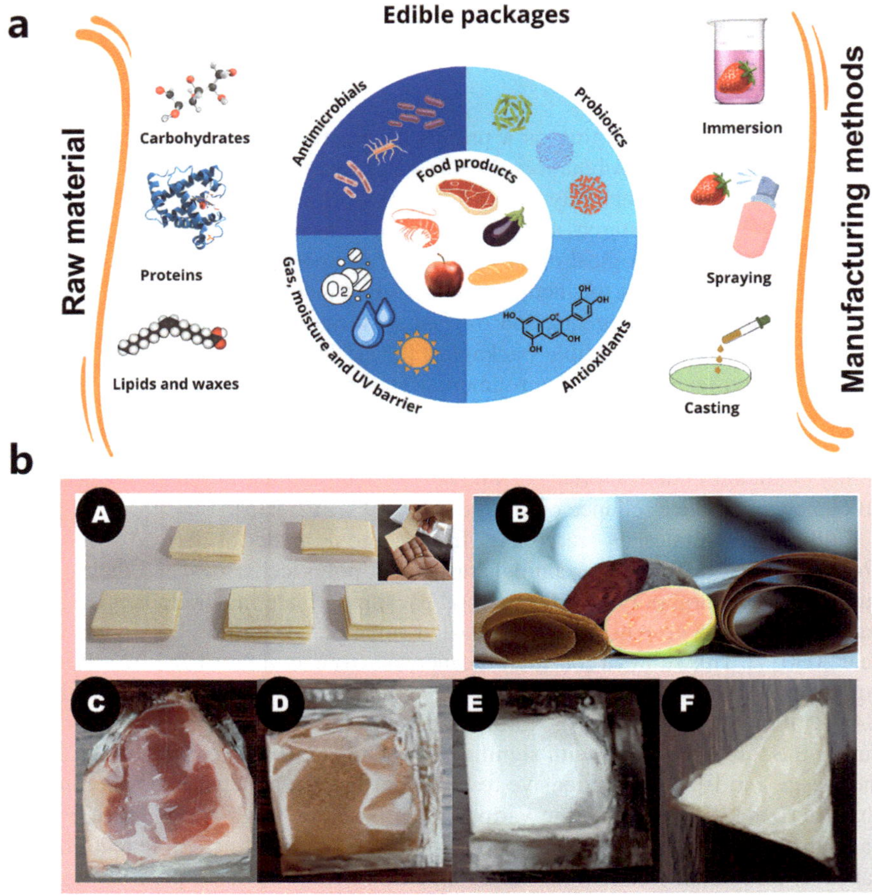

Fig. 5.3 **a** Schematic diagram showing the use of raw materials (natural polymers), fabrication methods, and benefits in the food industry, Reproduced with permission from Ribeiro et al. [204], Copyright Elsevier. **b** Illustrates the use of different types of edible films in some foods, including cheese slices (**a**), guava (left), and beetroot (right) purees (**b**). An edible film based on sodium alginate applied to meat slices (**c**), soluble coffee (**d**), powdered medicine (**e**), and cheese slices (**f**) [8, 78, 90, 176, 188]

exceptional effective blocks gases and physical attributes [135]. Protein-based packaging helps mitigate the oxidation of fried foods or high-fat or, a major contributor to quality deterioration and shortened shelf life, by inhibiting oxygen permeation (Fig. 5.3a) [260]. The protein's structure also influences packaging characteristics; for instance, the oxygen permeability of some animal proteins (e.g., WP) and plant proteins (such as soy protein (SPI), corn zein, etc.) is greater than that of collagen films due to globular structure and differences in molecular arrangement in these proteins [237]. Furthermore, protein-derived edible packaging provides significant application in individually packaging small food portions, especially for products like beans, nuts, and cashews, where individual packaging is impractical. Dry and wet procedures can be used in the construction of protein-driven film materials (Fig. 5.4) [188]. When paired with substances such as WP concentrate, ascorbic acid, bee waxes, and cysteine have shown effectiveness in reducing browning in freshly cut apples. WP doesn't substantially reduce the weight loss of fresh-cut apples. However, it's worth noting WP nanofibers (WPNF) can be created through the self-assembly of WP isolate with trehalose. These nanofibers exhibit increased hydrophobicity and enhanced antioxidant properties compared to conventional WP-based films. Additionally, WPNF-based films display improved smoothness and uniformity, potentially offering advantages in various applications beyond fresh-cut apple preservation [27].

5.2.1.1 Edible Packaging from Whey or Milk

Among milk proteins, WP, and casein contain a small amount of lactose, which contributes to the formation of edible films. This property further enhances their tendency to crystallize, resulting in packaging materials with high mechanical properties [241]. Natural polymers packaging is developed from WP fraction, including WP concentrate WP isolate by incorporating various plasticizers and emulsifiers [32, 79, 214, 223, 265]. Films from WP demonstrate outstanding Oil, oxygen, and aroma barrier properties in small to medium relative humidity environments, alongside suitable animal properties for diverse applications such as food coating, layer separation, and pouch formation. Additionally, recent efforts have focused on enhancing the functional attributes of WP films through the combination of prebiotics and probiotics [73, 265]. WP coatings/films present a compelling alternative to synthetic packaging materials, addressing both the disposal of whey and the adverse environmental impacts of packaging waste (Table 5.1 and Figs. 5.1, 5.2, 5.3 and 5.4).

5.2.1.2 Edible Packaging from Wheat Gluten

Wheat gluten comprises a major protein component in wheat flour. It is obtained by washing wheat flour to remove the starch granules from the flour [265]. This protein mainly consists of gliadins' monomers that impart its extensibility and glutenin polymers which are responsible for its elasticity. The cohesion and elastic behavior of gluten contribute to the structural integrity required for film formation [71]. Hence, it provides excellent gas (oxygen and/or carbon dioxide) barrier characteristics to packaging food materials [266]. Moreover, the thermal stability of the wheat gluten protein-carboxymethyl film surpassed that of the wheat gluten protein film significantly. Furthermore, the incorporation of egg white proteins led to a notable enhancement in the opacity of wheat gluten protein films, resulting in superior physical strength and improved light-blocking capabilities (Fig. 5.2). Additionally, the tartaric acid improved the thickness of the films, and its exceptional

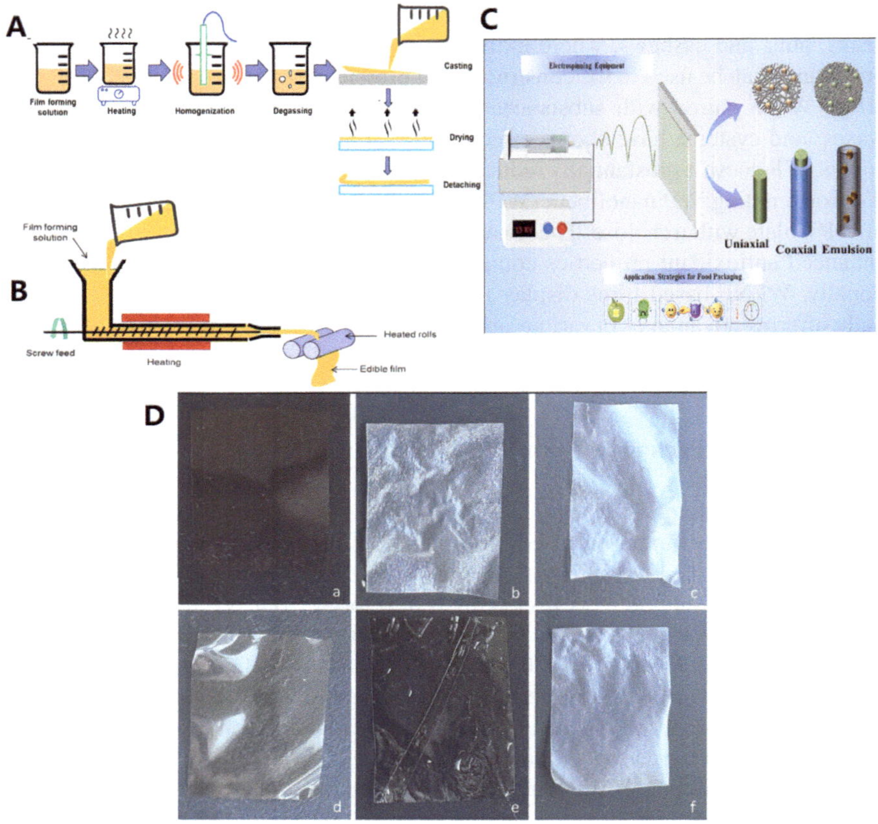

Fig. 5.4 Schematic diagram illustrating the fabrication methods of edible packaging including casting method (**a**), extrusion method (**b**) [7], and electrospinning method (**c**) Reproduced with permission from [248], Copyright Elsevier. Macrostructures of different films **a** pectin, **b** cellulose (HPMC), **c** gelatin protein, **d** pectin-HPMC film, **e** gelatin-pectin film, **f** gelatin-HPMC-based films for fresh fish fillets (**d**) (Athanasopoulou et al. [16])

UV resistance rendered the film suitable for use as a novel type of edible film. These films, produced through these approaches, were transparent, light-colored, possessed high mechanical strength, and exhibited good water resistance [61]. Moreover, its capacity for cross-linking under heat, low water solubility, viscoelastic properties, cost-effectiveness, and accessibility as a byproduct make it a better choice for edible film materials [13]. After processing wheat gluten displays brittleness and can absorb more water. Hence, various strategies such as incorporating additives, and plasticizers containing reactive groups (–SH, NH_2, –COOH), and mixing with biopolymers such as cassava starch, poly lactic acid, polycaprolactone, polyvinyl alcohol (PVC), polyesters, etc., can mitigate these effects (Table 5.1) [106]. Furthermore, gluten-derived packaging exhibits mechanical strength that is influenced by several factors, including pH and concentration. However, water vapor permeable characteristics in packaging materials can be associated with pH and ethanol contents [71, 98]. Nonetheless, the physicochemical characteristics of wheat gluten-based material are enhanced by incorporating various organic acids, polysaccharides, and proteins [59, 104].

5.2.1.3 Edible Packaging from Soybean Protein Isolate (SPI)

SPI stands out as a prominent source of edible film, extracted from defatted and dehulled soybeans (Table 5.1) [53]. The film formation from soy protein involves two main steps. The first step in the process is the destruction of the soy protein chain complex. This is achieved through the utilization of alkaline or thermal treatment methods. These methods rupture the soy protein structure, thereby exposing hydrophobic and sulfhydryl regions [82]. Additionally, these treatments cleave the native disulfide bonds (S–S) located in the soy protein chain complex [82]. Despite its inherent structural challenges, native soy protein can be modified through protein structure cross-linking using methods such as natural cross-linking, thermal treatment, and denaturation [77]. SPI-based films show several characteristics including low oxygen permeability, flexibility, and transparency, making them suitable for a wide range of applications. Additionally, SPI films offer advantages like low-density PE material having low costs but are easily available. The Gel-forming capacity of the SPI makes it a suitable candidate for creating the framework for composite films that incorporate lipids and essential bioactive components such as antioxidants and antimicrobial agents [24, 53, 104]. Hamed et al. [99] formed functional and homogeneous SPI material using Purslane seed oil. Results showed a smooth sheet-like structure without cracks or pores. Spectral analysis revealed consistent patterns, indicating minimal impact on the functional groups of soy protein isolates. These films displayed higher thermal stability due to increased protein concentration. Additionally, soy protein isolate-Purslane seed oil films effectively inhibited pH increase, TBARS, and peroxide content in microbial growth during cold storage and processed beef burger patties showing promise as food film for preservation [99].

5.2.1.4 Edible Film Derived from Sodium Caseinate (SC)

SC originates from casein, a protein naturally present in milk. While casein is typically insoluble in water, it can undergo a pH adjustment process to become soluble in water [258]. SC has a flexible, random, and coiled structure that facilitates the production of effective edible coating materials. The surface-modifying properties of SC can further enhance the barrier characteristics of the SC coating by associating it with other proteins, such as zein (Fig. 5.2 and Table 5.1) [258]. Conversely, the direct coating reduces oxygen barrier property, with evenly distributed zein nanospheres on the film surface (Fig. 5.2). Similarly, genipin chain cross-linking and its combination with vital compounds, such as antimicrobial agents and essential oils

(EOs) enhance the mechanical strength of SC coating [12, 92, 142, 191]. Sodium caseinate-based edible coating facilitates the delivery of ginger essential oil, acting as a carrier for antimicrobial agents. Applying this edible solution onto chicken breast and subsequent storage at 4 °C reduced aerobic bacteria counts for 12 days. Additionally, Carnauba wax combined with nano clay bio composite enhances the nutritional quality of Valencia oranges. Furthermore, the utilization of chitosan-based layer-by-layer coatings in combination with sodium alginate or tannic acid demonstrates improved antimicrobial properties, showcasing significant potential in the packaging industries (Parreidt et al. 2018).

5.2.1.5 Edible Film Derived from Corn Zein

Corn zein protein derived from corn, is crucial in the creation of edible materials for packaging, such as pouches, coating/ and or films [40, 262]. It is obtained in the form of by-product of bioethanol production and is easily available [67]. It is commonly utilized in conjugation with other biopolymers, such as polysaccharides and SPI, to increase barrier characteristics and impart a glossy surface finish [43, 246]. Coating hydrophilic protein-derived film with zein through targeted protein–protein interactions offers exciting prospects for improving their barrier properties. Zein features unique hydrophobic and hydrophilic regions on its surface, facilitating its self-assembly (Table 5.1). Moreover, the hydrophobicity of corn zein can be attributed to its high amount of non-polar amino acids, including alanine, proline, and leucine, resulting in the production of a strong barrier with higher mechanical properties [104]. Corn zein protein boasts numerous beneficial characteristics for film development, including biodegradability, biocompatibility, and exceptional film-forming capabilities. These films act as effective barriers against environmental factors such as oxygen and moisture to enhance the shelf-life stability of packaging materials. Moreover, corn zein-based films can be further enhanced by incorporating several essential components, such as antioxidants, antimicrobial agents, and flavor enhancers. Subsequently, by enriching corn zein films with these active ingredients, the efficacy in food preservation can be substantially amplified. Additionally, the renewable and abundant nature of corn as a raw material renders corn zein-based edible films a smart and sustainable alternative to conventional petroleum-based packaging materials [104, 137].

5.2.1.6 Edible Packaging from Collagen and Gelatin

Collagen, sourced from animals, is utilized in edible packaging and is characterized by its hydrophilic nature and richness in hydroxyproline, proline, and glycine, causing it to swell in polar liquids with high solubility parameters (Figs. 5.2, 5.3a, 5.4d and Table 5.1) [47]. Historical records trace the use of collagen casing for the meat industry [104]. Gelatin is produced through the hydrolysis of insoluble collagen that exists in the skins and bones of animals. Generally, it is considered a waste product when animals are slaughtered and/or processed [144]. Gelatin has gained considerable attention in recent years as a packaging material due to its several applications such as barrier against oil and gas, non-toxicity, affordability, and biodegradability. In food packaging, they effectively shield against moisture, oxygen, and other environmental factors, preserving freshness and extending shelf life. Moreover, these films can be customized with additives like antimicrobials, antioxidants, and flavorings, enhancing their effectiveness in food preservation and sensory enhancement. Collagen and gelatin films have achieved considerable attention in pharmaceutical industries and biomedical engineering [158, 196]. However, these proteins possess several limitations, which include rapid biodegradation, poor mechanical strength, as well as weak water resistance that need to be overcome before utilizing them for food packaging applications [42, 85, 144].

5.2.1.7 Edible Packaging from Alternative Protein Sources

Scientists have begun exploring proteins from various origins to promote the sustainable utilization of protein sources to evaluate their potential for forming edible films. For selecting protein sources, two main approaches have been selected either by utilizing by-product or leveraging the unique characteristics and nutritional benefits of different sources (Table 5.1). Proteins sourced from lentils [14], sesame isolates [217] fish muscle-derived proteins, such as myofibrillar proteins [123], pumpkin seeds [136, 253], egg white [100, 108], and rice [242] have been explored for edible film formation. However, the main challenge faced by these proteins is their limited ability to form a cohesive film matrix. One method to enhance the cohesion of protein film is employing transglutaminase (TG) cross-linking using the ultrasonication technique [49]. Nevertheless, such proteins are utilized in composite film form or directly for food packaging applications [104].

5.2.2 Smart and Sustainable Edible Packaging Derived from Polysaccharides

Recently, there has been intense interest in utilizing polysaccharides to create edible films and coatings [109]. Among the polysaccharides utilized for edible packaging production are cellulose, pectin, starch, chitosan, alginate, etc. (Fig. 5.4d and Table 5.1) [36, 232]. Films from Polysaccharides exhibit strong network formation and hydrogen bonding in their structures, enhancing their effective gas barrier properties. However, polysaccharides' strong hydrophilicity creates challenges for their packaging products resulting in less effective vapor barriers. Polysaccharide coatings, devoid of oil ratio and colorless in appearance, are utilized to enhance shelf-life stability without inducing anaerobic conditions [104, 158].

5.2.2.1 Edible Packaging from Cellulose and Its Derivatives

The most common natural polymer, cellulose, is the main structural component of plant cell walls.

It is made from the linear homopolysaccharide consisting of β-1,4 sugar (glucose) units (Fig. 5.4 and Table 5.1). Cellulose and its derivatives, including methylcellulose (MC), hydroxypropyl cellulose (HPC), and hydroxypropyl methylcellulose (HPMC) are applied to create different edible films utilized in the food sector (Fig. 5.4d and Table 5.1). These derivatives of cellulose are produced through structural modifications such as the attachment of small functional groups, including methyl, carboxyl, and hydroxyl, to cellulose molecules [104, 113, 169]. Methylcellulose-based coatings form a barrier that restricts the movement of lipids or oil making them suitable for use in confectionery food items. Furthermore, HPMC-derived films help inhibit oil absorption, making them prominent packaging materials for fried foods (Fig. 5.2).

5.2.2.2 Edible Packaging from Starch-Derived

Starch is a well-known homopolysaccharide that consists of amylose and amylopectin and contributes a significant role in developing edible films due to its continuous matrix formation and its renewable nature. Amylose comprises a linear chain α-(1 → 4) glucose units, while a branched polymer chain from α-(1 → 6) exists in an amylopectin chain. Amylose is greatly favored for edible filmmaking because of its solubility in water, low oxygen penetration, and high flexibility [36, 232]. The films produced from starch and its derivatives perform efficient barriers against oxygen, moisture, and other environmental factors resulting in enhanced shelf-life stability for several food items (such as cream, butter, and meat) (Fig. 5.2 and Table 5.1). Moreover, starch-based coating is conveniently employed for the surface of various foods providing transparent, pliable films that adhere well and enhance the visual appeal of products. Additionally, these coatings can be personalized with additives like antioxidants, antimicrobials, and flavorings to enhance their functionality and shelf stability. Furthermore, starch-based edible packaging materials are renewable, smart, and sustainable helping to reduce plastic materials in the environment [4, 74].

5.2.2.3 Edible Packaging from Pectin and Pectin Derivatives

Pectin and its derivatives are mainly composed of galacturonic acids. It is obtained from fruits, such as apples (pomace), orange peel, berries, etc. [161]. The galacturonic acids' esterification of pectin with methyl alcohol impacts its ability to form pectin's films and gel (Fig. 5.4d). The esterification degree classified pectin into low or high methoxy pectins [68]. Pectin is extensively used in the food sector for its distinct properties that make it suitable for the creation of edible materials. This pectin film offers diverse applications, such as edibility, biodegradability, and versatile physicochemical characteristics [45]. These films provide efficient protection against moisture and oxygen. Furthermore, the designed pectin films maintain the shelf-life and lipid oxidation of perishable foods (Fig. 5.4d). Furthermore, pectin-based films offer flexibility and transparency, rendering them appropriate for coating a variety of food surfaces and enhancing their aesthetic appeal. Additionally, these films can be enhanced with functional additives like antimicrobials, antioxidants, and flavorings to enhance their effectiveness and customize them for specific food preservation requirements [243].

5.2.2.4 Edible Packaging from Alginate

Edible films designed from alginate contain guluronic acid and mannuronic acid, with the ratio of these acids impacting the molecular weight of alginates and their physical characteristics [148]. Alginates are generally extracted from seaweed. It exhibits several physicochemical properties such as thickening, film-forming, and stabilization, rendering it suitable for the creation of edible packaging materials (Fig. 5.2 and Table 5.1) [104]. The sugar acids in alginate favor the attachment with cations (Ca^{+2}, Fe^{+2}, and Mn^{+2}), influencing the physical properties in alginate-based edible packaging, utilizing retention of moisture and color preservation [65, 213]. The coating and films prepared from alginate have garnered attention for their potential to enhance and preserve the quality of several foods, which include meat, seafood, eggs, cheese, fruits, and vegetables. They achieve this by acting as sacrificial moisture agents, reducing dehydration, and enhancing the stability and mechanical properties of foods [213]. There are several phenolic compounds, such as thymol and tea polyphenols, play significant roles in enhancing the functionalities of alginate packaging materials. These compounds not only provide support to edible films but also improve their antimicrobial and mechanical characteristics. Tea polyphenols consumption may also benefit health by providing essential bioactive substances. Alginate combination with other biopolymer solutions enhances the film strength as well as prolongs the stability of various foods and reduces waste from foods. Additionally, ultrasonic-assisted pretreatment can improve the edible materials containing grape leaf extract and sodium alginate, reducing thickness, elongation at break (EAB), and moisture content [245]. Nonetheless, alginate coatings or films demonstrate reduced resistance to water or moisture owing to the intrinsic hydrophilic properties of alginates (Figs. 5.2 and 5.3a) [104].

5.2.2.5 Edible Packaging from Pullulan

Pullulan, consisting of maltotriose molecules linked through α-(1 $\rightarrow$ 4) linkages and α-(1 $\rightarrow$ 6) bonds. It is considered a water-soluble polysaccharide obtained from Aureobasidum pullulan, a yeast-like fungus [186]. It presents an excellent option for creating edible films used in packaging because of pullulan's distinctive properties, including heat prevention, resistance to oxygen, and oil, and its colorless and transparent nature [81]. In edible film formation, pullulan functions as coatings and thickener pullulan is beneficial for enhancing the shelf life and stability of fruits/ and or vegetables [175]. The coating characteristics of pullulan can be improved by combining it with chitosan-oligosaccharide (an antibacterial agent) and glutathione (a reducing agent) [104]. However, pullulan shows weak mechanical and water barrier characteristics due to its hydrophilic characteristics. These issues can be mitigated by incorporating lipids, such as oleic acid, palmitic acid, and beeswax (Table 5.1).

5.2.2.6 Edible Packaging from Chitosan (CS)

Many organisms comprise chitin in their cell walls, such as invertebrates and fungi. CS is produced by the deacetylation of chitin in an alkaline solution. Edible films derived from chitosan display barrier properties against CO_2 and O_2 and possess inherent antimicrobial characteristics (Fig. 5.2). The degree of deacetylation chitin contributes significant impact on the physicochemical characteristics of CS-based films [131]. Chitosan-based product's properties are in line with natural gums, exhibiting high viscosity and possessing antimicrobial growth that can be attributed to exposure to active regions of amino acids [170]. Both chitin and chitosan produce clear films that improve the shelf life and quality of foods [203].

5.2.3 Edible Packaging from Lipids

Lipids, naturally occurring compounds found in plants, insects, and animals, exhibit a diverse range of functional groups, including mono-, di-, and triglycerides, fatty alcohols, fatty acids, cerebrosides, terpenes phospholipids, and phosphatides [158]. In contrast to lipids, polysaccharides or proteins are unable to create films due to the lack of repeating units connected via covalent bonds, which are necessary for standalone film formation. However, they can create coatings material. To increase the hydrophobic characteristic of emulsion-based edible material, various animal and plant-derived lipids are introduced into the solution of film material [104, 158]. These fats or oils, chemically similar mixtures predominantly composed of triglycerides, vary in their physical and origins aspect. Plant-extracted oils are liquid at room temperature, while fats sourced from animals are solid. Various ingredients such as animal-based waxes (wool grease, beeswax, lanolin), plant-based waxes (sugar cane waxes, candelilla, carnauba), vegetable oils (olive, sunflower, rapeseed oil), petroleum wax, and synthetic waxes like paraffin are added to form a film-forming solution (Figs. 5.3, 5.4 and 5.6) [158, 202]. Waxes, comprised of alcohols/ esters of long-chain acids, possess higher molecular weights and increased hydrophobicity. Vegetable oils are raw oily materials obtained from nuts, seeds, or freshly harvested flowers through cold pressing (Fig. 5.6). Conversely, EOs are highly aromatic compounds extracted via steam distillation from many plants (flowers, fruits, leaves, and wood), rich in volatile, hydrophobic, and aromatic compounds such as terpenoids and terpenes and exhibit strong antimicrobial behavior. Incorporating lipids such as oil (virgin coconut) into the material-forming solution has been shown to decrease the vapor permeability to water of edible material produced via SPI [34]. To impart antimicrobial and antioxidant characteristics, EOs extracted from various fragrant plants such as oregano, garlic, lemon, clove, cinnamon, and rosemary are included in the coating solution during the process of emulsification. Encapsulation of Eos within the smart material enhances volatility stability [10, 103, 180]. However, a notable drawback of lipid-based materials is their brittleness, resulting in a taste and texture that is greasy and waxy, which might not be suitable for applications in various instances. Attaining the anticipated characteristic for the film requires biocompatibility between the polymer matrix and lipid phase. Lipid-based self-supporting materials are typically produced using a film-forming solution containing polymers with higher molecular weight (polysaccharide or protein) via solvent-casting techniques [104].

5.2.4 Edible Packaging from Composite Films

Composite films comprise various hydrophobic and hydrophilic components merged to enhance practical characteristics. Often, a biopolymer composite designed by a single functional compound is sufficient to provide all required properties, including mechanical and barrier capabilities (Figs. 5.3a, 5.4d and Table 5.1) [58]. While polar biopolymer films like proteins/polysaccharides show stronger mechanical and barrier characteristics under low humidity. These materials tend to have reduced water barrier properties because of hydrophilic behavior in high humidity (Figs. 5.3a and 5.4d). In contrast, hydrophobic lipids suggest reasonable resistance to moisture migration but an absence of mechanical strength compared to hydrocolloid films [104]. By combining hydrophobic lipids with hydrophilic polysaccharides or proteins, films with improved properties can be achieved. For instance, Omar-Aziz et al. [175] demonstrated the development of a composite film by blending beeswax and pullulan, resulting in important enhancements in mechanical tensile and water vapor permeability than material made from pullulan alone [175]. Composite films can be fabricated either in layered form or as emulsions of smart materials. Composite films consist of layered are categorized as numbers binary/ternary polymers. Various arrangements such as protein–protein, and carbohydrate-carbohydrate are feasible in binary films, with extensive literature available on this topic. Emulsified complex coatings offer advantages over layered ones, as they are less prone to delamination over time and require fewer drying and casting procedures. Complex Films created by incorporating the lipid phase into a hydrocolloid structure through emulsification demonstrate improved barrier and functional characteristics. Numerous studies have explored composite films combining polysaccharides and lipids to improve water barrier characteristics. The mechanical properties of the composite film are considerably affected by the lipid particle size, owing to their larger surface area. Additionally, complex coating incorporating proteins/ minerals shows notable enhancements in physical characteristics. Two-layer coating, often seen in complex material, is formed by layering one over the other, but this method poses a threat of layer separation. One potential solution to this issue is to utilize a single-step mechanism, where one substance is to be dispersed into another film-forming solution and then cast to produce the film. A recent trend among researchers is the combination of ESo in nanoemulsions to enhance efficacy [104].

5.2.5 Edible Packaging Natural Polymer Nanoparticles

Nanotechnology has become increasingly prominent in recent years, as a novel approach for obtaining organic and inorganic compounds at the nanoscale, offering unique properties [17, 69]. Nanomaterials, defined by their particle sizes ranging from 1 to 100 nm, represent a growing field in the food packaging industry. Incorporating nanomaterials into polymer matrices has shown promise in enhancing mechanical and physical characteristics (Table 5.1) [25, 212]. This approach also enables the production of film for packaging materials through nanoencapsulation of biologically active natural substances, enhancing stability and solubility and resulting in superior performance compared to conventional films [177]. For example, nano clay is extensively used because of its ability to enhance mechanical and barrier characteristics, due to its surface-to-volume ratio and high aspect proportion [218]. Çağrı Mehmetoğlu et al. [29] investigated the impact of silver particles on WP-material, revealing an 84% increase in mechanical tensile (tensile strength, TS) and a 67% enhancement in barrier characteristics than that of control material [29]. Similarly, nanocellulose and zinc/titanium oxide have been extensively employed in packaging industries [51, 151]. The safety of biopolymers remains a contentious topic within the Academic researchers. These materials can elicit many harmful impacts determined by factors such as their particle size distribution, chemical composition, surface condition, and shape. Experimental studies commonly report oxidative stress, inflammatory responses, and genotoxic effects as potential outcomes [150]. Furthermore, the ability of nanomaterials to penetrate human cells is diverse in size; for instance, particles around 100 nm able to certainly enter cells, while those around 40 nm can enter nuclei, and those less than 35 nm may breach the blood–brain barrier. Lower-size particles typically exhibit greater Catalytic performance and higher potential for generating reactive oxygen species, with enhanced binding capabilities and adsorption rates compared to larger counterparts. Despite ongoing research, comprehensive data on the toxicological and safety impact of nanoparticles remains scarce in the public domain. Therefore, a thorough exploration of their effects on environmental microbiota and human health is imperative to mitigate potential adverse effects (Fig. 5.1) [28, 69, 118, 131].

5.3 Properties and Functionality of Natural Polymers-Based Edible Packaging

Careful consideration of the characteristics and functionality of natural polymer-derived packaging is vital in the food industry (Figs. 5.2a–c). These films present an appealing alternative to traditional plastic owing to their non-toxic components, renewability, and potential to minimize environmental impact (Table 5.1). Some key attributes encompass effective barrier properties, shielding against moisture, oxygen, light, and microbial contaminants, thereby ensuring the preservation of food quality and safety. Moreover, mechanical strength, flexibility, and stretchability are vital for enduring handling and distribution while conforming to various food product shapes. Transparency and clarity contribute to enhanced product visibility and presentation, while the eco-friendly nature of biodegradability and sustainability aligns with environmentally conscious packaging solutions. Including antioxidants and antimicrobial agents further bolsters the films' efficacy by prolonging shelf life and upholding food quality. Additionally, ensuring compliance with food safety standards and streamlining processing and manufacturing processes are crucial factors in the advancement of films, facilitating widespread acceptance in the industrial food sector as an environmentally sustainable wrapping choice [3, 209].

5.3.1 Mechanical Properties

The mechanical characteristics, which encompass Young's modulus, TS, and EAB constitute critical attributes for films employed in the food sector. Ensuring suitable mechanical characteristics is essential to guarantee the functional capacity of the material in safeguarding food items during transportation, encompassing distribution, handling, and storage before it is consumed. These characteristics predominantly hinge on the chemical/composition makeup of the biopolymers utilized in film preparation (Table 5.1). Polysaccharides, WP, and other biopolymers show commendable packing capabilities, but lipids fail to produce freestanding films. Nonetheless, pure film materials often demonstrate inferior mechanical characteristics to conventional polymers like PE, PP, and PET. Enhancing the mechanical characteristics of films, employing strategies such as composite and multilayer film construction [133, 209]. Additionally, methods such as nanoparticle or nanomaterial incorporation and the integration of nanofillers have gained prominence. Numerous research endeavors have explored nanoparticle reinforcement to increase the fabrication of films [134, 209, 201]. Recently, Tafa et al. [230] investigated edible films developed biodegradable starch exhibiting TS ranging from 17.97 to 24.25 MPa. According to Wu et al. [250] Films with higher concentrations of starch and agar demonstrated good tensile strength The increase in TS is attributed to the establishment of intermolecular hydrogen bonds among agar and tapioca starch, as well as the formation of compacted structures within the biodegradable films, leading to greater TS [183]. Furthermore, the addition of antimicrobial agents and plasticizers can alter the films, mechanical strength by inducing changes in the cross-linking or intermolecular arrangement within the molecules of biopolymer [20, 102].

As natural polymers typically display brittle mechanical characteristics, plasticizers are commonly introduced into film-forming solutions to enhance workability thereby increasing the EAB value. Plasticizers, characterized as smaller molecules with average molecular weights, work by reducing the deformation tension, hardness, and density of the biopolymer. They enhance chain flexibility by disrupting biopolymer-biopolymer interactions and fostering polymer-plasticizer interactions [22, 51, 127]. Consequently, the addition of plasticizers imparts softness and flexibility to films. Common food-grade plasticizers used in film formation include glycerol, polyethylene glycol (PEG), and sorbitol. Biopolymers plasticized with sucrose, fructose, and maltose demonstrated favored mechanical characteristics, containing EAB and TS, compared to those plasticized with glycerol, sorbitol, and glycols (Fig. 5.2) [3, 66, 211].

5.3.2 Gas Barrier Property

The gas barrier property of an edible film denotes its capacity to hinder the passage of gases like oxygen, carbon dioxide, and moisture. This attribute holds significant importance in food packaging, as it aids in increasing the durability of perishable items by obstructing the infiltration of gases that could hasten processes such as oxidation, microbial proliferation, and moisture depletion. Edible films possessing superior gas barrier properties play a pivotal role in shielding food items from environmental influences, thus preserving their quality and freshness over extended durations (Figs. 5.1 and 5.2). The concentration of gases significantly influences food quality attributes such as color, nutrition, flavor/aroma, and texture. Hence, the gas barrier characteristics of packing play a crucial role in food preservation. Unlike many plastic films, biopolymer-based films offer a distinct advantage due to their favorable gas barrier properties, attributed to the naturally polar nature of hydrocolloid polymers (proteins and polysaccharides) (Table 5.1). These polymers effectively impede the permeation of gases like oxygen (O_2) and carbon dioxide (CO_2). It has been reported that adding hydrophobic antimicrobials and lipids including ESo enhances the permeability of O_2 and CO_2. Conversely, the incorporation of certain oils, like carvacrol, citrus, and lemongrass essential oils, has resulted in contrasting outcomes (Figs. 5.1, 5.2 and 5.6) [3, 30, 64, 101, 167, 208].

The gas-blocking capabilities of films are critical for preserving various foods. When it comes to preserving vegetables or fruits, films with appropriate gas permeability help regulate the internal gas concentration of the packaging. This regulation, in turn, reduces the respiration rate of vegetables and fruits to a manageable level, thereby minimizing weight and loss of nutrition. Nonetheless, if the gas barrier of films is excessively low, it can lead to anaerobic respiration, resulting in the decay of texture and the presence of off-flavors in vegetables and fruits. Hence, fine-tuned adjustment of gas permeability in films is essential to retain the quality of food [3]. Sun et al. [228] introduced an innovative biopolymer-based film composed of cellulose nanocrystals (as the dispersed phase) and methylcellulose (as the continuous phase). This film demonstrated an enhanced CO_2/O_2 selectivity, as indicated by the permeability ratio of CO_2 to O_2. This enhancement was attributed to the film matrix's capability to differentiate between gas molecular sizes and the varied affinity of polymer matrices towards gases. Consequently, this selectivity proved advantageous in reducing the metabolic activity of vegetables/fruits. Zhou et al. [263] proposed an innovative biomimetic approach to craft plant leaf-mimetic packing capable of regulating gas permeation. This method involved utilizing chitosan porous microspheres or poly (L-lactic acid) (PLLA) with nanoscale pore structures, strategically designed as gas "switches" or "stomata" within a shellac film. Their findings revealed that by integrating varying quantities of porous microspheres tailored to the respiratory rate of fruits, the novel film could effectively control the permeability of oxygen (O_2), CO_2, and water vapor (H_2O), as well as the CO_2/O_2 selectivity. This controlled gas permeation exhibited exceptional preservation efficacy across five model fruits with distinct respiratory metabolisms [3].

Additionally, edible films were manufactured using proteins (corn zein and wheat gluten) and celluloses (methylcellulose and hydroxypropyl cellulose) and subjected to gas permeability testing (Table 5.1). Generally, the O_2 and CO_2 permeabilities of these materials were lower compared to those of plastic films. However, their water vapor permeabilities exceeded those of plastic films. The concentration of plasticizer significantly influenced gas permeability. In cellulose films, the permeabilities of O_2, CO_2, and water vapor increased with higher plasticizer concentrations. The addition of lipids to the hydroxypropyl cellulose (HPC) film reduced gas permeability. Linear relationships were observed between gas permeability and film thickness in protein films [179]. A potential drawback of gas barrier properties in edible films lies in their susceptibility to inadequacy, which may result in heightened gas permeability and decreased efficacy in preserving food freshness. This limitation can stem from various factors, including the selection of polymers with suboptimal gas barrier characteristics or insufficient film thickness. Nonetheless, numerous methods exist to surmount this challenge and bolster the gas barrier performance of edible films. Firstly, opting for polymers endowed with inherently superior gas barrier properties, such as those with higher molecular weights or denser molecular structures, can enhance overall performance. Additionally, integrating barrier additives like nanoparticles or nano clays into the film matrix creates intricate pathways for gas molecules, slowing down permeation (Fig. 5.2). Embracing multilayer structures and employing surface modifications, such as plasma treatment or the application of barrier coatings, also serves to augment gas barrier effectiveness. Finally, the cross-linking of polymer chains within the film matrix via chemical agents or physical means fortifies the film and diminishes gas permeability. Through the implementation of these strategies, the gas barrier challenges faced by edible films can be effectively mitigated, thus ensuring the enhanced preservation of food products.

5.3.3 Water Resistance

The water resistance of smart packing film plays a crucial part in inhibiting degradation reactions in vegetables/foods, making it one of the extensively studied film properties. The water resistance of smart packing film plays a crucial role in preventing deteriorative reactions in foods, making it one of the most extensively studied properties of films. Naturally, bio-macromolecules show high hydrophilic characteristics, which limit their suitability as packaging materials for foods with high water activity (Table 5.1). Efforts to enhance the water resistance of polymers typically involve strategies such as emulsified film formation, wherein waxes and lipids are added as humidity shields (Figs. 5.2 and 5.4), Additionally, methods like the creation of multi-structured composite films have been employed to increase water barrier characteristics, as explained in "Advanced Strategies to Achieve Functionality of Films." This enhancement is primarily due to the increased films with hydrophobic nature and the complexity of water flow into the matrix resulting from the inclusion of lipids [3]. The utilization of lipids in conjunction with biopolymers to create emulsified/multilayered films has appeared as a significant approach to enhancing water resistance in the resulting films (Fig. 5.2). In two-layer composite films, a protein or polysaccharide film coating is either covered or immersed in a wax solution or melted lipid to form another layer of lipid (Fig. 5.4). In emulsified films, wax or lipids are first dispersed within a protein or polysaccharide solution as an emulsion, then captured within the biopolymer matrix as the dispersed phase following casting (Fig. 5.4). Multilayered films demonstrate enhanced water resistance compared to emulsified films, mainly because the successive lipid layers provide perpendicular resistance against water molecules [54, 129].

In contrast, Tang et al. [233], employed the solution casting method to prepare edible composite films. Analysis of the functional properties revealed that these films exhibited good compatibility, water resistance ranging from 33.50 to 60.15°, and strong UV-blocking properties. The inclusion of shellac increased both the to resist the water and the mechanical strength of the films. Based on assessments of weight loss, decay incidence, and other cherry indicators, these composite films were found to effectively preserve Chinese cherries at room temperature for at least 7 days. Hence, composite films represent a promising environmentally friendly packaging material for applications in the fruit preservation industry (Table 5.1) [233]. Similarly, Lee et al. [138] reported that mung bean starch-based edible films incorporating guar gum exhibited excellent overall water resistance properties.

5.3.4 Optical Property

The optical characteristics of edible films pertain to how they interact with light, encompassing transparency, opacity, color, and gloss. These attributes hold significant importance in food packaging, as they impact consumer perception, product display, and the ease of examining packaged contents (Figs. 5.2 and 5.4). The optical characteristics of edible films can vary considerably depending on factors such as their composition, thickness, and manufacturing processes. Transparency, often quantified as a key optical property, holds significant importance for biopolymer-based films as it directly influences the appearance of the packaged product and, consequently, consumer acceptance (Table 5.1). Typically, biopolymer films composed primarily of hydrocolloids demonstrate high transparency, reflecting the uniformity of the biopolymer matrix [48, 146]. Following the incorporation of lipids, the visibility films reduced in correlation with the number of lipids introduced into the solution [107, 244]. The introduction of lipids leads to the formation of heterogeneous networks within the hydrocolloid polymers, causing light scattering and reducing transparency [86]. Essentially, the lipid phase's impact on the optical characteristics of films typically relies on factors such as the droplet size, and distribution, and lipid fraction within the polymers [86, 87]. Several research has confirmed that higher lipid droplets induce greater light scattering effects compared to smaller ones, resulting in increased opacity [3, 55, 104, 156].

Jiménez and colleagues [119] investigated the effects of four unsaturated and saturated fatty acids on films developed from HPMC. Their findings revealed that the transparency of the material declined when saturated fatty acids were added, correlating with the enlargement of lipid clusters. Additionally, other substances like antioxidants and additives may also affect the optical properties of films. Recently, Malik and Mitra [151] developed active packaging solutions using zinc oxide nanoparticles in HPMC film. Their findings indicated a notable reduction in film transparency upon the addition of ZnO nanoparticles to the HPMC. This decrease in transparency was attributed to the white color of the ZnO nanoparticles, which intensified by light scattering within the film matrix. Despite its advantages, edible films may encounter limitations related to the influence of certain additives or components on their transparency, resulting in decreased clarity and visual appeal. Furthermore, the inclusion of lipids or other substances can lead to the formation of heterogeneous networks within the film structure, causing light scattering and diminishing transparency. To tackle these challenges, various strategies can be implemented. A thoughtful selection of ingredients and additives is crucial to uphold or improve film transparency. Additionally, optimizing processing parameters such as temperature and mixing techniques can reduce the formation of heterogeneous networks and minimize light scattering effects. Advanced technologies such as nanotechnology offer potential solutions by enhancing film uniformity and mitigating the effect of additives on optical characteristics. In conclusion, addressing the limitations associated with the optical properties of edible films requires a comprehensive approach involving ingredient selection, process refinement, and technological advancements to ensure their efficacy in food packaging applications.

5.4 Methods for the Fabrication of Edible Food Packaging

The fabrication of natural polymers-based edible food packaging involves a series of methods that are essential in producing smart and sustainable packaging materials. These methods comprise casting, extrusion, compression molding, and electrospinning [248, 252]. Below are some common fabrication methods utilized for smart and sustainable food packaging materials explained in detail.

5.4.1 *Casting Method*

This method is widely utilized to produce smart edible food packaging (Figs. 5.3a, 5.4a and d) [212]. While this method is predominantly utilized on a small scale, such as in research laboratories, it has also been employed on a large scale through continuous casting methods [5]. The process of creating a film using the casting method involves dissolving proteins and/or polysaccharides in a solution (Figs. 5.4a and d and Table 5.1) [37, 184]. This solution is then augmented by the addition of various fillers such as plasticizers, emulsifiers, nanofibers, and nanoparticles. The combination of biopolymer solution and fillers is carefully cast into petri dishes. Subsequently, the cast mixture is placed in a vacuum oven set at 50 °C for 48 h. This step supports the removal of water and other unwanted solvents from the mixture, ensuring the desired composition and properties of the edible packaging [5, 212]. Before examining the characteristics, of the films used for food packaging materials, it is important to ensure their proper storage. This involves keeping them in a desiccator to maintain a consistent humidity level and prevent several other issues, which include bubble formation, weak adhesion, and decreased stability of the film. By doing desiccation in the casting method, not only the sensitive samples can be protected from moisture and other environmental conditions that can degrade the film properties, but also provide consistency, stability, and preservation to the film (Fig. 5.2) [212]. For instance, casting is considered a useful method for producing food packaging that is transparent, uniform, and consistent (Fig. 5.4a). In terms of consistency, the casting method produces consistent material that enhances the mechanical characteristics (strength, elasticity, strength, etc.) and nice appearance of the packaging [212]. Hence, the natural polymers-based packaging produced through the casting method has versatility and offers advantages (i.e., biodegradability, extended shelf life, etc.) over plastic packaging (Figs. 5.3a and 5.4a) [154]. Despite these applications, the casting method is time-consuming due to its involvement in multiple steps, such as heating–cooling cycles and complex structures and shapes. Additionally, it can be expensive and prone to defects and imperfections, such as gas bubbles. Furthermore, environmental issues may arise from the release of hazardous chemicals or solvents in cases not properly managed.

5.4.2 Extrusion Method

Extrusion is a manufacturing process that facilitates shaping and creating new materials by forcing them through a die. The process is commonly used to create products with a constant cross-section, such as pipes, tubing, and profiles [212]. The materials processed in extrusion are typically dry or semi-dried, with varying levels of moisture content (Fig. 5.4b). These materials are subjected to forces under specific barrel conditions, including temperature control, screw parameters such as speed and configuration, feeding rate, moisture content, and the use of a die to shape the material into the desired cross-section [154]. These parameters significantly influence the physicochemical and extended functional properties of packaging materials. The extrusion method has some advantages to the casting method in producing edible packaging at both low scale (in the laboratory) and large (industrial) scale [154, 212]. Whereas the former is mostly utilized at aona low scale. Among these advantages are flexibility, consistency, material modification, and reduced waste properties. Extrusion methods facilitate edible film packaging for peanuts and edible containers and bottles for food materials, including lipids, proteins, rice flour, and other dairy products (Table 5.1). In extrusion methods, the controlled combination of high pressure, temperature, and mechanical forces is employed to effectively mix the raw materials, such as proteins and/or polysaccharides with other components (additives). This process utilizes blending and plasticizing the ingredients using a strong plasticizer, especially glycerol [140, 212]. The mixture is then pushed through a die to achieve the required characteristics, including the shape and profile of the final product [8]. For this purpose, various die conformations, such as tubes, sheets, or rods, influence the texture and appearance of packaging materials. One of the best applications of extrusion methods is to provide different shapes and sizes based on edible packaging materials. The efficiency of the extrusion process is high due to its continuous application, which allows for the rapid production of a high amount of packaging materials. Hence, the extrusion method is relatively cheap, hygienic, and cost-saving production of edible packaging for food industries. Apart from these advantages it deals with, the extrusion method also has several drawbacks. One of the primary drawbacks is the high cost associated with the equipment, machinery, and setups required for the technique. However, when compared with the casting process, the extrusion method is generally more complex, often requires skilled operators for maintenance, and consumes more energy during the thermal and molding phases compared to the casting process (Fig. 5.4). Therefore, it is challenging for food researchers to decide the appropriate biopolymers for food packaging involves considering various factors such as nature, combination, concentration, cost, and other factors which include heat, pressure, temperature, and ionic strength that effect on the shape and size of the film created during the extrusion method [154, 212].

5.4.3 Compression Molding

Compression molding is another highly efficient manufacturing method used for producing synthetic plastics and edible packaging materials. Compression molding offers many advantages to producing food packaging materials at small and large scales [212]. This method facilitates the production of desired shapes and dimensions of the packaging materials when molded by compressing and squeezing the preheated film materials. Compression molding is recognized as a highly efficient, consistent, and sustainable manufacturing method. It is considered faster preparation time compared to other methods, such as the casting process for film formation [120]. During compression molding the production of edible films where a mixture of biopolymers, including proteins, polysaccharides, and additives is placed into an appropriate mold. The mixture is then compressed to facilitate curing, which occurs through the crosslinking of the biopolymers existing in the mixture. The curing process is typically conducted at room temperature, around 25 °C, for cold compression. In contrast, hot compression involves the utilization of thermal treatment on the mold. This method facilitates the formation of a robust and consistent film resulting in a wide ranges of food packaging applications. It was investigated in previously published studies [130] that compression molding resulted in a final film produced had greater thickness and flexibility compared to edible packaging produced from the casting method.

5.4.4 Electrospinning

This is a simple, versatile, and efficient technique to create edible packaging. This method can be effectively utilized on both small scale (such as in laboratories), and large scale (commercial products) for the creation of packaging materials (Figs. 5.3a, 5.4c, and 5.5) [212]. In this process, micro- and nano-capsules (fibers) having sizes of 10–1000 nm are created by placing biopolymers and additive solution (such as proteins, polysaccharides, or their mixtures) in a syringe following to employee an external high-voltage electrostatic field supported by a grounded plate (Fig. 5.4c) [249, 93, 212]. The solution is then pulled out and obtained from the tip of the syringe following the drying process by subjecting it to intervening air, resulting in electrospun fibers, which are deposited on the collection plate [249]. The resulting fibers are then dried by subjecting them to high temperatures [249]. This is a cutting-edge technique extensively studied by food researchers, leading to significant advancements in food packaging and delivery systems (Fig. 5.5). The food packaging materials (nanofibers) produced by electrospinning result in exceptional characteristics, including strong mechanical characteristics, porosity, higher surface area for antimicrobial and barrier characteristics against external chemicals, etc. [249]. Several studies have demonstrated that electrospun nanofibers offer significant advantages (Fig. 5.5) [249, 245]. These advantages stem from the higher biocompatibility, ability to encapsulate functional ingredients, and low toxicity of the resulting electrospun nanofibers [248 249, 93]. Nonetheless, challenges were faced during the electrospinning process when attempting to incorporate proteins and polysaccharides into electrospun devices (Fig. 5.4d) [249, 93]. It was observed that there was minimal spinning success with globular proteins, and certain polysaccharides exhibited poor water vapor barrier characteristics [243].

Sani et al. [212], highlighted the selection of diverse techniques for edible packaging and pointed out that the selection of method for edible packaging is critical and is closely linked to the type of biopolymers used in the packaging materials. The choice of method, whether it involves proteins, polysaccharides, lipids, or additives, plays a substantial part in determining the characteristics and effectiveness of the edible packaging. The amount of edible packaging material being prepared is influenced by numerous factors, including the type of material used and the desired characteristics of the packaging. Several common properties are considered essential for edible packaging materials, such as mechanical strength, biodegradability, safety standards, etc. Also, as stated earlier the casting process for the development of edible packaging can be utilized at a laboratory scale, while extrusion, compression molding, and electrospinning techniques are applicable at both small and large scales to produce packaging from some specific types of natural polymers. Some proteins

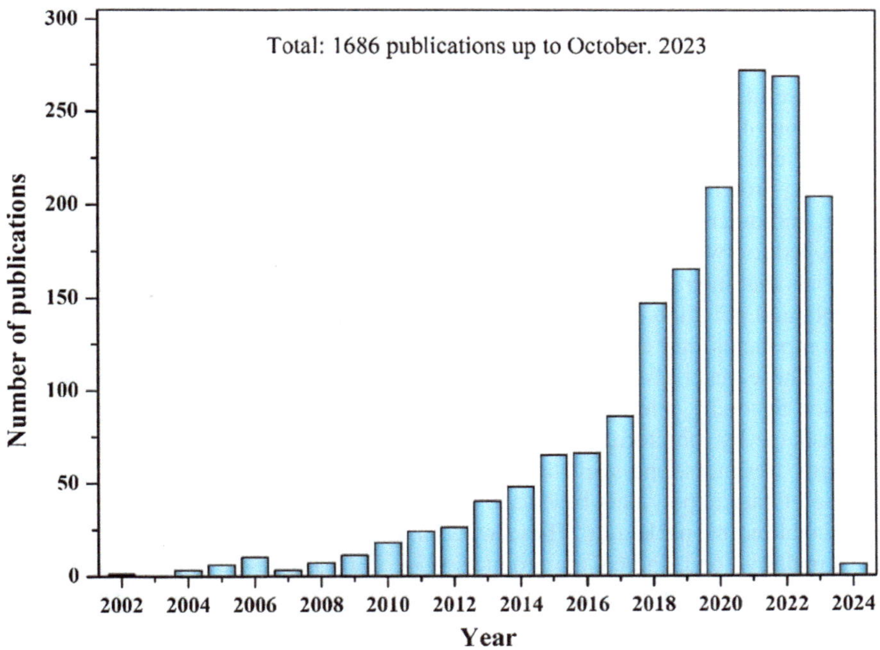

Fig. 5.5 The number of publications during the last 12 years (2002–2023) by searching the keywords foods and the impact of electrospinning on processing, storage, shelf-life, and protection of foods, Reproduced with permission from [248], Copyright Elsevier

and polysaccharides are sensitive to high temperature, pressure, or high applied shear rates, hence extrusion method is not suitable for such biopolymer solutions. Some issues can be seen in edible packaging materials with the compression method where biopolymers can be molded by compression or both compression and heating the system. Additionally, the electrospinning technique is only applicable to those natural polymers that have electrical charges and show electrostatic interactions, resulting in pulling through the nozzle.

5.5 Applications of Natural Polymers-Based Edible Packaging

Proteins and polysaccharides are utilized as emulsifying, gelling, and stabilizing agents in various food applications, leading to the production of film coatings and films that find extensive packaging use in dairy, bakery, and other food products (Table 5.1 and Figs. 5.2 and 5.3) [134, 154, 157]. Proteins and polysaccharides-based coatings and films derived from natural resources are considered innovative materials that are helping the production of smart and sustainable packaging for dairy and other food products (Figs. 5.1 and 5.3). Furthermore, these natural polymers protect and preserve the food materials from physicochemical and biological factors, which play a significant role in damaging packaging materials. These materials are classified into active and smart packaging for these food products, which can be useful in the improvement of food wastage, food quality, safety, and customer perception of sustainable materials practices (Fig. 5.2 and Table 5.1) [235]. In the realm of edible food packaging, active materials developed from natural polymers, including proteins and polysaccharides, enhance the stability, antioxidants, texture, flavor, and other properties of the final products [224]. Intelligent materials can be developed for two main purposes in food packaging. Firstly, they can serve as active materials that interact with the packaged food to preserve its quality and extend its life. Natural polymers-based materials are active in releasing and absorbing essential components that can preserve food, leading to higher stability, and shelf-life [224]. These materials release, absorb, or scavenge substances to preserve the food product. Secondly, intelligent materials can create a controlled environment within the package by monitoring and regulating factors such as pH, temperature, ionic strength, freshness, and shelf-life [224, 240]. This section explores and discusses the use of natural polymers-based active, intelligent, and smart materials to produce sustainable edible packaging. These materials are discussed because of their applications in dairy, food, and bakery products [154, 157]. The research provides insights into the existing trends and future forecasts of edible packaging to produce sustainable packaging materials.

5.5.1 Dairy Products

Edible packaging is crucial in the dairy sector. Various kinds of dairy products including cheese, ice cream, butter, and yogurt, need packaging to inhibit the detrimental effects and to enhance the stability and quality of these commercial products (Figs. 5.3, 5.4 and Table 5.1) [114, 154, 165]. These materials are designed from natural biopolymers, offering various benefits that contribute to the preservation of dairy products. The edible film contributes numerous characteristics such as interaction with water, emulsification, stabilization, and gelling agents, which help maintain the stability, mechanical properties, and texture of dairy products [83, 114]. The role of packaging materials in the protection and preservation of some of the dairy products are discussed below.

5.5.1.1 Cheese

Dairy products, such as cheese, undergo several processing steps before they reach our tables or supermarket. When consumers choose a brand of cheese products, they consider several important factors beyond just the price. Packaging is crucial in ensuring the quality, appearance, stability, and shelf-life of the cheese product. Edible packaging of cheese products has multiple functions that not only enhance the taste and texture of food products [114, 165]. These factors include exposure to light and air, moisture regulation, and contamination prevention (Fig. 5.3). Different kinds of cheeses are available in the market with different compositions of aqueous phase or moisture content, which influence on texture and durability of the cheese. For instance, hard cheese, like cheddar consisted of lowest moisture content (49–56%) than semisolid cheese, such as feta (54–69%), while soft cheese (mozzarella) developed with the highest moisture content (67%) [165].

Cheese is prone to spoilage because of various factors, including microbiological and biochemical changes that occur during its fabrication, growth, and delivery [165]. Previous studies show that cheese packaging fabricated from natural polymers offers superior barrier properties against gas and moisture permeability, and microbial activity compared to their synthetic counterparts [114, 154]. Doğan et al. [63] conducted a study where they developed natural polymers-based micro-nano fibers encasulated with lemon peel EOs (LPO)/gelatin for use in edible packaging applications. The authors [63] examined several important properties of fibers including physicochemical characteristics, encapsulation efficiency, and antimicrobial properties of cheese products. The results [63] indicated that the fibers had a greater impact on inhibiting the growth or viability of Staphylococcus aureus on the cheese than they did on *Escherichia coli*. The findings acquired from Doğan et al. [63] were found to be consistent with other factors, particularly contamination levels in cheese products. This suggests that centrifugation can be a favorable technology for enhancing packaging applications in the food sector. Gao et al. [83] prepared edible packaging for Mangolian cheese preservation utilizing protein (WP) and cellulose derivative

(HPMC) edible film incorporated with cinnamaldehyde. The authors [83] various mass ratios of WP and HPMC were prepared and analyzed. The ratios included 100/0, 75/25, 50/50, 25/75, and 0/100 of WP/HPMC. The study findings suggest that a 50/50 ratio of water-soluble polymer to HPMC results in a stronger barrier to moisture and vapor characteristics. The findings of the study show that a 50/50 ratio of WP/HM shows a stronger barrier to moisture and vapor characteristics, while a higher HPMC ratio was useful for releasing cinnamaldehyde. This study [83] presents valuable insights into the creation of edible films that possess antimicrobial properties, aiming to improve the lifecycle of various food products, particularly cheese.

There have been successful studies focusing on cheese packaging, aiming to improve its sustainability and storage duration (Fig. 5.3). However, further research is essential to advance the progress of green and sustainable edible film based on natural polymers. These advances include coatings and films that not only extend the shelf life of cheese but also positively impact consumer perception and marketing strategies related to foods, especially cheese packaging products.

5.5.1.2 Ice Cream

Ice cream is an oil-in-water (O/W) emulsion and is recognized as an important dairy product consumed by people worldwide, because of its creamy appearance, texture, and variation of flavors [2, 229]. However, during storage, ice cream faces several challenges that can affect its quality. These challenges include issues related to gases, vapors, and microbes, as well as stability concerns regarding color, lipid content, and water phases. Generally, it consisted of milk, sugars, cream, and some flavoring as well as emulsifying and/or stabilizing agents [229]. These ingredients come together to create the creamy texture and sweet flavor that ice cream is known for emulsifiers such as mono- and diglycerides, and polysorbate, and their mixture plays an key part in sustaining the stability, texture, and time frame of ice cream products (Fig. 5.1) [2, 126]. In recent years, the edible film has been subject of considerable research aimed at enhancing the preservation of ice cream during storage [2]. These materials are designed to inhibit gases, water, and microbial contact while also controlling and reducing the levels of thiobarbituric acid reactive substances (TBARS), free fatty acids (FFAs), and peroxide values (PV) in ice cream during storage [23]. There have therefore been some studies that have been performed by direct addition of plant extract materials or using plant extract to produce bioactive films. However, when incorporating plant extracts into bioactive films, it is essential to consider the method of addition to avoid potential detrimental effects. While some studies have explored the direct addition of plant extract materials or utilizing plant extracts in film production, certain drawbacks can arise from the direct incorporation of these extracts into the films [23]. Edible packaging made from a combination of proteins, polysaccharides, and lipids has emerged as a valuable innovation in food products, especially in ice cream products [2, 126]. It serves as a barrier against gases and water vapors, helps control microbial activity, and aids in preserving the bioactive components present in both full-fat and low-fat ice creams. Mahajan et al.

[149] prepared innovative edible packaging materials for ice cream (kulfi) using a combination of aloe vera, a medicinal plant known for its health applications, including antioxidant and antibacterial characteristics, carrageenan (a polysaccharide commonly used as a gelling agent in foods), and an additive, called glycerol. The results showed that edible film worked strongly as an antibacterial activity, especially against Escherichia with a lower level of lipid oxidation to kulfi product [149]. Furthermore, the findings of the published study [149] show that compared to the control system, the aloe vera-based edible film showed lower levels of TBARS, FFAs, and PV, especially for 15% aloe vera system. This study [149] provides valuable insights into developing commercial edible packaging suitable for various types of ice cream and other frozen products.

Nonetheless, future studies focusing on natural polymers-based edible packaging for ice cream, particularly those with bioactive surfaces, are essential for advancing sustainable packaging solutions in the food industry. Additionally, edible packaging enriched with natural polymers and bioactive agents represents an important development in food packaging knowledge [114, 154]. By offering improved texture and taste, inhibiting microbial activity, and providing essential barrier properties, these innovative solutions play a crucial role in ensuring the safety and quality of packaged food products [23, 126, 229].

5.5.1.3 Butter, Margarine, and Spreads

Water-in-oil (W/O) emulsions are a kind of emulsion where water is dispersed in oil (continuous phase) [89, 84, 222]. Common examples of W/O emulsions include butter, margarine, and spreads. These products are prepared by homogenizing water and oil phases to create a stable mixture [221, 113, 222]. In these systems, the oil phase typically constitutes a high amount of oil and/or fat, which serves as the continuous phase [110, 120]. W/O emulsion products always have a major issue called lipid oxidation [89]. Lipid oxidation further causes flavor problems, unsuitable colors, and nutrient degradation in the W/O emulsion products [89, 76]. Lipid oxidation becomes fast when fatty acids, especially unsaturated fatty acids react with oxygen gas encouraging the formation of toxic compounds, which contribute to rancidity, off-flavors, and color issues in butter and margarine products quality [89]. However, proper packaging is important for retaining the quality as well as physical and chemical stability of margarine and butter products [221, 120]. Food Packaging contributes a considerable function in shows a major part to protecting butter and margarine products from environmental factors including light, oxygen, and light, water vapors leading to lipid oxidation and deterioration in these products (Fig. 5.6) [89, 221]. In addition to lipid oxidation, several other factors can disturb the shelf longevity and quality preservation of butter and margarine, which include microbial growth, exposure to oxygen, moisture content, and changes in rheological properties [152].

Butter is known as bulk fat, enjoyed by consumers worldwide for centuries. It is obtained traditionally or by advanced methods using cream or milk, but mostly consists of saturated fats [110, 152]. However, saturated fats found in butter products have been linked to an increased risk of chronic diseases like obesity, diabetes, metabolic syndrome, etc. [83]. As a result, there is an increasing need for substitute materials in the food industry that can mimic the texture, rheology, creaminess, and flavor of traditional solid-like products while also meeting the body's energy requirements [83, 112]. Margarine and spreads are also known as W/O emulsions, which can replace bulk fat butter [89, 152]. These products encompass a variety of items such as table margarine, bakery margarine, vegetable spreads, and creams. Generally, these products consist of different proportions of aqueous and lipid phases, ranging from 10 to 90% [83, 222]. According to US legislation, margarine must contain at least 80% fat phase, while spreads are required to have a lipid phase ranging from 30 to 70% [221, 120]. Margarine and spreads undergo three distinct stages in their production process: emulsification, crystallization, and packaging. Traditional packaging methods for butter, margarine, and spreads are effective in protecting these products from physicochemical and environmental contaminants [221]. These non-biodegradable packaging in fat-rich products can cause a significant threat to the environment, leading to various detrimental effects such as increased landfill waste, water pollution, and serious health issues [152, 222]. There is a growing interest has emerged among food scientists in developing green and sustainable food packaging materials using natural biopolymers, like proteins and polysaccharides (Fig. 5.6 and Table 5.1) [83, 114, 152]. Edible materials from proteins and/or polysaccharides are used individually to create food packaging materials that are safe for consumption with food. On the other hand, natural polymers can also be combined with synthetic petroleum-based polymers to produce packaging materials that offer improved characteristics while reducing the negative health impacts associated with fully synthetic polymers [133, 154, 157].

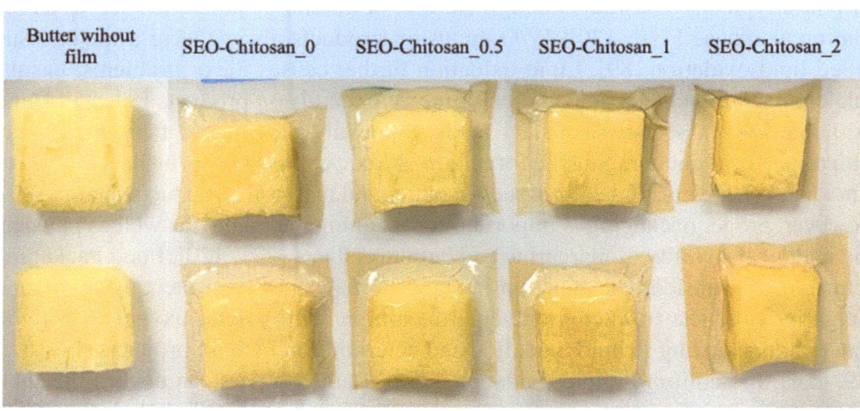

Fig. 5.6 Encapsulation of butter materials in sandalwood (Santalum album) essential oil (SEO)-chitosan films and stored at a constant temperature of 4 °C for 90 days [76]

Nor Adilah et al. [173] prepared polyethylene films using mango peel extract and fish gelatin to enhance the features of margarine products by preparing polyethylene films. These films were designed through a film-forming solution applied onto polyethylene film with varying thicknesses, including 10, 20, 40, and 60 μm. The findings of the study show the fabrication of colored layers using a mixture of polyethylene and gelatin. These films exhibited enhanced properties in terms of light barrier and DPPH radical scavenging capabilities. Supporting information was obtained from scanning electron microscopy, indicating the formation of compact microstructures in films. The study further suggested the stronger chemical stability of margarine over extended storage periods (28 days), suggesting the effectiveness of bilayer film in providing antioxidant properties. The authors concluded that the integration of bilayer film in the packaging of margarine significantly contributed to inhibiting lipid oxidation and preventing other forms of chemical contamination within the product. Another study conducted by Florez et al. [76] focused on designing active packaging film materials using chitosan and santalum album essential oil (SEO) to inhibit the lipid oxidation of butter. The production of chitosan films was explored using different concentrations of SEO at concentrations of 0.5, 1, and 2%. The authors [76] examined various characteristics of films, which included moisture content, mechanical properties, antioxidants, thermostability, solubility, and water vapor permeability (WVP). Except for puncture and equilibrium moisture content, the chitosan-SEO film showed a significant effect due to SEO. The film produced in this study [76] featured biodegradable butter sachets. These biodegradable sachets exhibited a significant reduction in thiobarbituric acid reactive substances (TBARS) and lipid oxidation than to the controlled trials used for comparison. The research findings suggest that the use of biodegradable materials in packaging, such as these butter sachets, can contribute to a decrease in lipid oxidation and possibly improve the durability of products stored within them.

Despite using of natural polymers in the creation of W/O emulsion products like butter and margarine, there is a growing demand for biodegradable films that can effectively protect these products from lipid oxidation and chemical contamination and maintain their rheological, mechanical, and physicochemical properties and stability during storage [84, 154].

5.5.2 *Meat, Seafood, and Other Animal-Based Foods*

Perishable foods, such as cheese, bulk fats like margarine and butter, meat (both animal and chicken), seafood (such as fish), and other animal and egg-derived products, undergo various changes in their appearance, texture, flavor, and lipid oxidation over because of biological origin [23, 154]. These alterations can lead to spoilage and potential food safety issues if not properly handled and stored. To prevent these changes and extend their shelf life, it's essential to package meat (fresh and frozen), fish, and other animal-based foods effectively (Figs. 5.2 and 5.3) [23]. Therefore, it is important to preserve such kinds of foods from various factors including light, oxygen, temperature, humidity, enzymatic activity, and microorganisms [154]. Petroleum-based packaging is widely used for meat, fish, and egg products. This type of packaging is favored for several reasons, including its low cost, widespread availability, and exceptional vapor barrier properties [62]. However, studies show that non-degradable and toxic packaging materials have considerable detrimental effects on the environment by causing water pollution, threatening ecosystems, and impacting human health (Fig. 5.2) [181, 23, 62]. The development of natural polymer-based edible packaging materials is gaining attention to protect meat, fish, and other animal foods (Fig. 5.3) [23]. Edible packaging developed from proteins, polysaccharides, and lipids offers a wide range of benefits such as biodegradability and shelf-life extension for meat, fish, eggs, etc. These edible materials play a significant role in reducing packaging waste, promoting green and sustainability in various industries leading applications of food packaging, and protecting meat, fish, and other animal-based foods [23, 154].

Puscaselu et al. [189] performed a study where they developed film meat products using natural polymers such as agar, water, sodium alginate, and glycerol. These materials were used to create 15 different film types for packaging purposes. The authors [189] examined various physicochemical properties of the films, including taste, aroma, color, microstructure, and mechanical characteristics. These properties are crucial in responsible the suitability of the packaging materials for diverse applications. Out of the 15 film materials developed, Puscaselu et al. [189] found three films that were considered the best for packaging raw salami slices. These films were suggested as potential substitutes for traditional plastic packaging materials due to their favorable properties.

Additionally, seafood, encompassing fish products, typically faces a reduced shelf life due to fast propagation of microbes, posing both health risks to consumers and economic losses [96, 145]. In recent years, edible packaging has emerged as a viable approach to improve the storage stability of fish (Figs. 5.1, 5.2 and Table 5.1) [97]. Several polysaccharides such as chitosan, and SA-CMC have been effectively utilized to package fish and other seafood (Figs. 5.3a, 5.6 and Table 5.1) [199, 201]. Chitosan–Gelatin-based edible packaging has emerged as a promising material for extending the shelf stability and retaining the quality of fish, seafood, and meat products [16] (Fig. 5.6). This innovative packaging material offers several benefits that play an important role in reducing spoilage, preserving texture, color, and flavor during storage and transportation [16]. Castro et al. [35] investigated the prospective of whey protein concentrate (WPC) film combined with the extract of green tea to delay lipid oxidation and oxidative properties in fresh salmon during a 14-day storage period.

Over the past two decades, there has been a marked surge in global interest regarding hen egg production and consumption [216]. The rise in production demand for egg and egg-based products is due to several factors such as their nutritional values, protein need, versatility in cooking, ease of digestibility, and lifestyle. The basic composition of a hen's egg is water (74%), protein (12%), lipids (11.6%), and a small amount of starch and sugars (1%) [231, 261]. Plastic packaging plays a vital part in continuing the freshness and quality of eggs during transportation and storage

[91]. It provides a protective barrier that protects the eggs from external chemical contaminants, physicochemical changes and damage, and microbial growth [23]. Liquid eggs, obtained after the removal of their shells, can be managed to produce egg-based products for the food and biomedical engineering industries [261]. The food products where a liquid egg is used include biscuits, cookies, cakes, mayonnaise, ice cream, and several other food products [198]. Egg products are commonly preserved using packaging methods, which can range from traditional plastic to more innovative edible-based coatings and films [23, 216]. The choice of film is essential not only for maintaining the quality and freshness of the eggs but also for making the environment green [216]. Conventional petroleum-derived plastic packaging presents notable challenges because of its adverse effects on various living organisms, including plants, animals, and humans [249, 154, 205]. Edible packaging prepared with natural polymers, including proteins, polysaccharides, and lipids, are promising materials that have a negligible effect on the environment [23, 154]. These innovative packaging solutions can be consumed along with processed eggs and their products [23, 216].

de Araújo Soares et al. [52] prepared edible packaging using a protein coating designed with WPI-sodium montmorillonite nanoparticles (MMT-Na), and sodium metabisulfite (SMB) to enhance the eggs' internal quality, such as yolk index, albumin pH, weight loss, foam stability, etc., were investigated during the storage period under the controlled conditions (25 ± 3 °C, 70 $70 \pm 5\%$ RH). For this purpose, initially, the eggs were immersed in and then dried at room temperature. The other part of the coating that was not cooked was considered a control to compare with the results of the cooked edible coating for eggs [52]. The results of the de Araújo Soares et al. [52] show that compared to the control (43%), more improvement was noted in foam stability of WPI-MMT-Na-SMB coating because of a little change in pH after five weeks of storage period. The authors [52] suggested that WPI-MMT-Na-SMB-based coating can be a useful material in the protection of egg products and reduce the requirements of refrigeration storage. Rachtanapun et al. [192], prepared edible coating materials for egg protection using polysaccharides (cassava starch, CS, and carboxymethylcellulose, CMC) blended with paraffin for *four* weeks of storage at different temperatures. The blends had different compositions (CS/CMC/paraffin), for instance, 6% CS, 1% CMC, and 0.5% paraffin [192]. The physicochemical properties, such as Haugh unit (HU), weight loss, pH, and microbes activity (load) of egg coating were investigated after the four weeks of storage. The findings [192] show that low temperature (4 °C) was considered the best HU grade, weight loss (2.14%) than at the higher temperature (at 30 °C storage, weight loss $\approx$3%). The pH of the egg protein (albumen) of coated and uncoated storage eggs at 4 °C increased, while no microbes were loaded at 4 °C. The coating developed from CS/CMC/paraffin led applications in the quality of eggs, inhibited microbial development, and enhanced the storage period of eggs at higher temperatures (25 and 30 °C) [192].

Numerous studies have been carried out in the field of edible packaging, focusing on the development of film and coating materials to encapsulate bioactive components and safeguard various perishable foods including meat, seafood, eggs, and egg-based products. However, it is essential to continuously enhance these materials to ensure they meet the necessary standards in terms of physicochemical, structural, mechanical, biological, rheological, and digestive properties [18, 132].

5.5.3 *Fruits and Vegetables*

Market demand for fresh fruits or vegetables has been a driving force behind the continuous advancements in preservation methods aimed at extending the durability and retaining the quality of these products [75]. However, fruits and vegetables are perishable materials that face challenges related to storage and preservation because of their elevated moisture levels and exposure to microbial degradation [175]. Additionally, using appropriate packaging, such as edible film or coating, able to enhances the durability of vegetables and fruits by creating barriers against gas, moisture, microbes, etc. (Fig. 5.3a and b). Numerous research endeavors have examined the application of biopolymers in preserving fruits and vegetables (Table 5.1). For instance, Zhang et al. [259] examined the use of a chitosan-derived film combined with TiO_2 for grapes packaging and storing [259], Reyes-Avalos et al. [200] investigated the impact of an alginate-derived composite film on the durability of figs. Moreover, the effects of beeswax content in an HPMC-derived material on the state of plums following postharvest coating application were examined [167, 200]. These films are essential for ensuring the quality, safety, and durability of products by hindering lipid oxidation, discoloration, retaining appearance, moisture loss, trapping essential flavors, and serving as mediums for healthy food nutrients [175, 204]. For instance, Ochoa et al. [174] used edible layers having *Euphorbia antisyphilitica*-derived wax and 0.01% ellagic acid (EA) to improve the quality and longevity of delicious golden apples through application. Another study conducted by Mehyar et al. [153] revealed that treating pine nuts and walnuts with carnauba wax and WPI notably decreased oxidative and hydrolytic rancidity, while also enhancing the sensory characteristics of the nuts. Similarly, Fagundes et al. [70] evaluated films fabricated by HPMC, beeswax, and other food preservatives having antifungal properties, such as sodium benzoate, sodium ethylparaben, and sodium methylparaben. These edible films were used for cherry tomatoes during cold preservation. Fagundes et al. [70] found that these edible films were significantly helpful in respiration rate, weight loss, and retaining the control of cherry tomatoes during storage time [70].

5.6 Challenges, Research Gap, and Future Perspectives

Conventional packaging resulting from petroleum such as PE, PP, PS, and PET are extensively utilized in the food sector because of their mechanical and barrier characteristics that help preserve food freshness and maintain product safety while providing adequate protection during shipping and management [181, 212]. However, petroleum-based packaging is a significant contributor to environmental problems like landfill waste, water pollution, and health issues due to its non-biodegradable nature. The edible film refers to the fabrication of alternative materials for petroleum-based packaging offering green and sustainable materials to reduce land and water waste and environmental issues [105, 193, 235]. These natural polymers-based edible innovative packaging materials are designed to be safe for consumption with foods, biodegradable, and deliver healthy components to the body during gastrointestinal tract (GIT) digestion [105]. However, food packaging industries face a significant challenge in creating edible packaging materials that possess the same mechanical strength and barrier characteristics as traditional petroleum-based and plastic packaging [212]. Consequently, it is essential to create smart films and coating effective formulations of smart material (natural polymers and emulsions based) and coatings for the utilization of natural polymers. To maximize the performance of smart edible coatings, it is important to optimize formulation within controlled conditions. Parameters like polymer concentration, heating temperature, pH level, polymer combination, processing temperature, ionic strength, and other environmental factors can significantly influence the mechanical, gas, and water barrier characteristics of the food packaging. Several factors, including heat, humidity, and temperature, can influence these properties. The permeability of oxygen is enhanced when the humidity is low, while it reduces when these humidity conditions are high. The CO_2 permeability of emulsion-based packaging is closely linked to the composition of the oil phase [5, 243]. The stability of emulsions is essential as it directly impacts the mechanical barrier and characteristics of emulsified material during different stages like emulsifying, packaging, storage, and digestion within the GIT. Therefore, the utilization of advanced emulsion technology is crucial for creating stable emulsions that effectively inhibit various physical and chemical instability processes [41, 178]. These processes include flocculation, coalescence, oxidation, and the oxidative properties of lipid-based materials utilized as food packaging [89]. Hence, by utilizing unique natural polymer characteristics and optimizing processing parameters, researchers can create innovative packaging materials in the future that address the increasing demands of the food sector.

5.7 Regulatory Considerations and Commercial Aspects

Food safety and regulatory standards vary greatly from one country to another [57]. In Europe and the USA, consumable food packaging (coatings and e.g., films) is classified on different regularity frameworks, based on food ingredients, food additives, and contact materials. To ensure the standards of the Food and Drug Administration (FDA), the components used in food packaging materials adhere to the Generally Recognized as Safe (GRAS) description [57]. The edible film packaging presents an environmentally friendly option compared to traditional packaging, but it comes with several essential regulatory and commercial challenges for effective market deployment. Regulatory, the films must meet stringent food safety standards, with all ingredients thoroughly listed to satisfy labeling requirements, and they must secure necessary certifications like those for organic or biodegradable products. Commercially, despite a growing consumer preference for green packaging, the increased cost of producing edible films over standard materials remains a significant hurdle. Companies need to assess how these films will fit into current production lines and their broader supply chain impact. Additionally, using edible film technology could give companies a market advantage by attracting eco-aware customers, though this requires careful strategic planning and possibly forming partnerships to enhance technological development [44, 117, 194].

Furthermore, consumer perception plays a major part in the acceptance and success of edible films [6, 236]. Unlike traditional petroleum-based food packaging, edible packaging materials offer numerous advantages by providing certain sensory expectations that are essential for consumer satisfaction [257]. Texture, taste, aroma, and mouthfeel are considered important when designing edible packaging materials [18]. Green and sustainable edible materials have gained considerable importance due to the increasing consumer perception and demand for biodegradable and renewable packaging [23, 154]. Encapsulation is a process where essential ingredients are trapped with a protective barrier material having a protective barrier [18]. Recently, there has been pressing interest in utilizing natural polymers for encapsulation purposes [154, 192, 205, 243]. Moreover, emulsion-based edible packaging materials offer a range of benefits beyond traditional packaging materials, such as enhancing the creaminess, texture, flavor, mouthfeel, and digestion of food materials [5]. Nonetheless, developing edible packaging materials that meet consumer demands requires an effective approach to utilizing natural polymers and emulsions under controlled conditions [5, 6, 236]. By considering the interplay between composition, concentration, processing parameters, and other factors involved in creating edible films and coatings, researchers and manufacturers can innovate sustainable packaging solutions that are associated with market requirements [18, 257]. To improve the success of commercial aspects related to edible packaging materials from natural polymers, companies must implement effective marketing trends and cultivate a positive consumer perception [6, 122]. Using social media platforms, opinion-maker cooperation, effective approaches, and wide availability, companies can create a solid market demand for edible packaging products [220, 257]. Moreover, addressing consumer concerns regarding pricing, potential benefits, and sustainability is vital in shaping consumer attitudes toward these green, smart, and sustainable alternatives. The Institute of Food Science and Technology (IFST), Europe recommends treating edible packaging as potentially hazardous unless proven safe [117]. In European countries, including Switzerland, regulations require the disclosure of any risks associated with edible packaging. In Europe, the size of edible packaging is used as a criterion for regulation [25]. The US Food and Drug Administration (USFDA) provides a comprehensive list of approved food components and contact substances, highlighting the importance of manufacturers compiling toxicological profiles for containers that incorporate edible packaging. However, enhanced research into the toxicological impact of edible packaging could help in establishing more robust safety standards and regulations for their use [117].

5.8 Conclusion

Recent advancements in the food and beverage industry have led to a significant increase in research and development endeavors to produce edible packaging materials using natural polymers. Natural polymers, such as proteins, polysaccharides, and lipids, are organic compounds derived from renewable resources like animals, plants, and microorganisms. These polymers offer several advantages when used in food packaging materials, primarily they exhibit smartness, biodegradability, and sustainability. Edible packaging materials enhance the physicochemical, gelling, structural, rheological, and sensory properties of food materials. These properties play a considerable role in preserving the freshness, stability, and quality of foods during extended periods of storage.

Utilizing edible packaging materials, such as films and coatings, presents a unique set of challenges for the food industry because these materials are consumed along with the foods they package. Several studies have been performed to explore the use of edible packaging materials as alternatives to petroleum-based packaging. These efforts aim to reduce or eliminate petroleum-based packaging materials by blending biopolymers with synthetic substances or producing them entirely from natural polymers such as proteins, polysaccharides, and lipids. However, challenges remain in enhancing various properties of these materials, such as texture stability, taste preservation, flavor retention, rheological properties, and mechanical strength. To achieve the desired outcomes in various applications, it is essential to consider and manipulate a range of physicochemical characteristics. These characteristics include the composition of natural polymers, the ratio of different components, ionic strength, pH ranges, thermal and non-thermal treatment methods, temperature control, and more. Subsequently, to enhance the taste, texture, and appearance of natural materials in the food industry, one effective strategy is the development of biopolymers-based emulsion materials. Hence, future studies can be made on designing smart, active, and sustainable natural polymers and emulsion-based edible films using advanced emulsion technology. These films should undergo a thorough examination of their physicochemical, mechanical, and digestive properties before, during, and after the GIT digestion stage by utilizing in vitro and in vivo trials.

Competing Interest The authors confirm that they have no conflicts of interest including financial interests and/or personal relationships to disclose for this chapter.

Declaration of AI-Assistant Tools We disclose that the AI tool was used to enhance the text in some parts of the chapter during the writing process, following which we performed a thorough review and editing of the final version of the chapter and took responsibility for the chapter publication.

References

1. Abbasi, A., Sabahi, S., Bazzaz, S., Tajani, A. G., Lahouty, M., Aslani, R., Hosseini, H.: An edible coating utilizing Malva sylvestris seed polysaccharide mucilage and postbiotic

from Saccharomyces cerevisiae var. boulardii for the preservation of lamb meat. Int. J. Biol. Macromol. **246**, 125660 (2023). https://doi.org/10.1016/j.ijbiomac.2023.125660
2. Abdulla, S.F., Shams, R., Dash, K.K.: Edible packaging as sustainable alternative to synthetic plastic: a comprehensive review. Environ. Sci. Pollut. Res. Int. (2024). https://doi.org/10.1007/s11356-024-32806-z
3. Abdullah, Cai, J., Hafeez, M.A., Wang, Q., Farooq, S., Huang, Q., Tian, W., Xiao, J.: Biopolymer-based functional films for packaging applications: a review. Front. Nutr. **9**, 1000116 (2022). https://doi.org/10.3389/fnut.2022.1000116
4. Abera, W.G., Kasirajan, R., Majamo, S.L.: Synthesis and characterization of bioplastic film from banana (Musa Cavendish species) peel starch blending with banana pseudo-stem cellulosic fiber. Biomass Convers. Biorefinery (2023). https://doi.org/10.1007/s13399-023-04207-8
5. Aguirre-Joya, J.A., De Leon-Zapata, M.A., Alvarez-Perez, O.B., Torres-León, C., Nieto-Oropeza, D.E., Ventura-Sobrevilla, J.M., Aguilar, M.A., Ruelas-Chacón, X., Romeo Rojas, R., Ramos-Aguiñaga, M.E., Aguilar, C.N.: Basic and applied concepts of edible packaging for foods. Food Packag. Preserv. **1–61**, 1–61 (2018). https://doi.org/10.1016/b978-0-12-811516-9.00001-4
6. Ahmed, M.W., Haque, M.A., Mohibbullah, M., Khan, M.S.I., Islam, M.A., Mondal, M.H.T., Ahmmed, R.: A review on active packaging for quality and safety of foods: Current trends, applications, prospects, and challenges. Food Packag. Shelf Life **33**, (2022). https://doi.org/10.1016/j.fpsl.2022.100913
7. Ajesh Kumar, V., Hasan, M., Mangaraj, S.M.P., Verma, D.K., Srivastav, P.P.: Trends in edible packaging films and its prospective future in food: a review. Appl. Food Res. **2**(1), 100118 (2022). https://doi.org/10.1016/j.afres.2022.100118
8. Ajesh Kumar, V., Pravitha, M., Srivastav, P.P., Mangaraj, S.R.P., Hasan, M.: Development of soy-based nanocomposite film: Modeling for barrier and mechanical properties and its application as cheese slice separator. J. Texture Stud. **53**(6), 809–819 (2022). https://doi.org/10.1111/jtxs.12636
9. Akhtar, H.M.S., Ahmed, S., Olewnik-Kruszkowska, E., Gierszewska, M., Brzezinska, M.S., Dembińska, K., Kalwasińska, A.:. Carboxymethyl cellulose based films enriched with polysaccharides from mulberry leaves (Morus alba L.) as new biodegradable packaging material. Int. J. Biol. Macromol. **253**, 127633 (2023). https://doi.org/10.1016/j.ijbiomac.2023.127633
10. Alexandre, E.M.C., Lourenço, R.V., Bittante, A.M.Q.B., Moraes, I.C.F., Sobral, P.J.D.A.: Gelatin-based films reinforced with montmorillonite and activated with nanoemulsion of ginger essential oil for food packaging applications. Food Packag. Shelf Life **10**, 87–96 (2016). https://doi.org/10.1016/j.fpsl.2016.10.004
11. Alipour, A., Rahaiee, S., Rajaei Litkohi, H., Jamali, S.N., Jafari, S.M.: Development and optimization of whey protein- Lepidium perfoliatum gum packaging films: an approach towards antimicrobial and biodegradable films. Ind. Crops Prod. **196**, 116447 (2023). https://doi.org/10.1016/j.indcrop.2023.116447
12. Alizadeh-Sani, M., Rhim, J.-W., Azizi-Lalabadi, M., Hemmati-Dinarvand, M., Ehsani, A.: Preparation and characterization of functional sodium caseinate/guar gum/TiO_2/cumin essential oil composite film. Int. J. Biol. Macromol. **145**, 835–844 (2020). https://doi.org/10.1016/j.ijbiomac.2019.11.004
13. Ansorena, M.R., Zubeldía, F., Marcovich, N.E.: Active wheat gluten films obtained by thermoplastic processing. LWT-Food Sci. Technol. **69**, 47–54 (2016). https://doi.org/10.1016/j.lwt.2016.01.020
14. Apodaca, E., Montanari, A., Castro, L., Umilta, E., Arroyo, L., Zurlini, C., Villaran, M.C.: Lentil by-products as a source of protein for food packaging applications. Am. J. Food Technol. **15**, 1–10 (2020). https://doi.org/10.3923/ajft.2020.1.10
15. Asiamah, E., Arthur, W., Kyei-Barfour, V., Sarpong, F., Ketemepi, H.K.: Enhancing the functional and physicochemical properties of tomato (Solanum lycopersicum L.) fruit through polysaccharides edible dipping technique coating under various storage conditions. Bioact. Carbohydr.S Diet. Fibre **30**, 100373 (2023). https://doi.org/10.1016/j.bcdf.2023.100373

16. Athanasopoulou, E., Bigi, F., Maurizzi, E., Karellou, E.I.E., Pappas, C.S., Quartieri, A., Tsironi, T.: Synthesis and characterization of polysaccharide- and protein-based edible films and application as packaging materials for fresh fish fillets. Sci. Rep. **14**(1), 517 (2024). https://doi.org/10.1038/s41598-024-51163-y
17. Ayyub, A., Li, B., Tahir, K., Chen, X.D., Iqbal, S., Bhutto, R.A., Rehman, W., Ullah, S., Munawar, J.: Impact of chitosan/metal oxide-based blends on the rheology, stability, and microstructures of oil-in-water emulsions. J. Food Eng. **375**, (2024). https://doi.org/10.1016/j.jfoodeng.2024.112076
18. Barbosa-Nuñez, J.A., Espinosa-Andrews, H., Cardona, A.A.V., Haro-González, J.N.: Polymer-based encapsulation in food products: a comprehensive review of applications and advancements. J. Future Foods **5**(1), 36–49 (2025). https://doi.org/10.1016/j.jfutfo.2024.01.003
19. Barzegar, H., Alizadeh Behbahani, B., Mirzaei, A., Ghodsi Sheikhjan, M.: Evaluation of the physicochemical and microbial properties of lamb meat coated with Shirazi balangu seed mucilage-based edible coating containing cell-free supernatant of Levilactobacillus brevis G145. Food Chem. Adv. **3**, 100456 (2023). https://doi.org/10.1016/j.focha.2023.100456
20. Bastarrachea, L., Dhawan, S., Sablani, S.S.: Engineering properties of polymeric-based antimicrobial films for food packaging: a review. Food Eng. Rev. **3**(2), 79–93 (2011). https://doi.org/10.1007/s12393-011-9034-8
21. Bayer, G., Shayganpour, A., Zia, J., Bayer, I.S.: Polyvinyl alcohol-based films plasticized with an edible sweetened gel enriched with antioxidant carminic acid. J. Food Eng. **323**, 111000 (2022). https://doi.org/10.1016/j.jfoodeng.2022.111000
22. Bertuzzi, M.A., Castro Vidaurre, E.F., Armada, M., Gottifredi, J.C.: Water vapor permeability of edible starch based films. J. Food Eng. **80**(3), 972–978 (2007). https://doi.org/10.1016/j.jfoodeng.2006.07.016
23. Bhat, Z.F., Bhat, H.F., Manzoor, M., Abdi, G., Aadil, R.M., Hassoun, A., Ait-Kaddour, A.: Enhancing the lipid stability of foods of animal origin using edible packaging systems. Food Chem. X **21**, 101185 (2024). https://doi.org/10.1016/j.fochx.2024.101185
24. Bhatia, S., Shah, Y.A., Al-Harrasi, A., Jawad, M., Koca, E., Aydemir, L.Y.: Novel applications of black pepper essential oil as an antioxidant agent in sodium caseinate and chitosan based active edible films. Int. J. Biol. Macromol. **254**, 128045 (2024). https://doi.org/10.1016/j.ijbiomac.2023.128045
25. Bizymis, A.-P., Tzia, C.: Edible films and coatings: properties for the selection of the components, evolution through composites and nanomaterials, and safety issues. Crit. Rev. Food Sci. Nutr. **62**(31), 8777–8792 (2022). https://doi.org/10.1080/10408398.2021.1934652
26. Bonilla, J., Sobral, P.J.A.: Investigation of the physicochemical, antimicrobial and antioxidant properties of gelatin-chitosan edible film mixed with plant ethanolic extracts. Food Biosci. **16**, 17–25 (2016). https://doi.org/10.1016/j.fbio.2016.07.003
27. Botalo, A., Inprasit, T., Ummartyotin, S., Chainok, K., Vatthanakul, S., Pisitsak, P.: Smart and UV-resistant edible coating and films based on alginate, whey protein, and curcumin. Polymers **16**(4), 447 (2024)
28. Bumbudsanpharoke, N., Choi, J., Ko, S.: Applications of nanomaterials in food packaging. J. Nanosci. Nanotechnol. **15**, 6357–6372(6316) (2015). https://doi.org/10.1166/jnn.2015.10847
29. Çağrı Mehmetoğlu, A., Sezer, E., Erol, S.: Development of antimicrobial whey protein-based film containing silver nanoparticles biosynthesised by Aspergillus Niger. Int. J. Food Sci. Technol. **56**(2), 965–973 (2021). https://doi.org/10.1111/ijfs.14749
30. Cai, J., Lu, W., Kan, Q., Chen, X., Cao, Y., Xiao, J.: Volatile composition changes of fruits in a biopolymer-coated polyethylene active packaging: effects of modified atmosphere and packaging-shaped bacterial community. Food Res. Int. **152**, 110843 (2022). https://doi.org/10.1016/j.foodres.2021.110843
31. Cai, L., Wang, Y., Cao, A.: The physiochemical and preservation properties of fish sarcoplasmic protein/chitosan composite films containing ginger essential oil emulsions. J. Food Process Eng. **43**(10), e13495 (2020). https://doi.org/10.1111/jfpe.13495

32. Çakmak, H., Özselek, Y., Turan, O.Y., Fıratlıgil, E., Karbancioğlu-Güler, F.: Whey protein isolate edible films incorporated with essential oils: antimicrobial activity and barrier properties. Polym. Degrad. Stab. **179**, 109285 (2020). https://doi.org/10.1016/j.polymdegradstab.2020.109285
33. Cardador, M.J., Gallego, M.: Origin of haloacetic acids in milk and dairy products. Food Chem. **196**, 750–756 (2016). https://doi.org/10.1016/j.foodchem.2015.10.011
34. Carpiné, D., Dagostin, J.L.A., Bertan, L.C., Mafra, M.R.: Development and characterization of soy protein isolate emulsion-based edible films with added coconut oil for olive oil packaging: barrier, mechanical, and thermal properties. Food Bioprocess Technol. **8**(8), 1811–1823 (2015). https://doi.org/10.1007/s11947-015-1538-4
35. Castro, F.V.R., Andrade, M.A., Sanches Silva, A., Vaz, M.F., Vilarinho, F.: The contribution of a whey protein film incorporated with green tea extract to minimize the lipid oxidation of salmon (Salmo salar L.). Foods **8**(8), 327 (2019)
36. Cazón, P., Velazquez, G., Ramírez, J.A., Vázquez, M.: Polysaccharide-based films and coatings for food packaging: a review. Food Hydrocoll. **68**, 136–148 (2017). https://doi.org/10.1016/j.foodhyd.2016.09.009
37. Channa, I.A., Ashfaq, J., Gilani, S.J., Chandio, A.D., Yousuf, S., Makhdoom, M.A., Jumah, M.N.B.: Sustainable and eco-friendly packaging films based on poly (vinyl alcohol) and glass flakes. Membranes **12**(7), 701 (2022). https://doi.org/10.3390/membranes12070701
38. Chen, J., Chai, J., Chen, X., Huang, M., Zeng, X., Xu, X.: Development of edible films by incorporating nanocrystalline cellulose and anthocyanins into modified myofibrillar proteins. Food Chem. **417**, 135820 (2023). https://doi.org/10.1016/j.foodchem.2023.135820
39. Chen, L., Jiang, X., Sun, Y., Gan, D., Liu, W., Wu, Y., Hao, X.: Composite optimization and characterization of dietary fiber-based edible packaging film reinforced by nanocellulose from grapefruit peel pomace. Int. J. Biol. Macromol. **253**, 127655 (2023). https://doi.org/10.1016/j.ijbiomac.2023.127655
40. Chen, Y., Ye, R., Liu, J.: Effects of different concentrations of ethanol and isopropanol on physicochemical properties of zein-based films. Ind. Crops Prod. **53**, 140–147 (2014). https://doi.org/10.1016/j.indcrop.2013.12.034
41. Chen, H., Iqbal, S., Wu, P., Pan, R., Wang, N., Bhutto, R. A., Rehman, W., Chen, X.D.: Enhancing rheology and reducing lipid digestion of oil-in-water emulsions using controlled aggregation and heteroaggregation of soybean protein isolate-peach gum microspheres. Int. J. Biol. Macromol. **273**(Pt 1), 132964 (2024). https://doi.org/10.1016/j.ijbiomac.2024.132964
42. Cheng, Y., Gao, S., Wang, W., Hou, H., Lim, L.-T.: Low temperature extrusion blown ε-polylysine hydrochloride-loaded starch/gelatin edible antimicrobial films. Carbohyd. Polym. **278**, 118990 (2022). https://doi.org/10.1016/j.carbpol.2021.118990
43. Cheng, S.-Y., Wang, B.-J., Weng, Y.-M.: Antioxidant and antimicrobial edible zein/chitosan composite films fabricated by incorporation of phenolic compounds and dicarboxylic acids. LWT-Food Sci. Technol. **63**(1), 115–121 (2015). https://doi.org/10.1016/j.lwt.2015.03.030
44. Chiralt, A., González-Martínez, C., Vargas, M., Atarés, L.: 18 - Edible films and coatings from proteins. In: Yada, R.Y. (ed.) Proteins in Food Processing, 2nd edn., pp. 477–500. Woodhead Publishing (2018)
45. Chodijah, S., Husaini, A., Zaman, M., Hilwatulisan.: Extraction of pectin from banana peels (Musa paradiasica fomatypica) for biodegradable plastic films. J. Phys. Conf. Ser. **1167**, 012061 (2019). https://doi.org/10.1088/1742-6596/1167/1/012061
46. Coimbra, P., Marona, B., Henriques, M.H., Campos, L., Gomes, D.M., Vitorino, C., Sousa, J.J., Braga, M.E., Gaspar, M.C.: Edible films based on potato and quince peels with potential for the preservation of cured cheese. Food Packag. Shelf Life **40**, 101176 (2023). https://doi.org/10.1016/j.fpsl.2023.101176
47. Coppola, D., Oliviero, M., Vitale, G.A., Lauritano, C., D'Ambra, I., Iannace, S., de Pascale, D.: Marine collagen from alternative and sustainable sources: extraction, processing and applications. Mar. Drugs **18**(4), 214 (2020)
48. Correa-Pacheco, Z.N., Bautista-Baños, S., Corona-Rangel, M.L., Ventura-Aguilar, R.I., Jiménez-Pérez, J.L., Cruz-Orea, A., Fonseca-García, A., López-Gamboa, G., Olvera-Cano,

L.I.: Morphological, optical and thermal properties of bioactive-chitosan nanostructured edible films for food packaging applications. Food Biophys. **19**(1), 207–218 (2024). https://doi.org/10.1007/s11483-023-09818-2
49. Cruz-Diaz, K., Cobos, Á., Fernández-Valle, M. E., Díaz, O., Cambero, M.I.: Characterization of edible films from whey proteins treated with heat, ultrasounds and/or transglutaminase. Application in cheese slices packaging. Food Packag. Shelf Life **22**, 100397 (2019). https://doi.org/10.1016/j.fpsl.2019.100397
50. Cui, H., Cheng, Q., Li, C., Chen, X., Lin, L.: Improving packing performance of lily polysaccharide based edible films via combining with sodium alginate and cold plasma treatment. Int. J. Biol. Macromol. **206**, 750–758 (2022). https://doi.org/10.1016/j.ijbiomac.2022.02.181
51. Dash, K.K., Ali, N.A., Das, D., Mohanta, D.: Thorough evaluation of sweet potato starch and lemon-waste pectin based-edible films with nano-titania inclusions for food packaging applications. Int. J. Biol. Macromol. **139**, 449–458 (2019). https://doi.org/10.1016/j.ijbiomac.2019.07.193
52. de Araújo Soares, R., Borges, S.V., Dias, M.V., Piccoli, R.H., Fassani, E.J., Silva, E.M.C.D.: Impact of whey protein isolate/sodium montmorillonite/sodium metabisulfite coating on the shelf life of fresh eggs during storage. LWT **139**, (2021). https://doi.org/10.1016/j.lwt.2020.110611
53. de Souza, K.C., Correa, L.G., da Silva, T.B.V., Moreira, T.F.M., de Oliveira, A., Sakanaka, L.S., Dias, M.I., Barros, L., Ferreira, I.C., Valderrama, P., Leimann, F.V.: Soy protein isolate films incorporated with Pinhão (Araucaria angustifolia (Bertol.) Kuntze) extract for potential use as edible oil active packaging. Food Bioprocess Technol. **13**(6), 998–1008 (2020). https://doi.org/10.1007/s11947-020-02454-5
54. Debeaufort, F., Voilley, A.: Effect of surfactants and drying rate on barrier properties of emulsified edible films. Int. J. Food Sci. Technol. **30**(2), 183–190 (1995). https://doi.org/10.1111/j.1365-2621.1995.tb01370.x
55. Devi, L.S., Jaiswal, A.K., Jaiswal, S.: Lipid incorporated biopolymer based edible films and coatings in food packaging: a review. Curr. Res. Food Sci. **8**, 100720 (2024). https://doi.org/10.1016/j.crfs.2024.100720
56. Devidas Meshram, B., Kisanrao Lule, V., Vyawahare, S., Rani, R.: Application of edible packaging in dairy and food industry. IntechOpen (Chapter 13). In: Tumuluru, J.S. (ed.) Food Processing and Packaging Technologies-Recent Advances (2023). https://doi.org/10.5772/intechopen.107850
57. Dhall, R.K.: Advances in edible coatings for fresh fruits and vegetables: a review. Crit. Rev. Food Sci. Nutr. **53**, 435–450 (2013)
58. Dhumal, C.V., Sarkar, P.: Composite edible films and coatings from food-grade biopolymers. J. Food Sci. Technol. **55**(11), 4369–4383 (2018). https://doi.org/10.1007/s13197-018-3402-9
59. Dong, Y., Li, Y., Ma, Z., Rao, Z., Zheng, X., Tang, K., Liu, J.: Effect of polyol plasticizers on properties and microstructure of soluble soybean polysaccharide edible films. Food Packag. Shelf Life **35**, 101023 (2023). https://doi.org/10.1016/j.fpsl.2022.101023
60. Dong, Y., Rao, Z., Liu, Y., Zheng, X., Tang, K., Liu, J.: Soluble soybean polysaccharide/gelatin active edible films incorporated with curcumin for oil packaging. Food Packag. Shelf Life **35**, 101039 (2023). https://doi.org/10.1016/j.fpsl.2023.101039
61. Dong, M., Tian, L., Li, J., Jia, J., Dong, Y., Tu, Y., Liu, X., Tan, C., Duan, X.: Improving physicochemical properties of edible wheat gluten protein films with proteins, polysaccharides and organic acid. LWT **154**, 112868 (2022). https://doi.org/10.1016/j.lwt.2021.112868
62. Donkor, L., Kontoh, G., Yaya, A., Bediako, J.K., Apalangya, V.: Bio-based and sustainable food packaging systems: relevance, challenges, and prospects. Appl. Food Res. **3**(2), (2023). https://doi.org/10.1016/j.afres.2023.100356
63. Doğan, N., Doğan, C., Eticha, A. K., Gungor, M., Akgul, Y.: Centrifugally spun micro-nanofibers based on lemon peel oil/gelatin as novel edible active food packaging: fabrication, characterization, and application to prevent foodborne pathogens E. coli and S. aureus in cheese. Food Control **139**, (2022). https://doi.org/10.1016/j.foodcont.2022.109081

64. Du, W.X., Olsen, C.W., Avena-Bustillos, R.J., McHugh, T.H., Levin, C.E., Friedman, M.: Storage stability and antibacterial activity against Escherichia coli O157:H7 of carvacrol in edible apple films made by two different casting methods. J. Agric. Food Chem. **56**(9), 3082–3088 (2008). https://doi.org/10.1021/jf703629s
65. Dursun Capar, T.: Characterization of sodium alginate-based biodegradable edible film incorporated with Vitis vinifera leaf extract: nano-scaled by ultrasound-assisted technology. Food Packag. Shelf Life **37**, 101068 (2023). https://doi.org/10.1016/j.fpsl.2023.101068
66. Edhirej, A., Sapuan, S.M., Jawaid, M., Zahari, N.I.: Effect of various plasticizers and concentration on the physical, thermal, mechanical, and structural properties of cassava-starch-based films. Starch-Stärke **69**(1–2), 1500366 (2017). https://doi.org/10.1002/star.201500366
67. Escamilla-García, M., Calderon-Dominguez, G., Chanona-Perez, J.J., Farrera-Rebollo, R.R., Andraca-Adame, J.A., Arzate-Vazquez, I., Mendez-Mendez, J.V., Moreno-Ruiz, L.A.: Physical and structural characterisation of zein and chitosan edible films using nanotechnology tools. Int. J. Biol. Macromol. **61**, 196–203 (2013). https://doi.org/10.1016/j.ijbiomac.2013.06.051
68. Espitia, P.J.P., Batista, R.A., Azeredo, H.M.C., Otoni, C.G.: Probiotics and their potential applications in active edible films and coatings. Food Res. Int. **90**, 42–52 (2016). https://doi.org/10.1016/j.foodres.2016.10.026
69. Espitia, P.J.P., Otoni, C.G.: Nanotechnology and edible films for food packaging applications. In: Ahmed, S. (ed.) Bio-based Materials for Food Packaging: Green and Sustainable Advanced Packaging Materials, pp. 125–145. Springer Singapore, Singapore (2018)
70. Fagundes, C., Palou, L., Monteiro, A.R., Pérez-Gago, M.B.: Hydroxypropyl methylcellulose-beeswax edible coatings formulated with antifungal food additives to reduce alternaria black spot and maintain postharvest quality of cold-stored cherry tomatoes. Sci. Hortic. **193**, 249–257 (2015). https://doi.org/10.1016/j.scienta.2015.07.027
71. Fakhouri, F.M., Martelli, S.M., Caon, T., Velasco, J.I., Buontempo, R.C., Bilck, A.P., Innocentini Mei, L.H.: The effect of fatty acids on the physicochemical properties of edible films composed of gelatin and gluten proteins. LWT **87**, 293–300 (2018). https://doi.org/10.1016/j.lwt.2017.08.056
72. Falguera, V., Quintero, J.P., Jiménez, A., Muñoz, J.A., Ibarz, A.: Edible films and coatings: Structures, active functions and trends in their use. Trends Food Sci. Technol. **22**(6), 292–303 (2011). https://doi.org/10.1016/j.tifs.2011.02.004
73. Fernandes, L.M., Guimarães, J.T., Pimentel, T.C., Esmerino, E.A., Freitas, M.Q., Carvalho, C.W.P., Cruz, A.G., Silva, M.C.: Chapter 7-edible whey protein films and coatings added with prebiotic ingredients. In: Barba, F.J., Putnik, P., Kovačević, D.B. (eds.) Agri-Food Industry Strategies for Healthy Diets and Sustainability, pp. 177–193. Academic Press (2020)
74. Fitch-Vargas, P.R., Aguilar-Palazuelos, E., Ruiz-Armenta, X.A., Delgado-Nieblas, C.I., Barraza-Elenes, C., Calderón-Castro, A.: Development of edible films based on reactive extrusion succinylated corn starch for the preservation of mango (Mangifera indica L. Cv. Kent). J. Food Meas. Charact. **18**(3), 2345–2358 (2024). https://doi.org/10.1007/s11694-024-02378-0
75. Flores-López, M.L., Cerqueira, M.A., de Rodríguez, D.J., Vicente, A.A.: Perspectives on utilization of edible coatings and nano-laminate coatings for extension of postharvest storage of fruits and vegetables. Food Eng. Rev. **8**(3), 292–305 (2016). https://doi.org/10.1007/s12393-015-9135-x
76. Flórez, M., Cazón, P., Vázquez, M.: Active packaging film of chitosan and Santalum album essential oil: Characterization and application as butter sachet to retard lipid oxidation. Food Packag. Shelf Life **34**, (2022). https://doi.org/10.1016/j.fpsl.2022.100938
77. Friesen, K., Chang, C., Nickerson, M.: Incorporation of phenolic compounds, rutin and epicatechin, into soy protein isolate films: mechanical, barrier and cross-linking properties. Food Chem. **172**, 18–23 (2015). https://doi.org/10.1016/j.foodchem.2014.08.128
78. Galus, S., Kadzińska, J.: Food applications of emulsion-based edible films and coatings. Trends Food Sci. Technol. **45**(2), 273–283 (2015). https://doi.org/10.1016/j.tifs.2015.07.011
79. Galus, S., Kadzińska, J.: Whey protein edible films modified with almond and walnut oils. Food Hydrocoll. **52**, 78–86 (2016). https://doi.org/10.1016/j.foodhyd.2015.06.013

80. Gamboni, J.E., Bonfiglio, G.V., Slavutsky, A.M., Bertuzzi, M.A.: Evaluation of edible films as single-serve pouches for a sustainable packaging system. Food Chem. Adv. **3**, 100547 (2023). https://doi.org/10.1016/j.focha.2023.100547
81. Gan, L., Jiang, G., Yang, Y., Zheng, B., Zhang, S., Li, X., Tian, Y., Peng, B.: Development and characterization of levan/pullulan/chitosan edible films enriched with ε-polylysine for active food packaging. Food Chem. **388**, 132989 (2022). https://doi.org/10.1016/j.foodchem.2022.132989
82. Gao, Z.-H., Zhang, Y.-H., Fang, B., Zhang, L.-P., Shi, J.: The effects of thermal-acid treatment and crosslinking on the water resistance of soybean protein. Ind. Crops Prod. **74**, 122–131 (2015). https://doi.org/10.1016/j.indcrop.2015.04.026
83. Gao, Y., Zhao, Y., Yao, Y., Chen, S., Xu, L., Wu, N., Tu, Y.: Recent trends in design of healthier fat replacers: type, replacement mechanism, sensory evaluation method and consumer acceptance. Food Chem. **447**, 138982 (2024). https://doi.org/10.1016/j.foodchem.2024.138982
84. Gao, Y., Lei, Y., Wu, Y., Liang, H., Li, J., Pei, Y., Li, Y., Li, B., Luo, X., Liu, S.: Beeswax: a potential self-emulsifying agent for the construction of thermal-sensitive food W/O emulsion. Food Chem. **349**, Article 129203 (2021). https://doi.org/10.1016/j.foodchem.2021.129203
85. Ge, L., Zhu, M., Xu, Y., Li, X., Li, D., Mu, C.: Development of antimicrobial and controlled biodegradable gelatin-based edible films containing nisin and amino-functionalized montmorillonite. Food Bioprocess Technol. **10**(9), 1727–1736 (2017). https://doi.org/10.1007/s11947-017-1941-0
86. Ge, A., Iqbal, S., Chen, X.D.: Alteration in rheology and microstructure of O/W emulsions using controlled soy protein isolate-polysaccharide aggregation in aqueous phases. J. Food Eng. **317**, (2022). https://doi.org/10.1016/j.jfoodeng.2021.110872
87. Ge, A., Iqbal, S., Kirk, T.V., Chen, X.D.: Modulating the rheological properties of oil-in-water emulsions using controlled WPI-polysaccharide aggregation in aqueous phases. J. Food Eng. **297**, (2021). https://doi.org/10.1016/j.jfoodeng.2021.110488
88. Gharibzahedi, S.M.T., Ahmadigol, A., Khubber, S., Altintas, Z.: Whey protein isolate/jujube polysaccharide-based edible nanocomposite films reinforced with starch nanocrystals for the shelf-life extension of banana: optimization and characterization. Int. J. Biol. Macromol. **222**, 1063–1077 (2022). https://doi.org/10.1016/j.ijbiomac.2022.09.232
89. Ghelichi, S., Hajfathalian, M., Yesiltas, B., Sørensen, A.-D.M., Garcia-Moreno, P.J., Jacobsen, C.: Oxidation and oxidative stability in emulsions. Compr. Rev. Food Sci. Food Saf. **22**, 1864–1901 (2023). https://doi.org/10.1111/1541-4337.13134
90. Gheorghita, R., Amariei, S., Norocel, L., Gutt, G.: New edible packaging material with function in shelf life extension: applications for the meat and cheese industries. Foods **9**(5), 562 (2020)
91. Giampietro-Ganeco, A., Borba, H., Scatolini-Silva, A.M., Boiago, M.M., de Souza, P.A., de Mello, J.L.M.: Quality assessment of eggs packed under modified atmosphere. Ciênc. Agrotec. Lavras **39**(1), 82–88 (2015)
92. Di Giuseppe, F.A., Volpe, S., Cavella, S., Masi, P., Torrieri, E.: Physical properties of active biopolymer films based on chitosan, sodium caseinate, and rosemary essential oil. Food Packag. Shelf Life **32**, 100817 (2022). https://doi.org/10.1016/j.fpsl.2022.100817
93. Goksen, G., Nisha, P., Ekiz, H.I.: Electrospinning technology: Its process conditions and food packaging applications. Nonthermal Process. Agri-Food-Bio Sci. Sustain. Futur. Goals 447–468 (2022). https://doi.org/10.1007/978-3-030-92415-7_12
94. Goswami, M., Mondal, K., Prasannavenkadesan, V., Bodana, V., Katiyar, V.: Effect of guar gum-chitosan composites edible coating functionalized with essential oils on the postharvest shelf life of Khasi mandarin at ambient condition. Int. J. Biol. Macromol. **254**, 127489 (2024). https://doi.org/10.1016/j.ijbiomac.2023.127489
95. Gul, O., Saricaoglu, F.T., Besir, A., Atalar, I., Yazici, F.: Effect of ultrasound treatment on the properties of nano-emulsion films obtained from hazelnut meal protein and clove essential oil. Ultrason. Sonochem. **41**, 466–474 (2018). https://doi.org/10.1016/j.ultsonch.2017.10.011

96. Gómez-Estaca, J., López de Lacey, A., López-Caballero, M.E., Gómez-Guillén, M.C., Montero, P.: Biodegradable gelatin–chitosan films incorporated with essential oils as antimicrobial agents for fish preservation. Food Microbiol. **27**(7), 889–896 (2010). https://doi.org/10.1016/j.fm.2010.05.012
97. Günlü, A., Koyun, E.: Effects of vacuum packaging and wrapping with chitosan-based edible film on the extension of the shelf life of sea bass (Dicentrarchus labrax) fillets in cold storage (4 °C). Food Bioprocess Technol. **6**(7), 1713–1719 (2013). https://doi.org/10.1007/s11947-012-0833-6
98. Hadnađev, M., Kalić, M., Krstonošić, V., Jovanović-Lješković, N., Erceg, T., Škrobot, D., Dapčević-Hadnađev, T.: Fortification of chocolate with microencapsulated fish oil: Effect of protein wall material on physicochemical properties of microcapsules and chocolate matrix. Food Chem. X **17**, 100583 (2023). https://doi.org/10.1016/j.fochx.2023.100583
99. Hamed, Y.S., Bakry, A.M., El-Sayed, A.M., Swailam, H.M., Ahsan, H.M., Wang, J., Zou, X.G., Bu, T., Zhang, H., Rayan, A.M., Yang, K.: Development of soy protein isolate-based films containing omega rich oil from purslane seed for beef burger patties packaging. J. Food Meas. Charact. (2024). https://doi.org/10.1007/s11694-024-02472-3
100. Han, K., Liu, Y., Liu, Y., Huang, X., Sheng, L.: Characterization and film-forming mechanism of egg white/pullulan blend film. Food Chem. **315**, 126201 (2020). https://doi.org/10.1016/j.foodchem.2020.126201
101. Han, Y., Yu, M., Wang, L.: Physical and antimicrobial properties of sodium alginate/carboxymethyl cellulose films incorporated with cinnamon essential oil. Food Packag. Shelf Life **15**, 35–42 (2018). https://doi.org/10.1016/j.fpsl.2017.11.001
102. Hashemi Gahruie, H., Ziaee, E., Eskandari, M.H., Hosseini, S.M.H.: Characterization of basil seed gum-based edible films incorporated with Zataria multiflora essential oil nanoemulsion. Carbohyd. Polym. **166**, 93–103 (2017). https://doi.org/10.1016/j.carbpol.2017.02.103
103. Hashemi, S.M.B., Mousavi Khaneghah, A.: Characterization of novel basil-seed gum active edible films and coatings containing oregano essential oil. Prog. Org. Coat. **110**, 35–41 (2017). https://doi.org/10.1016/j.porgcoat.2017.04.041
104. Hassan, B., Chatha, S.A.S., Hussain, A.I., Zia, K.M., Akhtar, N.: Recent advances on polysaccharides, lipids and protein based edible films and coatings: a review. Int. J. Biol. Macromol. **109**, 1095–1107 (2018). https://doi.org/10.1016/j.ijbiomac.2017.11.097
105. Hassoun, A., Boukid, F., Ozogul, F., Aït-Kaddour, A., Soriano, J.M., Lorenzo, J.M., Perestrelo, R., Galanakis, C.M., Bono, G., Bouyahya, A., Bhat, Z.: Creating new opportunities for sustainable food packaging through dimensions of industry 4.0: new insights into the food waste perspective. Trends Food Sci. & Technol. **142**, 104238 (2023). https://doi.org/10.1016/j.tifs.2023.104238
106. Hemsri, S., Asandei, A.D., Grieco, K., Parnas, R.S.: Biopolymer composites of wheat gluten with silica and alumina. Compos. A Appl. Sci. Manuf. **42**(11), 1764–1773 (2011). https://doi.org/10.1016/j.compositesa.2011.07.032
107. Hopkins, E.J., Chang, C., Lam, R.S.H., Nickerson, M.T.: Effects of flaxseed oil concentration on the performance of a soy protein isolate-based emulsion-type film. Food Res. Int. **67**, 418–425 (2015). https://doi.org/10.1016/j.foodres.2014.11.040
108. Huang, X., Luo, X., Liu, L., Dong, K., Yang, R., Lin, C., Song, H., Li, S., Huang, Q.: Formation mechanism of egg white protein/κ-Carrageenan composite film and its application to oil packaging. Food Hydrocoll. **105**, 105780 (2020). https://doi.org/10.1016/j.foodhyd.2020.105780
109. Imre, B., García, L., Puglia, D., Vilaplana, F.: Reactive compatibilization of plant polysaccharides and biobased polymers: review on current strategies, expectations and reality. Carbohyd. Polym. **209**, 20–37 (2019). https://doi.org/10.1016/j.carbpol.2018.12.082
110. Iqbal, S., Ayyub, A., Iqbal, H., Chen, X.D.: Protein microspheres as structuring agents in lipids: potential for reduction of total and saturated fat in food products. J. Sci. Food Agric. **101**(3), 820–830 (2021). https://doi.org/10.1002/jsfa.10645
111. Iqbal, S., Xu, Z., Huang, H., Chen, X.D.: Controlling the rheological properties of oil phases using controlled protein-polysaccharide aggregation and heteroaggregation in water-in-oil

emulsions. Food Hydrocoll. **96**, 278–287 (2019). https://doi.org/10.1016/j.foodhyd.2019.05.028

112. Iqbal, S., Zhang, P., Wu, P., Deng, R., Chen, X.D.: Impact of amylose from maize starch on the microstructure, rheology and lipolysis of W/O emulsions during simulated semi-dynamic gastrointestinal digestion. Int. J. Food Sci. Technol. **57**(6), 3578–3588 (2022). https://doi.org/10.1111/ijfs.15682
113. Iqbal, S., Baloch, M.K., Abbas, S., Khurram, M.J., Ahmed, R., Chen, X.D.: Rheological alteration in W/O emulsions using controlled aggregation and heteroaggregation of cellulose ethers with protein and polysaccharide in aqueous phases. Food Hydrocoll. **146**, (2024). https://doi.org/10.1016/j.foodhyd.2023.109261
114. Jafarzadeh, S., Salehabadi, A., Mohammadi Nafchi, A., Oladzadabbasabadi, N., Jafari, S.M.: Cheese packaging by edible coatings and biodegradable nanocomposites; improvement in shelf life, physicochemical and sensory properties. Trends Food Sci. Technol. **116**, 218–231 (2021). https://doi.org/10.1016/j.tifs.2021.07.021
115. Janjarasskul, T., Krochta, J.M.: Edible packaging materials. Annu. Rev. Food Sci. Technol. **1**, 415–448 (2010). https://doi.org/10.1146/annurev.food.080708.100836
116. Jayakumar, A., Radoor, S., Kim, J.T., Rhim, J.W., Nandi, D., Parameswaranpillai, J., Siengchin, S.: Recent innovations in bionanocomposites-based food packaging films – a comprehensive review. Food Packaging Shelf Life **33**, 100877 (2022). https://doi.org/10.1016/j.fpsl.2022.100877
117. Jeya Jeevahan, J., Chandrasekaran, M., Venkatesan, S.P., Sriram, V., Britto Joseph, G., Mageshwaran, G., Durairaj, R.B.: Scaling up difficulties and commercial aspects of edible films for food packaging: a review. Trends Food Sci. Technol. **100**, 210–222 (2020). https://doi.org/10.1016/j.tifs.2020.04.014
118. Jiang, J., Chen, X., Zhang, G.-L., Hao, H., Hou, H.-M., Bi, J.: Preparation of chitosan-cellulose-benzyl isothiocyanate nanocomposite film for food packaging applications. Carbohyd. Polym. **285**, 119234 (2022). https://doi.org/10.1016/j.carbpol.2022.119234
119. Jiménez, A., Fabra, M.J., Talens, P., Chiralt, A.: Effect of lipid self-association on the microstructure and physical properties of hydroxypropyl-methylcellulose edible films containing fatty acids. Carbohyd. Polym. **82**(3), 585–593 (2010). https://doi.org/10.1016/j.carbpol.2010.05.014
120. Jiménez-Carvelo, A.M., Arroyo-Cerezo, A., Bikrani, S., Jia, W., Koidis, A., Cuadros-Rodríguez, L.: Rapid and non-destructive spatially offset Raman spectroscopic analysis of packaged margarines and fat-spread products. Microchem. J. **178**, (2022). https://doi.org/10.1016/j.microc.2022.107378
121. Jridi, M., Boughriba, S., Abdelhedi, O., Nciri, H., Nasri, R., Kchaou, H., Kaya, M., Sebai, H., Zouari, N., Nasri, M.: Investigation of physicochemical and antioxidant properties of gelatin edible film mixed with blood orange (Citrus sinensis) peel extract. Food Packag. Shelf Life **21**, 100342 (2019). https://doi.org/10.1016/j.fpsl.2019.100342
122. Jurić, M., Maslov Bandić, L., Carullo, D., Jurić, S.: Technological advancements in edible coatings: Emerging trends and applications in sustainable food preservation. Food Biosci. **58**, (2024). https://doi.org/10.1016/j.fbio.2024.103835
123. Kaewprachu, P., Osako, K., Benjakul, S., Rawdkuen, S.: Effect of protein concentrations on the properties of fish myofibrillar protein based film compared with PVC film. J. Food Sci. Technol. **53**(4), 2083–2091 (2016). https://doi.org/10.1007/s13197-016-2170-7
124. Karkar, B., Patır, İ., Eyüboğlu, S., Şahin, S.: Development of an edible active chitosan film loaded with Nigella sativa L. extract to extend the shelf life of grapes. Biocatal. Agric. Biotechnol. **50**, 102708 (2023)
125. Kaur, J., et al.: Edible packaging: an overview. In: Poonia, A., Dhewa, T. (eds.) Edible Food Packaging. Springer, Singapore (2022). https://doi.org/10.1007/978-981-16-2383-7_1
126. Khan, I., Tango, C.N., Miskeen, S., Lee, B.H., Oh, D.-H.: Hurdle technology: a novel approach for enhanced food quality and safety–a review. Food Control **73**, 1426–1444 (2017). https://doi.org/10.1016/j.foodcont.2016.11.010

127. Khwaldia, K., Banon, S., Perez, C., Desobry, S.: Properties of sodium caseinate film-forming dispersions and films. J. Dairy Sci. **87**(7), 2011–2016 (2004). https://doi.org/10.3168/jds.S0022-0302(04)70018-1
128. Khwaldia, K., Perez, C., Banon, S., Desobry, S., Hardy, J.: Milk proteins for edible films and coatings. Crit. Rev. Food Sci. Nutr. **44**(4), 239–251 (2004). https://doi.org/10.1080/10408690490464906
129. Kowalczyk, D., Baraniak, B.: Effect of candelilla wax on functional properties of biopolymer emulsion films–a comparative study. Food Hydrocoll. **41**, 195–209 (2014). https://doi.org/10.1016/j.foodhyd.2014.04.004
130. Krishna, M., Nindo, C.I., Min, S.C.: Development of fish gelatin edible films using extrusion and compression molding. J. Food Eng. **108**(2), 337–344 (2012). https://doi.org/10.1016/j.jfoodeng.2011.08.002
131. Kumar, S., Mukherjee, A., Dutta, J.: Chitosan based nanocomposite films and coatings: emerging antimicrobial food packaging alternatives. Trends Food Sci. Technol. **97**, 196–209 (2020). https://doi.org/10.1016/j.tifs.2020.01.002
132. Kumar, A., Srivastav, P.P., Pravitha, M., Hasan, M., Mangaraj, S.V.P., Verma, D.K.: Comparative study on the optimization and characterization of soybean aqueous extract based composite film using response surface methodology (RSM) and artificial neural network (ANN). Food Packag. Shelf Life **31**, 100778 (2022). https://doi.org/10.1016/j.fpsl.2021.100778
133. Kumar, N., Tripathi, S., Pratibha, Mehra, M., Heena, Trajkovska Petkoska, A.: Whey protein based edible coatings: recent trends. In Poonia, A., Trajkovska Petkoska, A. (eds.) Whey Valorization: Innovations, Technological Advancements and Sustainable Exploitation, pp. 187–209. Springer Nature Singapore, Singapore (2023)
134. Kumar, N., Pratibha, Prasad, J., et al.: Recent trends in edible packaging for food applications—perspective for the future. Food Eng. Rev. **15**, 718–747 (2023). https://doi.org/10.1007/s12393-023-09358-y
135. Kumari, M., Mahajan, H., Joshi, R., Gupta, M.: Development and structural characterization of edible films for improving fruit quality. Food Packag. Shelf Life **12**, 42–50 (2017). https://doi.org/10.1016/j.fpsl.2017.02.003
136. Lalnunthari, C., Devi, L.M., Badwaik, L.S.: Extraction of protein and pectin from pumpkin industry by-products and their utilization for developing edible film. J. Food Sci. Technol. **57**(5), 1807–1816 (2020). https://doi.org/10.1007/s13197-019-04214-6
137. Lan, X., Zhang, X., Wang, L., Wang, H., Hu, Z., Ju, X., Yuan, Y.: A review of food preservation based on zein: the perspective from application types of coating and film. Food Chem. **424**, 136403 (2023). https://doi.org/10.1016/j.foodchem.2023.136403
138. Lee, J.-S., Lee, E.-S., Han, J.: Enhancement of the water-resistance properties of an edible film prepared from mung bean starch via the incorporation of sunflower seed oil. Sci. Rep. **10**(1), 13622 (2020). https://doi.org/10.1038/s41598-020-70651-5
139. Li, Y., Cao, C., Pei, Y., Liu, X., Tang, K.: Preparation and properties of microfibrillated chitin/gelatin composites. Int. J. Biol. Macromol. **130**, 715–719 (2019). https://doi.org/10.1016/j.ijbiomac.2019.03.014
140. Li, X.L., Shen, Y., Hu, F., Zhang, X.X., Thakur, K., Rengasamy, K.R., Khan, M.R., Busquets, R., Wei, Z.J.: Fortification of polysaccharide-based packaging films and coatings with essential oils: a review of their preparation and use in meat preservation. Int. J. Biol. Macromol. **242**(Pt 1), 124767 (2023). https://doi.org/10.1016/j.ijbiomac.2023.124767
141. Liang, W., Ge, X., Lin, Q., Niu, L., Zhao, W., Muratkhan, M., Li, W.: Ternary composite degradable plastics based on Alpinia galanga essential oil Pickering emulsion templates: a potential multifunctional active packaging. Int. J. Biol. Macromol. **257**, 128580 (2024). https://doi.org/10.1016/j.ijbiomac.2023.128580
142. Lin, H.-C., Wang, B.-J., Weng, Y.-M.: Development and characterization of sodium caseinate edible films cross-linked with genipin. LWT **118**, 108813 (2020). https://doi.org/10.1016/j.lwt.2019.108813

143. Loukri, A., Kyriakoudi, A., Oliinychenko, Y., Stratakos, A.C., Lazaridou, A., Mourtzinos, I.: Preparation and characterization of chitosan-citric acid edible films loaded with Cornelian cherry pomace extract as active packaging materials. Food Hydrocoll. **150**, 109687 (2024). https://doi.org/10.1016/j.foodhyd.2023.109687
144. López, D., Márquez, A., Gutiérrez-Cutiño, M., Venegas-Yazigi, D., Bustos, R., Matiacevich, S.: Edible film with antioxidant capacity based on salmon gelatin and boldine. LWT **77**, 160–169 (2017). https://doi.org/10.1016/j.lwt.2016.11.039
145. López de Lacey, A.M., López-Caballero, M.E., Montero, P.: Agar films containing green tea extract and probiotic bacteria for extending fish shelf-life. LWT-Food Sci. Technol. **55**(2), 559–564 (2014). https://doi.org/10.1016/j.lwt.2013.09.028
146. López-Ortiz, A.N.A.B.E.L., Pacheco Pineda, I.Y., Méndez-Lagunas, L.L., Balbuena Ortega, A., Guerrero Martinez, L., Pérez-Orozco, J.P., Del Río, J.A., Nair, P.K.: Optical and thermal properties of edible coatings for application in solar drying. Sci. Rep. **11**(1), 10051 (2021). https://doi.org/10.1038/s41598-021-88901-5
147. Ma, W., Tang, C.-H., Yin, S.-W., Yang, X.-Q., Qi, J.-R., Xia, N.: Effect of homogenization conditions on properties of gelatin–olive oil composite films. J. Food Eng. **113**(1), 136–142 (2012). https://doi.org/10.1016/j.jfoodeng.2012.05.007
148. Madsen, M., Westh, P., Khan, S., Ipsen, R., Almdal, K., Aachmann, F.L., Svensson, B.: Impact of alginate mannuronic-guluronic acid contents and pH on protein binding capacity and complex size. Biomacromol **22**(2), 649–660 (2021). https://doi.org/10.1021/acs.biomac.0c01485
149. Mahajan, K., Kumar, S., Bhat, Z.F., Singh, M., Bhat, H.F., Bhatti, M.A., Bekhit, A.E.-D.A.: Aloe vera and carrageenan based edible film improves storage stability of ice-cream. Appl. Food Res. **2**(1), (2022). https://doi.org/10.1016/j.afres.2022.100128
150. Malakar, A., Kanel, S.R., Ray, C., Snow, D.D., Nadagouda, M.N.: Nanomaterials in the environment, human exposure pathway, and health effects: a review. Sci. Total. Environ. **759**, 143470 (2021). https://doi.org/10.1016/j.scitotenv.2020.143470
151. Malik, G.K., Mitra, J.: Zinc oxide nanoparticle synthesis, characterization, and their effect on mechanical, barrier, and optical properties of HPMC-based edible film. Food Bioprocess Technol. **14**(3), 441–456 (2021). https://doi.org/10.1007/s11947-020-02566-y
152. Manzoor, M.F., Afraz, M.T., Yılmaz, B.B., Adil, M., Arshad, N., Goksen, G., Ali, M., Zeng, X.A.: Recent progress in natural seaweed pigments: Green extraction, health-promoting activities, techno-functional properties and role in intelligent food packaging. J. Agric. Food Res. **15**, (2024). https://doi.org/10.1016/j.jafr.2024.100991
153. Mehyar, G.F., Al-Ismail, K., Han, J.H., Chee, G.W.: Characterization of edible coatings consisting of pea starch, whey protein isolate, and carnauba wax and their effects on oil rancidity and sensory properties of walnuts and pine nuts. J. Food Sci. **77**(2), E52–E59 (2012). https://doi.org/10.1111/j.1750-3841.2011.02559.x
154. Meshram, D.B., Lule, K.V., Vyawahare, S., Rani, R.: Application of edible packaging. In: Dairy and Food Industry, pp. 1–43 (2023). IntechOpen. https://doi.org/10.5772/intechopen.107850
155. Miao, W., Fu, Y., Zhang, Z., Lin, Q., Li, X., Sang, S., McClements, D.J., Jiang, H., Ji, H., Qiu, C., Jin, Z.: Fabrication of starch-based oleogels using capillary bridges: potential for application as edible inks in 3D food printing. Food Hydrocoll. **150**, 109647 (2024). https://doi.org/10.1016/j.foodhyd.2023.109647
156. Milani, J.M., Nemati, A.: Lipid-based edible films and coatings: a review of recent advances and applications. J. Packag. Technol. Res. **6**(1), 11–22 (2022). https://doi.org/10.1007/s41783-021-00130-3
157. Moghadas, H.C., Chauhan, R., Smith, J.S.: Application of plant oils as functional additives in edible films and coatings for food packaging: a review. Foods **13**(7), (2024). https://doi.org/10.3390/foods13070997
158. Mohamed, S.A.A., El-Sakhawy, M., El-Sakhawy, M.A.-M.: Polysaccharides, protein and lipid -based natural edible films in food packaging: a Review. Carbohyd. Polym. **238**, 116178 (2020). https://doi.org/10.1016/j.carbpol.2020.116178

159. Mondéjar-López, M., Castillo, R., Jiménez, A.J.L., Gómez-Gómez, L., Ahrazem, O., Niza, E.: Polysaccharide film containing cinnamaldehyde-chitosan nanoparticles, a new eco-packaging material effective in meat preservation. Food Chem. **437**, 137710 (2024). https://doi.org/10.1016/j.foodchem.2023.137710
160. Moradi, M., Tajik, H., Razavi Rohani, S.M., Mahmoudian, A.: Antioxidant and antimicrobial effects of zein edible film impregnated with Zataria multiflora Boiss. Essential oil and monolaurin. LWT Food Sci. Technol. **72**, 37–43 (2016). https://doi.org/10.1016/j.lwt.2016.04.026
161. Morales-Contreras, B.E., Wicker, L., Rosas-Flores, W., Contreras-Esquivel, J.C., Gallegos-Infante, J.A., Reyes-Jaquez, D., Morales-Castro, J.: Apple pomace from variety "Blanca de Asturias" as sustainable source of pectin: composition, rheological, and thermal properties. LWT **117**, 108641 (2020). https://doi.org/10.1016/j.lwt.2019.108641
162. Mu, J., Hu, R., Tang, Y., Dong, W., Zhang, Z.: Microencapsulation of green coffee oil by complex coacervation of soy protein isolate, sodium casinate and polysaccharides: physicochemical properties, structural characterisation, and oxidation stability. Int. J. Biol. Macromol. **256**, 128064 (2024). https://doi.org/10.1016/j.ijbiomac.2023.128064
163. Nabeel Ahmad, H., Yong, Y., Wang, S., Munawar, N., Zhu, J.: Development of novel carboxymethyl cellulose/gelatin-based edible films with pomegranate peel extract as antibacterial/antioxidant agents for beef preservation. Food Chem. **443**, 138511 (2024). https://doi.org/10.1016/j.foodchem.2024.138511
164. Nair, S.S., Trafiałek, J., Kolanowski, W.: Edible packaging: a technological update for the sustainable future of the food industry. Appl. Sci. **13**(14), 8234 (2023). https://doi.org/10.3390/app13148234
165. Najera, A.I., Nieto, S., Barron, L.J.R., Albisu, M.: A review of the preservation of hard and semi-hard cheeses: quality and safety. Int. J. Environ. Res. Public Health **18**(18), (2021). https://doi.org/10.3390/ijerph18189789
166. Naseri, H.R., Beigmohammadi, F., Mohammadi, R., Sadeghi, E.: Production and characterization of edible film based on gelatin–chitosan containing Ferulago angulate essential oil and its application in the prolongation of the shelf life of turkey meat. J. Food Process. Preserv. **44**(8), e14558 (2020). https://doi.org/10.1111/jfpp.14558
167. Navarro-Tarazaga, M.L., Massa, A., Pérez-Gago, M.B.: Effect of beeswax content on hydroxypropyl methylcellulose-based edible film properties and postharvest quality of coated plums (Cv. Angeleno). LWT-Food Sci. Technol. **44**(10), 2328–2334 (2011). https://doi.org/10.1016/j.lwt.2011.03.011
168. Nazir, S., Wani, I.A.: Development and characterization of an antimicrobial edible film from basil seed (Ocimum basilicum L.) mucilage and sodium alginate. Biocatal. Agric. Biotechnol. **44**, 102450 (2022). https://doi.org/10.1016/j.bcab.2022.102450
169. Ngatirah, N., Ruswanto, A., Sunardi, S.: Effect of Hydroxypropyl methylcellulose from oil palm empty fruit bunch on oil uptake and physical properties of French fries. Food Sci. Technol. **42**, (2022). https://doi.org/10.1590/fst.110421
170. Nguyen, T.T., Thi Dao, U.T., Thi Bui, Q.P., Bach, G.L., Ha Thuc, C.N., Ha Thuc, H.: Enhanced antimicrobial activities and physiochemical properties of edible film based on chitosan incorporated with Sonneratia caseolaris (L.) Engl. leaf extract. Prog. Org. Coat. **140**, 105487 (2020). https://doi.org/10.1016/j.porgcoat.2019.105487
171. Ni, H., Li, H., Hou, W., Chen, J., Miao, S., Wang, Y., Li, H.: From sea to sea: Edible, hydrostable, and degradable straws based on seaweed-derived insoluble cellulose fibers and soluble polysaccharides. Carbohyd. Polym. **334**, 122038 (2024). https://doi.org/10.1016/j.carbpol.2024.122038
172. Nohara, N.M.L., Ariza-Tarazona, M.C., Triboni, E.R., Nohara, E.L., Villarreal-Chiu, J.F., Cedillo-Gonzalez, E.I.: Are you drowned in microplastic pollution? A brief insight on the current knowledge for early career researchers developing novel remediation strategies. Sci. Total Environ. **918**, 170382 (2024). https://doi.org/10.1016/j.scitotenv.2024.170382
173. Nor Adilah, A., Noranizan, M.A., Jamilah, B., Nur Hanani, Z.A.: Development of polyethylene films coated with gelatin and mango peel extract and the effect on the quality of margarine. Food Packag. Shelf Life **26**, (2020). https://doi.org/10.1016/j.fpsl.2020.100577

174. Ochoa, E., Saucedo-Pompa, S., Rojas-Molina, R., Garza, H.D.L., Charles-Rodríguez, A.V., Aguilar, C.N.: Evaluation of a candelilla wax-based edible coating to prolong the shelf-life quality and safety of apples. Am. J. Agric. Biol. Sci. **6**, 92–98 (2011)
175. Omar-Aziz, M., Khodaiyan, F., Yarmand, M.S., Mousavi, M., Gharaghani, M., Kennedy, J.F., Hosseini, S.S.: Combined effects of octenylsuccination and beeswax on pullulan films: water-resistant and mechanical properties. Carbohyd. Polym. **255**, 117471 (2021). https://doi.org/10.1016/j.carbpol.2020.117471
176. Otoni, C.G., Avena-Bustillos, R.J., Azeredo, H.M.C., Lorevice, M.V., Moura, M.R., Mattoso, L.H.C., McHugh, T.H.: Recent advances on edible films based on fruits and vegetables—a review. Compr. Rev. Food Sci. Food Saf. **16**(5), 1151–1169 (2017). https://doi.org/10.1111/1541-4337.12281
177. Pal, N., Dubey, P., Gopinath, P., Pal, K.: Combined effect of cellulose nanocrystal and reduced graphene oxide into poly-lactic acid matrix nanocomposite as a scaffold and its anti-bacterial activity. Int. J. Biol. Macromol. **95**, 94–105 (2017). https://doi.org/10.1016/j.ijbiomac.2016.11.041
178. Pan, R., Iqbal, S., Wang, N., Wu, P., Chen, H., Bhutto, R.A., Chen, X.D.: Alteration in rheological and digestive properties of O/W emulsions using controlled aggregation of konjac glucomannan and xanthan gum in aqueous phases. Food Hydrocolloids **160** (2025). https://doi.org/10.1016/j.foodhyd.2024.110795
179. Park, H.J., Chinnan, M.S.: Gas and water vapor barrier properties of edible films from protein and cellulosic materials. J. Food Eng. **25**(4), 497–507 (1995). https://doi.org/10.1016/0260-8774(94)00029-9
180. Perdones, Á., Escriche, I., Chiralt, A., Vargas, M.: Effect of chitosan–lemon essential oil coatings on volatile profile of strawberries during storage. Food Chem. **197**, 979–986 (2016). https://doi.org/10.1016/j.foodchem.2015.11.054
181. Perera, K.Y., Jaiswal, A.K., Jaiswal, S.: Biopolymer-based sustainable food packaging materials: challenges, solutions, and applications. Foods **12**(12) (2023). https://doi.org/10.3390/foods12122422
182. Pinto, J.P., D'souza, O.J., Hiremani, V.D., Dalbanjan, N.P., Kumar, S.P., Narasagoudr, S.S., Masti, S.P., Chougale, R.B.: Functional properties of taro starch reinforced polysaccharide based films for active packaging. Food Biosci. **56**, 103340 (2023). https://doi.org/10.1016/j.fbio.2023.103340
183. Prakash Maran, J., Sivakumar, V., Sridhar, R., Prince Immanuel, V.: Development of model for mechanical properties of tapioca starch based edible films. Ind. Crops Prod. **42**, 159–168 (2013). https://doi.org/10.1016/j.indcrop.2012.05.011
184. Prakoso, F.A.H., Indiarto, R., Utama, G.L.: Edible film casting techniques and materials and their utilization for meat-based product packaging. Polym. (Basel) **15**(13), 2800 (2023). https://doi.org/10.3390/polym15132800
185. Pratap Singh, D., Packirisamy, G.: Biopolymer-based edible coating for enhancing the shelf life of horticulture products. Food Chem.: Mol. Sci. **4**, 100085 (2022). https://doi.org/10.1016/j.fochms.2022.100085
186. Priyadarshi, R., Kim, S.M., Rhim, J.W.: Pectin/pullulan blend films for food packaging: effect of blending ratio. Food Chem. **347**, 129022 (2021). https://doi.org/10.1016/j.foodchem.2021.129022
187. Purewal, S.S., Kaur, A., Bangar, S.P., Singh, P., Singh, H.: Protein-based films and coatings: an innovative approach. Coatings **14**(1), 32 (2024)
188. Puscaselu, R., Gutt, G., Amariei, S.: Rethinking the future of food packaging: biobased edible films for powdered food and drinks. Molecules **24**(17), 3136 (2019)
189. Puscaselu, R.G., Anchidin-Norocel, L., Petraru, A., Ursachi, F.: Strategies and challenges for successful implementation of green economy concept: edible materials for meat products packaging. Foods **10**(12), (2021). https://doi.org/10.3390/foods10123035
190. Qi, W., Tong, X., Wang, M., Liu, S., Cheng, J., Wang, H.: Impact of soybean protein isolate concentration on chitosan-cellulose nanofiber edible films: focus on structure and properties. Int. J. Biol. Macromol. **255**, 128185 (2024). https://doi.org/10.1016/j.ijbiomac.2023.128185

191. Qiu, Y.-T., Wang, B.-J., Weng, Y.-M.: Preparation and characterization of genipin cross-linked and lysozyme incorporated antimicrobial sodium caseinate edible films. Food Packag. Shelf Life **26**, 100601 (2020). https://doi.org/10.1016/j.fpsl.2020.100601
192. Rachtanapun, P., Homsaard, N., Kodsangma, A., Phongthai, S., Leksawasdi, N., Phimolsiripol, Y., Seesuriyachan, P., Chaiyaso, T., Chotinan, S., Jantrawut, P., Ruksiriwanich, W.: Effects of storage temperature on the quality of eggs coated by cassava starch blended with carboxymethyl cellulose and paraffin wax. Poult. Sci. **101**(1), 101509 (2022). https://doi.org/10.1016/j.psj.2021.101509
193. Rahman, S., Gogoi, J., Dubey, S., Chowdhury, D.: Animal derived biopolymers for food packaging applications: a review. Int. J. Biol. Macromol. **255**, 128197 (2024). https://doi.org/10.1016/j.ijbiomac.2023.128197
194. Ramos, Ó.L., Fernandes, J.C., Silva, S.I., Pintado, M.E., Malcata, F.X.: Edible films and coatings from whey proteins: a review on formulation, and on mechanical and bioactive properties. Crit. Rev. Food Sci. Nutr. **52**, 533–552 (2012)
195. Ramos, Ó.L., Fernandes, J.C., Silva, S.I., Pintado, M.E., Malcata, F.X.: Edible films and coatings from whey proteins: a review on formulation, and on mechanical and bioactive properties. Crit. Rev. Food Sci. Nutr. **52**(6), 533–552 (2012). https://doi.org/10.1080/10408398.2010.500528
196. Ramos, M., Valdés, A., Beltrán, A., Garrigós, M.C.: Gelatin-based films and coatings for food packaging applications. Coatings **6**(4), 41 (2016)
197. Ranjbar, M., Azizi Tabrizzad, M.H., Asadi, G., Ahari, H.: Investigating the microbial properties of sodium alginate/chitosan edible film containing red beetroot anthocyanin extract for smart packaging in chicken fillet as a pH indicator. Heliyon **9**(8), e18879 (2023). https://doi.org/10.1016/j.heliyon.2023.e18879
198. Rehault-Godbert, S., Guyot, N., Nys, Y.: The golden egg: nutritional value, bioactivities, and emerging benefits for human health. Nutrients **11**(3), (2019). https://doi.org/10.3390/nu11030684
199. Remya, S., Mohan, C.O., Bindu, J., Sivaraman, G.K., Venkateshwarlu, G., Ravishankar, C.N.: Effect of chitosan based active packaging film on the keeping quality of chilled stored barracuda fish. J. Food Sci. Technol. **53**(1), 685–693 (2016). https://doi.org/10.1007/s13197-015-2018-6
200. Reyes-Avalos, M.C., Femenia, A., Minjares-Fuentes, R., Contreras-Esquivel, J.C., Aguilar-González, C.N., Esparza-Rivera, J.R., Meza-Velázquez, J.A.: Improvement of the quality and the shelf life of figs (Ficus carica) using an alginate-chitosan edible film. Food Bioprocess Technol. **9**(12), 2114–2124 (2016). https://doi.org/10.1007/s11947-016-1796-9
201. Rezaei, F., Shahbazi, Y.: Shelf-life extension and quality attributes of sauced silver carp fillet: a comparison among direct addition, edible coating and biodegradable film. LWT **87**, 122–133 (2018). https://doi.org/10.1016/j.lwt.2017.08.068
202. Rhim, J.W., Shellhammer, T.H.: 21-Lipid-based edible films and coatings. In: Han, J.H. (ed.) Innovations in Food Packaging, pp. 362–383. Academic Press, London (2005)
203. Ribeiro, A.M., Estevinho, B.N., Rocha, F.: Preparation and incorporation of functional ingredients in edible films and coatings. Food Bioprocess Technol. **14**(2), 209–231 (2021). https://doi.org/10.1007/s11947-020-02528-4
204. Ribeiro, I.S., Maciel, G.M., Bortolini, D.G., Fernandes, I.D.A.A., Volpato, W.M., Pedro, A.C., Rubio, F.T.V., Haminiuk, C.W.I.: Sustainable innovations in edible films and coatings: an overview. Trends Food Sci. & Technol. **143**, (2024). https://doi.org/10.1016/j.tifs.2023.104272
205. Ritchie, H., Samborska, V., Roser, M.: "Plastic Pollution" Published online at OurWorldInData.org. https://ourworldindata.org/plastic-pollution (2023)
206. Rodrigues, D.C., Cunha, A.P., Brito, E.S., Azeredo, H.M.C., Gallão, M.I.: Mesquite seed gum and palm fruit oil emulsion edible films: influence of oil content and sonication. Food Hydrocoll. **56**, 227–323 (2016). https://doi.org/10.1016/j.foodhyd.2015.12.018
207. Rofeal, M., Abdelmalek, F.: Evaluation of bioactive multilayer and edible gelatin/xanthan gels crosslinked with Echinacea purpurea for Ricotta cheese preservation. Sustain. Chem. Pharm. **36**, 101322 (2023). https://doi.org/10.1016/j.scp.2023.101322

208. Rojas-Graü, M.A., Avena-Bustillos, R.J., Olsen, C., Friedman, M., Henika, P.R., Martín-Belloso, O., Pan, Z., McHugh, T.H.: Effects of plant essential oils and oil compounds on mechanical, barrier and antimicrobial properties of alginate–apple puree edible films. J. Food Eng. **81**(3), 634–641 (2007). https://doi.org/10.1016/j.jfoodeng.2007.01.007
209. Râpă, M., Popa, E.E.: Biopolymers for edible films and coatings in food applications. In: Thomas, S., Ar, A., Jose Chirayil, C., Thomas, B. (eds.) Handbook of Biopolymers, pp. 1085–1115. Springer Nature Singapore, Singapore (2023)
210. Râpă, M., Popa, E.E.: Biopolymers for edible films and coatings in food applications. In: Thomas, S., AR, A., Jose Chirayil, C., Thomas, B. (eds.) Handbook of Biopolymers. Springer, Singapore (2022). https://doi.org/10.1007/978-981-16-6603-2_40-1
211. Saberi, B., Chockchaisawasdee, S., Golding, J.B., Scarlett, C.J., Stathopoulos, C.E.: Physical and mechanical properties of a new edible film made of pea starch and guar gum as affected by glycols, sugars and polyols. Int. J. Biol. Macromol. **104**(Pt A), 345–359 (2017). https://doi.org/10.1016/j.ijbiomac.2017.06.051
212. Sani, M.A., Azizi-Lalabadi, M., Tavassoli, M., Mohammadi, K., McClements, D.J.: Recent advances in the development of smart and active biodegradable packaging materials. Nanomater. (Basel) **11**(5), (2021). https://doi.org/10.3390/nano11051331
213. Senturk Parreidt, T., Müller, K., Schmid, M.: Alginate-based edible films and coatings for food packaging applications. Foods **7**(10), 170 (2018)
214. Seydim, A.C., Sarikus-Tutal, G., Sogut, E.: Effect of whey protein edible films containing plant essential oils on microbial inactivation of sliced Kasar cheese. Food Packag. Shelf Life **26**, 100567 (2020). https://doi.org/10.1016/j.fpsl.2020.100567
215. Shahidi, F., Hossain, A.: Preservation of aquatic food using edible films and coatings containing essential oils: a review. Crit. Rev. Food Sci. Nutr. **62**(1), 66–105 (2022). https://doi.org/10.1080/10408398.2020.1812048
216. Sharaf Eddin, A., Ibrahim, S.A., Tahergorabi, R.: Egg quality and safety with an overview of edible coating application for egg preservation. Food Chem. **296**, 29–39 (2019). https://doi.org/10.1016/j.foodchem.2019.05.182
217. Sharma, L., Singh, C.: Sesame protein based edible films: development and characterization. Food Hydrocolloids **61**, 139–147 (2016). https://doi.org/10.1016/j.foodhyd.2016.05.007
218. Shekarabi, A.S., Oromiehie, A.R., Vaziri, A., Ardjmand, M., Safekordi, A.A.: Investigation of the effect of nanoclay on the properties of quince seed mucilage edible films. Food Sci. Nutr. **2**(6), 821–827 (2014). https://doi.org/10.1002/fsn3.177
219. Shellhammer, T.H., Krochta, J.M.: Whey protein emulsion film performance as affected by lipid type and amount. J. Food Sci. **62**(2), 390–394 (1997). https://doi.org/10.1111/j.1365-2621.1997.tb04008.x
220. Siddiqui, S.A., Zannou, O., Bahmid, N.A., Fidan, H., Alamou, A.F., Nagdalian, A.A., Hassoun, A., Fernando, I., Ibrahim, S.A., Arsyad, M.: Consumer behavior towards nanopackaging-a new trend in the food industry. Future Foods **6**, (2022). https://doi.org/10.1016/j.fufo.2022.100191
221. Silva, T.J., Fernandes, G.D., Bernardinelli, O.D., Silva, E.C., da, R., Barrera-Arellano, D., Ribeiro, A.P.B.: Organogels in low-fat and high-fat margarine: a study of physical properties and shelf life. Food Res. Int. **140**, 110036 (2021). https://doi.org/10.1016/J.FOODRES.2020.110036
222. Smetana, S., Leonhardt, L., Kauppi, S.-M., Pajic, A., Heinz, V.: Insect margarine: processing, sustainability, and design. J. Clean. Prod. **264**, (2020). https://doi.org/10.1016/j.jclepro.2020.121670
223. Soazo, M., Pérez, L.M., Piccirilli, G.N., Delorenzi, N.J., Verdini, R.A.: Antimicrobial and physicochemical characterization of whey protein concentrate edible films incorporated with liquid smoke. LWT-Food Sci. Technol. **72**, 285–291 (2016). https://doi.org/10.1016/j.lwt.2016.04.027
224. Song, T., Qian, S., Lan, T., Wu, Y., Liu, J., Zhang, H.: Recent advances in bio-based smart active packaging materials. Foods (Basel, Switzerland) **11**(15), 2228 (2022). https://doi.org/10.3390/foods11152228

225. Soni, A., Gurunathan, K., Mendiratta, S.K., Talukder, S., Jaiswal, R.K., Sharma, H.: Effect of essential oils incorporated edible film on quality and storage stability of chicken patties at refrigeration temperature (4 ± 1 °C). J. Food Sci. Technol. **55**(9), 3538–3546 (2018). https://doi.org/10.1007/s13197-018-3279-7
226. Sudheer, S., Bandyopadhyay, S., Bhat, R.: Sustainable polysaccharide and protein hydrogel-based packaging materials for food products: a review. Int. J. Biol. Macromol. **248**, 125845 (2023). https://doi.org/10.1016/j.ijbiomac.2023.125845
227. Sun, Q., Sun, C., Xiong, L.: Mechanical, barrier and morphological properties of pea starch and peanut protein isolate blend films. Carbohyd. Polym. **98**(1), 630–637 (2013). https://doi.org/10.1016/j.carbpol.2013.06.040
228. Sun, B., Wang, W., Zhang, M., Sain, M.: Biomass-based edible film with enhanced mass barrier capacity and gas permeable selectivity. Cellulose **25**(10), 5919–5937 (2018). https://doi.org/10.1007/s10570-018-1976-z
229. Syed, Q.A., Anwar, S., Shukat, R., Zahoor, T.: Effects of different ingredients on texture of ice cream. J. Nutr. Health & Food Eng. **8**(6), (2018). https://doi.org/10.15406/jnhfe.2018.08.00305
230. Tafa, K.D., Satheesh, N., Abera, W.: Mechanical properties of tef starch based edible films: development and process optimization. Heliyon **9**(2), e13160 (2023). https://doi.org/10.1016/j.heliyon.2023.e13160
231. Tahergorabi, R., Jaczynski, J.: Egg as nutraceuticals. In: Hester, P. (ed.) Egg Innovations and Strategies for Improvements, pp. 271–279. Academic Press, Cambridge, MA, USA (2016)
232. Tang, H., Han, Z., Zhao, C., Jiang, Q., Tang, Y., Li, Y., Cheng, Z.: Preparation and characterization of aloe vera polysaccharide-based packaging film and its application in blueberry preservation. Prog. Org. Coat. **177**, 107445 (2023). https://doi.org/10.1016/j.porgcoat.2023.107445
233. Tang, B., Ma, J., Liu, L., Xu, J., Zhang, H., Li, K., Tu, X., Guan, Q.: Preparation of pullulan-shellac edible films with improved water-resistance and UV barrier properties for Chinese cherries preservation. J. Future Foods **5**(1), 107–118 (2025). https://doi.org/10.1016/j.jfutfo.2024.01.010
234. Tavassoli-Kafrani, E., Shekarchizadeh, H., Masoudpour-Behabadi, M.: Development of edible films and coatings from alginates and carrageenans. Carbohyd. Polym. **137**, 360–374 (2016). https://doi.org/10.1016/j.carbpol.2015.10.074
235. Taylor, S., Colonna, A., Jung, J., Gutierrez, J., Zhao, Y.: Consumer perception and acceptance of edible packaging for various food products. J. Food Sci. 1–15, (2024). https://doi.org/10.1111/1750-3841.16992
236. Teixeira-Costa, B.E., Andrade, C.T.: Natural polymers used in edible food packaging—history, function and application trends as a sustainable alternative to synthetic plastic. Polysaccharides **3**(1), 32–58 (2021). https://doi.org/10.3390/polysaccharides3010002
237. Thawien, W.: Protein-based edible films: characteristics and improvement of properties. In: Ayman Amer, E. (ed.) Structure and Function of Food Engineering, pp. Ch. 3. IntechOpen, Rijeka (2012)
238. Tian, Y., Yang, X., Cao, C., Lv, Z., Han, C., Guo, Q., Duan, Y., Zhang, J.: Improved antioxidant activities of edible films by curcumin-containing with zein/polysaccharide. Food Biosci. **57**, 103538 (2024). https://doi.org/10.1016/j.fbio.2023.103538
239. Tkaczewska, J., Jamróz, E., Zając, M., Guzik, P., Derbew Gedif, H., Turek, K., Kopeć, M.: Antioxidant edible double-layered film based on waste from soybean production as a vegan active packaging for perishable food products. Food Chem. **400**, 134009 (2023). https://doi.org/10.1016/j.foodchem.2022.134009
240. Versino, F., Ortega, F., Monroy, Y., Rivero, S., López, O.V., García, M.A.: Sustainable and bio-based food packaging: a review on past and current design innovations. Foods (Basel, Switzerland), **12**(5), 1057 (2023). https://doi.org/10.3390/foods12051057
241. Wagh, Y.R., Pushpadass, H.A., Emerald, F.M.E., Nath, B.S.: Preparation and characterization of milk protein films and their application for packaging of Cheddar cheese. J. Food Sci. Technol. **51**(12), 3767–3775 (2014). https://doi.org/10.1007/s13197-012-0916-4

242. Wang, L., Ding, J., Fang, Y., Pan, X., Fan, F., Li, P., Hu, Q.: Effect of ultrasonic power on properties of edible composite films based on rice protein hydrolysates and chitosan. Ultrason. Sonochem. **65**, 105049 (2020). https://doi.org/10.1016/j.ultsonch.2020.105049
243. Wang, Y., Khan, M.A., Chen, K., Zhang, L., Chen, X.: Electrospinning of natural biopolymers for innovative food applications: a review. Food Bioprocess Technol. **16**, 704–725 (2023). https://doi.org/10.1007/s11947-022-02896-z
244. Wang, L.-F., Rhim, J.-W.: Isolation and characterization of melanin from black garlic and sepia ink. LWT **99**, 17–23 (2019). https://doi.org/10.1016/j.lwt.2018.09.033
245. Wang, M., Wei, Z., Zhang, Z.: Antimicrobial edible films for food preservation: recent advances and future trends. Food Bioprocess Technol. (2023). https://doi.org/10.1007/s11947-023-03178-y
246. Wang, K., Wu, K., Xiao, M., Kuang, Y., Corke, H., Ni, X., Jiang, F.: Structural characterization and properties of konjac glucomannan and zein blend films. Int. J. Biol. Macromol. **105**, 1096–1104 (2017). https://doi.org/10.1016/j.ijbiomac.2017.07.127
247. Wang, F., Yu, G., Yang, Q., Yi, X., Fu, L., Wang, Y.: Antibacterial Gelidium amansii polysaccharide-based edible films containing cyclic adenosine monophosphate for bioactive packaging. Int. J. Biol. Macromol. **212**, 324–336 (2022). https://doi.org/10.1016/j.ijbiomac.2022.05.090
248. Wang, T., Su, E.: Electrospinning meets food packaging: a promising pathway towards novel opportunities in food preservation. Food Packaging Shelf Life **41**, 101234 (2024). https://doi.org/10.1016/j.fpsl.2023.101234
249. Williams, A.T., Rangel-Buitrago, N.: The past, present, and future of plastic pollution. Mar. Pollut. Bull. **176**, 113429 (2022). https://doi.org/10.1016/j.marpolbul.2022.113429
250. Wu, Y., Geng, F., Chang, P.R., Yu, J., Ma, X.: Effect of agar on the microstructure and performance of potato starch film. Carbohyd. Polym. **76**(2), 299–304 (2009). https://doi.org/10.1016/j.carbpol.2008.10.031
251. Wu, C., Jiang, H., Zhao, J., Humayun, M., Wu, S., Wang, C., Zhi, Z., Pang, J.: A novel strategy to formulate edible active-intelligent packaging films for achieving dynamic visualization of product freshness. Food Hydrocoll. **133**, 107998 (2022). https://doi.org/10.1016/j.foodhyd.2022.107998
252. Wu, Y., Wu, H., Hu, L.: Recent advances of proteins, polysaccharides and lipids-based edible films/coatings for food packaging applications: a review. Food Biophys. **19**(1), 29–45 (2023). https://doi.org/10.1007/s11483-023-09794-7
253. Xu, X., Liu, H., Duan, S., Liu, X., Zhang, K., Tu, J.: A novel pumpkin seeds protein-pea starch edible film: mechanical, moisture distribution, surface hydrophobicity, UV-barrier properties and potential application. Mater. Res. Express **6**, (2019). https://doi.org/10.1088/2053-1591/ab63f7
254. Yang, S., Wang, J., Wang, Y., Luo, Y.: Key role of collagen fibers orientation in casing-meat adhesion. Food Res. Int. **89**, 439–447 (2016). https://doi.org/10.1016/j.foodres.2016.08.035
255. Yasmeen, F., Karamat, H., Rehman, R., Akram, M., Ghfar, A.A., Abdelghani, H.T.M., Dar, A., Mitu, L.: Fabrication and testing of edible films incorporated with ZnO nanoparticles to enhance the shelf life of bread. Food Biosci. **56**, 103111 (2023). https://doi.org/10.1016/j.fbio.2023.103111
256. Yekta, R., Mirmoghtadaie, L., Hosseini, H., Norouzbeigi, S., Hosseini, S.M., Shojaee-Aliabadi, S.: Development and characterization of a novel edible film based on Althaea rosea flower gum: investigating the reinforcing effects of bacterial nanocrystalline cellulose. Int. J. Biol. Macromol. **158**, 327–337 (2020). https://doi.org/10.1016/j.ijbiomac.2020.04.021
257. Yin, Y., Woo, M.W.: Transitioning of petroleum-based plastic food packaging to sustainable bio-based alternatives. Sustain. Food Technol. (2024). https://doi.org/10.1039/d4fb00028e
258. Yin, Y.C., Yin, S.W., Yang, X.Q., Tang, C.H., Wen, S.H., Chen, Z., Xiao, B.J., Wu, L.Y.: Surface modification of sodium caseinate films by zein coatings. Food Hydrocolloids **36**, 1–8 (2014). https://doi.org/10.1016/j.foodhyd.2013.08.027
259. Zhang, X., Xiao, G., Wang, Y., Zhao, Y., Su, H., Tan, T.: Preparation of chitosan-TiO_2 composite film with efficient antimicrobial activities under visible light for food packaging

applications. Carbohyd. Polym. **169**, 101–107 (2017). https://doi.org/10.1016/j.carbpol.2017.03.073
260. Zhang, S., Zhao, H.: Preparation and properties of zein–rutin composite nanoparticle/corn starch films. Carbohyd. Polym. **169**, 385–392 (2017). https://doi.org/10.1016/j.carbpol.2017.04.044
261. Zhang, Y., Pham, H.M., Tran, S.D.: The chicken egg: an advanced material for tissue engineering. Biomolecules **14**(4), (2024). https://doi.org/10.3390/biom14040439
262. Zhang, S., Zhao, H.: Preparation and properties of zein-rutin composite nanoparticle/corn starch films. Carbohydr. Polym. **169**, 385–392 (2017). https://doi.org/10.1016/j.carbpol.2017.04.044
263. Zhou, Z., Ma, J., Li, K., Zhang, W., Li, K., Tu, X., Liu, L., Xu, J., Zhang, H.: A plant leaf-mimetic membrane with controllable gas permeation for efficient preservation of perishable products. ACS Nano **15**(5), 8742–8752 (2021). https://doi.org/10.1021/acsnano.1c00997
264. Zioga, M., Papantonopoulou, G., Evageliou, V.: High internal phase emulsions and edible films with high methoxyl pectin and pea protein isolate or sodium caseinate. Food Hydrocoll. **140**, 108605 (2023). https://doi.org/10.1016/j.foodhyd.2023.108605
265. Zoghi, A., Khosravi-Darani, K., Mohammadi, R.: Application of edible films containing probiotics in food products. J. Consum. Prot. Food Saf. **15**(4), 307–320 (2020). https://doi.org/10.1007/s00003-020-01286-x
266. Zubeldía, F., Ansorena, M.R., Marcovich, N.E.: Wheat gluten films obtained by compression molding. Polym. Testing **43**, 68–77 (2015). https://doi.org/10.1016/j.polymertesting.2015.02.001

Chapter 6
Replacement of Conventional Packaging Materials with Green Polymers

M. Prasanna Kumar, Arthi Vijayakumar, Priyanga Govindarasu, Jaymina Murdamootoo, Perraulta Lavanya, Ramachandran Ramkumar, and Narayanasamy Sangeetha

Abstract Synthetic polymers are considered a threat due to their impact on the environment via disposal and their dependence on fossil sources. The upcoming alternative technology in the packaging industry is green packaging materials obtained from natural sources like plants, animals, and microorganisms. These are considered to be environmentally safe and include carbohydrates like starch, lignin, and cellulose and also proteins like keratin, collagen, and gelatin, as well as microbial-origin biopolymers like polyhydroxy alkanoates and exo-polysaccharides (EPS). Among the various classes of bio-polymers, starch is considered non-toxic, available, and economical and is used in the form of active films in food packaging. Bio-polymeric polyphenolic compounds like gallic acid, which are extracted via supercritical extraction, are added to enhance the anti-oxidant properties of the chitin and gelatin-based films in active food packaging. Nanomaterials of biological origin are also incorporated in active food packaging inclusively bacterial nanocellulose, which is a hydrocolloid nanocellulose substrate that is synthesized from the non-pathogenic strains of Komagataeibacter due to its high crystallinity, thermal stability, and mechanical

M. Prasanna Kumar · A. Vijayakumar · P. Govindarasu · J. Murdamootoo · P. Lavanya · N. Sangeetha (✉)
Department of Food Science and Technology, Pondicherry University, Puducherry, India
e-mail: praba_sangeen@pondiuni.ac.in

M. Prasanna Kumar
e-mail: pk143957@pondiuni.ac.in

A. Vijayakumar
e-mail: arthivijayakumar1999@gmail.com

P. Govindarasu
e-mail: g.priyanga1998@pondiuni.ac.in

P. Lavanya
e-mail: perrtaultalavanya@gmail.com

R. Ramkumar
Institute of Animal Nutrition, Tamil Nadu Veterinary and Animal Sciences University, Chengalpattu, Tamil Nadu, India

S. ul Islam and M. Shahid (eds.), *Green Materials for Active Food Packaging*, Engineering Materials, https://doi.org/10.1007/978-981-96-0369-5_6

strength. It is necessary to conduct more basic research on applying biodegradable polymers in food packaging and to make environmental protection efforts.

Keywords Synthetic polymers · Bio-polymers · Active food packaging · Super-critical extraction · Nanomaterials · Nano-encapsulation

6.1 Introduction

The primary function of consumable packaging is to preserve and protect the food matrix. Conventional packaging materials, such as thermosets, have been extensively used in the food packaging industry due to their desirable properties like gas and vapor resistance, favorable physio-chemical characteristics, lightweight nature, global availability, and commercial viability. However, these materials are often discarded into the environment after use, contributing significantly to plastic pollution. Recent studies have highlighted the alarming rate of plastic production, estimating around 320 million tons per year, which results in considerable environmental degradation due to the accumulation of non-biodegradable plastics [54]. The primary types of thermoset plastics used in food packaging include polyolefins such as polypropylene (PP), high-density polyethylene (HDPE), low-density polyethylene (LDPE), and polystyrene (PS). These materials, once set, cannot be reprocessed through thermal treatment, rendering them unsuitable for sustainable packaging solutions. As global populations rise, the demand for consumable materials is expected to increase by 50% by 2050, further exacerbating the issue of plastic waste [31]. The accumulation of thermoset plastics presents significant challenges, including environmental pollution, economic burden, and health problems. The amplifying contribution of thermoset plastics to ecological degradation, aquatic pollution, and petrochemical waste necessitates a shift toward more sustainable plastic manufacturing frameworks [30, 43, 50]. Hazardous substances from thermosets containing antioxidants, coloring agents, monomers, and depolymerized compounds can leach into the ecosystem, raising significant concerns about food safety and environmental health. Fossil-based chemicals used in food packaging lack durability and soil degradability, presenting various challenges to ecosystem preservation [76]. Polyamides, important polymers in food packaging, are notable for their resistance to carbon dioxide, oxygen, flavors, and durability [38]. Despite their inherent hydrophilic properties, polyamides exhibit limitations as moisture barriers, which can adversely affect the texture and quality of food. However, they have excellent oleophobicity and vapor impermeability, which helps maintain the quality of packaged food. Therefore, the use of polymers derived from fossil fuels is limited due to their non-biodegradable nature.

Consequently, there is an essential need to shift towards new sources for plastic synthesis [33]. Fossil-based petrochemicals used in food packaging disrupt the ecosystem due to their resistance to biological decomposition. Regenerating traditional packaging materials like thermosets is economically impractical, prompting a shift towards green polymers as effective alternatives. Experts in polymer chemistry advocate for this transition, emphasizing the potential of biodegradable materials to replace commonly used petro-based thermosets [76]. These green materials are emerging as strong competitors in the biodegradable industry, offering significant benefits for individual wellness and the ecosystem.

6.2 Green Polymers in Modern Packaging

6.2.1 Green Materials in Active Food Packaging

Since the 1980s, "green" science has evolved within professional circles, intending to incorporate biological materials into food packaging to maintain food quality and promote environmental sustainability [67]. Biopolymer-based packaging adheres to the 12 principles of green chemistry and technology, emphasizing secure packaging with minimal environmental impact. By following these principles, green packaging manufacturing aims to reduce pollution and align packaging technology with the goals of green chemistry and technology. Over the past few decades, the implementation of advanced packaging methods has increased to achieve specific desirable outcomes [45]. Advanced packaging involves interactions between the biopolymer and food, either by incorporating compounds in the initial packaging matrix or by creating layers between the food and the container. Unlike conventional packaging that acts as inert barriers, these materials are designed to offer multiple benefits, including improved food safety, sustainability, and quality. These innovative packaging solutions are created to preserve food's nutritional value and quality, inhibit pathogen growth, and indicate any leakages, thus ensuring food authenticity. Most film packaging or coatings protect against various factors that contribute to food spoilage and maintain product texture and flavor during transport and storage. Green polymers help reduce food waste by extending the shelf life of food products. Green biomass, which consists of chemical substances derived from the ecosystem, is used in various packaging applications such as wraps and covers. Due to their sustainability, these materials effectively preserve the tastes and textures of food during storage and transport. Biopolymers are compatible with various chemicals, including probiotics, antioxidants, antibacterial agents, and nutraceuticals, making them highly effective in packaging applications and capable of incorporating beneficial ingredients to enhance food quality and safety. Biopolymers offer a viable approach to developing sustainable packaging, addressing environmental issues, and meeting the requirements of the food industry. Three categories of green materials have been identified based on their sources and methods of production [44]:

- Polymers like polysaccharides and polypeptides extracted from renewable biomass.
- Polymers like polylactic acid (PLA) created by chemically polymerizing monomers.

- Polymers like polyhydroxyalkanoates (PHA) produced through microbial fermentation.

Additionally, biopolymers can be categorized based on their behavior as thermoplastic or hydroplastic materials, outlining their diverse features and applications in biotechnology and sustainable materials science.

6.2.2 Replacement of Thermosets by Novel Green Materials

Bio-film technology in food packaging provides a sustainable alternative to traditional thermosets. This advanced technology helps reduce waste segregation, as biodegradable polymer films exhibit mechanical and barrier properties distinct from petroleum-based polymers commonly used in food packaging. Blending or compositing different biopolymers can further enhance the performance of these biopolymeric films [6]. Materials' specific uses and suitability are partly determined by their erosion mechanisms. For instance, in drug delivery systems, bulk erosion leads to a slower, longer-lasting release of active ingredients, while surface erosion results in a rapid release. The potential application of green polymer films in food packaging is a topic of ongoing discussion. Identifying the right combinations of polymeric compounds and additives is crucial for enhancing the biochemical properties of green polymer materials.

Further research is needed to improve the qualities of bioplastics to match the performance of petroleum-based plastics. The shift from synthetic plastics to bio-based alternatives remains an aspirational goal that requires deep understanding, increased awareness, and careful consideration in future research. Public awareness about reducing waste from non-renewable and non-biodegradable petroleum-based food packaging is gradually rising. According to the Food and Agriculture Organization, post-harvest losses account for 20–50% of the world's annual food output. To minimize waste, reduce post-harvest losses, and improve food safety, there is a continuous push for more effective and sustainable food packaging solutions. Compostable films combined with nanoparticles for intelligent packaging show great promise. These innovative packaging materials help maintain or improve food quality and extend shelf life. Bio-based polymers are environmentally friendly substitutes with excellent performance characteristics that can replace conventional thermosets. Utilizing green films with nano-sized particles in advanced packaging technology can support green policies in packaging and enhance the green economy [52]. Ongoing research in these areas can contribute to a pollution-free environment.

6.2.3 Green Materials Yielding Methodology

Chitin extraction complied with Duman's methodology [24]. De-ionized water was used to wash the crustacean shells, and they were subsequently subjected to water removal treatment like drying at the ambient temperature and pulverized. Minerals are removed followed by blended with sodium hydroxide aiming to get samples that are devoid of polypeptides. At the conclusive part, they are prepared in a 4:2:1 combination of de-ionized water, methanol, and chloroform to remove oil and pigments in the samples. After that, the yield obtained was dried for 72 h at 40 °C in an oven to extract chitin. After each procedure, they are cleaned with de-ionized water pH similar to neutrality. The material obtained will be acting as a source in making packaging films. Selection of the yielding technology will be optimized by digging into the wide concentration ranges, period and temperature. The yielding mechanism begins with preliminary procedures followed by the pulversing starch into powder form, that was soaked sodium sulfate solution. Now the mixture is homogenized at fifty degree celsius and then those materials wer filtered followed by the filtrate is cleaned with de-mineralized water. After separating the starchy material from the the residue, it is washed multiple times with normal water until it turns clear. Upon getting de-moisturized for 24 h at 50 °C, the starch that was extracted remains at room temperature for storage. This technique ensures the effective extraction of starch, with each stage painstakingly planned to exclude contaminants and maximize starch output for subsequent uses [82]. Many chemical based methodologies can also be utilized in getting a higher yield of starch. Nonthermal techniques encompassing of processing based on ultrasonic vibartions which can greatly influence yield of biopolymers like polypeptides. The ultrasonic apparatus used to extract protein from spirulina ran at lower frequencies (20 kHz) and 1000 W of power. Stainless-steel cylindrical ultrasonic chamber, a sonication horn, and a feed reactor designed to sustain the food matrix's flow were all part of the system [86]. A double layer of water circulation encircled the 250 mL treatment chamber and the 1 L feed reactor, with temperature regulation. Manometers attached to output portion of the pump was used in managing pressure in the ultrasonic chamber's upstream and a manual pressure valve was used to maintain pressure in ultrasonic reactor's output and the reactor feed. Yielding phenomena of green materials involved blending of algae and the solvent in addition to the vigourous mixing to prevent gravitational segregation of that algae. By adjusting or by controlling the flow rate of the pump, protein yield is increased. These bio-compatible polymers that have been successfully derived may additionally be employed for developing creative functional packaging films. This could possibly allow a replacement of traditional petrochemicals materials in modern food packaging with these bio-compatible materials.

6.2.4 Effectiveness in Consumable Packing

Using biodegradable plastic films within consumable limits provides a feasible path towards the development of green and compostable equivalents for existing petrochemical-derived plastics provided with packaging innovations. These innovations can potentially tackle the problem of trash build-up imposed by plastics that are composed of not sustainable petrochemical compounds. Even though the resilience and bending qualities of reusable films created from organic polymers may initially not be as effective compared to those conventional petroleum-based polymers employed for consumables packaging, continual attempts are being made to improve these qualities by combining different types of biological polymers in combined forms. When evaluating whether bio-polymeric materials are appropriate for a certain application, their erosion process is a crucial factor to consider. For example, in delivery of drugs, polymers with surface erosion properties enable the quick release of active ingredients. Bulk erosion-undergoing materials, on the other hand, are better suited for long-term, continuous drug release. The possible uses of bioplastic films for food packaging are a topic of debate, with a focus on maximizing polymer pairing with the aim of improving their physio-chemical attributes. Characteristics of bio-polymeric wrapping materials depend upon selecting adding materials and green polymeric materials to improve its attributes [22]. The public's growing desire for healthier and better-quality food alternatives has prompted researchers to investigate cutting-edge preservation techniques. Various approaches have been studied to reduce lipid oxidation, including adding antioxidants directly to food items or integrating them into packaging materials. Directly adding antioxidants to food presents difficulties, though, as chemical ingestion reduces their effectiveness and speeds up food deterioration. Adding antioxidants to the packaging matrix to improve the stability of food items that are susceptible to oxidation is known as antioxidant packaging, and it is a potentially effective way to solve this problem. Recent research has shown that natural antioxidants, especially plant essential oils, have attracted a lot of interest because of their strong antioxidant and antibacterial qualities [7]. Bioplastic films and antioxidants packaging design promises to enhance food preservation and satisfy customer demand for safer, healthier food alternatives while lowering environmental impact via continuous research and innovation.

6.3 Bio-Polymers in Active Packaging of Food

The main purpose of food processing industries maintaining food from spoilage is crucial in the modern era caused by food-borne pathogens and other environmental threats. The packaging material plays an important role from manufacturing to the time of consumption of the consumable products. Packaging the edible items is influenced due to various numbers of variables like electricity, the resource's economic viability, its disposability, recycling potential, and sustainability [98]. Owing to the environmental threats and concerns, the usage of traditional plastic packaging has become a major concern due to its non-biodegradable nature and eco-toxic behavior. Hence, effective packaging material that does not threaten the environment and mimics the properties of conventional packaging is needed. Biodegradable active packaging is making its way into the market to address the issues that conventional packaging faces due to increased environmental concerns and the need to maintain food quality. This biodegradable active packaging has several benefits such as biodegradability, and it also mimics the properties of conventional packaging, including reducing the moisture content, leakage, mechanical and oxidative stability, preserving the colour and aroma, prevents contamination, thereby increasing the keeping time of the edible material. [41].

Biopolymers are designed with the raw materials of renewable resources such as agricultural or marine sources [21]. The biopolymers are entirely derived from organic substances with higher biodegradability and composability behavior and enable the package to be disposed of in the soil, which serves as fertilizers and soil conditioners [26]. Different categories of bio-polymers based including polysaccharides such as gums, cellulose, chitosan/chitins, starch and protein sources like (gelatin, casein, whey and collagen), and plant sources including (soya, and gluten), lipids sources including waxes, fats and oils and dietary polyphenols. Other types of biopolymers developed from marine prokaryotes and the microbes based by-products, including exopolysaccharides derived from bacteria and fungi [15, 21]. Bio-polymer-based packaging showed extended shelf-life by reducing lipid oxidation, preventing microbial growth and contamination, and reducing water movement, exhibiting better physical and functional properties has promising aspects in food packaging [34].

6.3.1 Carbohydrates

Polysaccharides are the most abundant macromolecule on earth. The biodegradable packaging materials, films, edible coatings, and membranes have been produced from polysaccharides, and their derivatives utilized in several industries, including food, pharmaceutical, medical, and other industrial applications [27]. The polysaccharides-extracted films possess better gas permeability of gas characteristic; thereby, the desired modified atmosphere can be developed to extend the shelf life, while polysaccharides-derived membranes have been used in food industries for packaging and edible coatings. These membranes also showed better barrier against O_2 and CO_2 and exhibited improved mechanical properties [92]. The majority of these polymeric materials are rigid and hydrophilic, but showed reduced barrier against water vapor; however, several improvements and modifications have been studied to improve their properties for their utilization in packaging and other industrial applications [60]. The starch, cellulose, chitosan, and other polysaccharides originate from agricultural, underwater, and animal-based sources were used to develop biopolymer/biodegradable packaging materials.

6.3.1.1 Starch

Starches are one of the low-cost polysaccharides and an easily available group of biodegradable biopolymers. It is otherwise known as hydrocolloid biopolymer. It is a renewable resource that is easy to handle and can be used as thermoplastic. Starch is extracted from various sources, including sorghum, corn, wheat, and rice and from tuber sources such as cassava, potato, and tapioca, and from pseudocereals, including buckwheat, quinoa, amaranth and other non-conventional sources including kithul and talipot palm [1, 9, 79]. Starch is plasticized through a certain amount of water or plasticizers with the application of heat, followed by an extrusion process. As starch exhibits low barrier to moisture which restricts their application and also influence the mechanical properties. Hence, the starch is combined with other biopolymers and additives to improve their properties [53]. For instance, starch is blended with chitosan and gelatin through solution casting method which increase the density and reduce the moisture uptake [92]. Corn starch blended with polyvinyl alcohol (PVOH) through solution casting method which increase the biodegradability and composability of corn starch in soil [8]. This blending of starch brings biodegradable polymers with better mechanical strength, flexibility and suitable moisture barrier properties, which finds its way into commercial packaging of consumer products and non-food sectors [9].

6.3.1.2 Cellulose

Cellulose is a degradable, reusable, and most widely distributed natural polymer in the biosphere. β-D-glucopyranose units of glucose makesup the structure of cellulose. Due to thestructure and the presence of hydroxyl groups, which tend to form hydrogen interchain bonded crystalline microfibrils and fibers used in packaging for paper, cardboard, and corrugated paperboard [20]. Cellulose exhibits specific characteristics, including low density, good mechanical strength, durability, economic feasibility, toxic-free, biocompatibility, renewability, stability over chemicals, and better film-forming properties, increasing its usage as a packaging material [23]. However, in its natural form, it is difficult to utilize for packaging because of its hydrophilic behavior and crystalline properties, as it exhibits undesirable mechanical properties and is considered a hard polymer. Hence, it is treated with other chemicals, such as sodium hydroxide, sulphuric acid, and other chemicals, to produce cellophane, which exhibits desirable mechanical properties for its usage in packaging [98]. Cellophane has been used for packaging food products, which requires a longer shelf-life.

6.3.1.3 Chitosan/Chitin

The next predominant agro-polymer that exists naturally is chitosan/chitin. It seems to be an essential exoskeleton structural element of arthropods and present in the cell walls of fungi and yeast [75]. It is a commercially produced acetylated polysaccharide made of N-acetyl-D-glucosamine, obtained from the waste products of prawns and crabs by chemical extraction methods. It is an inelastic nitrogenous polysaccharide. Chitin applicability is limited because of its insoluble nature in solvents. The three processes separating chitin from other molecules are deproteination, demineralization, and decolorization, which eliminate protein, salts, and colour [73].

The biodegradable material identified as chitosan is formed by partially N-deacetylating chitin in the presence of 40–50% sodium hydroxide (alkaline deacetylation). Chitosan is hydrophobic but may dissolve in acidic solutions. The membranes derived from chitosan were biodegradable, biocompatible, non-toxicity and renewable. The chitosan with good mechanical properties were developed by adding glycerol through thermo-mechanical treatment, resulting in thermoplastic material.

6.3.1.4 Gums

Gums are naturally occurring polysaccharides which are revealed to have good gelling properties when water is added. Gums are biodegradable polysaccharides, which is cheap, sustainable and an edible packaging material. Gums have nutraceutical and antimicrobial effects that benefit human health. Gums also possess anti-cancer, anti-inflammatory properties and act as an antioxidant. Gums deliver widespread application in both food and pharmaceutical industries as stabilizers, thickeners, binders, emulsifiers and as suspending agent [81].

6.3.1.5 Alginates

Alginates are the linear polysaccharides derived from seaweeds and by some bacteria. Alginates are water soluble and an anionic in nature. It comprises segments of alternating mannuronic (M blocks) and glucuronic acids (G blocks) collectively called MG blocks. These three blocks determine alginates' physical properties directly with their extraction source [60]. Alginates were considered a better film forming agent because of its selective permeability, non-toxic, biodegradable, biocompatible nature which is also low in cost. The efficient extraction methods would produce eco-friendly biopolymer from seaweed, which has a wider application in food and non-food sectors.

6.3.1.6 Carrageenan

Carrageenan is an intricate blend of many types of polysaccharides. They can hydrophilic, anionic linear chain polysaccharides extracted from seaweeds. It is mainly used as emulsifiers and stabilizers in various food products. Edible films and coating were prepared from carrageenan and blended with other polymers to increase their application. Edible coatings prepared from carrageenan are used to preserve fresh cut fruit by reducing water loss and gas exchange, maintains the colour and organoleptic properties [57]. Multiple investigations have shown that in order to enhance the quality and antibacterial qualities of films, additives such as lecithin, salts, and antibiotics were used along with antioxidants such ascorbic acid and gallic acid. Carrageenan is widely employed as a biopolymer and in food and other industrial uses these days.

6.3.2 Proteins

Whey proteins, caseins, zein, gelatin and soy protein have been used as biodegradable biopolymers for packaging. These protein-based substances will adhere to hydrophilic surfaces, and provides barriers to O_2 and CO_2 but do not show resistance to water diffusion.

6.3.2.1 Whey Proteins

Whey proteins are considered as active milk protein that are added in infant formula and as a protein supplement in sports formula. Due to its good mechanical strength and gas barrier properties films also prepared from whey protein. Edible coatings are also prepared from whey thus exhibit good barrier effects on encapsulating aromas and lipids [19]. These films and coatings are edible, economically feasible, biodegradable, and renewable and find its potential markets for various categories of food products. Hence, edible packaging materials can be prepared from whey proteins.

6.3.2.2 Casein

Casein is also milk protein which comprises of 80% from total protein. Casein also shown better film-making properties. It can be used as emulsifiers and caseinates also showed stability over thermal and mechanical barriers. Because of the presence of 4 major subunits such as αs1, αs2, β and K-casein with greater number of amino acids [4]. Casein is also considered as an excellent encapsulating material for bioactive compounds and it has proven efficient edible packaging films [56].

6.3.2.3 Zein

Zein is storage protein in corn and considered as an excellent material for edible packaging and it also insoluble, thermoplastic compound provides good filming properties. Films prepared from zein provides thermal stability but showed reduced mechanical properties. This can be overcome by blending zein with other biopolymer.

The films were shown to be smooth in appearance showed selective permeability and also possess antibacterial and barrier properties [99].

6.3.2.4 Gelatin

Gelatin is denatured protein prepared by collagen hydrolysis. Collagen being insoluble fibrous polypeptide present within the extracellular matrix and in connective tissue. The utilization of gelatin relies on the good gelling behaviour. Nowadays in food industries applicability of gelatin increases due to its increased water vapour permeability and biodegradable film making properties [5]. And used as a foaming agents, stabilizers and emulsifiers [9]. Used in nanoencapsulation by incorporating the plasticiser to improve the properties and also delays oxidation and ripening of fresh produce [78].

6.3.2.5 Soy Protein

Soy protein is the economically feasible and biodegradable material obtained as the by-products from the oil-industries including isolates of soy protein (90% protein) and concentrates of soy protein (70% protein). Soy proteins are hydrophilic compounds and also shown sensitivity to environmental conditions hence its exhibits poor mechanical and other barrier properties comparing to synthetic polymers [85]. Hence the films prepared from them can be used for low moisture content food [77]. But these problems are improved through blending them other polymers. Other advantage of using soy protein in films, edible coatings and in packaging materials added nutritional benefits.

6.3.3 Polyphenolics and Other Polymers

Biopolymer based packaging films were mainly used as wrapping material for the food products and edible coatings were applied on the food stuff directly. Hence, both the films and edible coatings not brings the properties of protecting the food stuffs by providing the moisture supply, barriers to gas and also carriers for bioactive constituents like polyphenolic compounds [97]. Biopolymer based packaging materials can blended with bioactive compounds which retards oxidation, prevents browning and extends the shelf life by reducing the microbial spoilage through its anti-microbial and antioxidant activities [63]. Propolis is the bee glue which is the resin collected by worker-bees that posses antimicrobial and antioxidant properties was mixed with biopolymers for to develop edible packaging films and coatings [95]. The polyphenols extracted from thinned young apples blended with Making active packaging materials using chitosan which showed increased properties of being anti-oxidant aided by chlorogenic acid [80].

Chinese tea is good source of polyphenolic compounds including flavanols and phenolic acids. Green tea extract blended with chitosan for developing edible packaging films for pork sausages [72]. The results showed that green tea impregnated chitosan film showed enhanced antioxidant and antimicrobial properties which enhanced the shelf life of the pork sausages. The incorporation of essential oils in chitosan based films increases the anti-microbial properties increase shelf-life of fresh harvest produces [89]. Application of cinnamon oil on the chitosan based edible coating extends keeping quality of sweet pepper and also maintained the sensory properties [91]. As the presence of cinnamaldehyde which prevents the growth of microorganisms thereby extending the shelf life of the product.

Other biopolymers such as lipids also used for developing edible films and coatings. Due to its poor mechanical stability and transparency it is used with the combination of hydrophobic lipid layer followed by hydrophilic layer for developing emulsified films and coatings. Waxes are the hydrophobic substances used to develop edible films and coatings. The impregnation of waxes in composite films and coatings minimise the water loss in fresh produce improve the barrier properties and thermal stability [29].

6.4 Conversion of Edible Waste to Green Active Edible Packaging Material

Production of edible waste matrices, from agro, dairy, vegetables, fruits, cereals, meat, fish, and cooking oils are all commonplace in open marketplaces, homes, and processing companies. According to estimates from the Food and Agriculture Organization (FAO), during post-harvest treatment, waste from horticultural sources ranges up to 30% are dumped aside. In addition to encouraging trash recovery, turning these food wastes into value-added products can increase returns and provide commodities that are commercially viable. The production of biopolymers is becoming more and more popular due to its biobased origin, potential for biodegradation, biocompatibility, and biocompostability. The potential of food waste-derived green edible active bio-polymers, and other polymers are being highlighted by a comparison of their physicochemical properties with those of manufactured polymers. Compared to chemical synthesis techniques that depend on co-polymer blending or alkali and acid treatment, the creation of microbial synthesis and bio-based polymers derived from renewable resources are straightforward, scalable, and less harmful to the environment [11]. When improperly managed, waste materials from homes, canteens, and the food processing industry—such as seeds, stems, leaves, roots, pomace, and peels—pose a risk to public health and safety. Edible waste has a high biochemical profile, rich in oils, polyphenols, micro- and macronutrients, and bioactive chemicals, even after it is thrown out. A 1.3 billion metric tons, ends up being wasted every year globally; up to 60% of these food losses are caused by horticultural waste. The production of fruits and vegetables alone in India amounts to over 87 million metric tons per year, producing about 5.7 million tons of garbage, much of which is dumped on the outskirts of cities. There are serious ecological repercussions to using conventional waste management methods including composting, landfilling, cremation, and feeding livestock feed.

6.4.1 Biopolymers Based on Fruit Peels

Non Starch Polysaccharides like pectin which is well-known for its many uses in food manufacturing sector as a gelling agent thickening reagents, stabilizing agent, that has lately drawn interest due to its ability to make films that are edible from food processing byproducts. Though regarded as agricultural waste, peels from pineapple affords a natural supply of natural plasticizer and pectin. Peel of Pineapple based pectin extracts (PPS) has been suggested as a biological plasticizing agent to improve the characteristics of non-starch polysaccharide based films, specifically in terms of their ability to resist oxidation and function as a gas barrier, by using microwave heating extraction procedures [36]. In a comparable manner pectin from dragon fruit peels may be extracted with HCl solution, producing a yield of around 11% of non-starch polysaccharide like pectin. In manufacturing green plastics, this recovered pectin is frequently combined with an ethylene glycol plasticizer [42]. Furthermore, research has demonstrated that pineapple fruit peel extracts may change the crystallographic structure of chitosan, which in turn can change the mechanical strength of the films that are produced from chitosan. The antibacterial and antioxidant qualities of films made from chitosan have been discovered to be greatly supported by higher concentrations of pineapple peel extract [62]. Furthermore, it has been shown that jabuticaba peel flour is a useful resource for creating biodegradable films. It has been demonstrated that increasing the content of glycerol and jabuticaba peel flour improves the hydrophilicity and thickness of the film. Still, a peel of potato used in producing polymerized propelene (PP) films, adding curcumin and cellulose obtained through bacteria. Cellulose from bacteria interacts with biological polymers.

6.4.2 Grain Waste in the Production of Green Polymers in Active Food Packaging

Utilizing rice straw that has been enhanced with high-lignin nanofibers offers a chance to create films of cellulose acetate nanocomposite. When made utilizing the casting procedure, these films show improved hydrophilicity and acceptable transparency [28]. Moreover, lignocellulose byproducts derived from wheat, such bran and straw, have been used to produce enzymes and bio-based materials at the same time. These materials have potential uses in packaging, especially when combined with the compression molding process [74]. Additionally, the possibility of antimicrobial packaging, an aspect of active consumable packing, to interact with food products and extend shelf life has drawn attention [28]. The effectiveness of antioxidants derived from barley husks by acid hydrolysis and delignification has been demonstrated by their excellent dispersion over polyolefin in frozen swordfish packing.

6.5 Technology of Nanomaterials in Green Based Packing

It will be widely acknowledged that the use of nanotechnology is one of the most significant advances of the modern period, with enormous potential for producing inventive products with several benefits. Food processing and packaging are two areas where its use is gradually growing [70]. Since technology of nano particles provides solutions to a wide range of industrial difficulties, it is attracting a lot of interest in the edible packaging industry. Packaging materials may be made to have better rigidity, insulating qualities, microbial resistance, and other desirable attributes by incorporating nanoparticles into them. The development of environmentally friendly packaging materials has shown to be a viable use of nanotechnology. By using nanoparticles, materials that are robust, strong, and light-weight may be produced that retain their structural integrity and use less packing material overall. Moreover, nanoparticles can improve a package's resistant qualities, eventually lowering edible waste and increasing storage duration [17] (Fig. 6.1).

a. **Resistance property enhancement**

Particles of nano size can exhibit resistance characteristics of films that biodegrade and are utilized in the consumable packing sector. This attribute aids in prolonging the shelf life and maintaining the quality of the food particles by shielding them from moisture, fumes, and volatile substances.

b. **Increased physical characteristics**

A few nano-sized particles mixed with natural polymers enhance the physical attributes of packing polymers, allowing them to be more resilient and appropriate for a range of uses.

c. **Release kinetics of bioactive compounds**

To preserve food quality, active ingredients that are released gradually can be included into packaging materials using the nano-encapsulation approach. Antioxidants and antibacterial agents, for instance, can be added to the packaging to better protect the contents within.

d. **Nano-sensors for consumable item authenticity**

Nano-sensors have has been incorporated into packaging to offer pathogen, toxin, and spoilage indication and detection as well as a wide conclusion about the authenticity of the consumable item.

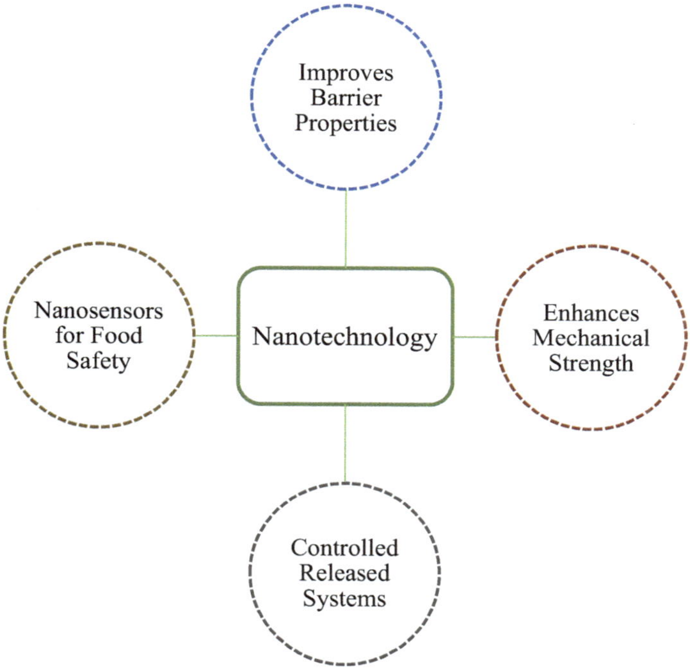

Fig. 6.1 Utilization of technology of nano-based materials in food packing

e. **Ecological methods**

Utilizing by encouraging the use of biodegradable alternatives and decreasing dependency on plastics generated from fossil fuels, nanomaterials created from other renewable sources can further assist in minimizing the environmental effect of packaging.

The importance of nanoparticles is growing dramatically across a range of industries, including cosmos, nutrition, animal feed, pharmaceuticals, and beauty products [39]. Because of the advancements in green chemistry and environmentally friendly approaches, the synthesis of particles of nano-size may be done in a sustainable and beneficial manner. These biological elements, such as plants, microbes, and other bio-regenerable materials, have made it possible to create techniques for the manufacture of green nano-particles [61]. These days, nanomaterials are widely employed in packaging applications based on technology of nano-compounds. Certain nanoparticles of nano-metals, such titanium dioxide and silver, have antibacterial qualities that allow them to be used as food packaging materials, inhibiting the development of microorganisms and enhancing food safety. Additionally, packaging materials may be strengthened and mechanically flexible with the help of nanoparticles, making them more resilient to rips and punctures.

6.5.1 Nanocomposites

In the previous decades, Nanocomposites are becoming more and more popular for use in food packaging. Low quantities of fillers designed with the aid of nanotechnology can significantly enhance the polymers' mechanical, thermal, and barrier properties [65]. Additionally, it is possible to incorporate antioxidant and antibacterial qualities, which are important for consumable packing utilization. Nanoparticles technology makes it easier to create composites of nano-size, which are made of a matrix of polymers that has been injected with nano sized particles or nano-fillers. Nano-fillers, such as carbon tubes of nano-size and nanoparticles of clay may efficiently restrict the passage of gases, moisture, and other contaminants by strengthening the resistance of packing materials. By increasing the shelf life of edible mart, packaging made of composites of nano size can reduce food waste and promote sustainability. A variety of polymeric materials exhibit various resistance qualities. As an example, high-density polyethylene (HDPE) has better vapor resistance than Polyethylene terephthalate, but Polyethylene terephthalate has better oxygen barrier qualities [93]. Resistance of a polymeric substance is impacted by a variety of elements, such as polarity, hydrogen bonding, branching extent, degree of crystallinity, and cross-linking [25]. Furthermore, the presence of another chemical might affect a drug's permeability. In particular, polymer swelling and plasticization cause ethylene vinyl alcohol's oxidative resistance to significantly decline in greater humidity environments (*The Wiley Encyclopedia of Packaging Technology*, 3rd ed., 2009).

Most of the consumable packaging applications, several layers of polymer compounds with significant characteristics are used to attain the appropriate levels of mechanical strength, barrier qualities, and sealing abilities. Polymer nanocomposites (PNCs) are a type of packaging material that may provide the required functions without sacrificing their capacity to be recycled [25]. An innocuous, nanoscale filler is incorporated into a polymeric matrix to generate PNCs. The common filler materials are silica nano-sized particles of silica, carbon nano-sized tubes, nano based crystals of starch, nano based fibers derived from cellulosic polymers, nano based particles derived from chitosan or chitin, clay, silicate nano-platelets, and minerals. Additionally, research from the literature suggest that Nanocomposites of Polymer (PNC) have improved thermal, flame-retardant, and strength characteristics. Currently, the nano based composites og green polymeric substances that have been investigated exclusively for packing applications which are thermal starch and its derivatives, as well as synthetic biopolymers such as polylactic acid, polybutyrate (PHB), and PBS [96]. For usage in consumable packaging and cellulose-based paper packaging materials, along with growing affinity for investigating utilization of technology of nanomaterials, improvising novel characteristics of more complex natural biopolymers, such as cellulose derivatives.

6.5.1.1 Resistance of Nano-Composite

Enhancing resistance characteristics of polymeric nano-composites (PNCs) was frequently explained in literature using the idea that the addition of filler causes an increase in tortuosity. When nano based fillers are added, a complex route is formed that alters the extent of diffusion within particles, improving their resistance. It is difficult to achieve the required improvement unless there is a consistent segregation of fillers throughout the green polymers. An further plausible phenomena pertains to the interplay between nanoparticles and polymeric material, which may influence resistant properties by delaying polymer molecules [48].

Silicates and Clay Based Polymeric Nano Based Composites

Other silicate materials in addition to clays is steady. Fillers are a very attractive option because they are widely accessible at affordable rates and are assumed to be non-toxic. Over the past 20 years, much research has been done on polymer nanocomposites (PNCs) based on nano clay for food contact applications. Numerous clay varieties, including saponite, kaolinite, hectorite, and montmorillonite, is thoroughly investigated for potential PNC uses. When manufacturing Melt compounding, in situ polymerization, and nano based composite materials which are usually the two main techniques used. Systems thorough exfoliating of the nano based composites, nano based clays made of hydrophilic polymeric compounds, including polyamide, have been reported regularly. Achieving intercalated morphologies in matrices of nonpolar polymers like LDPE and HDPE is tedious. Studies [35] states that employment of methodologies which can improve the rearranging the clay to create a contact between the non-polar polymer and the surface of nanoclays of aluminosilicate using bridging with alkyl quaternary ammonium ions and polyolefins using compatibilizers such as maleic anhydride. The shape and dispersion of nanoparticles determine the bulk characteristics of PNC materials. Regarding barrier qualities, monodispersed exfoliated systems outperform other maqterials.

6.5.2 *Other Nano Based Materials*

Materials which are active is defined as products and materials meant to improve the condition or extend the shelf life of wrapped consumable items. Their parts that allow substances to seep or leak into the food packaging or the surrounding environment are particularly designed into them. Furthermore, Regulation (EC) 1935/2004 Framework lays forth the specifications for goods and active materials. Oxidation and microbial resistant nanocomposites possess the ability in prolonging edible items' keeping time by preventing degradation caused by oxidation, microbes, and enzymes. The antimicrobial packaging uses of nanoparticles, such as Ag, Ti, Cu, Zn,

Zinc oxide, Magnesium oxide, and Titanium oxide, is studied extensively. Antioxidants can be added to food to prevent lipids and proteins from oxidizing, which can delay the emergence of bad tastes and improve food color stability. In addition to active packaging that uses oxygen scavengers, strategies to lessen food oxidation include limiting oxygen present using high-barrier packing materials and establishing an anaerobic environment. Furthermore, packaging techniques that release antioxidants and synthetic antioxidants like butylated hydroxytoluene (BHT), which is frequently used in food packaging, can both help reduce lipid oxidation [94].

Food packaging has a lot of potential for using nanotechnology, especially in the form of nanocomposites, which fits in with the increased focus on sustainability. When evenly dispersed, these nanocomposites augmented with nanofillers can improve the barrier qualities of polymers. The enhancement of barrier qualities in conventional and bio-based polymer composites is frequently attributed to nanoclays. The use of surface grafting and layer-by-layer assembly techniques guarantees even filler dispersion and gets over agglomeration problems. In addition to improving mechanical and thermal characteristics, silica nanoparticles in polymer matrices also increase gas barrier qualities. For particular applications, it is essential to optimize the physical–chemical characteristics of nanofillers, including surface chemistry, size, and form. Moreover, food items' shelf lives may be increased by nanocomposites with antioxidant and antibacterial qualities.

6.6 Packaging Authenticity

6.6.1 Efficiency on Resistance Against Diseases

In petro-based consumable packing materials discussed previously these are traditional thermosets, that are important due to their endurance, resistance to degradation, and performance/cost ratio. Also, these qualities provide a substantial environmental risk because plastics do not naturally decompose considerably. Moreover, the bulk of plastics have a relatively short life cycle and are typically only used once or for short periods of time. The phenomenon of environmental contamination from plastic garbage, known as "white pollution," has become increasingly problematic for the waste management process due to its increasing prevalence. Out of the 280 million tons of plastic waste produced in 192 nations and coastal regions, 8 million tons were thought to be entering the ocean in 2011. It is a significant drawback for package that is used later. It is recommended to use agricultural waste as the raw material for bio-film production. Technological developments have made it feasible to produce renewable films and non-polluting sources by reducing pollution. When different types of edible waste are used as an organic raw material for polymers, material films that are rapidly reintegrated into the environment are produced, and trash accumulation is reduced, both of which are good for the environment.

Another type of active packaging is compostable film, which is described as intentionally engaging with food that is being packed in order to improve it, especially in terms of its sensory and nutritional attributes. When applied in practice, active packaging stresses the addition of bioactive compounds and/or their immobilization in polymeric chains, allowing for interactions between the polymer, the food, and the environment. It also offers innovative methods for preserving or extending the useful life as well as improving and examining the security of food as well as reliability (Bhardwaj et al. 2019). The main characteristics of advanced packing are its capacity to absorb moisture, ethylene, oxygen, and smells; moreover, the presence of antioxidant, antibacterial, and aromatic substances defines it Barone et al. [11]. The importance of packaging's hygienic and sanitary qualities for food safety has received increased attention throughout time. The COVID-19 pandemic of 2019 has brought attention to how important packaging is for food safety, especially for the general population [11]. The inclination to use plastic packing is associated with an enhancement in processed food consumption and keeping time, as well as with the delivery of meals that call for the use of disposable packaging. Making investments to develop biodegradable and active packaging that can compete with traditional packaging.

6.6.2 Individual Protection

The preservation of the environment on a worldwide scale has gained popularity recently. Thus, encouraging sustainable growth and reducing the consequences of debris and environmental degradation requires use of bio-based packing. "Packing with green" signifies packing materials made of ecologically friendly materials that always prioritizes protecting the environment and public health. Green packaging may be classified into two main groups based on the viewpoint from which it is viewed: that of the company or that of the customer. Recall that customers and laws are the two main forces behind organizations' adoption of cutting-edge green packaging techniques. In reality, governments are pressuring businesses to promote sustainable packaging by enacting regulations, new laws, levies, and other steps to make packaging environmentally friendly and sustainable. This is in addition to consumers' increased concern for the environment. For example, in accordance with EU guidelines, member states have to implement legal measures for packaging waste management.

Though it poses to be novel, sustainable packaging recently attracted a lot of attention. In actuality, it is an important consideration that must be made in order to accomplish the SDGs, which have an effect on both the economy and society [88]. Numerous reviews have noted that the majority of studies on environmentally friendly packaging concentrate on the production method (technical issues) and the contents of the packaging (scientific aspects). Other research has concentrated on a few topics, such as consumer behavior, circular economy marketing, and the function of packaging in supply networks. As far as we are aware, no thorough literature review has been conducted on green packaging that takes into account several consumer and business factors.

6.6.3 Physio-Chemical Attributes

Polylactic acid is a polymer derived from its monomer lactic acid. Poly Lactic acid's monomer Lactic acid is obtained by fermenting sugarcane and other sources that are polymerized by condensation or by Lactide ring-opening polymerization. PLA is extremely biocompatible, fabricable, heat-moldable, and durable mechanically.

Researchers are very interested in polylactides because of their biocompatibility and biodegradability, which may find applications in biotech and medicine. The green polymer's ability to decompose into naturally existing metabolites through enzymatic or hydrolytic reactions is one of its attractive qualities. In recent decades, the degradation of polylactic acid materials for practical medical applications such as sutures has been studied, medication delivery systems, surgical implants. Polylactic acid based poly Anhydrides have been researched for implantation, with the possibility of substituting architecturally tailored stents for conventional metallic ones. Compared to conventional metallic or non-biodegradable polymers, biopolymers based on polylactides are simpler for the body to remove from the body and retain their form over time.

Polylactic acid was researched and found it out to be a possible replacement in alleviating the surrounding problems like segregation of thermoset trash, with a major focus on packaging. The ability of PLA to biodegrade suggests that it will ultimately change back into green, that offers an advantage in joining the eco-pathways. The use of PLA in food packaging has already garnered a lot of interest. At the moment, packaged goods containing PLA are found in stores across North America, Europe, and Japan. These items include yogurt, juice, and bottled water. PLA-based packaging materials are especially well-suited for usage in food applications since they are regarded as safe (GRAS).

Packaging as the main focus, Polylactic acidic substances is currently researched as a possible replacement for addressing the ecological problem of thermoset waste buildup. The fact that polylactic acid may get degraded by ecosystem suggests that it will eventually change back into biomass, which is a major advantage for joining the ecological pathway [2]. PLA' application in consumable packing attracted many positive investigations. At the moment, packaged goods containing polylactic acid are found in stores across European nations and Japan. These products include yogurt, juice, and bottled water. PLA-based packaging materials are especially well-suited for usage in applications since they are regarded as safe (GRAS) in food.

6.6.4 Influence in the Consumable Sector

Concerns concerning environmental contamination have arisen from Materials that are not biodegradable are being used in a range of packaging uses. Consumable packing is the market segment in the synthetic plastic packaging industry that is expanding at the quickest rate. The food sector alone accounts for more than two thirds of the materials required to produce packaging. Paper, glass, and plastics are among the many materials that are utilized extensively around the world. Due to changes in food preparation and consumption habits as well as the economic expansion of several areas and global markets, this amount has been rising steadily over time. The packaging industry uses the majority of plastics produced globally and is the main cause of waste plastics that are shockingly finding their way into the environment. This has to do with single-use plastics and the increase of prepared foods and snacks that need to be sometimes stored in durable plastic containers. As a result, there is a rising need for ecologically friendly, sustainable packaging materials with the right mechanical, barrier, and physical properties to be used in food packaging. According to a number of studies, PLA may eventually take the place of conventional materials with a wide range of packing. This research mostly concentrates on the function of polylactic acid-derived packing films and the implications of food-related packing waste due to anticipated worldwide development in the near future. The application of natural fiber composites and fillers filled with biopolymer is rapidly growing. This work describes, for the first time, the variety of PLA-derived composite substances that may be made by using Southern African plant waste. It also provides an overview of recent studies that demonstrate the biodegradability of virgin polylactic acid in oceanic settings.

Food packaging uses 50% of the substances derived from petroleum and petroleum products used in the packaging industry. These polymers made from petro-based products are persistent in the environment and decompose slowly. As they do so, they break down into tiny plastic particles that fish and other predators may readily consume, allowing them to contribute to bioaccumulation by entering the food chain. Large amounts of plastic waste are consistently generated by packaging, which has sparked environmental concerns and necessitated extensive research into renewable alternatives. The majority of thermosets utilized in modern packing come from petro by-products, also the amount produced annually is continually rising. Thus, it is vital to keep in mind that the greatest approach to lessen the quantity of plastic trash in the environment is to improve waste management, particularly in developing countries [47]. It is crucial to use environmentally friendly packaging, and there are three types of environmentally friendly packaging. Glass and other reusable packaging are easily cleaned and reused. Paper and other recyclable packaging can be processed and used again later. Packaging that decomposes naturally, like cotton bags, blends in with the surroundings without causing any damage.

6.7 Barriers in Green Material Utilization

Due to packing restrictions, a large amount of food is wasted annually around the world, with the United States accounting for about forty percent of this total. The current yearly output of plastics worldwide exceeds three twenty million tons. Plastics industry is shifting to alternative polymers including polymers of vinyl chloride, ethylene terephthalate, propylene, ethylene along with amide due to the rising costs of fossil fuels. Yet, using these materials that are not biodegradable results in emission of greenhouse gases in addition to medical issues. The cost of recycling is impacted by the use of conventional packaging materials, which makes recycling more difficult [16, 51]. The most sensible way to lessen the pollution non-biodegradable packaging materials contribute to the environment is to switch to green polymers. With the need of replaceable eco-friendly substances in switching to current polymers used for packing business is growing daily. Innovative packaging has garnered significant attention in recent years. This remains an essential factor in accomplishing the sustaining development. Goals, having ramifications for society and the economy. Numerous research studies have focused on certain areas, such the impact of packaging on the supply chain. The circularity of economy, techniques for advanced selling, in addition to the actions of consumers [71].

6.7.1 Financial Impact

European nations generated fifty-eight million tons of the three hundred and twenty-two million tons of plastic produced worldwide in 2015; forty percent of this plastic was used for packaging. Remarkably, the yearly plastic use was comprised of bio-based polymers [59]. The consumable packing sector is keen in advancement of eco-based polymers that can guarantee the edible matrix present inside is safe, wholesome, with its potential preservation in an effort to eliminate the use of non-biodegradable plastics and it was anticipated that the need for bio-based packaging materials would reach approximately nine and a half million tons [10].

The consumer market's current shift toward greener packaging and waste reduction has resulted in a considerable development in the applications of green materials in the active packaging business. Bio-based plastics are generally defined as organic macromolecules that come from biological components and are manufactured or processed by humans. The need for sustainability pushes companies to investigate circular and sustainable alternatives with the goal of preserving resources for coming generations, with an emphasis on biodegradable and bio-renewable materials. Kernel of mango starch based natural pouches poses to be a good substitute for retort thermoset pouches used to package red chili powder, are produced using a casting process. Compared to being stored in polyethylene pouches, the chili powder's color and pungency were more preserved in the bio-based mango kernel starch pouches [46].

6.7.2 Accessibility

The use of biopolymers for single films in modern packaging is expected to expand in Europe, according to market research data, from one thousand seven hundred and forty-three million in 2016 to two thousand four hundred and twenty thousand million in 2021. Specialists and food makers are also focusing on bio-based substitutes because of its many advantages, including expandability, non-toxic nature, and biodegradation ability. It may be argued that there are bio based substitutes for almost all fossil fuel-based products used in the packaging sector. However, the cost-effectiveness, renewability, and large-scale availability of these green materials goods present considerable obstacles. A renewable supply for the production of biobased and sustainable packaging materials is offered by various forms of bio-waste. The generation of feed stock from agricultural waste may be improved by optimizing biorefinery processes to meet industrial needs, such as lowering energy and labor intensiveness. Furthermore, it is becoming more widely acknowledged that the type of substrate and coating solution selected have an impact on the overall application, accessibility, and commercial viability of bio-based materials in food packaging. A number of variables, such as improvements in packaging materials with improved qualities, more availability, and lower costs, are driving market expansion. Further factors driving market development include rising consumer interactions in ecological problems along with the introduction of upcoming regulations. One general packaging trend that is supported by the use of barrier materials is the reduction of packing weight (thickness) in order to lessen the impact on the environment. This tactic promotes resource efficiency and makes it easier to save energy while traveling.

6.7.3 Acceptance of Green Materials Between Consumers and Industries

Thermoset bags are a symbol of contemporary society; however, restrictions are imposed by different public organizations in the form of bans to lessen the usage of thermoset bags [49]. Ten million multipurpose bags were utilized in the Union of European Nations (EU) in 2010. However, years later, single-use bags based on biodegradable choices were created and have become increasingly popular. While ten billion multipurpose bags were utilized annually in the Union of European Nations in 2010, shopping bags have changed in the last few years as most EU nations have outlawed single-use, non-biodegradable bags made of petrochemicals. Therefore, a more thorough study into bio-based materials that affect ecologically acceptable and sustainable alternatives is required. To completely or substantially replace plastic packaging materials derived from petroleum, industries are concentrating their efforts on creating a variety of green-based or green-degradable replacements utilizing green resources. It is known that the qualitative characteristics of food items, such as flavor, scent, shelf life, and market attractiveness, are significantly impacted by the use of barriers in food packaging made of green or bio-based materials. Resistance against the penetration of several elements, such as water, bacteria, aromatic compounds, carbon dioxide, and oils, is crucial when using environmentally friendly packaging materials [64] and Table 6.1 acknowledged the application of non-synthetic green resources. On the other hand, packaging made of plastic shows a wide range of mass transfer properties as well as poor to outstanding barrier qualities [12]. Although biopolymers are environmentally beneficial, their industrial uses are restricted because of a number of problems, including cost of manufacture, thermal resistance, water and oxygen vapour barriers, and thermal resistance. Packaging typically uses synthetic or non-biodegradable polymers because of the aforementioned factors. In order to surmount these constraints and facilitate the industrial utilization of biopolymers as packaging materials, more investigation is imperative to optimize their nutritional qualities, shelf-life, and resilience against microbiological deterioration. Manufacturers rank packaging choices according to consumer preferences and prices when it comes to environmentally friendly packaging [14]. According to some studies, there is a sizable market for environmentally friendly packaging made of biodegradable polymers, but there are obstacles to overcome, including high production prices, complicated processing steps, and performance problems [68]. Moreover, merchants and packaged products firms support cost-cutting measures for environmentally friendly packaging and energy costs [3]. Several studies enhanced the resistance characteristics of wrapping materials by adding green materials. These biopolymers have found application in mixed films, bio-nano composites, and coatings, among other forms [40]. Even though green polymers have many benefits, there are still several mechanical and chemical characteristics that need to be improved. These characteristics are considered major barriers to the commercial application of these polymeric materials. Those features allow green polymers to be used in various uses, including consumables, coloring agents, cosmetics, medicine of bio-origin, and engineering of tissues, in addition to medicines. Bioplastics still confront difficulties with important physicochemical and biological qualities, which prevents their broad use in the industry despite the advances in green materials packaging. However, as bioplastic manufacturing and application technologies evolve quickly, new avenues for academic and industrial collaboration to create creative and sustainable packaging systems are opened up.

6.8 Conclusion

The shift from conventional synthetic polymers to green polymers represents a significant advancement in packaging materials. This transition addresses the environmental concerns posed by synthetic polymers, such as their non-biodegradability and reliance on fossil fuels. Green polymers, derived from natural sources like plants, animals, and microorganisms, offer a sustainable alternative with minimal environmental impact. They include various materials, such as starch, lignin, cellulose, keratin, collagen, gelatin, polyhydroxyalkanoates, and exopolysaccharides.

Table 6.1 Biological based, non-synthetic materials for packaging

Bio-polymers	Types	Methodology of synthesis	Current studies of these materials in packaging domain	References
Polypeptides	Polypeptides	Nature based	Enzyme-based and physiological and chemical changes of polypeptide coatings and films	[13, 18, 66]
Wax and lipid	Fats	Nature based	Several layers of edible materials	[32]
Polyhydroxyalkanoates	Polyesters that are aliphatic	Fermentation by microbes	Manufacturing expenses	[58]
The cellulose fiber	Complex polysaccharides	Nature based	Film based on composites of cellulose crystals of nano-size and nano based fiber	[58]
Starch polymers	Simple polysaccharides	Green	Starches in film-based composites	[37, 58]

There are numerous advantages of green polymers in food packaging. They are non-toxic, economical, and possess desirable properties such as high crystallinity, thermal stability, and mechanical strength. By incorporating bio-polymeric polyphenolic compounds and nanomaterials, green polymers can enhance packaging films' antioxidant and antimicrobial properties, thereby improving food preservation and safety. Despite the promising potential, there is a need for further research to optimize the application of biodegradable polymers in food packaging. This includes understanding their physicochemical properties, interaction with food materials, and their impact on food quality. Developing composite and blended bio-polymers can help overcome the limitations of individual materials and achieve the desired attributes for packaging. Implementing green polymers aligns with green chemistry and technology principles, aiming to reduce pollution and support environmental sustainability. By replacing traditional petro-based thermosets with biodegradable alternatives, the packaging industry can contribute to a healthier ecosystem and promote individual wellness. As research and innovation continue, green polymers promise to revolutionize the packaging industry, providing eco-friendly solutions that meet the demands of modern consumers and the environment.

References

1. Aaliya, B., Sunooj, K. V., Ashina, S., Priyanka, K., Navaf, M., Akhila, P. P., Raviteja, P., Sinha, S. K., Yadav, V., Mir, S. A., Lackner, M., George, J., & Nemţanu, M. R.: Synthesis, characterization, and in-vivo application of non-thermal plasma-modified talipot starch-Aloe vera gel biocomposite films for enhancing the shelf life of guava fruits. Food Packag. Shelf Life, **40**, 101195 (2023). https://doi.org/10.1016/J.FPSL.2023.101195
2. Ahmed, J., Varshney, S.K.: Polylactides—chemistry, properties and green packaging technology: a review. Int. J. Food Prop. **14**(1), 37–58 (2011). https://doi.org/10.1080/109429109 03125284
3. Ahmed, J., Varshney, S.K.: Polylactides—chemistry, properties and green packaging technology: a review. Int. J. Food Prop. **14**, 37–58 (2011)
4. Aloui, H., Baraket, K., Sendon, R., Silva, A.S., & Khwaldia, K.: Development and characterization of novel composite glycerol-plasticized films based on sodium caseinate and lipid fraction of tomato pomace by-product. Int. J. Biol. Macromol. **139**, 128–138 (2019). https://doi.org/10.1016/J.IJBIOMAC.2019.07.156
5. Amin, U., Khan, M.U., Majeed, Y., Rebezov, M., Khayrullin, M., Bobkova, E., Shariati, M. A., Chung, I. M., & Thiruvengadam, M.: Potentials of polysaccharides, lipids and proteins in biodegradable food packaging applications. International J. Biol. Macromol. **183**, 2184–2198 (2021). https://doi.org/10.1016/J.IJBIOMAC.2021.05.182
6. Asgher, M., Qamar, S.A., Bilal, M., Iqbal, H.M.N.: Bio-based active food packaging materials: sustainable alternative to conventional petrochemical-based packaging materials. Food Res. Int. **137**, 109625 (2020). https://doi.org/10.1016/j.foodres.2020.109625
7. Asgher, M., Urooj, Y., Qamar, S.A., Khalid, N.: Improved exopolysaccharide production from Bacillus licheniformis MS3: optimization and structural/functional characterization. Int. J. Biol. Macromol. **151**, 984–992 (2020). https://doi.org/10.1016/j.ijbiomac.2019.11.094
8. Azahari, N.A., Othman, N., & Ismail, H.: Biodegradation studies of polyvinyl alcohol/corn starch blend films in solid and solution media. J. Phys. Sci. **22**(2), 15–31 (2011)
9. Babu, R.P., O'connor, K., & Seeram, R.: Current progress on bio-based polymers and their future trends. Prog. Biomater. **2**(8) (2013). https://doi.org/10.1186/2194-0517-2-8
10. Bajpai, P.: Biobased polymers. In: Properties and Applications in Packaging. Elsevier, Amsterdam, The Netherlands (2019). ISBN 0128184043
11. Barone, A.S., Matheus, J.R.V., de Souza, T.S.P., Moreira, R.F.A., Fai, A.E.C.: Green-based active packaging: opportunities beyond COVID-19, food applications, and perspectives in circular economy—a brief review. Compr. Rev. Food Sci. Food Saf. **20**(5), 4881–4905 (2021). https://doi.org/10.1111/1541-4337.12812
12. Beitzen-Heineke, E.F., Balta-Ozkan, N., Reefke, H.: The prospects of zero-packaging grocery stores to improve the social and environmental impacts of the food supply chain. J. Clean. Prod. **140**, 1528–1541 (2017)
13. Bugnicourt, E., Schmid, M., Nerney, O.M., Wildner, J., Smykala, L., Lazzeri, C., Cinelli, P.: Processing and validation of whey-protein-coated films and laminates at semi-industrial scale as novel recyclable food packaging materials with excellent barrier properties. Adv. Mater. Sci. Eng. **2013**, 1–10 (2013)
14. Burek, J., Kim, D., Nutter, D., Selke, S., Auras, R., Cashman, S., Sauer, B., Thoma, G.: Environmental sustainability of fluid milk delivery systems in the United States. J. Ind. Ecol. **22**, 180–195 (2018)
15. Cha, D.S., & Chinnan, M.S.: Biopolymer-based antimicrobial packaging: a review. Crit. Rev. Food Sci. Nutr. **44**(4), 223–237 (2004). https://doi.org/10.1080/10408690490464276
16. Chausali, N., Saxena, J., Prasad, R.: Recent trends in nanotechnology applications of bio-based packaging. J. Agric. Food Res. **7**, 100257 (2022)
17. Ciambelli, P., La Guardia, G., Vitale, L.: Nanotechnology for green materials and processes. In: Studies in Surface Science and Catalysis, vol. 179, pp. 97–116. Elsevier Inc. (2019). https://doi.org/10.1016/B978-0-444-64337-7.00007-0

18. Cinelli, P., Schmid, M., Bugnicourt, E., Wildner, J., Bazzichi, A., Anguillesi, I., Lazzeri, C.: Whey protein layer applied on biodegradable packaging film to improve barrier properties while maintaining biodegradability. Polym. Degrad. Stab. **108**, 151–157 (2014)
19. Costa, M.J., Maciel, L.C., Teixeira, J.A., Vicente, A.A., & Cerqueira, M.A.: Use of edible films and coatings in cheese preservation: opportunities and challenges. Food Res. Int. **107**, 84–92 (2018). https://doi.org/10.1016/J.FOODRES.2018.02.013
20. Cruz-Romero, M., & Kerry, J.P.: Crop-based biodegradable packaging and its environmental implications. CAB reviews: perspectives in agriculture, veterinary science, nutrition and natural resources, **3**, (2008). https://doi.org/10.1079/PAVSNNR20083074
21. Cutter, C.N.: Opportunities for bio-based packaging technologies to improve the quality and safety of fresh and further processed muscle foods. Meat Sci. **74**(1), 131–142 (2006). https://doi.org/10.1016/J.MEATSCI.2006.04.023
22. Deng, W., Liu, K., Cao, S., Sun, J., Zhong, B., Chun, J.: Chemical composition, antimicrobial, antioxidant, and antiproliferative properties of grapefruit essential oil prepared by molecular distillation. Molecules **25**(1), 217 (2020). https://doi.org/10.3390/molecules25010217
23. Duan, J., Reddy, K.O., Ashok, B., Cai, J., Zhang, L., & Rajulu, A.V.: Effects of spent tea leaf powder on the properties and functions of cellulose green composite films. J. Environ. Chem. Eng. **4**(1), 440–448 (2016). https://doi.org/10.1016/J.JECE.2015.11.029
24. Duman, F., Kaya, M.: Crayfish chitosan for microencapsulation of coriander (Coriandrum sativum L.) essential oil. Int. J. Biol. Macromol. **92**, 125–133 (2016). https://doi.org/10.1016/j.ijbiomac.2016.06.068
25. Duncan, T.V.: Applications of nanotechnology in food packaging and food safety: barrier materials, antimicrobials and sensors. J. Colloid Interface Sci. **363**(1), 1–24 (2011). https://doi.org/10.1016/j.jcis.2011.07.017
26. Ferreira, A.R.V., Alves, V.D., & Coelhoso, I.M.: Polysaccharide-based membranes in food packaging applications. Membr. **6**(2), 22 (2016). https://doi.org/10.3390/MEMBRANES6020022
27. Freitas, F., Alves, V.D., Reis, M.A., Crespo, J.G., & Coelhoso, I. M.: Microbial polysaccharide-based membranes: current and future applications. J. Appl. Polym. Sci. **131**(6), 40047 (2014). https://doi.org/10.1002/APP.40047
28. Ganesh, K.S., Sridhar, A., Vishali, S.: Utilization of fruit and vegetable waste to produce value-added products: conventional utilization and emerging opportunities-a review. Chemosphere, **287**, (2022). https://doi.org/10.1016/j.chemosphere.2021.132221
29. Goslinska, M., & Heinrich, S.: Characterization of waxes as possible coating material for organic aerogels. Powder Technol. **357**, 223–231 (2019). https://doi.org/10.1016/J.POWTEC.2019.08.096
30. Groh, K.J., Backhaus, T., Carney-Almroth, B., Geueke, B., Inostroza, P.A., Lennquist, A., Leslie, H.A., Maffini, M., Slunge, D., Trasande, L., Warhurst, A.M., Muncke, J.: Overview of known plastic packaging-associated chemicals and their hazards. Sci. Total. Environ. **651**, 3253–3268 (2019). https://doi.org/10.1016/j.scitotenv.2018.10.015
31. Guillard, V., Gaucel, S., Fornaciari, C., Angellier-Coussy, H., Buche, P., Gontard, N.: The next generation of sustainable food packaging to preserve our environment in a circular economy context. Front. Nutr. **5**, (2018). https://doi.org/10.3389/fnut.2018.00121
32. Hanstveit, A.O.: Biodegradability of petroleum waxes and beeswax in an adapted CO_2 evolution test. Chemosphere **25**, 605–620 (1992)
33. Hemavathi, A.B., Siddaramaiah, H.: Food packaging: polymers as packaging materials in food supply chains. In: Encyclopedia of Polymer Applications, pp. 1374–1397 (2023)
34. Huang, J., Hu, Z., Hu, L., Li, G., Yao, Q., & Hu, Y.: Pectin-based active packaging: a critical review on preparation, physical properties and novel application in food preservation. Trends Food Sci. Technol. **118**, 167–178 (2021). https://doi.org/10.1016/J.TIFS.2021.09.026
35. Jacquelot, E., Espuche, E., Gérard, J.F., Duchet, J., Mazabraud, P.: Morphology and gas barrier properties of polyethylene-based nanocomposites. J. Polym. Sci. Part B Polym. Phys. **44**(2), 431–440 (2006). https://doi.org/10.1002/polb.20707

36. Jatav, J., Bhattacharya, B.: Pineapple-peel waste to pineapple-peel based active food packaging film: paving the way for a sustainable environment. Biomass Convers. Biorefinery (2022). https://doi.org/10.1007/s13399-022-03104-w
37. Johansson, C., Bras, J., Mondragon, I., Nechita, P., Plackett, D., Simon, P., Svetec, D.G., Virtanen, S., Baschetti, M.G., Breen, C., et al.: Renewable fibers and bio-based materials for packaging applications-a review of recent developments. BioResources **7**, 2506–2552 (2012)
38. Kalendova, A., Smotek, J., Stloukal, P., Kracalik, M., Slouf, M., Laske, S.: Transport properties of poly(lactic acid)/clay nanocomposites. Polym. Eng. Sci. **59**(12), 2498–2501 (2019). https://doi.org/10.1002/pen.25251
39. Kannadasan, Dr., M., Bichala, P.K., Agrawal, A., Singh, S.: A review: nano particle drug delivery system. Int. J. Pharm. Sci. Med. **5**(12), 46–58 (2020). https://doi.org/10.47760/ijpsm.2020.v05i12.008
40. LakshmiBalasubramaniam, S., Howell, C., Tajvidi, M., Skonberg, D.: Characterization of novel cellulose nanofibril and phenolic acid-based active and hydrophobic packaging films. Food Chem. **374**, 131773 (2022)
41. López-Rubio, A., Almenar, E., Hernandez-Muñoz, P., Lagarón, J.M., Catalá, R., & Gavara, R.: Overview of active polymer-based packaging technologies for food applications. Food Rev. Int. **20**(4), 357–387 (2004). https://doi.org/10.1081/FRI-200033462
42. Maraveas, C.: Production of sustainable and biodegradable polymers from agricultural waste. Polymers **12**(5), (2020). https://doi.org/10.3390/POLYM12051127. (MDPI AG)
43. Mishra, R.K., Ha, S.K., Verma, K., Tiwari, S.K.: Recent progress in selected bio-nanomaterials and their engineering applications: an overview. J. Sci. Adv. Mater. Devices **3**(3), 263–288 (2018). https://doi.org/10.1016/j.jsamd.2018.05.003
44. Moeini, A., Germann, N., Malinconico, M., Santagata, G.: Formulation of secondary compounds as additives of biopolymer-based food packaging: a review. Trends Food Sci. Technol. **114**, 342–354 (2021). https://doi.org/10.1016/j.tifs.2021.05.040
45. Müller, P., Schmid, M.: Intelligent packaging in the food sector: a brief overview. Foods **8**(1), 16 (2019). https://doi.org/10.3390/foods8010016
46. Nawab, A., Alam, F., Haq, M.A., Haider, M.S., Lutfi, Z., Kamaluddin, S., Hasnain, A.: Innovative edible packaging from mango kernel starch for the shelf life extension of red chili powder. Int. J. Boil. Macromol. **114**, 626–631 (2018)
47. Ncube, L.K., Ude, A.U., Ogunmuyiwa, E.N., Zulkifli, R., Beas, I.N.: Environmental impact of food packaging materials: a review of contemporary development from conventional plastics to polylactic acid based materials. Materials **13**(21), 4994 (2020). https://doi.org/10.3390/ma13214994
48. Nielsen, L.E.: Models for the permeability of filled polymer systems. J. Macromol. Sci. Part A-Chem. **1**(5), 929–942 (1967). https://doi.org/10.1080/10601326708053745
49. Nielsen, T.D., Holmberg, K., Stripple, J.: Need a bag? A review of public policies on plastic carrier bags—where, how and to what effect? Waste Manag. **87**, 428–440 (2019)
50. Nielsen, T.D., Hasselbalch, J., Holmberg, K., Stripple, J.: Politics and the plastic crisis: a review throughout the plastic life cycle. WIREs Energy Environ. **9**(1), (2020). https://doi.org/10.1002/wene.360
51. Nilsen-Nygaard, J., Fernández, E.N., Radusin, T., Rotabakk, B.T., Sarfraz, J., Sharmin, N., et al.: Current status of biobased and biodegradable food packaging materials: impact on food quality and effect of innovative processing technologies. Compr. Rev. Food Sci. Food Saf. **20**, 1333–1380 (2021)
52. Omerović, N., Djisalov, M., Živojević, K., Mladenović, M., Vunduk, J., Milenković, I., Knežević, N.Ž, Gadjanski, I., Vidić, J.: Antimicrobial nanoparticles and biodegradable polymer composites for active food packaging applications. Compr. Rev. Food Sci. Food Saf. **20**(3), 2428–2454 (2021). https://doi.org/10.1111/1541-4337.12727
53. Özeren, H.D., Olsson, R.T., Nilsson, F., & Hedenqvist, M.S.: Prediction of plasticization in a real biopolymer system (starch) using molecular dynamics simulations. Mater. Des. **187**, 108387 (2020)

54. Paletta, A., Leal Filho, W., Balogun, A.-L., Foschi, E., Bonoli, A.: Barriers and challenges to plastics valorisation in the context of a circular economy: case studies from Italy. J. Clean. Prod. **241**, 118149 (2019). https://doi.org/10.1016/j.jclepro.2019.118149
55. Pawar, S.N., & Edgar, K.J.: Alginate derivatization: A review of chemistry, properties and applications. Biomate. **33**(11), 3279–3305 (2012). https://doi.org/10.1016/J.BIOMATERIALS.2012.01.007
56. Picchio, M.L., Linck, Y.G., Monti, G.A., Gugliotta, L.M., Minari, R.J., & Alvarez Igarzabal, C.I.: Casein films crosslinked by tannic acid for food packaging applications. Food Hydrocoll. **84**, 424–434 (2018). https://doi.org/10.1016/J.FOODHYD.2018.06.028
57. Plotto, A., Narciso, J.A., Rattanapanoneb, N., & Baldwina, E.A.: Surface treatments and coatings to maintain fresh-cut mango quality in storage. J. Sci. Food Agric. **90**(13), 2333–2341 (2010). https://doi.org/10.1002/JSFA.4095
58. Rameshkumar, S., Shaiju, P., O'Connor, K.E., Padamati, R.B.: Bio-based and biodegradable polymers—state-of-the-art, challenges, and emerging trends. Curr. Opin. Green Sustain. Chem. **21**, 75–81 (2020)
59. Reichert, C.L., Bugnicourt, E., Coltelli, M.B., Cinelli, P., Lazzeri, A., Canesi, I., Schmid, M.: Bio-based packaging: materials, modifications, industrial applications and sustainability. Polymers **12**(7), 1558 (2020)
60. Rhim, J.W., Park, H.M., & Ha, C.S.: Bio-nanocomposites for food packaging applications. Prog. Polym. Sci. **38**(10–11), 1629–1652 (2013). https://doi.org/10.1016/J.PROGPOLYMSCI.2013.05.008
61. Rincon, A.M.: Chapter 6-presence of nanomaterials on consumer products: food, cosmetics, and drugs. In: Exposure to Engineered Nanomaterials in the Environment, pp. 165–181. Elsevier (2019). https://doi.org/10.1016/B978-0-12-814835-8.00006-6
62. Rodsamran, P., Sothornvit, R.: Preparation and characterization of pectin fraction from pineapple peel as a natural plasticizer and material for biopolymer film. Food Bioprod. Process. **118**, 198–206 (2019). https://doi.org/10.1016/j.fbp.2019.09.010
63. Sahraee, S., Milani, J.M., Regenstein, J.M., & Kafil, H.S.: Protection of foods against oxidative deterioration using edible films and coatings: A review. Food Biosci. **32**, 100451 (2019). https://doi.org/10.1016/J.FBIO.2019.100451
64. Sangroniz, A., Zhu, J.-B., Tang, X., Etxeberria, A., Chen, E.Y.-X., Sardon, H.: Packaging materials with desired mechanical and barrier properties and full chemical recyclability. Nat. Commun. **10**, 3559 (2019). https://doi.org/10.1038/s41467-019-11525-x
65. Sarfraz, J., Gulin-Sarfraz, T., Nilsen-Nygaard, J., Pettersen, M.K.: Nanocomposites for food packaging applications: an overview. Nanomaterials **11**(1), 1–27 (2021). https://doi.org/10.3390/nano11010010. (MDPI AG)
66. Schmid, M., Dallmann, K., Bugnicourt, E., Cordoni, D., Wild, F., Lazzeri, C., Noller, K.: Properties of whey-protein-coated films and laminates as novel recyclable food packaging materials with excellent barrier properties. Int. J. Polym. Sci. 1–7 (2012)
67. Sdrolia, E., Zarotiadis, G.: A comprehensive review for green product term: from definition to evaluation. J. Econ. Surv. **33**(1), 150–178 (2019). https://doi.org/10.1111/joes.12268
68. Sharma, C., Bhardwaj, N.K., Pathak, P.: Static intermittent fed-batch production of bacterial nanocellulose from black tea and its modification using chitosan to develop antibacterial green packaging material. J. Clean. Prod. **279**, 123608 (2021)
69. Shen, M., Huang, W., Chen, M., Song, B., Zeng, G., Zhang, Y.: (Micro) plastic crisis: unignorable contribution to global greenhouse gas emissions and climate change. J. Clean. Prod. **254**, 120138 (2020)
70. Siegrist, M., Cousin, M.E., Kastenholz, H., Wiek, A.: Public acceptance of nanotechnology foods and food packaging: the influence of affect and trust. Appetite **49**(2), 459–466 (2007). https://doi.org/10.1016/j.appet.2007.03.002
71. Singh, R., Sharma, R., Shaqib, M., Sarkar, A., & Chauhan, K.D.: Biodegradable polymers as packaging materials. In Biopolymers and their industrial applications (pp. 245–259) (2021). Elsevier. https://doi.org/10.1016/B978-0-12-819240-5.00010-9

72. Siripatrawan, U., & Noipha, S.: Active film from chitosan incorporating green tea extract for shelf life extension of pork sausages. Food Hydrocoll. **27**(1), 102–108 (2012). https://doi.org/10.1016/J.FOODHYD.2011.08.011
73. Sjaifullah, A., & Santoso, A.B.: Autolytic isolation of chitin from white shrimp (Penaues Vannamei) waste. Procedia Chem. **18**, 49–52 (2016). https://doi.org/10.1016/J.PROCHE.2016.01.009
74. Souza Filho, P.F., Zamani, A., Ferreira, J.A.: Valorization of wheat byproducts for the co-production of packaging material and enzymes. Energies **13**(6), (2020). https://doi.org/10.3390/en13061300
75. Srinivasa, P.C., & Tharanathan, R.N.: Chitin/Chitosan — Safe, ecofriendly packaging materials with multiple potential uses. Food Rev. Int. **23**(1), 53–72 (2007). https://doi.org/10.1080/87559120600998163
76. Stoica, M., Marian Antohi, V., Laura Zlati, M., Stoica, D.: The financial impact of replacing plastic packaging by biodegradable biopolymers-a smart solution for the food industry. J. Clean. Prod. **277**, 124013 (2020). https://doi.org/10.1016/j.jclepro.2020.124013
77. Su, J.F., Huang, Z., Yuan, X.Y., Wang, X.Y., & Li, M.: Structure and properties of carboxymethyl cellulose/soy protein isolate blend edible films crosslinked by maillard reactions. Carbohyd. Polym. **79**(1), 145–153 (2010). https://doi.org/10.1016/J.CARBPOL.2009.07.035
78. Suderman, N., Isa, M. I.N., & Sarbon, N.M.: The effect of plasticizers on the functional properties of biodegradable gelatin-based film: A review. Food Biosci. **24**, 111–119 (2018). https://doi.org/10.1016/J.FBIO.2018.06.006
79. Sudheesh, C., Valiyapeediyekkal Sunooj, K., Aaliya, B., Navaf, M., Parambil Akhila, P., Ahmad Mir, S., Sabu, S., Sasidharan, A., Pulissery Sudheer, K., Kumar Sinha, S., Appukuttan Sajeevkumar, V., & George, J.: Effect of energetic neutrals on the kithul starch retrogradation; potential utilization for improving mechanical and barrier properties of films. Food Chem. **398**, 133881 (2023). https://doi.org/10.1016/J.FOODCHEM.2022.133881
80. Sun, L., Sun, J., Chen, L., Niu, P., Yang, X., & Guo, Y.: Preparation and characterization of chitosan film incorporated with thinned young apple polyphenols as an active packaging material. Carbohydr. Polym. **163**, 81–91 (2017). https://doi.org/10.1016/J.CARBPOL.2017.01.016
81. Tahir, H.E., Xiaobo, Z., Mahunu, G.K., Arslan, M., Abdalhai, M., & Zhihua, L.: Recent developments in gum edible coating applications for fruits and vegetables preservation: A review. Carbohydr. Polym. **224**, 115141 (2019). https://doi.org/10.1016/J.CARBPOL.2019.115141
82. Tesfaye, T., Ayele, M., Ferede, E., Gibril, M., Kong, F., Sithole, B.: A techno-economic feasibility of a process for extraction of starch from waste avocado seeds. Clean Technol. Environ. Policy **23**(2), 581–595 (2021). https://doi.org/10.1007/s10098-020-01981-1
83. Testa, F., Iovino, R., Iraldo, F.: The circular economy and consumer behaviour: the mediating role of information seeking in buying circular packaging. Bus. Strategy Environ. **29**, 3435–3448 (2020)
84. Thielen, M.: Bioplastics: Basics, Applications. Markets; Polymedia Publisher, Mönchengladbach, Germany (2012)
85. Tulamandi, S., Rangarajan, V., Rizvi, S.S.H., Singhal, R.S., Chattopadhyay, S.K., & Saha, N.C.: A biodegradable and edible packaging film based on papaya puree, gelatin, and defatted soy protein. Food Packag. Shelf Life, **10**, 60–71 (2016). https://doi.org/10.1016/J.FPSL.2016.10.007
86. Vernès, L., Abert-Vian, M., El Maâtaoui, M., Tao, Y., Bornard, I., Chemat, F.: Application of ultrasound for green extraction of proteins from spirulina. Mechanism, optimization, modeling, and industrial prospects. Ultrason. Sonochem. **54**, 48–60 (2019). https://doi.org/10.1016/j.ultsonch.2019.02.016
87. Vila-Lopez, N., Küster-Boluda, I.: A bibliometric analysis on packaging research: towards sustainable and healthy packages. Br. Food J. (2021)
88. Wandosell, G., Parra-Meroño, M.C., Alcayde, A., Baños, R.: Green packaging from consumer and business perspectives. Sustainability **13**(3), 1356 (2021). https://doi.org/10.3390/su13031356

89. Win, N.K.K., Jitareerat, P., Kanlayanarat, S., & Sangchote, S.: Effects of cinnamon extract, chitosan coating, hot water treatment and their combinations on crown rot disease and quality of banana fruit. Postharvest Biol. Technol. **45**(3), 333–340 (2007). https://doi.org/10.1016/J.POSTHARVBIO.2007.01.020
90. Wu, C.L., Zhang, M.Q., Rong, M.Z., Friedrich, K.: Tensile performance improvement of low nanoparticles filled-polypropylene composites. www.elsevier.com/locate/compscitech (n.d.)
91. Xing, Y., Li, X., Xu, Q., Yun, J., Lu, Y., & Tang, Y.: Effects of chitosan coating enriched with cinnamon oil on qualitative properties of sweet pepper (Capsicum annuum L.). Food Chem. **124**(4), 1443–1450 (2011). https://doi.org/10.1016/J.FOODCHEM.2010.07.105
92. Yadav, A., Mangaraj, S., Singh, R., Das, K., Kumar, N., & Arora, S.: Biopolymers as packaging material in food and allied industry. Int. J. Chem. Stud. **6**(2), 2411–2418 (2018)
93. Yam, K.L. (Ed.).: The Wiley encyclopedia of packaging technology. John Wiley & Sons. Bhardwaj, A., Alam, T., & Talwar, N. (2019). Recent advances in active packaging of agri-food products: a review . J.Postharvest Technol. **7**(1), 33–62 (2010). Retrieved from: https://acspublisher.com/journals/index.php/jpht/article/view/15377
94. Yildirim, S., Röcker, B., Pettersen, M.K., Nilsen-Nygaard, J., Ayhan, Z., Rutkaite, R., Radusin, T., Suminska, P., Marcos, B., Coma, V.: Active packaging applications for food. In Compr. Rev. Food Sci. Food Saf. **17**(1), 165–199 (2018). https://doi.org/10.1111/1541-4337.12322. (Blackwell Publishing Inc.)
95. Yong, H., & Liu, J.: Active packaging films and edible coatings based on polyphenol-rich propolis extract: a review. Compr. Rev. Food Sci. Food Safety, **20**(2), 2106–2145 (2021). https://doi.org/10.1111/1541-4337.12697
96. Youssef, A.M., El-Sayed, S.M.: Bionanocomposites materials for food packaging applications: concepts and future outlook. Carbohydr. Polym. **193**, 19–27 (2018). https://doi.org/10.1016/j.carbpol.2018.03.088. (Elsevier Ltd.)
97. Yuan, L., Wu, Y., Qin, Y., Yong, H., & Liu, J.: Recent advances in the preparation, characterization and applications of locust bean gum-based films. J. Renew. Mater. **8**(12), 1565–1579 (2020). https://doi.org/10.32604/JRM.2020.014562
98. Yuvaraj, D., Iyyappan, J., Gnanasekaran, R., Ishwarya, G., Harshini, R.P., Dhithya, V., Chandran, M., Kanishka, V., & Gomathi, K.: Advances in bio food packaging - An overview. Heliyon, e07998. (2017). https://doi.org/10.1016/j.heliyon.2021.e07998
99. Zhang, L., Liu, Z., Wang, X., Dong, S., Sun, Y., & Zhao, Z.: The properties of chitosan/zein blend film and effect of film on quality of mushroom (Agaricus bisporus). Postharvest Biol. Technol. **155**, 47–56 (2019). https://doi.org/10.1016/J.POSTHARVBIO.2019.05.013

Chapter 7
Advances in Polymer Nanocomposites for Active Food Packing

Vishaka Chauhan, Chinmay, and Anjaneyulu Bendi

Abstract Active food packaging has emerged as a potential technology in recent years, intending to increase the shelf life of food goods while maintaining quality and safety. Polymer nanocomposites have garnered significant attention among the many materials being researched for active packaging due to their versatile and adjustable properties. This book chapter thoroughly overviews the latest developments in polymer nanocomposites for active food packaging applications. It goes into the principles of active packaging, clarifies the function of polymer nanocomposites, and investigates how they might be used to solve problems with food safety and preservation. This chapter looks at the different kinds of nanoparticles mixed into polymer matrices, how they work, and how they affect the characteristics of food packaging. The chapter also covers new sensor developments that use nanocomposite technology and antimicrobial packaging options. Future perspectives and issues in the field are highlighted, and the debate offers insights into possible research routes and chances for innovation in active food packaging technology.

Keywords Active food packaging · Polymer nanocomposites · Nanoparticles · Food preservation · Anti-microbial packaging · Sensors

7.1 Introduction

The food industry continuously evolves, driven by consumer demands for convenience, safety, and sustainability. Packaging is essential in preserving food goods' safety and quality over their entire shelf life. Glass, metal, and plastic are traditional packaging materials that offer passive protection but frequently fall short when

V. Chauhan · Chinmay
Department of Chemistry, Faculty of Applied and Basic Sciences, Shree Guru Gobind Singh Tricentenary University, Gurugram 122505, Haryana, India

A. Bendi (✉)
Innovation and Translational Research Hub (iTRH) & Department of Chemistry, Presidency University, Itgalpura, Rajanukunte, Bangalore 560064, Karnataka, India
e-mail: anjaneyulu.bendi@gmail.com

S. ul Islam and M. Shahid (eds.), *Green Materials for Active Food Packaging*, Engineering Materials, https://doi.org/10.1007/978-981-96-0369-5_7

tackling new issues, including infection, food spoiling, and environmental effects. One potential way to address these issues is through active food packaging, which combines functional materials to actively interact with the food being packaged or its surroundings.

Polymer nanocomposites' distinctive qualities and versatility in active food packaging have garnered much interest. Researchers may modify packaging materials' mechanical, barrier, and functional properties by including nanoscale fillers in polymer matrices to meet specific requirements. Due to their high surface area-to-volume ratio and enhanced interactions with food components, nanoparticles are ideal for active packaging applications. This review covers the principles of active packaging and how nanocomposites might improve its functionality [1–4].

7.2 Fundamentals of Active Food Packaging

Active food packaging refers to packaging solutions that actively interact with the packaged food or its surroundings to enhance sensory elements, extend shelf life, or improve safety. This contrasts with protection and passive confinement. Unlike traditional packing materials, which primarily serve as barriers against external factors, active packaging materials are designed to release or absorb chemicals, inhibit the growth of germs, or continually monitor the food's quality [5–14].

The key functionalities of active packaging include:

- **Oxygen scavenging**: Oxygen plays an important role in the degradation of food by oxidizing proteins, lipids, and vitamins, which results in changes in color, taste, and nutritional loss. By adding oxygen scavengers to packing materials, food items that are sensitive to oxygen can have their shelf life extended by carefully removing oxygen from the headspace of the container.
- **Moisture control**: The microbiological stability, texture, and crispness of food items are all influenced by their moisture level. Incorporating materials that absorb moisture or act as desiccants into packaging can assist control of relative humidity and stave off moisture-related problems like caking, spoiling, and mold development.
- **Ethylene absorption**: Fruits and vegetables mature, and ethylene, a plant hormone, accelerates senescence. To extend the shelf life of perishable goods, ethylene scavengers—which absorb ethylene gas and slow the ripening process—can be added to packaging materials.
- **Antimicrobial activity**: Microbial contamination is a major issue in the food industry that leads to foodborne illnesses and spoiling. Food safety can be enhanced and shelf life extended using antimicrobial packaging materials that use active substances such as bacteriocins, essential oils, or nanoparticles to prevent the growth of dangerous and spoilage microorganisms.
- **Flavor and aroma preservation**: Food items' distinctive taste and scent are caused by volatile chemicals that can evaporate or deteriorate over time.

Packed meals' sensory qualities can be preserved using packaging materials that have barrier qualities against flavor migration or active scavenging of volatile chemicals.

7.3 Polymer Nanocomposites in Active Food Packaging

To accomplish these intended capabilities, polymer nanocomposites provide a flexible framework for incorporating active agents and functional fillers. By precisely controlling the concentration, dispersion, and interactions of nanoparticles inside the polymer matrix, scientists can tailor the properties of active packaging materials to specific food products and storage conditions. In the following sections, we will address the various nanoparticles used in polymer nanocomposites for active food packaging and how they affect the package's performance [15–20].

There are particular dispersion, compatibility, and processing difficulties when adding nanoparticles to polymer matrices. To meet these obstacles and enhance the qualities of active packaging materials, tactics including compatibilization agents, surface modification, and nanocomposite synthesis procedures are used. Food packaging polymer nanocomposites have varying mechanical, barrier, and functional properties depending on the quantity of nanoparticles included in the polymer.

Polymer matrices with nanoparticles added to them can significantly affect the mechanical, barrier, and functional properties of packaging materials [21–25]. Comprehending these impacts is vital in customizing the functionality of active packaging systems to fulfill distinct application demands.

Advances in synthesizing nanomaterials, processing methods, and packaging technologies are driving a rapid evolution in polymer nanocomposites for active food packaging. Despite great progress in fully utilizing nanocomposite-based active packaging systems, many opportunities and challenges must still be addressed.

A viable platform for creating active food packaging materials with improved functions and performance is provided by polymer nanocomposites. Nanoparticles added to polymer matrices allow researchers to tailor packaging materials' mechanical, barrier, and functional properties to specific application requirements. The latest developments in nanocomposite-based sensors and antimicrobial packaging systems could completely change how food is packaged, preserved, and consumed. However, realizing the full potential of nanotechnology in active food packaging requires tackling issues with nanoparticle dispersion, migration, safety, and scalability. Moving forward, research should concentrate on creating smart, multipurpose packaging that can guarantee food safety, quality, and sustainability in a food supply chain that is becoming more intricate and dynamic.

With an emphasis on active food packaging, this book chapter attempts to present a thorough summary of current developments in polymer nanocomposites. We then look at the many kinds of nanoparticles utilized in polymer matrices, how they work, and how they affect the packaging characteristics. This chapter also examines new advances in antimicrobial packaging technologies and nanocomposite-based sensors, emphasizing how these innovations may be used to preserve and safeguard food.

Qin et al. [26] synthesized polymer nanocomposite films for the active food packaging applications. Hydrogen bonding was the primary mechanism by which the polymer and PCE interacted (Scheme 7.1).

Because of the combination of PCE and silver nanoparticles, the film demonstrated the best barrier, mechanical, antioxidant, and antibacterial capabilities. Furthermore, because PCE contains a lot of anthocyanins, polymer/PCE and polymer nanocomposite films may exhibit color changes in various pH buffers. A few particles and spots showed up on the composite films' surfaces and cross-sections. To extend food shelf life, the films composed of synthetic polymer nanocomposite may be utilized as active food packaging materials (Fig. 7.1).

Brito et al. [27] fabricated polyethylene films using nanoparticles for active food packaging applications. The low-density polyethylene films' thermal, chemical, and physical characteristics were unaffected in the presence of silver nanoparticles. Although all of the examined microorganisms were resistant to the antibacterial properties of the nanostructured films, fungi and Gram-negative bacteria exhibited more resistance than Gram-positive bacteria. Moreover, the packages containing

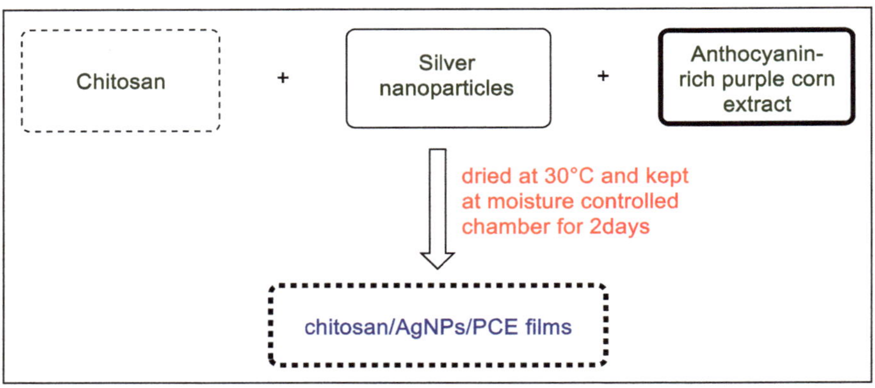

Scheme 7.1 Synthesis of polymer nanocomposite films

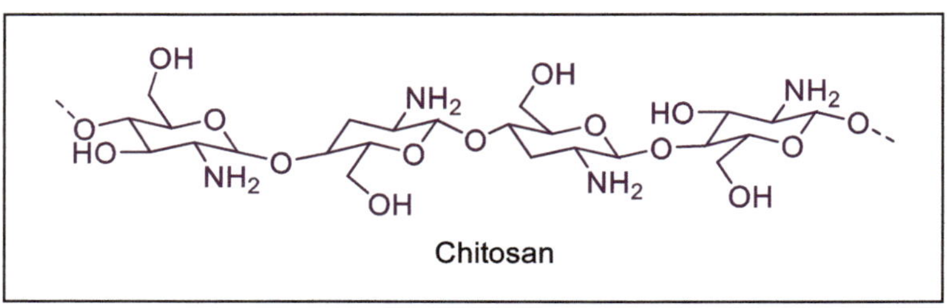

Fig. 7.1 Structure of chitosan

silver nanoparticles were shown to inhibit the growth and reproduction of early-stage bacterial cells. These results showed that extruded polyethylene sheets combined with silver nanoparticles might be used to make antimicrobial food packaging, especially for liquid meals.

For use in active food packaging, Chen et al. [28] produced polymer nanocomposite films. The nanoparticles were made using a simple method for the in-situ reduction of silver nitrate. The concurrent formation and cross-linking of silver nanoparticles with cellulose was made possible by the one-pot synthesis (Scheme 7.2). The nanoparticles were evenly distributed throughout the composite film's inside and outside. The composite material demonstrated exceptional antibacterial characteristics, clarity, thermal stability, and good tensile qualities up to 86.4 MPa. Composite film 0.04 has a sterilizing performance that is more than 99.9% against *Escherichia coli* and *Staphylococcus aureus*. The composite film may be used in food packaging and bacterial barriers because of its low residual LiCl/DMAc and low diffusion of silver nanoparticles (Fig. 7.2).

Chowdhury et al. [15] synthesized food packaging films based on PVA crosslinked with nanocomposites. Glyoxal was used as a crosslinking agent when making films (Scheme 7.3). The mechanical and physical qualities have been enhanced by the

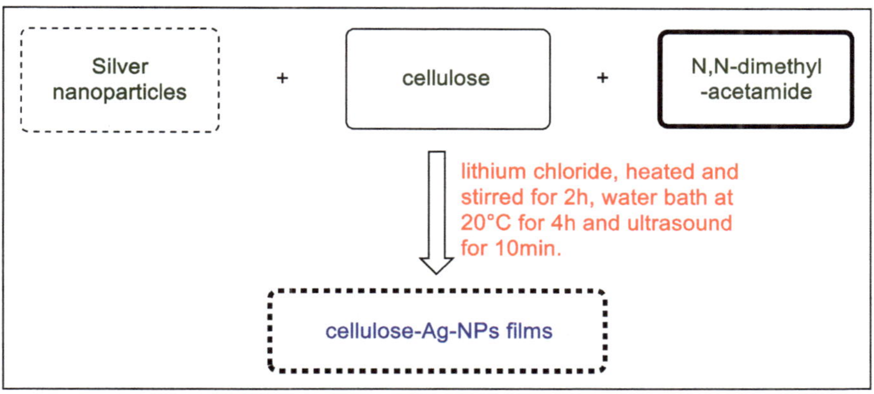

Scheme 7.2 Synthesis of polymer nanocomposite films

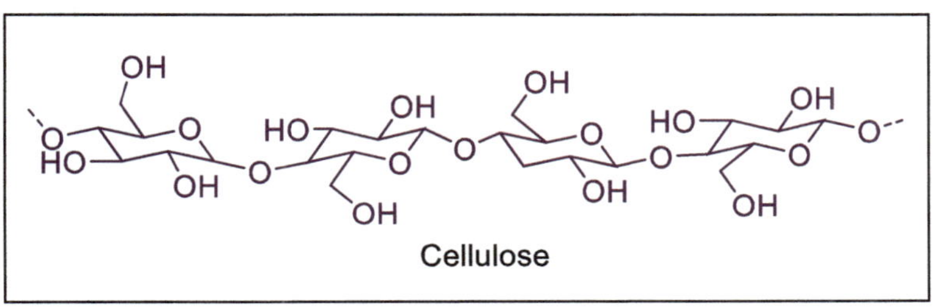

Fig. 7.2 Structure of cellulose

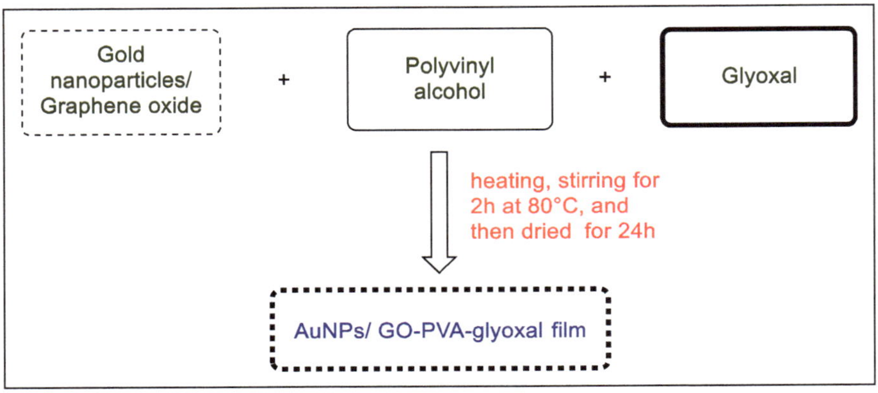

Scheme 7.3 Synthesis of food packaging composite films

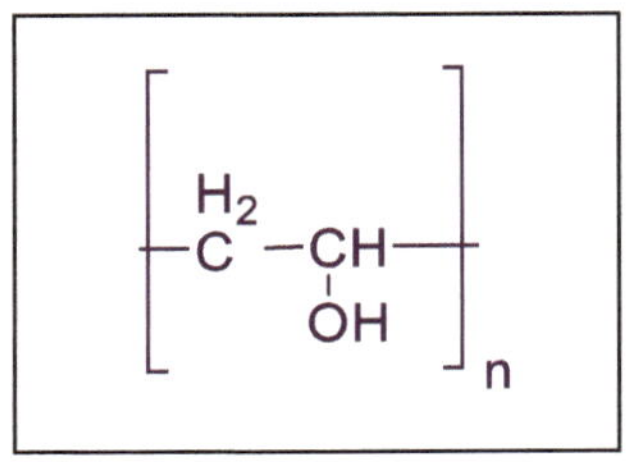

Fig. 7.3 Structure of polyvinyl alcohol

dispersion of nanocomposites within the polymer. When it comes to preventing microbiological contamination, the produced nanocomposite film outperforms the graphene/polymer crosslinked composite film due to its wider inhibitory zone creation. The composite film for food preservation has confirmed its applicability in food packaging applications and increased the quality of banana shelf life (Fig. 7.3).

Roy et al. [29] fabricated polymer-based nanocomposite films for active food packaging applications. Solution-casting was used to create the polymer films for the nanocomposite. Within the polymer matrix, the nanoparticles were dispersed uniformly (Scheme 7.4). The addition of nanoparticles improved the polymer-based film's mechanical strength, hydrophobicity, UV barrier, and water vapor barrier while having little to no effect on its thermal stability. Although the generated nanocomposite film exhibited substantial antibacterial activity against the Gram-negative (*E. Coli*) bacterium, its impact on the growth rate of the Garm-positive (*L. monocytogenes*) bacteria was negligible. Active food packaging applications might greatly benefit from the produced nanocomposite films, which have much enhanced physical and functional qualities (Fig. 7.4).

Shankar et al. [30] fabricated food packaging films based on carbohydrates crosslinked with nanocomposites. By combining nanoparticles with five distinct carbohydrates, multifunctional composite films were created. The nanocomposite

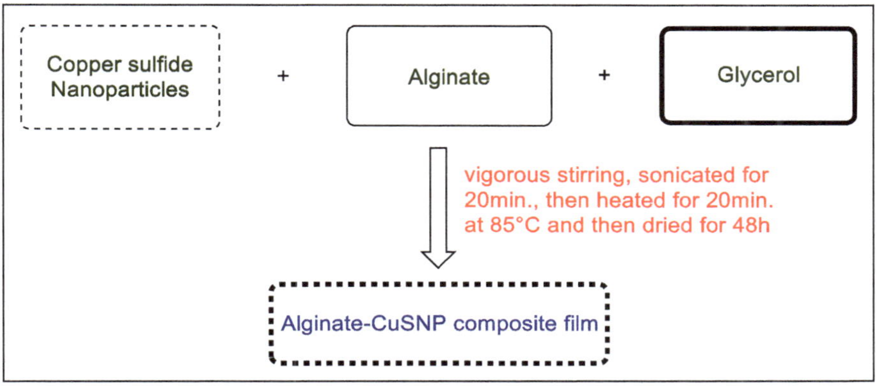

Scheme 7.4 Synthesis of food packaging composite films

Fig. 7.4 Structure of alginate

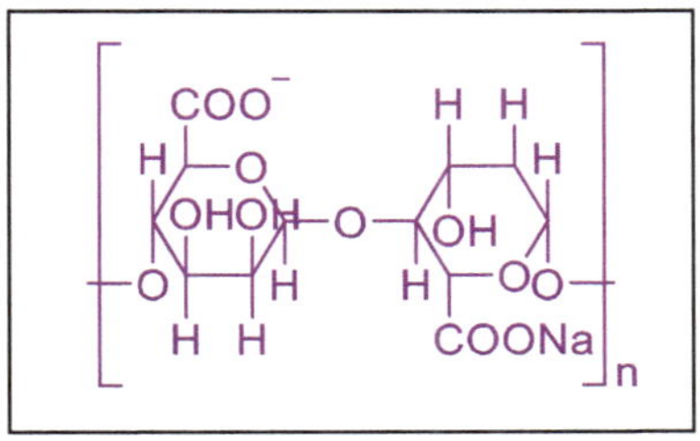

films were produced using a solvent-casting method (Scheme 7.5). Nanoparticles improved the material's thickness, tensile strength, relative humidity, UV barrier, and water vapor barrier properties. The composite films containing nanoparticles exhibited strong antibacterial action against Escherichia coli and Listeria monocytogenes. The resulting composite films might be used as UV-light-barrier antibacterial films for active food packaging.

Ahmadi et al. [31] produced composite sheets based on cellulose for improved food packaging uses. The composite films were made via the casting process. The incorporation of nanoparticles improved the water resistance and physico-mechanical characteristics of the composite films (Scheme 7.6). The composite films with 5% w/w zinc oxide nanoparticles and those with 0.1% w/w selenium nanoparticles had the highest tensile strength values. Additionally, composites with 3% w/w zinc oxide nanoparticles had the lowest water vapor permeability and water solubility of any film. Zinc oxide nanoparticles were shown to have a better antibacterial impact than selenium nanoparticles. The composite films that included selenium nanoparticles showed a greater ability to scavenge free radicals than films that included zinc oxide nanoparticles. The films' antioxidant impact was further amplified by the combination of the two nanoparticles. In conclusion, creating active food packaging items may use fabricated packaging films.

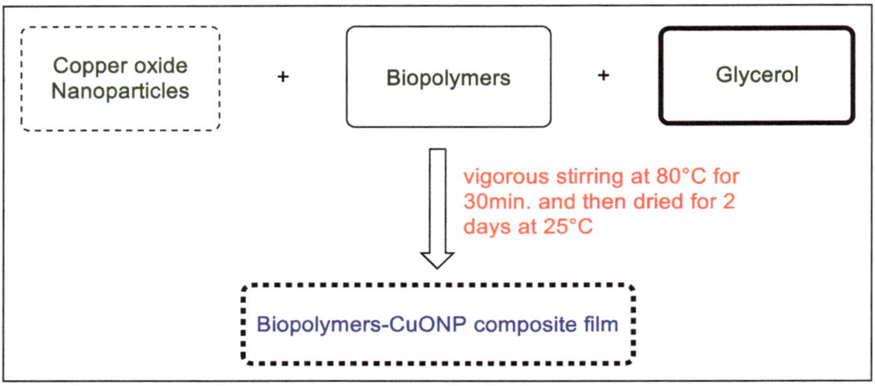

Scheme 7.5 Synthesis of biopolymers composite films

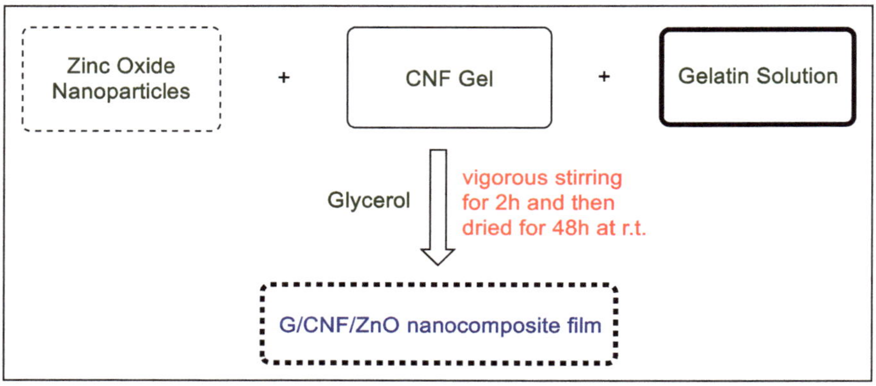

Scheme 7.6. Synthesis of food packaging composite films

Li et al. [32] fabricated cellulose-based composite films as active food packaging material. The composite films were made using the casting procedure. Nanoparticles mixed in the networks on the surface of the films showed an uneven arrangement, indicating that a smooth surface could be achieved with the optimal weight ratio of around 1:1 for the CMC to DOM (Scheme 7.7). The 2 g zinc oxide nanoparticle-containing films showed antibacterial activity against both Gram-positive (*S. aureus*) and Gram-negative (*E. coli*) bacteria. At 400 and 450 μL/mL, the antibacterial rate dropped as DOM increased. Higher zinc oxide nanoparticle concentration film-forming solutions behave like liquids and exhibit shear-thinning characteristics. Food packaging applications hold significant promise for the utilization of edible films.

Ojha et al. [33] fabricated food packaging films based on polymer nanocomposites. The composite films were made using the solution-casting method. The generated films of nanocomposite showed the intercalated hybrid interaction between the

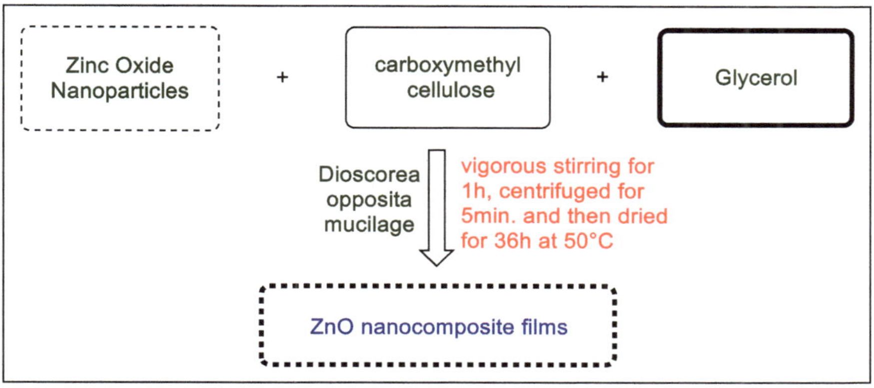

Scheme 7.7. Synthesis of composite films

polymer matrix and nanocomposites (Scheme 7.8). Strong interactions between the carbonyl of the ester groups in the polymer matrix and the surface hydroxyl moiety of the nanoparticles are observed. The composite films exhibited notable enhancements in physical, chemical, thermo-mechanical, and biodegradation properties in comparison to the bare polymer matrix. This was attributed to the uniform dispersion of nanoparticles and their robust interfacial adhesion with the matrix via hydrogen bonding interactions. Superb biocompatibility was found while using L929 murine fibroblast cells in the cell viability experiment. Regarding *E. Coli* and *S. aureus*, it was shown that the antibacterial activity of synthetic composite films rose as the concentration of nanoparticles increased. The composite films demonstrated exceptional biocompatible, biodegradable, and antibacterial qualities that may be very helpful in reducing the development of contaminating microorganisms, increasing food's shelf life, and enhancing its quality.

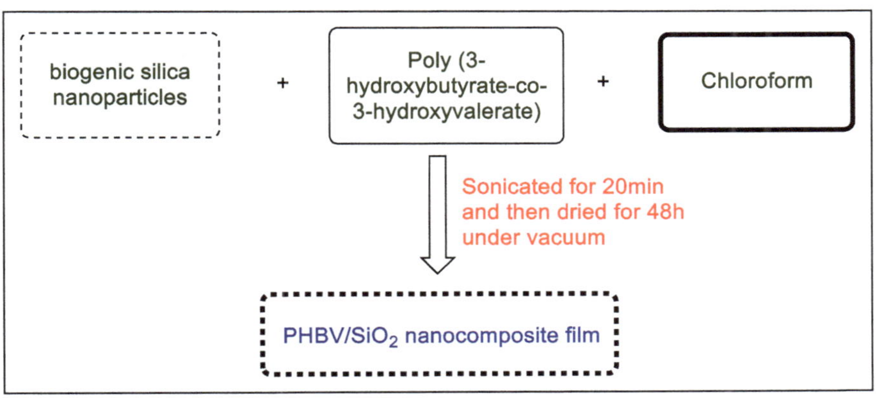

Scheme 7.8 Synthesis of bio-nanocomposite films

Mackiw et al. [34] fabricated plasma-modified food packaging films for the extension of the shell life of food. Films treated with argon and sulfur dioxide combined considerably reduced microbial growth in comparison to composites coated with argon and sodium oxide. The greatest levels of growth inhibition were produced by sulfur-containing film, which showed 86% inhibition for *Staphylococcus aureus* and 82% inhibition for *Escherichia coli.* Sulfur dioxide-containing antimicrobial surfaces may be used to package foods such as potato products, dried fruit and vegetables, and juices and fruit drinks.

Wu et al. [35] synthesized protein-based composite films as active food packaging films. The nanoparticles were produced by a simple wet-chemical method. The composite films were created using the liquid precipitation method (Scheme 7.9). The addition of nanoparticles to the polymer matrix affected the characteristics of the resulting composite films, including their oxygen permeability, storage modulus, and color. The synthetic composite film exhibited superior mechanical qualities, antifungal activities, and barrier capabilities. In contrast, the synthetic composite film's oxygen and water permeabilities are 66.1% and 54.8%, respectively, of those of ordinary SPI film. According to these studies, the created composite film may be used as the perfect packaging matrix to preserve food.

Amjadi et al. [36] fabricated betanin-based nanocomposite films for active food packaging applications. The electrospinning method proved effective in producing the synthetic nanofibers. The produced nanofibers were found to be of an average size of around 535 nm (Scheme 7.10). The manufactured sample has strong surface hydrophobicity and acceptable mechanical characteristics. The combination of betanin and titanium dioxide nanoparticles exhibited high antibacterial activity in nanofibers against two food-borne pathogenic microorganisms. The amount of DPPH scavenging activity in the synthetic nanofiber sample was 64.42%. The biocompatibility of the produced nanofibers was further validated by an in vitro cell cytotoxicity assay. So, using synthetically produced electrospun nanofiber would be a positive step toward increasing food quality and shelf life.

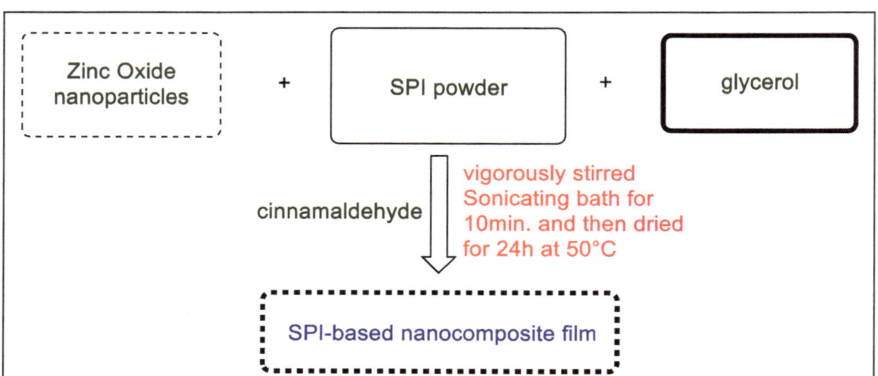

Scheme 7.9 Synthesis of SPI-based nanofibers

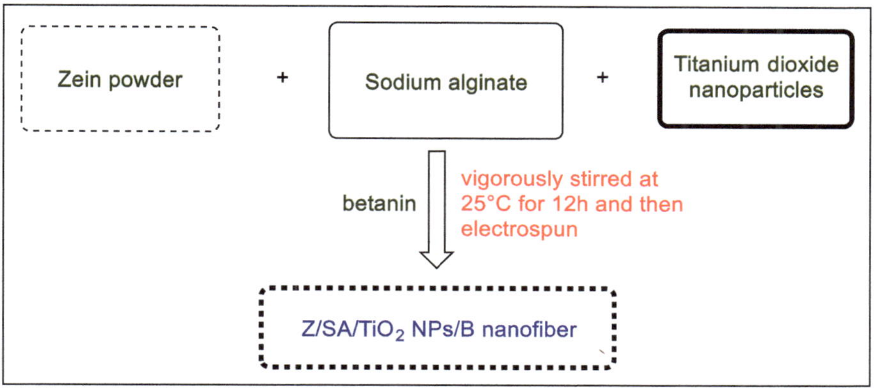

Scheme 7.10 Synthesis of nanofibers

Swaroop et al. [37] synthesized biofilms based on polylactic acid for food packaging applications. The films are synthesized using the solvent-casting method (Scheme 7.11). Compared to virgin PLA films, the 2-weight percent reinforced PLA films from the synthesized bio-composite showed the largest gains in tensile strength and oxygen barrier properties (up to 29% and 25%, respectively). However, free volumes close to magnesium oxide nanoparticles and interfacial behavior caused a roughly 25% drop in the water vapor barrier characteristics. Additionally, the synthetic films showed better antibacterial activity. After 12 h of treatment, roughly 46% of the *E. Coli* bacterial culture died due to increasing damage induced by the two wt. percent biofilms. The resulting films work well as food packaging materials since they are translucent, UV radiation-shielding, and have high antibacterial activity.

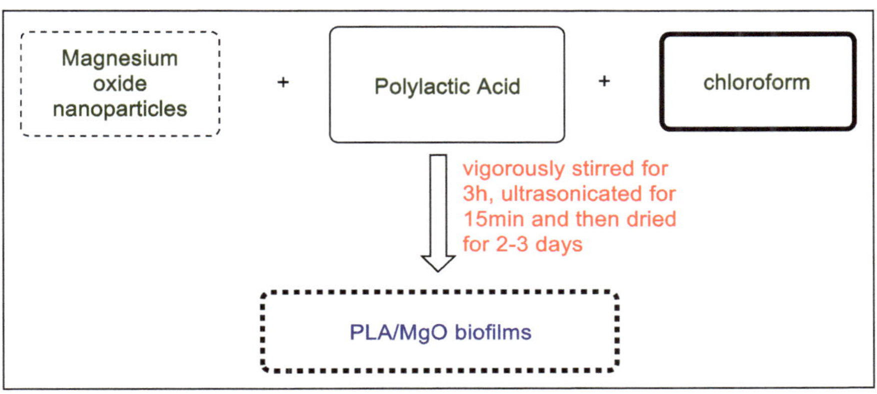

Scheme 7.11. Synthesis of food packaging biofilms

Kousheh et al. [38] fabricated nano cellulose films for active food packaging applications. The hydrothermal process was used to synthesize the films. To create antimicrobial/ultraviolet protective nano-paper, the as-prepared carbon dots are inserted into nanocellulose using an ex-situ technique. The high-hydroxylated groups and 2.8 nm average size of the photoluminescent carbon dots were produced. Both bacteria were susceptible to the antibacterial action of carbon dots at a dosage of 500 mg/mL. Furthermore, when exposed to UV light, nano-paper fluoresced. Compared to nanocellulose, nanocellulose with a carbon dots loading capacity of 71.74 mg/cm^2 demonstrated adequate stretchability and flexibility. Furthermore, the carbon dot-infused nano paper showed more potent UV blocking and inhibitory activity against Gram-positive bacteria than Gram-negative bacteria. As a novel fluorescence antimicrobial/ultraviolet protection material in the nano cellulose film, carbon dots may be used to create antibacterial packaging that is resistant to forgeries because of their fluorescence look.

Bi et al. [39] fabricated chitosan-based food packaging films. The simultaneous addition of TPGS and nano-silicon dioxide resulted in strong intermolecular hydrogen bond interactions and a tight inner microstructure in the chitosan film (Scheme 7.12). The inclusion of TPGS was found to slightly increase the crystallinity of the films containing chitosan and silicon dioxide. The highest tensile strength and elongation at break, as well as the lowest moisture content, water vapor, and oxygen permeability, were observed in the TPGS/chitosan/silicon dioxide film when compared to other films. Furthermore, TPGS/chitosan and TPGS/chitosan/silicon dioxide films performed better at scavenging free radicals than chitosan and chitosan/silicon dioxide films did. The films containing silicon dioxide and chitosan showed lower antibacterial activity than the films containing TPGS/chitosan and chitosan alone. TPGS/chitosan and TPGS/chitosan/silicon dioxide film packaging, in particular, markedly increased the oxidative stability of soybean oil. Our results suggest the potential use of TPGS/chitosan/silicon dioxide film as a novel antioxidant and antibacterial packaging material for the food industry.

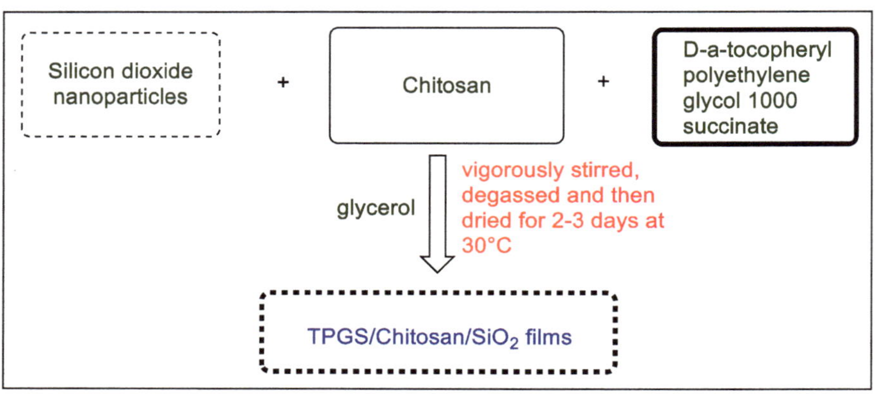

Scheme 7.12 Synthesis of chitosan-based nanocomposite food packaging films

Huang et al. [12] synthesized nanocomposite films by mixing GO and ZnO nanoparticles with PLA using the solution-blending method. Particles of graphene oxide/zinc oxide and polylactic acid were individually dissolved in chloroform and then completely combined over many hours at higher temperatures. To create the required blend films, the resulting mixture was now cast on coating machines and cured (Scheme 7.13). According to the results, the inclusion of graphene oxide/zinc oxide nanoparticles increased the tensile strength from 60.0 to 72 MPa and significantly improved the glass transition temperature and UV shielding properties. After testing the films' antibacterial activity against *S. aureus* and *E. coli*, it was discovered that exposure to light increased the films' effectiveness by 99.2 and 97.6%, respectively (Fig. 7.5).

Mahmoudi et al. [40] developed composite films by combining chitosan films with polyvinylpyrrolidone and graphene oxide. An aqueous polyvinylpyrrolidone solution was also created after a chitosan solution was initially made using acetic acid to synthesize the desired blend film. The two synthesized solutions were combined in varying ratios at this point. The mixture was then gradually mixed with graphene oxide suspension until the graphene oxide content was between 1 and 3 wt. percent (Scheme 7.14). When the produced films were tested for various functional characteristics, it was found that the PVP addition increased the water vapor permeability,

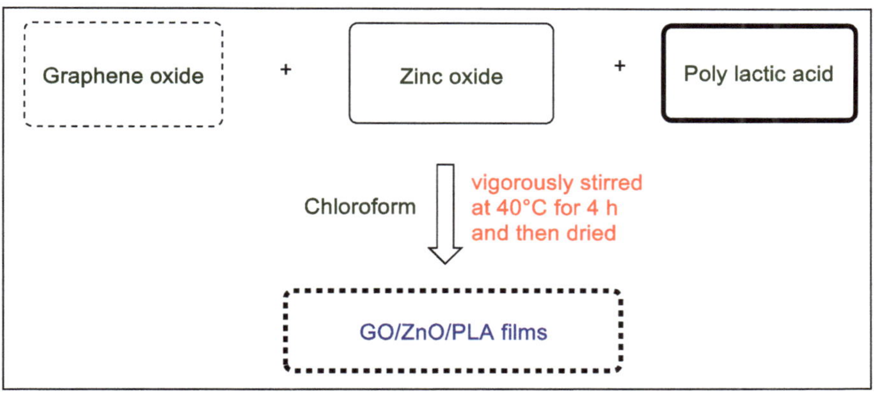

Scheme 7.13 Synthesis of polylactic-based nanocomposite films

Fig. 7.5 Chemical structure of polylactic acid

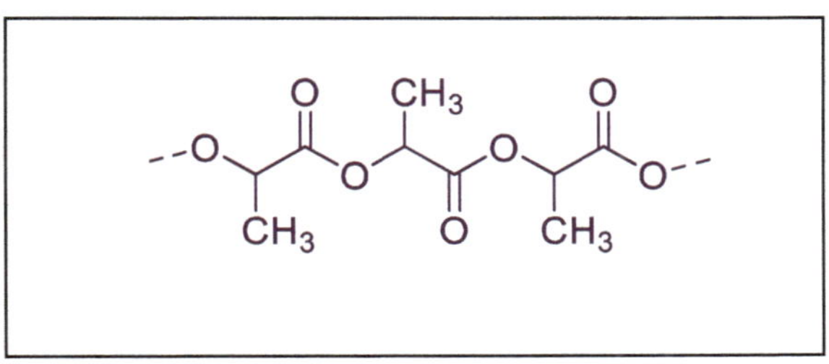

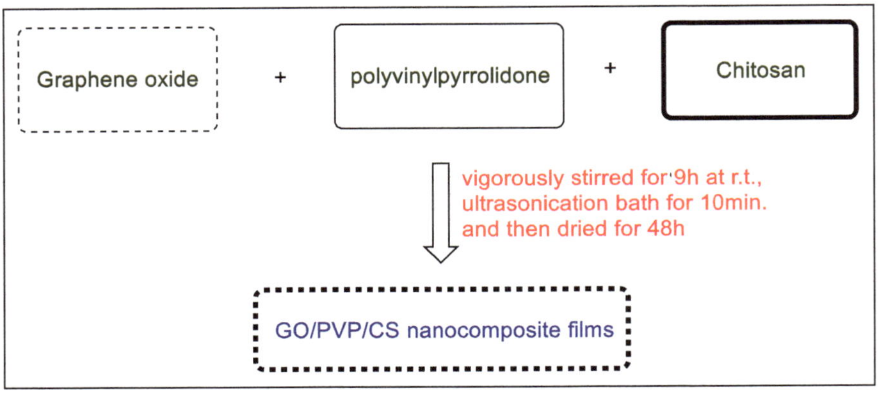

Scheme 7.14 Synthesis of food packaging films

Fig. 7.6 Chemical structure of polyvinylpyrrolidone

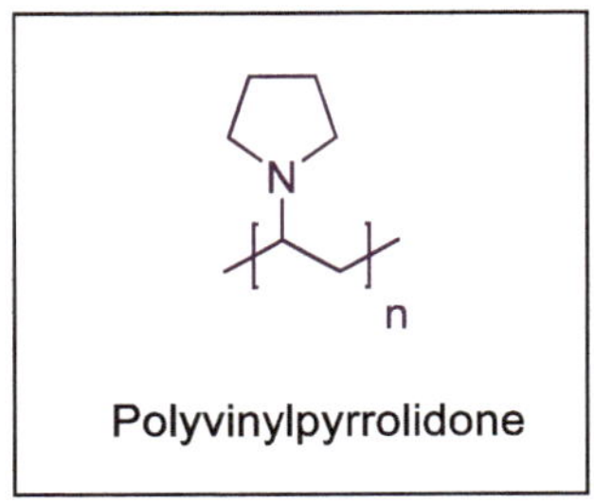

opacity, and flexibility of the films. The primary cause of the blend films' antibacterial activity was the combination of graphene oxide and chitosan (Fig. 7.6).

Usman et al. [41] formed nanocomposites by incorporating graphene oxide with polyvinyl alcohol, starch, and silver. Deionized water was used to create separate solutions of PVA and GO, which were then completely combined to create the target film. After that, the mixture was mixed with silver nitrate and starch, and the resulting solution was applied to cast plates to create the films (Scheme 7.15). The films' young modulus and tensile strength values increased by 145% and 26.81%, respectively, after being assessed for various qualities. In contrast, there was a discernible increase in anti-microbial activity, as seen by the zone of inhibition values for *S. aureus* and *E. Coli* to 17 and 20.5 mm, respectively (Fig. 7.7).

Xu et al. [42] created mix films of titanium dioxide, chitosan, and graphene oxide to lessen microbial food deterioration. Comparable techniques of formation were employed, in which solutions of graphene oxide and chitosan were fully combined, and then three types of solutions with varying ratios were formed by adding colloidal titanium dioxide (Scheme 7.16). After testing the films against *Bacillus subtillis* and *Aspergillus niger*, the results showed that the film with the highest activity had a ratio of 1:20:4 (graphene oxide: chitosan and titanium dioxide).

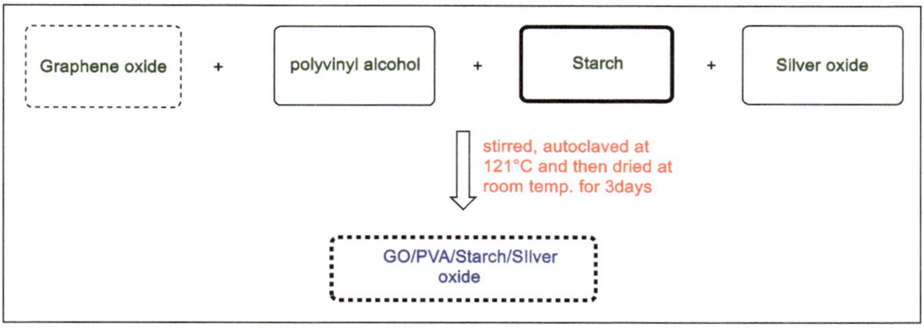

Scheme 7.15 Synthesis of nanocomposites by incorporating graphene oxide with polyvinyl alcohol, starch, and silver

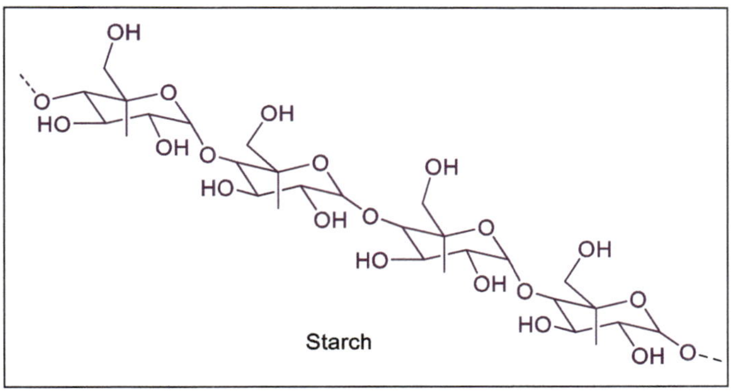

Fig. 7.7 Chemical structure of starch

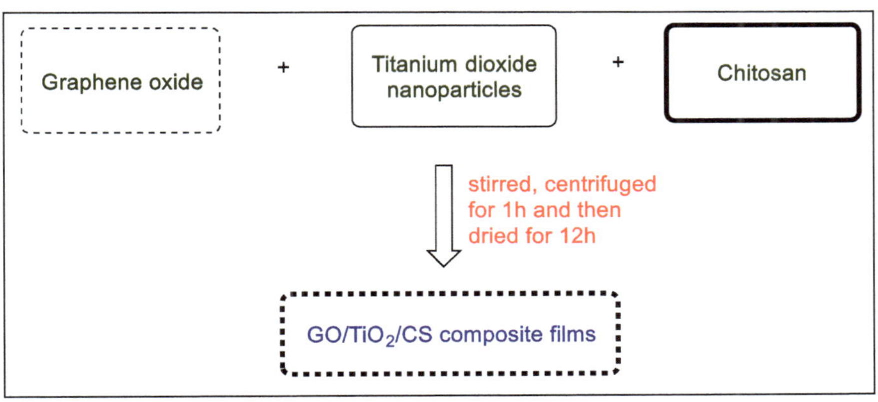

Scheme 7.16 Synthesis of active food packaging films

Arfat et al. [43] synthesized polylactic acid-based composite food packaging films. The solvent-casting process created the target films by adding PLA and polyethylene glycol in the presence of dichloromethane (Scheme 7.17). Dichloromethane was used to create the GO nanosheet suspension. Following a brief sonication, the two solutions were combined. The addition of CLO was followed by several more minutes of thoroughly mixing the mixture, after which it was placed onto glass Petri dishes to form the sheets or films. A functional examination of the sheets showed that the films were more flexible when clove oils were present and that the thermo-mechanical and barrier characteristics were significantly affected by graphene oxide. The movies also revealed notable anti-microbial activity against *E. Coli* and *S. aureus.*

Sharma et al. [44] developed PVA-based films by incorporating starch. Graphene oxide was applied to enhance the sheets' water barrier qualities. Solutions and melts were mixed to create the blend films (Scheme 7.18).

Lin et al. [45] prepared polyvinyl alcohol and graphene oxide-based food packaging films. After making an aqueous solution of graphene oxide and polyvinyl alcohol, glycerine and Tween-80 were added and carefully mixed (Scheme 7.19).

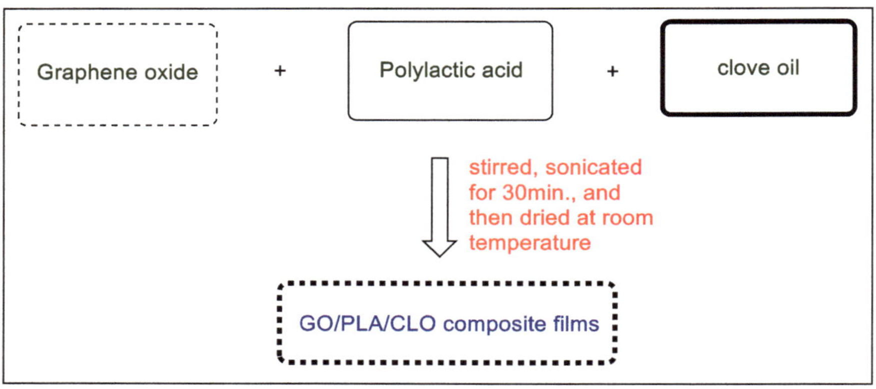

Scheme 7.17 Synthesis of food packaging films based on polylactic acid

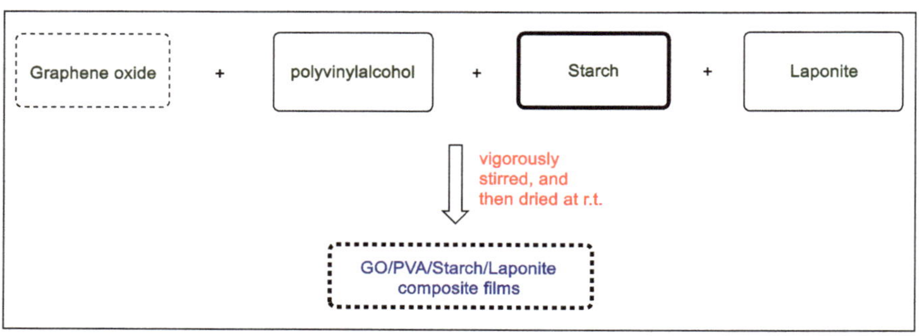

Scheme 7.18 Synthesis of polyvinyl alcohol-based films by incorporating starch and laponite RD

Subsequently, the solution was heated to a higher temperature and agitated continuously for several minutes while thyme oil was added. Finally, the mixture was poured onto plates, let to dry, and then formed into sheets or films. Test findings indicated that the composite film using 1% oregano oil had better qualities. Their zone of inhibition values were as high as 11.7 and 12.6 when tested against *E. Coli* and *L. monocytogenes*, indicating interesting antibacterial properties. Their zone of inhibition values reached 12.6 mm and 11.7 mm, respectively, when evaluated for antimicrobial activity against *L. monocytogenes* and *E. coli*. These results provided encouraging antibacterial properties. Additionally, the films demonstrated a tensile strength of 20.760 MPa and tensile fracture elongation of 25.9%, respectively.

Using the solution casting approach, Fang et al. [46] created high-quality food packaging materials by combining poly (3-hydroxybutyrate-co-3-hydroxyvalerate) with graphene oxide-cellulose nanocrystal hybrids (Scheme 7.20). After screening the synthetic films for a variety of functional characteristics, it was discovered that the elongation to break and tensile strength had increased to 170.2% and 52.1%, respectively. In a similar vein, the films demonstrated superior antibacterial and barrier qualities (Fig. 7.8).

Using mechanical and polymerization processes, Shen et al. [47] created blend films by combining polylactic acid with a hybrid of graphene oxide and silver. Lactic acid was distilled at 100 °C for a few hours to create a silver-graphene oxide solution, which was then used to create the mixed film (Scheme 7.21). Toluene sulfonic acid and stannous octate were added to the solution, and it was carefully agitated for a couple more hours in a nitrogen environment. The flexural strength of the film improved significantly and reached values of 942 K Pa due to the inclusion of a hybrid of silver and graphene oxide; in contrast, tests against *E. Coli* and *S. aureus* showed comparable increases, with values reaching as high as 99%.

Graphene oxide and carbon nanofibers were used by Rivera-Briso et al. [48] to create food packaging films. 1% of GO, or CNF nanoparticles, by weight, was utilized

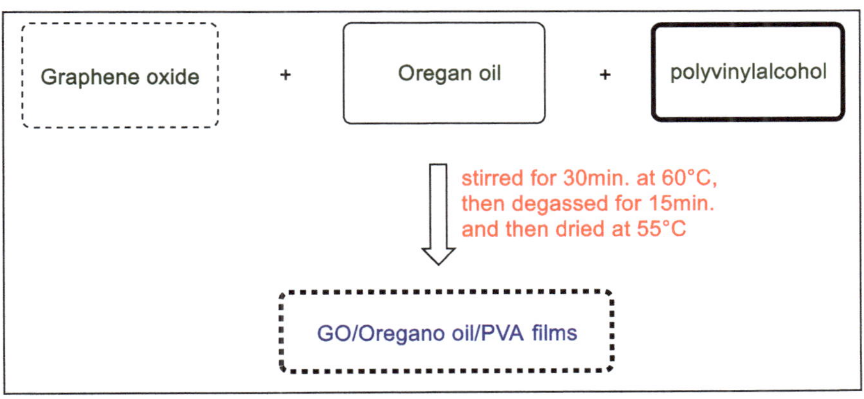

Scheme 7.19 Synthesis of graphene oxide/oregano oil/PVA films

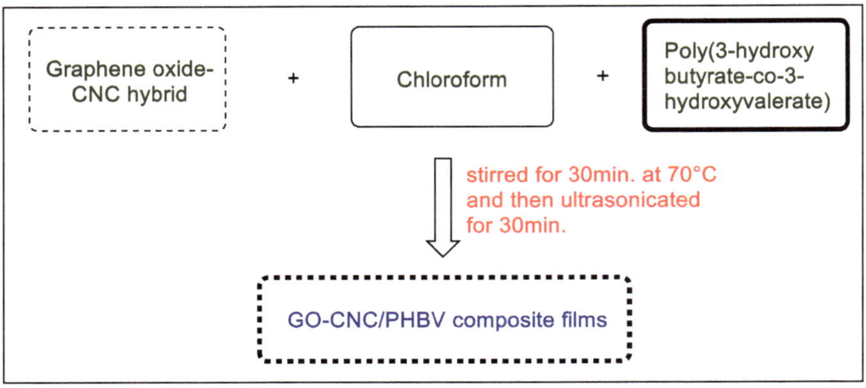

Scheme 7.20 Integration of GO-CNC hybrid with PHBV

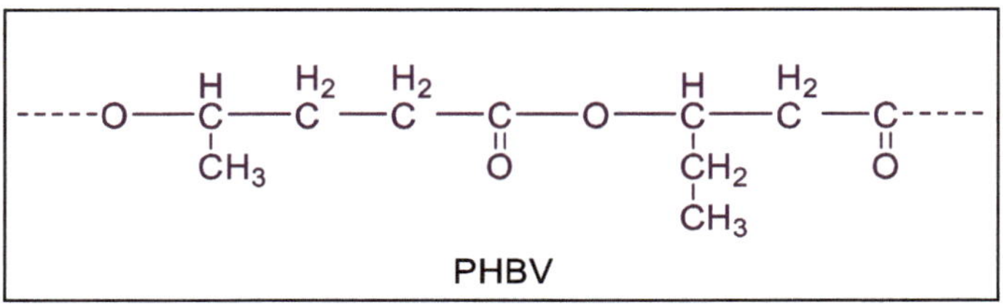

Fig. 7.8 Chemical structure of polylactic acid

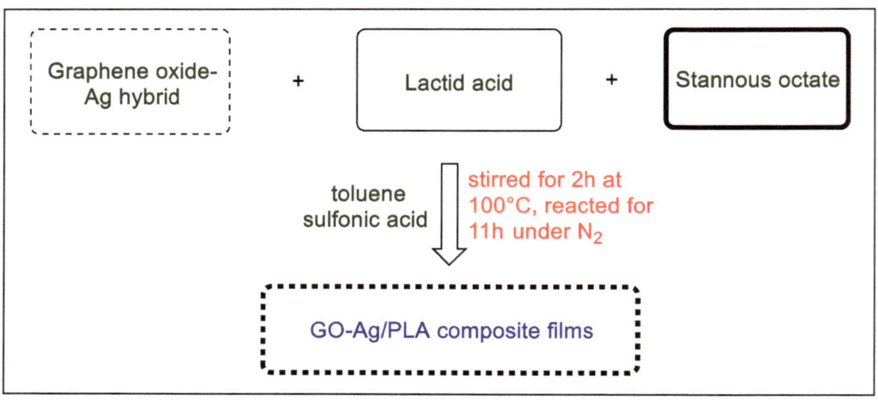

Scheme 7.21 Integration of silver-graphene oxide hybrid with polylactic acid

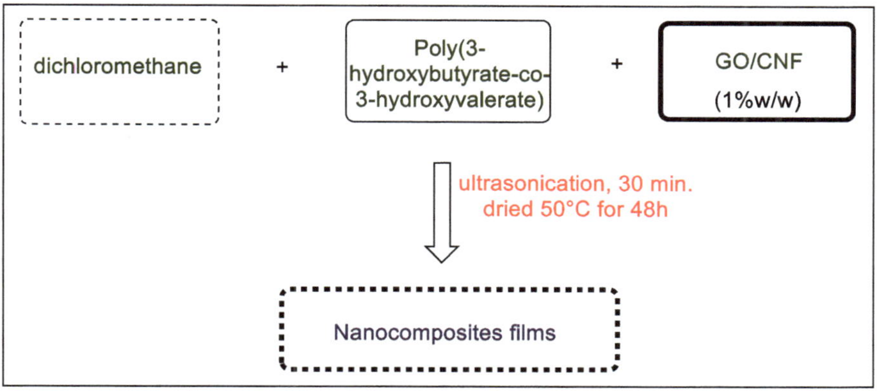

Scheme 7.22 Synthesis of nanocomposite films

to create the films. Similarly, both nanomaterials improved wettability, compression performance, wettability, and thermal behavior (Scheme 7.22). Both nanocomposites had antibacterial activity sufficient to inhibit the development of certain *S. aureus* bacteria when nanoparticles were added to the polymer matrix. Graphene oxide-based nanocomposites exhibited greater antibacterial activity, cell adhesion, and proliferative activity versus time compared to CNF-based nanocomposites. Nonetheless, both nanocomposite films have great promise for use in biomedical applications.

7.4 Conclusion and Future Perspectives

Food packaging has benefited immensely from the application of nanotechnology, especially with the creation of polymeric nanocomposites. As sustainability gains more importance, there is a growing need for materials made from renewable resources. Using nanotechnology as a support can help overcome some obstacles to using these materials in the industry. Despite the many benefits that polymer nanocomposites provide, the physicochemical discrepancies between polymeric nanocomposites and their macroscale counterparts make their use in food packaging concerning for human health. The first stages in detecting, characterizing, and quantifying these particles are crucial, and more research is required to comprehend nanoparticles' effects on human health fully. Migration studies should be carried out to assess how many nanoparticles may permeate food matrices.

References

1. Montes, S., Etxeberria, A., Mocholi, V., Rekondo, A., Grande, H., Labidi, J.: Effect of combining cellulose nanocrystals and graphene nanoplatelets on the properties of poly(lactic acid) based films. Express Polym. Lett. **12**, 543–555 (2018). https://doi.org/10.3144/expresspolymlett.2018.45
2. Song, K., Gao, A., Cheng, X., Xie, K.: Preparation of the superhydrophobic nano-hybrid membrane containing carbon nanotube based on chitosan and its antibacterial activity. Carbohydr. Polym. **130**, 381–387 (2015). https://doi.org/10.1016/j.carbpol.2015.05.023
3. Rezić, I., Haramina, T., Rezić, T.: Metal nanoparticles and carbon nanotubes—perfect antimicrobial nano-fillers in polymer-based food packaging materialsa. In: Food Packaging. pp. 497–532. Elsevier (2017)
4. Mohamed, N.A., Abd El-Ghany, N.A.: Novel aminohydrazide cross-linked chitosan filled with multi-walled carbon nanotubes as antimicrobial agents. Int. J. Biol. Macromol. **115**, 651–662 (2018). https://doi.org/10.1016/j.ijbiomac.2018.04.101
5. Kumar, S., Raj, S., Jain, S., Chatterjee, K.: Multifunctional biodegradable polymer nanocomposite incorporating graphene-silver hybrid for biomedical applications. Mater. Des. **108**, 319–332 (2016). https://doi.org/10.1016/j.matdes.2016.06.107
6. Gouvêa, R.F., Del Aguila, E.M., Paschoalin, V.M.F., Andrade, C.T.: Extruded hybrids based on poly(3-hydroxybutyrate-co-3-hydroxyvalerate) and reduced graphene oxide composite for active food packaging. Food Packag. Shelf Life **16**, 77–85 (2018). https://doi.org/10.1016/j.fpsl.2018.02.002
7. Sharma, S., Biswal, B.K., Kumari, D., Bindra, P., Kumar, S., Stobdan, T., Shanmugam, V.: Ecofriendly fruit switches: graphene oxide-based wrapper for programmed fruit preservative delivery to extend shelf life. ACS Appl. Mater. Interfaces **10**, 18478–18488 (2018). https://doi.org/10.1021/acsami.8b02048
8. Demitri, C., De Benedictis, V.M., Madaghiele, M., Corcione, C.E., Maffezzoli, A.: Nanostructured active chitosan-based films for food packaging applications: effect of graphene stacks on mechanical properties. Meas. (Lond.) **90**, 418–423 (2016). https://doi.org/10.1016/j.measurement.2016.05.012
9. Cui, R., Jiang, K., Yuan, M., Cao, J., Lin, L., Tang, Z., Qin, Y.: Antimicrobial film based on polylactic acid and carbon nanotube for controlled cinnamaldehyde release. J. Market. Res. **9**, 10130–10138 (2020). https://doi.org/10.1016/j.jmrt.2020.07.016
10. Liu, Y., Wang, S., Lan, W., Qin, W.: Fabrication of polylactic acid/carbon nanotubes/chitosan composite fibers by electrospinning for strawberry preservation. Int. J. Biol. Macromol. **121**, 1329–1336 (2019). https://doi.org/10.1016/j.ijbiomac.2018.09.042
11. Barra, A., Ferreira, N.M., Martins, M.A., Lazar, O., Pantazi, A., Jderu, A.A., Neumayer, S.M., Rodriguez, B.J., Enăchescu, M., Ferreira, P., Nunes, C.: Eco-friendly preparation of electrically conductive chitosan-reduced graphene oxide flexible bionanocomposites for food packaging and biological applications. Compos. Sci. Technol. **173**, 53–60 (2019). https://doi.org/10.1016/j.compscitech.2019.01.027
12. Huang, Y., Wang, T., Zhao, X., Wang, X., Zhou, L., Yang, Y., Liao, F., Ju, Y.: Poly(lactic acid)/graphene oxide-ZnO nanocomposite films with good mechanical, dynamic mechanical, anti-UV and antibacterial properties. J. Chem. Technol. Biotechnol. **90**, 1677–1684 (2015). https://doi.org/10.1002/jctb.4476
13. Zhang, R., Wang, Y., Ma, D., Ahmed, S., Qin, W., Liu, Y.: Effects of ultrasonication duration and graphene oxide and nano-zinc oxide contents on the properties of polyvinyl alcohol nanocomposites. Ultrason Sonochem. **59**, (2019). https://doi.org/10.1016/j.ultsonch.2019.104731
14. Pourmoslemi, S., Shokouhi, S., Mahjub, R.: Investigation of antibacterial activity of polyvinyl alcohol packaging films composed of silver oxide nanoparticles, graphene oxide and tragacanth gum using Box-Behnken design. Packag. Technol. Sci. **34**, 613–622 (2021). https://doi.org/10.1002/pts.2599

15. Chowdhury, S., Teoh, Y.L., Ong, K.M., Rafflisman Zaidi, N.S., Mah, S.K.: Poly(vinyl) alcohol crosslinked composite packaging film containing gold nanoparticles on shelf life extension of banana. Food Packag. Shelf Life. **24**, (2020). https://doi.org/10.1016/j.fpsl.2020.100463
16. Goh, K., Heising, J.K., Yuan, Y., Karahan, H.E., Wei, L., Zhai, S., Koh, J.X., Htin, N.M., Zhang, F., Wang, R., Fane, A.G., Dekker, M., Dehghani, F., Chen, Y.: Sandwich-architectured poly(lactic acid)-graphene composite food packaging films. ACS Appl. Mater. Interfaces **8**, 9994–10004 (2016). https://doi.org/10.1021/acsami.6b02498
17. Marques, C.S., Dias, M.V., de Soares, N.F.F., Borges, S.V., de Oliveira, I.R.N., Pires, A.C. dos S., Medeiros, E.A.A., Alves, E.: Ultrastructural and antimicrobial impacts of allyl isothiocyanate incorporated in cellulose, β-cyclodextrin, and carbon nanotubes nanocomposites. J. Vinyl Addit. Technol. **27**, 795–805 (2021). https://doi.org/10.1002/vnl.21850
18. Youssef, A.M., El-Naggar, M.E., Malhat, F.M., El Sharkawi, H.M.: Efficient removal of pesticides and heavy metals from wastewater and the antimicrobial activity of f-MWCNTs/PVA nanocomposite film. J. Clean. Prod. **206**, 315–325 (2019). https://doi.org/10.1016/j.jclepro.2018.09.163
19. Azizi, S., Azizi, M., Sabetzadeh, M.: The role of multiwalled carbon nanotubes in the mechanical, thermal, rheological, and electrical properties of pp/pla/mwcnts nanocomposites. J. Compos. Sci. **3**, (2019). https://doi.org/10.3390/jcs3030064
20. Pal, N., Banerjee, S., Roy, P., Pal, K.: Reduced graphene oxide and PEG-grafted TEMPO-oxidized cellulose nanocrystal reinforced poly-lactic acid nanocomposite film for biomedical application. Mater. Sci. Eng. C **104**, (2019). https://doi.org/10.1016/j.msec.2019.109956
21. Torres-Giner, S., Echegoyen, Y., Teruel-Juanes, R., Badia, J.D., Ribes-Greus, A., Lagaron, J.M.: Electrospun poly(ethylene-co-vinyl alcohol)/graphene nanoplatelets composites of interest in intelligent food packaging applications. Nanomaterials **8**, (2018). https://doi.org/10.3390/nano8100745
22. Bhathiwal, A.S., Bendi, A., Tiwari, A.: A study on synthesis of benzodiazepine scaffolds using biologically active chalcones as precursors. J. Mol. Struct. **1258**(5), 132649 (2022). https://doi.org/10.1016/j.molstruc.2022.132649
23. Singh, A., Tiwari, A.: A study on biologically active chalcone based benzodiazepines. IARS' Int. Res. J. **12**, (2022). https://doi.org/10.51611/iars.irj.v12i02.2022.216.
24. Anjaneyulu, B., Chinmay, Chauhan, V., Carabineiro, S.A.C., Afshari, M.: Recent advances on zinc ferrite and its derivatives as the forerunner of the nanomaterials in catalytic applications. J. Inorg. Organomet. Polym. Mater. **34**, 1887–1907 (2024). https://doi.org/10.1007/s10904-023-02952-x.
25. Rao, G.B.D., Anjaneyulu, B., Kaushik, M.P.: Greener and expeditious one-pot synthesis of dihydropyrimidinone derivatives using non-commercial β-ketoesters via the Biginelli reaction. RSC Adv. **4**, 43321–43325 (2014). https://doi.org/10.1039/c4ra06587e
26. Qin, Y., Liu, Y., Yuan, L., Yong, H., Liu, J.: Preparation and characterization of antioxidant, antimicrobial and pH-sensitive films based on chitosan, silver nanoparticles and purple corn extract. Food Hydrocoll. **96**, 102–111 (2019). https://doi.org/10.1016/j.foodhyd.2019.05.017
27. da Brito, S.C., Bresolin, J.D., Sivieri, K., Ferreira, M.D.: Low-density polyethylene films incorporated with silver nanoparticles to promote antimicrobial efficiency in food packaging. Food Sci. Technol. Int. **26**, 353–366 (2020). https://doi.org/10.1177/1082013219894202
28. Chen, Q.Y., Xiao, S.L., Shi, S.Q., Cai, L.P.: A one-pot synthesis and characterization of antibacterial silver nanoparticle-cellulose film. Polym. (Basel) **12**, (2020). https://doi.org/10.3390/polym12020440
29. Roy, S., Rhim, J.W.: Effect of CuS reinforcement on the mechanical, water vapor barrier, UV-light barrier, and antibacterial properties of alginate-based composite films. Int. J. Biol. Macromol. **164**, 37–44 (2020). https://doi.org/10.1016/j.ijbiomac.2020.07.092
30. Shankar, S., Wang, L.F., Rhim, J.W.: Preparation and properties of carbohydrate-based composite films incorporated with CuO nanoparticles. Carbohydr. Polym. **169**, 264–271 (2017). https://doi.org/10.1016/j.carbpol.2017.04.025
31. Ahmadi, A., Ahmadi, P., Sani, M.A., Ehsani, A., Ghanbarzadeh, B.: Functional biocompatible nanocomposite films consisting of selenium and zinc oxide nanoparticles embedded in gelatin/

cellulose nanofiber matrices. Int. J. Biol. Macromol. **175**, 87–97 (2021). https://doi.org/10.1016/j.ijbiomac.2021.01.135
32. Li, X., Ren, Z., Wang, R., Liu, L., Zhang, J., Ma, F., Khan, M.Z.H., Zhao, D., Liu, X.: Characterization and antibacterial activity of edible films based on carboxymethyl cellulose, Dioscorea opposita mucilage, glycerol and ZnO nanoparticles. Food Chem. **349**, (2021). https://doi.org/10.1016/j.foodchem.2021.129208
33. Ojha, N., Das, N.: Fabrication and characterization of biodegradable PHBV/SiO2 nanocomposite for thermo-mechanical and antibacterial applications in food packaging. In: IET Nanobiotechnology. pp. 785–795. Institution of Engineering and Technology (2020)
34. Maćkiw, E., Mąka, Ł.,Śiezyńska, H., Pawlicka, M., Dziadczyk, P., Rzanek-Boroch, Z.: The impact of plasma-modified films with sulfur dioxide, sodium oxide on food pathogenic microorganisms. Packag. Technol. Sci. **28**, 285–292 (2015). https://doi.org/10.1002/pts.2104
35. Wu, J., Sun, Q., Huang, H., Duan, Y., Xiao, G., Le, T.: Enhanced physico-mechanical, barrier and antifungal properties of soy protein isolate film by incorporating both plant-sourced cinnamaldehyde and facile synthesized zinc oxide nanosheets. Colloids Surf. B Biointerfaces **180**, 31–38 (2019). https://doi.org/10.1016/j.colsurfb.2019.04.041
36. Amjadi, S., Almasi, H., Ghorbani, M., Ramazani, S.: Preparation and characterization of TiO_2NPs and betanin loaded zein/sodium alginate nanofibers. Food Packag. Shelf Life. **24**, (2020). https://doi.org/10.1016/j.fpsl.2020.100504
37. Swaroop, C., Shukla, M.: Nano-magnesium oxide reinforced polylactic acid biofilms for food packaging applications. Int. J. Biol. Macromol. **113**, 729–736 (2018). https://doi.org/10.1016/j.ijbiomac.2018.02.156
38. Kousheh, S.A., Moradi, M., Tajik, H., Molaei, R.: Preparation of antimicrobial/ultraviolet protective bacterial nanocellulose film with carbon dots synthesized from lactic acid bacteria. Int. J. Biol. Macromol. **155**, 216–225 (2020). https://doi.org/10.1016/j.ijbiomac.2020.03.230
39. Bi, F., Zhang, X., Liu, J., Yong, H., Gao, L., Liu, J.: Development of antioxidant and antimicrobial packaging films based on chitosan, D-α-tocopheryl polyethylene glycol 1000 succinate and silicon dioxide nanoparticles. Food Packag. Shelf Life. **24**, (2020). https://doi.org/10.1016/j.fpsl.2020.100503
40. Mahmoudi, N., Ostadhossein, F., Simchi, A.: Physicochemical and antibacterial properties of chitosan-polyvinylpyrrolidone films containing self-organized graphene oxide nanolayers. J. Appl. Polym. Sci. **133**, (2016). https://doi.org/10.1002/app.43194
41. Usman, A., Hussain, Z., Riaz, A., Khan, A.N.: Enhanced mechanical, thermal and antimicrobial properties of poly(vinyl alcohol)/graphene oxide/starch/silver nanocomposites films. Carbohydr. Polym. **153**, 592–599 (2016). https://doi.org/10.1016/j.carbpol.2016.08.026
42. Xu, W., Xie, W., Huang, X., Chen, X., Huang, N., Wang, X., Liu, J.: The graphene oxide and chitosan biopolymer loads TiO_2 for antibacterial and preservative research. Food Chem. **221**, 267–277 (2017). https://doi.org/10.1016/j.foodchem.2016.10.054
43. Arfat, Y.A., Ahmed, J., Ejaz, M., Mullah, M.: Polylactide/graphene oxide nanosheets/clove essential oil composite films for potential food packaging applications. Int. J. Biol. Macromol. **107**, 194–203 (2018). https://doi.org/10.1016/j.ijbiomac.2017.08.156
44. Sharma, C., Manepalli, P.H., Thatte, A., Thomas, S., Kalarikkal, N., Alavi, S.: Biodegradable starch/PVOH/laponite RD-based bionanocomposite films coated with graphene oxide: preparation and performance characterization for food packaging applications. Colloid Polym. Sci. **295**, 1695–1708 (2017). https://doi.org/10.1007/s00396-017-4114-9
45. Lin, D., Wu, Z., Huang, Y., Wu, J., Li, C., Qin, W., Wu, D., Li, S., Chen, H., Zhang, Q.: Physical, mechanical, structural and antibacterial properties of polyvinyl alcohol/oregano oil/graphene oxide composite films. J. Polym. Environ. **28**, 638–646 (2020). https://doi.org/10.1007/s10924-019-01627-4
46. Li, F., Yu, H.Y., Wang, Y.Y., Zhou, Y., Zhang, H., Yao, J.M., Abdalkarim, S.Y.H., Tam, K.C.: Natural biodegradable poly(3-hydroxybutyrate- co-3-hydroxyvalerate) nanocomposites with multifunctional cellulose nanocrystals/graphene oxide hybrids for high-performance food packaging. J. Agric. Food Chem. **67**, 10954–10967 (2019). https://doi.org/10.1021/acs.jafc.9b03110

47. Shen, X.J., Yang, S., Shen, J.X., Ma, J.L., Wu, Y.Q., Zeng, X.L., Fu, S.Y.: Improved mechanical and antibacterial properties of silver-graphene oxide hybrid/polylactid acid composites by in-situ polymerization. Ind. Crops Prod. **130**, 571–579 (2019). https://doi.org/10.1016/j.indcrop.2019.01.018
48. Rivera-Briso, A.L., Aachmann, F.L., Moreno-Manzano, V., Serrano-Aroca, Á.: Graphene oxide nanosheets versus carbon nanofibers: Enhancement of physical and biological properties of poly(3-hydroxybutyrate-co-3-hydroxyvalerate) films for biomedical applications. Int. J. Biol. Macromol. **143**, 1000–1008 (2020). https://doi.org/10.1016/j.ijbiomac.2019.10.034

Chapter 8
Sustainable Coatings and Films for Active Food Packaging

Khalid Mehmood Wani, Ghumika Pandita, Athira R. S. Pillai, Ansu Sara Eapen, and Swarup Roy

Abstract Active food packaging incorporates the indulgence of novel technology to ensure food quality and safety. However, traditional petroleum-based materials significantly account for pollution in the form of a carbon footprint and resource depletion. Perhaps the significance of sustainability has given rise to the use of eco-friendly polymers as primary components for active food packaging. The chapter focuses on the study of numerous sustainable polymers, natural extracts, and bio-based compounds to be incorporated as coating and films for active food packaging systems to be potential alternatives to replace conventional packaging materials. The materials include bio-based compounds such as biopolymers and bioplastics. In this regard, biopolymers such as carbohydrates, protein, and polyester polymers could be useful. Adding active ingredients like antimicrobial agents, antioxidants, gas scavengers, etc., is useful for producing functional film and coatings for food packaging. The chapter also includes evaluating the physical and functional properties of such materials to establish compatibility with various food products. Furthermore, the application of sustainable packaging film in food applications is also discussed. The current study affirms a comprehensive understanding of the ecological footprint associated with these materials. Furthermore, the chapter focuses on revolutionizing the food packaging landscape to foster a more sustainable environment.

Keywords Active food packaging · Films & coatings · Biopolymers · Bioplastics · Sustainability

K. M. Wani (✉) · G. Pandita · A. R. S. Pillai · A. S. Eapen · S. Roy (✉)
Department of Food Technology and Nutrition, School of Agriculture, Lovely Professional University, Phagwara 144001, India
e-mail: khalid.iitg@gmail.com

S. Roy
e-mail: swaruproy2013@gmail.com

S. ul Islam and M. Shahid (eds.), *Green Materials for Active Food Packaging*, Engineering Materials, https://doi.org/10.1007/978-981-96-0369-5_8

8.1 Introduction

Consumer preference is an ever-changing cycle towards the product. Recent years have shown a paradigm shift in preference towards safe, convenient, and environmentally friendly products. Packaging, along with the actual products, is an important element of this cycle. Packaging serves as a critical linkage between manufacturer and consumer by playing a crucial role in modern life, ensuring product protection, the absence of containments, and allowing delivery across various parts of the world. However, conventional packaging systems are mainly developed with polymers obtained from petrochemical wastes. These petrochemical-derived polymers are often considered among non-renewable resources. The fossil-based resources, mainly petrochemicals, are finite, and the dependence of plastic on petrochemicals causes further strain on resources [12]. Though these polymers are nonrenewable, their very low price leads to a very high magnitude of utilization. The large utilization leads to the generation and, consequently, expelling of large amounts of waste. The global packaging industry generates an alarming amount of waste. Geyer et al. [39] estimated over 300 million tons of plastic end up in annual waste. Though generated in such huge quantities, very little attention is given to the effective disposal and recycling of these non-degradable, non-renewable materials. Non-degradable plastic materials are persistently applied in the landfills. Very minute degradation in the landfills leads to the leaching of harmful chemicals into the environment and causes a threat to land microflora and other organisms [43]. On the other hand, the majority of the plastic waste gets washed away into the oceans through different channels. The washed plastic further undergoes micritization and forms into microplastics. These microplastics end up in the food systems and finally in human plates [111]. These plastic wastes aggregate in soil, water, and air at an alarming rate.

The alarming environmental issue and globally elevating concerns have necessitated the paradigm shift towards eco-friendly solutions in many industries. The packaging industry is not untouched by such approaches and has experienced a primary focus on sustainable approaches, including packaging. This important transition helps to reduce the damage to the environment, encourage effective utilization of resources, and promote the circular economy with several closed-loop and open-loop principles [71]. The transition also demands innovative ideas for both the utilization of biomolecules for the development of sustainable packaging and the selection and addition of active ingredients, fillers, and additives [2].

Sustainable packaging utilizes a complete process that ponders the whole lifecycle of packaging material from obtaining the source, extraction, and purification of biopolymers, adding active ingredients, developing packaging, utilizing the packaging by the consumer, and end-of-life disposal. These sustainable approaches, coatings, and films have emerged as a vital component in making eco-friendly packaging solutions that meet consumer demand and satisfy the prescribed regulations [136]. This approach prioritizes the utilization of renewable, biodegradable, and compostable biomolecules with minimal resource utilization, energy consumption, and carbon footprints throughout the supply chain, curbing the concerns associated with conventional packaging [44]. The sustainable approach of packaging shares several advantages that mainly include lower dependency on non-renewable fossil resources and very high dependency on plant-based renewable polymers, enhanced biodegradability that can be achieved through the breakdown of biomolecules by natural microorganisms that are present in the soil, reducing landfills accommodating chemical leaching and plastic pollution in land and ocean systems and lower the energy requirement and environmental food prints [27].

The search for sustainable packaging material has led to the emergence of several promising polymers. Among these, bio-based polymers result from renewable biomass resources, including carbohydrate, protein, lipids, and their derivatives. Carbohydrates, mostly polysaccharides such as starch, chitosan, pectin, etc., are abundantly found in plant material and have the remarkable advantage of being renewable and biodegradable polymers [11]. These carbohydrate-based biopolymers have become one of the most widely used classes of polymer in sustainable packaging due to advancements in processing techniques. Proteins resulting from both plants and animals, for example, soybean, zein, whey, gelatin, etc., have generated prominence for the making of packaging systems [25]. These protein molecules show excellent film farming properties, biocompatibility, and biodegradability, making them suitable biomaterials to resolve a range of packaging needs. The derived polymers, such as polylactic acid (PLA), represent another category of sustainable polymers [35]. These are synthesized from renewable sources like cornstarch, sugarcane glucose, and vegetable oils, providing viable alternatives to traditional petroleum-based plastics. These derived biomolecules have comparable properties with plastic and other conventional materials. These biopolymers have the advantage of being utilized in the existing process facilities for the development of sustainable packaging materials [95].

The quality of coatings and films made from these bio-polymers can be measured in terms of physical properties, morphological properties, barrier properties, tensile properties, and thermal properties [122, 123]. These polymers show the ability to form a comprehensive structure and thin protective layer. The films developed with these biopolymers show good barrier properties, especially against oxygen. However, these films struggle to show resistance against moisture, oil, and even oxygen in some cases. Though these bio-polymers show good film-forming capacity, the tensile strength, flexibility, and elongation ratio do not serve the purpose and need either biopolymer modification or added active compounds, fillers, and additives to improve the quality [124]. In improving the quality, these sustainable films can be applied with active ingredients to enhance their applicability in protecting the food. This makes the biopolymer a great area of interest for achieving sustainability [121].

Therefore, this chapter looks into new innovations and developments focusing on the utilization of non-conventional polymers from renewable sources and the utilization of sustainable active ingredients and other additives. These innovations are expected to improve the performance of packaging with enhanced functionality and sustainability.

8.2 Polymers Used in Sustainable Coating and Films

The development of a sustainable packaging system is highly dependent on the biopolymer used. The selection of a biopolymer is an imperative feature for the long-term acceptability of the system. The ready availability of the biopolymer, the possibility of modification, and the versatility of application are some of the parameters that need to be considered for utilization in sustainable food packaging [137]. As we look into the compositional aspects of different sources, carbohydrates are the most abundantly found biomolecules in nature, followed by proteins and fats. Carbohydrates such as cellulose, chitosan, carrageenan, hemicellulose, and lignin have been utilized as the base material for various packaging materials in food and non-food systems due to their ready availability, easy bio-degradability, and minimal impact on the environment in terms of carbon footprint [135]. Other carbohydrates, such as starch and their modified forms and derivatives, have recently been emphasized for their utilization. However, having an affinity toward water molecules hinders the application of these biomolecules in sustainable packaging and reveals the need for modification [20].

Proteins are another major compositional biopolymer that is present in various plant and animal sources. However, the proteins applicability of the protein is highly dependent on their solubility, film forming ability, techno-functional properties and its interaction with other constituents off the enclosed food. Several attempts have been made to incorporate proteins in the packaging system; however, the limitation of nutritional security is also another concern in terms of protein when utilized as a base for packaging rather than as a food ingredient [61].

8.2.1 Carbohydrates

8.2.1.1 Starch

Starch is a naturally occurring sugar consisting of glucose units attached together by a glycosidic bond. Starch is among one of the most abundantly found polysaccharides in seeds, tubers, roots, and fruits (making it a highly potent bio-polymer for the development of sustainable packaging. It comprises of amylose and amylopectin. Amylose can be characterized by linear polysaccharides in which the glucose is attached with α-(1 $\rightarrow$ 4) glycosidic linkages [19]. Conversely, the amylopectin is observed in vastly branched polymer with α-(1 $\rightarrow$ 4) and α-(1 $\rightarrow$ 6) glycosidic bonds. The content of both amylose and amylopectin has a significant impact on the properties of starch. This unparalleled composition contributes to the diverse properties of starch, making it a suitable biopolymer for various packaging applications [11].

The conversion of starch into packaging film is achieved through a complex phenomenon of gelatinization that is mainly triggered by water and heat. The starch natively exists in the granular form that, when heated in the presence of water, absorbs a very high amount of water and gets swelled. This swelling disrupts the crystallinity of the starch granule, allowing the water to penetrate further and solubilize starch molecules [9]. The water molecules enter the starch granules; they form hydrogen bonds with amylose and amylopectin molecules, eventually reaching the elastic limit of the starch granule and finally bursting it. The bursting of granules releases the amylose and amylopectin molecules into the surrounding. This dispersed amylose and amylopectin interact with other starch molecules to form a viscous solution [24]. With further heating, the starch granules undergo further changes, resulting in gel formation. During gel formation, the starch molecules interact with water through hydrogen bonding and other molecular forces. This interaction led to the development of a three-dimensional network where starch molecules are held with water. This strength of the 3-D network provides gel texture and stability. The fully formed gel will then be cast and subjected to drying. The drying of the gel can be considered as the important step at which the gel gets stabilized in the form of packaging film. The gel undergoes retrogradation, the process in which the gel structure gets reorganized and reforms itself into a highly ordered arrangement [39]. The process of retrogradation progresses with the removal of heat, which leads to the breakage of hydrogen bonds and, consequently, the removal of water. This phenomenon leads to increasing the stability of the system with the formation of a firm gel or even the formation of insoluble complexes depending on specific conditions. This traditional technique of casting starch-based packaging film was being followed to enhance its large-scale production with higher moisture content, leading to large drying times and huge utilization of energy resources. Henceforth new techniques including foaming process in which starch-based foams are being prepared by subjecting starch under baking, molding, extrusion and supercritical foaming also addition of external components ultimately leading to heating [14]. According to Chen et al. [18] baking process evaporates the water and formation of foam take place in form of vapor bubbles which is directly dependent of the viscosity of starch. Furthermore, in the molding process the material is put under hydraulic press in which temperature controlled and due to temperature–pressure combination, foam is formed. Starch based foam in addition with grape stalk have successfully been formulated using this technique and has an application in food packaging industry. Supercritical fluid foaming employs supercritical carbon dioxide for gas bubble formation in which carbon dioxide is mixed up with the foaming material under pressure then consecutive depressurization to bubble formation and ultimately foamed product can be formed. Glutaraldehyde can be added into these foams to stabilize their structure. Extrusion is also another technique in which blowing, compression molding, and injection molding could be potentially done in which blowing involves blowing, pressing and cooling of material to produce packaging material. Similarly, compression involves hot pressing and cooling of the extruded product. Injection process utilizes the raw molten material subjected to the mold and obtains the packaging material post-cooling operation. However, Injection molding process of extrusion produces large number of complex plastic products at low costs making it a most viable process of packaging product formation [64].

The newer techniques are also employed for the strengthened packaging material formation with utilization of lower energy resources, less time, optimized process and cost effective. These techniques include Electrospinning, 3D-Printing, Reactive extrusion, Nanotechnology. Electrospinning process produces the nanometer or micron size polymers by electrostatic interaction [132]. Based on the intended use of the polymer with solid, hollow, flat, or ribbon like structures and interiors, respectively. Starch possesses great affinity towards nanofiber formation by electrospinning due its appropriate biocompatibility [77]. According to Munteanu and Vasile [28], numerous polymers such as polycaprolactone, polyvinyl alcohol, polyepoxyethane, polyglycolide and Poly-lactic acid can be used amalgamated with starch having numerous strengthened properties along with wide field of application in food industry. 3D-Printing involves designing of 3D models using computer software and printing them layer to layer [21]. As this technology is widely being applied in the food industry however in the formulation of starch-based packaging materials it exhibits various limitations such as direct dependence on printability, gelling behavior, mechanical properties as well as heat resistance based upon the variety of starch being used. Reactive extrusion is yet another technique which employs chemicals and /or enzymatic reactions leading to polymer modification by polycondensation, polymerization, grafting and crosslinking processes. In this technique the extruder acts as a bioreactor [143]. The technique subjects' starch to milder conditions providing more time for chemical reactions, modify in less time along with low cost and high efficiency. In a study by Fonseca-Florido et al. [25], plasticization and succinization of cassava starch can be achieved on reactive extrusion. Furthermore, nanotechnology can be also applicable in formulation of packaging material. The molecular change to nanoparticles leads to improvement on the mechanical and barrier properties of packaging material [115]. As per Falcoet al. [34], the inclusion of cellulose particles into the starch based films can boost their mechanical performance.

The starch-based films showed superior quality traits in terms of oxygen properties owing to compact molecular arrangement of amylose and amylopectin at the time of gelatinization and further reduction in the porosity at the during retrogradation [84]. Nevertheless, restriction for the starch-based films in properties like inferior elongation, flexibility, water vapor transmission rate makes it challenging for practical use. These limitations lead to derivation of different polymers from starch and utilization into packaging system.

8.2.1.2 Cellulose

Cellulose represents as one of the most appropriate polymers to be utilized as food packaging material along with its superior mechanical properties. It is as well a low cost, biodegradable and sturdy material with its derivation from Agricultural wastes, cotton, stalks and forest remnants etc. Cellulose is a homopolysaccharide comprising of β-1, 4-anhydro-D-glucopyranose unit [98]. This indicates the asymmetrical chair conformation structure with glucose units about the molecular axis whereas hydroxyl groups at equatorial position. Furthermore, each cellulose monomer unit consists of three hydroxy groups [48]. These hydroxyl bonds indicate the Hydrogen bond formation necessity in the formation of packaging materials. According to [110], the strength of this polysaccharide reliant on the number and alignment of hydrogen bonds. It includes the characteristics like little solubility and high crystallinity. Furthermore, Cellulose is always biosynthesized as a polymer chain having possession of both amorphous and crystalline regions. The H bonds within hemicelluloses and pectin indicates no penetration of solvents and reagents through the polymer chain [110]. However, in the packaging film formation various techniques such as periodate oxidation wherein aldehydes groups are utilized for enhancing cross linking within the cellulose chains; coupling agents including organo-silanes increase the adhesion properties by reducing the water intake capacity of the polymer; Esterification includes esterification of cellulose fibers for improving hydrophobicity of the polymers ultimately resulting in its resistance to the humidity. Furthermore, techniques such as isocyanate and polymer grafting could be used for the polymer modification and property enhancement [110].

The fabrication of cellulose based packaging materials could be done by numerous techniques. Solution casting involves amalgamation of two or more polymers in a solution for facilitation of cross linking within each other. In this technique, the polymer is soluble in the solution while as the nanoparticles may be scattered [61]. Cellulose triacetate and numerous other cellulose acylate films are fabricated by this technique. Layer by layer technique involves the addition of polymers along with the coatings in layer-by-layer fashion. This technique enhances the barrier, functional as well as physicochemical properties of the packaging film [50]. Qatarized cellulose along with CMC were mixed together to form an antifogging and anti-frosting behavior film. Extrusion involves the preparation of variable shape polymer at low cost, high efficiency and less time. Its majorly applicable in fabrication of tubular cellulose films [13]. Coating refers to direct application of the polymer material on to the surface of the food product by the process of dipping and/or spraying. Cellulose based coatings are generally preferred these days due to their ability to carry bioactive, moisture barrier and antimicrobial properties which maintains the aesthetics and quality of the food product [140]. Polymeric Hydrogels refer to the formation of 3D structure of polymer with and ability to encapsulate bioactive compounds within to enhance controlled release delivery of the compound along with the enhancement in the area application of cellulose polymers in the area of food packaging. Spray drying technique involves the spraying of polymer in order to obtain a packaging material [139]. Polymers along with the small size particles are dried under high temperature conditions [147]. Electrospinning is yet another technique which is one of the most versatile and robust techniques utilizing coaxial and emulsion electrospinning to obtain a biopolymer. In a study by Ohkawa [85], for regeneration of thin cellulose fibers cellulose acetate was subjected to two step electrospinning process for formulating a packaging material. Nano and micro-encapsulation are the other techniques in which the polymers are continuously utilized for encapsulation of the bioactive compounds which are highly sensitive towards environmental conditions. This technique enhances the application of cellulose as well as the quality of the food product [82]. Liposomes and nanoliposomes can as well be utilized in the cellulose derivative based films formation. These molecules can also be used for enhancing biodegradability, amphiphilic nature and biocompatibility of packaging material with the food product [37]. Nano micelles and Nano emulsion techniques also belong to the similar category of biocompatibility enhancement, solubility upgradation and carrier for bioactive compounds, respectively. Furthermore, Adsorption refers to the technique of cellulose acts as an adsorbent for holding liquid molecule in order to enhance the shelf life of the packed food product. However, in order to enhance the overall stability of the packaging material for better dispersion in polymer matrix, improved modifications in the form of derivates or nanoparticles formation would prove to be a new era for the utilization of cellulose fibers and its derivatives for development of green and sustainable environment [48].

8.2.1.3 Pectin

Pectin is one of the renowned names in renewable packaging materials and can be sourced from citrus peels. Pectin is one of the most complex polysaccharide majorly composed of galacturonic acid (Gal-A) which is linked by α-1,4-glycosidic bonds [51]. According to Kumar et al. [57], homogalacturonan, rhamnogalacturonan-I and rhamnogalacturonan-II are the conformation structure forming unit of pectin. It ultimately impacts the properties of pectin packaging films such as higher the linear forming units in pectin, higher will be the viscosity and the packaging film mechanical strength would be improved. Furthermore, the Gal-A C-6 carboxy group can be esterified and based on the degree of esterification high methoxyl pectin (HMP) (>50%) and low methoxyl pectin (LMP) (<50%) in which HMP forms gel in higher sugar and acidic conditions whereas LMP forms gel in contact with metal ions. Therefore, based on the intermolecular bonding of HMP and LMP the gelling performance diversity could be detected along with their film formation caliber. However, the pectin based packaging films could be formulated by the process of casting utilizing a suitable plasticizer including chlorine [124]. Spraying which includes addition of pectin with a suitable plasticizer such as chitosan using spray gun and further drying to obtain an acceptable packaging material [18]. Extrusion involves the blending of pectin and plasticizer while subjecting to different extrusion conditions to determine the quality of formulated pectin film [19]. Henceforth, the characteristics such as hydrophobicity, film components, gelling and thermal stability indicate the quality of the pectin films. However, the pectin films exhibit lower tensile strength which limits their utilization as the packaging material. Thus, the incorporation of additional strength enhancing compounds could be done to show synergistic impact on the mechanical performance of the packaging film [51].

8.2.1.4 Proteins

Proteins are complex molecules made of amino acid linked together with peptide linkage [149]. The proteins have a diverse range of structures and functions in organisms making them one of the abundant biomolecules. The alteration in distribution of amino acids determines the overall three-dimensional structure of protein known as tertiary structure which is crucial for its functionality [56]. The proteins can undergo various alteration and modifications to form a packaging film. A very common approach involves extraction of protein from suitable sources such as plant, animal or microorganisms. These proteins are then processed to form film through techniques such as casting extrusion, or compression molding. During film formation, the proteins molecules undergo structural changes driven by molecular interaction and are stabilized by the forces such as H bonding, hydrophobic and electrostatic interaction [51].

Initially, the proteins are dispersed insolvent or aqueous solution where they may undergo conformational changes, unfolding the native structure. The unfolding or denaturation led to expulsion of original hydrophilic region and exposes hydrophobic regions [104]. This exposure provides the potential sites for the binding of other protein molecules initiating and enhancing intermolecular and intramolecular crosslinking. Hydrogen bonding between polar groups on adjacent biomolecules helps in stabilizing the film structure while hydrophobic interaction contributes to cohesion within the film. The casted films further subjected to drying for removal solvents. This allows the solvent to evaporate and provide the opportunity to protein molecules to rearrange and interact to form a cohesive film structure [38]. In addition to these interactions, crosslinking agents may be added to the protein solution to enhance film stability and mechanical properties. Addition of the crosslinking agents promotes the formation of covalent bonds between protein molecules reinforcing film structure and improving its resistance to mechanical stress and environmental factors [17].

Similar wet and dry process of film formation as plastics is also involved in protein based packaging. The wet method includes solvent casting of the polymer whereas dry method involves the compression, molding or extrusion ultimately resulting into fabrication of thermoplastic [40]. The denaturation protein on exposure to heat, acid or base, and solvent is a characteristic feature of both the method. The greater the rate of denaturation greater is the extension of the of the molecule which enhances their interaction among the protein chains on intrusion of electrostatic, van der waals, hydrogen bond, covalent and disulfide bond interaction. The generation of the cohesive matrix of the protein takes place [45]. To improve performance of the protein biopolymers produced the techniques such as plasticization, cross-linking, composite film development and nanotechnology are employed. Plasticization involves the addition of plasticizer in order to enhance the rigidness and brittleness of protein polymer [29]. Plasticizers are the small size molecules having low volatility which reduce the protein interactions within themselves, enhance flexibility, dispensability and extensibility of the biopolymer film. Water, glycerol, sorbitol and xylitol are one of the most commonly used plasticizers [70]. Cross linking is yet another technique to enhancing number of bonds on linking of proteins with each other assuring stronger intermolecular covalent bonds leading to reduction in polymer mobility. Heat treatment, acid and/or base treatment, aldehydes addition glyoxal and phenolic compounds etc. induce cross linking in the bio polymers. Furthermore, composite films could be formulated through the process such as blending, laminating, extruding or coating with suitable polymers or polysaccharides. Numerous composite films could be fabricated based on their utilization such as protein–protein films, protein-polysaccharide films, lipid protein films, protein polymer films and nanocomposites [96].

The protein biopolymer films serve with many advantages for environment however; they have limitation on the barrier properties compared with synthetic polymers. The proteins are inherently having hydrophilic sites on the molecules that inevitably absorb the water molecules and further depriving the performance for oxygen barrier too. Other properties reliant on the source of protein, conditions of extraction, thermal and pH stability. These drawbacks can be overcome with the approaches such as blending, physical and chemical modification and application on fillers [8].

8.2.1.5 Derived Polymers

The biopolymers that are naturally present and utilized into the different packaging serve with several advantages such as renewable, biodegradable and lower environmental impact. However, when it comes to the flexibility, plasticizing effect and barrier properties, these biopolymers could not withstand the comparison with synthetic polymers. Considering these aspects, several natural biopolymers can be tweaked through physical and chemical processing to withstand the requirement. When incorporated into coating formulations, these can improve the strength, flexibility, and barrier performance, thereby reducing their environmental footprint [4]. The major derived biopolymers are provided in Table 8.1.

8.3 Active Ingredients Applied in Sustainable Packaging

Active compounds refer to the secondary metabolites produced by plants and animals and this section summarize the compounds that are having an application in the sustainable packaging. Active ingredients addition along with the biodegradable sustainable packaging material has lead to development of new concept called as responsive active packaging [40]. It states the addition of anti-microbial and antioxidative compound such as essential oils and compounds like anthocyanin into the packaging materials and final product not only to maintain the quality of the food product as well as to indicate the activity of certain microorganism or undesirable chemical reactions within food material. Therefore, the effectiveness of these compounds has been highlighted in recent years with respect to the reduction in food spoilage rate, stimuli-response of compounds and quality product production [55].

Table 8.1 Derived biopolymers with their application for improvement of sustainable packaging

Derived polymer	Process technology	Experimental design	Improvement	References
Carboxymethyl cellulose (CMC)	Cellulose when reacted with sodium chloroacetate at alkaline condition it produce CMC	CMC + ZnO	Improved UV-Barrier property	[105]
Carboxymethyl chitosan (CMC)	Chitosan when reacted with sodium chloroacetate at alkaline condition it produce CMC	Gelatin + CMC + Agar	Improve solubility, and biodegradability	[145]
Hydroxypropyl starch (HPS)	Starch when treated with Propylene oxide at alkaline condition gives HPS	Cellulose nanofiber + HPS	Excellent flexibility and mechanical properties	[144]
Alginate-gelatin	Alginate and gelatin crosslinked with calcium ions	Nanoclay + sodium alginate-gelatin	Improved antimicrobial activity and biodegradability	[90]
Soy protein isolate (SPI)	Alkali extraction of protein followed by acid precipitation and centrifugal separation	Nanocellulose + SPI + Green extract phenolic compound	Improved antioxidant, antibacterial, and mechanical properties	[78]
Low-methoxy pectin	Aqueous solubilization followed by separation, filtration, precipitation, and drying	Low-methoxyl pectin + chitosan	Lowered water absorption and better WVP	[146]
Lignosulfonates	Lignin allowed to react with sulfonic acid gives lignosulfonates	Chitosan + bentonite + lignosulfonate	Improved strength, antioxidant and UV-shielding properties	[116]
Acetylated gum Arabic	Acetylation of guar gum with acetic anhydride in presence of pyridine or sodium acetate	Chitosan + Arabic Gum Nanoparticles + Oregano Essential Oil	Enhanced biodegradability	[62]

(continued)

Table 8.1 (continued)

Derived polymer	Process technology	Experimental design	Improvement	References
Polylactic acid (PLA)	Lactic acid is polymerized in presence of tin(II) chloride or titanium(IV) isopropoxide, at elevated temperatures	PLA + plasticizers	Enhanced biodegradability	[117]
Polyvinyl alcohol (PVOH)	Polyvinyl acetate is hydrolysed in acidic or alkaline condition to acetate groups (-$COOCH_3$) into hydroxyl groups (-OH)	Poly (Vinyl Alcohol) + Glass Flakes	Maintain flexibility and barrier properties	[17]

With the growing era the treatments used to reduce down the action of microbes especially Gram positive bacteria have become nugatory due to their development of resistance over time such as benzylpenicillin [40]. However, as the issue became serious the development of novel techniques such as utilization of organic acids like sorbic acid was applied to inhibit the microbe and its spores by acting on its plasma membrane in a protonated form and kill the cell by accumulation of free radicals and oxidative stress. Development in science and technology has led to the production of a newer concept of addition of antimicrobial compounds such as organic acids, salts, aldehydes, phenol, terpenoids and chitosan etc. [5].

When it comes to mechanism of action of these active compounds, majorly two mechanism comes forward for giving the preservative action, namely microbial cell destruction upon ionization of the active compound and inactivation of essential proteins through molecular alteration and binding to active sites [121, 123]. Organic acid like sorbic acid serve as effective food preservative due to their ability to prevent the microbial contamination. Sorbic acid naturally occurring in fruit like berries and can synthetically be produced for commercial purpose. When applied in food, sorbic acid comes in contact with water and dissociate into sorbet ions in acidic environment of the food. These eyes penetrate microbial cell membrane disrupting the integrity and causing leakage of seller content comprising essential functions and osmotic balance of microbial cell. Furthermore, sorbic acids protonated form lowers intracellular PH enters in the microbial enzyme activity and interfering other biochemical and metabolic processes. These disruptions include Cellular stress triggering the production of periodical the damaged cellular component such as lipids proteins in cell organelles and DNA. To oxidative stress consequently weakens the microbial defense mechanism leading to eventual cell death due to accommodation of oxidative damage and membrane description. Micros exposed to sorbic acid cannot survive are reproduced effectively injuring the food preservation [134].

Considering the other mechanism, the organic compounds such as eugenol, potential antifungal compound exacts its effects through multiple mechanisms targeting various cellular structures and processes in fungal cells. Primarily eugenol interacts with the fungal cell membrane, a vital barrier responsible for maintaining cell integrity and regulating nutrient transport in microbial cells. By descriptive the lipid-protein bilayer structure, eugenol increases membrane permeability comprising the fungal cell's ability to maintain osmotic balance. Eugenol interferes with ergosterol synthesis a critical component of fungal cell membrane analogous to cholesterol in animal cells. Reduced ergosterol levels weakens the membrane making it more susceptible to damage and further compromising fungal cell integrity. Eugenol Interact with the proteins and disrupt fungal mitochondria, an essential organism involved in energy production (ATP synthesis) and other cellular processes. By disrupting mitochondrial functions, eugenol to energy depletion and impaired growth for the compromising fungal cell viability. Furthermore, eugenol Trigger the production of reactive oxygen species within fungal cell such as super oxide radicals and hydrogen peroxide causing oxidative stress consequently damaging cellular proteins, lipids, and DNA ultimately leading to Cell death. Moreover, usual inhibits specific fungal enzymes and metabolic Pathways such as Glucan synthase Involved in cell wall synthesis and adenosine triphosphate (ATP) production affecting energy metabolism and cellar integrity [53].

This concept delves into the incorporation of these compounds into the food packaging matrix to target certain undesirable microorganisms. In the sustainable packaging Industry Essential oils (EOs) are mostly used compounds in which terpenoids, aldehydes and phenol is of main interest which interact with the microbial cell due to their higher hydrophobicity that leads to their enhanced interaction with cell membrane and mitochondria which eventually results in the vital component loss of the cell [79]. According Khaneghah et al. [79], The membrane of *E. Coli* could be disintegrated by the action of EOs consisting of carvacrol and thymol by facilitation of eugenol entry into the cell and suffice interaction with the proteins.

Chitosan is a potent antimicrobial agent which is active against the broad spectrum microbes [119]. If the compound is given the required acidic conditions the glucosamine monomer at C-2 carries a positive charge which can easily interact with the negatively charged segment of cell membrane like proteins which can substantially reduce down the microbial load in the packaging material [112]. Furthermore, chitosan is also known to hinder the activity of baker's yeast by inhibiting the permeation of glucose inside the cell eventually leading to cell death [7].

Anthocyanins are also potential antimicrobial compounds which are highly effective against the *E. coli* [65]. These compounds are also potential inhibitors of proline in *L. monocytogenes* [9]. Anthocyanin's also potentially damage the bacterial cell by down-regulating the gene encoding, cell wall phospholipids synthesis etc.

Furthermore, the microbial activity of microorganisms can be reduced through various approaches which include protein structure alteration, cell membrane protein or lipid change, cell wall component blockage, replication, transcription and translation inhibition of nucleic acids, cellular metabolism disruption [52]. The activity of the antimicrobial agent is directly dependent on direct and indirect contact with the microorganism. The direct action can be achieved by diffusion and solubilization of non-volatile compounds into the surface of food where microorganisms could be mainly present [42]. However, the indirect contact can be achieved by interaction of volatile substance around the food and antimicrobial compound by the absorption phenomenon [79]. Therefore, the antimicrobial compounds could be added into the packaging or the material could have intrinsic antimicrobial activity such as chitosan. The antimicrobial compound adding could also be due to the chemical or physical modification by covalent or non-covalent bonds with the matrix [56]. Thus, the addition of active components in the packaging films makes the food spoilage indication easier, reduces food waste and spoilage, and enhances the approach to food security, sustainability, and biodegradability.

8.4 Properties of Sustainable Packaging

The environmental concerns related to polymer-based packaging film paved the path for the interest in sustainable packaging research [142]. The characteristics of the film, to withstand tear and stress during storage and transportation, barrier to protect from external factors such as gases, UV light and moisture and functional properties including antioxidant and antimicrobial are critical to maintaining food freshness [49]. Bio-based polymers exhibit high properties even though to modify certain properties and to enhance shelf life some substances, like active compounds, bio nanocomposites, and copolymers are added (Table 8.2) and it is schematically representation is given in Fig. 8.1.

8.4.1 Barrier Properties

Barrier properties are referred to as the packaging material's ability to restrict gases, moisture, light, and other elements that could harm the product. Sustainable packaging films endeavour to maintain or enhance these properties while minimizing ecological impact. They achieve this through material selection, using multilayer structures, applying barrier coatings, employing advanced technologies, and facilitating modified atmosphere packaging [10]. By optimizing these approaches, sustainable packaging films can effectively protect products from external factors while reducing their ecological footprint. Furthermore, the incorporation or blending of natural substances like herb extract, bio nanocomposites, or coating bee wax can obstruct the gas and water vapor thereby increasing the storage time of packaged food [1].

Commonly starch, cellulose, chitosan, and gelatine are used as the base material for the synthesis of biopolymers due to their cheaper cost, and less environmental impact. Starch derivatives are synthesized as a result of chemical and physical modification from agricultural residues, especially starch nanocrystalline can provide an excellent barrier to gases and moisture. The complex hierarchical arrangement

Table 8.2 Various biopolymers used for sustainable packaging and their properties and role in the shelf life of different food components

Biopolymer	Filler	Properties of the film	Shelf life	References
Carboxymethyl cellulose/starch/ alginate	Grape seed extract	Increase in thermal stability, ultra-violet resistance, antioxidant and antimicrobial	Increase in the shelf life of green chilies	[96]
Carboxymethyl cellulose	Curcumin + zinc nano oxide	Increase in mechanical strength, moisture barrier and UV barrier and also antioxidant antimicrobial properties		[105]
Cellulose nanofiber	Shikonin	Increase in ultra-violet (UV) barrier, increase in tensile strength, and volatile gas sensor	Extension in the shelf life of shrimp	[102]
Polylactic acid	Curcumin	High mechanical strength, high transparency, antioxidant, and antimicrobial action	Improvement in the shelf life of food	[106]
Polylactic acid	Piper betel leaf extract	Increase in structural, mechanical, antioxidant, and antimicrobial properties	The shelf life of tuna meat extended	[93]
Pullulan/ chitosan	Mushroom-mediated zinc nano oxide propolis	High barrier to UV and water vapour, high transparency, antimicrobial, and antioxidant properties	Betterment in the shelf life of meat	[104]
Corn starch + chitosan nanoparticles	Thymol	High mechanical, barrier, antioxidant, and antimicrobial activity	Higher shelf life of cherry tomato	[87]
Starch	*Adiantum capillus* extract	High antioxidant and antimicrobial function	Increase in the shelf life fresh cut pear	[47]
Gelatin/sodium alginate	Date pits extract	Better optical, physical, and functional properties	Improvement in the shelf life of raw minced beef	[30]
Gelatin	Citrus seed extract	Prevention against microbial growth and oxidation reaction	Higher shelf life of sea bass fillets	[129]

(continued)

Table 8.2 (continued)

Biopolymer	Filler	Properties of the film	Shelf life	References
Sturgeon gelatin	*Portulaca oleracea* extract	High antimicrobial, and sensory acceptance	Increased the shelf life of fish sausage	[23]
Pullulan	Sweet basil extract	High antioxidant, anti-bacterial (*Staphylococcus aureus, Bacillus subtilis*) and anti-fungal (*Rhizopus arrhizus*) properties	Increased the shelf life of apple	[118]
Chitosan	Pink pepper residue	High antioxidant antimicrobial property against psychotropic and lactic acid bacteria	Increase in the shelf life of salmon fillet	[72]
Chitosan	*Herba lophatheri* extract	Prevented the growth of yeast and mould	Increase in the shelf life of bighead carp fillets	[18]
Chitosan	Mango leaf extract	High tensile strength, and antioxidant property	Increase in the shelf life of cashew nut	[97]
Chitosan/ pullulan	Saffron	Good resistance against moisture and gases, prevented the growth microbes	Food preservation	[31]
Carrageenan	Olive leaf extract	Thickness and elongation at break increased, and water vapour permeability decreased	Increase in the shelf life of lamb meat	[70]
Zein	Pomegranate peel extract	Increase in mechanical, structural, and antimicrobial properties	Increase in the shelf life of Himalayan cheese	[80]

of starch granules is characterized by layers of crystalline and amorphous regions resulting in a tortuous pathway supporting the barrier property to the film [20].

Similar to starch, cellulose also exhibits a good barrier to oxygen and light. Currently, there has been a digression towards bio-nanocomposites which exhibit comparatively higher characteristics in regards to biopolymers. Recently research was conducted by [134] on the inclusion of attapulgite into cellulose, the composite film exhibited better characteristics compared to the synthetic polymer. The incorporation of attapulgite into the cellulose matrix provides a physical obstacle to oxygen,

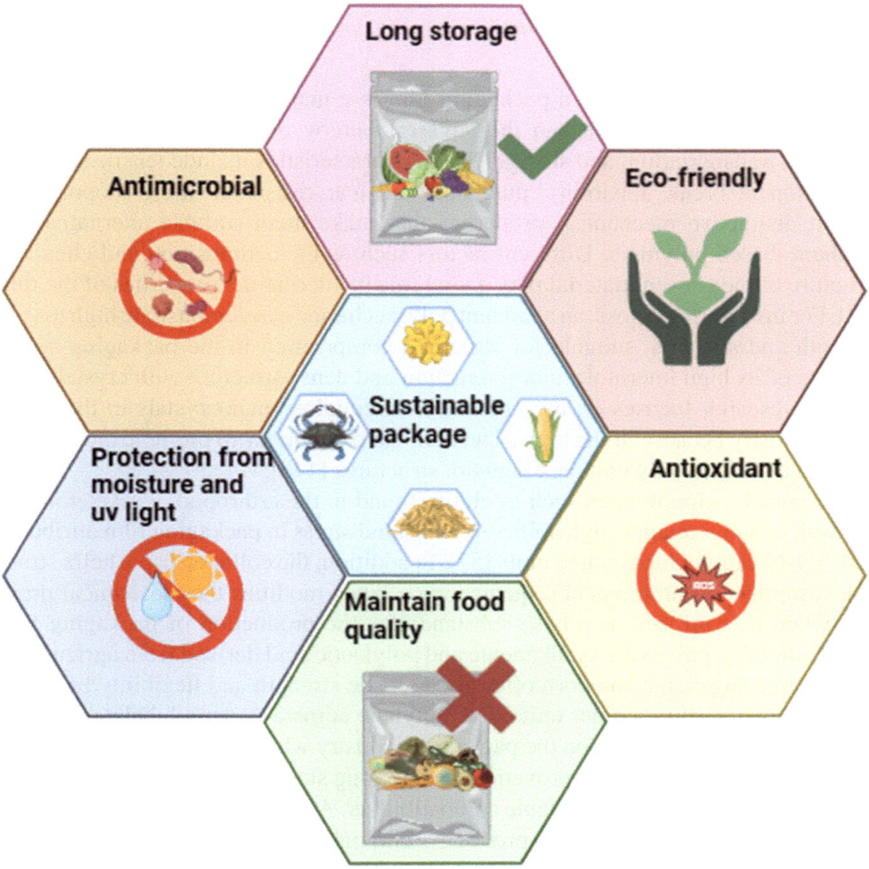

Fig. 8.1 Various properties of sustainable packaging

light, and water vapor. Attapulgite blended into the biopolymer resulting in the forms a composite material with modified structural integrity and reduced permeability. These substances act as fillers in the cellulose matrix, creating a network of interconnected particles that restrict the passage of moisture, light, and gases [134]. In addition, another preferred mechanism is to coat the surface to protect the food enhance shelf life and make it more sustainable. Coating with natural substances like essential oil and bee wax can enhance the barrier properties as well as additional benefits such as prevention of microbial growth. The pectin-based film was developed by using bee wax as a plasticizer, castor oil, and clove oil as additive enhanced antimicrobial and barrier properties in comparison to control pectin film [26].

8.4.2 Mechanical Properties

The mechanical property of a packaging film is a major attribute that defines the ability of the film to withstand the stresses thereby protecting the food during handling, transportation, and storage. These characteristics include tensile strength, elongation at break, flexibility, puncture, and tear resistance [86]. Biopolymers exhibit distinctive mechanical properties that make them suitable alternatives to synthetic-based polymers. Different factors such as the composition and chemical structure of the chosen material play a vital role in altering the properties of the film [92]. For instance, cellulose, an abundant polysaccharide can demonstrate high tensile strength and stiffness, suitable for structural composition in the packaging sector because of its high intermolecular interaction and dense structure with crystallinity. Current research focuses on the utilization of cellulose nanocrystals in the packaging industry because of the high aspect ratio and efficiency to promote mechanical strength contributed by complex network structures [148].

Conversely, biopolymers such as chitin, found in the arthropod exoskeleton and cell wall of fungi retain a high ability to withstand stress in packaging film attributed to β- 1,4-N-acetyl glucosamine units [57]. In addition, the collagen triple helix structure comprised of amino acid sequence in collagen modifies the mechanical property when it is utilized as a base substance for the production of packaging film [50]. Generally, polyhydroxy alkanoate and polylactic acid derived from agricultural sources like sugarcane and corn offer good tensile strength and flexibility based on the composition of monomer units. Polybutylene adipate-co-terephthalate (PBAT), has obtained a huge impact on the packaging industry which provides high mechanical properties. For further improvement, combining starch as a plasticizer is considered an effective way with a suitable compatibilizer. A research study conducted [22] discovered, that PBAT is the most productive compatibilizer for enhancing the tensile strength and elongation at break and can be widely exploited in the food industry. However, PBAT exhibits high flexibility and tear resistance have less demand in the market due to its cost compared to conventional packaging material [22]. Overall, biopolymers derived from renewable sources possess high mechanical properties but vary according to their structure and composition. Methods like blending with other biopolymers, bio nanocomposites, or natural substances can modify the properties and exploited in the packaging industry to a large extent. As research progresses, understanding and optimizing the mechanical properties of biopolymers for their application suitability, as they play a vital role in sustainable material development.

8.4.3 Thermal Properties

Thermal stability is an essential parameter to prevent the degradation of the substance due to temperature effect in short it can be explained as the ability of the packaging substance to resist temperature change to maintain its characteristics during various stages such as transportation, storage, and production [50]. Biodegradable packaging substances often derived from renewable sources including natural polymers or recycled materials are designed to maintain integrity and functionality when exposed to elevated temperature conditions. Assuring the thermal resilience of the biopolymer is important to prevent unacceptable consequences such as warping and chemical leaching which can affect the food quality and safety of the food product [141].

The current trend in film development involves utilizing a combination of biopolymers to enhance film properties. In particular, the synergistic application of cellulose and chitosan has been demonstrated to effectively modify the thermal stability of wrapping films, making them suitable for sustainable packaging. This enhancement is attributed to the incorporation of lignin-containing nanocellulose along with chitosan, forms covalent bonds between lignin and the polymer matrix, thereby contributing to elevated thermal resistance in the packaging film [143]. Similarly, the utilization of polylactic acid (PLA) in conjunction with nano-crystalline cellulose has shown outstanding properties in packaging materials, encompassing notable enhancements in thermal stability alongside other advantages such as biodegradability and widespread availability, thus establishing it as a superior option for packaging applications [114].

Gelatin, is recognized for its formidable barrier properties against oxygen and UV radiation, high thermal stability, and emulsifying attributes. Nevertheless, its susceptibility to mechanical strength has hindered its widespread adoption in packaging applications, though this limitation can be overcome through the addition of natural additives to enhance its performance [63]. The inclusion of soluble soybean polysaccharide (SSPS) into gelatin for edible film significantly enhanced the thermal stability of the film [61]. The strong interactions observed between gelatin and SSPS, characterized by hydrogen bonding and electrostatic interactions, likely resulted in greater steric hindrance, significantly impeding the mobility of macromolecules. These strong interactions may have minimized the available free volume and intermolecular spacing within the films, consequently elevating the glass transition temperature of gelatin.

8.4.4 Optical Properties

The optical characteristics of packaging are essential components that contribute significantly to both the functionality and visual appeal of such packaging. Sustainable packaging materials are carefully chosen to possess properties such as transparency, translucency, or opacity, that are suitable for the specific requirements of the product being packaged and the preferences of consumers [125]. Transparency, for instance, allows consumers to view the constituents inside the package, aids product visibility, and enhances the overall consumer experience by offering a clear vision of what they are purchasing. Moreover, sustainable packaging materials with controlled light transmission properties serve the multiple purposes of showcasing the product attractively while also protecting against potentially harmful ultraviolet radiation, thus helping to prevent food degradation. On the other hand, controlled opacity in packaging materials can offer privacy and protection for sensitive items while still maintaining an appealing aesthetic [45]. By effectively exploiting these optical properties, sustainable packaging solutions can meet the environmental objectives of minimizing waste and reducing ecological impact, while simultaneously fulfilling functional requirements such as protection, visibility, and consumer appeal. This balance between transparency, protection, and aesthetic appeal emphasises the significance of optical properties in the design and implementation of sustainable packaging solutions. In recent times, there has been a significant rise in augmenting biopolymer materials with natural substances or bionanomaterials to alter the optical characteristics of films. A study by Sadadekar et al. [109] revealed that blending essential oil derived from fennel seeds with potato peel extract into a nano chitosan film resulted in enhanced transparency and remarkable UV light prevention properties. Likewise, another composite film, formulated using chitosan, natural kaolinite clay, and extract from *Ficus carica* leaves, demonstrated outstanding optical attributes with good transparency [29]. Nano chitosan and pectin-based films, integrated with fennel essential oil and potato peel extract, exhibit remarkable transparency owing to various factors [109]. The diminutive size of chitosan and pectin particles aids in reducing light dispersion, while the inclusion of fennel essential oil and potato peel extract compounds further enhances transparency. Additionally, the even distribution of these additives throughout the film ensures uniformity, minimizing light obstruction and augmenting transparency. At times, the hue of an extract can directly impact the clarity of a film, as evidenced by findings indicating that the addition of blueberry extract into polysaccharides sourced from *Morinda citrifolia* resulted in increased red and yellow color, leading to a slight decrease in transparency relative to the control film; nevertheless, this alteration coincided with enhanced light barrier properties, potentially attributed by the pigments inherent in the extract [46]. The optical properties of packaging films are significantly affected when plant extract pigments are integrated into biopolymer matrices, primarily due to their interaction with light. These pigments, including chlorophylls, carotenoids, and anthocyanins, possess unique absorption spectra dictating their response to different light wavelengths [55]. Upon incorporation into biopolymers, these pigments absorb and scatter incident light, leading to alterations in film color, transparency, and light transmission properties. Moreover, pigments can impact the film's light barrier capabilities by absorbing specific light wavelengths, thereby reducing UV radiation penetration and providing photoprotection to the packaged contents. The mechanism underlying these changes involves the interaction between light and pigment molecules within the film matrix [138]. Upon light exposure, pigments absorb certain wavelengths, leading to electronic excitation and subsequent energy dissipation through processes like fluorescence or heat generation. Consequently, these interactions result in modifications to the film's optical characteristics, affecting its overall performance as a packaging material [127].

8.4.5 Structural Properties

The structural characteristics of sustainable packaging films are significantly influenced by their composition. Recycled plastics offer durability and barrier qualities, while bio-based polymers may deliver exceptional strength and flexibility [128]. Through careful selection and combination of these ingredients, producers can produce films that are tailored to meet specific needs for various products and applications. To modify the performance of these films, additional chemicals can be added, including UV stabilizers, antioxidants, and antibacterial agents [64].

Complex carbohydrates, which are the building blocks of polysaccharide-based films, are mostly derived from natural sources like microorganisms and plants. In food packaging, cellulose, starch, chitosan, gum arabic, agar, and their derivatives are the polysaccharides most frequently used [151]. Chitosan, which is made from chitin, has a semi-crystalline structure, and it can form films with strong mechanical and antibacterial properties [83]. Gum Arabic is a complex polysaccharide that can form amorphous films that possess good flexibility and moisture barrier properties. Agar is derived from the cell wall of red algae and is used in sustainable packaging applications because of its exceptional gelation and thickening abilities [101]. The structure, configuration, and characteristics of the amylose and amylopectin polymers found in starch molecules, along with the distribution of crystalline and amorphous regions inside starch granules, influence the structural properties of starch-based films [76]. The functionality and performance of starch-based films can be improved for a variety of applications by modifying their qualities in different ways [83]. The hydroxyl groups influences the structural properties of polysaccharide-based films, resulting in stronger intermolecular interactions, higher crystallinity, and higher melting temperatures. However, their use as packaging material is restricted due to their sensitivity to heat [24]. The disintegration into tiny fragments is caused by the breaking of hydrogen bonds, which is additionally supported by the hydroxyl groups [151]. Also, polysaccharide-based films have poor moisture resistance properties due to their high hydrophilicity. These films tend to absorb moisture at high relative humidity levels, which can compromise their moisture and gas barrier qualities by causing the matrix to swell and hydrogen bonds to break. Furthermore, some polysaccharide-based films like starch films might have reduced mechanical properties, which would restrict their application even more [64].

The focus on developing sustainable packaging materials is based on the structural properties of protein-based films. These films must possess strength, elasticity, and low permeability to be suitable for food packaging [5]. In the development of active packaging materials, proteins such as whey protein isolate, blood meal, gluten, peanut protein, sunflower protein, egg, soy protein, soybean, and casein have been utilized due to their antimicrobial properties and ability to restrict oxygen permeation [76]. These films serve as effective oxygen blockers, preventing food spoilage, and are transparent, odorless, and tasteless, making them suitable for a wide range of food products. The molecular properties of proteins, such as their charge, conformation, molecular mass, flexibility, and thermal stability, determine their potential to form films [5]. Various biopolymers-based packaging and their application in food is discussed in Table 8.2.

8.5 Application of Sustainable Packaging for Shelf-Life Improvement in Food

For reducing food waste and preventing food deterioration selection of suitable packaging materials is very important. Food stored inside the package should not react with the packaging material because it can cause toxic compounds production. The use of biobased films can extend the storage life by protecting the food from various external factors like gases, light, and moisture. In addition, use of modern technologies like active packaging or modification of packaging film can further enhance the storage time and food quality [4]. Sources of biopolymers and the addition of natural substances are illustrated in Fig. 8.2.

8.5.1 Different Biopolymers Use for the Synthesis of Sustainable Packaging

Due to the rising concern related to environmental issues created by synthetic polymers, there is a drastic demand for biobased packaging material which can provide advantages similar to petroleum film and cause no harm to the ecosystem. These materials which are known to be biopolymer or bioplastics are synthesized from living, microbial sources, or agricultural wastes thereby this can give multiple benefits like eco-friendly packaging film by undergoing degradation naturally without leaving any residue on the earth's surroundings and proper utilization of agricultural debris which can cause severe threat to the nature [60]. These films can exhibit considerable characteristics similar to petroleum polymers regarding structural, mechanical, and barrier properties. To modify the properties adding natural additives, fillers or nanomaterials can be adopted, respectively. The biopolymers are classified based on the origin which includes biopolymers such as cellulose, starch extracted from plant sources, bio polyolefins and polylactic acid prepared chemically from bio sources, and biopolymers processed from micro-organisms like polyhydroxyalkanoates and polyhydroxy butyrate [10].

Starch, which is categorized under carbohydrate, is a hydrocolloid polymer consisting of amylose and amylopectin obtained from wheat, corn, or tubers. The concentration of these constituents can result in the modification of film properties. An increase in the concentration of amylose varied the tensile strength of starch-based packaging film [10]. For the improvement of film characteristics, starch is blended with plasticizers for instance glycerol and sorbitol resulted bioplastics exhibiting high mechanical properties. Plasticizers can destroy the hydrogen bonds present in

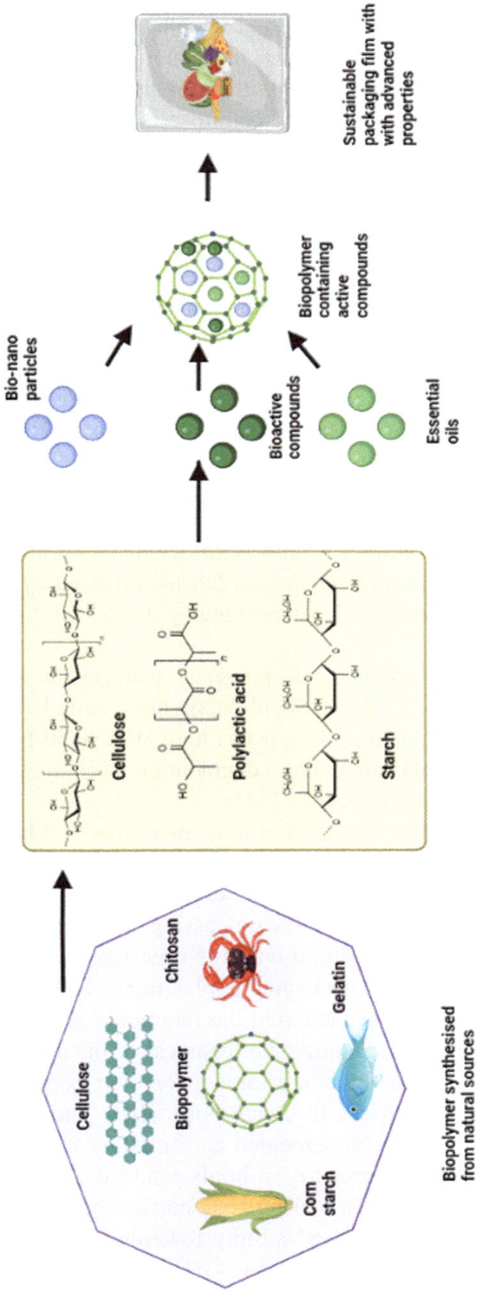

Fig. 8.2 Schematic representation of sources of biopolymer and its application in the food system

the starch. Starch extracted from tapioca, corn, and rice has provided good translucent and tensile strength and better biodegradability [81].

Cellulose is a natural polymer that present in cells of plants and can effectively utilized for the manufacture of green packaging materials. Cellulose is structurally composed of repeating glucose monomers with β-1,4 glycosidic bonds [134]. Rather than using it directly, a modified form of cellulose is found to be more effective in packaging because of its poor solubility. Among the various forms of cellulose including cellulose nanocomposites acquired wide acceptance in the packaging industry. In addition, combination with another biopolymer further enhances the properties. For example, the cellulose nanocomposite added with PBAT and PLA presented better mechanical and thermal stability than conventional plastics[6]. Besides, the inclusion of bio extract with cellulose nanoparticles can provide superior qualities to the film. Cellulose nanofiber incorporated with grape extract increased the functional characteristics of the film [103].

Pectin, a branched polysaccharide that is predominantly present in the plant cell wall and is widely utilized for the preparation of packaging film because of its flexibility and good resistance to quality-decreasing factors such as gases, moisture, light, aroma, and microbes [131]. Moreover, extracts from natural sources can be added as an active ingredient into the pectin-based film to improve the functional characteristics of the film and storage time. Gelatin-pectin composite film added with essential extracted from *Mentha pulegium* and *Lavandula angustifolia* showed high antimicrobial activity and was effective in prolonging the storage time of tilapia fillet [3]

Polylactic acid is a hydrophobic, bio-polyester that can be synthesized from fermentable sugar, corn starch, or lactic acid depolymerization [67]. PLA is highly selective for sustainable packaging since it has high stretchability, toughness, and mechanical characteristics and can undergo degradation through ester hydrolyses or microbes, and enzymes are not necessary [14].

Polyhydroxyalkanoates (PHA) consist of monomer units of 3-hydroxy fatty acid making it suitable in the packaging industry providing desirable properties including thermal and barrier characteristics. The synthesis of PHA is expensive therefore sustainable alternatives can be opted for the processing of PHA. Hydrothermal treatment of lignocellulose from agricultural waste is used for the production of PHA and can be effectively utilized for packaging applications [125]. The incorporation of phloretin into PHA film further increased the functional property of the film by hindering microbial growth and retaining the freshness of the apple [73].

Packaging films made from lipids are known for their barrier qualities. They are integrated into various matrices to create both edible and inedible films and coatings [5]. Such films can be implemented as a mixture of a hydrophobic and hydrophilic layer,. The nonpolar property of lipids enhances their capacity to lower water transfer rates and improve film moisture and barrier characteristics [152]. This improvement stems from the decreased affinity towards water or moisture which aids in reducing the water transmission rates [64]. However, utilizing lipids in these films can cause a reduction in mechanical strength and clarity. To overcome this, the infusion of essential oils and appropriate plasticizing agents in coating solutions has been explored by various studies. Essential oil extracts, for example, lemongrass, rosemary, and clove, have been employed as plasticizers, lending an antibacterial effect and improving mechanical properties [76]. Pure coconut oil has also been implemented, enhancing structural and moisture characteristics in edible composite materials [15]. Also, the blending of fats and oils within matrices, such as proteins and carbohydrate-based constituents, can create a uniform film-forming structure, adding to the overall attributes of the packing material [75].

8.5.1.1 Key Factors Influencing the Shelf Life of Food

The concept of shelf life can be concisely defined as the period during which food can be stored without experiencing deterioration or loss of quality by maintaining its desired characteristics in terms of nutrition, freshness, and sensory. The duration of food preservation is integrally linked to a variety of factors, including storage temperature, packaging conditions, processing techniques, and the presence of microbial contaminants. These elements collectively determine the extent to which food products can maintain their freshness, flavor, texture, and nutritional value over time [84]. Bacteria such *as Escherichia coli, Streptococcus,* and *Listeria monocytogenes* can induce food-borne illness, resulting in spoilage and decreased shelf life. These microbes can propagate on food surfaces and cause unpleasant odors and also can produce enzymes that degrade food constituents, leading to the alteration in texture and nutrient breakdown ultimately reducing shelf life [32]. In addition to microbial activity, oxygen serves a crucial function in preserving the shelf life of food. Its presence can facilitate the growth of aerobic microorganisms and initiate oxidative reactions within food, thereby contributing to quality deterioration through processes such as lipid oxidation and the development of off-flavors. Exposure to light poses another challenge as it can break down sensitive compounds present in food, leading to alterations in color and flavour. Similarly, water vapor acts as a catalyst for microbial growth within food products. Consequently, the interplay of oxygen, light, and water vapor alongside microbial activity significantly influences the longevity and overall quality of stored food items. To address this issue, food researchers have adopted novel innovations in food packaging that effectively preserve food quality and enhance shelf life by preserving its nutritional characteristics by maintaining sustainability[54].

The use of bio-based polymers alone is not sufficient to withstand the advanced properties of synthetic package, food industry has adopted modern technologies to enhance the film characteristics and to food shelf life. This innovative technique are known as active and intelligent packaging [68]. Incorporation of bio-active substance into the biopolymer substance can reshape the film mechanical, structural, barrier, and functional properties of the film and increase food shelf life. The active package is referred as the inclusion of natural or active substances into the base material and absorption or diffusion of these elements as in a controlled manner into the food system to scavenge free radicals or gases or provide some sensory attribute to the food in terms of nutrition and quality and can replace the synthetic antioxidants and antimicrobials which can cause a carcinogenic effect on human beings [4]. An intelligent package does not increase the food shelf life but can monitor and provide accurate information regarding food freshness or quality inside the package through the sensor. Simply put, active packaging maintains food superiority while intelligent packaging provides real information about the food product [59].

Natural Antimicrobials

Natural antimicrobials have greatly influenced the food industry due to their benefits over synthetic antimicrobials. This substance can be successfully incorporated into biopolymers to control the growth of harmful microbes. Antimicrobials derived from spices or herb sources that contain ample bioactive compounds can inhibit the growth of undesirable organisms through different mechanisms [32]. The bioactive substance, including polyphenols and other compounds present in the extract, can destroy the phospholipid layer of the microbes, thereby leaking out the organelles, which can result in the microbe's death. Furthermore, polyphenols obstruct essential cellular functions by interfering with DNA and microbial enzymes. By producing reactive oxygen species and oxidative stress in microbial cells, their antioxidant qualities also support antibacterial activity [69]. Essential oils obtained from aromatic plants also exhibit high anti-bacterial and anti-viral effects.

Oxygen Scavengers

Oxidation reactions in food occur when molecules within the food react with oxygen, leading to changes in taste, color, texture, and ultimately spoilage. For instance, fats and oils can become rancid, while fruits and vegetables can be subjected to browning due to reactive oxygen species generated as a result of oxidation. This can significantly affect the shelf life of food by hastening its deterioration through the promotion of microbial growth and the development of undesirable flavors and odours [150]. Antioxidants or oxygen scavengers are commonly added to food products to slow down oxidation and prolong shelf life by inhibiting or delaying the oxidation process.

In the packaging industry, there is a prevalent use of oxygen scavengers designed to counteract excessive oxygen in the headspace of packages. These scavengers are substances capable of reacting with surplus oxygen within the packaging material, thereby synthesizing harmless by-products without compromising the quality of the packaged product. The accumulation of excess oxygen within packages can occur due to various factors such as insufficient exhaustion mechanisms or chemical reactions [37]. One of the most commonly employed oxygen scavengers involves the utilization of iron salts, which react to form stable iron oxides within the package. This chemical process helps mitigate the detrimental effects of oxygen and ensures the preservation of product quality throughout the packaging lifecycle.

Antioxidants such as butylated hydroxy anisole (BHA) and butylated hydroxytoluene (BHT) are commonly used synthetic antioxidants, but due to their carcinogenic effect, natural compounds are incorporated into biopolymer films to maintain sustainability and reduce environmental hazards [41]. The most commonly used natural antioxidants are plant extracts containing phenolic compounds and flavonoids, essential oils, and natural pigments. The hydroxyl group present in the compound can react with the reactive oxygen species, resulting in the formation of stable molecules without affecting the food quality.

Moisture Absorbers

Moisture is a crucial factor that serves as the medium for the growth of microbes which contributes to food spoilage and reduces the shelf life. Therefore, the excess amount of moisture that should be removed from the packaging film is essential. The elevated relative humidity (RH) within the packaging of high-water activity foods like fish, poultry, meat, and fresh produce hastens the proliferation of bacteria and mold, consequently resulting in the degradation of the nutritional value and quality of the food. To resolve this issue, moisture absorbers can be added to packaging film as sachets or pads filled with hygroscopic substances[36]. The most common absorbent pads used in packaging film are natural clay, fructose, silica gel, modified starch, and sorbitol. Excess moisture in the food gets adsorbed into the surface of this substance thereby reducing the water content in the food [36]. Moisture absorbers can be further classified into organic (fructose, sorbitol) and inorganic (calcium oxide, silica gel, potassium carbonate). Organic absorbers like fructose and sorbitol are good moisture absorbers but their application is limited due to cost compared to silica gel and calcium oxide which are non-toxic with high moisture resistance [37].

8.5.2 Suitability of Different Materials for Various Food Products (Dairy, Fruits and Vegetables, Fish and Meat)

8.5.2.1 Dairy

Milk and dairy products are packed in various materials that are specifically chosen for their processing conditions, storage, handling, and eventual use. Shelf-life extension only matters if weight and loss of nutrients can be minimized by packaging. The selection of packing materials for dairy items is based on several factors like product compatibility, chemical inertness, absence of toxicity, impact resistance or tamper indication, sanitation maintenance, light and odor protection, etc. [100]. All these should be done within given costings while considering size, shape, weight, printability, marketing appeal, etc. The choice of suitable materials and methods is determined by the nature and characteristics of the packaged milk products [13].

Product integrity could, therefore, be maintained by ensuring that appropriate packaging materials are used while shelf life and consumer satisfaction could also be increased, thus enabling them to meet both regulatory standards and environmental sustainability objectives at the same time [100].

The existence of a lot of dairy product-related illnesses like human listeriosis, necessitates the need to prevent microbial contamination. The other bacteria that cause diseases in humans associated with dairy products are *Salmonella* which is usually found in spoiled cheese while the presence of molds during ripening and storage of different types of dairy products especially cheese is a big problem faced by the dairy industry [91]. Common fungal contaminants include species of *Aspergillus* and *Penicillium*. One way that has been used to solve this issue historically was by infusing spices, herbs, or their oils physically into the product. Several studies have shown that this type of edible packaging works well in the dairy industry. For example, research has investigated packaging cheese using chitosan-coated nisin-silica liposome and shown anti-listeria effects. Probiotic coatings as innovative strategies are also emerging as promising approaches for preserving dairy products. It has been observed that incorporating lactic acid bacteria like *Lactobacillus acidophilus* and *Lactobacillus helveticus* onto eatable cheese coverings can reduce the presence of bacteria within ten days [91]. These developments not only enhance food safety but also improve shelf life and general quality attributes for milk-based products. In cheese packaging, starch laminates, and nitrocellulose lacquered cellophane are commonly used because of their superior moisture and gas barrier properties, aiding in prolonging the shelf life and preserving the quality of cheese. As for yogurt and fermented foods, biodegradable plastics like PLA (polylactic acid) derived from fermented plant starch (typically corn), are frequently employed [99]. It is extensively utilized in dairy packaging, including yogurt cups and lids, to ensure mechanical strength while remaining environmentally sustainable. They serve as reliable barriers against moisture and gases, safeguarding the product's freshness and integrity. For ice creams acetylated monoglyceride coated paper cartons are found to be applied due to their moisture barrier properties [76]. Increased mechanical, structural, and antimicrobial properties of zein-based film with pomegranate peel extract enhanced the shelf life of Himalayan cheese.

8.5.2.2 Fruits and Vegetables

Most quality losses of fresh produce occur during storage and transport, including moisture loss, shrinkage, weight loss, microbial attack, chemical degradation, mechanical damage, etc. Food packaging acts as a barrier against free air exchange and prevents unwanted odors [53]. However, it should be noted that gaseous exchange protection with food coatings can alter the environment around fresh fruits/vegetables, potentially creating anaerobic conditions and causing an alcoholic fermentation process, which can alter the sensory characteristics of packaged foods [113]. Thus, careful selection of food materials for packaging fruits and vegetables is necessary to obtain the best balance of permeability properties. This selection ensures that the packaging retains moisture, exhibits adequate adhesion due to hydrophilic properties, prevents off-flavor development, maintains consistency and appearance (e.g., glossy), preserves fruit firmness, reduces weight loss, and minimizes solute leakage [53].

Carrageenan film blended with green tea extract has been utilized as coatings for blueberries and raspberries, while Alginate-oleic acid-based coatings supplemented with green tea have been applied to strawberries and raspberries [33]. These coatings offer antiviral activity, contributing to their preservation and quality enhancement. A rice starch edible coating blended with sucrose esters is found to be effective in delaying ethylene biosynthesis and reducing the respiration rate in bananas [120]. In cucumber, corn starch and mint extract edible coating is found to increase the shelf at room temperature [94]. The pectin-based coating on lime fruit was shown to prevent a progressive increase in shriveling or wilting and loss in green color during storage [66]. Carboxymethyl cellulose, starch or alginate with grape seed extract as filler increased the shelf life of green chilies by providing thermal stability, ultraviolet barrier, antioxidant property and antimicrobial property [96]. Corn starch and chitosan nanoparticles with thymol increased the shelf life of cherry tomato as it provided high mechanical, barrier, antioxidant, and antimicrobial properties [87]. The shelf life fresh cut pear was increased by the antioxidant and antimicrobial of starch film incorporated with *Adiantum capillus* extract [47].

8.5.2.3 Meat, Poultry, and Fish

When it comes to packaging meat, fish, and poultry, standard packaging often employs trays formed from materials like polystyrene, polypropylene, polyvinylchloride (PVC) with ethylene vinyl alcohol, Low-density polyethylene, or PVC coatings, enhanced with foil (PVC) or lids, bags for limited-duration storage, and wax-coated paper for coverings. Conversely, bio-derived packaging alternatives for such items feature multiple layers of packaging substances endowed with practical biobased coatings, such as modified starches and antimicrobial and anti-fogging systems [130]. Even though these bio-derived packaging alternatives are accessible for pilot packaging in specific markets, they generally carry a heavier price tag than regular options [130].

Lignin, a natural polymer found in wood, can be used for meat packaging. It can prevent UV light in packaging films and possesses antioxidant properties that prevent oxidative degradation and antimicrobial properties that can help inhibit bacterial growth. The edible packing technique efficiently reduces moisture loss, maintains the texture, and reduces the unappealing leak of product fluids [130]. This effectively decreases food spoilage and wastage. Moreover, edible packaging plays a part in slowing down the biochemical degradation of the product, protects against fat oxidation, fends off rancidity, and prevents unwanted color changes. The inclusion of antimicrobials further provides safety against microbes and product spoilage. In sliced ham, sodium alginate edible films can be used as vehicles for delivering probiotic bacteria [88], and oregano essential oil incorporated in sodium alginate edible film inhibits the growth of *Listeria* [89]. Reduced moisture loss, minimized lipid oxidation, prevented discoloration, and reduced dripping were found in muscle foods coated with collagen edible films [88]. The storage period of shrimp is improved when packed in cellulose nanofiber incorporated with shikonin, as it acts as an ultra-violet barrier and has increased in tensile strength [102]. Polylactic acid biopolymer combined with piper betel leaf extract has structural, mechanical, antioxidant, and antimicrobial properties, which will increase the shelf life of tuna meat [93]. Mushroom-mediated zinc nano oxide propolis in pullulan/chitosan film act as a barrier to UV and water vapor, and has high transparency, antimicrobial, and antioxidant properties. Antimicrobial characteristics of *Herba lophatheri* extract incorporated chitosan films enhanced the shelf life of bighead carp fillets [19].

Thus, using sustainable polymers in coating and film applications can be advantageous over synthetic, non-biodegradable polymers. There are multiple choices for the polymers, and selecting appropriate polymers is key to specific applications. The biobased, sustainable polymers added with functional ingredients are promising for active packaging purposes. Much research has been done on improving the physical and functional properties of biobased packaging film incorporating suitable active fillers. Various studies suggested that bio-based degradable polymers showed encouraging results in preserving and monitoring the safety and quality of packed food. Even though biopolymers have the potential to replace conventional packaging material, there is still a need for more research to make it on a commercial scale.

8.6 Conclusions

The transition to sustainable packaging materials marks a significant advancement in addressing environmental challenges posed by conventional plastic packaging. This chapter has explored various bio-based polymers, such as carbohydrates, proteins, and polyesters, highlighting their potential to replace traditional petrochemical-based materials. Integrating these biopolymers with active ingredients, such as antimicrobial and antioxidant agents, has demonstrated the capability to enhance the functionality and applicability of sustainable films and coatings in food packaging. A key aspect of sustainable packaging is the comprehensive approach that considers the entire lifecycle of packaging materials—from sourcing and extraction to consumer use and end-of-life disposal. This lifecycle perspective ensures minimal environmental impact by promoting renewable, biodegradable, and compostable materials, reducing carbon footprints, energy consumption, and dependency on non-renewable resources. Developing bio-based polymers offers several advantages, including biodegradability, reduced environmental pollution, and the potential for existing processing facilities to adapt to these new materials. Despite these benefits, challenges remain, particularly in improving the physical properties, such as moisture and oxygen barrier performance, tensile strength, and flexibility of these materials. Ongoing research and innovation in polymer modification and the addition of active compounds are crucial to overcoming these challenges and ensuring that sustainable packaging materials meet industry standards and consumer expectations. Adopting sustainable packaging materials is crucial to a more eco-friendly and resource-efficient future. By continuing to develop and refine bio-based polymers, the packaging industry can significantly reduce its environmental impact while providing effective and safe packaging solutions. The shift towards sustainable materials supports environmental conservation and fosters innovation and economic growth within the sector, paving the way for a circular economy.

References

1. Adibi, A., Trinh, B.M., Mekonnen, T.H.: Recent progress in sustainable barrier paper coating for food packaging applications. Prog. Org. Coat. **181**, 107566 (2023)
2. Accorsi, R., Baruffaldi, G., Manzini, R.: A closed-loop packaging network design model to foster infinitely reusable and recyclable containers in food industry. Sustain. Prod. Consumpt. **24**, 48–61 (2020)
3. Aitboulahsen, M., Chairi, H., Laglaoui, A., Zantar, S., Abdelhay, A., Bakkali, M., Zerrouk, M.H.: Gelatin/pectin-based film incorporated with essential oils: Functional characteristics and shelf life extension of tilapia fillets under refrigeration. J. Food Saf. **40**(3), e12774 (2020)
4. Amin, U., Khan, M.K.I., Maan, A.A., Nazir, A., Riaz, S., Khan, M.U., Sultan, M., Munekata, P.E.S., Lorenzo, J.M.: Biodegradable active, intelligent, and smart packaging materials for food applications. Food Packag. Shelf Life **33**, 100903 (2022)
5. Amin, U., Khan, M.U., Majeed, Y., Rebezov, M., Khayrullin, M., Bobkova, E., Shariati, M.A., Chung, I.M., Thiruvengadam, M.: Potentials of polysaccharides, lipids and proteins in biodegradable food packaging applications. Int. J. Biol. Macromol. **183**, 2184–2198 (2021)
6. Andrade, M.S., Ishikawa, O.H., Costa, R.S., Seixas, M.V.S., Rodrigues, R.C.L.B., Moura, E.A.B.: Development of sustainable food packaging material based on biodegradable polymer reinforced with cellulose nanocrystals. Food Packag. Shelf Life **31**, 100807 (2022)
7. Colangelo, A.C.M.D., Chinnici, F.: Relevance and perspectives of the use of chitosan in winemaking: a review. Crit. Rev. Food Sci. Nutr. **61**(20), 3450–3464 (2021). https://doi.org/10.1080/10408398.2020.1798871
8. Anushikha, Gaikwad, K.K.: Lignin as a UV blocking, antioxidant, and antimicrobial agent for food packaging applications. Biomass Convers. Biorefinery 1–13 (2023)
9. Apostolidis, E., Kwon, Y.-I., Shetty, K.: Inhibition of Listeria monocytogenes by oregano, cranberry and sodium lactate combination in broth and cooked ground beef systems and likely mode of action through proline metabolism. Int. J. Food Microbiol. 128(2), 317–324 (2008). https://doi.org/10.1016/j.ijfoodmicro.2008.09.012
10. Basnur, J., Putra, M.F.F., Jayusman, S.V.A., Zulhilmi, Z.: Sustainable packaging: Bioplastics as a low-carbon future step for the sustainable development goals (SDGs). ASEAN J. Sci. Eng. Mater. **3**(1), 51–58 (2024)
11. Barhoum, A., Jeevanandam, J., Rastogi, A., Samyn, P., Boluk, Y., Dufresne, A., Danquah, M.K., Bechelany, M.: Plant celluloses, hemicelluloses, lignins, and volatile oils for the synthesis of nanoparticles and nanostructured materials. Nanoscale **12**(45), 22845–22890 (2020)
12. Barreto, R.A.: Fossil fuels, alternative energy and economic growth. Econ. Model. **75**, 196–220 (2018)
13. Biegańska, M.: Packaging of Dairy Products: Emerging Strategies. In: Food packaging: the smarter way (pp. 127–164). Springer (2022)
14. Boey, J.Y., Mohamad, L., Khok, Y.S., Tay, G.S., Baidurah, S.: A review of the applications and biodegradation of polyhydroxyalkanoates and poly (lactic acid) and its composites. Polymers **13**(10), 1544 (2021)

15. Carpena, M., Nuñez-Estevez, B., Soria-Lopez, A., Garcia-Oliveira, P., Prieto, M.A.: Essential oils and their application on active packaging systems: a review. Resources **10**(1), 7 (2021)
16. Ceylan, Z., Sengor, G.F.U., Yilmaz, M.T.: Nanoencapsulation of liquid smoke/thymol combination in chitosan nanofibers to delay microbiological spoilage of sea bass (*Dicentrarchus labrax*) fillets. J. Food Eng. **229**, 43–49 (2018)
17. Channa, I.A., Ashfaq, J., Gilani, S.J., Chandio, A.D., Yousuf, S., Makhdoom, M.A., & Jumah, M.N. bin.: Sustainable and eco-friendly packaging films based on poly (vinyl alcohol) and glass flakes. Membranes 12(7) (2022). https://doi.org/10.3390/membranes12070701
18. Chen, Y., Xu, L., Wang, Y., Chen, Z., Zhang, M., Chen, H.: Characterization and functional properties of a pectin/tara gum based edible film with ellagitannins from the unripe fruits of Rubus chingii Hu. Food Chemistry 325, 126964 (2020). https://doi.org/10.1016/j.foodchem.2020.126964
19. Chen, H., Wang, J., Jiang, G., Guo, H., Liu, X., & Wang, L.: Active packaging films based on chitosan and Herba lophatheri extract for the shelf life extension of fried bighead carp fillets. Int. Food Res. J. 27(4), (2020)
20. Chi, K., Wang, H., Catchmark, J.M.: Sustainable starch-based barrier coatings for packaging applications. Food Hydrocolloids **103**, 105696 (2020)
21. Chuanxing, F., Qi, W., Hui, L., Quancheng, Z., Wang, M.: Effects of pea protein on the properties of potato starch-based 3D printing materials. Int. J. Food Eng. **14**(3), 20170297 (2018)
22. Dammak, M., Fourati, Y., Tarrés, Q., Delgado-Aguilar, M., Mutjé, P., & Boufi, S.: Blends of PBAT with plasticized starch for packaging applications: mechanical properties, rheological behaviour and biodegradability. Ind. Crops Prod. 144, 112061 (2020). https://doi.org/10.1016/j.indcrop.2019.112061
23. Dehghan Tanha, L., Khoshkhoo, Z., Azizi, M.H.: Application of edible coating made of sturgeon gelatin and *Portulaca oleracea* extract for improving the shelf life of fish sausages. J. Food Measure. Character. **15**(5), 4306–4313 (2021). https://doi.org/10.1007/s11694-021-01013-6
24. Deng, J., Zhu, E.-Q., Xu, G.-F., Naik, N., Murugadoss, V., Ma, M.-G., Guo, Z., Shi, Z.-J.: Overview of renewable polysaccharide-based composites for biodegradable food packaging applications. Green Chem. **24**(2), 480–492 (2022)
25. de Matos Fonseca, J., Koop, B.L., Trevisol, T.C., Capello, C., Monteiro, A.R., Valencia, G.A.:. An overview of biopolymers in food packaging systems. Nanotechnol. Enhanced Food Packag 19–53 (2022)
26. Dharhshini, A., Sistla, Y.S.: Beeswax and castor oil to improve the moisture barrier and tensile properties of pectin based edible films for food packaging applications. Eng. Proc. 5 (2023)
27. Drishya, C.M., Wani, K.M.: Exploring muskmelon peel as a novel source of pectin: Extraction methods and applications. Food. Human. **3**, 100331 (2024)
28. Dumitriu, R.P., Stoleru, E., Mitchell, G.R., Vasile, C., Brebu, M.: Bioactive electrospun fibers of poly (ε-caprolactone) incorporating α-tocopherol for food packaging applications. Molecules **26**(18), 5498 (2021)
29. El Mouzahim, M., Eddarai, E.M., Eladaoui, S., Guenbour, A., Bellaouchou, A., Zarrouk, A., Boussen, R.: Food packaging composite film based on chitosan, natural kaolinite clay, and Ficus. carica leaves extract for fresh-cut apple slices preservation. Int. J. Biol. Macromol. 233, 123430 (2023). https://doi.org/10.1016/j.ijbiomac.2023.123430
30. Elhadef, K., Chaari, M., Akermi, S., Ennouri, K., Hlima, H.B., Fourati, M., Mtibaa, A.C., Ennouri, M., Sarkar, T., Shariati, M.A., et al.: Gelatin-sodium alginate packaging film with date pits extract: an eco-friendly packaging for extending raw minced beef shelf life. Meat Sci. **207**, 109371 (2024)
31. Ezati, P., Roy, S., & Rhim, J.-W.: Effect of saffron on the functional property of edible films for active packaging applications. ACS Food Sci. Technol. 2(8), 1318–1325 (2022)
32. Fadiji, T., Rashvand, M., Daramola, M.O., Iwarere, S.A.: A review on antimicrobial packaging for extending the shelf life of food. Processes **11**(2), 590 (2023)

33. Falcó, I., Flores-Meraz, P.L., Randazzo, W., Sánchez, G., López-Rubio, A., Fabra, M.J.: Antiviral activity of alginate-oleic acid based coatings incorporating green tea extract on strawberries and raspberries. Food Hydrocolloids **87**, 611–618 (2019)
34. Falcó, I., Randazzo, W., Sánchez, G., López-Rubio, A., Fabra, M.J.: On the use of carrageenan matrices for the development of antiviral edible coatings of interest in berries. Food Hydrocolloids **92**, 74–85 (2019)
35. Fathima, P.E., Panda, S.K., Ashraf, P.M., Varghese, T.O., Bindu, J.: Polylactic acid/chitosan films for packaging of Indian white prawn (*Fenneropenaeus indicus*). Int. J. Biol. Macromol. **117**, 1002–1010 (2018)
36. Gaikwad, K.K., Singh, S., Ajji, A.: Moisture absorbers for food packaging applications. Environ. Chem. Lett. **17**(2), 609–628 (2019)
37. Gaikwad, K.K., Singh, S., Lee, Y.S.: Oxygen scavenging films in food packaging. Environ. Chem. Lett. **16**, 523–538 (2018)
38. Gaur, S. S., Dhar, P., Wani, K. M., Srivastava, M., Sakurai, S., Kumar, A., Katiyar, V.: Ion transfer channel network formed by flower and rod shape crystals of hair hydrolysate in poly (vinyl alcohol) matrix and its application as anion exchange membrane in fuel cells. J. Colloid Interface Sci. **587**, 214–228 (2021)
39. Geyer, R., Jambeck, J.R., Law, K.L.: Production, use, and fate of all plastics ever made. Sci. Adv. **3**(7), e1700782 (2017)
40. Gouveia, T.I.A., Biernacki, K., Castro, M.C.R., Gonçalves, M.P., Souza, H.K.S.: A new approach to develop biodegradable films based on thermoplastic pectin. Food Hydrocoll 97, 105175 (2019). https://doi.org/10.1016/j.foodhyd.2019.105175
41. Gupta, P.: Role of oxygen absorbers in food as packaging material, their characterization and applications. J. Food Sci. Technol. 1–11 (2023)
42. Gulin-Sarfraz, T., Kalantzopoulos, G.N., Kvalvåg Pettersen, M., Wold Åsli, A., Tho, I., Axelsson, L., Sarfraz, J.: Inorganic nanocarriers for encapsulation of natural antimicrobial compounds for potential food packaging application: a comparative study. Nanomaterials, 11(2) (2021). https://doi.org/10.3390/nano11020379
43. Gunaalan, K., Fabbri, E., Capolupo, M.: The hidden threat of plastic leachates: a critical review on their impacts on aquatic organisms. Water Res. **184**, 116170 (2020)
44. Gururani, P., Bhatnagar, P., Dogra, P., Joshi, H.C., Chauhan, P.K., Vlaskin, M.S., Joshi, N.C., Kurbatova, A., Irina, A., Kumar, V.: Bio-based food packaging materials: a sustainable and holistic approach for cleaner environment-a review. Curr. Res. Green Sustain. Chem. 100384 (2023)
45. Guzman-Puyol, S., Ben\'\itez, J.J., & Heredia-Guerrero, J.A.: Transparency of polymeric food packaging materials. Food Res. Int. 111792 (2022)
46. Han, H.-S., Song, K. Bin.: Noni (*Morinda citrifolia*) fruit polysaccharide films containing blueberry (*Vaccinium corymbosum*) leaf extract as an antioxidant packaging material. Food Hydrocolloids 112, 106372 (2021). https://doi.org/10.1016/j.foodhyd.2020.106372
47. Hashemi, S.M.B., Zahabi, N., Rezaee, Z., Maherani, Z., Boghori, P., Keshavarz, Z.: Evaluation of a starch-based edible film as carrier of A diantum capillus-veneris extract to improve the shelf life of fresh-cut pears. J. Food Saf. **36**(3), 291–298 (2016)
48. Heinze, T.: Cellulose: structure and properties. In O. J. Rojas (Ed.): Cellulose chemistry and properties: fibers, nanocelluloses and advanced materials (pp. 1–52). Springer International Publishing (2016). https://doi.org/10.1007/12_2015_319
49. Helanto, K.E., Matikainen, L., Talja, R., Rojas, O.J.: Bio-based polymers for sustainable packaging and biobarriers: a critical review. BioResources **14**(2), 4902–4951 (2019)
50. Hou, C., Gao, L., Wang, Z., Rao, W., Du, M., Zhang, D.: Mechanical properties, thermal stability, and solubility of sheep bone collagen–chitosan films. J. Food Process Eng. **43**(1), e13086 (2020)
51. Huang, J., Hu, Z., Hu, L., Li, G., Yao, Q., & Hu, Y.: Pectin-based active packaging: a critical review on preparation, physical properties and novel application in food preservation. Trends Food Sci. Technol. 118, 167–178 (2021). https://doi.org/10.1016/j.tifs.2021.09.026

52. Huang, Y., Huang, J., Chen, Y.: Alpha-helical cationic antimicrobial peptides: relationships of structure and function. Protein Cell **1**(2), 143–152 (2010). https://doi.org/10.1007/s13238-010-0004-3
53. Kadzińska, J., Janowicz, M., Kalisz, S., Bryś, J., Lenart, A.: An overview of fruit and vegetable edible packaging materials. Packag. Technol. Sci. **32**(10), 483–495 (2019)
54. Kanatt, S.R.: Development of active/intelligent food packaging film containing Amaranthus leaf extract for shelf life extension of chicken/fish during chilled storage. Food Packag. Shelf Life **24**, 100506 (2020)
55. Kanha, N., Osiriphun, S., Rakariyatham, K., Klangpetch, W., Laokuldilok, T.: On-package indicator films based on natural pigments and polysaccharides for monitoring food quality: a review. J. Sci. Food Agric. **102**(15), 6804–6823 (2022)
56. Keppler, J. K., Schwarz, K., & van der Goot, A. J.: Covalent modification of food proteins by plant-based ingredients (polyphenols and organosulphur compounds): A commonplace reaction with novel utilization potential. Trends Food Sci. Technol. 101, 38–49 (2020). https://doi.org/10.1016/j.tifs.2020.04.023
57. Kumar, S., Mukherjee, A., Dutta, J.: Chitosan based nanocomposite films and coatings: Emerging antimicrobial food packaging alternatives. Trends Food Sci. Technol. 97, 196–209 (2020)
58. Kumar, M., Tomar, M., Saurabh, V., Mahajan, T., Punia, S., del Mar Contreras, M., Rudra, S.G., Kaur, C., & Kennedy, J.F. (2020). Emerging trends in pectin extraction and its anti-microbial functionalization using natural bioactives for application in food packaging. *Trends in Food Science & Technology*, *105*, 223–237. https://doi.org/10.1016/j.tifs.2020.09.009
59. Kuswandi, B., others.: Active and intelligent packaging, safety, and quality controls. Fresh-Cut Fruits Vegetables 243–294 (2020)
60. Lawal, U., Valapa, R.B.: Bioplastics: an introduction to the role of eco-friendly alternative plastics in sustainable packaging. In: Bio-Based Packaging: Material, Environmental and Economic Aspects, 319–334 (2021)
61. Liu, C., Huang, J., Zheng, X., Liu, S., Lu, K., Tang, K., Liu, J.: Heat sealable soluble soybean polysaccharide/gelatin blend edible films for food packaging applications. Food Packag. Shelf Life **24**, 100485 (2020)
62. Lobato-Guarnido, I., Luzón, G., Ríos, F., Fernández-Serrano, M.: Synthesis and characterization of environmentally friendly chitosan–arabic gum nanoparticles for encapsulation of oregano essential oil in pickering emulsion. Nanomaterials 13(19), (2023). https://doi.org/10.3390/nano13192651
63. Luo, Q., Hossen, M.A., Zeng, Y., Dai, J., Li, S., Qin, W., Liu, Y.: Gelatin-based composite films and their application in food packaging: a review. J. Food Eng. 313, 110762 (2022). https://doi.org/10.1016/j.jfoodeng.2021.110762
64. Luzi, F., Torre, L., Kenny, J. M., & Puglia, D.: Bio- and fossil-based polymeric blends and nanocomposites for packaging: structure–property relationship. Materials 12(3), (2019). https://doi.org/10.3390/ma12030471
65. Ma, Y., Ding, S., Fei, Y., Liu, G., Jang, H., Fang, J.: Antimicrobial activity of anthocyanins and catechins against foodborne pathogens Escherichia coli and Salmonella. Food Control 106, 106712 (2019). https://doi.org/10.1016/j.foodcont.2019.106712
66. Maftoonazad, N., Ramaswamy, H.S.: Application and evaluation of a pectin-based edible coating process for quality change kinetics and shelf-life extension of lime fruit (*Citrus aurantifolium*). Coatings **9**(5), 285 (2019)
67. Malek, N.S.A., Faizuwan, M., Khusaimi, Z., Bonnia, N.N., Rusop, M., Asli, N.A.: Preparation and characterization of biodegradable polylactic acid (PLA) film for food packaging application: a Review. J. Phys: Conf. Ser. **1892**(1), 12037 (2021)
68. Mane, K.A.: A review on active packaging: an innovation in food packaging. Int. J. Environ., Agric. Biotechnol. **1**(3), 238566 (2016)
69. Manso, T., Lores, M., de Miguel, T.: Antimicrobial activity of polyphenols and natural polyphenolic extracts on clinical isolates. Antibiotics **11**(1), 46 (2021)

70. Martiny, T.R., Raghavan, V., de Moraes, C.C., da Rosa, G.S., Dotto, G.L.: Bio-based active packaging: Carrageenan film with olive leaf extract for lamb meat preservation. Foods **9**(12), 1759 (2020)
71. Meherishi, L., Narayana, S.A., Ranjani, K.S.: Sustainable packaging for supply chain management in the circular economy: a review. J. Clean. Prod. **237**, 117582 (2019)
72. Merlo, T.C., Contreras-Castillo, C.J., Saldana, E., Barancelli, G.V., Dargelio, M.D.B., Yoshida, C.M.P., Junior, E.E.R., Massarioli, A., Venturini, A.C.: Incorporation of pink pepper residue extract into chitosan film combined with a modified atmosphere packaging: Effects on the shelf life of salmon fillets. Food Res. Int. **125**, 108633 (2019)
73. Mirpoor, S.F., Patanè, G.T., Corrado, I., Giosafatto, C.V.L., Ginestra, G., Nostro, A., Foti, A., Gucciardi, P.G., Mandalari, G., Barreca, D., et al.: Functionalization of polyhydroxyalkanoates (pha)-based bioplastic with phloretin for active food packaging: characterization of its mechanical, antioxidant, and antimicrobial activities. Int. J. Mol. Sci. **24**(14), 11628 (2023)
74. Mirza Alizadeh, A., Masoomian, M., Shakooie, M., Zabihzadeh Khajavi, M., Farhoodi, M.: Trends and applications of intelligent packaging in dairy products: a review. Crit. Rev. Food Sci. Nutr. **62**(2), 383–397 (2021)
75. Mohamed, S.A.A., El-Sakhawy, M., Nashy, E.-S.H.A., Othman, A.M.: Novel natural composite films as packaging materials with enhanced properties. Int. J. Biol. Macromol. **136**, 774–784 (2019)
76. Mondal, K., Ghosh, T., Bhagabati, P., Katiyar, V.: Chapter 8—sustainable nanostructured materials in food packaging. In: Karak, N.: (Ed.) Dynamics of advanced sustainable nanomaterials and their related nanocomposites at the bio-nano interface (pp. 171–213). Elsevier (2019). https://doi.org/10.1016/B978-0-12-819142-2.00008-2
77. Monika, Mehmood, K., Katiyar, V.: Effect of dicumyl peroxide on biodegradable poly (lactic acid)/functionalized gum arabic based films. J. Appl. Polym. Sci. **138**(45), 51341 (2021)
78. Mostafa, H., Airouyuwaa, J. O., Hamed, F., Wang, Y., Maqsood, S.: Structural, mechanical, antioxidant and antibacterial properties of soy protein isolate (SPI)-based edible food packaging films as influenced by nanocellulose (NC) and green extracted phenolic compounds from date palm leaves. Food Packaging Shelf Life 38, 101124 (2023). https://doi.org/10.1016/j.fpsl.2023.101124
79. Mousavi Khaneghah, A., Hashemi, S. M. B., & Limbo, S.: Antimicrobial agents and packaging systems in antimicrobial active food packaging: An overview of approaches and interactions. Food Bioproducts Process. 111, 1–19 (2018). https://doi.org/10.1016/j.fbp.2018.05.001
80. Mushtaq, M., Gani, A., Gani, A., Punoo, H.A., Masoodi, F.A.: Use of pomegranate peel extract incorporated zein film with improved properties for prolonged shelf life of fresh Himalayan cheese (Kalari/kradi). Innov. Food Sci. Emerging Technologies 48, 25–32 (2018)
81. Navasingh, R.J.H., Gurunathan, M.K., Nikolova, M.P., Królczyk, J.B.: Sustainable bioplastics for food packaging produced from renewable natural sources. Polymers **15**(18), 3760 (2023)
82. Nayana, P., Wani, K.M.: Unlocking the green potential: sustainable extraction of bioactives from orange peel waste for environmental and health benefits. J. Food Meas. Charact. **18**(10), 8145–8162 (2024)
83. Nešić, A., Cabrera-Barjas, G., Dimitrijević-Branković, S., Davidović, S., Radovanović, N., Delattre, C.: Prospect of polysaccharide-based materials as advanced food packaging. Molecules **25**(1), 135 (2019)
84. Nethra, P.V., Sunooj, K.V., Aaliya, B., Navaf, M., Akhila, P.P., Sudheesh, C., Mir, S.A., Shijin, A., George, J.: Critical factors affecting the shelf life of packaged fresh red meat--a review. Measurement: Food, 100086 (2023)
85. Ohkawa, K., Hayashi, S., Nishida, A., Yamamoto, H., Ducreux, J.: Preparation of pure cellulose nanofiber via electrospinning. Text. Res. J. **79**(15), 1396–1401 (2009)
86. Othman, S.H., Kahar, N.S., Nordin, N., Ahmad Shapi'i, R.: Properties and food packaging applications of solvent casting-made starch-based films incorporated with essential oils: a review. Starch-Stärke 2200152 (2023).
87. Othman, S.H., Othman, N.F.L., Shapi'i, R.A., Ariffin, S.H., Yunos, K.F.M.: Corn starch/chitosan nanoparticles/thymol bio-nanocomposite films for potential food packaging applications. Polymers **13**(3), 390 (2021)

88. Pandhi, S., Kumar, A., Alam, T.: Probiotic edible films and coatings: concerns, applications and future prospects. J Packag Technol Res **3**, 261–268 (2019)
89. Pavli, F., Argyri, A. A., Skandamis, P., Nychas, G.-J., Tassou, C., Chorianopoulos, N.: Antimicrobial activity of oregano essential oil incorporated in sodium alginate edible films: control of listeria monocytogenes and spoilage in ham slices treated with high pressure processing. Materials 12(22) (2019). https://doi.org/10.3390/ma12223726
90. Perera, K. Y., Jaiswal, A. K., Jaiswal, S.: Extending cheese shelf-life using eco-friendly sodium alginate-gelatin films reinforced with nanoclay. Food Biosci. 56, 103304 (2023). https://doi.org/10.1016/j.fbio.2023.103304
91. Petkoska, A.T., Daniloski, D., D'Cunha, N.M., Naumovski, N., Broach, A.T.: Edible packaging: sustainable solutions and novel trends in food packaging. Food Res. Int. **140**, 109981 (2021)
92. Priyadarshi, R., Roy, S., Purohit, S. D., & Ghosh, T.: Biopolymers for food packaging and biomedical applications: Options or obligations? In: Coatings (Vol. 12, Issue 9, p. 1261). MDPI (2022)
93. Radhalakshmi, V., Raman, M., Joy, M.R.: Development of active packaging film based on poly (lactic acid) incorporated with Piper betel leaf ethanolic extract and its application in the shelf-life extension of tuna meat. Int. J. Biol. Macromol. **246**, 125751 (2023)
94. Raghav, P., Saini, M.: (2018). Development of mint (*Mentha viridis* L.) herbal edible coating for shelf life enhancement of cucumber (*Cucumis sativus*). Int. J. Green Herbal Chem. 7. https://doi.org/10.24214/IJGHC/GC/7/2/37991
95. Rahman, S., Gogoi, J., Dubey, S., Chowdhury, D.: Animal derived biopolymers for food packaging applications: a review. Int. J. Biol. Macromol. 128197
96. Ramakrishnan, R., Kulandhaivelu, S.V., Roy, S.: Alginate/carboxymethyl cellulose/starch-based active coating with grapefruit seed extract to extend the shelf life of green chilli. Ind. Crops Prod. **199**, 116752 (2023)
97. Rambabu, K., Bharath, G., Banat, F., Show, P.L., Cocoletzi, H.H.: Mango leaf extract incorporated chitosan antioxidant film for active food packaging. Int. J. Biol. Macromol. **126**, 1234–1243 (2019)
98. Raza, M., Abu-Jdayil, B.: Cellulose nanocrystals from lignocellulosic feedstock: a review of production technology and surface chemistry modification. Cellulose **29**(2), 685–722 (2022). https://doi.org/10.1007/s10570-021-04371-y
99. Reichert, C. L., Bugnicourt, E., Coltelli, M.-B., Cinelli, P., Lazzeri, A., Canesi, I., Braca, F., Martínez, B. M., Alonso, R., Agostinis, L., Verstichel, S., Six, L., Mets, S. De, Gómez, E. C., Ißbrücker, C., Geerinck, R., Nettleton, D. F., Campos, I., Sauter, E., Schmid, M.: Bio-based packaging: materials, modifications, industrial applications and sustainability. Polymers 12(7) (2020). https://doi.org/10.3390/polym12071558
100. Rejeesh, C. R., Anto, T.: Packaging of milk and dairy products: approaches to sustainable packaging. Mater. Today Proc. 72, 2946–2951 (2023). https://doi.org/10.1016/j.matpr.2022.07.467
101. Roy, S., Chawla, R., Santhosh, R., Thakur, R., Sarkar, P., Zhang, W.: Agar-based edible films and food packaging application: a comprehensive review. Trends Food Sci. Technol. 104198 (2023)
102. Roy, S., Ezati, P., Biswas, D., Rhim, J.-W.: Shikonin functionalized packaging film for monitoring the freshness of shrimp. Materials **15**(19), 6615 (2022)
103. Roy, S., Kim, H.C., Panicker, P.S., Rhim, J.-W., Kim, J.: Cellulose nanofiber-based nanocomposite films reinforced with zinc oxide nanorods and grapefruit seed extract. Nanomaterials **11**(4), 877 (2021)
104. Roy, S., Priyadarshi, R., Rhim, J.-W.: Development of multifunctional pullulan/chitosan-based composite films reinforced with ZnO nanoparticles and propolis for meat packaging applications. Foods **10**(11), 2789 (2021)
105. Roy, S., Rhim, J.-W.: Carboxymethyl cellulose-based antioxidant and antimicrobial active packaging film incorporated with curcumin and zinc oxide. Int. J. Biol. Macromol. **148**, 666–676 (2020)

106. Roy, S., Rhim, J.-W.: Preparation of bioactive functional poly (lactic acid)/curcumin composite film for food packaging application. Int. J. Biol. Macromol. **162**, 1780–1789 (2020)
107. Roy, S., Rhim, J.-W.: Starch/agar-based functional films integrated with enoki mushroom-mediated silver nanoparticles for active packaging applications. Food Biosci. **49**, 101867 (2022)
108. Roy, S., Rhim, J.-W.: Carboxymethyl cellulose-based antioxidant and antimicrobial active packaging film incorporated with curcumin and zinc oxide. Int. J. Biol. Macromol. 148, 666–676 (2020). https://doi.org/10.1016/j.ijbiomac.2020.01.204
109. Sadadekar, A.S., Shruthy, R., Preetha, R., Kumar, N., Pande, K.R., Nagamaniammai, G.: Enhanced antimicrobial and antioxidant properties of Nano chitosan and pectin based biodegradable active packaging films incorporated with fennel (*Foeniculum vulgare*) essential oil and potato (*Solanum tuberosum*) peel extracts. J. Food Sci. Technol. **60**(3), 938–946 (2023). https://doi.org/10.1007/s13197-021-05333-9
110. Saedi, S., Garcia, C.V., Kim, J.T., Shin, G.H.: Physical and chemical modifications of cellulose fibers for food packaging applications. Cellulose **28**(14), 8877–8897 (2021). https://doi.org/10.1007/s10570-021-04086-0
111. Setälä, O., Lehtiniemi, M., Coppock, R., Cole, M.: Microplastics in marine food webs. In: Microplastic contamination in aquatic environments (pp. 339–363). Elsevier (2018)
112. Shahidi, F., Hossain, A.: Preservation of aquatic food using edible films and coatings containing essential oils: a review. Crit. Rev. Food Sci. Nutr. **62**(1), 66–105 (2022). https://doi.org/10.1080/10408398.2020.1812048
113. Sharma, H., Shami, V., Samsher, Chaudhary, V., Sunil, Kumar, M.: Importance of edible coating on fruits and vegetables: a review. 8, 4104–4110 (2019)
114. Sucinda, E. F., Abdul Majid, M. S., Ridzuan, M. J. M., Cheng, E. M., Alshahrani, H. A., Mamat, N.: Development and characterisation of packaging film from Napier cellulose nanowhisker reinforced polylactic acid (PLA) bionanocomposites. Int. J. Biol. Macromol. 187, 43–53 (2021). https://doi.org/10.1016/j.ijbiomac.2021.07.069
115. Suderman, N., Isa, M. I. N., Sarbon, N. M.: The effect of plasticizers on the functional properties of biodegradable gelatin-based film: A review. Food Biosci. **24**, 111–119 (2018)
116. Sun, M., Liu, N., Ni, S., Xu, Q., Gao, H., Dai, H., Fu, Y., Qin, M.: Enhanced strength, antioxidant and barrier properties of chitosan-based film by bentonite and lignosulfonate. Food Packag Shelf Life 42, 101246 (2024). https://doi.org/10.1016/j.fpsl.2024.101246
117. Swetha, T. A., Bora, A., Mohanrasu, K., Balaji, P., Raja, R., Ponnuchamy, K., Muthusamy, G., Arun, A.: A comprehensive review on polylactic acid (PLA)—synthesis, processing and application in food packaging. Int. J. Biol. Macromol. 234, 123715 (2023). https://doi.org/10.1016/j.ijbiomac.2023.123715
118. Synowiec, A., Gniewosz, M., Kraśniewska, K., Przybył, J. L., B\kaczek, K., & W\keglarz, Z.: Antimicrobial and antioxidant properties of pullulan film containing sweet basil extract and an evaluation of coating effectiveness in the prolongation of the shelf life of apples stored in refrigeration conditions. Innov. Food Sci. Emerging Technol. 23, 171–181 (2014)
119. Tachaboonyakiat, W.: 9—Antimicrobial applications of chitosan. In: J. A. Jennings & J. D. Bumgardner (Eds.), *Chitosan Based Biomaterials Volume 2* (pp. 245–274). Woodhead Publishing (2017). https://doi.org/10.1016/B978-0-08-100228-5.00009-2
120. Thakur, R., Pristijono, P., Bowyer, M., Singh, S.P., Scarlett, C.J., Stathopoulos, C.E., Vuong, Q.V.: A starch edible surface coating delays banana fruit ripening. Lwt **100**, 341–347 (2019)
121. Thivya, P., Reddy, B.P., N., Bhosale Yuvraj, K., & Sinija, V. R.: Recent advances and perspectives for effective utilization of onion solid waste in food packaging: a critical review. Rev. Environ. Sci. Bio/Technol. **22**(1), 29–53 (2023)
122. Thivya, P., Bhosale, Y.K., Anandakumar, S., Hema, V., Sinija, V.R.: Development of active packaging film from sodium alginate/carboxymethyl cellulose containing shallot waste extracts for anti-browning of fresh-cut produce. Int. J. Biol. Macromol. **188**, 790–799 (2021)
123. Thivya, P., Bhosale, Y. K., Anandakumar, S., Hema, V., Sinija, V. R.: Exploring the effective utilization of shallot stalk waste and tamarind seed for packaging film preparation. Waste Biomass Valoriz. 1–16 (2021b)

124. Thivya, P., Bhosale, Y.K., Anandakumar, S., Hema, V., Sinija, V.R.: Study on the characteristics of gluten/alginate-cellulose/onion waste extracts composite film and its food packaging application. Food Chem. **390**, 133221 (2022)
125. Trinetta, V., Cutter, C.N., Floros, J.D.: Effects of ingredient composition on optical and mechanical properties of pullulan film for food-packaging applications. LWT-Food Sci. Technol. **44**(10), 2296–2301 (2011)
126. Tripathi, A.D., Mishra, P.K., Darani, K.K., Agarwal, A., Paul, V.: Hydrothermal treatment of lignocellulose waste for the production of polyhydroxyalkanoates copolymer with potential application in food packaging. Trends Food Sci. Technol. 123, 233–250 (2022)
127. Tripathi, S., Kumar, L., Deshmukh, R. K., Gaikwad, K. K.: Ultraviolet blocking films for food packaging applications. Food Bioprocess Technol. 1–20 (2023)
128. Tyagi, P., Salem, K.S., Hubbe, M.A., Pal, L.: Advances in barrier coatings and film technologies for achieving sustainable packaging of food products—a review. Trends Food Sci. Technol. 115, 461–485 (2021)
129. Ucak, I., Abuibaid, A.K.M., Aldawoud, T.M.S., Galanakis, C.M., Montesano, D.: Antioxidant and antimicrobial effects of gelatin films incorporated with citrus seed extract on the shelf life of sea bass (*Dicentrarchus labrax*) fillets. J. Food Process. Preserv. **45**(4), e15304 (2021)
130. Umaraw, P., Munekata, P. E. S., Verma, A. K., Barba, F. J., Singh, V. P., Kumar, P., Lorenzo, J. M.: Edible films/coating with tailored properties for active packaging of meat, fish and derived products. Trends Food Sci. Technol. 98, 10–24 (2020)
131. Vanitha, T., & Khan, M.: Role of pectin in food processing and food packaging. In: Pectins-extraction, purification, characterization and applications (2019)
132. Vasile, C., Baican, M.: Progresses in food packaging, food quality, and safety—controlled-release antioxidant and/or antimicrobial packaging. Molecules **26**(5), 1263 (2021)
133. Varghese, S.M., Bhosale, Y.K., Aruna Nair, U.K., Hema, V., Sinija, V.R.: Valorisation and Characterization of *Allium cepa* var. aggregatum biowastes for the production of microcrystalline cellulose. Waste Biomass Valoriz. 13(4), 1931–1944 (2022)
134. Wang, C., Shi, J., He, M., Ding, L., Li, S., Wang, Z., Wei, J.: High strength cellulose/ATT composite films with good oxygen barrier property for sustainable packaging applications. Cellulose **25**(7), 4145–4154 (2018). https://doi.org/10.1007/s10570-018-1855-7
135. Wani, K.M., Uppaluri, R.V.: Characterization of pectin extracted from pomelo peel using pulsed ultrasound assisted extraction and acidic hot water extraction process. Appl. Food Res. **3**(2), 100345 (2023)
136. Wani, K.M., Uppaluri, R.V.: Comparative efficacy of ultrasound-assisted and hot water extraction of papaya leaves. J. Herb. Med. **42**, 100809 (2023)
137. Wani, K.M., Uppaluri, R.V.: Continuous and pulsed ultrasound-assisted extraction of pectin from pomelo fruit peel using citric acid. Biomass Convers. Biorefin. 1–16 (2022)
138. Wani, K.M., Uppaluri, R.V.: Efficacy of ionic gelation based encapsulation of bioactives from papaya leaf extract: characterization and storage stability. Biomass. Convers. Biorefin. **14**(16), 19911–19928 (2024)
139. Wani, K.M., Uppaluri, R.V.: Efficacy of ultrasound-assisted extraction of bioactive constituents from Psidium guajava leaves. Appl. Food Res. **2**(1), 100096 (2022)
140. Wani, K.M., Uppaluri, R.V.: Recovery of bioactive constituents from unripe papaya peel and pulp using ultrasound assisted extraction: Optimization of process parameters. In North-East research conclave (pp. 119–139). Singapore: Springer Nature Singapore (2022)
141. Westlake, J.R., Tran, M.W., Jiang, Y., Zhang, X., Burrows, A.D., Xie, M.: Biodegradable biopolymers for active packaging: demand, development and directions. Sustain. Food Technol. **1**(1), 50–72 (2023)
142. Wu, F., Misra, M., Mohanty, A.K.: Challenges and new opportunities on barrier performance of biodegradable polymers for sustainable packaging. Prog. Polym. Sci. **117**, 101395 (2021)
143. Xu, K., Li, Q., Xie, L., Shi, Z., Su, G., Harper, D., Tang, Z., Zhou, J., Du, G., Wang, S.: Novel flexible, strong, thermal-stable, and high-barrier switchgrass-based lignin-containing cellulose nanofibrils/chitosan biocomposites for food packaging. Ind. Crops Products 179, 114661 (2022). https://doi.org/10.1016/j.indcrop.2022.114661

144. Yao, F., Wu, Z., Gu, Y., Di, Y., Liu, Y., Srinivasan, V., Lian, C., Li, Y.: Acetylated nanocellulose reinforced hydroxypropyl starch acetate realizing polypropylene replacement for green packaging application. Carbohydrate Polymers 331, 121886 (2024). https://doi.org/10.1016/j.carbpol.2024.121886
145. Yaradoddi, J.S., Banapurmath, N.R., Ganachari, S.V., Soudagar, M.E.M., Mubarak, N.M., Hallad, S., Hugar, S., Fayaz, H.: Biodegradable carboxymethyl cellulose based material for sustainable packaging application. Sci. Rep. **10**(1), 21960 (2020)
146. Ye, Y., Zeng, F., Zhang, M., Zheng, S., Li, J., Fei, P.: Hydrophobic edible composite packaging membrane based on low-methoxyl pectin/chitosan: effects of lotus leaf cutin. Food Packag Shelf Life 26, 100592 (2020). https://doi.org/10.1016/j.fpsl.2020.100592
147. Liu, C., Huang, J., Zheng, X., Liu, S., Lu, K., Tang, K., Liu, J.: Heat sealable soluble soybean polysaccharide/gelatin blend edible films for food packaging applications. Food Packaging Shelf Life 24, 100485 (2020). https://doi.org/10.1016/j.fpsl.2020.100485
148. Yu, T.-Y., Tseng, Y.-K., Lin, T.-H., Wang, T.-C., Tseng, Y.-H., Chang, Y.-H., Wu, M.-C., Su, W.-F.: Effect of cellulose compositions and fabrication methods on mechanical properties of polyurethane-cellulose composites. Carbohyd. Polym. **291**, 119549 (2022)
149. Yu, Y.M., Fukagawa, N.K.: Protein and amino acids. In Present knowledge in nutrition (pp. 15–35). Academic Press (2020)
150. Zhang, H., Tsao, R.: Dietary polyphenols, oxidative stress and antioxidant and anti-inflammatory effects. Curr. Opin. Food Sci. 8, 33–42 (2016). https://doi.org/10.1016/j.cofs.2016.02.002
151. Zhao, Y., Li, B., Li, C., Xu, Y., Luo, Y., Liang, D., Huang, C.: Comprehensive review of polysaccharide-based materials in edible packaging: a sustainable approach. Foods **10**(8), 1845 (2021)
152. Zubair, M., Pradhan, R.A., Arshad, M., Ullah, A.: Recent advances in lipid derived bio-based materials for food packaging applications. Macromol. Mater. Eng. **306**(7), 2000799 (2021)

Chapter 9
Advances in Bio-based Barrier Materials for Active Food Packaging

Payel Dasgupta and Deepak Subramani

Abstract The present chapter aims to give a multifaceted view of the biobased barrier materials suitable for active food packaging. To begin with, barrier functions for mass migration of moisture, gas, odor, and microbes and controlled release of active constituents from package to food are addressed along with energy barrier features like light/irradiation trap and heat insulation. It is followed by exploring various biomaterials used for delivering barrier functions. This context explores polysaccharides, proteins, lipid coatings, fermentation-derived polylactic acid and polyhydroxyalkanoates, food wastes, their derivatives, and green nanomaterials. The factors affecting the features of barrier materials and the latter's suitability for active packaging are critically evaluated from consumer acceptability, health and safety, food properties, and active material compatibility perspectives. Moreover, multilayering techniques to overcome poor barrier properties of biopolymers are elaborated. With this, this chapter provides a solid understanding of the challenges and potential of biobased barrier materials for active packaging.

Keywords Active packaging · Barrier · Bio-based · Coating · Nanomaterial

9.1 Introduction

The primary function of a food packaging material is to protect the food from the external environment. Hence, the packaging material must offer a superior barrier to water, water vapor, gas, grease, oils, and microorganisms. It should also prevent the loss of aroma compounds from foods and the uptake of off-flavors from the external environment [1]. In active packaging, the barrier material should also ensure a sustained release of active ingredients (e.g., phenolics and oils) loaded on the film [2].

P. Dasgupta (✉)
Agricultural and Food Engineering Department, Indian Institute of Technology, Kharagpur, West Bengal 721302, India
e-mail: payeldasgupta.30@gmail.com

D. Subramani
Asserpark, 6706 HC Wageningen, The Netherlands

S. ul Islam and M. Shahid (eds.), *Green Materials for Active Food Packaging*, Engineering Materials, https://doi.org/10.1007/978-981-96-0369-5_9

Many petroleum-derived polymers like low-density polyethylene (LDPE), ethylene vinyl alcohol (EVOH), polystyrene (PS), polypropylene (PP), and polyvinyl chloride (PVC) have long been used to offer barrier functions. For their non-biodegradability and environmental impact, there is a greater need to replace synthetic barrier materials with natural, bio-based, and biodegradable materials [3]. Consumers today are also willing to pay more for products packed in eco-friendly material [4].

The bio-based barrier material can be organic biopolymers like polysaccharides, proteins, and lipids. The production and usage of fermentation-derived biomaterials like polylactic acid and polyhydroxyalkanoates are also rising [1]. Fruits and vegetable wastes like peel, pomace, and trimmings, animal and aquatic by-products like bones and skin, and cereals/legume wastes like bran, hull, and pods can be suitable raw materials for extraction of these biopolymers [5]. The derived polysaccharides like Pectin, alginate, galactomannan, cellulose, starch, and proteins have moderate to high hydrophilicity and offer poor water barrier function. However, chemical or physical modification of biopolymers by esterification, etherification, self-assembly, acid-heat treatment, and blending with essential oils or waxes are reported to improve water barrier properties [6–8]. Furthermore, nanomaterials like cellulose nanocrystals and zein nanoparticles obtained from biomaterials can further improve the barrier function [9, 10].

The material with a good barrier function alone does not make it an eligible candidate for active packaging of food applications. The material should be non-toxic, non-allergenic, recyclable, available, affordable, biodegradable, non-flammable, offer good mechanical strength and thermal stability, and should be easy to process for film development and food packaging in industries [1]. Moreover, one efficient barrier function of the biomaterial is simply not sufficient. This is because food products often demand more than one barrier function. Hence, multilayering techniques that bring the barrier functions of different biopolymers are desirable [11]. Bridging the gap between film and product characteristics is essential to designing barrier materials for active packaging.

Therefore, this book chapter briefly discusses the different biobased barrier materials used for active packaging while critically evaluating the factors affecting the barrier material features and their suitability for specific food applications. The multilayering techniques for barrier materials are also discussed.

9.2 Barrier Functions in Active Food Packaging

The active food packaging should offer a barrier to mass and heat transfer to reduce the quality deterioration/spoilage rate and extend the shelf life. The mass transfer includes water, water vapor, gases like O_2 and CO_2, non-polar lipids like fats, oils, and grease, aroma compounds, biological/live organisms like bacteria and insects, colorants/printing dyes and active constituents added in active packaging [12, 13]. From the energy barrier category, the packaging film should resist heat transfer and UV rays [14].

9.2.1 Moisture Barrier

Moisture barrier is the most important barrier property for a packaging material. The packaging material may come in direct contact with food having high water content or may get exposed to water vapor when the food is stored in high relative humidity conditions. High water activity makes the product perishable due to the more pronounced growth of bacteria, yeast, and mold [15]. This also promotes oxidative or hydrolytic reactions in food, causing off-flavor generation and loss of sensitive vitamins (vitamin C) and pigments. The high water activity activates/accelerates the enzyme activity, increases the mobility of precursors, and makes the product unstable [16]. The moisture barrier can also be a requirement for the effective functioning of active ingredients used in packaging [13]. The different indicators measured to understand the moisture barrier property of a film are the contact angle, water absorptivity, swelling ratio, water vapor permeability (WVP, and water vapor transmission rate (WVTR) [1, 17]. The contact angle reflects the surface hydrophobicity of the material. The higher the contact angle, the higher the surface hydrophobicity. A film with a highly hydrophobic surface (contact angle) has a significantly lower water permeation [18]. By dividing the maximum quantity of allowed water/water vapor by the water/water vapor transmission rate, the product's shelf life can be estimated (Fig. 9.1).

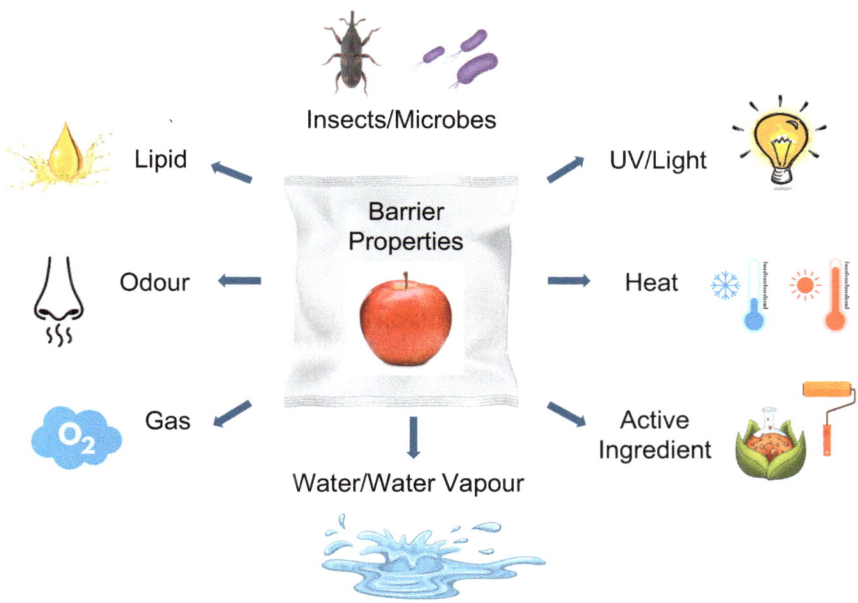

Fig. 9.1 Barrier properties for active packaging of food

9.2.2 Gas Barrier

Unlike liquids, gases can penetrate the packaging material more easily. Oxygen in the headspace or as dissolved oxygen in lipids (fat/oil) and water can initiate oxidation reactions. The presence of enzymes like catalase, peroxidase, lipoxygenase, and polyphenol oxidase intensifies the oxidation rate. The oxidation of fats and oils causes rancidity and off-flavor generation. Beyond affecting flavor and taste, oxidation can also influence the texture of foods (refrigerated fresh dough products). This can change the color of foods by enzymatic browning and ascorbic acid browning. Water-soluble vitamins like vitamin C and fat-soluble vitamins like tocopherols and antioxidant pigments naturally present in foods or added as an active ingredient in active packaging become unavailable upon oxidation [19, 20].

Perishable commodities like fruits and vegetables, meat, and seafood demand modified gas composition (CO_2, O_2, N_2, or other inert gases) for higher shelf life. This is employed in modified atmospheric packaging. In this case, it is necessary for the packaging to offer a barrier to gas escaping from the package [21]. The selectivity of the package for specific gases prevents the accumulation of specific gases (ethylene) while ensuring moderate oxygen for enhancing shelf life [22]. Storage conditions like relative humidity and temperature largely influence the gas barrier. If the packaging material absorbs water in a high relative humidity condition, it also increases oxygen permeability [1]. This is because water is a plasticizer, and this loosens the polymer interactions in the network, facilitating the oxygen barrier. Since oxygen can dissolve in non-polar solvents better than in aqueous medium, fat and oil products demand a greater gas barrier [19]. Oxygen transmission rate (OTR) and oxygen permeability (OP) are the key indicators studied for the gas barrier property of a packaging material [23].

9.2.3 Lipids Barrier

Like water, the barrier to lipid transfer is also important for certain foods, especially when they are rich in fats, oils, and waxes. The transfer of fats/waxes into the outer cardboard cover is typically encountered in the bakery box. This is due to the capillary transfer of lipids through corrugated/porous packaging materials [24]. Plant oils and waxes are also incorporated into active packaging to improve the hydrophobicity and water barrier properties of the package [25]. However, this should not get leaked/leached to the other side of the laminate. Since oils and fats could be carriers of natural pigments and flavors, their transfer could also leach the colors and flavor compounds [26]. This may leave colored or uncolored (wet grease-like) stains on the package surface, making consumers displeased about the product's perceived quality. Since lipids can highly absorb external odors, the presence of lipids on the outer surface may also give an objectionable odor to food [27]. Unlike liquid foods demanding highly hydrophobic surfaces, oily foods demand hydrophilic surfaces and a minimal porosity to hinder lipid permeation [28]. Kit test is the most commonly used method for determining the effectiveness of lipid barrier. In this, Kits solvents (organic non-polar type) with a number from 1 to 12 are used. A higher kit number for a packaging material reflects the lipid barrier nature [29].

9.2.4 Odor Trap

Foods contain many volatile flavor compounds. This constitutes hundreds to thousands of aromatic hydrocarbons, aldehydes, ketones, acids, and alcohols, naturally present in a food or generated from oxidation during processing and storage [30]. Their volatility and low molecular weight (or size) allow these compounds to permeate the packaging material [26]. Loss of volatiles makes the product flavorless. This is particularly relevant for hydrophilic foods. Permeation of water from food to the periphery of packaging can take away volatile aroma compounds. Moreover, water or high humidity loosening the film matrix also allows the loss of aroma. Since most aromatic compounds are hydrophobic, their nature to avoid water allows their greater chance for permeation through the package [31–33]. Odor trap material is also expected to prevent external odor permeation into food or adsorb to package [34]. With food stimulants of varying polarity, the diffusion coefficient of aroma compounds (alkanes) of varying molecular weight, representative of food, are determined for a packaging material. In another way, the odor permeated through the package is measured by storing the packed food in an air-tight box and feeding the permeant to gas chromatography [29].

9.2.5 Microbial/Insect Barrier

Active packaging, beyond inhibiting the growth of microbes in food, should also offer a physical barrier to the entry of microbes through the package. Since microbes are of sub-micron to micron ranges, highly porous packaging material can allow the permeation of microbes. This contamination can happen from air or close contact with other objects (food or non-food surfaces) carrying microbes [35]. This becomes relevant for biobased materials as they provide nutrients for the growth of microbes. Since biomaterials are hydrophilic, the water absorption by package in a high relative humidity environment can greatly support the activity of microbes on the surface [36]. When the initial growth of fungi-producing extracellular enzymes can break the biopolymers, the subsequent growth of bacterial colonies from the broken polymers can support microbial growth well. This may also enter the food eventually due to loss of structure or increased film permeability. Next to microbes, other predators like insects are also important [37]. Insects or their larvae can penetrate the food through available pores in packaging or bore through it. Regardless, the packaging material should resist the insects by paralyzing or killing effect. The material can also be strong enough physically to resist the puncture of the film by insects [38, 39].

9.2.6 Controlled Release of Active Constituents in Food

In active packaging, the active ingredients like phenolic extracts, essential oils, etc., are loaded into the package. These active constituents are expected to be released into food in a sustained manner. A rapid release may not give a prolonged antioxidant or antimicrobial function. This affects the effectiveness of active packaging material and product shelf life [40]. Depending on the product characteristics, such as polarity, viscosity, density, surface tension, surface-to-volume ratio of food, headspace volume, and air-food interface, the release kinetics of one or more active ingredients may vary. Overall, the packaging should offer a controlled or minimal barrier to active constituents in the direction of food while being impermeable for their loss to the external environment through the package [41, 42].

9.2.7 UV Barrier

Photooxidation by exposure to light, particularly UV wavelength (200–400 nm), is a major issue, and this may happen at any point in the supply chain when food is exposed to light. This is expected during distribution or storage in retail outlets or even post-purchase storage/handling conditions at households. This can induce rancidity and off-flavor generation in foods with high oil content. Sensitive vitamins and pigments are also lost by oxidation [43]. Though sunlight constitutes only 5% UV, it is still significant to break the stable covalent bonds in the food compounds. The active ingredient may lose its function by UV exposure [44]. For these reasons, the barrier material must hinder UV rays while ensuring film transparency, which is expected in many products. The UV barrier property is determined by measuring the percentage of transmittance of light at the UV wavelength. A barrier material that can absorb the UV and reduce the transmittance is highly preferred [45].

9.2.8 Heat Insulation

Though not used very much in bio-based food packaging context, it is also an important energy barrier function. The heat transfer through the package may affect the sustained release of active ingredients in the package, and when food is heated, this can cause many deleterious effects. This can accelerate microbial growth, melt fat to recrystallize later for fat bloom development, affect the structure of food, and accelerate enzymatic oxidation and hydrolysis reactions, leading to off-flavor generation and vitamin loss [46, 47]. This is also suitable for frozen food wrapping. Materials with a lower thermal conductivity, high porosity, and low bulk density (e.g., aerogels, cardboard) are suitable for imparting heat insulation [48].

9.3 Bio-based Barrier Materials for Active Food Packaging

See Fig. 9.2.

9.3.1 Cellulose

Cellulose is a linear polysaccharide of β-glucose units and is the most abundant renewable and biodegradable polymer in plants and bacteria. Though cellulose is largely utilized for making paper and paperboards, it has poor barrier properties due to its hydrophilic nature, poor aqueous solubility, crystallinity (40–60%), and worse film-formation capacity [49]. However, the properties of cellulose can be modified by either physical or chemical treatments. In chemical treatment, esterification and etherification are the most common practice. The etherification of cellulose includes carboxymethyl cellulose (CMC), ethyl cellulose (EC), hydroxypropyl cellulose (HPC), hydroxypropyl methylcellulose (HPMC) and methylcellulose (MC). These materials have been widely recognized for use as paper coatings to offer water barrier properties. These compounds are produced by substituting the position of hydroxyl groups of cellulose with alkyl groups. The type of alkyl groups attached and the carbon chain length vary the water-barrier property of the material [50].

Since cellulose is highly crystalline, the alkyl substitution mainly happens at amorphous regions of cellulose where hydroxyl groups are easily accessible. With pre-treatments, the crystallinity degree can be modified, and the degree of alkyl substitution can be enhanced to vary the barrier property of a material to a great

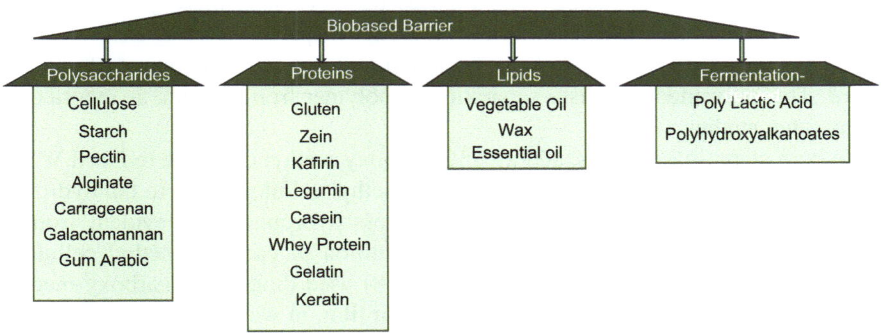

Fig. 9.2 Bio-based materials offering barrier properties to active food packaging

extent [49]. Many studies have reported the better potential of cellulose derivatives to offer barrier properties in active packaging. This can be further improved by adding essential oils and other additives. da Silva et al. [51] reported the barrier properties of curcumin-loaded HPMC films. The WVP of the HPMC film was not affected by curcumin presence. However, the permeability of oxygen and carbon dioxide was significantly reduced. A lower curcumin crystal size given by the antisolvent precipitation method gave better dispersibility in the matrix and better released (>50%) into the food stimulant when it came in contact. HPMC, made by reacting cellulose with methylene chloride and propylene oxide, is an inexpensive, biodegradable thermoplastic with high oil resistance. Though HPMC has water vapor permeability lower than Cellophane, the surface is still hydrophilic and requires more improvements. Adding hydrophobic ingredients such as fatty acids, waxes, surfactants, resins, and zein nanoparticles to its film can reduce water vapor permeability [52, 53]. Lee et al. [39] found that oregano essential oil added HPMC film exhibited a strong antioxidant and antibacterial effect against *Salmonella typhimurium* and improved UV and water vapor barrier properties.

Yang et al. [54] prepared a nanofibrous active packaging film from ethyl cellulose, tea polyphenols, and polyvinyl alcohol using an electrospinning blending technique. Compared to polyvinyl alcohol film, the prepared composite film exhibited a lower wettability, water solubility, and water vapor permeability. At the same time, the film also showed antimicrobial (90% inhibition of *E.coli* and *Streptococcus aureus*) and antioxidant effects (90% DPPH radical scavenging rate) from tea polyphenols (5%). Since tea polyphenols are present in minimal amounts, the pronounced water barrier property of the film was attributed to ethyl cellulose with minimal presence of hydroxyl groups for water binding. The film was found to improve the shelf life of meat.

Methylcellulose is a cellulose derivative obtained by treating cellulose with methylene chloride or dimethyl sulfide. Its solubility is evident only up to 50 °C, and higher temperatures result in thermoreversible gels. It exhibits high stability at a wide pH range (2–12) [55]. Erdohan and Turhan [56] developed a methylcellulose and whey protein-based film whose water vapor permeability was significantly reduced by methylcellulose. Unlike proteins with bulky side chains, the linear polymer methylcellulose forms a tight packing and reduces the pore size, thereby bringing the material's water vapor permeability down. Better dispersibility was ensured since methylcellulose was mixed into heat-denatured proteins and subsequently homogenized. This prevented the phase separation of polymer from protein, as reported in many other studies.

In corn starch-based films, adding 15% carboxy methyl cellulose reduced WVP. Since the carboxylic acid groups of carboxy methyl cellulose bind to the hydroxyl groups in starch, the film surface becomes less hydrophilic, as evident from a decreased contact angle. However, further addition of carboxy methyl cellulose caused no improvements [57]. Zheng et al. [58] used dialdehyde carboxy methyl cellulose as a cross-linking agent in soy-protein film, in which the aldehyde group cross-links with the amino group of soy protein to form a compact and tightly structured film exhibiting higher water barrier properties. However, carboxyl methyl cellulose itself is capable of high water affinity, dense cross-linking, and amino acid binding to aldehyde, reducing potential water binding sites that contribute to a low WVP. The film also had a low light (200–280 nm wavelength) transmission, indicating the effective UV-barrier property. This is because the free aromatic amino acids in protein structure absorb UV rays.

In active packaging, cellulose acetate and cellulose sulfate are the most commonly used cellulose esters. Cellulose acetate is made by substituting hydroxyl groups in cellulose with acetate groups by reacting cellulose with acetic anhydride and acetic acid. This non-toxic, low-cost, and bio-degradable material is sparingly soluble in water and exhibits great thermal and chemical resistance [49]. Laroque et al. [59] prepared cellulose acetate film loaded with the active ingredient carvacrol, an antimicrobial compound in oregano essential oil. Carvacrol was found to have a minimal effect on the water barrier property of cellulose acetate film. The carvacrol was released into the vapor phase gradually for ten weeks. The active constituent was bactericidal to *Weissella viridescens* and *Pseudomonas fluorescens*, which increased the shelf life of ham by three times. Santos et al. [60] prepared an active packaging material by incorporating essential oil derived from oregano and fennel into cellulose acetate. The film exhibited superior water vapor barrier properties and was active against *Escherichia coli, Staphylococcus aureus, and Penicillium spp.* Active packaging material with UV-resistant and antimicrobial qualities was developed by adding phenyl salicylate into cellulose acetate and polylactic acid-based film [61]. It exhibited two folds of lower water vapor permeability. Since cellulose acetate filled the gaps in polylactic acid, the film structure was very compact for the water to diffuse. The film absorbed 95% of UV rays and had an antibacterial effect against *E.coli.* In the case of cellulose sulfate, a sulfate group replaces hydrogen in cellulose [62]. The presence of the sulfate group provides high water solubility and antibacterial properties to the film. It has a better ability to form a film. Chen and Liu [63] developed cellulose sulphate-based active packaging film added with mustard essential oil and β-cyclodextrin. Including mustard essential oil reduced the film's moisture sorption without influencing the WVTR. The film inhibited *E. coli, S. Aureus, B. subtilis*, and *Aspergillus niger* as β-cyclodextrin ensured a sustained release of oil constituents.

Microcrystalline cellulose (MCC) is another native cellulose derivative that has garnered attention largely in the past two decades [64]. It is produced by dissolving the amorphous regions of cellulose with hydrochloric acid to leave behind only a crystalline portion. Since the crystalline cellulose is compact and hydrogen bonding can happen only minimally at the surface, it has strong water and water vapor barrier properties. Adding microcrystalline cellulose as filler in tapioca and pea starch-based films significantly improved their barrier properties [65, 66]. At lower concentrations (3%), microcrystalline cellulose showed better dispersibility, and this impermeable filler makes water and oxygen take a tortuous path through starch film rather than a straight or perpendicular path for permeation. Also, if dispersed well, microcrystalline cellulose interacts with starch by hydrogen bonding. This polymer–polymer interaction avoids polymer-water interactions, thereby reducing water affinity. By these mechanisms, MCC at a low concentration showed better barrier properties.

However, at a high concentration (10%), the filler is poorly dispersed into the starch matrix and exists as agglomerates. This poor dispersibility at high concentrations gave higher water and oxygen permeability.

Oxygen and oil barrier traits of microfibrillated cellulose films were studied by Aulin et al. [67]. The film had randomly assembled fibers of 5–10 nm thickness. The double coating of the paper base with microfibrillated cellulose significantly reduced the oxygen permeability due to reduced sheet porosity, with effectiveness similar to ethylene vinyl alcohol packaging. However, at relative humidity greater than 70 and 80%, oxygen permeability and water uptake increased as some polymer–polymer interactions were replaced by polymer-water bonds, especially at amorphous regions. Moreover, the films showed better resistance to oil penetration.

9.3.2 Starch

Starch is the most commonly used biopolymer for packaging due to its cheap cost, availability, and film-forming ability. Derived from many plant-based sources, starch is used alone or with other polymers for making films and coatings. The films made from starch are oil-resistant, devoid of color and odor, and show semi-permeability to CO_2 and impermeability to O_2. Compared to LDPE films, which do not show selectivity between the permeability of different gases, starch exhibits a greater gas selectivity. It is suitable for preserving fruits and vegetables that continue respiration post-harvest [49]. Although starch has superior gas barrier properties, it is hydrophilic and, therefore, has a poor WVP. Films made from high amylose starch exhibit a lower WVP than normal corn starch-based film. Blending with other suitable polymers/additives or derivatizing the starch similar to cellulose, can improve the water barrier property [68]. Since starch-based film has more pores and cracks on the surface, the addition of plasticizers like sorbitol reduces the gas permeability [69]. Active ingredients loaded into starch-based active packaging can also modify the barrier properties of film. In a study by Cheng et al. [70], the addition (<20%) of anthocyanin extract from red cabbage into acetylated distarch phosphate (ADSP) film reduced its moisture absorption and oxygen permeability. This is because hydroxyl groups in anthocyanin interact with starch to make a dense and cohesive film with improved barrier properties.

Films made by hydrophobically modified potato starch have high surface hydrophobicity (contact angle), and they offer low oxygen and water vapor permeability [71]. Similarly, Phinainitisatra and Harnkarnsujarit [72] found that the addition of hydroxypropylated cassava starch into rice starch increased the contact angle and thereby reduced the water vapor permeability. Shaikh et al. [8] reported that acetylation of pearl millet starch reduced the water vapor permeability and water solubility when the level of acetylation was kept at 4–8%. However, the same study has reported that hydropyrrolation (10–30%) of starch offers higher water vapor permeability as its introduction weakens the inter and intra-molecular hydrogen bonding, allowing increased water permeation. The differences between studies could be due to the difference in the level of hydroxy propylation, crystallinity, molecular weight, amylose-amylopectin ratio, starch granule morphology and film preparation method. Acetylated starch offers better barrier properties than hydroxypropylated starch. Therefore, it has great potential to act as a bio-based barrier material. Paperboard coated with starch-based polyelectrolytic complexes gave a defect-free coated film surface that dropped water vapor transmission by 40%, compared to uncoated paperboard [73]. Moreover, the coating also offered grease and oil barrier properties. The ionic cross-linked and entangled starch polymer network is highly dense, compact, and homogenous. The effectiveness of starch-based polyelectrolytic complexes highly depends on molecular weight and charge density. For instance, the starch-based polyelectrolytic complex had a grease barrier comparable to polyethylene-coated paperboard only when the molecular weight of the starch derivative was high. Benzylation of mango kernel starch dropped water vapor permeability by 58% and increased UV absorption by 80%. Thus, benzylated starch can offer both water vapor and UV barrier [74]. According to [75], hydroxypropylation drops the water and oxygen barrier properties of native starch. However, when starch was subjected to debranching before hydroxypropylation, the resulting film exhibited a superior water and oxygen barrier. This is because debranching avoids steric hindrance and improves the interaction of starch chains (Table 9.1).

9.3.3 Chitosan

Chitosan, derived from the chitin of crustaceans, is the second most abundant biopolymer, next to cellulose. It is obtained by deacetylation of chitin (N-acetyl-D-glycosamine). It also originated from insects and many fungal species. Chitosan is semi-crystalline in nature and is soluble in dilute organic acid solutions. This polysaccharide naturally possesses antimicrobial activity against bacteria, mold, and yeast [91]. For this reason, it is largely explored in active packaging. However, the functionality of chitosan does not stop with antimicrobial properties, it also possesses barrier properties that are useful in packaging design. In a study by Ren et al. [92], adding chitosan (21–81%) into corn starch film significantly reduced the WVP. Though chitosan possesses amino and hydroxyl groups, which might favor water binding, it however did not happen. It is possible that the intermolecular hydrogen bonding between the amino group of chitosan and the hydroxyl group of starch reduces the hydrophilicity of the composite polymer, thereby improving water barrier properties.

Xu et al. [76] prepared composite film from bacterial cellulose, chitosan, and curcumin for active packaging application. It was found that chitosan of a higher molecular weight gave a higher contact angle (hydrophobicity), mechanical strength, and lower WVTR and OP. The hydrogen bonding between the film components might have reduced the sites available for water adsorption, thereby offering restriction to water vapor. However, the molecular weight of chitosan is too high, affecting WVTR and OP due to the chitosan chain undergoing self-aggregation, thereby weakening the

Table 9.1 Barrier properties and surface characteristics of films from various bio-based barrier materials

Barrier material	Film/Film constituents	WVP	WVTR	OP	OTR	Contact angle (θ°)	Other	References
Chitosan	Curcumin, Cellulose	–	$50\ g\ m^{-2}\ d^{-1}$	$0.05\ g\ cm^{-2}\ Pa^{-1}\ h^{-1}$	–	106	0.13% WS	Xu et al. [76]
	propolis extract	$0.50\ g\ mm\ Pa^{-1}\ m^{-2}\ d^{-1}$	–	$0.2\ cm^3\ mm\ atm^{-1}\ m^{-2}\ d^{-1}$	–	–	–	Siripatrawan and Vitchayakitti [77]
	Cardboard Paper	–	$280.7\ g\ m^{-2}\ d^{-1}$	–	–	–	$49.74\ g\ m^{-2}$ WA	Gatto et al. [78]
	Paper	–	$594\ g\ m^{-2}\ d^{-1}$	–	$1.1\ cm^3\ m^{-2}\ d^{-1}$	–	–	Vartiainen et al. [79]
Beeswax	Cassava starch, whey protein	$1.55\ g\ mm\ Pa^{-1}\ m^{-2}\ d^{-1}$	–	–	–	–	–	Cortés-Rodríguez et al. [80]
Carnauba wax	Gelatin	$0.03\ g\ mm\ Pa^{-1}\ m^{-2}\ d^{-1}$	–	–	–	–	26.5% WS	Romani et al. [81]
Palm wax	Gelatin	$1.02 \times 10^{-10}\ g\ Pa^{-1}\ m^{-1}\ s^{-1}$	–	–	–	116	21% UV trasnmision, 25% WS	Syahida et al. [82]
Casein	Tannic acid	$4.6\ g\ Pa^{-1}\ m^{-1}\ s^{-1}$	–	–	–	–	32% WS	Picchio et al. [83]
Kafirin	Gelatin	$6.70 \times 10^{-10}\ g\ Pa^{-1}\ m^{-1}\ s^{-1}$	–	–	–	105	<1% UV transmission	Wang et al. [84]
PLA	Gelatin	$2.65 \times 10^{-11}\ g\ Pa^{-1}\ m^{-1}\ s^{-1}$	–	$5.88 \times 10^{-18}\ mol\ m^{-1}\ Pa^{-1}\ s^{-1}$	–	–	30–40% UV transmission	Nilsuwan et al. [85]
Gluten	Pectin	$1.2 \times 10^{-10}\ g\ Pa^{-1}\ m^{-1}\ s^{-1}$	–	–	–	–	31% WS	Sartori et al. [86]

(continued)

Table 9.1 (continued)

Barrier material	Film/Film constituents	WVP	WVTR	OP	OTR	Contact angle (θ°)	Other	References
Zein	Paper	–	–	400 g m^{-2} d^{-1}	–	80	45° oil contact angle	Hamdani et al. [87]
	Oleic acid	0.37 g mm m^{-2} h^{-1} kPa^{-1}	–	–	–	–	2.8% WS	Masamba et al. [88]
Soy protein amyloid fibrils	PVA, clove essential oil	0.6 × 10^{-3} g KPa $^{-1}$ m^{-1} h^{-1}	–	0.88 × 10^{-5} g m^{-1} kPa^{-1} h $^{-1}$	–	90	25% WS	Yang et al. [89]
Cellulose nanocrystals	Chitosan	1.39 × 10^{-10} g Pa $^{-1}$ m^{-1} s^{-1}	63,239 g cm^{-1} d^{-1}	–	3 cm^3 μm^{-1} m^{-2} kPa^{-1} d^{-1}	108	36% WA	Lavrič et al. [10]
	Alginate	9.36 × 10^{-10} g Pa $^{-1}$ m^{-1} s^{-1}	239 g cm^{-1} d^{-1}	–	1 cm^3 μm^{-1} m^{-2} kPa^{-1} d^{-1}	58	50% WA	
Starch nanocrystals	Starch	1.5 × 10^{-5} g Pa $^{-1}$ m^{-1} h^{-1}	–	–	–	–	35%	Mukurumbira et al. [90]
Zein nanoparticles	Starch, Olive oil	1.28 × 10^{-7} g Pa $^{-1}$ m^{-1} h^{-1}	–	–	–	97	<1% UV transmission	Farajpour et al. [9]

WVP—Water Vapour Permeability; WVTR—Water Vapour Transmission Rate; OP—Oxygen Permeability; OTR—Oxygen Transmission Rate; WS—water solubility; WA—Water Absorption hydrogen bonds and barrier properties. Chitosan of moderate molecular weight (500–850 kDa) had a stronger reducing power than low (15–50 kDa) and high molecular weight (1700–4150 kDa) chitosan. Depending on the binding strength of chitosan with curcumin, the release characteristics of curcumin vary, and therefore, the antioxidant activity of the film is affected. Therefore, it is necessary to choose chitosan of appropriate molecular weight. A similar study by Cazón et al. [93] found that bacterial cellulose, chitosan and polyvinyl alcohol-based film had an excellent UV barrier property. Grafting chitosan on the regenerative cellulose film significantly reduced the OP as the chitosan cemented the highly porous surface of the cellulose matrix. Chitosan in the film also inhibited pathogens *E.coli* and *Streptococcus aureus* by 99% in a dose-dependent manner [94].

Chitosan film added with propolis extract (2.5–20%) was tested for its potential as an antioxidant and antimicrobial packaging by Siripatrawan and Vitchayakitti [77]. The OP of the film significantly reduced with an increase in the concentration of extract in the film. A similar result was also noted for chitosan film added with cinnamon oil. A strong cross-linking effect between phenolic compounds (polar phenolic acids and esters) and chitosan increased the interfacial adhesion, which gave a compact polymer structure that affected gas transmission through the material. Like OP, propolis extract added chitosan film also exhibited low WVP because the phenolics occupy the position of the functional group of chitosan by hydrogen and covalent bonding, thereby reducing the hydrogen groups present for binding to water. The film exhibited profound antibacterial and antioxidant effects. The ability of this combination to offer better barrier properties with no compromise on active packaging function makes this a suitable material [95].

Since chitosan possesses an amino group, it can interact with the carboxylic acid group of fatty acids differently, depending on pH. If pH is maintained at 5.5 to 6, chitosan can hinder lipid transfer. Chitosan (at 2.2% coating level) was known to function as an effective oil-barrier coating on kraft paper, with oil-resistance efficiency as high as fluorinated resins. The incorporation of alginate further raised the oil barrier while dropping the coating treatment cost [96]. Chitosan and lemongrass essential oil coating on a paperboard made the latter impermeable to microbes. The citral and other terpenes in the essential oil are known to paralyze insects and protect the packed pasta and wheat grains from insect infestation. The film also had a superior water and water vapor barrier and fat-barrier properties [38].

9.3.4 Gums

Gums represent hydrocolloids from various origins. This includes seaweed polysaccharides (e.g., alginate, carrageenan), galactomannans (guar gum, tara gum, fenugreek gum), pectin, plant exudates (gum arabic) and microbial gums (gellan, curdlan). Pectin is a heterogeneous polysaccharide. It comprises α-(1,4)-D-galacturonic acid [97]. Meerasri and Sothornvit [98] developed a pectin film incorporated with γ-amino butyric acid. The film exhibited strong antioxidant activity and lower film solubility and water vapor permeability. Active packaging made from pectin film added with curcumin showed results similar to previous findings [99]. Sistla and Mehraj [100] reported the molecular dynamics of CO_2, O_2 and water in pectin film. The surface adsorption and diffusion of water through linear polygalacturonic acid chains is greater than for O_2 and CO_2. This is because of the strong hydrogen bonding of water with pectin. Due to poor surface adsorption of O_2, the intrinsic diffusion of gases relies on the pores and free volume channels in the film matrix. Though pectin offers a greater oxygen barrier, the water barrier properties can be improved by adding lipophilic essential oils [101]. Pectin-based nanocomposite aerogels were found to have thermal conductivity lower than air. This can serve as an effective insulated food packaging [48].

Galactomannan is a non-charged polysaccharide found in many dicot seeds, like legumes. It consists of a backbone made of mannose, to which a galactose branch is substituted. The properties of different galactomannans like Tara gum, locust bean gum, and guar gum differ depending on the ratio of galactose and mannose. Galactomannans have poor moisture barrier properties due to the presence of many hydroxyl groups in sugar units [102]. The galactomannan-based film offers very low water vapor and oxygen permeability when mixed with corn oil and glycerol. Its application over cheese reduced weight loss during storage [103]. UV treatment is also found to improve the barrier properties of galactomannan [104]. Mikkonen et al. [105] developed spruce galactomannan film using sorbitol as a plasticizer. The film exhibited oxygen permeability comparable to polyethylene and ethylene vinyl alcohol-based films. When tested for aroma permeability, it was found that the diffusion coefficient of tested aroma compounds, D-limonene and carvone was very low. The film was stated to have a superior aroma barrier than starch or even synthetic polyethylene, polycaprolactone and polyhydroxybutyrate-based films. While providing the needed film transparency, the film also acted as a UV barrier. The selective permeability of galactomannan film for CO_2 and O_2 can be modulated by adjusting the concentration of galactomannan and adding glycerol and oil levels. While oil addition promoted the permeability of both gases, galactomannan reduced the same. Commercial packagings for fruits and vegetables have a CO_2 to O_2 permselectivity ratio of 8–10. Since galactomannan packaging has permselectivity ratio of 3.5–11.5, the film can offer an optimum barrier to specific gases, and is highly suitable for fruit and vegetable packaging [106]. Chemical modification of galactomannan with acetic anhydride esterification is another technique to improve the barrier properties of packaging. Esterification of galactomannan increased the surface hydrophobicity and reduced the water vapor permeability of the film made from modified galactomannan. Although this modification slightly increased the oxygen permeability, it is still lower than low-density polyethylene, high-density polyethylene and linear low-density polyethylene [50]. Galactomannan film added with antimicrobial sodium acetate decreased O_2 permeability while elevating CO_2 permeability. The chitosan, sodium acetate and galactomannan-based film also exhibited superior water barrier properties. An antimicrobial film with improved water and gas barrier properties can be produced from this combination [107].

Alginate is a seaweed polysaccharide obtained from brown algae. It constitutes β-D-mannuronic acid linked with α-L-guluronic acid by a glycoside bond. The characteristics of alginate vary depending on the proportion of mannuronic acid and guluronic acid. For its gel-forming ability by interaction with cations, it is largely studied for its film-forming properties. However, alginate is highly hydrophilic and has poor water or water-vapour barrier properties [108]. For their hydrophilicity and ability to cement the holes and fractures in the paper, the alginate coating dropped gas permeability from 809 to 0 ml/min [29]. Alginate-based films were more effective than chitosan films for air barriers. Although it offered a greater gas barrier, water uptake, and surface hydrophilicity were higher. The water vapor transmission rate was only 35% lower than the uncoated sample. The permeation of aroma compounds (alkanes of molecular weight) was 18–50% lower for alginate-coated paper. Moreover, the alginate-coated film also gave a minimal migration of mineral oil aromatic hydrocarbons and mineral oil-saturated hydrocarbons. Overall, alginate film exhibits a superior barrier for gas, UV, aroma, and other lipids. However, water or water vapor transmission is still a challenge. However, the co-use of lipophilic substances can compensate for this disadvantage. Adding lipophilic thymol into Sodium-alginate films reduced swelling ratio, water solubility, and water vapor permeability. Also, the film had almost zero transmittance of UV rays [109]. Active packaging films made from alginate and castor oil had superior water vapor permeability and showed inhibition of gram-positive bacteria [110].

Similar to alginate, carrageenan, a sulfated polysaccharide from marine algae, is also hydrophilic and has a poor water barrier property. This is due to the presence of sulfur and hydroxyl groups capable of binding water [111]. However, the film still has a superior oxygen barrier and mechanical strength [112]. Zhou et al. [113] developed a film from emulsion made of camelia oil, carrageenan, and konjac glucomannan. The film had a highly hydrophobic surface and offered greater resistance to water vapor permeation. Esterification of kappa carrageenan is also reported to improve UV and water barrier properties [113]. Gum Arabic is an exudate from the acacia tree. Since it is a polysaccharide-protein conjugate, it has both hydrophobic and hydrophilic properties. The addition of gum arabic into almond protein film improved the latter's compactness and reduced surface porosity, thereby also improving the water barrier property of the film [114]. As an emulsifier, the gum arabic can influence the distribution of nanocellulose filler in starch film. When nanocellulose filler can reduce the water vapor permeability of starch film by 1.4 times, nanocellulose in the presence of gum arabic brings water permeability down twofold [115]. In an active packaging system, the gum arabic-chitosan complex ensured a sustained release of cinnamon essential oil, and this improved the antimicrobial effect against *E.coli.* However, the ratio of chitosan to gum Arabic is crucial. At a 1:4 ratio of chitosan and gum arabic, a strong binding caused the formation of an insoluble complex, which was desirable for the controlled release of oil [76]. Similar to plant gums, composite films from microbial polysaccharides like gellan are also reported to make films with superior UV and water barrier properties [116].

9.3.5 Lipids

Lipids are non-molar molecules originating from plants, animals, or microbial sources. Lipid-based films and coating have been used for a long time to avoid moisture loss from fruits and vegetables. Essential oils, waxes and fatty acids, resins, triglyceride-rich vegetable oils, hydrogenated fats, emulsifiers like lecithin, and mono and di-glycerides are groups used as barrier materials [117]. The homogenization of sunflower oil in high amylose starch and regular film formulations significantly reduced the water vapor permeability of the prepared film due to increased surface hydrophobicity. However, it did not influence the CO_2 and O_2 permeability of starch film [69]. Similarly, [118] found that coating of calcium alginate film with linseed and sunflower oil in submicron thickness (0.45 μm) improved the surface hydrophobicity and reduced wettability and water vapor permeability. The oil-coated film in a micrometer thickness (4.55 μm) also had a strong UV absorption below 320 nm.

Essential oils contain aldehydes, phenols, terpenes, alcohol and other aromatic volatile compounds known for their antioxidant and antimicrobial properties. It is the key active ingredient in many active packaging formulations. Including essential oil-derived carvacrol and citral into sago starch and guar gum-based bi-phasis edible films dropped the WVTR as the hydrophobic nature of these active constituents reduces the water vapour transfer rate. These lipophilic components might interact with the polymer/proteins and reduce their interaction with water. Thus, these essential oils could offer antimicrobial activity alongside improving water barrier properties [119]. Moreover, the phenolic compounds in clove essential oil can absorb UV rays by 80% and prevent their transmission through the material. Therefore, essential oils can also provide UV resistance in active packaging [120].

Ghasemlou et al. [121] stated that an increase in added plant essential oils level to 3% dropped the water vapor permeability of corn starch film. However, although the oxygen permeability of the film was unaffected at 1–2%, the 3% level gave a higher oxygen permeability. Poor oxygen permeability was also noted for starch-gelatin films mixed with fatty acid esters. Since oxygen has a higher dissolution ability in non-polar solvents, the oxygen permeability of lipid-based bio-barrier material must be carefully optimized. This can be done by adjusting the lipid concentration, multilayer coating and inclusion of nanoparticles.

Waxes are esters of long-chain fatty acids and have a high molecular weight. Plant waxes such as candelilla wax, sunflower wax, carnauba wax, rice bran wax, and insect waxes such as beeswax and shellac were known to improve barrier properties to biobased hydrophilic packaging [117]. Cortés-Rodríguez et al. [80] stated that the addition of beeswax into cassava starch and whey protein-based film improved their water barrier, suitable even for highly humid environments. Carnauba wax coating combined with cold plasma can effectively improve the barrier properties of film made from fish myofibrillar protein [81]. The wax coating was applied over fish protein film after plasma treatment. With or without plasma application on fish protein, carnauba wax gave lower water vapor permeability. However, plasma treatment further improvised by causing a 65% reduction in water vapor permeability.

This is because plasma functionalizes the protein surface by forming oxygen or nitrosyl radicals that can effectively cross-link the proteins or improve the adhesion of protein with wax upon coating of wax in subsequent steps. Although adding palm wax into fish gelatin-based film affected film transparency, it improved the UV resistance of film and is suited for light-sensitive foods [82].

Unsaturated fatty acids are less efficient than saturated fatty acids in offering resistance to moisture. This is because of their high polarity and very different crystallization behavior. Therefore, incorporating stearic acid and palmitic acid offers a better barrier function than oleic acid [122]. Moreover, fatty alcohols like stearyl alcohol give a lower water vapor permeability than fatty acids or triglyceride oils. This is because the hydroxyl group in fatty alcohol has less affinity for water than carboxyl groups in fatty acids. The higher the molecular weight or carbon chain length of fatty acids, the greater their moisture barrier property. Stearic and palmitic acids have lower water vapor permeability than lauric and myristic acids. Although arachidic and behenic acids have a higher carbon chain length than stearic acids, they still give a lower water vapor permeability due to the formation of a heterogenous film structure. The physical state of fat/oil is another variable that should be considered. Solid fats, for their firmness and high density, offer greater moisture resistance. Therefore, hydrogenated vegetable oils and waxes of high melting point characteristics are more suitable. However, above a critical value, solid fat also impairs the structure by affecting polymer interactions, increasing film heterogeneity, porosity and surface defects [123].

9.3.6 Proteins

Protein-based films are better known for their gas barrier property, compared to polysaccharide or lipid-based bio packaging. The oxygen permeability of soy protein film is 250–700 times lower than methylcellulose, polyethylene, starch and pectin. Plant proteins such as wheat gluten, soy protein and corn zein, as well as animal proteins like casein, whey protein and gelatin, have been largely explored for their barrier properties [124]. Casein, a milk protein of random coil structure, effectively forms a cohesive film by hydrogen, hydrophobic and electrostatic interactions. Casein films are highly hydrophilic, but their cross-linking by the addition of 4% tannic acid significantly reduced the water vapor permeability of the film. However, further addition of tannic acid did not make significant improvements as the hydroxyl group in tannic acid at concentration could bind water. Although unmodified casein films were fully solubilized upon soaking in water for 3 h, cross-linked casein maintained structural integrity and was not solubilized for seven days [83]. Polar amino acids distributed on casein random coils or micelles provide the film with better oxygen barrier properties. Adjustment of pH to the isoelectric point and ionic crosslinking of casein with calcium and their combined effects reduced water vapor permeability of the film by 36–43%. This is due to increased protein–protein interaction, which makes the film structure highly cross-linked and dense. Calcium caseinate-beeswax emulsion-based film exhibited a moisture barrier 90% more effective than pure caseinate film due to a higher hydrophobicity given by beeswax [125].

The film-forming ability of whey protein is determined by its extent of denaturation by heat, UV exposure, or alkaline pH. Since whey proteins are globular proteins, denaturation is essential to bring out the buried hydrophobic and sulfhydryl groups for better covalent/non-covalent interaction, aiding film formation and better barrier properties [12]. In the presence of plasticizers, whey proteins act as a better barrier to oxygen, oil and aroma. However, to improve the moisture barrier, lipids are often added to it in studies [126, 127]. Gelatin is a water-soluble polymeric protein obtained from hydrolysis of collagen. It originated from animal skin, bone and fish by-products. Active packaging made from gelatin exhibited a lower water solubility and increased water vapor permeability after the addition of citrus and lemongrass essential oil [128, 129]. Wang et al. [84] found that Kafirin and gelatin-based tri-layer composite film offered better UV barrier while also maintaining the transparency. However, a higher water barrier property of this film is not associated with gelatin but is due to the hydrophobic Kafirin protein (prolamin). A bilayer film made from fish gelatin and polylactic acid exhibited lower water vapor permeability than fish gelatin film due to the relatively higher hydrophobicity of polylactic acid. However, for oxygen permeability, fish gelatin gave lower values than polylactic acid. Moreover, fish gelatin film had UV transmittance of only 25%, and therefore, it also exhibits a very good UV barrier. This is because of the presence of aromatic amino acids capable of absorbing UV. Though oxygen and UV barrier property of gelatin was asserted by a few studies, it has a very poor water barrier property, but the latter can be improved when used as a composite or bilayer or in the presence of additives [85, 130].

Derived from wheat, gluten comprises two proteins: alcohol-soluble gliadin and insoluble glutenin. For its cohesiveness and elasticity, gluten exhibits excellent film-forming ability. The addition of gluten to soy protein film reduced the water vapor permeability [131]. This is because of dense networking by gluten and high amounts of non-polar amino acids in it. The pH significantly influences the water vapor permeability of wheat-gluten and soy protein-based films. Compared to pH 7 and 3, pH 9 had a higher water vapor permeability because a higher electrostatic repulsion between proteins affects the protein cross-linking. The less-dense network becomes more permeable to water at a higher pH. The water vapour permeability is inversely correlated with the oxygen permeability, which increases at a higher ratio of gluten. While pH played a significant role in influencing the water barrier property of gluten film, the other film processing variables, such as dispersion temperature, stirring time, and drying temperature, did not influence the water vapor barrier [132]. Sartori et al. [86] explored the combination of gluten with polysaccharide pectin in making a film. A gluten content reduced the water vapor permeability and solubility of the film. Moreover, the film also exhibited a UV barrier comparable to polyethylene film. At pH below 6, the oppositely charged pectin and gluten could have a strong electrostatic interaction. Since film immersion in calcium chloride was one of the steps in film preparation, calcium can also bind two proteins with pectin or even protein with protein. This enhanced protein–protein interaction and protein complexation with polysaccharides could have been the reason for improved barrier properties.

Zein is a hydrophobic prolamin obtained from corn. For its film-forming ability and excellent gas barrier properties, zein-based films are largely studied in the context of barrier properties. Since zein has a very minimal amount of cysteine amino acid, the disulfide bonding for protein interactions is minimal [124]. Transglutaminase enzyme-assisted cross-linking is preferred to strengthen the protein network. Masamba et al. [88] evaluated the dual effect of transglutaminase crosslinking and lipid addition on zein films. The cross-linking of zein with 1% transglutaminase and the addition of 3% oleic acid improved the film barrier properties by reducing the water solubility and water vapor permeability.

Globular proteins in legumes such as soy and peas are explored these days for their film-forming ability and barrier properties. Kamada et al. [6] tailored the soy proteins at a micro and nanoscale level by natural self-assembly to acquire the desired film barrier properties. The procedure involved heating of soy proteins and sonication in an acetic acid solution. This unfolded the proteins and improved the intermolecular interactions. A controlled cooling process self-assembled the proteins into β-sheet-rich fibrillar aggregates, forming a hydrogel. After drying, the water-insoluble film with a higher crystallinity and surface hydrophobicity was obtained. When coated over a paperboard, soy protein film significantly reduced the water uptake. The self-assembled soy protein film had a higher moisture barrier than a film made from commercial soy protein isolate. The addition of zein by only 0.5% further improved the water barrier properties of this film. Yang et al. [89] prepared an active packaging film using soy protein-derived amyloid fibrils (by acid/heat treatment) as a stabilizer for incorporating clove oil emulsion into polyvinyl alcohol film. The film preserved the strawberries by effectively declining the decay rate. A highly exposed non-polar amino acids in soy protein amyloid fibrils increased the surface hydrophobicity of the film lowering the water solubility and water vapour permeability of the film. A homogenously distributed emulsion droplets in the film and strong hydrogen bonding give a dense structure, effectively reducing the oxygen permeability. The phenolics in clove oil offered a UV barrier to the film. Thus, amyloid fibrils of soy proteins and their use with clove oil produced a film of water, oxygen and light barrier. The encapsulation of clove oil with amyloid fibrils ensured a sustained release of oil rich in antimicrobials. The use of this was found to preserve the quality of strawberries by dropping the decay rate. Native or denatured plant proteins can also be cross-linked by transglutaminase enzyme to improve the barrier property of the film. Studies on proteins from beans, hempseeds, grass peas, barley, quinoa and coconut reported that transglutaminase treatment increased the compactness and hydrophobicity of film and improved water and gas barrier properties [133–138]. Prior sonication or heating to denature the proteins, pectin addition, enzyme incubation conditions (pH, incubation temperature and period, transglutaminase level) and film drying temperature can significantly influence the transglutaminase treatment effect on film properties [139].

9.3.7 Green Nanomaterials

Cellulose nanocrystals have high thermal stability (up to 300 °C) [7]. It is also known to reduce the water vapor permeability of films. The physiological weight loss observed with mangoes during storage was significantly reduced (<30% weight loss after 20 days) if wrapped with cellulose nanocrystal film [140, 141]. When cellulose nanocrystals were added less than 3% into polylactic acid film, WVP and OP of the film were reduced by 82% and 90%, respectively. However, when cellulose microfibers were added to the polylactic acid film, the barrier properties did not improve. With nanofabrication, the cellulose nanocrystals retain higher crystallinity and also have better dispersibility into the matrix, thereby offering improved barrier properties [142]. Cellulose nanoparticles isolated from potato peel, when added into polyvinyl alcohol and fennel oil-based active packaging material, significantly reduced the OTR of the nanocomposite film. The film also exhibited significant free radical scavenging and antimicrobial activity against all the tested pathogens [143]. Similar results have been noted by Lavrič et al. [10], in which incorporation of cellulose nanocrystals (3%) in chitosan film gave the highest contact angle and reduced WVTR and OP by 45%, compared to only chitosan or alginate films. Riahi et al. [144] added chitosan-based quantum dots of 7.8 nm average size into carboxymethyl cellulose film. This gave the film a UV-blocking function without affecting the film's transparency. The film was active against bacteria, *E.coli* and *L. Monocytogens* and mold *Aspergillus niger* and *Penicillium chrysogenum.* The film extended the shelf life of lemons by inhibiting surface mold growth even after 21 days.

When crystalline nanocellulose of up to 10% was added into ion-bonded chitosan and carboxy-methyl cellulose-based matrix, WVTR dropped by 40% [145]. The WVTR for the nanocellulose-reinforced film is found to be lower than plastics, low-density polyethylene, polypropylene and polyethylene terephthalate. This is attributed to the crystallinity of nanocellulose hindering the diffusion of water vapor condensed on the side of the film. Moreover, the ionic complexation between cationic chitosan and anionic CMC gives a tight and compact packaging, offering resistance to water transport. However, further additions (>10%) caused agglomeration of nanocellulose, negatively affecting the barrier properties. When coated on a paperboard, nanocomposite film exhibited a better barrier against grease, oil, and water.

Starch nanocrystals are crystalline platelets produced by breaking the amorphous regions in native starch. The addition of starch nanocrystals (2.5–10%) into native starch film reduced water vapour permeability by 40%, while at higher levels, nanocrystals agglomerated similar to cellulose nanocrystals and affected the water barrier property. While some have reported that starch nanocrystals do not affect the solubility of starch film, a drop in solubility has also been noted. The difference could be due to the difference in the level of the hydrogen bonding interactions between starch nanocrystals and native starch. Since the nanocrystals increase the opacity of the film, light transmittance is reduced and is suited for oily foods sensitive to photo-oxidation [90, 146].

Zein, a protein derived from corn has been largely explored to develop nanoparticles or bioactive compound loaded nanoparticles. Being a hydrophobic protein, it is difficult to homogeneously distribute zein in the film matrix. There are techniques like liquid–liquid dispersion to produce zein nanoparticles. The interaction of zein with sodium caseinate introduces hydrophilic moieties on the surface of zein and prevents aggregate formation by hydrophobic interaction. The zein nanoparticles produced this way reinforced potato starch and olive oil-based edible film and improved the mechanical strength and water barrier properties of the film [9]. In whey-protein isolate-based film reinforced with zein nanoparticles, the casein binding creates an effective interface between the hydrophobic zein and hydrophilic whey proteins and causes a homogenous distribution of zein in the film matrix. The produced film exhibited a higher surface hydrophobicity and a lower fractional free volume water vapor permeability [147]. Similar results have also been noted for zein nanoparticle-loaded hydroxy propyl methyl cellulose film [148].

9.3.8 Food Wastes/By-products

Food industries generate immense food waste or by-products, which are discarded or used as animal feed. The presence of valuable antimicrobial and antioxidant compounds like phenolics and other fibers and biopolymers to reinforce the structure of films and improve mechanical and barrier properties makes food wastes/ by-products a highly valued material for active packaging [149]. While extraction of biopolymers from food wastes for use as barrier material is one approach, the complete food waste can also be used as a filler material. The letter can achieve zero carbon footprint in the food industry. Sunflower hull is a by-product of the edible oil industry. De'Nobili et al. [150] added micronized sunflower hull (50 μm) into pectin-based composite film as a filler material. Its addition at a 12% level improved the water vapor barrier and mechanical strength of the film. This is due to the presence of lipophilic lignin (26%) and micron-sized filler occupying the voids in the pectin matrix. Moreover, the film also exhibited excellent free radical scavenging behavior due to the presence of proanthocyanidins. Fruits and vegetable wastes like citrus peel are generated in huge tons every year. In a study by Merino et al. [151], the hydrolyzed orange peel was used along with chitosan to develop an active packaging film for dry foods. A strong electrostatic interaction between carboxylic acid groups of pectin in hydrolyzed orange peel and chitosan strengthened the film matrix and reduced water solubility, water vapor, and gas permeability. Also, the presence of phenolics gave strong antioxidant activity and UV-barrier to the film. However, the film lacked significant antimicrobial properties due to the complexation of phenolics with chitosan and other polymers.

9.3.9 Fermentation-derived Biomaterials

Polylactic acid is a biobased thermoplastic obtained from resources like cassava roots/chips, potato/corn starch, and sugarcane bagasse. These materials can act as a substrate for the production of lactic acid by fermentation. Polylactic acid is made from lactic acid by a condensation polymerization reaction. Because of its popularity as a biodegradable and bio-based material, it is largely produced these days and is increasingly replacing petroleum-based packaging materials [152]. Coating polylactic acid over paper drops the water absorption and water vapor permeability. This is due to the increased surface hydrophobicity of the material. However, polylactic acid is not an effective oxygen or carbon dioxide barrier. Similarly, the polylactic acid-based coating or film also poorly hinders the penetration of oil and grease [13]. However, the addition of biobased nanomaterials, like cellulose nanocrystals, improves the barrier function of the material [153]. Qin et al. [154] developed an active packaging film from a polylactic acid matrix to which essential oil was added as an active ingredient. While the addition of essential oil inhibited microbes like *E. coli* and *Bacillus subtills*, it also improved the water vapor barrier properties of the film. The lower water vapor permeability of the essential oil-loaded polylactic acid film extended the shelf life of meat to 18 days, during which the color and firm texture of meat was not largely affected [155].

Polyhydroxyalkanoates are polyesters produced naturally by *Bacillus megaterium*. It is non-toxic, bio-degradable, and has mechanical and barrier characteristics similar to synthetic polymer polypropylene. It is highly hydrophobic in nature, and its water vapor permeability is similar to polyvinyl chloride (PVC) and polyethylene terephthalate (PET) [156]. Its high degree of crystallinity offers poor water sorption and water vapor barrier properties to the material [157]. Though the material is rigid and has poor heat stability, it can be blended with polylactic acid or modified with the addition of nanomaterials [13]. The material also has a moderate barrier for O_2, CO_2 and non-polar solvents like hexane [156]. Double coating of polyhydroxyalkanoates reduces moisture, aroma and oxygen permeability of paper [158]. The barrier properties of the film can be further improved by sandwiching other film layers or by producing composite films. The addition of the zein layer into polyhydroxyalkanoate film improved the aroma and oxygen barrier properties of the film [159]. Similar to zein, keratin addition into polyhydroxyalkanoates also reduced the water, oxygen and aroma barrier properties of the film to half [160]. The incorporation of lignin into polyhydroxyalkanoate film reduced O_2 and CO_2 permeability by 77% and 91%, respectively [161].

9.4 Multilayering for Enhanced Barrier Function

A food product may require more than one barrier function. However, one biobased material cannot offer all the solutions. The problem can be overcome by multilayering or lamination techniques in which multiple layers made from different biopolymers of varying functionality are brought together as a single laminate with multiple layers. When designing multilayered packaging for a food product, it is important to know the product's characteristics. The packaging material that comes in immediate contact with food should be chemically inert, whereas the outer layer should offer mechanical strength and stiffness, abrasion resistance and printability. Between these two or more layers, there can be an adhesive layer to ensure the surface adhesion of films [162]. The viscosity and density of these different layer dispersions are important as the layers should be distinct even upon co-extrusion closely. These layers should have strong surface interactions but should not become homogenous/inseparable. Traditional coating/film formation methods use spray coating, dip coating, knife coating and roller coating. If multiple layers are to be generated through these conventional approaches, the coated film can go through multiple rounds of coating with new biomaterials. To ensure a surface adhesion of multiple layers and overcome defects in layers, biopolymer solutions of different charges can be used so that the strong electrostatic interaction between the layers can withhold the structure [163] (Fig. 9.3).

With the co-extrusion technique, many layers with more precise layer thickness can be obtained. In co-extrusion, the point at which different film layers come in direct contact can happen before or at the extrusion die. As an advancement of co-extrusion technology, which can produce 2–17 layers, thousands of layers with only a micron thickness can also be produced with the micro-layer coextrusion technique. This is achieved with a vein-splitting kind of design installed in the block before the extrusion die [163] (Fig. 9.4).

Extrusion lamination involves feeding a primary web along with a secondary web through rollers. These webs are then drawn into a sheet of molten polymeric material,

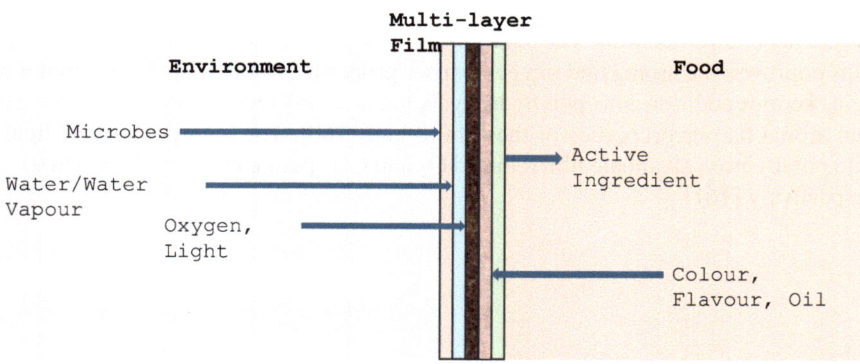

Fig. 9.3 Schematic representation of the barrier function of multi-layer film

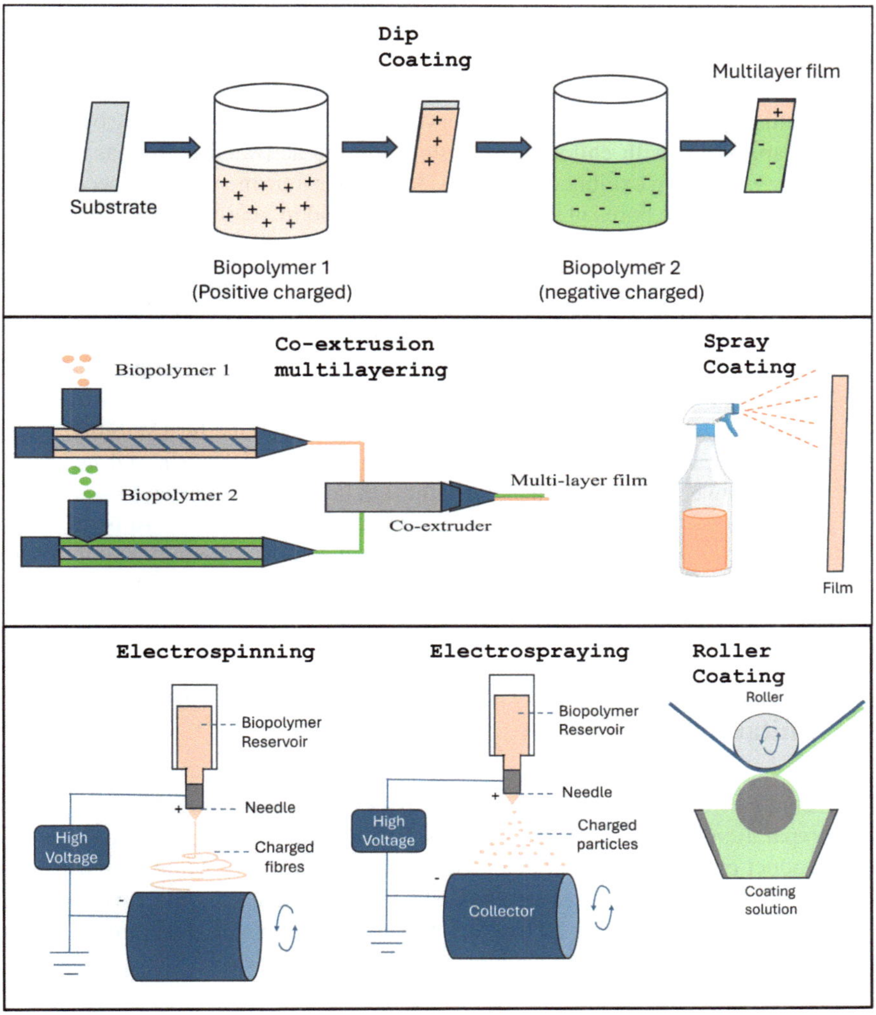

Fig. 9.4 Multilayering techniques for barrier film/coating formation

which is extruded from a die. The three layers are pressed together by a chilled drum and impression roller, bonding the two webs as the molten polymeric material solidifies in the center. Finally, the edges are trimmed, and the prepared laminate is wound [164]. Recent advancements such as electrospinning and electrospraying have the potential to produce nano-structured multilayered film. With electrostatic force, electrospinning produces a continuous jet of ultrathin polymer ejected from the spinneret to the collector, where it is deposited in randomly oriented nano-fibers of high surface area to volume ratio. Unlike electrospinning, electrospraying by atomization produces non-continuous beads. Both techniques produce materials with a high surface area-to-volume ratio. The overlay of fibers produces a dense and strong entangled network in the film [165, 166].

The multilayer packages for food packaging typically have 3 to 7 different layers of varying functions. dos Santos et al. [167] developed a multilayer packaging by surface coating paperboard with chitosan, palmitic acid, and activated carbon as separate layers. The packaging exhibited a lower water vapour transmission rate because the coating cemented the pores in the cellulose matrix and increased surface hydrophobicity given by activated carbon palmitic acid. Moreover, the coating increased the grease resistance while also improving the stiffness and elongation capacity. With only a minimal drop in biodegradation kinetics from uncoated paperboard, the three-layer coated paperboard degraded by 55% within 188 days [168]. Zhang et al. [169] developed a bilayer film from antimicrobial chitosan and zein. The two films had strong interfacial interaction. Overall, the film offered an excellent barrier to water, oxygen and UV while also improving the physical strength and heat stability. Similarly, [84] fabricated a tri-layered film in which a blended Kafirin and gelatin is sandwiched between pure Kafirin and gelatin skin layers. The multilayered film was transparent but also had a superior UV barrier property. Compared to hydrophilic gelatin, hydrophobic Kafirin was found effective in significantly reducing moisture transfer. Therefore, the film had a one-way barrier to the water transfer.

9.5 Factors Affecting Barrier Coating/Film and Their Suitability for Food Applications

The features of barrier material film/coatings depend on the biopolymer type and film preparation methods, whereas their suitability for food depends on specific food product characteristics, storage and handling conditions. Moreover, it is also necessary to know if the barrier material establishes immediate contact with food or the external environment. The factors affecting the barrier material features and their suitability for specific food applications are discussed below.

9.5.1 Barrier Coating and Film Processing

The barrier material of choice should be non-toxic, non-allergenic and non-flammable, and the barrier component should not migrate into food. Furthermore, affordability, availability, biodegradability, eco-friendliness, processability, re-cyclability and compatibility with food and other packaging constituents is also important [170]. From an affordability perspective, fermentation-derived polyhydroxyalkanoates have a minimal material cost. The input raw material for fermentation can be obtained from food wastes/by-products, and the formulation can be optimized to lower costs [171]. Other biopolymers like cellulose and starch may require chemical or physical modification to improve barrier properties, but this increases the material cost. The packaging should not only pass barrier properties but also mechanical properties and thermal stability. Therefore, a general compromise should be made with different properties expected in packaging [50]. The ease of molding/demoulding is another relevant criterion. The degree of crystallinity of biomaterial can influence the solubility and homogeneity in dispersion and film casting [172]. Also, the type of barrier material can affect the gloss, opacity, coating thickness, GSM, smoothness, printability and color of the film. Glass transition temperature of barrier material comes into the picture during film drying and storage. An increase in the storage temperature of biomaterial beyond the glass transition temperature increases the molecular mobility and free fractional volume of the film. This not only increases the barrier property but also may increase the migration of film components to food [173, 174]. Biopolymers (e.g., starch) have a partly crystalline structure, which makes the film brittle. Biopolymers often require plasticizers like sorbitol and glycerol to improve the film-forming ability and mechanical properties. However, its addition can compromise the water barrier property of the material [175].

The coating technique is determined based on the characteristics of the barrier material. The type of coating technique (e.g., dip/ spray/roller/brush coating) used, viscosity, density, and surface adhesion quality of barrier coating, and the thickness of the coating on the film may vary. Since coating thickness is affected, this also influences the barrier properties. The higher the coating thickness, the greater the barrier due to the increased mean free path for permeants [176]. The barrier coating may require more than one component, such as binders, thickeners, lubricants, cross-linking agents, etc. However, not all of these additives are compatible with one solvent in which the biopolymer is dissolved [177]. The porosity of the barrier film is largely influenced by the type of barrier material used, concentration, film processing conditions, and the process-induced surface defects. Since porosity directly equates with permeation rate, this is another variable to consider in the context of barrier materials specifications [178] (Fig. 9.5).

9.5.2 Consumer Acceptability

Unlike in the past when consumer purchase decisions were solely dependent on the intrinsic quality of food, today, consumers also evaluate packaging from eco-friendliness, safety, animal-friendliness, aesthetics, and convenience points of view. For example, animal-waste/by-product-derived proteins, collagen, and keratin could be unattractive to a few consumer segments [179]. Moreover, consumers are also concerned about the potential health threat from nanomaterials used in barrier coatings. Since products packed in transparent synthetic films of versatile designs/dimensions offer a higher standard of convenience (e.g., microwave processability of packed food), the effectiveness of biobased materials from this perspective is also considered [180]. Consumers expect packages to be transparent for many foods, to

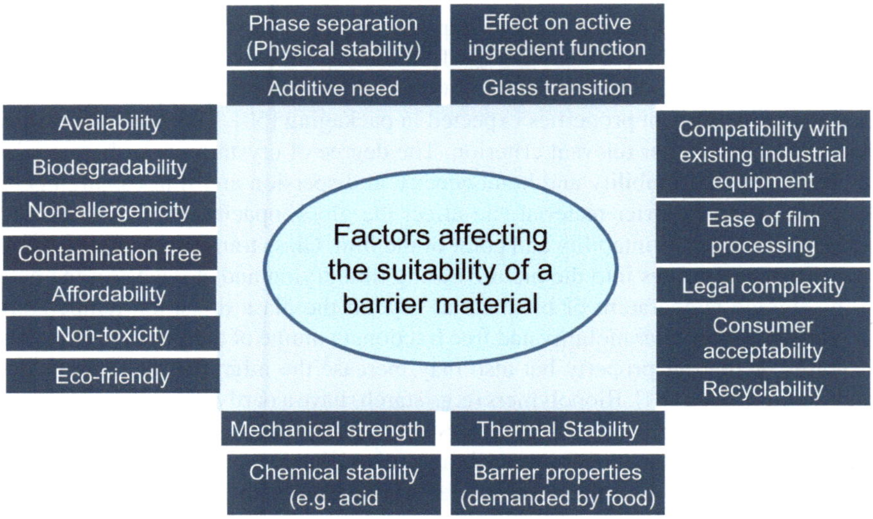

Fig. 9.5 Factors affecting the suitability of a barrier material for active packaging

inspect the quality, and to gain trust. However, with some biobased barrier materials, the transparency of the film is reduced, or the color got changed [181].

9.5.3 Food Characteristics and Storage Conditions

The barrier material must have high surface hydrophobicity for beverages and dairy products with high water content to contain and prevent leakage and deterioration. Since dairy products are stored in refrigerated conditions, water condensation on film can also happen. Therefore, lower water and water vapor permeability are key traits of barrier material [182]. These products can also be acidic due to fermentation-produced lactic acid, organic acids in fruit juices, or acidulants/preservatives added into the formulation. Therefore, the barrier material should also be resistant to acidic conditions. Since dairy products like butter are prone to the adsorption of flavors from the external environment, the packaging should also restrain the permeation of external flavors. Oxygen permeation can initiate the oxidation of vitamins, flavors, and pigments in fruit juices. This also changes the color of the product by enzymatic browning. Hence, oxygen permeability is also a crucial factor. Since the product contains high moisture content, the flavor or color compounds can also migrate to the packaging material. If the barrier material does not resist these compounds, the external package may look stained, or the product may go flavorless or color fading. Light-induced oxidation is another variable to consider, as it can induce rancidity in fatty dairy products and cause the loss of pigments/flavors in beverages [183]. From a product characteristics point of view, the barrier material for dairy products is expected to offer water, oxygen, light, and aroma/pigment barriers. Since polylactic acid has water vapor permeability greater than synthetic polymers like polystyrene and polyethylene cups, is resistant to pH (2–6), and offers an excellent barrier to aroma and gas, it can be considered for further scrutiny in this product segment [184].

Bakery products encompass foods of low to high water activity. While bread has a high water activity, cookies have a lower water activity. Spoilage of bread is mainly associated with fungal growth on the surface and staling due to starch retrogradation and loss of moisture. The use of active packaging (i.e., chitosan) can inactivate surface mold growth, while the barrier coating applied on it can overcome moisture desorption/adsorption. Since products like cookies, muffins and cake contain high-fat content, an oil/grease barrier is an important attribute of barrier materials [183].

For chocolates, fat bloom, sugar bloom and fat oxidation are the main quality changes affecting consumer acceptance of the product [185]. Fat bloom is difficult to control with packaging as it is a consequence of temperature changes or tempering defects. Packaging with heat insulation properties may help, especially when chocolates need to be transported to arid, hot regions [46]. Sugar bloom can be avoided by using strong water barrier packaging. However, the package should also be resistant to fat permeation if in direct contact with the chocolate. Dry products like breakfast cereals, powders (milk/coffee powder), crystals (sugar/salt), candies and snacks (chips/crisps) need to be stored in a low relative humidity environment due to the low water activity and hygroscopicity of the materials. The water permeation through the package causes caking in powders, loss of crispiness of snacks and also increased microbial growth [186, 187]. For such products, the water barrier property of the film is the most important criterion. Since dried products are prone to pick up off-flavors from the external environment, especially when they have a high surface area (powders), the aroma barrier becomes quite an important criterion. Snack foods have high oil content (if fried). Therefore, it is important for the barrier material to also provide an oil barrier and gas barrier to avoid rancidity and off-flavor generation. It is also necessary to offer a light barrier to avoid rancidity [188].

Fruits and vegetables continue to respire even after the harvest. There is a higher intake of oxygen and liberation of CO_2 and also ethylene, depending on the fruit. The fruit also transpires (moisture loss) during storage. With a few exceptions (e.g., onion), fruits and vegetables generally require storage under high humid (85–98% RH) and low-temperature environments. The effective water barrier property, coupled with a desirable CO_2/O_2 permeation selectivity of the biobased barrier material, is important. This is because the presence of oxygen promotes respiration and continued spoilage by senescence or microbial deterioration, but complete removal also causes spoilage by anaerobic respiration [189, 190]. Similar to fruits, meat also requires oxygen to maintain its red color, given by oxymyoglobin. The removal of oxygen affects the color due to the conversion of oxymyoglobin to metmyoglobin. The anaerobic condition also promotes the growth of toxic microbes like *Clostridium botulinum.* Therefore, meat is often packed under modified atmospheric conditions using different gas compositions. Nevertheless, it is necessary for meat packaging to offer an effective gas barrier and also a water barrier [191]. Similar to meat, seafood containing high amounts of polyunsaturated fatty acids and the ability to acquire warmed-over flavor by oxidation demand effective gas barrier properties to the film [192].

9.5.4 Safety, Health, and Legal Considerations

Health and safety are the most important concerns for any packaging material that comes in direct or indirect contact with food. The barrier material should neither have genotoxic nor cytotoxic effects. Since biobased packaging material is increasingly made from allergenic proteins like soy, peanut, gluten and casein, it is also important to consider the accidental consumption of these compounds [193]. Using biomaterial also increases the risk of microbial contamination during production and handling. Many in vivo and in vitro studies on cellulose nanocrystals done to date have confirmed their safety, up to 200 mg/kg/day body intake [194]. Similarly, polylactic acid has also received 'Generally Regarded As Safe' (GRAS) status from the FDA [195]. All the discussed biopolymers are used either as an additive (e.g., Pectin, gums, modified cellulose) or as an ingredient (Starch) in foods globally. These biopolymers exhibit no adverse toxicity. For their biodegradability, there is no concern about bioaccumulation in the food chain [14]. Packaging material that comes in direct/indirect contact with foods should comply with the legal standards and specifications. The use of biomaterials requires testing for heavy metals, pesticide residues, and other toxins (e.g., aflatoxin) [196]. The food packaging film should also meet the overall and specific migration limits [197]. The legal complexities and safety studies required around active packaging delay the full-scale commercialization of biobased materials used in active packaging.

9.5.5 Compatibility with Active Packaging

Active packaging incorporates functional ingredients like phenolic extract or oils whose release characteristics significantly influence the intended antimicrobial or antioxidant or other functions in food. Adding biobased materials as fillers could reduce the effectiveness of active packaging, for instance, the inability to suppress microbial growth. This is because of the complexation of phenolics (or another active constituent) with barrier materials added to the film [42, 198]. A sustained release of active ingredients is desirable to improve the shelf life of a product with no significant quality deteriorations. For certain materials, the release kinetics are very rapid, especially at the early stage, and it did not have a prolonged effect [199]. Thus, the effect of barrier material on the release kinetics of active ingredients, depending on storage conditions, is an important attribute to consider when choosing the barrier material. Since the active ingredients used in active packaging are often volatile, the barrier material should also prevent permeation and loss of active components [198].

9.6 Summary

- The bio-based barrier materials for active packaging offer a barrier to mass transfer (water vapor, gas) and also energy transfer (UV, heat).
- The polysaccharides, lipids, and phenolics naturally offer good barriers to gas, water, and UV, respectively.
- The poor water barrier properties of hydrophilic starch, gums, and pectin can be improved by esterification, etherification, and other physical or chemical modification methods.
- Novel green nanomaterials like cellulose nanocrystals are safe and non-toxic, and their addition into biopolymeric films improves barrier function and mechanical strength.
- Fermentation-derived polyhydroxyalkanoates have a superior barrier function compared to other biopolymers.
- With multilayering techniques, the barrier function of packaging can be improved.
- Barrier properties demanded by fruits and vegetables, dry foods, dairy and beverages, and meat products differ due to differences in water activity, fat content, surface area, and product storage conditions.
- There is great scope for exploring new processing routes for making green nanomaterials and their film deposition techniques. The loss of effectiveness of active packaging caused by barrier material-active ingredient complexation demands more studies looking into molecular dynamics.

References

1. Kunam, P.K., Ramakanth, D., Akhila, K., Gaikwad, K.K.: Bio-based materials for barrier coatings on paper packaging. Biomass Convers. Biorefinery 1–16 (2022)
2. Firouz, M.S., Mohi-Alden, K., Omid, M.: A critical review on intelligent and active packaging in the food industry: research and development. Food Res. Int. **141**, 110113 (2021)
3. Piergiovanni, L., Limbo, S., Piergiovanni, L., Limbo, S.: Plastic packaging materials. Food Packag. Mater. 33–49 (2016)
4. Gaffey, J., McMahon, H., Marsh, E., Vehmas, K., Kymäläinen, T., Vos, J.: Understanding consumer perspectives of bio-based products—A comparative case study from Ireland and The Netherlands. Sustainability **13**(11), 6062 (2021)
5. Varghese, S.A., Pulikkalparambil, H., Promhuad, K., Srisa, A., Laorenza, Y., Jarupan, L., Harnkarnsujarit, N., et al.: Renovation of agro-waste for sustainable food packaging: a review. Polymers **15**(3), 648 (2023)
6. Kamada, A., Rodriguez-Garcia, M., Ruggeri, F.S., Shen, Y., Levin, A., Knowles, T.P.: Controlled self-assembly of plant proteins into high-performance multifunctional nanostructured films. Nat. Commun. **12**(1), 3529 (2021)

7. Shaghaleh, H., Xu, X., Wang, S.: Current progress in production of biopolymeric materials based on cellulose, cellulose nanofibers, and cellulose derivatives. RSC Adv. **8**(2), 825–842 (2018)
8. Shaikh, M., Haider, S., Ali, T.M., Hasnain, A.: Physical, thermal, mechanical and barrier properties of pearl millet starch films as affected by levels of acetylation and hydroxypropylation. Int. J. Biol. Macromol. **124**, 209–219 (2019)
9. Farajpour, R., Djomeh, Z.E., Moeini, S., Tavakolipour, H., Safayan, S.: Structural and physico-mechanical properties of potato starch-olive oil edible films reinforced with zein nanoparticles. Int. J. Biol. Macromol. **149**, 941–950 (2020)
10. Lavrič, G., Oberlintner, A., Filipova, I., Novak, U., Likozar, B., Vrabič-Brodnjak, U.: Functional nanocellulose, alginate and chitosan nanocomposites designed as active film packaging materials. Polymers **13**(15), 2523 (2021)
11. Wang, Q., Chen, W., Zhu, W., McClements, D.J., Liu, X., Liu, F.: A review of multilayer and composite films and coatings for active biodegradable packaging. NPJ Sci. Food **6**(1), 18 (2022)
12. Schmid, M., Müller, K.: Whey protein-based packaging films and coatings. In: Whey proteins, pp. 407–437. Academic Press (2019)
13. Tyagi, P., Salem, K.S., Hubbe, M.A., Pal, L.: Advances in barrier coatings and film technologies for achieving sustainable packaging of food products–a review. Trends Food Sci. Technol. **115**, 461–485 (2021)
14. Reichert, C.L., Bugnicourt, E., Coltelli, M.B., Cinelli, P., Lazzeri, A., Canesi, I., Schmid, M., et al.: Bio-based packaging: materials, modifications, industrial applications and sustainability. Polymers **12**(7), 1558 (2020)
15. Gaikwad, K.K., Singh, S., Ajji, A.: Moisture absorbers for food packaging applications. Environ. Chem. Lett. **17**(2), 609–628 (2019)
16. Bell, L.N.: Moisture effects on food's chemical stability. In: Water activity in foods: fundamentals and applications, pp. 227–253 (2020)
17. Archaviboonyobul, T., Jinkarn, T., Sane, S., Chariyachotilert, S., Kongcharoenkiat, S.: Water resistance and barrier properties improvement of paperboard by poly (lactic acid) electrospraying. Packag. Technol. Sci. **27**(5), 341–352 (2014)
18. Balasubramaniam, S.L., Patel, A.S., Nayak, B.: Surface modification of cellulose nanofiber film with fatty acids for developing renewable hydrophobic food packaging. Food Packag. Shelf Life **26**, 100587 (2020)
19. Johnson, D.R., Decker, E.A.: The role of oxygen in lipid oxidation reactions: a review. Annu. Rev. Food Sci. Technol. **6**, 171–190 (2015)
20. Kołakowska, A., Bartosz, G.: Oxidation of food components. Food Oxid. Antioxid. Chem. Biol. Funct. Prop. **1** (2013)
21. Opara, U.L., Caleb, O.J., Belay, Z.A.: Modified atmosphere packaging for food preservation. In: Food quality and shelf life, pp. 235–259. Academic Press (2019)
22. Al-Ati, T., Hotchkiss, J.H.: The role of packaging film permselectivity in modified atmosphere packaging. J. Agric. Food Chem. **51**(14), 4133–4138 (2003)
23. Siracusa, V.: Food packaging permeability behaviour: a report. Int. J. Polym. Sci. (2012)
24. Twede, D., Selke, S.E., Kamdem, D.P., Shires, D.: Cartons, crates and corrugated board: handbook of paper and wood packaging technology. DEStech Publications, Inc. (2014)
25. Carpena, M., Nuñez-Estevez, B., Soria-Lopez, A., Garcia-Oliveira, P., Prieto, M.A.: Essential oils and their application on active packaging systems: a review. Resources **10**(1), 7 (2021)
26. Dekker, M., Van Willige, R.W.G., Linssen, J.P.H., Voragen, A.G.J.: Modelling the effect of oil/fat content in food systems on flavour absorption by LLDPE. Food Addit. Contam. **20**(2), 180–185 (2003)
27. Huber, M., Ruiz, J., Chastellain, F.: Off-flavour release from packaging materials and its prevention: a foods company's approach. Food Addit. Contam. **19**(S1), 221–228 (2002)
28. Khwaldia, K., Arab-Tehrany, E., Desobry, S.: Biopolymer coatings on paper packaging materials. Compr. Rev. Food Sci. Food Saf. **9**(1), 82–91 (2010)

29. Kopacic, S., Walzl, A., Zankel, A., Leitner, E., Bauer, W.: Alginate and chitosan as a functional barrier for paper-based packaging materials. Coatings **8**(7), 235 (2018)
30. Cheng, H.: Volatile flavor compounds in yogurt: a review. Crit. Rev. Food Sci. Nutr. **50**(10), 938–950 (2010)
31. de Roos, K.B.: Understanding and controlling the behaviour of aroma compounds in thermally processed foods. Trends Food Sci. Technol. **17**(5), 236–243 (2006)
32. Dury-Brun, C., Chalier, P., Desobry, S., Voilley, A.: Multiple mass transfers of small volatile molecules through flexible food packaging. Food Rev. Intl. **23**(3), 199–255 (2007)
33. Winotapun, C., Phattarateera, S., Aontee, A., Junsook, N., Daud, W., Kerddonfag, N., Chinsirikul, W.: Development of multilayer films with improved aroma barrier properties for durian packaging application. Packag. Technol. Sci. **32**(8), 405–418 (2019)
34. Sajilata, M.G., Savitha, K., Singhal, R.S., Kanetkar, V.R.: Scalping of flavors in packaged foods. Compr. Rev. Food Sci. Food Saf. **6**(1), 17–35 (2007)
35. Suominen, I., Suihko, M.L., Salkinoja-Salonen, M.: Microscopic study of migration of microbes in food-packaging paper and board. J. Ind. Microbiol. Biotechnol. **19**, 104–113 (1997)
36. Siroli, L., Patrignani, F., Serrazanetti, D.I., Chiavari, C., Benevelli, M., Grazia, L., Lanciotti, R.: Survival of spoilage and pathogenic microorganisms on cardboard and plastic packaging materials. Front. Microbiol. **8**, 2606 (2017)
37. Banu, J.R., Kavitha, S., Kannah, R.Y., Devi, T.P., Gunasekaran, M., Kim, S.H., Kumar, G.: A review on biopolymer production via lignin valorization. Biores. Technol. **290**, 121790 (2019)
38. de Fátima Silva, M., Maciel, V.B.V., Noletto, A.P.R., Venturini, A.C., de Carvalho, R.A., Yoshida, C.M.P.: Chitosan active coating on paperboard surface forming an anti-insect grain-based food packaging. Packag. Technol. Sci. **35**(4), 361–372 (2022)
39. Lee, J.Y., Garcia, C.V., Shin, G.H., Kim, J.T.: Antibacterial and antioxidant properties of hydroxypropyl methylcellulose-based active composite films incorporating oregano essential oil nanoemulsions. Lwt **106**, 164–171 (2019)
40. Almasi, H., Jahanbakhsh Oskouie, M., Saleh, A.: A review on techniques utilized for design of controlled release food active packaging. Crit. Rev. Food Sci. Nutr. **61**(15), 2601–2621 (2021)
41. Bahmid, N.A., Siddiqui, S.A.: Factors influencing release, absorption, and concentration of natural compounds in antimicrobial packaging. Food Rev Int 1–27 (2023)
42. Benbettaïeb, N., Mahfoudh, R., Moundanga, S., Brachais, C.H., Chambin, O., Debeaufort, F.: Modeling of the release kinetics of phenolic acids embedded in gelatin/chitosan bioactive-packaging films: influence of both water activity and viscosity of the food simulant on the film structure and antioxidant activity. Int. J. Biol. Macromol. **160**, 780–794 (2020)
43. Kwon, S., Orsuwan, A., Bumbudsanpharoke, N., Yoon, C., Choi, J., Ko, S.: A short review of light barrier materials for food and beverage packaging. Korean J. Packag. Sci. Technol. **24**(3), 141–148 (2018)
44. Roy, S., Ramakrishnan, R., Goksen, G., Singh, S., Łopusiewicz, Ł.: Recent progress on UV-light barrier food packaging films–a systematic review. Innov. Food Sci. Emerg. Technol. 103550 (2023)
45. Vilela, C., Pinto, R.J., Coelho, J., Domingues, M.R., Daina, S., Sadocco, P., Freire, C.S., et al.: Bioactive chitosan/ellagic acid films with UV-light protection for active food packaging. Food Hydrocolloids **73**, 120–128 (2017)
46. Chen, C., Ding, R., Yang, S., Wang, J., Chen, W., Zong, L., Xie, J.: Development of thermal insulation packaging film based on poly (vinyl alcohol) incorporated with silica aerogel for food packaging application. Lwt **129**, 109568 (2020)
47. Laorenza, Y., Chonhenchob, V., Bumbudsanpharoke, N., Jittanit, W., Sae-Tan, S., Rachtanapun, C., Harnkarnsujarit, N., et al.: Polymeric packaging applications for seafood products: packaging-deterioration relevance, technology and trends. Polymers **14**(18), 3706 (2022)
48. Nešić, A., Gordić, M., Davidović, S., Radovanović, Ž, Nedeljković, J., Smirnova, I., Gurikov, P.: Pectin-based nanocomposite aerogels for potential insulated food packaging application. Carbohyd. Polym. **195**, 128–135 (2018)

49. Nechita, P., Roman, M.: Review on polysaccharides used in coatings for food packaging papers. Coatings **10**(6), 566 (2020)
50. Liu, W., Gu, J., Huang, C., Lai, C., Ling, Z., Yong, Q.: Fabrication of hydrophobic and high-strength packaging films based on the esterification modification of galactomannan. Int. J. Biol. Macromol. **167**, 1221–1229 (2021)
51. da Silva, M.N., de Matos Fonseca, J., Feldhaus, H.K., Soares, L.S., Valencia, G.A., de Campos, C.E.M., Monteiro, A.R., et al.: Physical and morphological properties of hydroxypropyl methylcellulose films with curcumin polymorphs. Food Hydrocolloids **97**, 105217 (2019)
52. Ghadermazi, R., Hamdipour, S., Sadeghi, K., Ghadermazi, R., Khosrowshahi Asl, A.: Effect of various additives on the properties of the films and coatings derived from hydroxypropyl methylcellulose—a review. Food Sci. Nutr. **7**(11), 3363–3377 (2019)
53. Ghanbarzadeh, B., Almasi, H.: Physical properties of edible emulsified films based on carboxymethyl cellulose and oleic acid. Int. J. Biol. Macromol. **48**(1), 44–49 (2011)
54. Yang, Y., Shi, Y., Cao, X., Liu, Q., Wang, H., Kong, B.: Preparation and functional properties of poly (vinyl alcohol)/ethyl cellulose/tea polyphenol electrospun nanofibrous films for active packaging material. Food Control **130**, 108331 (2021)
55. Nasatto, P.L., Pignon, F., Silveira, J.L., Duarte, M.E.R., Noseda, M.D., Rinaudo, M.: Methylcellulose, a cellulose derivative with original physical properties and extended applications. Polymers **7**(5), 777–803 (2015)
56. Erdohan, Z.Ö., Turhan, K.N.: Barrier and mechanical properties of methylcellulose–whey protein films. Packag. Technol. Sci Int. J. **18**(6), 295–302 (2005)
57. Ghanbarzadeh, B., Almasi, H., Entezami, A.A.: Improving the barrier and mechanical properties of corn starch-based edible films: effect of citric acid and carboxymethyl cellulose. Ind. Crops Prod. **33**(1), 229–235 (2011)
58. Zheng, T., Yu, X., Pilla, S.: Mechanical and moisture sensitivity of fully bio-based dialdehyde carboxymethyl cellulose cross-linked soy protein isolate films. Carbohyd. Polym. **157**, 1333–1340 (2017)
59. Laroque, D.A., de Aragao, G.M.F., de Araújo, P.H.H., Carciofi, B.A.M.: Active cellulose acetate-carvacrol films: antibacterial, physical and thermal properties. Packag. Technol. Sci. **34**(8), 463–474 (2021)
60. Santos, D.C., Ribeiro-Santos, R., Ventura, L.A.F., Melo, N.R., Costa, B.D.S., Rojas, E.E.G., Salgado, N.L.: Antimicrobial activity studies and characterization of cellulose acetate films containing essential oils. Ital. J. Food Sci. **28**(2), 248 (2016)
61. Liang, L., Chen, H.: Development and characterization of biodegradable ultraviolet protective and antibacterial polylactic acid-cellulose acetate film modified by phenyl salicylate. Int. J. Biol. Macromol. **211**, 85–93 (2022)
62. Casalini, S., Giacinti Baschetti, M.: The use of essential oils in chitosan or cellulose-based materials for the production of active food packaging solutions: a review. J. Sci. Food Agric. **103**(3), 1021–1041 (2023)
63. Chen, G., Liu, B.: Cellulose sulfate based film with slow-release antimicrobial properties prepared by incorporation of mustard essential oil and β-cyclodextrin. Food Hydrocolloids **55**, 100–107 (2016)
64. Bangar, S.P., Esua, O.J., Nickhil, C., Whiteside, W.S.: Microcrystalline cellulose for active food packaging applications: a review. Food Packag. Shelf Life **36**, 101048 (2023)
65. Ma, X., Chang, P.R., Yu, J.: Properties of biodegradable thermoplastic pea starch/ carboxymethyl cellulose and pea starch/microcrystalline cellulose composites. Carbohyd. Polym. **72**(3), 369–375 (2008)
66. Othman, S.H., Majid, N.A., Tawakkal, I.S.M.A., Basha, R.K., Nordin, N., Shapi'i, R.A.: Tapioca starch films reinforced with microcrystalline cellulose for potential food packaging application. Food Sci. Technol. **39**, 605–612 (2019)
67. Aulin, C., Gällstedt, M., Lindström, T.: Oxygen and oil barrier properties of microfibrillated cellulose films and coatings. Cellulose **17**, 559–574 (2010)
68. Colussi, R., Pinto, V.Z., El Halal, S.L.M., Biduski, B., Prietto, L., Castilhos, D.D., Dias, A.R.G., et al.: Acetylated rice starches films with different levels of amylose: mechanical,

water vapor barrier, thermal, and biodegradability properties. Food Chem. **221**, 1614–1620 (2017)
69. Garcia, M.A., Martino, M.N., Zaritzky, N.E.: Lipid addition to improve barrier properties of edible starch-based films and coatings. J. Food Sci. **65**(6), 941–944 (2000)
70. Cheng, M., Yan, X., Cui, Y., Han, M., Wang, Y., Wang, J., Wang, X., et al.: Characterization and release kinetics study of active packaging films based on modified starch and red cabbage anthocyanin extract. Polymers **14**(6), 1214 (2022)
71. Jonhed, A., Andersson, C., Järnström, L.: Effects of film forming and hydrophobic properties of starches on surface sized packaging paper. Packag. Technol. Sci. Int. J. **21**(3), 123–135 (2008)
72. Phinainitisatra, T., Harnkarnsujarit, N.: Development of starch-based peelable coating for edible packaging. Int. J. Food Sci. Technol. **56**(1), 321–329 (2021)
73. Chi, K., Wang, H., Catchmark, J.M.: Sustainable starch-based barrier coatings for packaging applications. Food Hydrocolloids **103**, 105696 (2020)
74. Singaram, V.A.J., Guruchandran, S., Bakshi, A., Muninathan, C., Ganesan, N.D.: Study on enhanced mechanical, barrier and optical properties of chemically modified mango kernel starch films. Packag. Technol. Sci. **34**(8), 485–495 (2021)
75. Hu, X., Jia, X., Zhi, C., Jin, Z., Miao, M.: Improving properties of normal maize starch films using dual-modification: combination treatment of debranching and hydroxypropylation. Int. J. Biol. Macromol. **130**, 197–202 (2019)
76. Xu, Y., Liu, X., Jiang, Q., Yu, D., Xu, Y., Wang, B., Xia, W.: Development and properties of bacterial cellulose, curcumin, and chitosan composite biodegradable films for active packaging materials. Carbohyd. Polym. **260**, 117778 (2021)
77. Siripatrawan, U., Vitchayakitti, W.: Improving functional properties of chitosan films as active food packaging by incorporating with propolis. Food Hydrocolloids **61**, 695–702 (2016)
78. Gatto, M., Ochi, D., Yoshida, C.M.P., da Silva, C.F.: Study of chitosan with different degrees of acetylation as cardboard paper coating. Carbohyd. Polym. **210**, 56–63 (2019)
79. Vartiainen, J., Motion, R., Kulonen, H., Rättö, M., Skyttä, E., Ahvenainen, R.: Chitosan-coated paper: effects of nisin and different acids on the antimicrobial activity. J. Appl. Polym. Sci. **94**(3), 986–993 (2004)
80. Cortés-Rodríguez, M., Villegas-Yépez, C., González, J.H.G., Rodríguez, P.E., Ortega-Toro, R.: Development and evaluation of edible films based on cassava starch, whey protein, and bees wax. Heliyon **6**(9) (2020)
81. Romani, V.P., Olsen, B., Collares, M.P., Oliveira, J.R.M., Prentice, C., Martins, V.G.: Cold plasma and carnauba wax as strategies to produce improved bi-layer films for sustainable food packaging. Food Hydrocolloids **108**, 106087 (2020)
82. Syahida, S.N., Ismail-Fitry, M.R., Ainun, Z.M.A.A., Hanani, Z.A.N.: Effects of palm wax on the physical, mechanical and water barrier properties of fish gelatin films for food packaging application. Food Packag. Shelf Life **23**, 100437 (2020)
83. Picchio, M.L., Linck, Y.G., Monti, G.A., Gugliotta, L.M., Minari, R.J., Igarzabal, C.I.A.: Casein films crosslinked by tannic acid for food packaging applications. Food Hydrocolloids **84**, 424–434 (2018)
84. Wang, W., Xiao, J., Chen, X., Luo, M., Liu, H., Shao, P.: Fabrication and characterization of multilayered kafirin/gelatin film with one-way water barrier property. Food Hydrocolloids **81**, 159–168 (2018)
85. Nilsuwan, K., Benjakul, S., Prodpran, T.: Physical/thermal properties and heat seal ability of bilayer films based on fish gelatin and poly (lactic acid). Food Hydrocolloids **77**, 248–256 (2018)
86. Sartori, T., Feltre, G., do Amaral Sobral, P.J., da Cunha, R.L., Menegalli, F.C.: Properties of films produced from blends of pectin and gluten. Food Packag. Shelf Life **18**, 221–229 (2018)
87. Hamdani, S.S., Li, Z., Rabnawaz, M., Kamdem, D.P., Khan, B.A.: Chitosan–graft–poly (dimethylsiloxane)/zein coatings for the fabrication of environmentally friendly oil-and water-resistant paper. ACS Sustain. Chem. Eng. **8**(13), 5147–5155 (2020)

88. Masamba, K., Li, Y., Hategekimana, J., Zehadi, M., Ma, J., Zhong, F.: Evaluation of mechanical and water barrier properties of transglutaminase cross-linked zein films incorporated with oleic acid. Int. J. Food Sci. Technol. **51**(5), 1159–1167 (2016)
89. Yang, S., Ban, Z., Jin, L., Chen, C., Li, L., Yi, G., Liu, L., et al.: Polyvinyl alcohol films incorporated with clove essential oil emulsions stabilized by soy protein isolate-derived amyloid fibrils: fabrication, characterization, and its application for active packaging. Food Chem. **440**, 138245 (2024)
90. Mukurumbira, A.R., Mellem, J.J., Amonsou, E.O.: Effects of amadumbe starch nanocrystals on the physicochemical properties of starch biocomposite films. Carbohyd. Polym. **165**, 142–148 (2017)
91. Priyadarshi, R., Rhim, J.W.: Chitosan-based biodegradable functional films for food packaging applications. Innov. Food Sci. Emerg. Technol. **62**, 102346 (2020)
92. Ren, L., Yan, X., Zhou, J., Tong, J., Su, X.: Influence of chitosan concentration on mechanical and barrier properties of corn starch/chitosan films. Int. J. Biol. Macromol. **105**, 1636–1643 (2017)
93. Cazón, P., Vázquez, M., Velazquez, G.: Environmentally friendly films combining bacterial cellulose, chitosan, and polyvinyl alcohol: effect of water activity on barrier, mechanical, and optical properties. Biomacromol **21**(2), 753–760 (2019)
94. Wu, Y., Luo, X., Li, W., Song, R., Li, J., Li, Y., Liu, S., et al.: Green and biodegradable composite films with novel antimicrobial performance based on cellulose. Food Chem. **197**, 250–256 (2016)
95. Ojagh, S.M., Rezaei, M., Razavi, S.H., Hosseini, S.M.H.: Development and evaluation of a novel biodegradable film made from chitosan and cinnamon essential oil with low affinity toward water. Food Chem. **122**(1), 161–166 (2010)
96. Ham-Pichavant, F., Sèbe, G., Pardon, P., Coma, V.: Fat resistance properties of chitosan-based paper packaging for food applications. Carbohyd. Polym. **61**(3), 259–265 (2005)
97. Sharma, V., Sharma, S., Sharma, L., Bebartta, R.P., Sehrawat, R.: Gum-based food packaging. In: Biodegradable and edible food packaging, pp. 217–245. Academic Press (2024)
98. Meerasri, J., Sothornvit, R.: Characterization of bioactive film from pectin incorporated with gamma-aminobutyric acid. Int. J. Biol. Macromol. **147**, 1285–1293 (2020)
99. Ezati, P., Rhim, J.W.: PH-responsive pectin-based multifunctional films incorporated with curcumin and sulfur nanoparticles. Carbohyd. Polym. **230**, 115638 (2020)
100. Sistla, Y.S., Mehraj, S.: Molecular simulations to understand the moisture, carbon dioxide, and oxygen barrier properties of pectin films. J. Mol. Model. **28**(4), 83 (2022)
101. Nastasi, J.R., Kontogiorgos, V., Daygon, V.D., Fitzgerald, M.A.: Pectin-based films and coatings with plant extracts as natural preservatives: a systematic review. Trends Food Sci. Technol. **120**, 193–211 (2022)
102. Zhao, N., Chai, Y., Wang, T., Wang, K., Jiang, J., Yang, H.Y.: Preparation and physical/chemical modification of galactomannan film for food packaging. Int. J. Biol. Macromol. **137**, 1060–1067 (2019)
103. Cerqueira, M.A., Lima, A.M., Souza, B.W., Teixeira, J.A., Moreira, R.A., Vicente, A.A.: Functional polysaccharides as edible coatings for cheese. J. Agric. Food Chem. **57**(4), 1456–1462 (2009)
104. Atta, O.M., Manan, S., Shahzad, A., Ul-Islam, M., Ullah, M.W., Yang, G.: Biobased materials for active food packaging: a review. Food Hydrocoll. **125**, 107419 (2022)
105. Mikkonen, K.S., Heikkilä, M.I., Helén, H., Hyvönen, L., Tenkanen, M.: Spruce galactoglucomannan films show promising barrier properties. Carbohyd. Polym. **79**(4), 1107–1112 (2010)
106. Cerqueira, M.A., Souza, B.W., Teixeira, J.A., Vicente, A.A.: Utilization of galactomannan from Gleditsia triacanthos in polysaccharide-based films: effects of interactions between film constituents on film properties. Food Bioprocess Technol. **6**, 1600–1608 (2013)
107. Almeida, L.B.S., Figueiredo, E.A.T., Dias, F.G.B., Santos, F.M.S., Fernandes, B.D., Vicente, A.A., Souza, B.W.S., et al.: Antimicrobial properties of chitosan and galactomannan composite coatings and physical properties of films made thereof. Future Foods **3**, 100028 (2021)

108. Nechita, P., Roman, M.: Review on polysaccharides used in coatings for food packaging papers. Coatings **10**, 566 (2020)
109. Chen, J., Wu, A., Yang, M., Ge, Y., Pristijono, P., Li, J., Mi, H., et al.: Characterization of sodium alginate-based films incorporated with thymol for fresh-cut apple packaging. Food Control **126**, 108063 (2021)
110. Aziz, M.S.A., Salama, H.E., Sabaa, M.W.: Biobased alginate/castor oil edible films for active food packaging. Lwt **96**, 455–460 (2018)
111. Sedayu, B.B., Cran, M.J., Bigger, S.W.: A review of property enhancement techniques for carrageenan-based films and coatings. Carbohyd. Polym. **216**, 287–302 (2019)
112. Suryanto, H., Yanuhar, U., Aminnudin, Pradana, Y.R.A., Bintara, R.D.: Starch/Carrageenan blend-based biocomposites as packaging materials. In: Composites from the aquatic environment, pp. 139–161. Springer Nature Singapore, Singapore (2023)
113. Zhou, X., Zong, X., Wang, S., Yin, C., Gao, X., Xiong, G., Mei, L., et al.: Emulsified blend film based on konjac glucomannan/carrageenan/camellia oil: physical, structural, and water barrier properties. Carbohyd. Polym. **251**, 117100 (2021)
114. Tahsiri, Z., Mirzaei, H., Hosseini, S.M.H., Khalesi, M.: Gum arabic improves the mechanical properties of wild almond protein film. Carbohyd. Polym. **222**, 114994 (2019)
115. Vigneshwaran, N., Ammayappan, L., Huang, Q.: Effect of Gum arabic on distribution behavior of nanocellulose fillers in starch film. Appl. Nanosci. **1**, 137–142 (2011)
116. Maurya, A.K., Yadav, M., Maurya, P., Fatima, A., Yadav, D.: UV-resistant gellan gum film reinforced with chitosan nanoparticle for eco-friendly packaging. Emergent Mater. 1–15 (2024)
117. Yousuf, B., Sun, Y., Wu, S.: Lipid and lipid-containing composite edible coatings and films. Food Rev. Intl. **38**(sup1), 574–597 (2022)
118. Nehchiri, N., Amiri, S., Radi, M.: Improving the water barrier properties of alginate packaging films by submicron coating with drying linseed oil. Packag. Technol. Sci. **34**(5), 283–295 (2021)
119. Dhumal, C.V., Ahmed, J., Bandara, N., Sarkar, P.: Improvement of antimicrobial activity of sago starch/guar gum bi-phasic edible films by incorporating carvacrol and citral. Food Packag. Shelf Life **21**, 100380 (2019)
120. Sharma, S., Barkauskaite, S., Duffy, B., Jaiswal, A.K., Jaiswal, S.: Characterization and antimicrobial activity of biodegradable active packaging enriched with clove and thyme essential oil for food packaging application. Foods **9**(8), 1117 (2020)
121. Ghasemlou, M., Aliheidari, N., Fahmi, R., Shojaee-Aliabadi, S., Keshavarz, B., Cran, M.J., Khaksar, R.: Physical, mechanical and barrier properties of corn starch films incorporated with plant essential oils. Carbohyd. Polym. **98**(1), 1117–1126 (2013)
122. Morillon, V., Debeaufort, F., Blond, G., Capelle, M., Voilley, A.: Factors affecting the moisture permeability of lipid-based edible films: a review. Crit. Rev. Food Sci. Nutr. **42**(1), 67–89 (2002)
123. Debeaufort, F., Voilley, A.: Lipid-based edible films and coatings. In: Edible films and coatings for food applications, pp. 135–168 (2009)
124. Chen, H., Wang, J., Cheng, Y., Wang, C., Liu, H., Bian, H., Han, W., et al.: Application of protein-based films and coatings for food packaging: a review. Polymers **11**(12), 2039 (2019)
125. Avena-Bustillos, R.J., Krochta, J.M.: Water vapor permeability of caseinate-based edible films as affected by pH, calcium crosslinking and lipid content. J Food Sci 58(4):904–907 (1993)
126. Anker, M., Berntsen, J., Hermansson, A.M., Stading, M.: Improved water vapor barrier of whey protein films by addition of an acetylated monoglyceride. Innov. Food Sci. Emerg. Technol. **3**(1), 81–92 (2002)
127. Kokoszka, S., Debeaufort, F., Lenart, A., Voilley, A.: Liquid and vapour water transfer through whey protein/lipid emulsion films. J. Sci. Food Agric. **90**(10), 1673–1680 (2010)
128. Ahmad, M., Benjakul, S., Prodpran, T., Agustini, T.W.: Physico-mechanical and antimicrobial properties of gelatin film from the skin of unicorn leatherjacket incorporated with essential oils. Food Hydrocolloids **28**(1), 189–199 (2012)

129. Tongnuanchan, P., Benjakul, S., Prodpran, T.: Properties and antioxidant activity of fish skin gelatin film incorporated with citrus essential oils. Food Chem. **134**(3), 1571–1579 (2012)
130. Hosseini, S.F., Javidi, Z., Rezaei, M.: Efficient gas barrier properties of multi-layer films based on poly (lactic acid) and fish gelatin. Int. J. Biol. Macromol. **92**, 1205–1214 (2016)
131. Were, L., Hettiarachchy, N.S., Coleman, M.: Properties of cysteine-added soy protein-wheat gluten films. J. Food Sci. **64**(3), 514–518 (1999)
132. Lens, J.P., De Graaf, L.A., Stevels, W.M., Dietz, C.H.J.T., Verhelst, K.C.S., Vereijken, J.M., Kolster, P.: Influence of processing and storage conditions on the mechanical and barrier properties of films cast from aqueous wheat gluten dispersions. Ind. Crops Prod. **17**(2), 119–130 (2003)
133. Giosafatto, C.V.L., Al-Asmar, A., D'angelo, A., Roviello, V., Esposito, M., Mariniello, L.: Preparation and characterization of bioplastics from grass pea flour cast in the presence of microbial transglutaminase. Coatings **8**(12), 435 (2018)
134. Mariniello, L., Giosafatto, C.V.L., Di Pierro, P., Sorrentino, A., Porta, R.: Swelling, mechanical, and barrier properties of albedo-based films prepared in the presence of phaseolin cross-linked or not by transglutaminase. Biomacromol **11**(9), 2394–2398 (2010)
135. Mirpoor, S.F., Giosafatto, C.V.L., Di Girolamo, R., Famiglietti, M., Porta, R.: Hemp (Cannabis sativa) seed oilcake as a promising by-product for developing protein-based films: effect of transglutaminase-induced crosslinking. Food Packag. Shelf Life **31**, 100779 (2022)
136. Qazanfarzadeh, Z., Kadivar, M., Shekarchizadeh, H., Di Girolamo, R., Giosafatto, C.V.L., Porta, R.: Secalin enzymatically cross-linked by either papain and N-acetyl-DL-homocysteine thiolactone or transglutaminase: improving of protein functional properties and film manufacturing. Food Hydrocolloids **120**, 106912 (2021)
137. Sorde, K.L., Ananthanarayan, L.: Effect of transglutaminase treatment on properties of coconut protein-guar gum composite film. Lwt **115**, 108422 (2019)
138. Vera, A., Tapia, C., Abugoch, L.: Effect of high-intensity ultrasound treatment in combination with transglutaminase and nanoparticles on structural, mechanical, and physicochemical properties of quinoa proteins/chitosan edible films. Int. J. Biol. Macromol. **144**, 536–543 (2020)
139. Zhang, W., Hedayati, S., Tarahi, M., Karaca, A.C., Hadidi, M., Assadpour, E., Jafari, S.M.: Advances in transglutaminase cross-linked protein-based food packaging films: a review. Int. J. Biol. Macromolecules 127399 (2023)
140. Dey, D., Dharini, V., Selvam, S.P., Sadiku, E.R., Kumar, M.M., Jayaramudu, J., Gupta, U.N.: Physical, antifungal, and biodegradable properties of cellulose nanocrystals and chitosan nanoparticles for food packaging application. Mater. Today Proc. **38**, 860–869 (2021)
141. Paralikar, S.A., Simonsen, J., Lombardi, J.: Poly (vinyl alcohol)/cellulose nanocrystal barrier membranes. J. Membr. Sci. **320**(1–2), 248–258 (2008)
142. Ding, W.D., Pervaiz, M., Sain, M.: Cellulose-enabled polylactic acid (PLA) nanocomposites: recent developments and emerging trends. Funct. Biopolymers 183–216
143. Ramesh, S., Radhakrishnan, P.: Cellulose nanoparticles from agro-industrial waste for the development of active packaging. Appl. Surf. Sci. **484**, 1274–1281 (2019)
144. Riahi, Z., Rhim, J.W., Bagheri, R., Pircheraghi, G., Lotfali, E.: Carboxymethyl cellulose-based functional film integrated with chitosan-based carbon quantum dots for active food packaging applications. Prog. Org. Coat. **166**, 106794 (2022)
145. Chi, K., Catchmark, J.M.: Improved eco-friendly barrier materials based on crystalline nanocellulose/chitosan/carboxymethyl cellulose polyelectrolyte complexes. Food Hydrocolloids **80**, 195–205 (2018)
146. Li, X., Qiu, C., Ji, N., Sun, C., Xiong, L., Sun, Q.: Mechanical, barrier and morphological properties of starch nanocrystals-reinforced pea starch films. Carbohyd. Polym. **121**, 155–162 (2015)
147. Oymaci, P., Altinkaya, S.A.: Improvement of barrier and mechanical properties of whey protein isolate based food packaging films by incorporation of zein nanoparticles as a novel bionanocomposite. Food Hydrocolloids **54**, 1–9 (2016)

148. Garavand, F., Khodaei, D., Mahmud, N., Islam, J., Khan, I., Jafarzadeh, S., Cacciotti, I., et al.: Recent progress in using zein nanoparticles-loaded nanocomposites for food packaging applications. Crit. Rev. Food Sci. Nutr. 1–21 (2022)
149. Bayram, B., Ozkan, G., Kostka, T., Capanoglu, E., Esatbeyoglu, T.: Valorization and application of fruit and vegetable wastes and by-products for food packaging materials. Molecules **26**(13), 4031 (2021)
150. De'Nobili, M.D., Bernhardt, D.C., Basanta, M.F., Rojas A.M.: Sunflower (Helianthus annuus L.) seed hull waste: composition, antioxidant activity, and filler performance in pectin-based film composites. Front. Nutr. **8**, 777214 (2021)
151. Merino, D., Bellassi, P., Paul, U.C., Morelli, L., Athanassiou, A.: Assessment of chitosan/pectin-rich vegetable waste composites for the active packaging of dry foods. Food Hydrocolloids **139**, 108580 (2023)
152. Swetha, T.A., Bora, A., Mohanrasu, K., Balaji, P., Raja, R., Ponnuchamy, K., Arun, A., et al.: A comprehensive review on polylactic acid (PLA)–synthesis, processing and application in food packaging. Int. J. Biol. Macromol. **234**, 123715 (2023)
153. Gan, I., Chow, W.S.: Antimicrobial poly (lactic acid)/cellulose bionanocomposite for food packaging application: a review. Food Packag. Shelf Life **17**, 150–161 (2018)
154. Qin, Y., Li, W., Liu, D., Yuan, M., Li, L.: Development of active packaging film made from poly (lactic acid) incorporated essential oil. Prog. Org. Coat. **103**, 76–82 (2017)
155. Mohamad, N., Mazlan, M.M., Tawakkal, I.S.M.A., Talib, R.A., Kian, L.K., Fouad, H., Jawaid, M.: Development of active agents filled polylactic acid films for food packaging application. Int. J. Biol. Macromol. **163**, 1451–1457 (2020)
156. Masood, F.: Polyhydroxyalkanoates in the food packaging industry. In: Nanotechnology applications in food, pp 153–177. Academic Press (2017)
157. Follain, N., Chappey, C., Dargent, E., Chivrac, F., Cretois, R., Marais, S.: Structure and barrier properties of biodegradable polyhydroxyalkanoate films. J. Phys. Chem. C **118**(12), 6165–6177 (2014)
158. Hernández-García, E., Freitas, P.A., Zomeño, P., González-Martínez, C., Torres-Giner, S.: Multilayer sheets based on double coatings of poly (3-hydroxybutyrate-co-3-hydroxyvalerate) on paper substrate for sustainable food packaging applications. Appl. Sci. **13**(1), 179 (2022)
159. Fabra, M.J., Lopez-Rubio, A., Lagaron, J.M.: Nanostructured interlayers of zein to improve the barrier properties of high barrier polyhydroxyalkanoates and other polyesters. J. Food Eng. **127**, 1–9 (2014)
160. Pardo-Ibáñez, P., Lopez-Rubio, A., Martínez-Sanz, M., Cabedo, L., Lagaron, J.M.: Keratin–polyhydroxyalkanoate melt-compounded composites with improved barrier properties of interest in food packaging applications. J. Appl. Polymer Sci. **131**(4) (2014)
161. Kovalcik, A., Machovsky, M., Kozakova, Z., Koller, M.: Designing packaging materials with viscoelastic and gas barrier properties by optimized processing of poly (3-hydroxybutyrate-co-3-hydroxyvalerate) with lignin. React. Funct. Polym. **94**, 25–34 (2015)
162. Eissenberger, K., Ballesteros, A., De Bisschop, R., Bugnicourt, E., Cinelli, P., Defoin, M., Schmid, M., et al.: Approaches in sustainable biobased multilayer packaging solutions. Polymers **15**(5), 1184 (2023)
163. Anukiruthika, T., Sethupathy, P., Wilson, A., Kashampur, K., Moses, J.A., Anandharamakrishnan, C.: Multilayer packaging: advances in preparation techniques and emerging food applications. Compr. Rev. Food Sci. Food Saf. **19**(3), 1156–1186 (2020)
164. Ajitha, A.R., Aswathi, M.K., Maria, H.J., Izdebska, J., Thomas, S.: Multilayer polymer films. In: Multicomponent Polymeric Materials, pp. 229–258. Springer, Dordrecht (2016)
165. La Fuente Arias, C.I., Kubo, M.T.K.N., Tadini, C.C., Augusto, P.E.D.: Bio-based multilayer films: a review of the principal methods of production and challenges. Crit. Rev. Food Sci. Nutr. **63**(14), 2260–2276 (2023)
166. Lim, L.T.: Electrospinning and electrospraying technologies for food and packaging applications. In: Electrospun polymers and composites, pp. 217–259. Woodhead Publishing (2021)

167. dos Santos, J.W.S., Noletto, A.P.R., de Azeredo, H.M.C., de Carvalho, R.A., da Silva, C.F., Yoshida, C.M.P.: Bio-based multilayer paperboard for sustainable packaging application. J. Appl. Polym. Sci. **140**(37), e54415 (2023)
168. Dos Santos, J.W.S., Garcia, V.A.D.S., Venturini, A.C., Carvalho, R.A.D., da Silva, C.F., Yoshida, C.M.P.: Sustainable coating paperboard packaging material based on chitosan, palmitic acid, and activated carbon: water vapor and fat barrier performance. Foods **11**(24), 4037 (2022)
169. Zhang, L., Li, K., Yu, D., Regenstein, J.M., Dong, J., Chen, W., Xia, W.: Chitosan/zein bilayer films with one-way water barrier characteristic: physical, structural and thermal properties. Int. J. Biol. Macromol. **200**, 378–387 (2022)
170. Pandit, P., Nadathur, G.T., Maiti, S., Regubalan, B.: Functionality and properties of bio-based materials. In: Bio-based materials for food packaging: green and sustainable advanced packaging materials, pp. 81–103 (2018)
171. Koller, M., Maršálek, L., de Sousa Dias, M.M., Braunegg, G.: Producing microbial poly-hydroxyalkanoate (PHA) biopolyesters in a sustainable manner. New Biotechnol. **37**, 24–38 (2017)
172. Rojas, A., Velásquez, E., Patiño Vidal, C., Guarda, A., Galotto, M.J., López de Dicastillo, C.: Active PLA packaging films: effect of processing and the addition of natural antimicrobials and antioxidants on physical properties, release kinetics, and compostability. Antioxidants **10**(12), 1976 (2021)
173. Chalier, P., Ben Arfa, A., Guillard, V., Gontard, N.: Moisture and temperature triggered release of a volatile active agent from soy protein coated paper: effect of glass transition phenomena on carvacrol diffusion coefficient. J. Agric. Food Chem. **57**(2), 658–665 (2009)
174. Rocca-Smith, J.R., Chau, N., Champion, D., Brachais, C.H., Marcuzzo, E., Sensidoni, A., Debeaufort, F., et al.: Effect of the state of water and relative humidity on ageing of PLA films. Food Chem. **236**, 109–119 (2017)
175. Yang, L., Paulson, A.T.: Mechanical and water vapour barrier properties of edible gellan films. Food Res. Int. **33**(7), 563–570 (2000)
176. Hirvikorpi, T., Vähä-Nissi, M., Harlin, A., Karppinen, M.: Comparison of some coating techniques to fabricate barrier layers on packaging materials. Thin Solid Films **518**(19), 5463–5466 (2010)
177. Marturano, V., Marotta, A., Salazar, S.A., Ambrogi, V., Cerruti, P.: Recent advances in bio-based functional additives for polymers. Progr. Mater. Sci. 101186 (2023)
178. de Paula Herrmann, P.S., Yoshida, C.M.P., Antunes, A.J., Marcondes, J.A.: Surface evaluation of whey protein films by atomic force microscopy and water vapour permeability analysis. Packag. Technol. Sci. Int. J. **17**(5), 267–273 (2004)
179. Ahsan, S., Liaqat, A., Khaliq, A., Iqbal, R., Chughtai, M.F.J., Mehmood, T., Sameed, N., et al.: Current trends and prospects of transforming animal waste into food. In: Climate Changes Mitigation and Sustainable Bioenergy Harvest Through Animal Waste: Sustainable Environmental Implications of Animal Waste, pp. 469–503. Springer Nature Switzerland, Cham (2023)
180. Thanakkasaranee, S., Sadeghi, K., Seo, J.: Packaging materials and technologies for microwave applications: a review1. Crit. Rev. Food Sci. Nutr. **63**(23), 6464–6483 (2023)
181. Wang, H., Wang, L.: Developing a bio-based packaging film from soya by-products incorporated with valonea tannin. J. Clean. Prod. **143**, 624–633 (2017)
182. Ščetar, M., Barukčić, I., Kurek, M., Jakopović, K.L., Božanić, R.: Packaging perspective of milk and dairy products. Dairy/Mljekarstvo **69**(1) (2019)
183. Petersen, K., Nielsen, P.V., Bertelsen, G., Lawther, M., Olsen, M.B., Nilsson, N.H., Mortensen, G.: Potential of biobased materials for food packaging. Trends Food Sci. Technol. **10**(2), 52–68 (1999)
184. Naser, A.Z., Deiab, I., Darras, B.M.: Poly (lactic acid)(PLA) and polyhydroxyalkanoates (PHAs), green alternatives to petroleum-based plastics: a review. RSC Adv. **11**(28), 17151–17196 (2021)

185. Subramaniam, P.: The stability and shelf life of confectionery products. In: The stability and shelf life of food, pp. 545–573. Woodhead Publishing (2016)
186. Primo-Martín, C., Van Vliet, T.: Sensory crispness of crispy rolls: effect of formulation, storage conditions, and water distribution in the crust. J. Food Sci. **74**(8), S377–S383 (2009)
187. Stapelfeldt, H., Nielsen, B.R., Skibsted, L.H.: Effect of heat treatment, water activity and storage temperature on the oxidative stability of whole milk powder. Int. Dairy J. **7**(5), 331–339 (1997)
188. Bauer, A.S., Leppik, K., Galić, K., Anestopoulos, I., Panayiotidis, M.I., Agriopoulou, S., Krauter, V., et al.: Cereal and confectionary packaging: Background, application and shelf-life extension. Foods **11**(5), 697 (2022)
189. Al-Ati, T., Hotchkiss, J.H.: Application of packaging and modified atmosphere to fresh-cut fruits and vegetables. In: Fresh-Cut Fruits and Vegetables: Science, Technology, and Market, pp. 305–338 (2002)
190. Sandhya. Modified atmosphere packaging of fresh produce: current status and future needs. LWT-Food Sci. Technol. **43**(3), 381–392 (2010)
191. Kandeepan, G., Tahseen, A.: Modified atmosphere packaging (MAP) of meat and meat products: a review. J. Packag. Technol. Res. **6**(3), 137–148 (2022)
192. DeWitt, C.A.M., Oliveira, A.C.: Modified atmosphere systems and shelf life extension of fish and fishery products. Foods **5**(3), 48 (2016)
193. Hadidi, M., Jafarzadeh, S., Forough, M., Garavand, F., Alizadeh, S., Salehabadi, A., Jafari, S.M., et al.: Plant protein-based food packaging films; recent advances in fabrication, characterization, and applications. Trends Food Sci. Technol. **120**, 154–173 (2022)
194. Ede, J.D., Ong, K.J., Mulenos, M.R., Pradhan, S., Gibb, M., Sayes, C.M., Shatkin, J.A.: Physical, chemical, and toxicological characterization of sulfated cellulose nanocrystals for food-related applications using in vivo and in vitro strategies. Toxicol. Res. **9**(6), 808–822 (2020)
195. Jamshidian, M., Tehrany, E.A., Imran, M., Jacquot, M., Desobry, S.: Poly-lactic acid: production, applications, nanocomposites, and release studies. Compr. Rev. Food Sci. Food Saf. **9**(5), 552–571 (2010)
196. Silva, F.A., Dourado, F., Gama, M., Poças, F.: Nanocellulose bio-based composites for food packaging. Nanomaterials **10**(10), 2041 (2020)
197. Panou, A., Karabagias, I.K.: Migration and safety aspects of plastic food packaging materials: need for reconsideration? Coatings **14**(2), 168 (2024)
198. Yildirim, S., Röcker, B., Pettersen, M.K., Nilsen-Nygaard, J., Ayhan, Z., Rutkaite, R., Coma, V., et al.: Active packaging applications for food. Compr. Rev. Food Sci. Food Saf. **17**(1), 165–199 (2018)
199. Min, T., Zhou, L., Sun, X., Du, H., Bian, X., Zhu, Z., Wen, Y.: Enzyme-responsive food packaging system based on pectin-coated poly (lactic acid) nanofiber films for controlled release of thymol. Food Res. Int. **157**, 111256 (2022)

Chapter 10
Bioactive Compounds in Active Food Packaging (ACP): Health and Safety Considerations

Aeyaz Ahmad Bhat, Anjuman Ayub, and Atif Khurshid Wani

Abstract Food packaging is critical in assuring product safety, ease of handling, and shelf life extension, all contributing to increased consumer convenience. Packaging made from various materials, including plastics, glass, metals, papers, and composites, must also address growing health concerns about the possible migration of hazardous substances into food. This chapter reviews bioactive compounds integrated into Active Food Packaging (ACP) systems, focusing on various compounds and technologies used. It investigates the functions of essential bioactive compounds such as antioxidants, antimicrobials, and oxygen scavengers, demonstrating their efficacy in improving food quality, safety, and shelf life. Furthermore, it emphasizes the importance of balancing technology advancements in ACP with thorough health and safety assessments. It also seeks to provide significant insights into the responsible development and deployment of ACP technologies by investigating allergenicity, toxicity, and environmental effects, ensuring that they meet consumer health requirements and promote sustainability.

Keywords Active packaging · Bioactive compounds · Coating technologies · Food safety · Health and safety considerations

10.1 Introduction

The food packaging industry is an important global sector, with developing countries playing a major role in its expansion. Packaging is vital in food processing because it contains, protects, and preserves items to suit consumer expectations [44].

A. A. Bhat
School of Chemical Engineering and Physical Sciences, Lovely Professional University, Phagwara 14411, Punjab, India

A. Ayub · A. K. Wani (✉)
School of Bioengineering and Biosciences, Lovely Professional University, Phagwara 14411, Punjab, India
e-mail: atifkhurshid61200216@gmail.com

S. ul Islam and M. Shahid (eds.), *Green Materials for Active Food Packaging*, Engineering Materials, https://doi.org/10.1007/978-981-96-0369-5_10

It performs a variety of roles, including product protection, information dissemination, and transportation facilitation, all of which influence consumer impressions and purchasing decisions [27]. Consumer demand for food packaging options that provide both convenience in handling and longer shelf life has recently increased. This tendency has considerably increased the prominence of scientific research and studies focusing on advancements in food packaging technology, especially active and intelligent packaging [34]. Active packaging alters packaging to preserve or elevate food's sensory attributes, safety, and quality. Intelligent packaging offers information on historical data, quality, and food safety through indicators [54]. Despite the potential benefits of active and intelligent packaging, standardization of their rationale has not been achieved yet. Active packaging systems, developed since the 1980s, offer a wide range of product applications using extracts from plant and food sources and biodegradable materials [39]. These systems can be categorized into non-migratory and active-releasing packaging, serving functions such as oxygen and moisture scavenging, ethylene absorption, carbon dioxide emission, antimicrobial protection, and antioxidant release [55]. While technological advancements have led to the development of new active packaging systems, their commercial implementation remains limited. Safety and legislative concerns contribute to the relatively low adoption rate of active packaging, accounting for only about one percent of total packaging in the United States. However, recent studies emphasize the potential of active packaging to ensure food safety during transportation and storage [17]. This study aims to explore active packaging systems and their functions concerning food products, focusing on commercial prospects and industry challenges. It seeks to address concerns regarding packaging costs, acceptability, safety, and legislation, providing insights into the opportunities and barriers associated with active ACP in the industry and consumer contexts [13].

Active packaging has emerged as a dynamic field of study within the broader packaging area, garnering significant attention from researchers in recent years. This heightened interest is driven by the substantial potential of active packaging technology to revolutionize the packaging industry, introducing innovative concepts in both theory and practice [48]. In contrast to conventional packaging, which mainly serves passive roles like containing and protecting products, active packaging integrates packaging materials with dynamic functionalities. These functions actively engage with the packaged product to improve its quality, safety, and longevity [57]. In contrast to the static nature of traditional packaging, active packaging engages dynamically with its surrounding environment, including the food system and external conditions. Thus, active packaging can be more accurately described as an integrated "environment/package/food" system, wherein the packaging material interacts dynamically with its surroundings to improve various aspects of food quality, safety, and shelf life [7]. This chapter explores the diverse landscape of ACP, focusing on incorporating bioactive compounds into packaging materials to enhance functionality and performance. Through a comprehensive examination of various bioactive compounds and their integration into packaging materials, this study aims to provide insights into ACP technologies' potential applications, benefits, and challenges [42]. By highlighting key examples of ACP technologies, such as antimicrobials, flavor enhancers, and nutraceuticals, as well as discussing innovative integration techniques like coating technologies, nanotechnology applications, and encapsulation techniques, this chapter aims to offer a comprehensive overview of the field [21]. Additionally, it discusses the environmental impact and sustainability considerations associated with active packaging materials, exploring eco-friendly alternatives and strategies for minimizing environmental harm [22]. Through this exploration, we aim to contribute to a deeper understanding of ACP and its role in enhancing food quality, safety, and sustainability in the modern food industry [20]. By leveraging the potential of active packaging technologies and embracing sustainable practices, we can work towards a future where packaging plays a proactive role in preserving food quality, reducing waste, and promoting environmental stewardship [50].

10.2 Types of Bioactive Compounds in Active Food Packaging (ACP)

Active packaging has evolved as a key research topic within the broader packaging science, receiving significant interest from scientists in recent years [3]. This increased attention arises from the significant potential impact that this technology may have. Active packaging can potentially transform the packaging business by introducing novel concepts in theory and practice [78]. Unlike traditional packaging, which is often considered "passive" because it performs static functions such as containment, protection, and communication, active packaging incorporates dynamic functionalities that interact with the packaged product to improve its quality, safety, and shelf life. Indeed, in the field of packaging, barrier qualities such as resistance to gases and UV light are often regarded as passive attributes inherent to the material itself [39]. In contrast, "active" refers to packaging that may interact with its surroundings, such as the food system and the outside world. As a result, active packaging is better defined as an integrated "environment/package/food" system in which the packaging material interacts dynamically with its surroundings to improve food quality, safety, and shelf life. Below are some examples of ACP technologies and how they work.

10.2.1 Antimicrobials and Antioxidants

Active packaging works toward the overarching goal of improving food quality, safety, and convenience. It includes absorbing and releasing mechanisms, with the former including scavengers for oxygen, carbon dioxide, and humidity, among others [40]. However, much of the research on active packaging for food and beverage applications primarily focuses on antibacterial and antioxidant systems, classified as releasing systems [78].

Controlled release packaging (CRP) is a key development in active packaging technology. CRP materials serve as reservoirs for active chemicals, releasing them into food at regulated rates throughout storage [80]. Achieving this controlled release requires careful encapsulation of active chemicals within the packaging material, considering chemical compatibility, possible interactions, and physical qualities [17]. Historically, efforts were mostly focused on directly integrating active chemicals into plastic films. However, issues such as the stability of active chemicals during processing, particularly under severe shear and heat stress, have [40]. Coatings present a promising alternative, providing an additional layer for embedding and releasing active compounds without compromising other essential properties like barrier and mechanical strength [60]. Coatings have the advantage of potentially responding to environmental stimuli, such as pH, temperature, or humidity, releasing active chemicals. Coating procedures are softer than extrusion, addressing concerns about high temperatures and manufacturing conditions [31]. Bio-macromolecule-based coatings produced from proteins, carbohydrates, and lipids may provide safety and applicability benefits, particularly regarding migration into food [49, 85]. Overall, developments in active packaging, particularly coating technology innovations, are projected to substantially impact the continued development and deployment of these systems in the food packaging industry.

10.2.2 Flavor Enhancers

Flavor enhancers may be included as ingredients in food packaging, especially in processed or pre-packaged goods, where they are frequently employed to increase taste and palatability. Manufacturers must identify all ingredients on food labels, including flavor enhancers, so consumers know what they eat [40]. Some food packaging may also incorporate labels or marketing messaging emphasizing the absence of specific taste enhancers, appealing to customers who prefer products free of artificial additives [3]. Furthermore, food packaging may include claims or certifications linked to flavor enhancers, such as "No MSG Added" or "Natural Flavors Only," to appeal to certain customer preferences or dietary restrictions [66].

Flavor enhancers are used in various applications, including culinary and food processing. Chefs and food scientists frequently use these ingredients to create balanced and savory dishes that ensure consumer consistency and pleasure [87]. Furthermore, flavor enhancers can disguise unpleasant smells or scents in food goods, enhancing palatability. This is especially essential in producing reduced-fat or low-sodium food formulations because taste augmentation helps preserve sensory appeal despite the absence of specific constituents [32]. In recent years, there has been an increasing interest in discovering natural sources of flavor improvement, such as herbs, spices, and fermentation products [1]. This trend reflects broader consumer demand for clean-label products and a preference for ingredients that are believed to be more nutritious and authentic. Overall, food packaging plays an important role in expressing information regarding the presence or absence of flavor enhancers, assisting customers to make informed choices about the items they purchase and consume [35].

10.2.3 Nutraceuticals

Nutraceuticals significantly impact food packaging by emphasizing health benefits through clear labeling, vibrant design, and educational content. Packaging often includes specific health claims, certifications, and detailed nutritional information, with designs that use health-associated imagery and colors. Functional packaging protects the integrity of nutraceuticals, while eco-friendly materials appeal to environmentally conscious consumers. Additionally, smart packaging technologies and regulatory compliance ensure consumers are well-informed and trust the product's health claims. This comprehensive approach helps attract health-conscious consumers and promotes the product's efficacy and benefits [23].

10.2.3.1 Highlighting Health Benefits

Food packaging commonly highlights nutraceutical elements with claims like "fortified with omega-3 fatty acids" or "rich in antioxidants." These statements are intended to inform consumers about the product's possible health benefits, making it more enticing to people seeking foods with increased nutritional value and health benefits. This method appeals to health-conscious consumers by emphasizing the functional features of the meal [17].

10.2.3.2 Functional Claims

Nutraceutical-rich foods frequently use functional claims on their packaging to emphasize the specific health advantages of their components. For example, a cereal box might say "supports heart health" or "promotes digestive wellness," stressing the functional benefits of the added nutrients or bioactive substances. These claims are intended to inform consumers about the product's health-promoting features, making it more enticing to people looking for foods with additional health advantages [4].

10.2.3.3 Clear Labeling

Food packaging labels play an important role in providing specific information about the presence and quantity of nutraceuticals in a product, maintaining transparency, and building customer trust. Transparency is critical for individuals to make educated dietary decisions. Knowing exactly what they're consuming and how much of a specific nutraceutical they're getting can help health-conscious customers manage their health and wellness objectives. For example, someone trying to improve their immune system may deliberately seek foods high in antioxidants or vitamins C and E. Similarly, someone looking to improve their cardiovascular health may opt for foods supplemented with omega-3 fatty acids [11].

10.2.3.4 Certifications and Seals

Certain nutraceutical-rich products frequently include certificates or seals on the packaging to verify compliance with specified quality standards or regulatory requirements. For example, a vitamin D-fortified product may bear the "USDA Organic" seal or the "Non-GMO Project Verified" certification. These certifications demonstrate that the product met high requirements for ingredient purity, manufacturing methods, and sourcing policies. Such endorsements reassure consumers about the product's quality and integrity, which boosts their trust in its health advantages and safety. By displaying these symbols, firms may successfully communicate their dedication to high standards and transparency, making their products more appealing to health-conscious consumers [56].

Similarly, the ACP includes Oxygen scavengers, which are incorporated into packaging materials to absorb oxygen from the surrounding atmosphere. This helps prevent oxidation and spoilage of oxygen-sensitive foods, such as meats, cheeses, and snack foods. Ethylene absorbers remove ethylene gas, produced naturally by fruits and vegetables, and accelerate their ripening and decay [39]. By reducing ethylene levels inside the packaging, these absorbers can extend the shelf life of fresh produce. Packaging materials with moisture-absorbing or moisture-releasing properties help regulate humidity levels inside the package, preventing moisture-related issues such as mold growth or soggy textures in foods [41]. In addition to these safety packaging bioactives, ACP utilizes various technologies to extend the shelf life, enhance safety, and maintain the quality of perishable foods during storage and transportation [75]. Phase change materials and thermal insulation layers are incorporated into the packaging to regulate temperature, minimizing the risk of spoilage caused by fluctuations in environmental conditions [43]. Additionally, time–temperature indicators are integrated into the packaging to provide visual or electronic cues about the product's temperature history, allowing consumers and retailers to assess freshness and safety [69]. Some active packaging systems also release bioactive compounds such as flavorings, antioxidants, and additives into the food to improve taste, aroma, and nutritional value, further enhancing the overall quality of packaged foods [74]. Table 10.1 summarizes different types of ACP technologies with their function, mechanisms, examples, and uses.

Table 10.1 Food packaging technologies with their function, mechanism, examples, and uses

Food packaging technology	Functionality	Mechanism of action	Examples of use
Oxygen scavengers	Absorb oxygen to prevent oxidation	Chemical reaction with oxygen to form harmless compounds	Packaging for snacks, meats, bakery products
Ethylene absorbers	Remove ethylene gas to delay ripening	Adsorption or oxidation of ethylene gas molecules	Packaging for fruits, vegetables
Antimicrobial packaging	Inhibit microbial on growth surface	Release of antimicrobial agents or creation of hostile environment	Packaging for meat, poultry, cheese
Moisture control	Absorb or release moisture to maintain humidity	Absorption of excess moisture or release to prevent dryness	Packaging for dried fruits, cereals, baked goods
Temperature control	Maintains desired temperature to prevent spoilage	Insulation or phase change materials to regulate temperature	Packaging for frozen foods, seafood, dairy
Time–temperature indicators	Provide visual or electronic cues about temperature history	Color change, digital display, or data logging	Packaging for perishable goods with sensitive temperature requirements
Flavor release systems	Release flavorings or additives into food	Encapsulation or controlled release mechanisms	Packaging for snacks, beverages, convenience foods
pH modifiers	Regulate acidity or alkalinity of food environment	Release of acid or base to maintain desired pH	Packaging for acidic or alkaline foods
Antioxidant releasers	Release antioxidants to prevent oxidation	Controlled release of antioxidants to inhibit oxidation	Packaging for oils, nuts, and other oxidative-prone foods
Gas emitters	Release gases to modify atmosphere inside package	Controlled release of gases to slow ripening or inhibit microbial growth	Packaging for fresh produce, baked goods

10.3 Integration of Bioactive Compounds into Packaging Materials

10.3.1 Coating Technologies

At the moment, the key technical challenge is the creation of high oxygen barrier materials that can work efficiently, especially in high humidity circumstances. Researchers are especially interested in discovering lighter alternatives to pricey multi-layer structures, such as coatings, to address the growing demand for decreased packing materials [38]. Plastic bottles and jars with improved gas barrier, mechanical, chemical, and heat-resistant qualities, made possible by developments in coating technology, are becoming increasingly popular, particularly for gas-sensitive beverages [2]. The next generation of coatings is expected to be multifunctional, offering advantages such as barrier characteristics, sealability, and scratch resistance in thin layers. Furthermore, there is a major emphasis on cost-effectiveness, with increasing line speeds being the preferred option. This necessitates packaging materials that can tolerate greater strains without sacrificing performance [53]. Responding to these trends, new technologies are emerging, promising highly engineered coatings for the future.

10.3.2 Nanotechnology Applications

Nanotechnology is possibly the most significant revolution in polymer research, which has continued for the past two decades. While some have questioned the tangible impact of nanotechnology on daily life, its significance in numerous research endeavors is obvious [30, 64]. The burgeoning interest in nanotechnology is justified by its vast potential within its intricate domain, which is only beginning to be fully comprehended and harnessed [15, 24]. A thorough investigation of nanotechnology and its applications in polymer science would be impractical here. This section presents an initial investigation of the main tendencies associated with nanotechnology, notably in the context of inventing new materials, specifically coatings suited for packaging applications in the food and beverage industry [73, 76]. In polymer science, nanotechnology often refers to creating novel materials with nano-sized properties (nanostructures) or inserting microscopic inorganic particles into polymer matrices to generate nanocomposites. These distinctions correspond closely to the two basic nanotechnology definitions commonly accepted in the field.

In the first scenario, nanotechnology entails the development of novel materials whose properties change as the structure approaches the nanoscale. Essentially, the 'micro' portion of the system plays a significant role in achieving novel qualities that are otherwise unachievable [51, 68]. The ability to design, produce, and manipulate nanosystems will be critical in developing novel materials with remarkable properties, such as coatings. This method of scaling down is known as top-down processing. The 'assembly' technique, particularly at the atomic, molecular, and supramolecular levels (bottom-up processing), has shown promise in building multifunctional, hierarchical structures suited for integration into mesoscale and macroscale systems [36].

10.3.3 Encapsulation Techniques

The growing need for innovation has prompted food and beverage packaging researchers to design and develop unique architectures. These technologies offer improved functional features and correspond with current environmental regulations while being economically feasible [45]. While significant progress has been made, particularly in academic circles, most of these innovations remain limited to laboratory settings. One key hurdle to their wider deployment is the sometimes weak connection between academics and industry, which hinders the transition from research findings to real-world applications [61]. Thus, fostering collaboration between academics and industry appears to be a critical requirement for the widespread adoption of novel concepts. This section will look at particular examples of improvements in advanced coatings and encapsulating processes designed for food and beverage packaging applications.

10.3.3.1 Oxygen and Water Vapor Barrier Coatings

The food business constantly looks for new ways to combat food spoilage, mostly caused by oxidation and water vapor transfer within packaging. Thus, effectively controlling oxygen permeability through packaging materials is critical [89]. There is a rising interest in creating highly effective barrier coatings that maintain a stable interior atmosphere, particularly in modified atmosphere packaging (MAP) goods [81]. These coatings must have superior oxygen barrier qualities while staying transparent, resistant to relative humidity changes, and cost-effective compared to conventional options. Researchers have developed a unique high oxygen barrier coating that meets these parameters. Scientists used a physicochemical approach to address the moisture susceptibility of polyvinyl alcohol (PVOH), a polymer recognized for its high oxygen barrier qualities, by creating a hybrid organic–inorganic coating. This method used a sol–gel approach to blend PVOH with tetraethoxysilane, an ethyl ester of orthosilicic acid. As a result of subsequent hydrolysis and condensation events, hydrogen and covalent bonds were formed between the two components, forming a unique hybrid three-dimensional structure [25]. The sol–gel solution was put onto several polymer substrates using a dip coating approach, including polyethylene terephthalate (PET), orientated polypropylene (OPP), cast polypropylene (CPP), linear low-density polyethylene (LLDPE), and blends with co-extruded low-density [16]. [89] found that coatings as thin as 1–2 μm significantly reduced oxygen transmission rate by around a 100-fold compared to untreated surfaces. Furthermore, even after prolonged exposure to water, the hybrid coatings maintained excellent oxygen barrier characteristics. It is worth mentioning, however, that these results were only seen on PET substrates. In a similar study, a hybrid coating composed of organic and inorganic phases reduced the oxygen transmission rate on polylactic acid films by approximately 48%, making it suitable for certain food packaging applications, particularly for products with a medium shelf life when used alongside modified atmosphere packaging [81]. Despite its efficiency against oxygen, this coating did not perform well against water vapor due to its hydrophilic nature [36]. Addressing challenges such as adherence to polyolefinic substrates and enhancing water vapor barrier qualities should make these organic–inorganic coatings a viable alternative to metalized films, particularly for oxygen and water vapor protection.

10.3.3.2 Sealant Coatings

Effective seals are essential for preserving the quality of food and drinks throughout transport and storage. Hermetic seals are critical for guaranteeing the package's integrity throughout the supply chain [70]. These seals can be created in two ways: by employing a thermoplastic polymer such as polyethylene in a multilayer structure or by applying a coating with remarkable sealing qualities. Heat-sealing, which includes applying heat over 100 °C then cooling, forms a bond by forcing heated interfaces together [47]. On the other hand, cold seals rely entirely on pressure or, in some cases, solvent to dissolve surfaces, removing the necessity for heat. Cold-sealing is used when high temperatures may harm packages or heat-sensitive goods, such as chocolate-based items [88]. Furthermore, because these coatings are obtained from natural sources, any contact with packaged food would pose no safety risk, as all constituents are legally recognized as food components or additives [79]. This allows for using these eco-friendly coatings as a substitute for typical cold sealants, removing the need for costly and time-consuming methods to apply the sealant just to the package edges [29].

10.3.3.3 Antimicrobial Coatings

In recent years, there has been an increasing demand to create antimicrobial packaging, although there are not many studies on antimicrobial coatings applied to packaging materials [40]. The examples below encourage further exploration in this area, demonstrating how incorporating active compounds into coatings can address different obstacles associated with directly integrating them into plastic substrates [72]. In 2007, Nerín's team made a significant addition to the subject by introducing the notion of "active coatings." They tested the antibacterial properties of four essential oils—clove, cinnamon, oregano, and cinnamon essential oils fortified with cinnamaldehyde—that were mixed into paraffin-based coatings for food-contact paper packaging [38]. These coatings were tested for fungal and bacterial pollutants often seen in food goods. Among these oils, cinnamaldehyde-rich cinnamon essential oil had the strongest antibacterial action [19]. Initial testing on two types of strawberries treated with this essential oil exhibited continuous freshness and quality over a seven-day refrigeration period at 4 °C, meeting the market's shelf life criteria without any microbiological contamination [28].

In the same year, researchers looked at developing antimicrobial coatings for plastic films used in food packaging. They presented a hybrid coating made with the sol–gel process that included silver salts such as $AgNO_3$ and AgCl as less expensive alternatives to silver nanoparticles [17]. Active thin coatings (0.6–1.1 μm thick) were applied to polyethylene and polyvinyl chloride films. The coated films' efficiency against both Gram-negative (*Escherichia coli*) and Gram-positive (*Staphylococcus aureus*) bacteria was assessed. Notably, the findings revealed that the antibacterial activities of the coated films were predominantly attributable to Ag+ ion diffusion within the coating's organic phase, providing a viable direction for antimicrobial packaging solutions [64]. They created a novel coating for polypropylene surfaces that provides antimicrobial protection for post-harvest crops without using active chemicals such as essential oils or enzymes. Their breakthrough entailed developing a covering with chitosan and pectin layers, leveraging chitosan's intrinsic antibacterial characteristics [59]. This covering efficiently inhibited fungus and bacteria, increasing the shelf life of packaged tomatoes by up to 13 days. Furthermore, increasing the chitosan content improved its antibacterial properties. A recent research cooperation between Clemson University and Sealed Air Corporation sought to address Listeria monocytogenes infection in fresh meat. They devised a unique method, mixing nisin with either lysozyme or rosemary extract in a cellulose coating applied to a barrier film [71]. Unlike traditional procedures, this study used a combination of antimicrobial drugs to increase efficacy and shelf life. The controlled release of these compounds provided continued protection against infections, demonstrating a cascade mechanism for antibacterial action.

10.4 Environmental Impact

10.4.1 Life Cycle Analysis of Active Packaging Materials

The life cycle study of active packaging materials entails evaluating their environmental impact across their entire lives, from raw material extraction to disposal. Researchers assess resource consumption, energy use, emissions, and waste formation at each production stage, use, and disposal [39, 84]. This analysis seeks to identify areas for improvement in sustainability and to lead the creation of more environmentally friendly packaging options. Companies may make educated decisions to reduce environmental harm and support circular economy concepts by considering the entire life cycle, including manufacturing, transportation, and end-of-life scenarios [8, 83, 86]. In a life cycle analysis (LCA) of active packaging materials, several key steps are typically undertaken to assess their environmental impact comprehensively.

10.4.1.1 Goal Definition and Scope

When performing an LCA research, it is critical to establish its goal and boundaries clearly. The purpose defines the goals and target audience, whereas the functional unit serves as a benchmark for comparison, such as per unit of packing material or product. System boundaries, such as cradle-to-gate or cradle-to-grave, define the scope, which includes multiple life cycle stages from raw material extraction to disposal. Finally, specific environmental impact categories, such as global warming potential and resource depletion, are chosen to evaluate the most important environmental implications. This organized technique assures that the outcomes are meaningful, accurate, and comparative [67].

10.4.1.2 Inventory Analysis

To collect data for an LCA, quantify the inputs and outputs at each stage of the packaging material's life cycle, from raw material extraction to production, distribution, use, and disposal. Inputs include raw materials, energy, and water, while outputs include emissions into the air, water, and land, as well as solid waste. This data collection necessitates extensive measurement or calculation of resource consumption, energy use, emissions, and trash creation at each stage, resulting in a full assessment of the environmental impacts associated with the packaging material's entire life cycle [14].

10.4.1.3 Impact Assessment

The environmental repercussions of inventory data in an LCA are evaluated using known impact assessment methodology and characterization models. This method converts basic data on resource consumption and emissions into measurable environmental impacts. Common effect categories include greenhouse gas emissions (measured by global warming potential), energy use, water consumption, air, and water contamination (such as acidification and eutrophication), land use, and toxicity (both human and ecological). By categorizing and quantifying these impacts, the evaluation provides a thorough understanding of the environmental implications of each stage of the product's life cycle, allowing for more informed decision-making [82].

10.4.1.4 Interpretation

Interpreting the impact assessment results in an LCA entails identifying major environmental hotspots, trade-offs, and possibilities for improvement across the active packaging material's lifecycle. This step involves determining which stages or procedures influence the environment most, comparing the performance of various packaging materials or situations, and identifying potential environmental benefits. It also entails examining uncertainties and limitations in the data and methodology to guarantee that the results are sound and reliable. This comprehensive interpretation directs decision-making toward more sustainable packaging options [17].

10.4.1.5 Sensitivity Analysis

Sensitivity assessments in LCAs involve determining how changes in key factors or assumptions affect the overall results. This includes modifying variables such as material composition, manufacturing technologies, and end-of-life situations to evaluate how these affect environmental results. Sensitivity analyses assist in discovering which factors have the greatest influence on the results, highlighting crucial areas where assumptions or data uncertainties can alter the overall assessment. This technique improves the robustness of the LCA by offering insights on the reliability and stability of the results under various scenarios, allowing for more resilient decision-making [10].

10.4.1.6 Improvement Strategies

Strategies for reducing the environmental impact of active packaging materials can be created based on LCA insights. These strategies could include optimizing material selection for more sustainable or less resource-intensive materials, improving manufacturing processes to increase efficiency and reduce emissions, increasing packaging recyclability to reduce waste, and promoting alternative end-of-life options such as composting or reusing materials. By focusing on the major areas identified in the LCA, these strategies seek to reduce the most important environmental consequences while improving the packaging material's overall sustainability [51]. This may involve collaboration with suppliers, manufacturers, distributors, and other stakeholders in the packaging supply chain.

10.4.2 Sustainability Considerations

In the world of active packaging materials, sustainability is critical. Ensuring optimal resource utilization across the full lifecycle of the packaging material is critical, including resource extraction, manufacturing processes, and disposal techniques [39]. Emphasizing renewable and recyclable materials coincides with circular economy concepts while also reducing dependency on scarce resources. Furthermore, promoting biodegradability and compostability promotes environmentally acceptable disposal techniques, reducing the pressure on landfills and ecosystems [18]. Comprehensive life cycle evaluations are critical in evaluating environmental consequences from manufacturing to end-of-life scenarios, guiding sustainable decision-making. Using active packaging technology to lengthen shelf life and reduce food waste strengthens sustainability efforts, improving product quality while lowering environmental impact [52]. Regulatory compliance and standard adherence promote agreement with recognized sustainability benchmarks, which fosters trust among consumers and stakeholders. Collaboration throughout the packaging value chain, including involvement with suppliers, manufacturers, retailers, and waste management organizations, fosters innovation in sustainable packaging solutions [9]. Social and ethical factors, such as labor standards and human rights, highlight the broader implications of sustainable packaging techniques, emphasizing the importance of equitable outcomes. By incorporating these sustainability factors into active packaging material design and implementation, organizations may minimize environmental impact, improve resource efficiency, and contribute to a more sustainable future [12].

10.4.3 Eco-Friendly Alternatives

Exploring eco-friendly alternatives to conventional packaging materials is essential for mitigating environmental impact. Some promising options include.

10.4.3.1 Biodegradable Plastics

Bioplastics, produced from renewable resources such as corn starch, sugarcane, and cellulose, present a promising biodegradable alternative to conventional petroleum-based plastics. Unlike traditional plastics that can persist in the environment for hundreds of years, bioplastics are designed to decompose more rapidly under natural conditions, thereby mitigating the accumulation of waste and reducing pollution. This biodegradability stems from their organic origins, allowing microorganisms to break them down into harmless by-products like water, carbon dioxide, and compost. As a result, bioplastics not only help to lower the carbon footprint associated with plastic production but also offer an eco-friendly solution to the growing global plastic pollution crisis. Their adoption can significantly contribute to more sustainable waste management practices and a healthier environment [10].

10.4.3.2 Compostable Packaging

Compostable packaging materials, such as films and containers made from plant-based polymers like polylactic acid (PLA), offer a sustainable solution by integrating into a closed-loop system. These materials can be industrially composted along with organic waste, meaning they break down in controlled composting environments that provide the right conditions of heat, moisture, and microbial activity. This process transforms the compostable packaging into nutrient-rich compost, which can be used to enhance soil health and fertility. By diverting waste from landfills and converting it into valuable compost, this system reduces the volume of waste, supports sustainable agriculture, and reduces the environmental impact of packaging waste [5].

10.4.3.3 Recycled Content

Using packaging materials that incorporate recycled content significantly lessens the demand for new raw materials and reduces the energy consumption and greenhouse gas emissions associated with manufacturing. Products such as recycled paper, cardboard, and plastics are crucial in promoting a circular economy, where materials are continually repurposed and reused rather than discarded. By extending the lifespan of these materials, we can minimize waste, conserve natural resources, and decrease the environmental footprint of packaging production [46].

10.4.3.4 Bio-Based Materials

Bio-based packaging materials derived from renewable sources such as bamboo, hemp, and algae provide eco-friendly alternatives to standard packaging. These materials typically have lower environmental footprints and may be biodegradable or recyclable, helping to promote sustainable packaging practices. These materials are made from renewable resources, so they may be replaced over time. Bamboo, for example, grows quickly, takes little water, and contains no chemicals, making it highly sustainable. Hemp is another rapidly growing plant that may be grown with little environmental impact.

Algae-based materials utilize microorganisms that can be cultivated using sunlight and CO_2, making them renewable and environmentally friendly [5, 58, 65].

10.4.3.5 Mushroom Packaging

Mycelium packaging, also known as mushroom-based packaging, leverages the natural root structure of mushrooms, called mycelium, to create sustainable and eco-friendly packaging solutions. Unlike traditional materials such as Styrofoam, mycelium packaging is fully biodegradable and compostable, breaking down naturally without leaving harmful residues. This innovative material not only reduces waste and environmental pollution but also offers a renewable alternative that aligns with the growing demand for greener, more sustainable packaging options [26].

10.4.3.6 Edible Packaging

Edible packaging materials made from natural substances such as seaweed, starches, or fruit skins provide innovative options for reducing packaging waste. These materials are not only biodegradable but also edible, providing an environmentally beneficial alternative to traditional packaging. They help retain freshness and reduce environmental impact by adding an extra layer of protection to food products. This technique is consistent with sustainable practices and meets the growing demand for ecologically responsible packaging solutions [40].

10.4.3.7 Reusable Packaging

Implementing reusable packaging technologies, such as refillable containers and durable tote bags, contributes greatly to a circular economy by reducing single-use packaging waste, preserving resources, and lowering overall waste creation. These systems enhance resource efficiency by using fewer resources and less energy to produce while also lowering environmental pollutants and greenhouse gas emissions. Encouraging consumers to return and refill packaging promotes sustainable behaviors, raises knowledge of environmental implications, and supports new business models based on sustainability. Overall, reusable packaging technologies help to ensure a more sustainable, resilient, and resource-efficient future [33].

10.5 Health and Safety Considerations

The interactions between food and packaging can negatively affect quality and safety. When foods are stored in polymer-based materials, changes in flavor can occur due to aroma absorption and the transfer of unwanted flavors from the packaging to the food. To address these issues, it's essential to carefully choose packaging materials and develop active or intelligent packaging solutions. These innovations aim to interact intentionally with the food to either enhance its quality and safety directly or prevent undesirable changes, thereby optimizing product quality, safety, and shelf life. When discussing health and safety considerations for bioactive compounds in ACP, it's important to address several key points.

10.5.1 Toxicity

Evaluating the safety of bioactive chemicals in packaging materials is critical to protecting consumer health and well-being. To detect possible dangers related with their migration into food, comprehensive toxicity tests are performed, including short-term (acute) and long-term (chronic) exposure examinations. Trace metals, which are commonly discovered in packing materials or as contaminants in food, have prompted serious concerns about their potential health effects. Heavy metal toxicity symptoms, whether acute or chronic, can range from moderate to severe, and may include dizziness, nausea, gastrointestinal disorders, sleep abnormalities, and reproductive problems. Furthermore, prolonged trace metal exposure has been associated to major health problems such as cardiovascular disease, poor growth and development, neurological abnormalities, weakened immunological function, reproductive concerns, higher rates of spontaneous abortions, and higher infant mortality rates. Thus, comprehensive assessment and control of bioactive chemicals and trace metals in food packaging are vital to protect public health and prevent undesirable health impacts among consumers [90].

Table 10.2 Compilation of typical food simulants employed in migration testing

Solvent used for migration testing	Simulant category
Distilled water	Simulant A
Aqueous acetic acid (3% w/v)	Simulant B
Aqueous ethanol (15% v/v)	Simulant C
Sunflower oil, rectified olive oil	Simulant D

10.5.1.1 Migration

Understanding the migratory behavior of bioactive chemicals is critical for determining their potential transfer from packaging materials to food. Migration testing should occur under various situations to emulate real-world scenarios. The nature and composition of the food heavily influence migration judgments. Foods with extra fats have been shown to have increased levels of migration. Studies have used several food simulants to investigate the impact of food kinds on migration [77]. These studies seek to understand the mass transfer of substances between packaging materials and food, frequently using solubility factors to estimate the extent of migration in real-time food production settings. Various European and United authorities prescribe specific food simulants for such investigations (Table 10.2).

10.5.1.2 Regulatory Compliance

Compliance with regulatory requirements and recommendations is critical to ensuring the safety of active and intelligent packaging (ACP) materials containing bioactive substances. Regulatory bodies such as the FDA (Food and Drug Administration) in the United States and the EFSA (European Food Safety Authority) in the European Union impose stringent safety standards to safeguard consumers from potential health concerns linked with food contact materials. These standards often include precise criteria for using bioactive substances in packaging, such as maximum permissible migration levels, acceptable materials, and adequate safety testing techniques. Compliance with these laws guarantees that ACP materials are thoroughly examined for potential dangers and fulfil specified safety standards before being released to the market [62].

10.5.1.3 Allergenicity

Assessing the allergenic potential of bioactive compounds, particularly those obtained from natural sources, is critical for protecting consumer health and avoiding harmful responses in people who have allergies or sensitivities to specific substances. This examination includes extensive testing to detect any allergies in the bioactive chemicals or their source materials. Understanding the allergenic characteristics of these chemicals allows manufacturers to make informed decisions about product formulation, labeling, and usage guidelines. Clear and precise labeling assists consumers in identifying products containing potential allergens, allowing them to make safe choices while avoiding allergens that may cause allergic responses. Furthermore, regulatory bodies frequently impose labeling requirements for allergenic compounds, assuring consumer safety [6].

10.5.1.4 Environmental Impact

It is critical to consider the ecological ramifications of bioactive compounds, such as their ability to accumulate in the environment and disrupt ecosystems. These compounds can persist in soil, water, and air after being discharged into the environment through various pathways, including industrial waste, consumer goods, and agricultural runoff [63]. This tenacity may cause organisms to bioaccumulate, harming ecosystems and species. Furthermore, some bioactive compounds may have unanticipated ecological implications that disrupt ecosystem equilibrium or alter organisms' behavior [37]. Therefore, a thorough examination of bioactive compounds' environmental implications is required to minimize any potential negative consequences and guarantee the sustainable use of these substances.

10.6 Conclusion

Including bioactive compounds in active food packaging (ACP) successfully improves food quality, safety, and shelf life. Packaging materials incorporating diverse bioactive chemicals and using sophisticated technologies such as nanotechnology and encapsulation can efficiently suppress microbial growth, minimize oxidative degradation, and alleviate other variables contributing to food deterioration. These improvements help to keep food fresh and nutritious for longer periods. However, health and safety must be prioritized when developing and implementing these new technologies. Comprehensive allergenicity tests are required to identify potential allergens that could cause unpleasant reactions in sensitive individuals. Toxicity tests are required to guarantee that the bioactive substances and materials utilized in ACP do not cause health concerns to users, both in the short and long run. Furthermore, environmental impact evaluations are critical to ensuring that the materials and technologies utilized in ACP do not harm the environment, thereby fostering sustainability. Migration issues must also be carefully considered to avoid dangerous compounds migrating from packaging materials to food products. This entails extensive testing to verify that any migrating contaminants remain within the safe limits set by regulatory agencies. Balancing technological innovation with health and safety considerations guarantees that ACP increases food shelf life and quality, protects consumer health, and promotes environmental sustainability. ACP can achieve its objectives while adhering to the highest safety and environmental requirements through thorough assessment and regulation.

Declaration The authors declare that they have no known competing financial interests or personal relationships that could have appeared to influence the work reported in this chapter. Generative AI and AI-assisted technologies have been used in the writing process to improve the readability and language of the work only. After using this tool, the authors reviewed and edited the content as needed and take full responsibility for the publication's content.

References

1. Acharya, S., Gunjal, M., Birwal, P.: Role of nanomaterials in edible packaging systems. In: Advances in Sustainable Food Packaging Technology, pp. 77–96. Apple Academic Press (2024)
2. Adeyemi, J.O., Fawole, O.A.: Metal-based nanoparticles in food packaging and coating technologies: a review. Biomolecules **13**, 1092 (2023)
3. Arroyo, B.J., Santos, A.P., de Melo, E. de A., Campos, A., Lins, L., Boyano-Orozco, L.C.: Bioactive compounds and their potential use as ingredients for food and its application in food packaging. In: Bioactive Compounds, pp. 143–156. Elsevier (2019)
4. Aryee, A.N., Ofori, K.F., Otchere, E.K., Dare, K.O., Akanbi, T.O.: Food-derived bioactive components and health claimsthe role of regulatory agencies. In: Plant Food Phytochemicals and Bioactive Compounds in Nutrition and Health, pp. 498–529. CRC Press (2024)
5. Asgher, M., Qamar, S.A., Bilal, M., Iqbal, H.M.: Bio-based active food packaging materials: sustainable alternative to conventional petrochemical-based packaging materials. Food Res. Int. **137**, 109625 (2020)
6. Arvanitoyannis, I.S., Kotsanopoulos, K.V.: Migration phenomenon in food packaging. Food-package interactions, mechanisms, types of migrants, testing and relative legislation—a review. Food Bioproc. Tech. **7**, 21–36 (2014)
7. Barkhordari, M., Mohseni-Shahri, F., Moeinpour, F.: Bio-based gelatin–TiO_2–purple basil extract nanocomposite films for monitoring fish freshness. J. Food Meas. Charact. 1–12 (2024)
8. Barone, G.D., Cernava, T., Ullmann, J., Liu, J., Lio, E., Germann, A.T., Nakielski, A., Russo, D.A., Chavkin, T., Knufmann, K.: Recent developments in the production and utilization of photosynthetic microorganisms for food applications. Heliyon (2023)
9. Betoret, E., Calabuig-Jiménez, L., Barrera, C., Dalla Rosa, M.: Sustainable Drying Technologies for the Development of Functional Foods and Preservation of Bioactive Compounds. IntechOpen (2016)
10. Bhargava, N., Sharanagat, V.S., Mor, R.S., Kumar, K.: Active and intelligent biodegradable packaging films using food and food waste-derived bioactive compounds: a review. Trends Food Sci. Technol. **105**, 385–401 (2020)
11. Bhatlawande, A.R., Ghatge, P.U., Shinde, G.U., Anushree, R., Patil, S.D.: Unlocking the future of smart food packaging: biosensors, IoT, and nano materials. Food Sci. Biotechnol. **33**, 1075–1091 (2024)
12. Borah, M.S., Tiwari, A., Sridhar, K., Narsaiah, K., Nayak, P.K., Stephen Inbaraj, B.: Recent trends in valorization of food industry waste and by-products: encapsulation and in vitro release of bioactive compounds. Foods **12**, 3823 (2023)
13. Carnaval, L.S., Jaiswal, A.K., Jaiswal, S.: Agro-food waste valorization for sustainable bio-based packaging. J. Compos. Sci. **8**, 41 (2024)
14. Castro-Muñoz, R., Conidi, C., Cassano, A.: Membrane-based technologies for meeting the recovery of biologically active compounds from foods and their by-products. Crit. Rev. Food Sci. Nutr. **59**, 2927–2948 (2019)

15. Chausali, N., Saxena, J., Prasad, R.: Recent trends in nanotechnology applications of bio-based packaging. J. Agric. Food Res. **7**, 100257 (2022)
16. Cruz, R.M.: Food Packaging (2020)
17. Deshmukh, R.K., Gaikwad, K.K.: Natural antimicrobial and antioxidant compounds for active food packaging applications. Biomass Convers. Biorefinery **14**, 4419–4440 (2024)
18. Dos Santos, M.S., Wancura, J.H., Oro, C.E., Dallago, R.M., Tres, M.V.: Opportunities and challenges of plant bioactive compounds for food and agricultural-related areas. Phyton **91**, 1105 (2022)
19. Duda-Chodak, A., Tarko, T., Petka-Poniatowska, K.: Antimicrobial compounds in food packaging. Int. J. Mol. Sci. **24**, 2457 (2023)
20. Falah, F., Zulfiana, D., Septiano, M., Nawawi, D.S., Sari, F.P., Fatriasari, W., Solihat, N.N.: Antimicrobial activity of food packaging biofilms derived from lignin-starch-poly (lactic acid). In: Presented at the AIP Conference Proceedings. AIP Publishing (2024)
21. Fan, X., Yang, T., Li, Y., Liu, M., Wang, H., Chang, L.: Zein–natamycin functional films with ethylene adsorption and fungi stasis for active food packaging applications. J. Coat. Technol. Res. 1–9 (2024)
22. Farhan, A.: Active packaging applications for dairy-based hygroscopic foods. In: Active Packaging for Various Food Applications, pp. 101–118. CRC Press (2021)
23. Gómez-Gaete, C., Avendaño-Godoy, J., Escobar-Avello, D., Campos-Requena, V.H., Rogel-Castillo, C., Estevinho, L.M., Martorell, M., Sharifi-Rad, J., Calina, D.: Revolutionizing fruit juice: exploring encapsulation techniques for bioactive compounds and their impact on nutrition, flavour and shelf life. Food Prod. Process. Nutr. **6**, 8 (2024)
24. Grevtseva, N., Gorodyska, O., Brykova, T., Gubsky, S.: The use of wine waste as a source of biologically active substances in confectionery technologies. In: Bioconversion of Wastes to Value-Added Products, pp. 69–111. CRC Press (2023)
25. Gupta, P.: Role of oxygen absorbers in food as packaging material, their characterization and applications. J. Food Sci. Technol. 1–11 (2023)
26. Hanula, M., Pogorzelska-Nowicka, E., Pogorzelski, G., Szpicer, A., Wojtasik-Kalinowska, I., Wierzbicka, A., Półtorak, A.: Active packaging of button mushrooms with zeolite and açai extract as an innovative method of extending its shelf life. Agriculture **11**, 653 (2021)
27. Hashem, A.H., El-Naggar, M.E., Abdelaziz, A.M., Abdelbary, S., Hassan, Y.R., Hasanin, M.S.: Bio-based antimicrobial food packaging films based on hydroxypropyl starch/polyvinyl alcohol loaded with the biosynthesized zinc oxide nanoparticles. Int. J. Biol. Macromol. **249**, 126011 (2023)
28. Jafarzadeh, S., Jafari, S.M., Salehabadi, A., Nafchi, A.M., Kumar, U.S.U., Khalil, H.A.: Biodegradable green packaging with antimicrobial functions based on the bioactive compounds from tropical plants and their by-products. Trends Food Sci. Technol. **100**, 262–277 (2020)
29. Jia, W., Wu, X., Kang, X.: Integrated the embedding delivery system and targeted oxygen scavenger enhances free radical scavenging capacity. Food Chem. X **17**, 100558 (2023)
30. Khezerlou, A., Tavassoli, M., Alizadeh Sani, M., Mohammadi, K., Ehsani, A., McClements, D.J.: Application of nanotechnology to improve the performance of biodegradable biopolymer-based packaging materials. Polymers **13**, 4399 (2021)
31. Kuai, L., Liu, F., Chiou, B.-S., Avena-Bustillos, R.J., McHugh, T.H., Zhong, F.: Controlled release of antioxidants from active food packaging: a review. Food Hydrocoll. **120**, 106992 (2021)
32. Kuroda, M.: Kokumi substance as an enhancer of Koku: its definition. In: Kokumi Substance as an Enhancer of Koku: Biochemistry, Physiology, and Food Science, pp. 15–22. Springer (2024)
33. Leichtweis, M.G., Oliveira, M.B.P., Ferreira, I.C., Pereira, C., Barros, L.: Sustainable recovery of preservative and bioactive compounds from food industry bioresidues. Antioxidants **10**, 1827 (2021)
34. Li, C., Liu, J., Li, W., Liu, Z., Yang, X., Liang, B., Huang, Z., Qiu, X., Li, X., Huang, K.: Biobased intelligent food-packaging materials with sustained-release antibacterial and real-time monitoring ability. ACS Appl. Mater. Interfaces **15**, 37966–37975 (2023)

35. Liu, M., Zhang, X., Wei, A., Li, H., Zhang, H., Zheng, L., Xia, N., Wang, J.: Protein-based active films: raw materials, functions, and food applications. Compr. Rev. Food Sci. Food Saf. **23**, e13302 (2024)
36. Luo, Y., Wang, Q., Zhang, Y.: Biopolymer-based nanotechnology approaches to deliver bioactive compounds for food applications: a perspective on the past, present, and future. J. Agric. Food Chem. **68**, 12993–13000 (2020)
37. Lau, O., Wong, S.: Contamination in food from packaging material. J. Chromatogr. A **882**, 255–270 (2000)
38. Manzoor, A., Yousuf, B., Pandith, J.A., Ahmad, S.: Plant-derived active substances incorporated as antioxidant, antibacterial or antifungal components in coatings/films for food packaging applications. Food Biosci. 102717 (2023)
39. Messinese, E., Pitirollo, O., Grimaldi, M., Milanese, D., Sciancalepore, C., Cavazza, A.: By-products as sustainable source of bioactive compounds for potential application in the field of food and new materials for packaging development. Food Bioprocess Technol. **17**, 606–627 (2024)
40. Moeini, A., Pedram, P., Fattahi, E., Cerruti, P., Santagata, G.: Edible polymers and secondary bioactive compounds for food packaging applications: antimicrobial, mechanical, and gas barrier properties. Polymers **14**, 2395 (2022)
41. Muñoz-Tebar, N., Pérez-Álvarez, J.A., Fernández-López, J., Viuda-Martos, M.: Chitosan edible films and coatings with added bioactive compounds: antibacterial and antioxidant properties and their application to food products: a review. Polymers **15**, 396 (2023)
42. Mushtaq, Z., Islam, M., Saeed, S., Bellitürk, K., Ilić, P., Parnian, A.: Nanoclay Packaging materials as biodegradable. Innov. Eng. Food Sci. 278–298 (2024)
43. Nabi, L., Nourani, M.: Biodegradable form stable phase change material for cold storage packaging of meat. Meat Sci. **201**, 109188 (2023)
44. Nayi, P., Kumar, N., Chen, H.-H.: Bio-Based Materials for Food Packaging. Biopolymer-Based Films and Coatings. Trends Chall (2023)
45. Nogueira, G.F., de Oliveira, R.A., Velasco, J.I., Fakhouri, F.M.: Methods of incorporating plant-derived bioactive compounds into films made with agro-based polymers for application as food packaging: a brief review. Polymers **12**, 2518 (2020)
46. Nour, V., Panaite, T.D., Ropota, M., Turcu, R., Trandafir, I., Corbu, A.R.: Nutritional and bioactive compounds in dried tomato processing waste. CyTA-J. Food **16**, 222–229 (2018)
47. Olawore, O., Ogunmola, M., Desai, S.: Engineered nanomaterial coatings for food packaging: design, manufacturing, regulatory, and sustainability implications. Micromachines **15**, 245 (2024)
48. Oluba, O.M., Ojeaburu, S.I., Bayo-Olorunmeke, O.A., Erifeta, G., Josiah, S.J.: Quality attributes and shelf-life of freshly cut beef coated with waste feather keratin-ginger starch composite enriched with avocado peel polyphenolic-rich extract. Food Sci. Preserv. **31**, 1–14 (2024)
49. Onyeaka, H., Passaretti, P., Miri, T., Al-Sharify, Z.T.: The safety of nanomaterials in food production and packaging. Curr. Res. Food Sci. **5**, 763–774 (2022)
50. Pan, Z., Zhong, W., Xu, J., Li, D., Lin, J., Wu, W., Pang, J., Wu, C.: Effects of oregano essential oil Pickering emulsion and ZnO nanoparticles on the properties and antibacterial activity of konjac glucomannan/carboxymethyl chitosan nanocomposite films. RSC Adv. **14**, 6548–6556 (2024)
51. Pateiro, M., Gómez, B., Munekata, P.E., Barba, F.J., Putnik, P., Kovačević, D.B., Lorenzo, J.M.: Nanoencapsulation of promising bioactive compounds to improve their absorption, stability, functionality and the appearance of the final food products. Molecules **26**, 1547 (2021)
52. Patil, B.S., Jayaprakasha, G.K., Chidambara Murthy, K., Vikram, A.: Bioactive compounds: historical perspectives, opportunities, and challenges. J. Agric. Food Chem. **57**, 8142–8160 (2009)
53. Patil, V., Shams, R., Dash, K.K.: Techno-functional characteristics, and potential applications of edible coatings: a comprehensive review. J. Agric. Food Res. 100886 (2023)

54. Perera, K.Y., Hopkins, M., Jaiswal, A.K., Jaiswal, S.: Nanoclays-containing bio-based packaging materials: properties, applications, safety, and regulatory issues. J. Nanostructure Chem. **14**, 71–93 (2024)
55. Pillai, B.P., Balasubramaniam, B., Gupta, R.K., Tyagi, A.: Bio-based materials for antimicrobial films in food applications: beyond the COVID-19 pandemic era. Oxf. Open Mater. Sci. **3**, (2024)
56. Pirsa, S.: Cellulose-based cartons: production methods, modification, and smart/active packaging. Cellulose 1–25 2024
57. Ponnusamy, A., SR, R.R., Rajan, R., Ashraf, F.: Chitosan silver nanoparticle inspired seaweed (Gracilaria crassa) biodegradable films for seafood packaging. Algal. Res. **78**, 103429 (2024)
58. Rahayu, F., Wani, A.K., Murianingrum, M., Marjani, M., Suhara, C., Hariyono, B.: Studies on dew retting process of kenaf by formulation of indigenous consortium bacteria. In: Presented at the AIP Conference Proceedings. AIP Publishing 2022
59. Ribeiro, I.S., Maciel, G.M., Bortolini, D.G., de Fernandes, I.A.A., Volpato, W.M., Pedro, A.C., Rubio, F.T.V., Haminiuk, C.W.I.: Sustainable innovations in edible films and coatings: an overview. Trends Food Sci. Technol. 104272 (2023)
60. Rojas, A., Velásquez, E., Patiño Vidal, C., Guarda, A., Galotto, M.J., López de Dicastillo, C.: Active PLA packaging films: effect of processing and the addition of natural antimicrobials and antioxidants on physical properties, release kinetics, and compostability. Antioxidants **10**, 1976 (2021)
61. Rostamabadi, H., Falsafi, S.R., Boostani, S., Katouzian, I., Rezaei, A., Assadpour, E., Jafari, S.M.: Design and formulation of nano/micro-encapsulated natural bioactive compounds for food applications. In: Application of Nano/Microencapsulated Ingredients in Food Products, pp. 1–41. Elsevier (2021)
62. Robertson, G.L.: Safety and legislative aspects of packaging. In: Food Packaging Principles and Practice, vol. 3, pp. 473–502. CRC Press, US (2006)
63. Raloff, J.: New concerns about phthalates: ingredients of common plastics; may harm boys as they develop. Sci. News **10**, 152–158 (2000)
64. Safakas, K., Giotopoulou, I., Giannakopoulou, A., Katerinopoulou, K., Lainioti, G.C., Stamatis, H., Barkoula, N.-M., Ladavos, A.: Designing antioxidant and antimicrobial polyethylene films with bioactive compounds/clay nanohybrids for potential packaging applications. Molecules **28**, 2945 (2023)
65. Selwal, N., Rahayu, F., Herwati, A., Latifah, E., Suhara, C., Suastika, I.B.K., Mahayu, W.M., Wani, A.K.: Enhancing secondary metabolite production in plants: exploring traditional and modern strategies. J. Agric. Food Res. 100702 (2023)
66. Shahid, S., Leghari, A.A., Farid, M.S., Saeed, M., Anwar, S., Anjum, R., Saeed, N., Abbas, Z.: Role of active food packaging developed from microencapsulated bioactive ingredients in quality and shelf life enhancement: a review. J. Am. Sci. **17**, 12–28 (2021)
67. Shashirekha, M., Mallikarjuna, S., Rajarathnam, S.: Status of bioactive compounds in foods, with focus on fruits and vegetables. Crit. Rev. Food Sci. Nutr. **55**, 1324–1339 (2015)
68. Shetta, A., Ali, I.H., Elshishiny, F., Mamdouh, W.: Different approaches for the inclusion of bioactive compounds in packaging systems. In: Releasing Systems in Active Food Packaging: Preparation and Applications, pp. 151–185. Springer (2022)
69. Siddiqui, S.A., Singh, S., Bahmid, N.A., Mehany, T., Shyu, D.J., Assadpour, E., Malekjani, N., Castro-Muñoz, R., Jafari, S.M.: Release of encapsulated bioactive compounds from active packaging/coating materials and its modeling: a systematic review. Colloids Interfaces **7**, 25 (2023)
70. Silvia, M.G., Leena, M.M., Bharathi, V.S., Moses, J., Anandharamakrishnan, C.: Minimal Processing and Edible Coating, in: Emerging Technologies for the Food Industry, pp. 85–129. Apple Academic Press (2024)
71. Singaram, A.J.V., Guruchandran, S., Ganesan, N.D.: Review on functionalized pectin films for active food packaging. Packag. Technol. Sci. (2024)
72. Singh, A.K., Kim, J.Y., Lee, Y.S.: Phenolic compounds in active packaging and edible films/coatings: natural bioactive molecules and novel packaging ingredients. Molecules **27**, 7513 (2022)

73. Singh, S., Angurana, R., Kaur, H., Katoch, V., Dhanjal, D.S., Aguilar-Marcelino, L., Ramamurthy, P.C., Singh, J.: Nanoencapsulation of fertilizers: delivery and applications. In: Nanofertilizer Delivery, Effects and Application Methods, pp. 97–110. Elsevier (2024)
74. Tavassoli, M., Khezerlou, A., Bangar, S.P., Bakhshizadeh, M., Haghi, P.B., Moghaddam, T.N., Ehsani, A.: Functionality developments of pickering emulsion in food packaging: principles, applications, and future perspectives. Trends Food Sci. Technol. **132**, 171–187 (2023)
75. Tkaczewska, J., Jamróz, E., Zając, M., Guzik, P., Gedif, H.D., Turek, K., Kopeć, M.: Antioxidant edible double-layered film based on waste from soybean production as a vegan active packaging for perishable food products. Food Chem. **400**, 134009 (2023)
76. Trajkovska Petkoska, A., Daniloski, D., Kumar, N., Pratibha, Broach, A.T.: Active edible packaging: a sustainable way to deliver functional bioactive compounds and nutraceuticals. Sustain. Packag. 225–264 (2021)
77. Triantafyllou, V.I., Akrida-Demertzi, K., Demertzi, P.G.: A study on the migration of organic pollutants from recycled paperboard packaging materials to solid food matrices. Food Chem. **101**(4), 1759–2176 (2007)
78. Valdés, A., Mellinas, A., Ramos, M., Burgos, N., Jiménez, A., del Garrigós, M.C.: Use of herbs, spices and their bioactive compounds in active food packaging. RSC Adv. **5**, 40324–40335 (2015)
79. Vasile, C., Baican, M.: Lignins as promising renewable biopolymers and bioactive compounds for high-performance materials. Polymers **15**, 3177 (2023)
80. Vasile, C., Baican, M.: Progresses in food packaging, food quality, and safety—controlled-release antioxidant and/or antimicrobial packaging. Molecules **26**, 1263 (2021)
81. Viacava, G.E., Marcovich, N.E., Ansorena, M.R.: Studies on nanostructure functionalized materials for active and intelligent food packaging. In: Nanostructured Materials for Food Packaging Applications, pp. 477–511. Elsevier (2024)
82. Vilas-Boas, A.A., Pintado, M., Oliveira, A.L.: Natural bioactive compounds from food waste: toxicity and safety concerns. Foods **10**, 1564 (2021)
83. Wani, A.K., Ahmad, S., Américo-Pinheiro, J.H.P., Tizro, N., Singh, R.: Building the taxonomic profile of the Riniaie Marwah hot spring of Kishtwar in Jammu and Kashmir: the first high-throughput sequencing-based metagenome study. Iran. J. Microbiol. **15**, 723–733 (2023)
84. Wani, A.K., Akhtar, N., Rahayu, F., Suhara, C., Anjli, A., Chopra, C., Singh, R., Prakash, A., El Messaoudi, N., Fernandes, C.D.: Eco-friendly and safe alternatives for the valorization of shrimp farming waste. Environ. Sci. Pollut. Res. 1–30 (2023b)
85. Wani, A.K., Akhtar, N., Sher, F., Navarrete, A.A., Américo-Pinheiro, J.H.P.: Microbial adaptation to different environmental conditions: molecular perspective of evolved genetic and cellular systems. Arch. Microbiol. **204**, 144 (2022)
86. Wani, A.K., Chopra, C., Singh, R., Ahmad, S., Américo-Pinheiro, J.H.P.: Mining microbial tapestry using high-throughput sequencing and in silico analysis of Trehalose synthase (TreS) derived from hot spring metagenome. Biocatal. Agric. Biotechnol. **52**, 102829 (2023). https://doi.org/10.1016/j.bcab.2023.102829
87. Weisany, W., Yousefi, S., Soufiani, S.P., Pashang, D., McClements, D.J., Ghasemlou, M.: Mesoporous silica nanoparticles: A versatile platform for encapsulation and delivery of essential oils for food applications. Adv. Colloid Interface Sci. 103116 (2024)
88. Yao, Q., Huang, F., Lu, Y., Huang, J., Ali, M., Jia, X.-Z., Zeng, X.-A., Huang, Y.: Polysaccharide-based food packaging and intelligent packaging applications: a comprehensive review. Trends Food Sci. Technol. 104390 (2024)
89. Yue, S., Zhang, T., Wang, S., Han, D., Huang, S., Xiao, M., Meng, Y.: Recent progress of biodegradable polymer package materials: nanotechnology improving both oxygen and water vapor barrier performance. Nanomaterials **14**, 338 (2024)
90. Yüzbaşi, N., Sezgin, E., Yıldırım, M., Yıldırım, Z.: Survey of lead, cadmium, iron, copper and zinc in Kaşar cheese. Food Addit. Contam. **20**, 464–469 (2003)

Chapter 11
Smart Sensors for Quality Monitoring in Green Packaging Solutions

Gazi Farhan Ishraque Toki, Anik Das, Rezaul Karim Khan Alave, and Rony Mia

Abstract There has been a lot of research into chemical and biosensor development over the last several decades, and that has led to some fascinating new sensor devices that could be useful in many fields, including food technology. A new kind of environmentally friendly food packaging called "smart" or "intelligent" has emerged by integrating these sensors. Integrating knowledge from several disciplines, these systems draw on food technology, science, chemistry, biochemistry, electronics, and physics. Smart packaging keeps tabs on food quality and safety throughout the supply chain, from farmers to consumers, using chemical or bio-type sensors. Many other kinds of sensors that measure things like time, temperature, carbon dioxide, oxygen, pathogens, freshness, and leakage can be created using this technology. Online assurance of quality and security for consumers, authorities, and food producers is a must, and this technology shows enormous promise for expanding the capabilities of green packaging with sensors that go beyond the existing state of the art in areas such as controlling the food's volume, color, and appearance.

Keywords Green packing · Intelligent packaging · Smart packing · Smart sensors · Quality monitoring · Food quality indicators

---

G. F. I. Toki
Nanotechnology Center, School of Fashion and Textiles, The Hong Kong Polytechnic University, Kowloon 999077, Hong Kong

A. Das
School of Textile Science and Engineering, Tiangong University, Tianjin 300387, People's Republic of China

R. K. K. Alave
Shanghai Frontier Science Research Center for Modern Textiles, College of Textiles, Donghua University, Shanghai 201620, People's Republic of China

R. Mia (✉)
Wilson College of Textiles, North Carolina State University, 1020 Main Campus Dr, Raleigh 27606, NC, United States
e-mail: mroni_mia@yahoo.com

Department of Textile Engineering, National Institute of Textile Engineering and Research, University of Dhaka, Dhaka 1000, Bangladesh

S. ul Islam and M. Shahid (eds.), *Green Materials for Active Food Packaging*, Engineering Materials, https://doi.org/10.1007/978-981-96-0369-5_11

11.1 Introduction

Food packaging technology is quite advanced, drawing on many different branches of science to accomplish many goals that go well beyond the original intent of preventing spoilage and damage to food as it travels from agriculture to the table. Aesthetics play a significant role in motivating consumers to purchase food products, for instance. Increasing the variety and quantity of meals given to consumers is one strategy employed by the food industry to boost profits. Because of this, there is a greater need for foods that can be stored for extended periods without significantly compromising their quality [1]. To ensure that food products maintain their quality throughout their entire lifecycle, from storage to transit to delivery and consumption, a complex system is in place, the most basic part of which is packaging. Preventing or slowing down chemical, biochemical, and microbiological degradation as well as mechanical damage to the payload is the primary goal of packing [2]. Not only does the package show the product's shape, size, color, and consistency, but it also provides information about the product's nutritional value and calorie content [3]. Many now consider plastics made from petroleum to be one of the utmost serious alarms to our surrounding environment. Businesses in this sector are working on solutions, such as new packaging technologies and biodegradable food packaging [4, 5]. Figure 11.1 shows the life cycles of various environmentally friendly packaging materials.

Innovative packaging solutions like smart food packaging can not only make sense, but also record fluctuations in quality, both inside and externally about the product's surroundings, and then communicate this information to the consumer effectively. One famous specimen of smart "interactive" food packaging is active and intelligent packaging (IP). IP includes active components that help enhance the shelf lifespan and preserve product eminence, while intelligent systems use indicators, sensors, and data carriers to convey information about the product's freshness during transit [6, 7]. The capability to perceive deviations in storage conditions (for instance changes in temperature or humidity) and prevent quality degradation is a remarkable feature. Intelligent electronics were integrated with active packaging components like antioxidants, humidity, carbon dioxide (CO_2), antibacterial mediators, emitters, and ethylene, along with oxygen scavengers to create the concept of "smart packaging" [8, 9]. Various devices can be unified into such packaging to regulate the temperature of the food within and provide real-time nutritional data on an electronic screen [10].

There are several sophisticated methods available for detecting harmful bacteria and other pollutants. Some of these devices necessitate the extraction of samples to determine whether the target contaminant is present. Detecting pollutants or pathogens is still a restricted application of these devices due to technological limitations in incorporating them into commodity packaging [11]. The perfect sensor would have these features: Superior selectivity, Extreme sensitivity to variations in bacterial concentrations, Quick response time, Extended shelf life, and Compact and affordable [2, 12]. One kind of sensor is the biosensor, which can detect biological reactions. It combines an electrochemical, optical, or acoustic transducer with a bioreceptor for

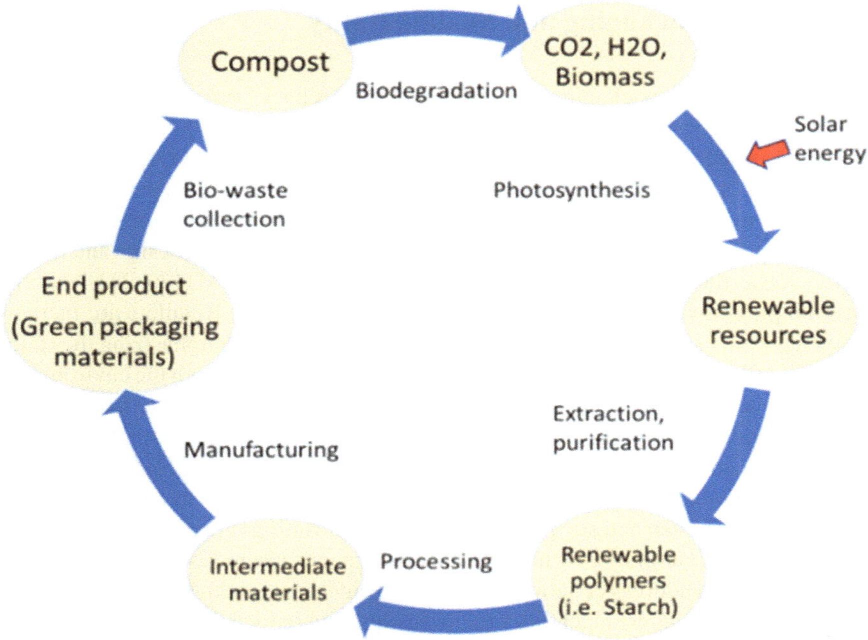

Fig. 11.1 The eco-friendly influence of packaging throughout time. Republished with the permission of Elsevier [5]. Copyright Elsevier, 1998

an analyte—an antigen, microbe, enzyme, nucleic acid, or hormone—to alter biological signals into quantifiable and visible responses [13, 14]. The parameters that need to be analyzed dictate the type of necessary transducer. Among a biosensor's most critical qualities are its specificity, sensitivity, longevity, mobility, and use [15, 16]. Optical sensors that specifically target natural amines in beer have also been developed; these work by utilizing the covalent contact between the tryptamine molecule and its active vinyl groups [17]. One type of biosensor can detect biogenic amines produced by amino acid decarboxylation or the amination and the transformation of ketones and aldehydes. Another type of biosensor can detect xanthine, a byproduct of adenine nucleotide destruction in bodily tissue, thru immobilizing xanthine oxide onto transducer electrodes [13].

This chapter aims to deliver a concise synopsis of the "Smart Sensors for Quality Monitoring in Green Packaging Solutions" from the outset. In this context, first, explain Green Packaging and its Sustainability. Next, the categories of quality monitoring, for instance, O_2, CO_2, Chemical, pH, humidity, time and temperature, and so on, are discussed. Since this chapter focuses on smart sensors for quality monitoring, it will include a quick brief on quality monitoring in packing and a smart sensor overview. Finally, the challenges and prospects of using smart sensors for quality monitoring in green packaging are summarized.

11.2 Green Packaging and Sustainability

The concept of green packaging, characterized by its reliance on renewable resources and sustainable design, is experiencing a surge in popularity. However, skepticism surrounds products labeled as "green" due to doubts about their true ecological integrity [18, 19]. In green packaging design, functionality is just one consideration among many. It seeks to minimize energy consumption and package size while maximizing effectiveness, often aided by tools like computer-aided design and frameworks such as the Sustainable Packaging Design Model (SPkDM) [20, 21]. Material selection plays a vital role in reducing environmental impact. Retailers are increasingly integrating sustainability into their packaging strategies, emphasizing the status of energy-efficient manufacture and the safety of ingredients for both the environment and human health [63]. Studies explore the risks associated with recycled materials in food packaging and investigate alternatives like bio-based and novel materials such as cellulose nanofibrils and bioplastics [22, 23]. Various sustainable materials are being explored, including recyclable options like paper-based packaging and novel materials like chitosan and soy protein isolate films. Printing directly onto packaging surfaces is also emerging as a method to reduce paper usage [24–26]. Sustainable packaging should address economic, social, and environmental concerns. It can lead to cost savings, stronger stakeholder relationships, and reduced energy consumption and waste. However, challenges persist, including the prioritization of aesthetics over sustainability and limitations due to insufficient knowledge or funding [27, 28]. Measuring sustainability through Key Performance Indicators (KPIs) ensures that packaging meets its purpose efficiently and safely while minimizing environmental harm [29].

The escalating consumption of packaged food, increasing annually by 5%, has resulted in a notable surge in packaging waste. Despite packaging offering advantages such as providing information, safeguarding products, and prolonging their shelf life, its environmental repercussions across its lifecycle are alarming [30, 31]. Conventional packaging, often dependent on non-renewable materials like plastics, contributes significantly to environmental issues: the extraction of raw materials for packaging consumes extensive energy and exhausts finite resources [32, 33]. The manufacturing process of packaging emits substantial greenhouse gases, exacerbating climate change. A considerable amount of packaging ends up in landfills, where it takes centuries to degrade, exacerbating landfill overflow. Plastic packaging is a primary culprit, polluting oceans and ecosystems, harming wildlife, and infiltrating the food chain. Some conventional packaging materials contain harmful chemicals that can seep into soil and water, potentially contaminating ecosystems and affecting human health [34, 35]. These adverse effects underscore the urgent necessity for transitioning towards sustainable packaging practices prioritizing renewable materials, waste reduction, and minimizing the environmental impact of the packaging industry [36, 37]. Plastic is extensively utilized due to its versatility and cost-effectiveness. However, its widespread adoption has resulted in; global plastic production has soared, reaching 381 million tonnes in 2015. Most plastic packing is discarded within a year, posing a noteworthy solid leftover topic, mostly when not adequately managed. Consumers often associate plastics with adverse environmental consequences, notably plastic pollution in marine environments. While Life Cycle Assessments (LCA) can quantify the ecological influence of various packing ingredients, disparities in methodology and system boundaries complicate comparisons. International policies such as bans on single-use plastics and initiatives for minimizing plastic waste are imperative. Bioplastics derived from renewable sources offer promise, although constraints exist. Approximately half of bioplastics are non-biodegradable and pose similar environmental risks as conventional plastics. Biodegradable bioplastics often necessitate specific composting conditions, which many consumers are unaware of leading to improper disposal [38].

Sustainable packaging is a crucial aspect of various industries, aiming to minimize environmental impact while ensuring product protection. This concept encompasses three foremost scopes, for instance, the economy, the environment, and human well-being. Ideally, sustainable packaging should contribute to a fairer distribution of resources and maintain a healthy ecosystem [39, 40]. From an ecological perspective, materials like glass are often considered more sustainable than plastic or aluminum cans due to their reusability. However, plastic remains the most extensively utilized packaging material as a result of its lightweight nature, lower rate, and moldability [41]. A shift towards material recycling, reuse, and the development of bio-based as well as ecological materials is necessary to achieve true sustainability. Packaging is vital in product protection, storage, and delivery, but its production and disposal significantly affect the environment. Studies have shown that packaging contributes to greenhouse gas emissions, resource depletion, and pollution. Single-use plastics, for instance, pose a particular threat due to their extensive use and long lifespan in landfills [42]. Many countries and organizations are implementing regulations and initiatives to promote sustainable packaging. The European Union (EU), for example, has restricted the usage of only-usage plastics, while country like Pakistan is progressively shifting towards executing the UN's Sustainable Development Goals, focusing on waste collection, recycling, and infrastructure development [43, 44]. Despite its benefits like extended shelf life, multi-layered packaging presents challenges in recycling due to the complexity of separating its various materials. Regulations are being enforced to limit its use and encourage the adoption of more recyclable alternatives like polyolefins [45, 46]. The concept of green packaging emphasizes the use of eco-friendly materials that minimize waste and pollution. This approach is gaining traction due to growing consumer awareness and regulations imposed by authorities. However, implementing green packaging strategies often requires overcoming obstacles related to cost, technological advancements, and human resource skillsets. Sustainable Supply Chain Management (SSCM) integrates ecological considerations throughout the supply chain, encompassing production, packaging, and logistics. Green supply chain management (GSCM) takes this a step further, directing to lessen the ecological impact of an entire organization's operations [47]. Reusable packaging, particularly plastic packaging, is being increasingly utilized for its efficiency and cost-effectiveness in reverse logistics. Solid waste management (SWM) remains a critical challenge, requiring collaborative efforts from governments, communities, and industries to implement effective solutions [46]. Sustainable packaging is a complex concept that necessitates a holistic approach considering its environmental, economic, and social impacts throughout the supply chain. Continuous research and development in material science, recycling technologies, and consumer education are crucial for achieving a sustainable packaging future [48, 49].

11.3 Quality Monitoring in Packaging

11.3.1 O_2 Detectors

Modified atmosphere packaging (MAP) has emerged as an extensively utilized methodology for prolonging the shelf life of diverse foodstuffs [50]. This technique entails enclosing the product within a meticulously regulated gas adsorption, typically displacing most atmospheric air with a blend of O_2, N_2, and CO_2. Precise monitoring of O_2 concentrations within MAPs is imperative, prompting the development of various sensor technologies (Fig. 11.2) [51]. One variant of the O_2 sensor employs luminescent probe molecules incorporated into a gas-permeable material, forming thin film structures. Despite reacting to O_2 exposure by modulating their photoluminescence, the resultant alterations often evade direct visual detection, necessitating specialized and costly analytical instrumentation [51].

Conversely, colorimetric redox-based indicator sensors offer a more straightforward solution grounded in reversible redox chemistry. These configurations, exemplified by the Ageless Eye, typically include a redox dye (e.g., methylene blue) in addition to a potent reducing agent (for instance, glucose in an alkaline milieu) [52–54]. These sensors change color with even slight fluctuations in O_2 levels, transitioning from purple to pink at O_2 concentrations as low as 0.1% and reverting to purple in the presence of elevated O_2 levels [55]. However, these systems mandate anaerobic storage and handling, as their efficacy rapidly diminishes upon exposure to ambient air [55]. A separate category of oxygen sensors utilizes phosphorescent dyes encapsulated within a permeable polymeric matrix, facilitating interaction with O_2 molecules. Although capable of detecting O_2 depletion within the enclosure, these sensors exhibit constraints akin to their luminescent counterparts [56].

Researchers have endeavored to mitigate the drawbacks of colorimetric systems by devising light-triggered, redox dye-based sensors. These setups incorporate photosensitive dyes necessitating UV or visible light activation for functionality [51]. Nonetheless, a significant drawback of this approach is the potential migration of redox dyes from the water-insoluble polymer film into the food, posing contamination and discoloration risks [57, 58]. To counter this concern, one study have fruitfully employed alginate polymers, which bind to the redox dyestuff, forestalling their release upon exposure to water [59]. MAP technology assumes a pivotal role in

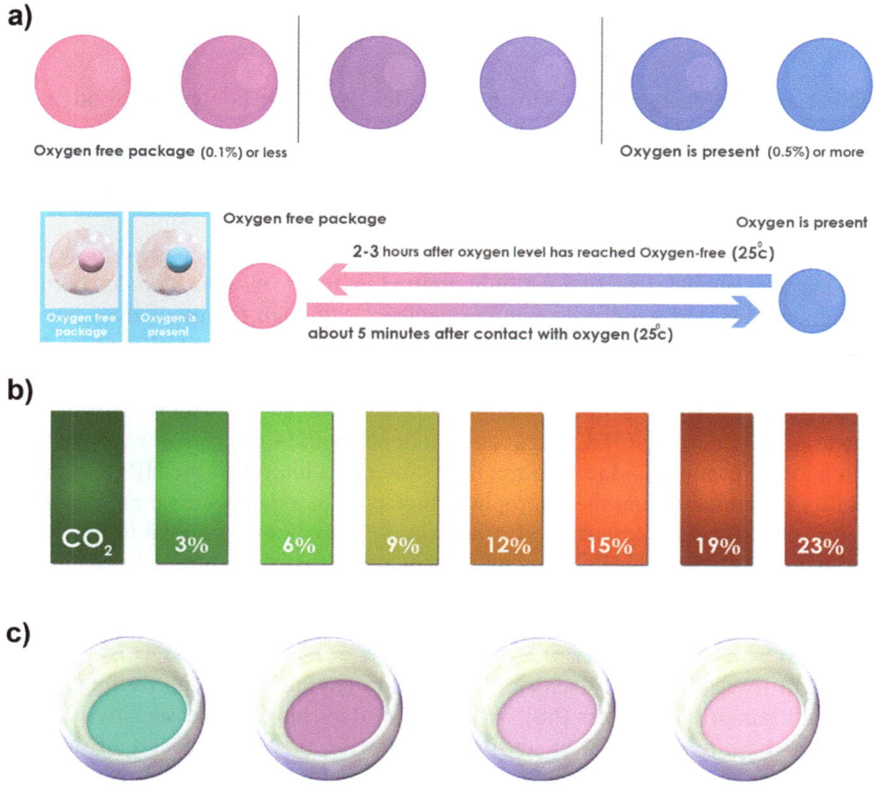

Fig. 11.2 **a** A colorimetric redox dye-based O_2 sensor, **b**)Color changes for various CO_2 levels from 0 to 23%, **c** a colorimetric CO_2 sensor utilizing a food-grade combination composed of an amino acid (l-lysine), a polypeptide (poly l-lysine, EPL), and naturally occurring colors (anthocyanins). Republished with permission from ACS [52]. Copyright © 2019 American Chemical Society

upholding food quality [60, 61]. Various sensor types have been devised for monitoring O_2 levels within MAPs, each harboring distinct advantages and constraints. While luminescent sensors offer heightened sensitivity, their detection necessitates specialized apparatus [51, 62–64]. Colorimetric redox sensors furnish a simpler alternative but mandate specific storage and handling protocols. Light-activated systems and phosphorescent sensors entail additional limitations concerning dye migration [65, 66]. Recent strides employing alginate polymers hold promise in addressing dye migration concerns, thereby enhancing O_2 monitoring within MAPs [53].

11.3.2 CO_2 Detectors

Optical sensors designed for sensing CO_2 levels in the context of food packaging can be considered into two basic kinds. For instance, luminescent dye-based sensors undergo a fluorescence change when they come into contact with CO_2, and colorimetric indicators rely on pH indicators to generate significant color changes [67–69].

To effectively utilize luminous dyes, sophisticated equipment is necessary for both stimulating and detecting the light emitted by the dyes. This includes portable instrumentation and handheld scanning devices. Although luminous dyes provide excellent analytical performance, their deployment is not without difficulties [70–72]. Researchers pioneered the development of this technology. The addition of HPTS (*8-Hydroxypyrene-1,3,6-trisulfonic acid*) onto a plastic sheet was the method they used to achieve it [73]. Consequently, a fluorometric CO_2 sensor was developed with a rapid response and restoration time. Scientists improved the HPTS system's stability by immobilizing the fluorophore molecule in a hydrophobic progressively altered silica (ormosil) matrix and enhancing sensitivity using a dual luminophore referencing system [71]. It is crucial to remember that the chemical constituents utilized in sensor production are not suitable for food consumption. Consequently, it is essential to take strict procedures to avoid the sensors from contacting food items [70].

Recent research on CO_2 sensors for food packaging has primarily focused on using colorimetric indicators [68]. While bright dyes have higher sensitivity, these indicators have relatively lower sensitivity. However, they have several advantages, including using more affordable components and eliminating the requirement for additional instrumentation instruments. Colorimetric pH dyestuff, for instance, m-cresol purple, have produced thin, vibrant plastic films that exhibit color changes when protonated or deprotonated [74]. Water-based systems provide excellent durability, especially for products that prerequisite to be kept at a humidity and temperature level that can be accurately regulated [70]. Furthermore, empirical designs utilizing food-grade blends have been validated, rendering them appropriate for mass production of labels that offer customers clear indications about food preservation.

However, CO_2 leak indicators often exhibit less reliability than O_2 gas sensors. This occurs when food undergoes CO_2 absorption in the initial days of storage, resulting in a reduced CO_2 content in the headspace. This behavior is particularly relevant to active pharmaceutical administration (MAP), where the package is initially flushed. Conversely, the metabolic activity of microbes could lead to a rise in CO_2 levels, potentially masking signs of leaks. Commercially accessible CO_2-sensitive MAP indicators, including plastic optical fluorescent films, may demonstrate irregular reversibility due to oxygen consumption caused by microbial growth. This can lead to compromised package integrity and the indication color remaining unaltered, creating the perception that the package is still intact even if it has spoiled [55].

The fluctuations in CO_2 levels during storage are the main factor contributing to the perception that CO_2 readings are often less reliable than O_2 indicators. However,

CO_2 indicators are still utilized to detect leaks. Furthermore, the metabolic activity of bacteria might lead to misleading signals, adding complexity to the assessment of detection precision. Several factors must be considered when using O_2 and CO_2 indicators to identify leaks in MAP [52]. The challenges associated with O_2 indicators are their poor sensitivity, the complexity of handling, the lack of clear color change in the presence of acidic CO_2, and the undesired reversibility [52, 75]. Although many advanced platforms are now technologically advanced for detecting contaminants and pathogens, most of these platforms necessitate the extraction of samples to identify specific molecules. Regarding food packaging applications, the main emphasis is on identifying the development of microbiological contamination. This necessitates incorporating integration within the package, clear-cut reactions that are readily distinguishable, and cost-effective manufacturing. Tracking alterations in the gas composition of a package due to microbial growth can provide as an indirect means of identifying microbial contamination. Gas sensors can be employed to achieve this objective [76]. The rise in CO_2 concentration can function as a signal of contamination or the presence of pathogens in packages that do not employ CO_2 as a protective gas. However, the detection of CO_2 has challenges in accurately detecting the proliferation of microbes in MAPs [76].

11.3.3 Specific Chemicals and pH Changes that Indicate Spoilage

Researchers have developed colorimetric assays to detect spoilage in fish, an alternative to relying solely on our senses. Researchers engineered an indicator comprising a specialized polymeric film that changes color upon contact with amines, indicating fish spoilage [77]. Another study used 16 color-changing compounds to detect spoilage chemicals, including trimethylamine and cadaverine [78]. Negative controls like hexanal in fresh fish were included to differentiate between fresh and spoiled products. While certain dyes exhibited stronger reactions to specific spoilage chemicals, additional dyes accommodate the variability in food spoilage, enhancing the sensor system's capacity to detect a comprehensive variety of spoilage mixes and bolstering quality control and food safety measures [79].

In tandem with colorimetric assays, electrochemical sensors offer another avenue for detecting volatile amines [79]. A sensor employing a hydrogel-coated pH electrode proficient of nursing the amount of essential volatile compounds in the air was developed. This wireless, cost-effective sensor relies on measuring changes in resonance frequency influenced by the volatile amine concentration [80]. Similar sensors are being developed for chicken, pork, and beef by directly incorporating pH-sensitive materials into packaging. These sensors exploit the alteration in air composition resulting from spoilage, detectable through pH changes [2, 81]. Spoilage in fruits, like fish and meat, can also be discerned using chemical sensors, particularly pH sensors, which enable real-time monitoring of fruit freshness without compromising the product. Sealed fruit packages undergo pH changes over time, detectable by pH sensors utilizing indicator dyes responsive to acids or bases. As fruits spoil, they release distinct volatile organic compounds, detectable by pH sensors, offering a profitable, subtle, and user-friendly technique for monitoring packaged fruit freshness [82]. Certain studies have employed pH sensors to detect specific chemicals like glutaraldehyde and formaldehyde emitted by spoiled fruits [51, 83, 84].

Chemical sensors, or receptors, are engineered to detect specific chemicals or gases through surface adsorption, translating their presence into measurable signals. Depending on their power requirements, sensors are classified as active or passive [85, 86]. Carbon-based nanomaterials, counting graphene, nanoparticles, in addition, nanotubes, are frequently employed in chemical sensors owing to their exceptional mechanical and electrical properties and large surface area [85]. These nano-based sensors facilitate the detection of pathogens, contaminants, spoilage, and product tampering, with recent advancements incorporating optical transducers, which eschew electricity and can be remotely read using visible or infrared light [87–89]. While the silicon (Si)-based optical transducers are integrated circuits embedded in Si semiconducting material [90].

Similar to Time–Temperature Indicators (TTIs) for fish and meat, inexpensive and straightforward chemical sensors have been devised to monitor fish freshness without product damage. These sensors exploit the increase in pH within sealed fish packages as spoilage occurs, detectable by pH indicator sensors employing dyes responsive to acids or bases. As fish spoil, they release volatile amines, detected by pH sensors typically incorporating a pH-sensitive dyestuff in a polymer medium. This dye changes color in response to volatile compounds emitted during spoilage, jointly acknowledged as total volatile basic nitrogen (TVB-N) [91].

11.3.4 Humidity Sensors

Humidity sensors are utilized to detect the moisture amount in food wrapping in real time. Their purpose is to maintain optimal conditions and prolong the shelf life of the food. Ensuring the humidity levels in food packing are at their best is crucial for preventing rotting and preserving the eminence of the product. The demand for extended shelf life in packaged goods has been directed to the creation of advanced humidity sensors for the purpose of real-time observation (Fig. 11.3) [92–95]. Tan et al. investigated using an economical substance for a wireless sensor intended for on-site measurement of moisture levels in dry food packaging [95]. The sensor incorporates a flat coil and capacitor onto a paper base strategically positioned taking place the inside surface of the packaging. As the level of humidity rises, which is frequently an indication of a decline in food quality, the paper material absorbs water vapor. The absorption modifies the capacitance of the capacitor and, consequently, the sensor's resonant frequency. The resonant frequency is determined via wireless measurement techniques by examining the impedance detected in the linked coil. Feng et al.

suggested including a humidity sensor in a completely printed Radio Frequency Identification (RFID) tag [77]. A related study examined the effects of different humidity levels (ranging from 20 to 70%) on five types of paper substrates [52]. The paper substrates provide a cost-effective medium while retaining good sensitivity and appropriate reaction times. The sensor tag utilizes two planar inductor-capacitor resonators that exploit inductive coupling for wireless functionality. One resonator utilizes distinct frequency spectrum characteristics to encode the identification data, while the other is the humidity-based sensor. The sensing device corresponds to the principle elucidated, in which alterations in the dielectric constant of the substrate caused by fluctuations in humidity are manifested in the resonant frequency [96, 97]. Measurements are acquired by wireless means by utilizing a copper (Cu) loop antenna that is coupled to a vector network analyzer (VNA) which functions as the reader antenna. Polyethylene terephthalate (PET) was found to be the most suited substrate among those examined, mostly because of its reduced water absorption [93]. Alternatively, another resonant structure served as a precise humidity sensor by detecting variations in resonant frequency caused by humidity [98, 99]. These sensor systems are especially suitable for dry packaged goods such as cereals and fried/baked snacks, where it is vital to maintain low moisture levels to preserve the product's quality. Using cost-effective manufacturing methods indicates that these sensors are suitable for commercial use. Nevertheless, additional investigation is required to enhance the efficiency of humidity sensors and achieve the highest possible shelf-life extension. Recently, there has been a significant increase in research efforts to develop humidity sensors for the purpose of real-time monitoring in food packaging [100]. These sensors possess the unique benefits of being simple and easily integrated into packaging systems. By means of a planar capacitor and inductor on a paper substratum created a cost-effective wireless humidity sensor. This sensor may be easily integrated into packages. Paper substrates provide a cost-efficient medium while maintaining a high level of sensitivity and a fair amount of time for reaction [92]. Once the moisture inside the wrapping increases, the paper substrate absorbs water vapor, which in turn alters the capacitance of the capacitor [96, 97].

Photonic crystal-based humidity sensing materials have become a subject of considerable investigation due to their notable sensitivity, affordability, and efficient manufacture, making them appealing for real-time food-packing monitoring [101–103]. The humidity sensors stated above display huge possibility for usage in dry food packaging, where preventing product degradation relies on avoiding moisture and maintaining extremely low humidity levels. These systems are very appropriate for extensive commercial applications because of their ease of manufacture and cost-effectiveness.

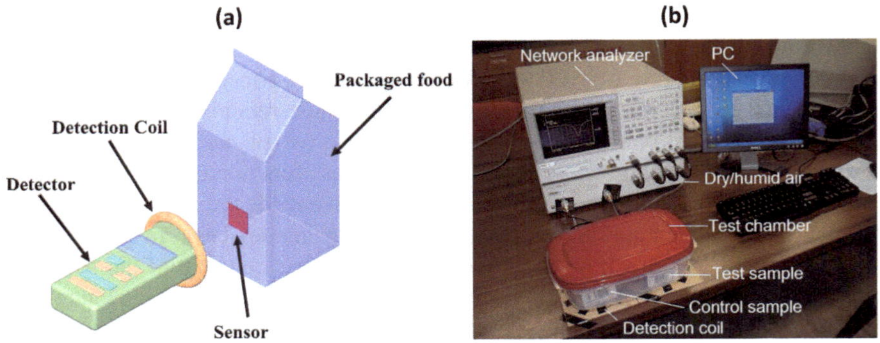

Fig. 11.3 **a** An LC-type humidity sensor positioned on the inner wall of the packaging offers a wireless response detected by a dedicated coil connected to the sensor reader unit, **b** Wireless in-package humidity monitoring system. Reproduced from Ref. [92], Copyright CC-BY terms, 2007

11.4 Time–Temperature Sensors

TTIs are an essential technology in the food sector. They deliver a visual representation of a product's temperature times gone by (Fig. 11.4). These devices operate by converting temperature exposure into at least one detectable and lasting alteration, accomplished through several methods such as physical changes, chemical reactions, electrochemical responses, enzymatic processes, or even microbial activity [104, 105]. The "Fresh Check" indication (TempTime Corporation, USA) is a commercially accessible TTI system with a solid-state polymerization process. This technique utilizes a thin layer of acetylenic monomer, which is transparent and lacks color initially. However, with time, it undergoes a chemical change and transforms into a polymer that is highly colored and opaque. The rate of this polymerization reaction is directly affected by the temperature to which the sensor is subjected [106].

TTI sensors are classified according to their level of contact with the food product. Temperature fluctuations can cause many alterations in food, such as mechanical, chemical, electrochemical, enzymatic, or microbiological changes. Therefore, these sensors provide essential information about patterns of food spoilage [107–109]. TTIs can be classified into three core divisions, such as critical time–temperature indicators, complete history indicators, and critical temperature indicators [106]. Critical temperature indicators exclusively give evidence about the degradation of a product and changes in its sensory characteristics resulting from temperature variations beyond a predetermined threshold within a specified time. These sensors usually depend on irreversible color changes that are visually noticeable in the indication (Fig. 11.4a). Figure 11.4a illustrates a temperature sensor system that relies on an enzyme activity. Thermal exposure causes a sensitive patch to change color, and if it turns orange, it indicates that the product has been compromised [110].

Critical time–temperature indicators provide comprehensive information about the combined influence of time and temperature on the excellence or safety of a

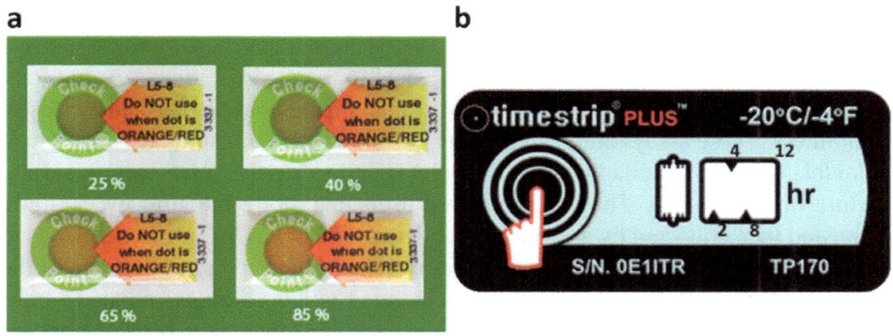

Fig. 11.4 **a** Biosensor for critical temperature monitoring with spoilage detection. **b** Time-integrating critical temperature indicator for threshold breach duration. Republished with permission from the American Chemical Society [52]. Copyright American Chemical Society, 2019

product. This information is particularly relevant when the product is exposed to temperatures over a specific threshold for a specified period of time. Figure 11.4b displays a commercially accessible crucial TTI device used to observe frozen food goods. If the threshold temperature is exceeded, a blue hue will be displayed in the viewing window of this system. The evolution of the blue hue along the time markers represents the duration of one or more breaches [111].

Ultimately, comprehensive historical indicators consistently observe and document variations in temperature that a sample experiences. Subsequently, this data is transformed into permanent alterations that convey information on the quality and safety of food to consumers [112–114].

11.4.1 Biosensors for Bacteria Detection

Microbial proliferation has a double impact on food quality. Firstly, the abundance of naturally existing bacteria affects the microbiological quality of the product and is linked to its deterioration. Furthermore, the proliferation of harmful bacteria, which are frequently introduced due to improper handling or contamination at diverse points in the food manufacturing procedure, presents substantial health hazards (such as pathogenic strains of *E. coli*) [115]. Today, Bacteria are one of the main reasons for foodborne losses and diseases, emphasizing their importance as a focus for monitoring food safety [116]. Conventional procedures for detecting bacteria mostly depend on advanced laboratory technologies such nucleic acid amplification, bacterial culture, colony counting, and immunological analysis. Although these technologies provide outstanding sensitivity, they are typically laborious and not easily accessible for mass utilization [117–119]. Lately, scientists have concentrated on creating bacterial detection technologies that are user-friendly and simply understandable, with low need for user involvement. Nevertheless, most of these gadgets are harmful and need the food item to be opened for analysis [120].

Lateral-flow immunoassays (LFIAs) are a method used to achieve user-friendliness. These assays entail combining bacterial cells with antibody-conjugated colloidal gold (Au) or palladium nanoparticles (PdNPs) and then putting the resulting mixture on a test strip. The complexes are propelled along the strip by capillary action and then collected by another immobilized antibody. This interaction leads to the formation of a discernible colored line or spot, which serves as evidence of the existence of bacteria (Fig. 11.5a). This technique provides a quick (about 15 min) and equipment-free way for identifying bacteria, specifically *Listeria monocytogenes* and *Klebsiella bacterium* [121–123]. Portable microfluidic devices provide an additional means for detecting germs. These devices, as demonstrated [124], consist of a surface coated with gold that has been modified with antibodies capable of reacting to certain antigens of *E. coli.* Bacterial aggregation can then be observed using bright-field microscopy. The used the method of attaching *E. coli*-specific antibodies to Au surfaces. It also included sensor chips to measure electrochemical signals in real-time (Fig. 11.5b) [125].

Basu et al. have established a sensor for detecting *E. coli* that is made from flexible acetate sheets containing graphene. This sensor employs electrodes for detection, utilizing graphene nanosheets that may capture *E. coli* bacteria and induce a noticeable alteration in impedance spectroscopy [126]. DuVall et al. [127] created a smartphone app that examines the clustering of paramagnetic silica beads to identify the existence of bacteria. This technique entails rupturing bacterial cells and subsequently utilizing loop-mediated isothermal amplification (LAMP) to amplify specific DNA strands.

Following the amplification of DNA, there is a following effect where the DNA causes the beads to come together and form aggregates, as seen in Fig. 11.5c [128]. Yoon, Lee, and Park investigated the application of laser speckle decorrelation to display the attendance of microbes in tissues. A static speckle pattern is observed when a laser is reflected off uncontaminated food samples in a static turbid medium. Nevertheless, the existence of living microbes generates a fluctuating speckle pattern that may be measured to assess the extent of contamination (Fig. 11.5d) [129]. Although these methods show advancements in user-friendly bacterial identification and their potential utility in monitoring food quality, they frequently necessitate user involvement and the employment of intricate laboratory-based methodologies. Methods for example microcontact printing, automated printers, and surface functionalization, have been intensively studied to create patterns and add functionality to surfaces to detect certain cells [130].

In a recent study, Yousefi et al. [130] introduced a significant breakthrough: a straightforward device capable of detecting *E. coli* that can be simply merged into food packing (Fig. 11.5e). This device shows great promise in identifying specific bacteria in food models without the introduction of the box. This is a significant achievement in advancing smart packaging sensors [130].

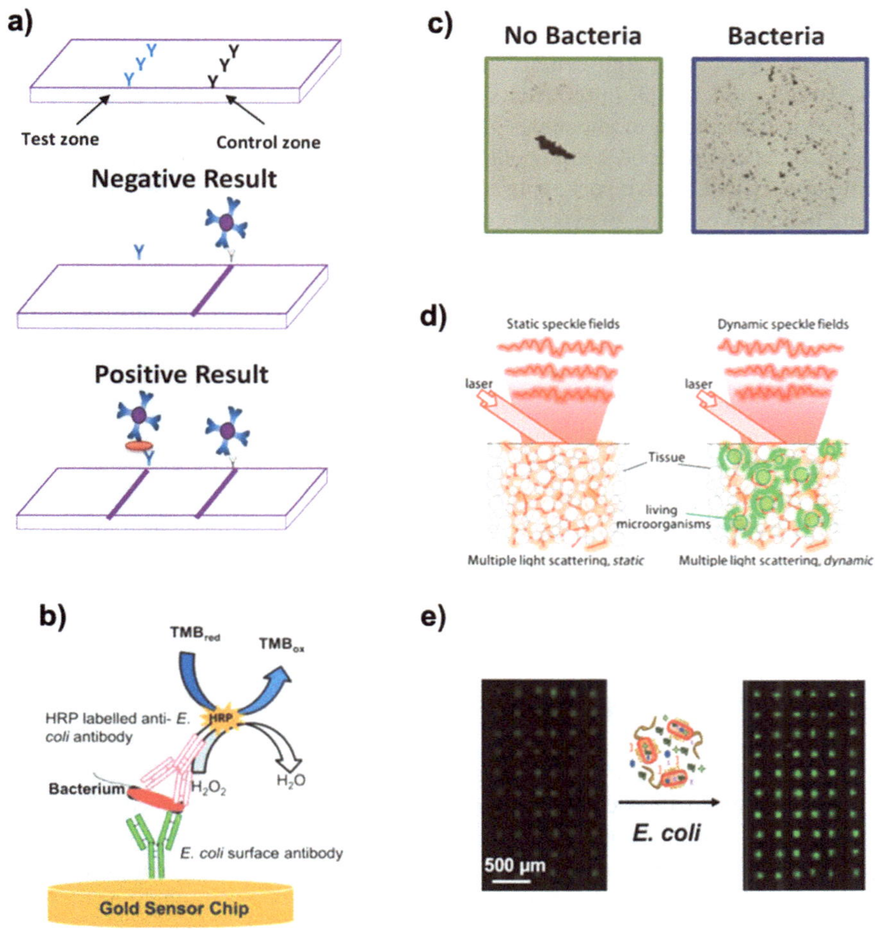

Fig. 11.5 **a** Lateral-flow test strip (LFT) schematic: A visible line in the testing zone specifies target bacteria, **b** Microfluidic device with a gold sensor chip for real-time electrochemical detection of E. coli using immobilized specific antibodies, **c** Paramagnetic silica bead aggregation intensifies upon target bacteria DNA detection, **d** Laser speckle imaging detects microscopic movement. A static, unmoving sample (like tissue) produces a static speckle pattern, while the presence of live microorganisms causes a dynamic, fluctuating pattern, **e** A DNAzyme sensor embedded in a cyclo-olefin polymer (COP) food wrap detects *E. coli* through a fluorescence signal increase. Republished with permission from American Chemical Society [52]. Copyright American Chemical Society, 2019

11.5 Smart Sensors: An Overview

The development of food packaging, shifting from basic safeguarding to sophisticated monitoring, represents a notable advancement in meeting customers changing needs in the food sector. In the past, packaging materials were mostly selected based on their capacity to provide convenience and prevent harmful interactions with the food contents. In the 20th century, significant advancements in packaging technology occurred, which introduced systems that incorporated antibacterial agents and oxygen-absorbing components. These advancements establish higher benchmarks for extending food's storage duration and protecting it from external factors, sometimes called "active packaging" [131]. However, the food and beverage packaging industry continues to experience ongoing challenges due to global trends such as the increasing use of industrialized food processing, the rise in import and export operations related to food products, and the growing desire for faster meal preparation times. These factors encourage the investigation of advanced and complex packaging options designed to satisfy the increasing customer demand for healthier, safer, functional, and economically sustainable processed meals while still prioritizing convenience. While the traditional perception of food packaging focused on protecting and preserving the product, promoting convenience has quickly become just as important. In modern food packaging design, factors such as traceability, tamper-proof features, and sustainability have become crucial considerations [132–134].

This discussion explores the idea and foundations of "smart" or "intelligent" packaging, with specific importance on incorporating sensor technology to enhance the safety and quality of food [135, 136].

11.5.1 Smart Packaging Principle

Smart packaging represents a substantial advancement in the packaging sector, incorporating various features customized to suit the product being packaged. This technology is utilized in many industries such as food & beverage, pharmaceuticals, health products, and household items. Although the future holds great promise for even more complex capabilities, the current and upcoming functionalities of smart packaging can be broadly classified as follows:

1. *Preserving Integrity and Prolonging Shelf-Life*: Intelligent packaging actively preserves the quality of products and inhibits food spoilage, thereby increasing the duration for which they can be stored. This can entail using oxygen scavengers to restrict oxidation or antimicrobial coatings to counteract detrimental microbes.
2. *Improving Product Characteristics*: Smart packaging has the ability to enhance various product attributes, including color, taste, flavor, scent, texture, and viscosity. Controlled-release methods can potentially govern the ripening process of fruits or vegetables.
3. *Adaptable to Environmental Changes*: Intelligent packaging has the capability to actively react to fluctuations in the product or its surrounding atmosphere. This could entail the use of thermochromic indicators that undergo a color change in the presence of high temperatures, so guaranteeing the safety of consumers.
4. *Expository Communication*: Intelligent packaging has the ability to convey vital information to consumers regarding the product, its past, or its present state. One

possible approach is to utilize printed signs that visually represent the freshness levels of the products, or to incorporate QR codes that can be scanned to get comprehensive information about the products.

5. *Assistance with Opening and Ensuring Seal Integrity*: Intelligent packaging can aid in the process of opening the package and offer visual indicators to verify the integrity of the seal, so ensuring the security and freshness of the goods.
6. *Measures to prevent counterfeiting and theft*: Intelligent packaging can be integrated with functionalities to authenticate products, impede counterfeiting, and discourage theft. This could entail the utilization of embedded radio frequency identification (RFID) tags or tamper-evident seals [137].

Nevertheless, it is crucial to acknowledge that not all intelligent packaging solutions will encompass every individual characteristic. In practical scenarios, the primary emphasis is frequently placed on detecting and transmitting information pertaining to the safety (indicating if food is safe or harmful) and quality (displaying freshness, ripeness, or firmness) of a product [138, 139]. This focus on the status of the product results in intelligent packaging that has the capability to: (*a*) *Product Tracking*: The packaging has the capability to track the precise position of the product at every stage of the supply chain, (*b*) *Environmental Sensing*: Sensors embedded in the packaging can detect and measure both internal and external environmental conditions, (*c*) *Notify stakeholders*: This data can be transmitted to manufacturers, merchants, and customers, offering immediate and accurate knowledge on the safety and quality of the product [140, 141].

The increasing need for extensive product information is fueling the expansion of intelligent packing, namely in the food sector. Consumers are becoming more inclined to comprehend component listings, storage suggestions, and instructions for proper usage. Smart labels and stickers that employ thin-film technology can fulfill this requirement by immediately conveying visual information to consumers [89]. Smart packaging can incorporate disposal instructions and visual protection for pharmaceutical and healthcare products, providing consumers with clear guidance on how to use and dispose of them correctly. In addition, there is great potential for developing drug delivery systems incorporated into intelligent packaging that may transmit patient data to healthcare facilities. Wireless and electronic communication for medicine delivery and patient data are still in the initial phases of expansion. However, there is a possibility that they could be integrated into packaging in the near future [142].

Another important aspect to consider is consumer confidence, especially concerning perishable food items. The existing dependence on "best by" dates frequently neglects to consider possible temperature variations during storage or shipping, resulting in consumer uncertainty regarding the safety of a product [143]. In the future, there is potential for incorporating microbial growth indicators and TTIs that rely on the chemical, physical, or enzymatic movement present in the food itself. These developments will offer precise and unequivocal signals of product quality, safety, and shelf-life. Scientists are currently engaged in the active development of novel TTIs, which provide a broader selection of choices for monitoring the deterioration of food [142]. Smart packaging is a dynamic field that has the potential to completely transform the way items are packaged, monitored, and delivered to consumers. In this dynamic field, innovation will be driven by a continued emphasis on features that improve product safety, quality, and consumer information.

11.5.2 Applications of Smart Sensors in Food Industries

11.5.2.1 Time–Temperature Indicators

Smart packaging refers to the usage of TTIs to monitor and control the temperature of products over time. TTIs are instruments that provide real-time info on the temperature conditions that a product has been exposed to. This technology allows for precise monitoring and ensures that products are transported and stored under optimal temperature circumstances [142]. An innovation currently available in the market is the integration of sensors into containers, enabling them to self-regulate temperature by either heating or cooling. This technology transforms the containers into "smart packaging" that informs consumers about the ideal temperature conditions for the product. TTIs, which are commonly used sensors, can be exemplified with thermochromic ink dots. These dots undergo a change in color when they reach certain temperatures after being refrigerated or heated in a microwave. Illustrations of this concept can be seen in pancake syrup containers that have dots to indicate the optimal temperature for heating in a microwave, as well as in orange juice packaging that visually shows when the refrigerated product is sufficiently cold to be consumed [144]. The TTIs that are now accessible utilize various methods, which can be classified into three categories: chemical, physical, or biological. Chemical and physical systems depend on reactions or alterations that are activated by time and temperature, such as acid–base reactions, melting, or polymerization. Biological responses occur when the biological activity of microbes, spores, or enzymes changes according to the effects of time and temperature [142]. For example, 3 M's MonitorMark™ provides two options: an industrial threshold indicator and a consumer smart label. The industrial variant serves as a gauge of misuse, lying dormant until a specific temperature limit is surpassed. This technique utilizes a material that has a specified melting point and is combined with a blue dye. A film strip is used to create a division between the wick and a reservoir. When activated, the wick becomes visible. When exposed to temperatures higher than its critical point, the substance will melt and spread into the wick, causing it to turn blue. These indicators are offered in a range of critical temperatures, spanning from −15 to 26 °C. The consumer version functions as a semi-historical indicator, undergoing a color change when subjected to temperatures that exceed the acceptable storage settings, and once more when the product reaches the end of its shelf-life. The operational mechanism closely resembles that of the industrial iteration, relying on the process of melting and diffusion of the blue dye [145].

Commercial products exhibit an additional illustration of intelligent labels that monitor the length of time a product has been opened or used. These labels have the capability to measure the passage of time ranging from minutes to more than a year, and can operate effectively under different temperature conditions. Activation is achieved by exerting pressure on a certain button, which triggers the interaction between a safe liquid for consumption and a unique permeable membrane inside the Timestrip. The diffusion of the liquid through the membrane is governed by physical rules, ensuring consistency and control. The upper surface showcases indicators that convey the amount of time that has passed after activation, as well as an area designated for displaying branding and other visual elements. Timestrips provide a range of adhesive tapes on the bottom side to stick to products or packaging, meeting specific customer requirements [142].

11.5.2.2 Sensors for Fish and Meat Freshness

While TTIs like Fresh-Check and CheckPoint offer valuable tools for estimating fish and meat freshness, researchers are exploring alternative methods for real-time, non-invasive monitoring. One promising approach utilizes simple and inexpensive chemical sensors that detect changes in fish pH. As fish spoils within a sealed package, its pH level gradually increases due to the release of basic volatile amines. These amines can be detected by a pH-sensitive sensor, offering a simple yet effective way to track freshness. These sensors rely on the fundamental property of pH indicator dyes, which change color based on the surrounding environment's acidity or alkalinity. In practice, these sensors can be constructed by entrapping a pH-sensitive dye (such as bromocresol green) within a polymer matrix. This dye reacts visibly to the spoilage-related volatile compounds, contributing to a measure known as TVB-N. This system facilitates rapid and non-invasive detection of spoilage compounds in fish through a colorimetric (color-based) method.

It's important to note that while pH-based sensors offer a promising approach, researchers are exploring various freshness indicator concepts. These concepts explore the detection of different spoilage indicators like CO_2, amines, ammonia, ethanol, and even hydrogen sulfide (H_2S). Integrating these indicators directly into food packaging creates visible tags that change color in response to specific analytes, providing a convenient and easy-to-understand method for consumers to assess food freshness [146–149].

11.5.2.3 Leak Indicators

Active packaging methods encompass MAP and equilibrium MAP, which substitute air with precisely regulated atmospheres. Typically, this involves reducing the amount of O_2 and increasing the level of CO_2. The ideal MAP for non-respiring foods entails of 20–80% CO_2 and 0–2% O_2 [150–152]. Nevertheless, detecting MAP leaks poses a challenging task. Gas leaks alter the composition by increasing the quantity of

O_2 and decreasing the quantity of CO_2. The modified environment encourages the proliferation of aerobic bacteria, leading to food spoilage. In extreme situations, the presence of microorganisms can effectively seal breaches by sustaining elevated amounts of carbon dioxide. [153, 154].

Hence, the algorithms used for detecting MAP leaks must be sophisticated and give priority to detecting O_2 levels over CO_2 levels. Commercial O_2-sensitive indicators mostly concentrate on the functionality of O_2 absorbers. Mitsubishi Gas Chemical markets oxygen-absorbing sachets designed to decrease the presence of oxygen. Furthermore, Cryovac-Sealed Air Ltd. has created an indicator that may be used to confirm the gas composition of a package [142, 154, 155]. Visual O_2 indicators consist of redox dyes for example methylene blue, reducing agents such as sugars, and alkaline substances such as sodium hydroxide. These items can take the form of tablets, printed layers, or laminated film package inserts [142].

Despite its appropriateness, CO_2 has downsides and absorbed by the product once it is packaged and progressively accumulates in the space above the product until it reaches a stable level. A significant decrease in CO_2 emissions following this initial period indicates the presence of a leak. The phenomenon known as the Microbial Masking Effect: Leaks can result in the proliferation of microorganisms and the generation of CO_2. The presence of CO_2 may obscure the decrease in the leak, resulting in a defective product that has normal levels of CO_2 [156].

Although O_2 indicators offer advantages, they also have disadvantages. Reduced Sensitivity: highly sensitive O_2 indicators may interact with small amounts of remaining O_2 (0.5–2.0%) that are occasionally present in properly sealed MAPs. This might make the handling process more difficult and necessitate the use of anaerobic preparation and packaging. Acidity Impact: MAPs containing CO_2 gas with an acidic nature may not exhibit a change in color of the oxygen indicator. Reversible indicators are ineffective for detecting leaks. Microbial growth has the ability to consume oxygen from a leak, which results in a change in the color of the indicator and masks the process of deterioration.

Ultimately, the effectiveness of MAP leak detection is constrained by the O_2 and CO_2 indicators. Scientists enhance the sensitivity, reliability, and clarity of response to enhance the effectiveness of these food safety measures [157].

11.5.2.4 Sensors for Food Pathogens and Contaminants

The realm of food safety is witnessing a surge in innovative platforms designed to detect pathogens and contaminants. However, the current focus leans towards standalone devices requiring sample extraction for analysis. In the context of food packaging, these systems primarily target microbial contaminants [158].

The challenge lies in seamlessly integrating these detection mechanisms within the packaging itself. This ideal solution necessitates a readily noticeable response, most likely a color change, while remaining cost-effective to manufacture. The most probable approach involves indirectly detecting microbial growth by measuring gas composition changes within the package, a consequence of microbial activity. This can be achieved using gas sensors as previously mentioned [159, 160]. Unfortunately, the number of practical packaging indicators for contaminants or pathogens remains scarce. While monitoring CO_2 levels can signal microbial growth, it becomes a challenge in MAP where CO_2 is already present in high concentrations. Therefore, CO_2 monitoring as a sole indicator is only viable for non-CO_2-containing packages [158].

For broader detection capabilities, conducting polymer biosensors offers another promising avenue. These sensors function by sensing gases released throughout microbial metabolism [161, 162]. They are designed by incorporating conducting nanoparticles into an insulating matrix, with changes in resistance correlating to the amount of gas released. These sensors have successfully detected foodborne pathogens by analyzing bacterial cultures [163]. Additionally, coupling such sensors with neural networks has confirmed potential in evaluating meat freshness. The continuous development of these innovative biosensor technologies holds immense promise for the future of food packaging, offering a more integrated and efficient approach to food safety.

11.5.2.5 Identification, Authentication, and Tracking

RFID tags have been used for many years to identify high-value products like electronics and clothing. These tags comprise two parts, like a microchip for processing and storing information, and an antenna for transmitting and receiving data. A separate reader device retrieves information from the tag, often at distances of numerous meters liable on the radio frequency utilized. Notably, RFID tags in packaging are passive, meaning they lack their own power source and rely on the reader's radio waves to transmit information [164]. The key benefit of RFID tags lies in their ability to monitor multiple items throughout the entire supply chain. This significantly enhances distribution speed and efficiency, especially crucial in today's world where raw materials travel from various regions to be processed in one location and then distributed globally. RFID technology is envisioned to replace the widely used barcodes of today.

Currently, silicon semiconductors are the primary material for RFID tags. However, this may change to facilitate cheaper and easier production on various materials. Developing printed electronics using conductive polymers and metallic inks offers a promising alternative to traditional methods. Research institutions and companies are actively exploring desktop inkjet printing and other high-volume production techniques for printed electronics [161, 165]. Additionally, some advancements are investigating the use of carbon nanotubes for RFID tag antennas [166, 167], although this technology remains less developed than conductive inks.

Interestingly, researchers are exploring the integration of RFID tags with chemical sensing functionalities. Prototypes for ethylene sensing (fruit ripeness) and moisture detection demonstrate the potential of this technology [168, 169]. Nanotechnology offers a clear path for further advancements as these systems are inherently microelectronic. Nanoscale barcodes, quantum dots, and magnetic nanoparticles are among the various systems being developed within the realm of nanotechnology. Nanotechnology, which studies and manipulates materials at the molecular level, has the latent to significantly reduce food spoilage and improve security throughout production, processing, and shipping [132]. However, the widespread adoption of such technologies in food packaging remains unclear due to cost and ease-of-use considerations.

For RFID tags, the future holds promise beyond simple tracking. Integration with functionalities like product authentication, anti-counterfeiting, and even theft prevention could become commonplace.

11.6 Challenges and Future Trends

More processing involving food is possible now that modern smart green packaging innovation has laid a firm groundwork. Although this technology has a lot of promising future uses, it might also face significant difficulties as it evolves. In this part, we will go over a few of these obstacles and the possibilities for food-smart green packaging.

The lack of advancements in sensory technology is a major roadblock to smart packaging's further growth. Incorporating sensors into large-scale packaged containers is challenging due to their size, cost, and the fact that numerous sensors could provide reliable readings for various freshness characteristics seen in food goods. Technologists have always been interested in finding the sweet spot where size, mobility, and functionality meet. The food business is particularly affected by this issue because of the demand for affordable, widely available, high-quality goods. The expansion of sensor-related nanotechnology research might provide a solution to this problem. These sensors may pick up on a wide range of environmental factors, including air pollutants, temperature, and humidity. Plus, they can do more intrusive tests on food quality with little to no harm to the item. Although this technique is now unfeasible for mass-produced food packaging, it has the potential to be commercialized for extremely precise food quality sensing with additional study [170]. Improving food quality sensors with the ability to use active packaging technology to deploy spoiling inhibitors is a major motivation. Antimicrobial packaging is one area that could use some improvement in this area of research. To ensure the safety of the product, food sanitization involves eliminating any microorganisms. Currently, most antimicrobial packaging methods rely on inorganic chemicals that are easily accessible. Concerns about potential responses or long-term impacts arise when artificial chemical components are used in food processing. Incorporating bioactive substances generated from natural sources can significantly enhance these systems [171]. Smart packaging is a system that incorporates multiple types of technology. Despite their importance for food safety, the disposal phase of packaging is complicated by monitoring and smart packaging technologies. Neither biodegradable nor readily recyclable is the majority of the material used; in fact, many package types containing particular chemical compounds require stringent disposal regulations. Research on biodegradable or recyclable materials is ongoing for use in this technology; while it may not be possible to entirely replace all components, doing so will greatly lessen the systems environmental effect [171]. Possibilities exist for cutting-edge smart packaging technology to connect with third-party processors via the Internet of Things (IoT). IoT systems rely on networks to facilitate communication and data sharing among various system components. A key feature of IoT infrastructure is the ability for "tag" and "reader" components to transmit data in both directions, building upon the concept of external communication offered by RFID systems. This would allow an external unit to control the package's circuits. Prospective growth areas in the food business also include analyzing the data received by the sensing devices [172]. Future smart packaging solutions may incorporate robust external computing units with processing power and performance capabilities well beyond those of traditional RFID readers. These auxiliary computing units would receive data transmitted by sensing devices in packaging. Extracting useful information from this data could help in forecasting food breakdown tendencies and finding the optimal active packaging technology combination for each product. Integration with IoT technology opens the door to consumer-oriented applications in addition to data transmission to industrialized computers. Whether you're at home or in the shop, you may use an app to keep tabs on food quality and circumstances, get storage tips, and receive specific information on packages and products. Integrating smart sensing systems with smart computing and meeting the needs of consumers and retailers is another potential future approach. By enhancing the evaluation of fruit quality, the combined systems will decrease food waste and promote customer well-being [173].

11.7 Conclusion

To boost food business profits, smart package solutions have been applied in recent years. This is accomplished by minimizing food waste caused by improper storage or handling and increasing the amount of time that food stays fresh. Additionally, shoppers gain since their food will always be in the best possible state of freshness and overall quality when they take it out of the package. Due to increased consumer expectations and longer food supply chains brought about by globalization, retail operations have been significantly impacted. These two forces have contributed to the development of novel packaging technologies that aim to measure and optimize the impact of various elements on food safety and quality, as well as to track the food's shelf life. The technology behind smart packaging is still in its early stages of development, and there is a lot of room for additional commercial adoption. Put another way, there still exists a technological chasm between what could be possible and what could be commercially viable. The use of smart packaging is significant in three ways. TTIs and similar tools do a great job of informing buyers about the quality, microbiological safety, and freshness of food as well as its shelf life. Data systems like barcodes and RFID tags also improve supply chain efficiency and effectiveness by providing real-time information on goods verification and tracking. Smart packaging materials have little commercial application while providing benefits that outweigh the cost of service. Further acceptance can only be achieved if the gap between lab results and practical, marketable food uses is reduced. Smart packaging has undoubtedly shaken up the packaging business, and we can expect to see more innovative approaches and advancements in food packaging soon.

Declaration The authors declare that they have no known competing financial interests or personal relationships that could have appeared to influence the work reported in this paper. Generative AI and AI-assisted technologies have been used in the writing process to improve the readability and language of the work only. After using this tool, the author(s) reviewed and edited the content as needed and take(s) full responsibility for the content of the publication.

7. References

1. Silvestre, C., Duraccio, D., Cimmino, S.: Food packaging based on polymer nanomaterials. Prog. Polym. Sci. **36**(12), 1766–1782 (2011). https://doi.org/10.1016/j.progpolymsci.2011.02.003
2. Ghaani, M., Cozzolino, C.A., Castelli, G., Farris, S.: An overview of the intelligent packaging technologies in the food sector. Trends Food Sci. Technol. **51**, 1–11 (2016). https://doi.org/10.1016/j.tifs.2016.02.008
3. Ghosal, K., et al.: Electrospinning over solvent casting: tuning of mechanical properties of membranes. Sci. Rep. **8**(1), 5058 (2018). https://doi.org/10.1038/s41598-018-23378-3
4. Rabnawaz, M., Wyman, I., Auras, R., Cheng, S.: A roadmap towards green packaging: the current status and future outlook for polyesters in the packaging industry. Green Chem. **19**(20), 4737–4753 (2017). https://doi.org/10.1039/C7GC02521A
5. Lörcks, J.: Properties and applications of compostable starch-based plastic material. Polym. Degrad. Stab. **59**(1), 245–249 (1998). https://doi.org/10.1016/S0141-3910(97)00168-7
6. Bhardwaj, A., Alam, T., Talwar, N.: Recent advances in active packaging of agri-food products: a review. J. Postharvest Technol. **7**(1), 33–62 (2019)
7. Mia, R., Sultana, S.: Fabrication and properties of silver nanowires (AgNWs) functionalized fabric. SN Appl. Sci. **2**(12), 2052 (2020). https://doi.org/10.1007/s42452-020-03845-1
8. Latos-Brozio, M., Masek, A.: The application of natural food colorants as indicator substances in intelligent biodegradable packaging materials. Food Chem. Toxicol. **135**, 110975 (2020). https://doi.org/10.1016/j.fct.2019.110975
9. Schaefer, D., Cheung, W.M.: Smart packaging: opportunities and challenges. Procedia CIRP **72**, 1022–1027 (2018). https://doi.org/10.1016/j.procir.2018.03.240
10. Kalpana, S., Priyadarshini, S.R., Maria Leena, M., Moses, J.A., Anandharamakrishnan, C.: Intelligent packaging: trends and applications in food systems. Trends Food Sci. & Technol. 93, 145–157 (2019). https://doi.org/10.1016/j.tifs.2019.09.008
11. Kuswandi, B., Wicaksono, Y., Jayus, Abdullah, A., Heng, L.Y., Ahmad, M.: Smart packaging: sensors for monitoring of food quality and safety. Sens. Instrum. Food Qual. Saf. **5**(3), 137–146 (2011). https://doi.org/10.1007/s11694-011-9120-x
12. Zannat, A., Uddin, M.N., Mahmud, S.T., Prithu, P.S.S., Mia, R.: Textile-based soft robotics for physically challenged individuals. J. Mater. Sci. **58**, 12491–12536 (2023). https://doi.org/10.1007/s10853-023-08799-4
13. Biji, K.B., Ravishankar, C.N., Mohan, C.O., Srinivasa Gopal, T.K.: Smart packaging systems for food applications: a review. J. Food Sci. Technol. **52**(10), 6125–6135 (2015). https://doi.org/10.1007/s13197-015-1766-7

14. Müller, P., Schmid, M.: Intelligent Packaging in the food sector: a brief overview. Foods **8**(1). https://doi.org/10.3390/foods8010016
15. Pereira de Abreu, D.A., Cruz, J.M., Paseiro Losada, P.: Active and intelligent packaging for the food industry. Food Rev. Int. 28(2), 146–187 (2012). https://doi.org/10.1080/87559129.2011.595022
16. Al Fahad, M.A., et al.: Pharmaceutical applications of natural dyes and pigments. In: Renewable Dyes and Pigments, pp. 165–175. Elsevier (2024)
17. Ramon-Marquez, T., Medina-Castillo, A.L., Fernandez-Gutierrez, A., Fernandez-Sanchez, J.F.: Novel optical sensing film based on a functional nonwoven nanofibre mat for an easy, fast and highly selective and sensitive detection of tryptamine in beer. Biosens. Bioelectron. **79**, 600–607 (2016). https://doi.org/10.1016/j.bios.2015.12.091
18. Salwa, H., Sapuan, S., Mastura, M., Zuhri, M.: Green bio composites for food packaging. Int. J. Recent Technol. Eng **8**(2), 450–459 (2019)
19. Turjo, M.N.A., et al.: Analysis of buying perception of textile engineering students to identify the scope of local brands in Bangladeshi apparel market. Clean. Eng. Technol. **12**, 100597 (2023). https://doi.org/10.1016/j.clet.2023.100597
20. Olsmats, C., Dominic, C.: Packaging scorecard–a packaging performance evaluation method. Packag. Technol. Sci. Int. J. **16**(1), 9–14 (2003)
21. Bucci, D.Z., Forcellini, F.A.: Sustainable packaging design model. In: Complex Systems Concurrent Engineering: Collaboration, Technology Innovation and Sustainability, pp. 363–370. Springer (2007)
22. Saxena, P., Bissacco, G., Meinert, K.Æ., Bedka, F.J.: Mold design and fabrication for production of thermoformed paper-based packaging products. J. Manuf. Process. **58**, 311–321 (2020)
23. Baranitharan, P., Mahesh, G.: Alkali treated maize fibers reinforced with epoxy poly matrix composites. Magnesium **15**(30), 150 (2014)
24. Wang, H., Wang, L.: Developing a bio-based packaging film from soya by-products incorporated with valonea tannin. J. Clean. Prod. **143**, 624–633 (2017)
25. Garrido, T., Etxabide, A., Leceta, I., Cabezudo, S., De La Caba, K., Guerrero, P.: Valorization of soya by-products for sustainable packaging. J. Clean. Prod. **64**, 228–233 (2014)
26. Ahmed, T., et al.: New trends in printing applications of natural dyes and pigments. Renew. Dye. Pigment. 139–163 (2024). https://doi.org/10.1016/B978-0-443-15213-9.00003-X
27. Sonneveld, K., James, K., Fitzpatrick, L., Lewis, H.: Sustainable packaging: how do we define and measure it. In: 22nd IAPRI Symposium, pp. 1–9 (2005)
28. Alliance, S.P. (ed.): Towards Sustainable Packaging–A Discussion Paper. Citeseer (2002)
29. Alliance, S.P.: Principles, Strategies & KPIs for Packaging Sustainability. Dandenong, Australia (2010)
30. Kumar, S., Deshmukh, R.: Packaged Food Market by Type (Dairy Products, Confectionery, Packaged Produce, Bakery & Snacks, Meat, Poultry & Seafood, Ready Meals, and Others), Sales Channel (Supermarket/Hypermarket, Specialty Stores, Grocery Stores, Online Stores, and Others): Global Opportunity Analysis and Industry Forecast 2021–2030. Allied Market Research (2021)
31. Mahmud, S.T., et al.: Recent developments of tin (II) sulfide/carbon composites for achieving high-performance lithium ion batteries: a critical review. Nanomaterials **12**(8), 1246 (2022). https://doi.org/10.3390/nano12081246
32. Geyer, R., Jambeck, J.R., Law, K.L.: Production, use, and fate of all plastics ever made. Sci. Adv. **3**(7), e1700782 (2017)
33. Banna, B.U., Mia, R., Hasan, M.M., Ahmed, B., Shibly, M.A.H.: Ultrasonic-assisted sustainable extraction and dyeing of organic cotton fabric using natural dyes from dillenia indica leaf. Heliyon **9**(8), e18702 (2023). https://doi.org/10.1016/j.heliyon.2023.e18702
34. Boesen, S., Bey, N., Niero, M.: Environmental sustainability of liquid food packaging: is there a gap between Danish consumers' perception and learnings from life cycle assessment? J. Clean. Prod. **210**, 1193–1206 (2019)

35. Lindh, H., Olsson, A., Williams, H.: Consumer perceptions of food packaging: contributing to or counteracting environmentally sustainable development? Packag. Technol. Sci. **29**(1), 3–23 (2016)
36. Narayan, R.: Carbon footprint of bioplastics using biocarbon content analysis and life-cycle assessment. MRS Bull. **36**(9), 716–721 (2011)
37. de Vargas Mores, G., Finocchio, C.P.S., Barichello, R., Pedrozo, E.A.: Sustainability and innovation in the Brazilian supply chain of green plastic. J. Clean. Prod. **177**, 12–18 (2018)
38. Neufeld, L., Stassen, F., Sheppard, R., Gilman, T.: The new plastics economy: rethinking the future of plastics. In: World Economic Forum, vol. 7, p. 158 (2016)
39. Reichert, C.L., et al.: Bio-based packaging: materials, modifications, industrial applications and sustainability. Polymers **12**(7), 1558 (2020)
40. Álvarez-Chávez, C.R., Edwards, S., Moure-Eraso, R., Geiser, K.: Sustainability of bio-based plastics: general comparative analysis and recommendations for improvement. J. Clean. Prod. **23**(1), 47–56 (2012)
41. De Feo, G., Ferrara, C., Minichini, F.: Comparison between the perceived and actual environmental sustainability of beverage packagings in glass, plastic, and aluminium. J. Clean. Prod. **333**, 130158 (2022)
42. Walther, B.A., Kusui, T., Yen, N., Hu, C.-S., Lee, H.: Plastic pollution in East Asia: Macroplastics and microplastics in the aquatic environment and mitigation efforts by various actors. Plast. Aquat. Environ.-Part I Curr. Status Chall. 353–403 (2020)
43. Nourmohamadi Shalke, P., Paydar, M.M., Hajiaghaei-Keshteli, M.: Sustainable supplier selection and order allocation through quantity discounts. Int. J. Manag. Sci. Eng. Manag. **13**(1), 20–32 (2018)
44. Nazam, M., Xu, J., Tao, Z., Ahmad, J., Hashim, M.: A fuzzy AHP-TOPSIS framework for the risk assessment of green supply chain implementation in the textile industry. Int. J. Supply Oper. Manag. **2**(1), 548–568 (2015)
45. Gardas, B.B., Raut, R.D., Narkhede, B.: Identifying critical success factors to facilitate reusable plastic packaging towards sustainable supply chain management. J. Environ. Manage. **236**, 81–92 (2019)
46. Azevedo, B.D., Scavarda, L.F., Caiado, R.G.G.: Urban solid waste management in developing countries from the sustainable supply chain management perspective: a case study of Brazil's largest slum. J. Clean. Prod. **233**, 1377–1386 (2019)
47. Nazam, M., Hashim, M., Randhawa, M.A., Maqbool, A.: Modeling the barriers of sustainable supply chain practices: A Pakistani perspective. In: Proceedings of the Thirteenth International Conference on Management Science and Engineering Management, vol. 2 13, pp. 348–364. Springer (2020)
48. Akmal, T., Jamil, F.: Testing the role of waste management and environmental quality on health indicators using structural equation modeling in Pakistan. Int. J. Environ. Res. Public Health **18**(8), 4193 (2021)
49. Kashif, M., Rani, T.: Customers' attitude towards green packaging: a case of sapphire, Pakistan. Journal of Marketing Strategies **3**(1), 1–28 (2021)
50. Sivertsvik, M., Rosnes, J.T., Bergslien, H.: Modified atmosphere packaging. Minimal Process. Technol. Food Ind. 61–86 (2002)
51. Mills, A.: Oxygen indicators and intelligent inks for packaging food. Chem. Soc. Rev. **34**(12), 1003–1011 (2005)
52. Yousefi, H., Su, H.-M., Imani, S.M., Alkhaldi, K., Filipe, C.D.M., Didar, T.F.: Intelligent food packaging: a review of smart sensing technologies for monitoring food quality. ACS Sens. **4**(4), 808–821 (2019)
53. Smolander, M.: The use of freshness indicators in packaging. Nov. Food Packag. Tech. 127–143 (2003)
54. Silva-Pereira, M.C., Teixeira, J.A., Pereira-Júnior, V.A., Stefani, R.: Chitosan/corn starch blend films with extract from Brassica oleraceae (red cabbage) as a visual indicator of fish deterioration. LWT-Food Sci. Technol. **61**(1), 258–262 (2015)

55. Eaton, K.: A novel colorimetric oxygen sensor: dye redox chemistry in a thin polymer film. Sens. Actuators B Chem. **85**(1–2), 42–51 (2002)
56. Banerjee, S., Kelly, C., Kerry, J.P., Papkovsky, D.B.: High throughput non-destructive assessment of quality and safety of packaged food products using phosphorescent oxygen sensors. Trends Food Sci. Technol. **50**, 85–102 (2016)
57. Gillman, P.K.: CNS toxicity involving methylene blue: the exemplar for understanding and predicting drug interactions that precipitate serotonin toxicity. J. Psychopharmacol. **25**(3), 429–436 (2011)
58. Paul, P., Kumar, G.S.: Toxic interaction of thionine to deoxyribonucleic acids: elucidation of the sequence specificity of binding with polynucleotides. J. Hazard. Mater. **184**(1–3), 620–626 (2010)
59. Vu, C.H.T., Won, K.: Novel water-resistant UV-activated oxygen indicator for intelligent food packaging. Food Chem. **140**(1–2), 52–56 (2013)
60. Oliveira, M., Abadias, M., Usall, J., Torres, R., Teixidó, N., Viñas, I.: Application of modified atmosphere packaging as a safety approach to fresh-cut fruits and vegetables–a review. Trends Food Sci. Technol. **46**(1), 13–26 (2015)
61. Zhang, M., Meng, X., Bhandari, B., Fang, Z., Chen, H.: Recent application of modified atmosphere packaging (MAP) in fresh and fresh-cut foods. Food Rev. Intl. **31**(2), 172–193 (2015)
62. Zhu, R., Desroches, M., Yoon, B., Swager, T.M.: Wireless oxygen sensors enabled by Fe(II)-polymer wrapped carbon nanotubes. ACS Sens. 2(7), 1044–1050 (2017). https://doi.org/10.1021/acssensors.7b00327
63. Kelly, C., et al.: Extruded phosphorescence based oxygen sensors for large-scale packaging applications. Sens. Actuators B Chem. **304**, 127357 (2020)
64. Kelly, C.A., Cruz-Romero, M., Kerry, J.P., Papkovsky, D.B.: Stability and safety assessment of phosphorescent oxygen sensors for use in food packaging applications. Chemosensors **6**(3), 38 (2018)
65. Wolfbeis, O.S.: Luminescent sensing and imaging of oxygen: Fierce competition to the Clark electrode. BioEssays **37**(8), 921–928 (2015)
66. Urriza-Arsuaga, I., Ielasi, G., Bedoya, M., Orellana, G.: Luminescence-based sensors for bioprocess applications. Fluoresc. Ind. 1–38 (2019)
67. Nopwinyuwong, A., Trevanich, S., Suppakul, P.: Development of a novel colorimetric indicator label for monitoring freshness of intermediate-moisture dessert spoilage. Talanta **81**(3), 1126–1132 (2010)
68. Rukchon, C., Nopwinyuwong, A., Trevanich, S., Jinkarn, T., Suppakul, P.: Development of a food spoilage indicator for monitoring freshness of skinless chicken breast. Talanta **130**, 547–554 (2014)
69. Jung, J., Puligundla, P., Ko, S.: Proof-of-concept study of chitosan-based carbon dioxide indicator for food packaging applications. Food Chem. **135**(4), 2170–2174 (2012)
70. Saliu, F., Della Pergola, R.: Carbon dioxide colorimetric indicators for food packaging application: applicability of anthocyanin and poly-lysine mixtures. Sens. Actuators B Chem. **258**, 1117–1124 (2018)
71. von Bültzingslöwen, C., et al.: Sol–gel based optical carbon dioxide sensor employing dual luminophore referencing for application in food packaging technology. Analyst **127**(11), 1478–1483 (2002)
72. Wang, J., Wen, Z., Yang, B., Yang, X.: Optical carbon dioxide sensor based on fluorescent capillary array. Results Phys. **7**, 323–326 (2017)
73. Mills, A., Chang, Q.: Fluorescence plastic thin-film sensor for carbon dioxide. Analyst **118**(7), 839–843 (1993)
74. Mills, A.: Optical sensors for carbon dioxide and their applications. In: Sensors for Environment, Health and Security: Advanced Materials and Technologies, pp. 347–370. Springer (2009)
75. McMillin, K.W.: Where is MAP going? A review and future potential of modified atmosphere packaging for meat. Meat Sci. **80**(1), 43–65 (2008)

76. Ammor, S., Tauveron, G., Dufour, E., Chevallier, I.: Antibacterial activity of lactic acid bacteria against spoilage and pathogenic bacteria isolated from the same meat small-scale facility: 1—screening and characterization of the antibacterial compounds. Food Control **17**(6), 454–461 (2006)
77. Kuswandi, B., Restyana, A., Abdullah, A., Heng, L.Y., Ahmad, M.: A novel colorimetric food package label for fish spoilage based on polyaniline film. Food Control **25**(1), 184–189 (2012)
78. Morsy, M.K., et al.: Development and validation of a colorimetric sensor array for fish spoilage monitoring. Food Control **60**, 346–352 (2016)
79. Rawat, S.: Food spoilage: microorganisms and their prevention. Asian J. Plant Sci. Res. **5**(4), 47–56 (2015)
80. Bhadra, S., Narvaez, C., Thomson, D.J., Bridges, G.E.: Non-destructive detection of fish spoilage using a wireless basic volatile sensor. Talanta **134**, 718–723 (2015)
81. Chen, H.-Z., Zhang, M., Bhandari, B., Yang, C.-H.: Development of a novel colorimetric food package label for monitoring lean pork freshness. LWT **99**, 43–49 (2019)
82. Taoukis, P., Koutsoumanis, K., Nychas, G.: Use of time–temperature integrators and predictive modelling for shelf life control of chilled fish under dynamic storage conditions. Int. J. Food Microbiol. **53**(1), 21–31 (1999)
83. Mia, R., Mofasser, A., Bhat, M.A., Howlader, M.I., Sayed, N.I., Bakar, M.A.: Utilizing neem wood waste as a natural dye in the development of bio-based functionalized colored jute fabric. Sustain. Chem. Pharm. **38**, 101489 (2024). https://doi.org/10.1016/j.scp.2024.101489
84. Zannat, A., Uddin, M.N., Mahmud, S.T., Mia, R., Ahmed, T.: Natural dyes and pigments in functional finishing. In: Renewable Dyes and Pigments, pp. 271–287. Elsevier (2024)
85. Vanderroost, M., Ragaert, P., Devlieghere, F., De Meulenaer, B.: Intelligent food packaging: the next generation. Trends Food Sci. Technol. **39**(1), 47–62 (2014)
86. Khan, M.T., Al Mamun, M.A., Rony, M., Anchang, X., Rashid, M.M.: Effect of different solvent systems on fiber morphology and property of electrospun PCL nano fibers. Tekstil ve Mühendis 28(122), 61–76 (2021). https://doi.org/10.7216/1300759920212812201
87. K. Nachay, "Analyzing nanotechnology," Food technology, vol. 61, no. 1, 2007.
88. De Azeredo, H.M.: Nanocomposites for food packaging applications. Food Res. Int. **42**(9), 1240–1253 (2009)
89. Liu, Y., Chakrabartty, S., Alocilja, E.C.: Fundamental building blocks for molecular biowire based forward error-correcting biosensors. Nanotechnology **18**(42), 424017 (2007)
90. Yebo, N.A., et al.: Selective and reversible ammonia gas detection with nanoporous film functionalized silicon photonic micro-ring resonator. Opt. Express **20**(11), 11855–11862 (2012)
91. Lee, S.Y., Lee, S.J., Choi, D.S., Hur, S.J.: Current topics in active and intelligent food packaging for preservation of fresh foods. J. Sci. Food Agric. **95**(14), 2799–2810 (2015)
92. Tan, E.L., Ng, W.N., Shao, R., Pereles, B.D., Ong, K.G.: A wireless, passive sensor for quantifying packaged food quality. Sensors **7**(9), 1747–1756 (2007)
93. Feng, Y., Xie, L., Chen, Q., Zheng, L.-R.: Low-cost printed chipless RFID humidity sensor tag for intelligent packaging. IEEE Sens. J. **15**(6), 3201–3208 (2014)
94. Harrey, P., Ramsey, B., Evans, P., Harrison, D.: Capacitive-type humidity sensors fabricated using the offset lithographic printing process. Sens. Actuators B Chem. **87**(2), 226–232 (2002)
95. Courbat, J., Kim, Y., Briand, D., De Rooij, N.: Inkjet printing on paper for the realization of humidity and temperature sensors. In: 2011 16th International Solid-State Sensors, Actuators and Microsystems Conference, pp. 1356–1359. IEEE (2011)
96. Uematsu, M., Franck, E.: Static dielectric constant of water and steam. J. Phys. Chem. Ref. Data **9**(4), 1291–1306 (1980)
97. Amin, Y., Chen, Q., Zheng, L., Tenhunen, H.: "Green" wideband log-spiral antenna for RFID sensing and wireless applications. J. Electromagn. Waves Appl. **26**(14–15), 2043–2050 (2012)
98. Virtanen, J., Ukkonen, L., Björninen, T., Sydänheimo, L.: Printed humidity sensor for UHF RFID systems. In: 2010 IEEE Sensors Applications Symposium (SAS), pp. 269–272. IEEE (2010)

99. Amin, E.M., Bhuiyan, M.S., Karmakar, N.C., Winther-Jensen, B.: Development of a low cost printable chipless RFID humidity sensor. IEEE Sens. J. **14**(1), 140–149 (2013)
100. Galstyan, V., Bhandari, M.P., Sberveglieri, V., Sberveglieri, G., Comini, E.: Metal oxide nanostructures in food applications: quality control and packaging. Chemosensors **6**(2), 16 (2018)
101. Burratti, L., De Matteis, F., Casalboni, M., Francini, R., Pizzoferrato, R., Prosposito, P.: Polystyrene photonic crystals as optical sensors for volatile organic compounds. Mater. Chem. Phys. **212**, 274–281 (2018)
102. Potyrailo, R.A., et al.: Towards outperforming conventional sensor arrays with fabricated individual photonic vapour sensors inspired by Morpho butterflies. Nat. Commun. **6**(1), 7959 (2015)
103. Yang, H., et al.: A visual water vapor photonic crystal sensor with PVA/SiO_2 opal structure. Appl. Surf. Sci. **423**, 421–425 (2017)
104. Ndraha, N., Hsiao, H.-I., Vlajic, J., Yang, M.-F., Lin, H.-T.V.: Time-temperature abuse in the food cold chain: review of issues, challenges, and recommendations. Food Control **89**, 12–21 (2018)
105. Göransson, M., Nilsson, F., Jevinger, Å.: Temperature performance and food shelf-life accuracy in cold food supply chains–insights from multiple field studies. Food Control **86**, 332–341 (2018)
106. Han, J.W., Ruiz-Garcia, L., Qian, J.P., Yang, X.T.: Food packaging: A comprehensive review and future trends. Compr. Rev. Food Sci. Food Saf. **17**(4), 860–877 (2018)
107. Zhu, R., Desroches, M., Yoon, B., Swager, T.M.: Wireless oxygen sensors enabled by Fe (II)-polymer wrapped carbon nanotubes. ACS Sens. **2**(7), 1044–1050 (2017)
108. Tsironi, T., Stamatiou, A., Giannoglou, M., Velliou, E., Taoukis, P.S.: Predictive modelling and selection of time temperature integrators for monitoring the shelf life of modified atmosphere packed gilthead seabream fillets. LWT-Food Sci. Technol. **44**(4), 1156–1163 (2011)
109. Nuin, M., et al.: Modelling spoilage of fresh turbot and evaluation of a time–temperature integrator (TTI) label under fluctuating temperature. Int. J. Food Microbiol. **127**(3), 193–199 (2008)
110. V. S. Freshtag™.: Temperature monitoring made simple. Vitsab™ International AB, vitsab.com
111. T. PLUS.: https://timestrip.com/indicator-labels/temperature-indicators/timestrip-plus. Last accessed May 10 2024
112. Wang, S., Liu, X., Yang, M., Zhang, Y., Xiang, K., Tang, R.: Review of time temperature indicators as quality monitors in food packaging. Packag. Technol. Sci. **28**(10), 839–867 (2015)
113. Pennanen, K., et al.: European consumers' perceptions of time–temperature indicators in food packaging. Packag. Technol. Sci. **28**(4), 303–323 (2015)
114. Choi, D.Y., Jung, S.W., Lee, D.S., Lee, S.J.: Fabrication and characteristics of microbial time temperature indicators from bio-paste using screen printing method. Packag. Technol. Sci. **27**(4), 303–312 (2014)
115. Kuchenmüller, T., Hird, S., Stein, C., Kramarz, P., Nanda, A., Havelaar, A.: Estimating the global burden of foodborne diseases-a collaborative effort. Eurosurveillance **14**(18), 19195 (2009)
116. Legese, M.H., Weldearegay, G.M., Asrat, D.: Extended-spectrum beta-lactamase-and carbapenemase-producing Enterobacteriaceae among Ethiopian children. Infect. Drug Resist. 27–34 (2017)
117. Gupta, V., Gulati, P., Bhagat, N., Dhar, M., Virdi, J.: Detection of Yersinia enterocolitica in food: an overview. Eur. J. Clin. Microbiol. Infect. Dis. **34**, 641–650 (2015)
118. Gokduman, K., Avsaroglu, M.D., Cakiris, A., Ustek, D., Gurakan, G.C.: Recombinant plasmid-based quantitative real-time PCR analysis of Salmonella enterica serotypes and its application to milk samples. J. Microbiol. Methods **122**, 50–58 (2016)
119. Dzantiev, B.B., Byzova, N.A., Urusov, A.E., Zherdev, A.V.: Immunochromatographic methods in food analysis. TrAC, Trends Anal. Chem. **55**, 81–93 (2014)

120. Fang, Z., Zhao, Y., Warner, R.D., Johnson, S.K.: Active and intelligent packaging in meat industry. Trends Food Sci. Technol. **61**, 60–71 (2017)
121. Tominaga, T.: Enhanced sensitivity of lateral-flow test strip immunoassays using colloidal palladium nanoparticles and horseradish peroxidase. LWT **86**, 566–570 (2017)
122. Tominaga, T.: Rapid detection of Klebsiella pneumoniae, Klebsiella oxytoca, Raoultella ornithinolytica and other related bacteria in food by lateral-flow test strip immunoassays. J. Microbiol. Methods **147**, 43–49 (2018)
123. Ramos, A.C., et al.: Evaluation of a rapid immunochromatographic test for detection of distinct variants of Klebsiella pneumoniae carbapenemase (KPC) in Enterobacteriaceae. J. Microbiol. Methods **142**, 1–3 (2017)
124. Tokel, O., et al.: Portable microfluidic integrated plasmonic platform for pathogen detection. Sci. Rep. **5**(1), 9152 (2015)
125. Altintas, Z., Akgun, M., Kokturk, G., Uludag, Y.: A fully automated microfluidic-based electrochemical sensor for real-time bacteria detection. Biosens. Bioelectron. **100**, 541–548 (2018)
126. Basu, P.K., et al.: Graphene based E. coli sensor on flexible acetate sheet. Sens. Actuators B Chem. **190**, 342–347 (2014)
127. DuVall, J.A., et al.: Optical imaging of paramagnetic bead-DNA aggregation inhibition allows for low copy number detection of infectious pathogens. PLoS ONE **10**(6), e0129830 (2015)
128. Yoon, J., Lee, K., Park, Y.: A simple and rapid method for detecting living microorganisms in food using laser speckle decorrelation. https://doi.org/10.48550/arXiv.1603.07343 (2016)
129. Didar, T.F., et al.: Improved treatment of systemic blood infections using antibiotics with extracorporeal opsonin hemoadsorption. Biomaterials **67**, 382–392 (2015)
130. Yousefi, H., Ali, M.M., Su, H.-M., Filipe, C.D., Didar, T.F.: Sentinel wraps: real-time monitoring of food contamination by printing DNAzyme probes on food packaging. ACS Nano **12**(4), 3287–3294 (2018)
131. Rooney, M.: Overview of active food packaging. In: Active Food Packaging, pp. 1–37. Springer (1995)
132. Mahalik, N.P.: Processing and packaging automation systems: a review. Sens. Instrum. Food Qual. Saf. **3**, 12–25 (2009)
133. Robertson, G.L.: Food Packaging: Principles and Practice. CRC Press (2005)
134. Andaleeb, S.S., Hasan, K.: Strategic Marketing Management in Asia: Case Studies and Lessons Across Industries. Emerald Group Publishing Limited (2016)
135. Han, J.H., Ho, C.H., Rodrigues, E.T.: Intelligent packaging. Innov. Food Packag. 138–155 (2005)
136. Butler, P.: Smart packaging–intelligent packaging for food, beverages, pharmaceuticals and household products. Mater. World **9**(3), 11–13 (2001)
137. Yezza, I.: Printed intelligence in packaging: current and potential applications of nanotechnology. In: Proc Symp Nanomaterials for Flexible Packaging, Columbus, Code, vol. 80863 (2009)
138. Rani, D.N., Abraham, T.E.: Kinetic study of a purified anionic peroxidase isolated from Eupatorium odoratum and its novel application as time temperature indicator for food materials. J. Food Eng. **77**(3), 594–600 (2006)
139. Yan, S., Huawei, C., Limin, Z., Fazheng, R., Luda, Z., Hengtao, Z.: Development and characterization of a new amylase type time–temperature indicator. Food Control **19**(3), 315–319 (2008)
140. Galagan, Y., Su, W.-F.: Fadable ink for time–temperature control of food freshness: novel new time–temperature indicator. Food Res. Int. **41**(6), 653–657 (2008)
141. Vaikousi, H., Biliaderis, C.G., Koutsoumanis, K.P.: Applicability of a microbial time temperature indicator (TTI) for monitoring spoilage of modified atmosphere packed minced meat. Int. J. Food Microbiol. **133**(3), 272–278 (2009)
142. Kuswandi, B., Wicaksono, Y., Jayus, Abdullah, A., Heng, L.Y., Ahmad, M.: Smart packaging: sensors for monitoring of food quality and safety. Sens. Instrum. Food Qual. Saf. **5**, 137–146 (2011)

143. Toki, G.F.I., et al.: Single minute exchange die (SMED): a sustainable and well-timed approach for Bangladeshi garments industry. Clean. Eng. Technol. **12**, 100592 (2023). https://doi.org/10.1016/j.clet.2022.100592
144. Islam, M.R., Mia, R., Habib, M.A.B., Lotif, M.A.: A comparative analysis for the cleaner production of vortex spun yarn based on the nozzle position. Heliyon **8**(10), e10963 (2022). https://doi.org/10.1016/j.heliyon.2022.e10963
145. Taoukis, P., Labuza, T.: Time-temperature indicators (TTIs). Nov. Food Packag. Tech. 103–126 (2003)
146. Wallach, D., Novikov, A.: Methods and devices for detecting spoilage in food products. PCT International Patent Application WO 98y20337. Biodetect Corporation (1998)
147. Pacquit, A., et al.: Development of a smart packaging for the monitoring of fish spoilage. Food Chem. **102**(2), 466–470 (2007)
148. Horan, T.J. (ed.): Method for determining bacteria contamination in food package. Google Patents (1998)
149. Cameron, A.: T. Talasila in modified-atmosphere packaging of fresh fruits and vegetables. In: IFT Annual Meeting, Book of Abstracts, p. 254 (1995)
150. Shen, Q., Kong, F., Wang, Q.: Effect of modified atmosphere packaging on the browning and lignification of bamboo shoots. J. Food Eng. **77**(2), 348–354 (2006)
151. Smolander, M., Hurme, E., Ahvenainen, R.: Leak indicators for modified-atmosphere packages. Trends Food Sci. Technol. **8**(4), 101–106 (1997)
152. Abe, Y.: Active packaging with oxygen absorbers. In: VTT Symposium, vol. 142, pp. 209–209. Valtion Teknillinen Tutkimuskeskus (1994)
153. Hurme, E., Ahvenainen, R.: Active and smart packaging of ready-made foods. In: Proceedings of International Symposium Minimal Process, pp. 1–7. Ready Made Foods (1996)
154. Kuswandi, B.: Real-time quality assessment of fruits and vegetables: sensing approaches. In: Sensor-Based Quality Assessment Systems for Fruits and Vegetables, pp. 1–30. Apple Academic Press, (2020)
155. Nakamura, H., Nakazawa, N., Kawamura, Y.: Food oxidation indicating material-comprises oxygen absorbtion agent containing indicator composed of methylene blue. Reducing Agent Resin Bind. JP 62–183834 (1987)
156. Mattila-Sandholm, T., Ahvenainen, R., Hurme, E., Järvi-Kääriänen, T.: Oxygen sensitive colour indicator for detecting leaks in gas-protected food packages. Eur. Pat. EP 666977 (1998)
157. Davies, E., Garner, C. (ed.): Oxygen indicating composition. UK Patent Application GB 2 298273. The Victoria University of Manchester Manchester (1996)
158. Mattila, T., Tawast, J., Ahvenainen, R.: New possibilities for quality control of aseptic packages: microbiological spoilage and seal defect detection using head-space indicators (1990)
159. DeCicco, B.T., Keeven, J.K. (ed.): Detection system for microbial contamination in health-care products. Google Patents (1995)
160. Kress-Rogers, E.: The marker concept: frying oil monitor and meat freshness sensor. In: Kress-Rogers, E. (ed.) Butterworth-Heinemann, vol. 523, p. 1993. Stoneham, Mass (1993)
161. Rubio Retama, J., Mecerreyes, D., Lopez-Ruiz, B., Lopez-Cabarcos, E.: Synthesis and characterization of semiconducting polypyrrole/polyacrylamide microparticles with GOx for biosensor applications. Colloids Surf. A Physicochemical Eng. Asp. **270**, 239–244 (2005)
162. Ahuja, T., Mir, I.A., Kumar, D.: Biomolecular immobilization on conducting polymers for biosensing applications. Biomaterials **28**(5), 791–805 (2007)
163. Arshak, K., Adley, C., Moore, E., Cunniffe, C., Campion, M., Harris, J.: Characterisation of polymer nanocomposite sensors for quantification of bacterial cultures. Sens. Actuators B Chem. **126**(1), 226–231 (2007)
164. Subramanian, V., et al.: Progress toward development of all-printed RFID tags: materials, processes, and devices. Proc. IEEE **93**(7), 1330–1338 (2005)
165. Tentzeris, M.: Novel paper-based inkjet-printed antennas and wireless sensor modules. In: 2008 IEEE International Conference on Microwaves, Communications, Antennas and Electronic Systems, pp. 1–8. IEEE (2008)

166. Loh, K.J., Lynch, J.P., Kotov, N.A.: Passive wireless sensing using SWNT-based multifunctional thin film patches. Int. J. Appl. Electromag. Mech. **28**(1–2), 87–94 (2008)
167. Demoustier, S., Minoux, E., Le Baillif, M., Charles, M., Ziaei, A.: Review of two microwave applications of carbon nanotubes: nano-antennas and nano-switches. C R Phys. **9**(1), 53–66 (2008)
168. Jedermann, R., Behrens, C., Westphal, D., Lang, W.: Applying autonomous sensor systems in logistics—combining sensor networks, RFIDs and software agents. Sens. Actuators A **132**(1), 370–375 (2006)
169. Potyrailo, R.A., Mouquin, H., Morris, W.G.: Position-independent chemical quantitation with passive 13.56-MHz radio frequency identification (RFID) sensors. Talanta **75**(3), 624–628 (2008)
170. Mahajan, P.V., Caleb, O.J., Singh, Z., Watkins, C.B., Geyer, M.: Postharvest treatments of fresh produce. Philos. Trans. R. Soc. A Math. Phys. Eng. Sci. **372**(2017), 20130309 (2014). https://doi.org/10.1098/rsta.2013.0309
171. Boerman, J.K., Bauersfeld, M.L., Schmitt, K., Wöllenstein, J.: Detection of gaseous ethanol by the use of ambient temperature platinum catalyst. Procedia Eng. **168**, 201–205 (2016). https://doi.org/10.1016/j.proeng.2016.11.217
172. Hu, L.-Y., et al.: Hydrogen sulfide prolongs postharvest shelf life of strawberry and plays an antioxidative role in fruits. J. Agric. Food Chem. **60**(35), 8684–8693 (2012). https://doi.org/10.1021/jf300728h
173. Alam, A.U., Rathi, P., Beshai, H., Sarabha, G.K., Deen, M.J.: Fruit quality monitoring with smart packaging. Sensors **21**(4). https://doi.org/10.3390/s21041509

Chapter 12
Edible Packaging: Novel Approaches for Food Protection and Sustainability

Vishal Srivastava and Hema Garg

Abstract In the chapter, we explored the innovative concept of edible packaging, a transformative approach in the food industry with extensive sustainability and environmental protection implications. Edible packaging emerges as a compelling alternative to traditional packaging materials, offering a dual function of protecting food while being consumable and biodegradable. This chapter delves into the current landscape of edible packaging, examining its composition, types, and the technological advancements that have made it a viable option. The chapter emphasizes the role of natural and renewable resources in creating edible coatings and films, highlighting their properties and use in various food products. Furthermore, the chapter explores the regulatory and safety aspects of edible packaging, considering consumer acceptance and legal frameworks across different regions. The prospect of using edible packaging for reducing food waste and its role in sustainability is also examined, offering insights into the environmental impact and lifecycle analysis. Finally, the chapter concludes with a forward-looking perspective, identifying future trends and research areas in edible packaging. It underscores the importance of interdisciplinary collaboration and innovation in developing edible packaging solutions that align with global sustainability goals and consumer demands.

Keywords Edible packaging · Sustainability · Biopolymers · Food protection · Shelf-life

---

V. Srivastava (✉)
Department of Design, Indian Institute of Technology, Delhi, New Delhi 110016, India
e-mail: vishal304444@gmail.com; vishal.srivastava@design.iitd.ac.in

H. Garg
School of Interdisciplinary Research, Indian Institute of Technology, Delhi, New Delhi 110016, India
e-mail: hema.garg16@gmail.com

S. ul Islam and M. Shahid (eds.), *Green Materials for Active Food Packaging*, Engineering Materials, https://doi.org/10.1007/978-981-96-0369-5_12

12.1 Introduction

Food packaging is essential in the food supply chain, playing a crucial role in the final preparation stages within food industries. It protects food from damage and degradation, enhancing safety, hygiene, and significant food waste reduction. The importance of effective food packaging for reducing food wastage is highlighted because nearly 30% of produced food is wasted due to spoilage while transporting and harvesting [1]. Proper packaging selection is essential to prevent negative impacts on food quality, such as color changes, microbial growth, lipid oxidation, or nutrient degradation [2]. Traditional packaging, often used once and discarded, includes materials like paper, plastic, glass, steel, aluminum, and various alloys. Although some materials like paper have relatively high recycling rates, others, particularly plastics, are less frequently recycled and pose significant environmental challenges due to their non-sustainable, non-biodegradable nature, often ending up in landfills or oceans [3]. These mostly petroleum-derived materials contribute to environmental pollution through CO_2 emissions and other toxicants during incineration. The increasing environmental concerns and consumers' desire for naturally processed, first-rate quality, and safe foods have spurred interest in developing sustainable, biodegradable, and edible packaging. Such packaging incorporates materials like starches, cellulose derivatives, chitosan, gums, and proteins, offering advantages like biocompatibility and moisture barrier properties [4]. Packaging materials that are biodegradable and edible have become a priority in the food industry, with ongoing research focused on making these solutions commercially viable despite challenges like low resistance to gases and difficulty in production. Awareness of edible packaging's benefits could significantly reduce landfill waste, decrease pollution, and aid in addressing climate change.

12.1.1 Overview of Edible Packaging: Advantages and Limitations

Edible packaging is emerging as an environmentally friendly and biodegradable option in active food packaging, offering enhancements over traditional packaging by optimizing food quality [5]. They also act as carriers for antioxidants and antimicrobial agents, which bolster their functionality. This versatility has spurred a considerable uptick in research over the past decade, although several challenges must be addressed to ensure safe and effective scaling to industrial production levels. Edible packaging materials are sourced from natural polymers that pose no health risks and are completely safe to consume. Coatings are directly applied to the food's surface and are meant to be integral to the food product, not removed before consumption. The choice of edible packaging solutions is largely influenced by the specific food product being packaged, the packaging material's characteristics, and the processing methods [6]. Furthermore, it is crucial that the packaging is sensorially compatible with the food it encases, ensuring that it does not alter the flavor or fragrance of the product. Edible materials offer several advantages over synthetic ones, yet some of their benefits are not fully understood. Edible packaging can replace and enhance the outer layers of packaged products, preventing moisture, aroma, and ingredient loss while permitting the regulated permeation of gases required for food respiration, such as ethylene, oxygen, and carbon dioxide. These packages can also enhance the sensory qualities of food by adding flavorings and colorings and modifying surface characteristics like hydrophobicity or hydrophilicity [4]. They serve as carriers for functionally active components that enhance health or well-being. While polysaccharides and proteins offer good oxygen barriers and mechanical strength, lipids provide superior moisture barrier properties but have poor mechanical strength and adhesion to hydrophilic food surfaces, affecting their practical application in food packaging.

12.2 Types of Edible Packaging Materials

See Fig. 12.1.

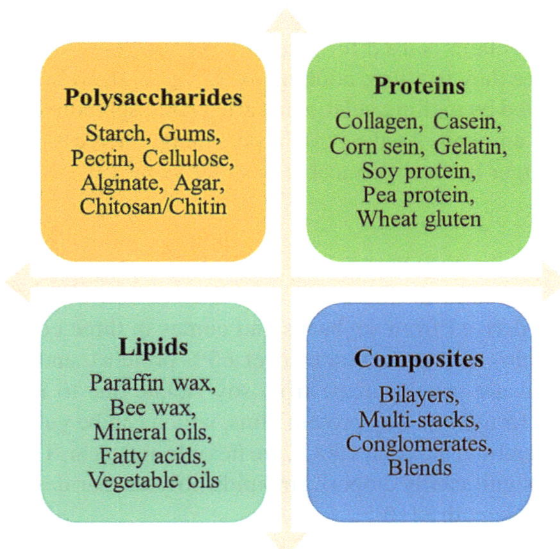

Fig. 12.1 Classification of edible materials [4]

12.2.1 Protein-Based Edible Films and Coatings

Proteins are biological macromolecules composed of amino acid monomers linked by peptide bonds. Each amino acid consists of a central carbon atom bonded to an amino group, a carboxyl group, a hydrogen atom, and a variable side chain. This variable side chain gives each amino acid its unique chemical properties. Protein structure is categorized into four levels: primary, secondary, tertiary, and quaternary [7]. Proteins differ from polysaccharides in being heteropolymers with at least 20 amino acids, whereas polysaccharides are often homopolymers like glucose. Protein films have superior mechanical properties compared to polysaccharides but show higher permeability to water vapor and have lesser mechanical strength than synthetic polymers. Proteins are classified as either fibrous, which are water-soluble, or globular, which are water-insoluble and require denaturation before film formation.

12.2.1.1 Gelatin

Collagen, a fibrous protein, is found in the skin, bones, tendons, vascular systems, and connective tissues of both vertebrates and invertebrates, including shellfish and fish. It is commonly extracted from pig skins, cow hides, and fish wastes. Gelatin is transparent, tasteless, and brittle and is derived from collagen through hydrolysis [8]. This gelatin is used for creating edible films, particularly in the livestock industry, where they facilitate cooking by acting as an edible skin when heated. The films are created by melting gelatin in warm water, spreading it onto a surface, and then drying it in an oven. As the protein percentage increases, these films become thicker and stronger but less permeable to water vapor [9].

12.2.1.2 Soy Protein

It is derived from soybeans and comes in three commercial forms: soy flour (56% protein), soy concentrate (over 65% protein), and soy isolate (over 90% protein). Films are usually made from soy isolate due to its higher protein content, aiding film formation. Soy protein films, which can be yuba (boiled and air-dried) or baked (spread and baked at 100C), are flexible and clear. They serve as excellent gas barriers and significantly outperform lipids/polysaccharides films in permeability for oxygen when kept dry [10].

12.2.1.3 Milk and Myofibrillar

Casein and whey proteins are primary components of the protein content of cow's milk, which has a protein content of 33 g/L of milk. 80% of the milk protein comprises Casein, featuring α, β, and κ-casein components. Caseinates, commercially produced by skimming milk at 20 °C and pH 4.6 and then precipitating with an alkali, exhibit superior emulsification properties due to their high proline content. These casein-based films are stable across various pH levels, salt concentrations, and temperatures. Whey protein is derived after precipitating casein at the same pH. Whey protein isolates, typically 90% protein, form excellent oxygen barriers in films. The relative humidity is not very high but has inadequate water vapor permeability. Edible films are also created using heat treatment or enzymes like transglutaminase to induce gelation and enhance film properties. Plasticizers improve film flexibility but can increase water vapor permeability, negatively impacting food quality [11]. Alternatively, hydrolyzing whey protein isolate reduces the need for plasticizers, thus maintaining film flexibility with less impact on water vapor permeability and reducing milk protein allergenicity. Fish is a common source of myofibrillar proteins like myosin and actin, used to create edible films.

12.2.2 Polysaccharides-Based Edible Films and Coatings

It is the most abundant natural polymer from plants, utilized in various ways, including wood for shelter and fuel, food from fruits, vegetables, and seeds, and fibers for paper and clothing. These polymers are present in all organisms and extracted using methods like anion exchange chromatography. Polysaccharides, which, when hydrated, have high viscosity and can be used as thickeners, are also employed to create edible films. They include cellulose, starch, chitosan, and alginate, with treatments to improve solubility or gel formation. While effective as carbon dioxide and oxygen barriers, their vapor barrier properties are limited, affecting their use in food packaging [12]. Enhancements such as crosslinking and multi-layered films improve their water resistance and overall barrier effectiveness.

12.2.2.1 Cellulose Derivatives

Cellulose, the most found polysaccharide, comprises repeated D-glucose units connected by β-1,4 glycosidic bonds. Naturally water-insoluble due to strong hydrogen bonds, cellulose is chemically modified to create soluble derivatives for edible films. Recent developments in cellulose derivatives for edible food packaging focus on enhancing functionality and sustainability [13]. Nanocellulose coatings have increased fresh-cut vegetables' shelf life and created bacterial cellulose-lactoferrin composites with antimicrobial properties suitable for edible packaging [14]. Particular emphasis has been placed on modifying cellulose's physical properties to improve its applicability in food packaging. For instance, hydroxypropyl cellulose, being thermoplastic, has been exploited for its suitability in processes such as extrusion, which are beneficial for industrial-scale production [15]. The ability of modified cellulose films to incorporate various functional additives has assisted in developing active and packaging solutions that can interface with food to increase freshness, monitor conditions, or improve safety [16].

12.2.2.2 Chitosan

After Cellulose, Chitin is the second most commonly found polysaccharide. It is found in invertebrate exoskeletons and consists of 2-deoxy-2-acetamido-b-D-glucose units. Chitosan, a derivative made soluble through de-N-acetylation, is used in anti-fungal preservative coatings for fruits and vegetables, forming transparent, strong, and flexible films with adequate resistance to oxygen and fats but are moisture sensitive [17]. Often combined with other biopolymers like gelatin, chitosan has been utilized to create composite films that exhibit improved mechanical strength, surface hydrophobicity, and barrier properties against oxygen and water vapor [18]. Applications of these advanced chitosan-based films include active food packaging embedded with antimicrobial and antioxidant agents to increase the shelf-life of spoilage-prone foods like meats, fruits, and vegetables [19]. These films effectively protect food from spoilage and pathogen contamination and maintain the sensory and nutritional quality of the food products [20]. Developing these chitosan-based films signifies an important step forward in food packaging, aligning with global sustainability trends and reducing plastic waste.

12.2.2.3 Starch

Starch, a natural polymer produced by plants like potatoes, corn, and rice, is colorless, edible, and tasteless, allowing for easy flavor or color additions. Commercially, it is often sourced from potatoes, corn, cassava, wheat, and quinoa. Starch is mainly comprised of a linear polymer (20–25% amylase) and a branched polymer (75–80% amylopectin) [21]. Developments in edible food packaging made out of starch primarily focus on improving their functional properties, most importantly regarding mechanical strength, flexibility, and barrier characteristics [22]. Researchers have made significant strides in enhancing the mechanical properties of starch-based films by incorporating various additives, such as plasticizers (Amino, water, glycerol), which reduce the rigidity and increase flexibility, making them more practical for food packaging applications [23]. These active films are designed to extend the shelf life of perishable items by suppressing microbial growth and oxidation processes [24]. Starch-based films have also been developed for specific applications, such as UV-protective packaging and improved gas barrier properties, to maintain the quality and safety of food products [25]. These advancements are significant because they support creating environmentally friendly packaging options, aligning with global environmental goals, and addressing consumer demands for safer, higher-quality packaged foods.

12.2.2.4 Alginate

Alginate, derived from brown algae and seaweed cell walls, forms edible gels through calcium ions interacting with polyuronides [26]. Unless treated with chelating agents at alkaline pH, these gels are irreversible. Cooling hot gels results in weak Ca++ alginate associations, enhancing cross-linking. Initially poor in water vapor barriers, alginate films become water-insoluble with added calcium, reducing water vapor permeability [27]. Alginate-based edible food packaging has improved mechanical properties and barrier functions. The innovations focus on creating composite films by incorporating synthetic and natural polymers or nanoparticles to leverage specific characteristics that improve overall performance [28]. One key advancement involves the integration of calcium ions, which interact with the alginate's carboxylate groups to form a stable, insoluble gel, significantly enhancing the film's structural integrity [29]. Additionally, these films are being developed for use in pharmaceutical and medical applications due to their biocompatibility and non-toxic nature [30]. This integration of alginate with other biodegradable materials highlights a move towards more sustainable packaging solutions that reduce reliance on synthetic polymers and support environmental conservation efforts.

12.2.2.5 Pullulan

Pullulan, a polysaccharide produced by the microbe Aureobasidium pullulans from starch, comprises maltotriose units connected by α (1,6) glycosidic links. It is notable for its high water solubility and ability to create edible films [31]. These films are heat-sealable, odorless, colorless, tasteless, and transparent, with low permeability to oil and oxygen, yet they allow water to pass through. Pullulan, a polysaccharide synthesized from starch by Aureobasidium pullulans, offers excellent film-forming abilities, making it ideal for food packaging [32]. Innovations include incorporating nanoparticles and bioactive compounds into pullulan films to improve mechanical strength, thermal stability, and antimicrobial activity, thus enhancing food preservation. For instance, pullulan films have been developed with enhanced barrier properties against oxygen and moisture, which is crucial for maintaining food quality [33]. These films are also being explored for pharmaceutical and medical applications due to their biocompatibility and non-toxic nature. The application of pullulan in food packaging is diversified, including its use as coatings for freshness preservation and as a base for active packaging materials [34].

12.2.2.6 Gellan

Gellan, a microbial polysaccharide from Sphingomonas elodea, forms hard, brittle gels used in the food industry as edible gels. Its deacylated form consists of linear tetrasaccharide units, gels with cations like K^+, Na^+, Ca^{++}, and Mg^{++} [35]. Edible coatings from alginate, pectin, and gellan extended fresh-cut melon's shelf-life by reducing water vapor transmission and maintaining firmness. Still, they didn't effectively block gas diffusion or stop microbial growth over 15 days at 4 °C [36]. The study found that gellan-based coatings with added lipids were more effective at reducing water loss than alginate ones [37, 38].

12.2.2.7 Carrageenan

Carrageenan, sulfated polysaccharides from red seaweed cell walls, are classified into three types—κ (20% sulfate), ι (33%), and λ (41%)—based on sulfate content. These form thermoreversible gels in water [39]. These polysaccharides, extracted from brown and certain red algae, are chosen for their gelling properties [40]. Recent developments in carrageenan-based edible food packaging have emphasized improving mechanical properties and barrier functions and incorporating active ingredients [41]. Innovations have focused on combining carrageenan with other natural polymers and functional substances like polyphenols and enzymes, creating composite films with improved performance [42]. These films are particularly useful for products requiring controlled freshness and extended shelf lifein the food industry. Moreover, the potential for carrageenan-based films to monitor food freshness through intelligent packaging has been explored [43]. These smart films can indicate food quality changes by responding to environmental factors, thus adding value to food safety and consumer information.

12.2.2.8 Pectin

Pectin, derived mainly from fruits with high concentrations of citric acid, consists of α-(1,4) D-galactopyranosyluronic acid units linked by (1,2) L-rhamnopyranosyl units. Edible films made from Pectin are initially rigid and brittle, with high modulus but low elongation. However, adding plasticizers enhances their flexibility [44]. These films possess strong mechanical properties and are effective in protecting low-moisture foods. Pectin, a plant-derived polysaccharide, is increasingly utilized in creating bio-based coatings and packaging films mainly because of its excellent film-forming capabilities and environmental friendliness [45]. Innovations include the development of pectin-based composite films that improve mechanical strength and barrier functions against moisture and gases, which are crucial for maintaining food quality [46]. A significant focus has been on integrating pectin with other natural substances like proteins and nanoparticles to form composite materials with high-quality barriers and mechanical properties [47].

12.2.2.9 Agar

Agar is found naturally as a gel within red algae, and it does not dissolve in water at low temperatures but becomes soluble at high temperatures. It acts as a gelling agent in producing candies and desserts [48]. Recent advancements in agar-based edible food packaging have leveraged agar's natural properties for improved film fabrication. Innovations aim to enhance agar film's functional properties by incorporating bio-active materials for active and smart packaging applications [49]. The integration of bio-active materials, plasticizers, and other biopolymers has been examined to address issues like brittleness and low barrier properties typically associated with agar films [50]. Applications of these advanced agar films include packaging fresh produce, meat, and bakery items, where the extended shelf life and maintenance of product quality are critical [51]. Furthermore, the high flexibility of these films allows them to be shaped into different shapes and sizes, allowing for customized packaging solutions to meet specific market needs.

12.2.3 Fruit Purees

Fruit purees like apple, peach, apricot, tomato, pear, mango, and banana are utilized to form edible films primarily composed of cellulose and pectin. These films typically lack flexibility, but adding plasticizers can enhance their toughness and processability [52]. Composite films with fiber or particles improve mechanical and barrier properties. Bananas, rich in starch and pectin, are ideal for creating biodegradable and cost-effective packaging with good oxygen barrier and mechanical properties, suitable for dried products [53]. These biodegradable films, derived from fruit purees like apple, banana, and mango, reduce environmental impact and maximize fruit utilization, aligning with the circular economy [54]. These films are advantageous because they retain the natural color and flavor of the fruit, adding sensory appeal compared to other films [55]. The primary component of these films includes natural polymers like pectin and cellulose, and the inclusion of sugars acts as natural plasticizers. Significant advancements include the usage of nanoparticles and nanoemulsions to enhance the films' mechanical and barrier properties [56]. For instance, it has been demonstrated that adding cellulose nanofibers and chitosan nanoparticles increases the tensile strength and decreases oxygen permeability. Furthermore, these films also act as vehicles for active substances, including antimicrobials and antioxidants, which migrate to the food, improving safety and increasing shelf life [57].

12.2.4 Lipid-Based Edible Films and Coatings

Lipids are diverse substances extracted from natural waxes like candelilla, carnauba, and beeswax and petroleum-based waxes such as paraffin. They also include mineral and vegetable oils, acetoglycerides, and oleic acids. Solid waxes are particularly effective in reducing water vapor transmission, making them ideal for applications that minimize dehydration, such as coatings for fruits and vegetables to reduce abrasion during handling. However, lipids do not form large macromolecules like polysaccharides and proteins and thus cannot create self-supporting film structures. This results in lipid films that are brittle, opaque, and sensitive to oxidation [58]. Bio-based waxes are preferred over mineral-based for food applications due to safety concerns. Carnauba wax, found in the leaves of the Brazilian Carnauba palm, is notable for its use in increasing the shelf life of fruits by decreasing moisture loss [59]. The effectiveness of wax in moisture resistance depends on its composition, with saturated fatty acids generally providing better barriers than unsaturated ones.

12.2.5 Composite Edible Packaging Materials

Significant advancements have been made in developing composite edible coatings and films, also known as multicomponent coatings. In the domain of multicomponent systems, also recognized as composite edible films and coatings, there is considerable scope for improving food quality [60]. These composites, crafted from various biopolymer combinations, leverage the synergistic effects of these polymers to enhance the thermal, mechanical, barrier, and other physicochemical characteristics of packaging [61]. Moreover, such packaging has enhanced the physicochemical and mechanical properties of fruits, increasing the shelf life of perishables, including fruits, vegetables, and bakery products. These edible films are categorized into binary and ternary systems—the former consisting of two hydrocolloids and the latter of three, tailored to meet specific food preservation needs [62]. The incorporation of active agents like essential oils and plant extracts boosts antimicrobial properties and also enhances mechanical, barrier, and thermal characteristics, contributing to reduced lipid oxidation and improved consumer acceptability [63].

12.2.6 Active Edible Packaging

Incorporating natural antimicrobial and antioxidant agents in developing edible packaging is gaining attention due to its ability to enhance food safety and extend the shelf life of products by minimizing oxidation and preserving quality attributes in foods [1]. The use of additives like plasticizers, emulsifiers, and biobased active agents, including essential oils and natural extracts, significantly enhances the properties of these edible packages. Commercially available active packaging solutions like Active-Film™, PEAKfresh, and others control moisture, reduce water vapor transmission, and inhibit ethylene synthesis while boosting antimicrobial and antioxidant activities [64]. These natural agents prevent oxidative stress and microbial contamination and disrupt microbial cell membranes, leading to cell death [65]. These components serve dual functions as antimicrobial and antioxidant agents and as colorimetric pH indicators, allowing consumers to monitor the freshness of meat

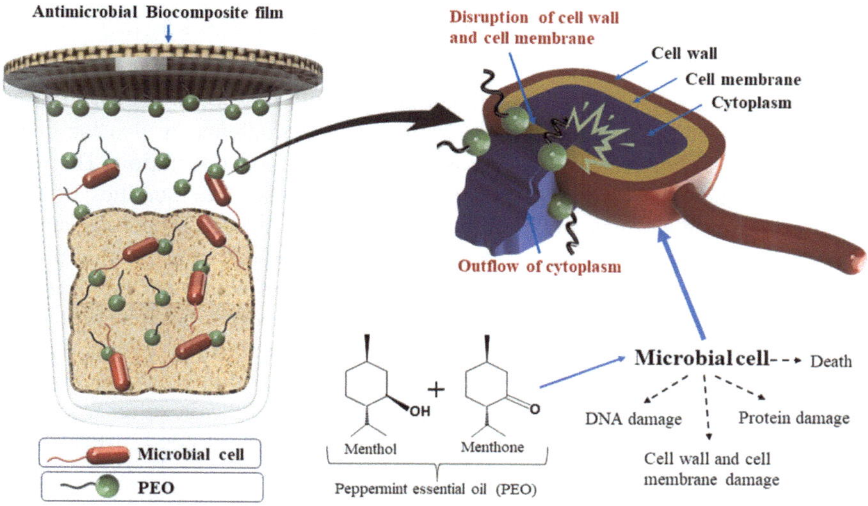

Fig. 12.2 Possible food shelf-life enhancement mechanism by using active food packaging [1]

through simple visual cues and smartphone applications, enhancing food safety and consumer convenience [66]. The mechanism of active food packaging has been shown in Fig. 12.2.

12.2.7 Nanomaterials for Edible Packaging

Nanotechnology is revolutionizing the food packaging industry by developing nanopackaging with enhanced properties. This technology is expected to account for 25% of all food packaging soon, significantly enhancing food safety by extending product shelf life, reducing waste, and repairing packaging tears, promoting better consumer health. Various nanomaterials, including nanoparticles, polymer nanocomposites, nanoemulsions, nanoformulations, and nanofillers like nanostarch, nanocellulose, nanoproteins, and nanolipids, are used to create active packaging [67]. Elements such as silver, gold, titanium dioxide, zinc, and nanoclay are incorporated to enhance these properties. Nanotechnology contributes to the development of active, smart, and intelligent packaging systems that guard against mechanical, physical, microbial, and UV damage, and it supports the controlled release of preservatives and other active agents, further prolonging the shelf life of food products [68].

12.3 Properties and Characteristics of Edible Packaging

Edible packaging acts as a multifunctional layer on the exterior of packaged products, preventing the loss of moisture, aromas, and ingredients. It also facilitates the controlled exchange of essential gases critical for food respiration. This type of packaging not only preserves but can enhance the sensory qualities of food by incorporating flavorings and colorings and adjusting surface properties such as hydrophobicity or hydrophilicity. Additionally, it can deliver functional additives that offer health or well-being benefits. The properties of edible packaging materials vary. Polysaccharides offer good oxygen barrier properties but lower moisture resistance; proteins provide considerable mechanical strength suitable for protecting fruits and vegetables during transport; and lipids, while providing excellent moisture barrier properties and maintaining the integrity of color and flavor, suffer from poor mechanical strength and often do not adhere well to hydrophilic food surfaces.

12.3.1 Barrier Properties

Oil and moisture absorption properties, the transfer of oxygen, and changes in flavor called migration properties are crucial for maintaining food quality [69]. These elements also affect the mass transfer processes amongst the food (including its ingredients) and packaging or the transfer amongst the surroundings and the packaged content. Edible films and coatings are designed to hinder these migration phenomena, thus enhancing packaged food product standards [70]. The performance of these edible barriers, particularly in terms of their transport properties, is influenced by the conditions in the surroundings (pressure, temperature, etc.) and the materials' composition. [71].

12.3.2 Mechanical Strength

The mechanical properties of food packaging are of critical significance in securing the packaged content from the external environment and impact. The mechanical properties of food packaging are largely governed by the interactions between fillers and the matrix, as well as physical, chemical, and thermal conditions that affect their stability and stiffness. As per the ASTM D882-91 method, the mechanical properties of packaging film are assessed by deformation at the break, percent elongation at break (EAB), elastic modulus, and tensile strength (TS) [72]. Enhanced mechanical resistance has been observed in edible film composites, particularly with the incorporation of citral essential oil, which gives favorable food wrapping. Furthermore, adopting sustainable bionanostructured materials like nanocellulose, nanochitosan, nanostrach, and various inorganic nano-fillers has improved packaging properties. For example, a study demonstrated that microfibrillated cellulose nanofibers (mgCNF) in a chitosan (CS) coating significantly enhanced Young's modulus and tensile strength compared to pure CS, underscoring the potential of nano-enhanced edible coatings in food packaging [73].

12.3.3 Optical Properties

The color and optical properties of edible films are crucial for applications in packaging. These properties, which include transparency and color coordinates (L* for lightness, a* for red-green spectrum, and b* for yellow-blue spectrum), significantly enhance the visual appeal of packaged foods, influencing consumer acceptance. Studies show that essential oil (EO) based films notably impact these color parameters, with lemongrass and sago nanoemulsions displaying higher lightness values, while alginate and thyme films register lower on the lightness scale [74]. The transparency of food packaging materials, crucial for shielding contents from harmful UV and visible light, is typically higher in edible films like those made from whey protein isolate, compared to conventional plastics like polyethylene and polypropylene. This property is measured using UV–Vis spectrophotometry, factoring in film thickness. Moreover, color properties are defined using L*, a*, b* values, providing a gauge for the food product's quality and safety [65]. These parameters also help detect adulteration and monitor pigment stability and browning effects, which are affected by polymer matrix used and the different types of fillers in it. Moreover, sensory criteria such as touch, texture, appearance, aroma, color, and overall perception are crucial parameters for assessing the quality of food and determining consumer acceptability of packaged food [75].

12.3.4 Sensory Evaluation

The sensory assessment of a given product requires analyzing along with interpreting how acceptable a product is to consumers using the human senses. This process employs experimental designs and statistical analysis to gauge consumer acceptance based on attributes like smell, taste, sight, and touch. Sensory evaluations, conducted by personnel ranging from skilled to unskilled, can be categorized into analytical testing, affective testing, and perception testing. Analytical testing focuses on identifying and describing differences between products, while affective testing assesses consumer preferences and likings through methods like hedonic and preference tests. Various sensory tests include the difference test (such as paired comparison, duo-trio, and triangle tests), rating tests (including ranking and hedonic rating tests), sensitivity tests, and descriptive tests. These tests help determine the most liked product features and measure specific attributes like flavor intensity or smoothness. This thorough sensory analysis helps ensure that food products meet consumer standards before they reach the market.

12.3.5 Morphological Properties

The morphological characteristics are crucial for understanding the size and structure of the polymer and filler to be used in coatings and edible films when considering packaging. The characterization typically uses FESEM (Field Emission Scanning Electron Microscopy) analysis. This method not only reveals the physical structure but also aids in assessing tensile and other physical properties. Research on edible casein films with zein hydrolysate showed that the addition of wax maintained a smooth and uniform texture similar to films where no wax was used in preparation, attributed to the hydrophobicity of the zein hydrolysate filler [76]. Further, FESEM studies demonstrated varied morphologies, such as fiber-like and aligned particle dispersions in CS-CNF and CS-mgCNF films, while analyses of neat pea starch and peanut protein isolate films revealed swollen starch granules and cavities, respectively, with blended films exhibiting a flexible network morphology.

12.3.6 Thermal Properties

Thermal characteristics are an important consideration for the broader application of edible film materials, typically analyzed using Thermogravimetric Analysis (TGA) and Differential Scanning Calorimetry (DSC). DSC measures the heat flow difference between a sample and a reference, while TGA assesses the composition and mass changes of the films as they are heated to various degradation temperatures, tracking the weight fraction and changes in mass based on the initial sample mass and the quantity of nanoparticles used along with their type [77]. Studies like the DSC analysis of kefiran films have highlighted the interaction between polymers and plasticizers, which are essential for food packaging. TGA analysis has shown that micro fibrillated cellulose nanofiber (mgCNF) reinforced chitosan (CS) nanocoatings exhibit improved thermal stability, with mgCNF providing greater heat dimensional stability than CNF.

12.3.7 Physicochemical Properties for Food Preservation

The essential physicochemical properties of coating materials and edible films for food packaging are moisture content, viscosity, and water solubility. Moreover, the stability of the film-forming dispersions is also assessed. Water solubility is crucial, defined by the percentage of total soluble matter (%TSM), indicating how much film dissolves in water. Glycerol-plasticized films, for instance, show increased solubility due to glycerol being hydrophilic, enhancing the film's water attraction. Viscosity measurements are vital for application methods like spray or immersion coating, with nanoemulsions' viscosity being particularly important for stability. The content of moisture that is calculated by the weight of moisture absorbed over time by the film, also called moisture content, impacts the film's mechanical properties [78]. Plasticizers can also increase a film's flexibility and decrease its moisture sensitivity, though a high concentration may compromise its structural integrity.

12.3.8 Migration Properties for Packaging Application

To ensure food safety, risk assessments and migration assays of food packaging are necessary [79]. Packaging materials have found an increase in their usage in the food contact materials (FCMs), leading to new innovations in intelligent packaging such as enhanced barrier properties, active with nanosensors, and based on biocomposites. Despite these advances, some issues like chemical migrations from the FCMs into the food are still faced, which involves the diffusion of low molecular weight components influenced by factors like the type of packaging material, temperature, and duration of exposure. Migration studies, guided by regulations such as Commission Regulation EU No. 10/2011, typically employ tests that simulate conditions for the assessment of overall and specific migration. These tests are crucial for understanding how substances like food additives, monomers, and residues transfer from packaging to food, potentially affecting food safety.

12.4 Applications of Edible Packaging in the Food Industry

See Fig. 12.3.

12.4.1 Fruits and Vegetables Packaging

A wholesome meal must contain the right number of vegetables and fruits due to their rich content of nutrients and phytochemicals. Consuming these foods reduces the risk of health disorders and diseases like cardiovascular disorders, diabetes, and stroke. An intake of at least 400 g of different vegetables and fruits is recommended by the World Health Organization (WHO) for optimal health [80]. However, managing these perishable items post-harvest during storage and transportation poses significant hurdles in the food industry, often leading to losses that contribute to landfills and

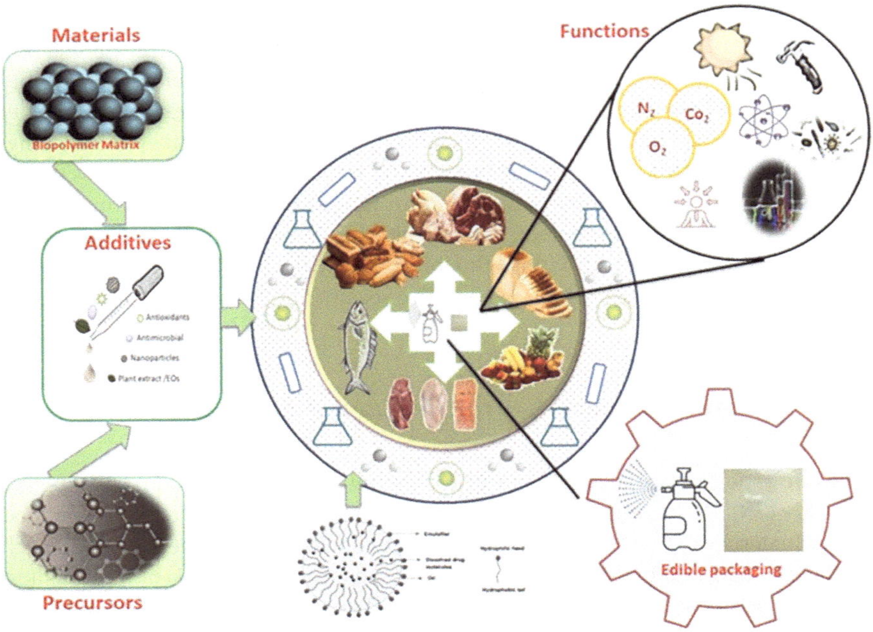

Fig. 12.3 Application, mechanism, and functions of edible coating on food product [62]

environmental issues. An elegant and sustainable solution is offered in the form of edible packaging, which helps keep the contents safe and fresh by improving their shelf life [81]. This packaging is crafted from materials such as polysaccharides, lipids, and proteins, serving as a barrier against gases and water vapor. During storage and transportation, it protects against physical, chemical, and biological damage.

12.4.2 Meat and Meat Products Packaging

The global market for meat and its derivative products is surging, with consumption levels rising worldwide except in Europe. In 2018, global meat consumption reached approximately 346.16 million tons and is projected to increase by 44%, reaching 453 million tons by 2030 [82]. Meat provides essential nutrients in the form of proteins, fatty acids, minerals, vitamins, and amino acids, along with essential micronutrients. However, meat quality can degrade due to factors like light exposure, fluctuating pH, temperature changes, and microbial contamination, leading to spoilage through processes that may be either microbial growth, lipid oxidation, or rancidity[83]. These factors have a negative effect on shelf life, and it also degrades the nature of meat products by affecting their texture, color, and nutritional content. Traditional non-biodegradable materials like plastics are replaced by eco-friendly, edible packaging from renewable resources [3]. This type of packaging preserves meat quality by controlling moisture loss and gas exchange and preventing microbial growth, thereby extending shelf life. These alternatives offer dietary advantages along with helping towards sustainability by helping lower harmful emissions and thus contributing positively towards climate [84].

12.4.3 Dairy, Bakery, and Other Food Products

Milk-based products are essential in diets globally and rich in nutrients. Maintaining their shelf life is challenging due to their perishability, with mold spoilage being a significant concern [85]. Edible packaging, particularly biopolymer-based coatings, has proven effective in extending shelf life by reducing microbial growth and oxidation. Research has shown that coatings using materials like chitosan and carrageenan can significantly inhibit mold growth and improve longevity. It was found that chitosan-based coatings were more potent than carrageenan in improving the shelf life of brown seaweed [62]. Other studies have explored coatings made from agar, whey protein with antimicrobial agents, and sweet whey on cheeses, finding them beneficial in preserving acidity, pH, taste, color, and texture while also reducing pathogen growth and weight loss [86].

12.5 Challenges and Future Perspectives

Edible packaging is emerging as a viable solution in sustainable and eco-friendly food packaging, answering the global call for greener practices. These coatings help reduce the environmental burden of traditional packaging materials, but their adoption requires addressing certain limitations. To fully leverage the benefits of edible coatings, a thorough exploration of these challenges and the development of innovative solutions are necessary. This collaborative effort between researchers and industry could facilitate their integration into mainstream packaging practices. As environmental concerns gain priority worldwide, overcoming these obstacles becomes crucial for transitioning to sustainable packaging norms. However, issues such as the production, storage, and commercial use of edible coatings remain, particularly with consumer acceptability and food safety concerning nutritional quality and shelf-life extension.

12.5.1 Technical Challenges in Edible Packaging Development

Edible food packaging is promising for extending food shelf life but often falls short in durability compared to conventional plastics. To address this, researchers are investigating methods to enhance their mechanical properties. This includes combining different biopolymers and adding strengthening elements such as nanomaterials. These efforts are directed at boosting edible coatings' structural integrity and resilience, making them more effective and broadening their use across various industries. Derived from natural or edible materials, these coatings offer environmental benefits and potential health advantages but may allow higher permeability due to their structural composition. In contrast, synthetic plastics provide more effective barriers but raise environmental concerns. Balancing environmental impact, health benefits and food protection is crucial, driving ongoing research to further improve the barrier properties of edible coatings. Researchers are exploring nanocomposites, incorporating nanoscale materials into coatings to improve their performance. By integrating nanoparticles or nanofibers, these advanced coatings could significantly enhance gas and vapor resistance, revolutionizing food preservation through nanotechnology, improving food safety and sustainability, and reducing waste.

12.5.2 Regulatory Considerations and Food Safety Issues

Developing and enforcing safety standards for edible packaging and nanomaterials in food packaging necessitates detailed regulatory guidance from bodies like the FDA and the EU [87]. Before commercial use, edible packaging must be certified as safe due to their direct contact with food, with materials needing to be recognized as Generally Recognized As Safe (GRAS) by the FDA. Plant extracts and essential oils used in these coatings can also trigger allergic reactions depending on the dosage. It is crucial to test for toxicity and allergenic potential of these substances regularly. According to regulations, edible coatings are categorized as food additives, coating ingredients, or packaging materials. They must adhere to specific safety standards established by US and EU regulations [88]. Food preservation can be strengthened by the incorporation of nanomaterials in the manufacturing of packaging, but it would require strict international regulation to ensure safety.

12.5.3 Challenges of Compatibility with Various Foods

Edible packaging shows promise but faces challenges in broad application due to the diverse textures, moisture levels, and acidity of different food products. The interaction between coatings and various food characteristics can affect their effectiveness in preserving quality and appearance. Researchers are working on customized formulations of edible coatings to meet the unique needs of each food type. This customization ensures that the coatings perform optimally, enhancing preservation and maintaining quality across various food products by considering their specific characteristics.

12.5.4 Environmental Impact of Biobased Food Packaging

The environmental impact of edible packaging is significantly positive compared to conventional plastic packaging. Edible packaging, derived from biopolymers like polysaccharides, proteins, and lipids, offers an eco-friendly alternative that can biodegrade naturally and reduce the accumulation of waste in landfills and oceans [89]. These materials are often sourced from agricultural by-products, which minimize waste and utilize renewable resources. Furthermore, the production of edible packaging generally involves less energy and lower emissions of greenhouse gases compared to the synthesis of synthetic polymers. Replacing plastic food packaging with edible, sustainable alternatives can significantly reduce environmental pollution and advance more sustainable packaging solutions [90]. A recent study showed that the environmental impact of the active food packaging made from agriculture waste showed a high sustainability score compared to the conventional packaging materials [91].

12.5.5 Future Directions and Emerging Trends

Edible food packaging and coatings are key strategies in keeping the quality and sensory parameters of various packaged products intact, helping to reduce weight loss, and serving as barriers against gas transmission and water vapor. Composite packaging, which utilizes multiple polymers, is preferred over single-polymer coatings due to its superior barrier properties. Various methods, like high-pressure homogenization technologies, including microfluidization and ultrasonication, are employed to improve the stability of biopolymer-based edible product packaging. However, ongoing research is necessary to identify compatible biopolymers and understand their interactions with additives like plasticizers and emulsifiers. This research aims to increase the quality of food and prolong shelf life. A significant research focus is the recovery and valorization of biopolymers from food waste and by-products, turning them into valuable resources for biodegradable packaging. This aligns with sustainable development goals, promoting environmental health and reducing waste. There's also potential in combining edible coatings with non-thermal preservation techniques to enhance food preservation. Moreover, developing cost-effective coating application technologies, especially for perishable foods like fruits and vegetables, remains a priority for further research and innovation.

12.6 Conclusion

Edible packaging, derived from natural biopolymers like polysaccharides, proteins, and lipids, is increasingly recognized within the sustainable packaging system for its capacity to extend food product shelf life and maintain quality attributes like freshness, color, and nutrition. Advanced formulations, including binary and ternary composites, enhance these properties by providing effective barriers against moisture and gas. Additionally, integrating natural antimicrobials like essential oils enhances food safety by inhibiting microbial growth. Technological advancements refine the material properties at the nanoscale. Examples are high-pressure homogenization techniques like micro-fluidization and ultrasonication, which enhance stability. As global demand shifts towards substitutes for fossil-based substances and products, the potential of edible packaging is expanding. This shift is driven by consumer preference for recyclable and biodegradable options. It is supported by ongoing research into incorporating bioactive compounds, which offer controlled release functionalities, contributing to food safety and reducing waste. Nanotechnology further promises to enhance the nutritional quality and sensory attributes of contents, aligning along the future directions in food packaging.

Declaration The authors declare that they have no known competing financial interests or personal relationships that could have appeared to influence the work reported in this paper. Generative AI and AI-assisted technologies have been used in manuscript preparation to improve the readability and language of the work only. After using this tool, the authors reviewed and edited the content as needed and take full responsibility for the contents of the publication.

References

1. Srivastava, V., Singh, S., Das, D.: Development and characterization of peppermint essential oil/ rice husk fibre/ corn starch active biocomposite film and its performance on bread preservation. Ind. Crops Prod. **208**, 117765 (2024). https://doi.org/10.1016/j.indcrop.2023.117765
2. Jeevahan, J., Anderson, A., Sriram, V., Durairaj, R. B., Britto Joseph, G., Mageshwaran, G.: Waste into energy conversion technologies and conversion of food wastes into the potential products: a review. Int. J. Ambient Energy **42**(9), pp. 1083–1101 (2021). https://doi.org/10.1080/01430750.2018.1537939

3. Srivastava, V., Singh, S., Das, D.: Biodegradable fibre-based composites as alternative materials for sustainable packaging design. Smart Innov. Syst. Technol. 262 SIST, pp. 87–98 (2022). https://doi.org/10.1007/978-981-16-6128-0_9/COVER/
4. Trajkovska Petkoska, A., Daniloski, D., D'Cunha, N. M., Naumovski, N., Broach, A. T.: Edible packaging: Sustainable solutions and novel trends in food packaging. Food Res. Int. **140**, p. 109981 (2021). https://doi.org/10.1016/J.FOODRES.2020.109981
5. Aguirre-Joya, J. A., et al.: Basic and Applied Concepts of Edible Packaging for Foods. Food Packag. Preserv. pp. 1–61 (2018). https://doi.org/10.1016/B978-0-12-811516-9.00001-4
6. Restrepo, A. E., Rojas, J. D., García, O. R., Sánchez, L. T., Pinzón, M. I., Villa, C. C.: Mechanical, barrier, and color properties of banana starch edible films incorporated with nanoemulsions of lemongrass (*Cymbopogon citratus*) and rosemary (*Rosmarinus officinalis*) essential oils. **24**(8), 705–712 (2018). https://doi.org/10.1177/1082013218792133
7. Nur Hanani, Z. A., Roos, Y. H., Kerry, J. P.: Use and application of gelatin as potential biodegradable packaging materials for food products. Int. J. Biol. Macromol. 71, 94–102 (2014). https://doi.org/10.1016/J.IJBIOMAC.2014.04.027
8. Hernandez-Izquierdo, V. M., Krochta, J. M.: Thermoplastic processing of proteins for film formation--a review. J. Food Sci. **73**(2) (2008). https://doi.org/10.1111/J.1750-3841.2007.00636.X
9. Jongjareonrak, A., Benjakul, S., Visessanguan, W., Prodpran, T., Tanaka, M.: Characterization of edible films from skin gelatin of brownstripe red snapper and bigeye snapper. Food Hydrocoll. **20**(4), 492–501 (2006). https://doi.org/10.1016/j.foodhyd.2005.04.007
10. Ou, S., Kwok, K.C., Kang, Y.: Changes in in vitro digestibility and available lysine of soy protein isolate after formation of film. J. Food Eng. **64**(3), 301–305 (2004). https://doi.org/10.1016/J.JFOODENG.2003.10.013
11. Fernández, L., De Apodaca, E.D., Cebrián, M., Villarán, M.C., Maté, J.I.: Effect of the unsaturation degree and concentration of fatty acids on the properties of WPI-based edible films. Eur. Food Res. Technol. **224**(4), 415–420 (2007). https://doi.org/10.1007/S00217-006-0305-1/FIGURES/5
12. Alves, V.D., Costa, N., Coelhoso, I.M.: Barrier properties of biodegradable composite films based on kappa-carrageenan/pectin blends and mica flakes. Carbohydr. Polym. **79**(2), 269–276 (2010). https://doi.org/10.1016/J.CARBPOL.2009.08.002
13. Liu, Y., et al.: A review of cellulose and its derivatives in biopolymer-based for food packaging application. Trends Food Sci. Technol. **112**, 532–546 (2021). https://doi.org/10.1016/J.TIFS.2021.04.016
14. Padrão, J., et al.: Bacterial cellulose-lactoferrin as an antimicrobial edible packaging. Food Hydrocoll. **58**, 126–140 (2016). https://doi.org/10.1016/J.FOODHYD.2016.02.019
15. Vedove, T.M.A.R.D., Maniglia, B.C., Tadini, C.C.: Production of sustainable smart packaging based on cassava starch and anthocyanin by an extrusion process. J. Food Eng. **289**, 110274 (2021). https://doi.org/10.1016/J.JFOODENG.2020.110274
16. Lu, P., et al.: Preparation of sugarcane bagasse nanocellulose hydrogel as a colourimetric freshness indicator for intelligent food packaging. Carbohydr. Polym. **249**, 116831 (2020). https://doi.org/10.1016/J.CARBPOL.2020.116831
17. Kittur, F.S., Kumar, K.R., Tharanathan, R.N.: Functional packaging properties of chitosan films. Eur. Food Res. Technol. **206**(1), 44–47 (1998). https://doi.org/10.1007/S002170050211/METRICS
18. Xu, J., Wei, R., Jia, Z., Song, R.: Characteristics and bioactive functions of chitosan/gelatin-based film incorporated with ε-polylysine and astaxanthin extracts derived from by-products of shrimp (*Litopenaeus vannamei*). Food Hydrocoll. **100**, 105436 (2020). https://doi.org/10.1016/J.FOODHYD.2019.105436
19. Wang, H., Ding, F., Ma, L., Zhang, Y.: Edible films from chitosan-gelatin: Physical properties and food packaging application. Food Biosci. **40**, 100871 (2021). https://doi.org/10.1016/J.FBIO.2020.100871
20. Haghighi, H., et al.: Comprehensive characterization of active chitosan-gelatin blend films enriched with different essential oils. Food Hydrocoll. **95**, 33–42 (2019). https://doi.org/10.1016/J.FOODHYD.2019.04.019

21. Basiak, E., Galus, S., Lenart, A.: Characterisation of composite edible films based on wheat starch and whey-protein isolate. Int. J. Food Sci. Technol. **50**(2), 372–380 (2015). https://doi.org/10.1111/IJFS.12628
22. Dang, K.M., Yoksan, R.: Morphological characteristics and barrier properties of thermoplastic starch/chitosan blown film. Carbohydr. Polym. **150**, 40–47 (2016). https://doi.org/10.1016/J.CARBPOL.2016.04.113
23. Sun, Q., Gong, M., Li, Y., Xiong, L.: Effect of dry heat treatment on the physicochemical properties and structure of proso millet flour and starch. Carbohydr. Polym. **110**, 128–134 (2014). https://doi.org/10.1016/J.CARBPOL.2014.03.090
24. Jiménez, A., Fabra, M.J., Talens, P., Chiralt, A.: Edible and Biodegradable Starch Films: A Review. Food Bioproc Tech **5**(6), 2058–2076 (2012). https://doi.org/10.1007/S11947-012-0835-4
25. Goudarzi, V., Shahabi-Ghahfarrokhi, I., Babaei-Ghazvini, A.: Preparation of ecofriendly UV-protective food packaging material by starch/TiO_2 bio-nanocomposite: characterization. Int. J. Biol. Macromol. **95**, 306–313 (2017). https://doi.org/10.1016/J.IJBIOMAC.2016.11.065
26. Sirviö, J.A., Kolehmainen, A., Liimatainen, H., Niinimäki, J., Hormi, O.E.O.: Biocomposite cellulose-alginate films: Promising packaging materials. Food Chem. **151**, 343–351 (2014). https://doi.org/10.1016/J.FOODCHEM.2013.11.037
27. Wong, D.W.S., Gregorski, K.S., Hudson, J.S., Pavlath, A.E.: Calcium alginate films: thermal properties and permeability to sorbate and ascorbate. J. Food Sci. **61**(2), 337–341 (1996). https://doi.org/10.1111/J.1365-2621.1996.TB14189.X
28. Wang, H., et al.: Preparation and characterization of multilayer films composed of chitosan, sodium alginate and carboxymethyl chitosan-ZnO nanoparticles. Food Chem. **283**, 397–403 (2019). https://doi.org/10.1016/J.FOODCHEM.2019.01.022
29. Ma, X., et al.: Biopolymer composite fibres composed of calcium alginate reinforced with nanocrystalline cellulose. Compos. Part A Appl. Sci. Manuf. **96**, 155–163 (2017). https://doi.org/10.1016/J.COMPOSITESA.2017.02.021
30. Abdul Khalil, H. P. S., et al.: Seaweed based sustainable films and composites for food and pharmaceutical applications: a review. Renew. Sustain. Energy Rev. **77**, 353–362 (2017). https://doi.org/10.1016/J.RSER.2017.04.025
31. Kanmani, P., Lim, S.T.: Development and characterization of novel probiotic-residing pullulan/starch edible films. Food Chem. **141**(2), 1041–1049 (2013). https://doi.org/10.1016/J.FOODCHEM.2013.03.103
32. Priyadarshi, R., Kim, S.M., Rhim, J.W.: Pectin/pullulan blend films for food packaging: effect of blending ratio. Food Chem. **347**, 129022 (2021). https://doi.org/10.1016/J.FOODCHEM.2021.129022
33. Unalan, I. U., Boyacı, D., Trabattoni, S., Tavazzi, S., Farris, S.: Transparent pullulan/mica nanocomposite coatings with outstanding oxygen barrier properties. Nanomaterials **7**, 281(2017). https://doi.org/10.3390/NANO7090281
34. Kraśniewska, K., Pobiega, K., Gniewosz, M.: Pullulan-biopolymer with potential for use as food packaging. Int. J. Food Eng. **15**(9), (2019). https://doi.org/10.1515/IJFE-2019-0030/DOWNLOADASSET/SUPPL/IJFE-2019-0030AD.ZIP
35. Lee, K.Y., Shim, J., Lee, H.G.: Mechanical properties of gellan and gelatin composite films. Carbohydr. Polym. **56**(2), 251–254 (2004). https://doi.org/10.1016/J.CARBPOL.2003.04.001
36. Oms-Oliu, G., Soliva-Fortuny, R., Martín-Belloso, O.: Using polysaccharide-based edible coatings to enhance quality and antioxidant properties of fresh-cut melon. LWT Food Sci. Technol. **41**(10), 1862–1870 (2008). https://doi.org/10.1016/J.LWT.2008.01.007
37. Tapia, M.S., Rojas-Graü, M.A., Carmona, A., Rodríguez, F.J., Soliva-Fortuny, R., Martin-Belloso, O.: Use of alginate- and gellan-based coatings for improving barrier, texture and nutritional properties of fresh-cut papaya. Food Hydrocoll. **22**(8), 1493–1503 (2008). https://doi.org/10.1016/J.FOODHYD.2007.10.004
38. Cazón, P., Velazquez, G., Ramírez, J.A., Vázquez, M.: Polysaccharide-based films and coatings for food packaging: a review. Food Hydrocoll. **68**, 136–148 (2017). https://doi.org/10.1016/J.FOODHYD.2016.09.009

39. Karbowiak, T., Debeaufort, F., Champion, D., Voilley, A.: Wetting properties at the surface of iota-carrageenan-based edible films. J. Colloid Interface Sci. **294**(2), 400–410 (2006). https://doi.org/10.1016/J.JCIS.2005.07.030
40. Jayakody, M.M., Vanniarachchy, M.P.G., Wijesekara, I.: Seaweed derived alginate, agar, and carrageenan based edible coatings and films for the food industry: a review. J. Food Measure. Characteriz. **16**(2), 1195–1227 (2022). https://doi.org/10.1007/S11694-021-01277-Y/METRICS
41. Cheng, C., et al.: Recent advances in carrageenan-based films for food packaging applications. Front. Nutr. **9**, 1004588 (2022). https://doi.org/10.3389/FNUT.2022.1004588/BIBTEX
42. Zia, K.M., et al.: A review on synthesis, properties and applications of natural polymer based carrageenan blends and composites. Int. J. Biol. Macromol. **96**, 282–301 (2017). https://doi.org/10.1016/J.IJBIOMAC.2016.11.095
43. Chen, S., et al.: Fabrication, characterization, stability and re-dispersibility of curcumin-loaded gliadin-rhamnolipid composite nanoparticles using pH-driven method. Food Hydrocoll. **118**, 106758 (2021). https://doi.org/10.1016/J.FOODHYD.2021.106758
44. Liu, L.S., Liu, C.K., Fishman, M.L., Hicks, K.B.: Composite films from pectin and fish skin gelatin or soybean flour protein. J. Agric. Food Chem. **55**(6), 2349–2355 (2007). https://doi.org/10.1021/JF062612U
45. Roy, S., Priyadarshi, R., Łopusiewicz, L., Biswas, D., Chandel, V., Rhim, J.W.: Recent progress in pectin extraction, characterization, and pectin-based films for active food packaging applications: a review. Int. J. Biol. Macromol. **239**, 124248 (2023). https://doi.org/10.1016/J.IJBIOMAC.2023.124248
46. Almasi, H., Azizi, S., Amjadi, S.: Development and characterization of pectin films activated by nanoemulsion and Pickering emulsion stabilized marjoram (*Origanum majorana* L.) essential oil. Food Hydrocoll. **99**, 105338 (2020). https://doi.org/10.1016/J.FOODHYD.2019.105338
47. Zeng, F., Ye, Y., Liu, J., Fei, P.: Intelligent pH indicator composite film based on pectin/chitosan incorporated with black rice anthocyanins for meat freshness monitoring. Food Chem. X **17**, 100531 (2023). https://doi.org/10.1016/J.FOCHX.2022.100531
48. Phan, T.D., Debeaufort, F., Luu, D., Voilley, A.: Functional properties of edible agar-based and starch-based films for food quality preservation. J. Agric. Food Chem. **53**(4), 973–981 (2005). https://doi.org/10.1021/JF040309S
49. Roy, S., Chawla, R., Santhosh, R., Thakur, R., Sarkar, P., Zhang, W.: Agar-based edible films and food packaging application: a comprehensive review, 924–2244 (2023). https://doi.org/10.1016/j.tifs.2023.104198
50. Yu, Z., et al.: Synthesis and evaluation of physical properties of Agar biopolymer film coating—an alternative for food packaging industry. Mater Res. Express **7**(9), 095307 (2020). https://doi.org/10.1088/2053-1591/ABB7AC
51. Mostafavi, F.S., Zaeim, D.: Agar-based edible films for food packaging applications—a review. Int. J. Biol. Macromol. **159**, 1165–1176 (2020). https://doi.org/10.1016/J.IJBIOMAC.2020.05.123
52. Martelli, M. R., Barros, T. T., De Moura, M. R., Mattoso, L. H. C., Assis, O. B. G.: Effect of chitosan nanoparticles and pectin content on mechanical properties and water vapor permeability of banana puree films. J. Food Sci. **78**(1), (2013). https://doi.org/10.1111/J.1750-3841.2012.03006.X
53. Sothornvit, R., Pitak, N.: Oxygen permeability and mechanical properties of banana films. Food Res. Int. **40**(3), 365–370 (2007). https://doi.org/10.1016/J.FOODRES.2006.10.010
54. Rabelo Vaz Matheus, J., Fontanive Miyahira, R., Elizabeth Cavalcante Fai, A.: Biodegradable films based on fruit puree: a brief review (2020). https://doi.org/10.1080/10408398.2020.1772715.
55. Du, W.X., et al.: Sensory evaluation of baked chicken wrapped with antimicrobial apple and tomato edible films formulated with cinnamaldehyde and carvacrol. J. Agric. Food Chem. **60**(32), 7799–7804 (2012). https://doi.org/10.1021/JF301281A/ASSET/IMAGES/LARGE/JF-2012-01281A_0001.JPEG

56. de Barros-Alexandrino, T.T., Tosi, M.M., Assis, O.B.G.: Comparison between chitosan nanoparticles and cellulose nanofibers as reinforcement fillers in papaya puree films: effects on mechanical, water vapor barrier, and thermal properties. Polym. Eng. Sci. **59**(S1), E287–E292 (2019). https://doi.org/10.1002/PEN.24938
57. Morsy, M. K., Khalaf, H. H., Sharoba, A. M., El-Tanahi, H. H., Cutter, C. N.: Incorporation of essential oils and nanoparticles in pullulan films to control foodborne pathogens on meat and poultry products. J. Food Sci. **79**(4) (2014). https://doi.org/10.1111/1750-3841.12400
58. Zahedi, Y., Ghanbarzadeh, B., Sedaghat, N.: Physical properties of edible emulsified films based on pistachio globulin protein and fatty acids. J. Food Eng. **100**(1), 102–108 (2010). https://doi.org/10.1016/J.JFOODENG.2010.03.033
59. Jo, W. S., Song, H. Y., Song, N. B., Lee, J. H., Min, S. C., Bin Song, K.: Quality and microbial safety of 'Fuji' apples coated with carnauba-shellac wax containing lemongrass oil. LWT—Food Sci. Technol. **55**(2), 490–497 (2014). https://doi.org/10.1016/J.LWT.2013.10.034
60. Dhumal, C.V., Sarkar, P.: Composite edible films and coatings from food-grade biopolymers. J. Food Sci. Technol. **55**(11), 4369–4383 (2018). https://doi.org/10.1007/S13197-018-3402-9/TABLES/1
61. Cortés-Rodríguez, M., Villegas-Yépez, C., Gil González, J. H., Rodríguez, P. E., Ortega-Toro, R.: Development and evaluation of edible films based on cassava starch, whey protein, and bees wax. Heliyon **6**(9), e04884 (2020). https://doi.org/10.1016/J.HELIYON.2020.E04884
62. Kumar, N., et al.: Recent trends in edible packaging for food applications—perspective for the future. Food Eng. Rev. **15**(4), 718–747 (2023). https://doi.org/10.1007/S12393-023-09358-Y
63. Heristika, W., Ningrum, A., Supriyadi, H., Munawaroh, S. H., Show, P. L.: Development of composite edible coating from gelatin-pectin incorporated garlic essential oil on physicochemical characteristics of red chili (*Capsicum annnum* L.). Gels **9**(1), 49 (2023). https://doi.org/10.3390/GELS9010049
64. Drago, E., Campardelli, R., Pettinato, M., Perego, P.: Innovations in smart packaging concepts for food: an extensive review. Foods **9**(11), 1628 (2020). https://doi.org/10.3390/FOODS9111628
65. Srivastava, V., Singh, S., Das, D.: Rice husk fiber-reinforced starch antimicrobial biocomposite film for active food packaging. J. Clean Prod. 138525 (2023). https://doi.org/10.1016/J.JCLEPRO.2023.138525
66. Kong, J., et al.: Multi-functional pH-sensitive active and intelligent packaging based on highly cross-linked zein for the monitoring of pork freshness. Food Chem. **404**, 134754 (2023). https://doi.org/10.1016/J.FOODCHEM.2022.134754
67. Cao, L., Ge, T., Meng, F., Xu, S., Li, J., Wang, L.: An edible oil packaging film with improved barrier properties and heat sealability from cassia gum incorporating carboxylated cellulose nano crystal whisker. Food Hydrocoll **98**, 105251 (2020). https://doi.org/10.1016/J.FOODHYD.2019.105251
68. Adeyeye, S.A.O.: Food packaging and nanotechnology: safeguarding consumer health and safety. Nutr. Food Sci. **49**(6), 1164–1179 (2019). https://doi.org/10.1108/NFS-01-2019-0020/FULL/PDF
69. Dubey, N. K., Dubey, R.: Edible films and coatings: An update on recent advances. Biopolym.-Formul.: Biomed. Food Appl. 675–695 (2020). https://doi.org/10.1016/B978-0-12-816897-4.00027-8
70. Zhang, R., et al.: Effects of preparation conditions on the properties of agar/maltodextrin-beeswax pseudo-bilayer films. Carbohydr. Polym. **236**, 116029 (2020). https://doi.org/10.1016/J.CARBPOL.2020.116029
71. Jeevahan, J., Chandrasekaran, M.: Nanoedible films for food packaging: a review. J. Mater. Sci. **54**(19), 12290–12318 (2019). https://doi.org/10.1007/S10853-019-03742-Y/TABLES/3
72. García, M. A., Pinotti, A., Martino, M. N., Zaritzky, N. E.: Characterization of starch and composite edible films and coatings. Edible Films Coatings Food Appl. 169–209 (2009). https://doi.org/10.1007/978-0-387-92824-1_6
73. Ghosh, T., Teramoto, Y., Katiyar, V.: Influence of nontoxic magnetic cellulose nanofibers on chitosan based edible nanocoating: a candidate for improved mechanical, thermal, optical, and

texture properties. J. Agric. Food Chem. **67**(15), 4289–4299 (2019). https://doi.org/10.1021/ACS.JAFC.8B05905/ASSET/IMAGES/LARGE/JF-2018-05905P_0009.JPEG
74. Mchugh1, T. H., Krochta, J. M.: Agricultural and food chemistry & copy; sorbitol-vs glycerol-plasticized whey protein edible films: integrated oxygen permeability and tensile property evaluation. **42** (1994). Accessed 30 April 2024. [Online]. Available: https://pubs.acs.org/sharingguidelines
75. Srivastava, V., Singh, S., Das, D.: Decision support system for assessing quality and consumer acceptance of active biocomposite based food packaging. Environ Dev Sustain (2024). https://doi.org/10.1007/s10668-024-05782-0
76. Sohail, S.S., Wang, B., Biswas, M.A.S., Oh, J.H.: Physical, morphological, and barrier properties of edible casein films with wax applications. J. Food Sci. **71**(4), C255–C259 (2006). https://doi.org/10.1111/J.1750-3841.2006.00006.X
77. Mourdikoudis, S., Pallares, R.M., Thanh, N.T.K.: Characterization techniques for nanoparticles: comparison and complementarity upon studying nanoparticle properties. Nanoscale **10**(27), 12871–12934 (2018). https://doi.org/10.1039/C8NR02278J
78. Katiyar, V., Ghosh, T.: Nanotechnology in edible food packaging (2021). https://doi.org/10.1007/978-981-33-6169-0
79. Sanches Silva, A., Cruz, J. M., Sendón García, R., Franz, R., Paseiro Losada, P.: Kinetic migration studies from packaging films into meat products. Meat Sci. **77**(2), 238–245 (2007). https://doi.org/10.1016/J.MEATSCI.2007.03.009
80. Appleton, K.M., et al.: The value of facial attractiveness for encouraging fruit and vegetable consumption: analyses from a randomized controlled trial. BMC Public Health **18**(1), 1–12 (2018). https://doi.org/10.1186/S12889-018-5202-6/TABLES/5
81. Árabe, K., et al.: Current advances on the development and application of probiotic-loaded edible films and coatings for the bioprotection of fresh and minimally processed fruit and vegetables. Foods **10**(9), 2207 (2021). https://doi.org/10.3390/FOODS10092207
82. Song, D.-H., et al.: Edible films on meat and meat products. Coatings **11**(11), 1344 (2021). https://doi.org/10.3390/COATINGS11111344
83. Sánchez-Ortega, I., García-Almendárez, B. E., Santos-López, E. M., Amaro-Reyes, A., Barboza-Corona, J. E., Regalado, C.: Antimicrobial edible films and coatings for meat and meat products preservation. Scientif World J **2014** (2014). https://doi.org/10.1155/2014/248935
84. Poore, J., Nemecek, T.: Reducing food's environmental impacts through producers and consumers. Science **360**(6392), 987–992 (2018). https://doi.org/10.1126/SCIENCE.AAQ0216/SUPPL_FILE/AAQ0216_DATAS2.XLS
85. Garnier, L., Valence, F., Mounier, J.: Diversity and control of spoilage fungi in dairy products: an update. Microorganisms **5**, 42 (2017). https://doi.org/10.3390/MICROORGANISMS5030042
86. Cerqueira, M.A., et al.: Use of galactomannan edible coating application and storage temperature for prolonging shelf-life of 'Regional' cheese. J. Food Eng. **97**(1), 87–94 (2010). https://doi.org/10.1016/J.JFOODENG.2009.09.019
87. Trajkovska Petkoska, A., Daniloski, D., Kumar, N., Pratibha, Broach, A. T.: Active edible packaging: a sustainable way to deliver functional bioactive compounds and nutraceuticals. Environmental Footprints and Eco-Design of Products and Processes 225–264 (2021). https://doi.org/10.1007/978-981-16-4609-6_9
88. Paidari, S., et al.: Edible coating and films as promising packaging: a mini review. J. Food Measure. Characteriz. **15**(5), 4205–4214 (2021). https://doi.org/10.1007/S11694-021-00979-7/METRICS
89. Srivastava, V., Singh, S., Das, D.: Enhancing packaging sustainability with natural fiber reinforced biocomposites: an outlook into the future. E3S Web of Conferences **436**, 08016 (2023). https://doi.org/10.1051/E3SCONF/202343608016
90. Ncube, L. K., Ude, A. U., Ogunmuyiwa, E. N., Zulkifli, R., Beas, I. N.: Environmental impact of food packaging materials: a review of contemporary development from conventional plastics to polylactic acid based materials. Materials **13**, 4994 (2020). https://doi.org/10.3390/MA13214994

91. Srivastava, V., Singh, S., Das, D.: Environmental impact assessment of active biocomposite packaging and comparison with conventional packaging for food application. In; DS 130: Proceedings of NordDesign 2024, Reykjavik, Iceland, 12th–14th August 2024, pp. 402–410. https://doi.org/10.35199/norddesign2024.43

Chapter 13
Eco-Friendly Active Packaging for Fresh and Minimally Processed Fruits and Vegetables

Devanampriyan Rajan, Deepak Subramani, and Manonmani Kumaraguruparaswami

Abstract The perishable nature of fruits and vegetables causes substantial post-harvest losses compared to other food commodities. This is attributed to their high moisture content, respiration rate, enzyme activity, ethylene generation, senescence, vulnerability to physical damage during handling, and other inherent nutrients that favor the growth of bacteria and mold. Active packaging made from eco-friendly materials provides a sustainable solution for preserving and extending the shelf life of fresh and minimally processed fruits and vegetables. The scavengers/adsorbers of ethylene, CO_2, O_2, and moisture, together with packaging designs aiding surface immobilization or release of bio-based antioxidants and antimicrobials, allow better preservation of physicochemical and sensory quality of fresh produce while also ensuring the safety from pathogens. This chapter briefly discusses eco-friendly active packaging solutions for fruit and vegetable preservation. The effect of active packaging on physical attributes, chemical stability, microbial growth, and organoleptic quality of the fresh produce are discussed. The integration of active packaging with minimal processing of fruits and vegetables is explored. Furthermore, the strengths, weaknesses, opportunities, and threats of active packaging of fruits and vegetables are critically reviewed.

Keywords Active packaging · Antimicrobial · Essential oils · Fruits and vegetables · Shelf life

D. Rajan (✉)
Department of Food Technology, Kongu Engineering College, Tamil Nadu, Erode, India
e-mail: devanampriyanr@gmail.com

D. Subramani
Asserpark, 6706 HC Wageningen, the Netherlands

M. Kumaraguruparaswami
Department of Food Technology, Rajalakshmi Engineering College, Chennai, India

S. ul Islam and M. Shahid (eds.), *Green Materials for Active Food Packaging*, Engineering Materials, https://doi.org/10.1007/978-981-96-0369-5_13

13.1 Introduction

The growing demand for fruits and vegetables is largely driven by their health advantages. Several factors, such as oxygen levels, storage temperatures, microbial decay, respiration rates, and the presence of ethylene, can significantly impact the freshness and quality of these perishable items [75]. In today's scenario, preserving fresh and minimally processed fruits and vegetables presents both a challenge and an opportunity. Innovative packaging technologies have emerged to address the need for longer shelf life and improved quality, especially due to increased international trade and the demand for fresh, safe, and nutritious produce. Packaging must adhere to specific criteria to prolong the shelf life of fruits and vegetables. Conventional packaging methods are no longer adequate to fully satisfy the demands of today's market [19]. The quest for eco-friendliness and sustainability has become essential in modern food packaging.

Active Packaging is an innovative packaging technology characterized by the interaction between the package, the product, and the surrounding environment. Its primary aim is to extend the shelf life, ensure safety, or enhance the product's sensory attributes, all while preserving its quality. This is especially crucial for fresh produce and foods with extended shelf life. Active-packaging systems typically aim to alter food packages' environmental or physiological conditions. This often involves removing or absorbing unwanted compounds like oxygen, carbon dioxide, and ethylene, as well as undesirable flavors or odors and excess moisture [115]. Some active-packaging systems achieve this by introducing compounds such as carbon dioxide, antioxidants, or preservatives into the airspace of the package using sachets, labels, or films. Bioactive substances can be added to films, including plant extracts, peptides, essential oils, and bacteriocins, which contain antioxidants and antimicrobial properties. These additives protect the quality of food products [26].

This chapter explores various aspects of eco-friendly active packaging tailored specifically for fresh and minimally processed fruits and vegetables. From understanding the principles underlying active packaging to exploring the incorporation of active substances, we aim to provide a comprehensive overview of sustainable packaging solutions. It deals with principles and mechanisms of active packaging, active packaging components and their mode of action, integration with minimal processing technologies, consumer acceptability, safety, legal compliance, and market-available active packaging solutions.

Ultimately, this chapter offers a comprehensive guide to eco-friendly active packaging for fresh and minimally processed fruits and vegetables, providing practical recommendations and a glimpse into the future of sustainable food preservation.

13.2 Factors Affecting the Quality and Shelf Life of Fruits and Vegetables

The quality and shelf-life of fruits and vegetables are influenced by various factors, encompassing pre-harvest conditions, post-harvest handling, and storage practices. These factors include temperature, humidity, packaging methods, and the ripeness/maturity of produce at harvest. Effectively managing these factors can prolong shelf-life and maintain quality, ensuring that fruits and vegetables reach consumers in optimal condition. This also minimizes the post-harvest losses. Understanding these variables is important in designing suitable active packaging for fresh produce.

13.2.1 Nature of Fruit/vegetable

The fruits and vegetables could be of tropical, sub-tropical, and temperate origins, and their properties significantly vary. The fruits and vegetables, after the harvest, are still alive and conduct regular physiological functions such as respiration, transpiration, and development changes in chemical components that result in color, flavor, texture, and nutritional value loss. The respiration rate of harvested commodities is inversely related to their shelf life. Higher respiration rates reduce shelf life. Transpiration is how fresh fruits lose water, affecting their appearance, saleable weight, firmness, crispness, or juiciness [6]. Based on their nature, fruits can be broadly classified into climacteric and non-climacteric fruits. The climacteric fruits continue to ripen by an accelerated ethylene generation and respiration rate after the harvest, whereas non-climacteric fruits cease to ripen after harvest but continue to degrade. While apples, tomatoes, bananas, papaya, and mango are a few examples of climacteric fruits, citrus fruits, cucumbers, and pineapple are examples of non-climacteric fruits [77]. It is highly important to choose the right physiological maturity of the fruit, which should consider the classification of the fruit and post-harvest storage life. Climacteric fruits are often harvested early to allow ripening during storage and distribution so that they reach consumers in a safe and highly edible state. While fruits, in general, are acidic and prone to spoilage by resistant yeast and mold, vegetables are neutral or slightly alkaline, and this supports bacterial spoilage. The skin or peel of fruit and vegetables offers natural protection from the external environment. Depending on the fruit or vegetable, the peel's toughness, porosity, and surface hydrophilicity may vary [88]. This also influences the vulnerability of fruits or vegetables to external factors influencing quality. The fruits and vegetables are a rich source of antioxidants and antimicrobials like pigmented carotenoids, phenolics, and alkaloids, influencing oxidation inhibition and the type of microbes that can grow on the surface. The chemical composition of fruits and vegetables within and between the both could be very different. While berries like tomato have a very high moisture content and sugars, root vegetables like potato and yam have a relatively low moisture content

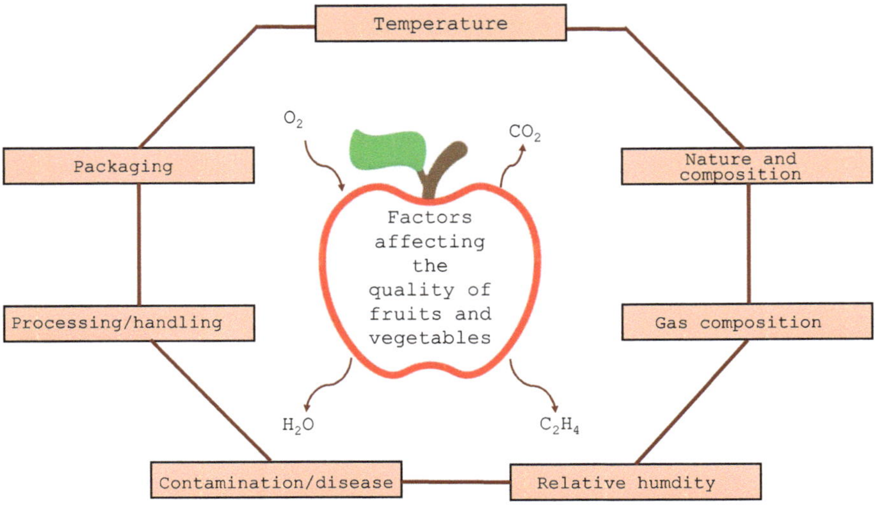

Fig. 13.1 Factors affecting the quality of fruits and vegetables

but high starch content. Both moisture content and the ability to readily utilize the compounds as nutrients influence the microbe's growth and spoilage (Fig. 13.1).

13.2.2 Contamination and Disease

Biological deterioration and illness are the leading causes of postharvest losses of fruits and vegetables in underdeveloped nations. Biodeterioration may be caused by insects, rodents, fungi, and microorganisms. The contamination can come from air, water, soil, equipment, and biological mediums like parasites, human touch, etc. Microorganisms easily contaminate and spread through fresh produce due to poor defensive mechanisms in tissues and the availability of nutrients and moisture. Contact with bruised or spoiled vegetables can affect a whole lot of vegetables. Moreover, mixing or close contact with different fruits and vegetables also causes cross-contamination as a microbe that expresses poor growth in fruit, if transferred to a vegetable, may exhibit accelerated growth depending on the substrate [42].

Since most fungi's optimum growth temperatures range from 25 °C to 30 °C, growth above 40 °C is difficult, and some occasions for bacterial mortality may occur. Tropical fresh fruits are easily contaminated by pathogens since the ideal temperature for contamination by bacteria is the average temperature in tropical regions. Mango is a tropical fruit with a shelf life of more than ten days if not affected by fungi. Even in affluent countries, infections cause approximately 20–25% of produced fruits and vegetables to deteriorate during postharvest processing. Physiological and storage circumstances after the harvesting stage, particularly in tropical environments, hasten the spread of the aforementioned illnesses [5].

13.2.3 Temperature

Proper temperature control between harvesting and consumption is the ideal technique to maintain the quality of fruits and vegetables. Keeping the fruit under refrigerated conditions can slow down several metabolic actions that cause ripening, giving more time to handle the produce after harvest. In general, an hour of delay from harvesting and cooling the crop results in a one-day reduction of shelf life. The surrounding environment's temperature directly impacts the respiration and metabolism of harvested climacteric fruits such as tomatoes [37]. High temperatures can accelerate the rate of respiration (CO_2 production) for stored fruit products.

13.2.4 Relative Humidity

Fruits and vegetables continue to transpire (lose moisture) after harvest. The relative humidity of storage conditions highly influences the loss of water from harvested fruit produce. Most fruits retain nutritional content, appearance, weight, and flavor at extremely high relative humidity while slowing wilting, softening, and juiciness. Tomato fruits are high in moisture and prone to shrinking after harvest. Fruit withering may appear with even a minor percentage of water loss. For mature green tomatoes, ideal relative humidity ranges from 85 to 95% (v/v), but firmer ripe fruits require 90–95% (v/v). The relative humidity and temperature of storage go hand in hand in influencing weight loss. Evapotranspiration increases when temperatures fall below the ideal range, resulting in shriveled fruits. A few exceptions, like onions, require a low relative humidity for a higher shelf life [97].

13.2.5 Gases Composition

The mixture of different gases in the storage environment is critical for extending the shelf life of tomato fruits. The presence of oxygen promotes the oxidation of nutrients and enzymatic browning by polyphenol oxidase and also supports aerobic bacterial growth. In the packaging context, modified atmospheric packaging that uses a different gas composition inside the sealed package extends the shelf life of fruits and vegetables. To limit senescence in matured green and ripe tomato fruit, the ideal atmosphere is 3–5% (v/v) of oxygen, 1–3% (v/v) of carbon dioxide, and 94–96% (v/v) of nitrogen gas. However, a high oxygen level is not desirable, as suspending aerobic respiration can promote anaerobic respiration. This leads to off-flavor generation and eventual structure loss due to catabolic reactions [85].

13.2.6 Processing/handling of Produce

Physical handling may significantly impact the post-harvest quality and longevity of fruits. Rough handling during hand or mechanical harvesting, packaging into crates, sorting, washing, or long-distance transportation might cause mechanical injuries like bruising, damage, scuffing, slicing, or piercing of fruits [78]. If fruits are to be processed, tissue damage happens during operations like cutting, shredding, and slicing. The damage inflicted on the tissues accelerates the respiration rate, ethylene synthesis, microbial growth, and oxidative damage to color and nutrients.

Sanitation of fruits often begins with a rinse in tap water to remove pesticide residues, plant debris, and other potential contaminants, then a dip in chlorinated water to effectively minimize the microbial count on the fruit surface. Chlorine is commonly used to disinfect fruit surfaces by adding sodium hypochlorite (NaOCl) to the water used for washing. The removal of surface contaminants improves both the safety and shelf life of the product. Blanching of fruits and vegetables, done to inactivate enzymes, avoids quality changes. Heat treatment also causes the destruction of endogenous microbes to some extent. However, if not careful, the same step can go wrong in multiple ways [41]. For instance, an inefficient blanching step may allow resistant enzymes to retain their activity, and now the heat-digested cell walls allow better accessibility of enzymes with substrates, thereby causing more quality deterioration. Furthermore, the presence of bruises or cuts on the surface of fruits allows the diffusion of washing water to fruit, accelerating microbial spoilage.

13.2.7 Packaging Conditions

Unlike many other foods, fruits and vegetables require a special packaging material design as they continue to respire even after harvest. Using highly porous packaging can cause greater weight loss by transpiration. On the other hand, the use of non-perforated packaging and continued transpiration causes an elevation of relative humidity in the headspace of the package. This causes fogging and condensation of moisture on the surface of fruits and vegetables, thereby promoting microbial spoilage. The porosity or micro-perforation and the type of packaging material can largely influence both water and gas exchange. Packing fruits and vegetables in a gas impermeable packaging can cause an eventual deficit of oxygen inside the package, promoting anaerobic respiration [72]. However, if the gas permeability is too high, this promotes the growth of aerobic bacteria and oxidation in fruits. For all these reasons, the packaging material needed for storing fruit and vegetables varies based on their nature, composition, and the conditions under which the packaging must be stored. The internal atmosphere of fruit and vegetables is often modulated with modified atmospheric packaging for increased shelf life.

13.3 Active Packaging for Preserving Fresh Produce: Technology, Properties, and Mode of Action

13.3.1 Oxygen Absorbers/scavengers

Oxygen deteriorates food quality by inducing fat oxidation and rancidity, enzymatic browning, and supporting microbial spoilage. Reducing the oxygen levels in the headspace of fruits and vegetables is desirable to reduce the respiration rate but should not be too low so that anaerobic respiration can be avoided. Though oxygen levels in the headspace can be modified by modified atmospheric packaging or vacuum packaging, these techniques have no control over the oxygen permeating the packaging during storage and transport. Therefore, using oxygen scavengers/absorbers can overcome the disadvantages in the barrier properties of the packaging. The present oxygen scavenger technology often uses mechanisms like oxidation of iron fillings, ascorbic acid, light-sensitive pigments, and unsaturated fatty acids. Moreover, oxidoreductive enzymes like glucose oxidase, alcohol oxidase, or live yeast are immobilized in the packaging to consume oxygen [20, 24].

Iron powders are the most common form of oxygen scavenger, which often comes as sachets. The reactions given below describe the mechanism of iron oxidation.

$$Fe \rightarrow Fe^{2+} + e^-$$

$$\tfrac{1}{2} O_2 + H_2O + 2e^- \rightarrow 2OH^-$$

$$\tfrac{1}{2} O_2 + H_2O + 2e^- \rightarrow 2OH^-$$

$$Fe^{2+} + 2OH^- \rightarrow Fe(OH)_2.$$

$$Fe(OH)_2 + \tfrac{1}{4} O_2 + \tfrac{1}{2} H_2O \rightarrow Fe(OH)_3.$$

As a thumb rule, 1 g of iron can react with 300 ml of oxygen. Escobar et al. [29] used the iron fillings-based oxygen scavenger and sodium acrylate-based moisture absorber to preserve fresh lulo (*Solanum quitoense*) fruit packed in biodegradable polylactic acid film. The oxygen scavenging followed a first-order reaction kinetics. The co-presence of sodium acrylate improved the oxygen-scavenging effect of iron fillings by allowing greater moisture absorption to convert iron to iron hydrogen. With a lowering of oxygen to 16.8%, the scavenger extended the shelf life of the fruit to 25 days. Iron fillings containing sachets with 210 ml oxygen scavenging capacity were used in a study by [52] to improve the storage stability of strawberries packed in a microperforated film. The oxygen scavenger reduced the carbon dioxide accumulation in the package by reducing the respiration rate due to depleted oxygen levels. After three weeks of storage, the gas composition inside the package was 15–20% CO_2 and 4–5% O_2. The microperforated film with the oxygen scavenger gave the lowest physicochemical changes in strawberries during storage. It should be noted that the sachet should be permeable to oxygen and water vapor to initiate the cascade of reactions but should not become wet or lose its structural integrity so that iron contamination into food can be avoided.

Phenolic antioxidants like gallic acid can scavenge oxygen. These molecules get oxidized in the presence of oxygen yet can also stabilize themselves without them becoming unstable free radicals. Consequently, these interactions between the scavenger and oxygen lead to a reduction in O_2 concentration. Ahn et al. [3] studied the effectiveness of gallic acid as an oxygen scavenger. The starting oxygen level in the headspace of the package was 21%. After seven days of storage with a film containing an oxygen scavenger, the oxygen levels decreased to 16.6% at 4 °C and 14.6% at room temperature. This organic oxygen scavenger exhibited an oxygen scavenging capacity of 0.709 mL/cm^2 at 23 °C. Taechutrakul et al. [98] used a multibranched polylactic acid and gallic acid film strip as an oxygen scavenger for bananas. The film exhibited a higher oxygen scavenging effect (20%). Within 30 h, the film caused a greater oxygen scavenging capacity but thereafter leveled off due to saturation of gallic acid moieties. The film preserved the quality of fresh banana for up to 14 days, with no significant changes to its initial green peel color or firmness.

Although less studied in the context of fruits and vegetables, there is a greater scope for using other oxygen scavengers. In enzyme-based oxygen scavenger systems, in the presence of oxygen and moisture, enzymes like glucose oxidase can oxidize glucose and deplete the oxygen levels in the headspace. Glucose oxidase is often combined with catalase enzyme to nullify hydrogen peroxide produced in the former reaction. The rate of reaction of glucose oxidase and catalase can be controlled to deplete oxygen in the headspace of the packed fruits and vegetables. Gluconic acid, formed as a product of glucose oxidation, could reduce the pH of food if it is not properly immobilized in packaging. If pH reduction is affecting the organoleptic quality of the product, calcium carbonate can be incorporated to neutralize the acids. This also generates CO_2, which can nullify the oxygen depletion-mediated pressure reduction within the package. Another possibility is using Laccase, an oxidoreductive enzyme with copper in its active site. This enzyme functions effectively only in a relatively high humidity (>84% RH) environment. An oxygen scavenger must be selected based on size, shape, oxygen scavenging capacity, rate of reaction, toxic or odorous by-products formed, price, duration to arrive at equilibrium, storage temperature, and properties of food, especially water activity [19, 24].

The scavenging ability of ascorbic acid is attributed to its ability to oxidize into dehydroascorbic acid. Although the scavenging potential of ascorbic acid is low, using metal catalysts like iron and copper can improve the oxygen scavenging function by 16%. Since ascorbic acids can exist in different states by protonation and

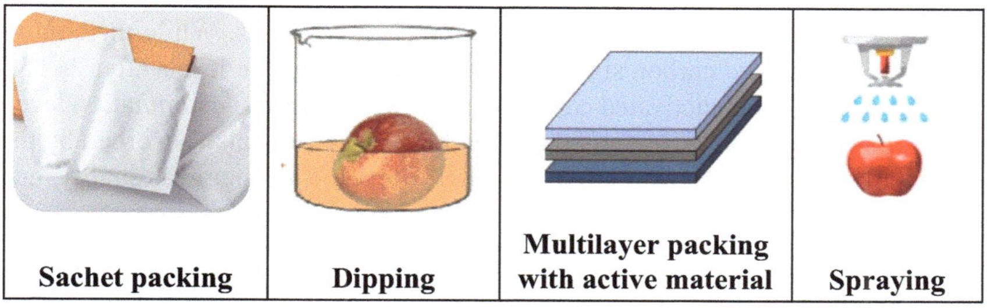

Fig. 13.2 Different ways of active food packaging

their stability or reactivity to oxygen varies, it is important to control the pH of the film. The oxidation of polyunsaturated fatty acids like linoleic acid and linolenic acid or even PUFA-rich oils like linseed oil can be used as an oxygen scavenger [4]. However, only primary oxidation of PUFA should be allowed. Secondary oxidation of PUFA generates low molecular weight volatile, which could produce a rancid off-flavor. A barrier material is required to avoid getting into fruits and vegetables if such odorous compounds are produced. Dried yeast or bacteria immobilized in a gel, wax, or other solid matrix can be an active biological oxygen scavenger [96]. For its activity, the organism requires both oxygen and moisture. The yeast and bacteria can consume oxygen and respire with moisture absorption from fruits and vegetables. Ensuring the viability of microbes is important for effective oxygen scavenging function (Fig. 13.2).

13.3.2 Ethylene Adsorbers/Scavengers

Ethylene is a plant growth-regulating hormone that accelerates the progression of fruits and vegetables toward senescence. Thus, control over ethylene synthesis by fruits and vegetables is highly important to extend the shelf life of fresh produce. Activated carbon and zeolites are the most commonly used ethylene adsorbers. Produced from biomaterials like food industrial wastes or lignocellulosic biomass, activated carbon contains a low inorganic constituent and exhibits high stability during storage [84]. The ability to adsorb ethylene by activated carbon greatly depends on the latter's pore volume and surface area. The adsorption kinetics fits with the Langmuir model, with direct effect from the partial pressure of ethylene in the headspace atmosphere and surface characteristics of activated carbon. Eventually, during the storage, an equilibrium is reached between adsorption and desorption of ethylene. It can regenerate the activated carbon after use if it can be collected. Activated carbon comes in powder, granule, and fiber formats. While the powder form has the maximum adsorption capacity (80%), fiber has the poor adsorption ability (40%). A pore size greater than 3.9 Å is needed for ethylene adsorption. A co-presence of palladium catalyst can further improve the ethylene scavenging ability [110]. Bailén et al. [14]

used activated carbon in modified atmosphere-packed tomatoes. While the control sample without activated carbon significantly increased ethylene levels after 24 h of storage, the package with activated carbon had a lower ethylene accumulation level. Moreover, the volatile odors generated during the storage were also adsorbed by the activated carbon. This improved tomatoes' shelf life, physicochemical quality, and sensory quality.

Unlike activated carbon, zeolite is an inorganic microporous crystalline aluminosilicate. They have negative charges in their framework. Zeolite, by molecular sieving, cation exchange, and adsorption, can remove ethylene from the headspace of the package. Natural zeolites with a large pore diameter of 12 Å facilitate the movement of ethylene through these openings and get entrapped in the zeolite structure [84]. In a study by Coloma et al. [23], a film with 10% Zeolite was shown to reduce ethylene by 37%. The addition of Zeolite into the packaging of Mangoes has been shown to reduce the ethylene level and extend shelf life to 40 days, while the shelf life of tomatoes was only 20 days. Zeolite-based Active packaging film for broccoli packing reduced ethylene by 0.33 ppm after 20 days of storage at 4°C, while the control sample had 62 ppm ethylene. The ethylene adsorbing capacity of zeolite can be further improved by modification with other metal cations. Using HCl to replace the cation exchange sites in the zeolite resulted in an elevation of ethylene adsorption capacity from 0.619 to 1.219 mmol/g [10]. Similar to zeolites, clays such as bentonite, kaolinite, and sepiolite have also been explored for active ethylene adsorbers. The differences originating from structure, surface area, and lattice substitution influence the ethylene adsorption capacity. Halloysite clay usage in the film has been demonstrated to reduce ethylene levels, thereby delaying the ripening of tomatoes and bananas [100].

Photocatalytic oxidation with Titanium dioxide catalyst is another method to scavenge ethylene. Titanium dioxide gets activated when exposed to UV light, which has a wavelength of around 380 nm. Water and oxygen in the immediate environment generate reactive oxygen species, which oxidize ethylene and convert the latter into H_2O and CO_2. A drop in particle size to elevate the surface area and doping with metal ions and non-metals are other strategies to improve the ethylene scavenging effect. Films incorporated with nano-sized Titanium dioxide when used to store tomatoes, the ethylene level reduced by 76–88% after exposure to UV light. This has also extended the shelf life of tomatoes. Increasing titanium dioxide from 1 to 10% in the packaging also improves ethylene degradation [61, 62].

Novel packaging has incorporated 1-methyl cyclopropane or other biochemicals into the packaging to block the binding site of ethylene. This alternative approach has also been proven to significantly delay tomatoes' ripening and mushroom browning [73].

13.3.3 Moisture Absorbers

Fruits and vegetables continue to transpire even after harvest. It is important to regulate the loss of moisture or moisture condensation on the surface to avoid shrinking and microbial spoilage. In general, moisture absorbers are highly hygroscopic materials that exhibit a greater affinity for water and the ability to hold the water. Moisture absorbers can absorb moisture in a vapor state from the headspace and the water exuded from the fruit skin. By controlling the relative humidity inside the packaging, moisture absorbers can preserve the quality of fruits and vegetables during storage. Moisture absorbers often come in either pad or sheet form [32].

Fructose is known for its hygroscopic nature, and therefore, it can draw moisture from its surroundings. It starts absorbing water vapor from the air when the relative humidity exceeds approximately 55%. Due to its good humectant properties, fructose can maintain moisture for extended periods, even at low relative humidity levels [32]. Bovi et al. [21] used Fruitpad (containing fructose) as a desiccant for packaging fresh strawberries. The Fruitpad containing fructose (30%) demonstrated the highest moisture absorption, with a capacity of 0.94 g of water per gram of pad when stored at 20 °C and 100% relative humidity. This study found that strawberries packaged with the fructose-infused Fruitpad experienced a minimal weight loss of $< 0.92\%$. Sorbitol, an organic sugar alcohol, has been proven to function as an effective desiccant for mushrooms, broccoli, and tomatoes [32]. Shirazi and Cameron [91] evaluated the effectiveness of sorbitol and xylitol in controlling the in-package relative humidity for storing tomatoes. While the relative humidity in the control package was 98%, the package with 10 g moisture absorbers, sorbitol, and xylitol stabilized relative humidity to 78% and 74%, respectively, over 48 days of storage. Other biopolymers, like cellulose or its derivatives, such as carboxy methyl cellulose, alginate, and other gums, make hydrogels suitable for biodegradable moisture absorbents.

The limitation of organic moisture absorbers is that they have a limited water-absorbing capacity compared to inorganic materials. Using silica gel as a moisture absorber significantly reduced the weight loss observed in guava fruit during storage for a month. In another study (Wang et al., 2017), sachets and pads containing carbonate were found to preserve the quality of mushrooms. Mahajan et al. [60] mixed rapid moisture adsorbers such as calcium chloride (0.55 g) and sorbitol (0.25 g) with slow moisture absorbers like bentonite (0.2 g) to prepare a composite material of a greater moisture absorption power (0.9 h/g). Relative humidity significantly influenced the absorber's ability to hold moisture. A rise in relative humidity from 76 to 96% increased the moisture-holding capacity of the absorber from 0.5 to 0.9 g/g. The moisture absorption kinetics followed a Weibull distribution model. Eggshell particles have a high surface area and porosity. Its addition into bacterial cellulose film significantly improved water adsorption. Increased bacterial cellulose to eggshell ratio from 1:1 to 1:5 increased water adsorption from 0.45 g to 1.05 g per gram of composite material [103]. Rux et al. (2016) developed humidity-regulating trays for fresh produce. The film had a hygroscopic ionomer lining to which sodium chloride was added up to 12%. A higher salt level improved the water absorption when the tray capacity was tested with tomatoes and strawberries. After two days of rapid water uptake, a steeper decline in water uptake was noted from 4th day. An equilibrium was achieved eventually by 12 days due to the unavailability of active sites for moisture absorption. An increase in relative humidity from 76 to 100% increased the water holding capacity from 1.6 to 7.3 g on day 1 and 4.3 to 16.7 g on day 12. Furthermore, calcium sulfate, sodium hexametaphosphate, silica gel, and sodium chloride have also been successful in use as moisture controllers for breba fruit, mushrooms, and avocado [43, 105, 113].

13.3.4 Carbon Dioxide Emitters/absorbers

Fruits and vegetables generate carbon dioxide during respiration, which continues after harvest. This causes accumulation of CO_2 in the package over time if the package is impermeable to gas escape. A higher level of CO_2 causes physiological injury to fresh produce and may cause discoloration, structural breakdown, and off-flavor development by anaerobic respiration. CO_2 absorbers are most useful in this context.

In chemical absorber-type systems, CO_2 absorption is commonly achieved with calcium hydroxide, which reacts with CO_2 to produce stable and non-toxic calcium carbonate and water as products.

$$Ca(OH)_2 + CO_2 \rightarrow CaCO_3 + H_2O.$$

The above reaction can happen spontaneously in the package without other additives or reactants. In its solid state, it is safe for having food contact. Alkaline salts such as sodium carbonate can be transformed into sodium bicarbonate by consuming carbon dioxide and water [53].

$$Na_2CO_3 + CO_2 + H_2O \rightarrow 2NaHCO_3.$$

Since this reaction requires a water molecule, water or water vapor uptake from headspace or food influences the reaction kinetics. Thus, regulating the water vapor permeability of sachets containing sodium bicarbonate can, in turn, regulate the rate of absorption of CO_2. An optimum in-package headspace gas composition can be obtained with this technique. Similarly, calcium oxide and iron fillings can also function as CO_2 absorbers [54].

CO_2 can also be physically adsorbed by microporous materials like activated carbon and Zeolite. Packed within a sachet, they can be used in powder, granule, bead, block, or sheet form. The surface area, pore volume, and pore size distribution influence CO_2 adsorption capacity. Moreover, surface hydrophilicity and selectivity for CO_2 also influence the effectiveness of adsorbents. Since these physical adsorbents can also adsorb moisture, the competition with moisture affects CO_2 adsorption. Zeolite preferentially adsorb moisture over CO_2. The adsorption of large amounts of water into the framework of these adsorbents also causes the desorption of already adsorbed CO_2. Therefore, designing a physical adsorbent for fruits and vegetables requires a plot of CO_2 adsorption isotherm as influenced by water vapor and partial pressure of CO_2 [20, 53].

With recent advancements, fruits and vegetables are packed in modified atmospheric packaging with an optimum CO_2 concentration to extend shelf life. If the package has poor gas barrier properties, the package may lose the CO_2 in the in-package environment. A drop in CO_2 levels in the headspace and their displacement with air can accelerate ripening and microbial growth. CO_2 in chilled conditions can have greater solubility in the product and affect the pH and microbial growth. Moreover, using oxygen adsorbers may reduce the pressure inside the package. A huge pressure difference may cause package collapse. To avoid that, CO_2 emitters are used proportionally with O_2 adsorbers to maintain the pressure. In such a scenario, carbon dioxide emitters are most useful in releasing CO_2 into the headspace.

Ferrous carbonate can consume O_2 and release CO_2 into the package. It follows the following reaction.

$$4FeCO_3 + 6H_2O + O_2 \rightarrow 4Fe(OH)_3 + 4CO_2.$$

For its poor stability, it often goes together with a metal catalyst and an oxygen scavenger. Similarly, a reaction between sodium bicarbonate and organic acids like ascorbic acid and citric acid can also produce CO_2 in the presence of moisture. Ascorbic acid is a well-known organic CO_2 emitter. With oxidation, ascorbic acid consumes O_2 and releases CO_2. The emitters of CO_2 are most commonly used as pads or sachets, along with moisture adsorbers, as these reactions often require a wet condition [54].

13.3.5 Surface Immobilized/release Type Antimicrobial and Antioxidant Agents

Microbial spoilage is a leading cause of post-harvest loss of fruits and vegetables. Moreover, enzymatic browning and discoloration are common in fresh produce due to oxidation. Fruits and vegetables are also increasingly associated with food-borne diseases. Mitigating microbes and oxidation can improve the quality and safety of produce while reducing post-harvest loss. Many plant, animal, marine, and microbial derivatives can effectively function as antioxidants and antimicrobials for active packaging. These substances can be immobilized permanently onto the film surface having close contact with food or be released in a sustained manner into headspace and food. The antimicrobial effect of these compounds can arise from cell wall disruption, increased membrane permeabilization, generation of reactive oxygen species, DNA or protein damage, and interference with the activity of intracellular enzymes

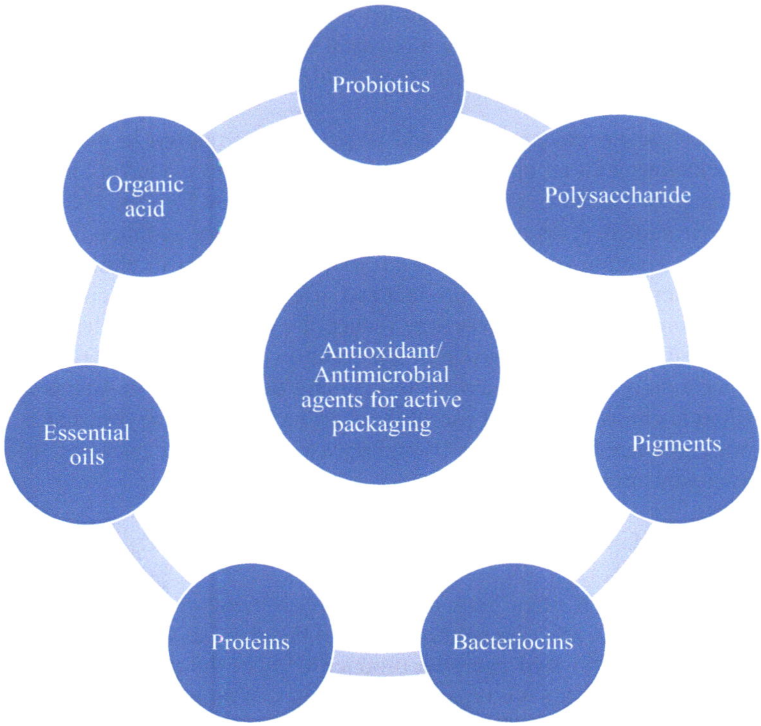

Fig. 13.3 Antioxidant/antimicrobial agents for active packaging

or other metabolic functions. This can delay the lag phase and decrease the viability of cells [22, 90] (Fig. 13.3).

13.3.5.1 Polysaccharides

A few Polysaccharides derived from plants, marine species, and microbes have been reported to have antimicrobial functions. Chitosan (N-acetyl-D-glycosamine) is a natural and biodegradable polysaccharide derived from chitin present in crustaceans and fungi. Chitosan naturally exhibits antifungal and antimicrobial activities. The effectiveness of chitosan is influenced by the pH. At a low pH (below pH 6), chitosan exists in a fully protonated state and becomes a positively charged polyelectrolytic cation. When chitosan with a cationic amino group comes into contact with the negatively charged bacterial surface, it can disrupt the cell wall permeability of bacteria. While some studies report that chitosan is more active against gram-positive than gram-negative bacteria, contradicting findings have also been reported. This is possibly due to a great variation in chitosan's structure and molecular weight [27]. Regardless, chitosan-based films have been proven to function against spoilage or pathogenic microbes like lactic acid bacteria, Aspergillus niger, Pseudomonas spp, E. coli, Listeria monocytes, Candida albicans, and many other yeast and mold species [107]. Like chitosan, Pullulan, a water-soluble exopolysaccharide produced by the fungi Aureobasidium pullulans, also possesses antibacterial and antifungal properties [101]. Both chitosan and pullulan are used along with plant-derived essential oils to have an effect on a wide spectrum of microbes with a greater degree of inhibition. Lentinan, a polysaccharide derived from the Shitake mushroom (Lentinus edodes), has been found to have an antimicrobial effect. The gelatin film incorporated with lentinan disrupted the cells of *E. coli* and *Streptococcus aureus*, thereby causing their death. The lentinan incorporated film was found to preserve the quality of mushrooms from microbial spoilage [36].

13.3.5.2 Protein Hydrolysates

Low to high-molecular-weight peptides obtained from enzymatic hydrolysis of proteins possess antioxidant and antimicrobial functions. Filho et al. [26] developed an alginate-based active packaging film with cottonseed protein hydrolysates. The film exhibited inhibitory activity against Staphylococcus aureus, *Colletotrichum gloeosporioides*, and *Rhizopus oligosporus*. It is release-type active packaging. Migration studies revealed that 60% of peptides get released into an aqueous medium in 30 min. However, adding a lipid layer can reduce the release kinetics and have a prolonged effect. The hydrolysates were also found to have potent free radical scavenging potential due to their electron donation and chelating effects. Similarly, the incorporation of gelatin hydrolysates into biodegradable gelatin film was also found to improve the antioxidant activity of the film. This was due to free radical scavenging and iron-reducing power [34].

13.3.5.3 Bacteriocins

Bacteriocins produced by lactic acid bacteria can inhibit gram-positive bacteria with no significant changes to the organoleptic quality of fresh produce. Of all bacteriocin tested for its efficacy as an antimicrobial film, nisin, an approved food additive in many countries, has been studied in an active packaging context. Barbosa et al. [15] incorporated nisin in a packaging film for storing mangoes. Nisin inhibited *Streptococcus aureus*, *Listeria monocytogenes*, *A. Acidoterrestris,* and *Bacillus cereus* spores. After four days of storage, the L. monocytogen count went below the detection limits. A six-log reduction was observed for Streptococcus aureus after six days of storage. The film also prolonged the shelf life of mangoes. Scannel et al. [87] found that nisin, after incorporating into a film as a release-type antimicrobial, could retain its activity over three months. It was active at both refrigerated and ambient temperature conditions. The edible film added with pediocin has been found effective in preserving the minimally processed papayas by exhibiting a 4-log reduction in total aerobic count. The diffusion coefficient of nisin from hydroxypropyl methylcellulose (HPMC) is higher than sodium caseinate, polylactic acid, and chitosan-based films. HPMC film loaded with nisin also had a higher antimicrobial effect. However, HPMC, sodium caseinate, and polylactic acid were suitable mediums for a controlled release of nisin [44].

13.3.5.4 Organic Acids

Organic acids such as sorbic acid, benzoic acid, lactic acid, acetic acid, citric acid, and ascorbic acid have long been approved food additives. Because they are generally considered safe, they have a great potential for use in antimicrobial packaging [39]. Fasihnia et al. [31] developed an active packaging film with different sorbic acid concentrations (2–6%). The film was found to be active against bacteria like *E. coli* and *Streptococcus aureus* and fungi like *Aspergillus niger*. Jipa et al. (2–12) developed a bacterial cellulose-based antimicrobial film incorporated with sorbic acid. Compared to a monolayer film, which had (a 10–12 m2 s-1 release rate), a multilayer film has a lower release rate (10^{-13} m^2 s^{-1} release rate) of sorbic acid. This lower release rate with multilayer packaging could give a prolonged antimicrobial effect. At a higher concentration of sorbic acid, poorly distributed crystals were formed in the film, and their dissolution also controlled the acid release kinetics. Sorbic acid can also be added in its salt form. The release rate of sorbic acid is 10–100 times lower than potassium sorbate due to the latter's higher solubility. Matrix characteristics can also play a role since the film's bacterial cellulose concentration influenced the release rate of sorbic acid. Furthermore, pH can influence the release rate and antimicrobial efficacy by influencing protonation. An increase in pH from 3 to 7 improved the release rate, and more than 99% of sorbic acid was in an active dissociated state at a higher pH [79].

13.3.5.5 Probiotics

Recent advancements have seen the use of antimicrobial-producing live bacteria in active packaging to extend the shelf life of fruits and vegetables, which is highly explored. Generally, probiotic bacteria produce bacteriocins, peptides, organic acids (acetic and lactic acid), hydrogen peroxide, and diacetyl (a protein-derived compound), all of which are capable of controlling pathogens, spoilage microbes, and the progression of rot and disease in fruits and vegetables [25]. These metabolites are released onto the foods from the probiotic active layer film when the food is in close surface contact. For example, adding *Lactococcus lactis* in a gellan gum-based film improved the antibacterial activity of the film. Although the cell viability remained unchanged for the first four days, a steady decline was noted up to 24 days due to nutrient depletion. The co-use of prebiotic cranberry extract in the film improved the viability of Lactococcus lactis during storage, making it suitable for preserving fresh-cut apples and potatoes [114]. Similarly, many other lactic acid bacteria have shown potential for use in antimicrobial packaging. The survivability of probiotic bacteria and their ability to preserve fruits and vegetables depend on various factors, including the type of selected culture or strain, the physiological state of the cells, food properties such as pH and water activity, storage conditions, and processing steps [25].

13.3.5.6 Essential Oils

Essential oils derived from plants are loaded with many plant secondary metabolites. This includes monoterpenes, terpenoids, phenolics, esters, aromatic hydrocarbons, aldehydes, lactones, and ketones. They are known to have both antioxidant and antimicrobial effects. Most natural antioxidants are phenolic compounds that inhibit oxidation by neutralizing free radicals such as superoxide and hydroxyl radicals. This process involves the phenolic compound donating a proton, which leads to its own oxidation. The compound then stabilizes itself through electron delocalization. When incorporating essential oils (EOs) into films, they are often added to a hydrophilic matrix to create an aqueous dispersion. During the film drying, lipid droplets become embedded within the matrix, causing structural rearrangements of the hydrophilic components. However, it is possible that essential oil, by coalescence or flocculation, may get separated. The stability of the film's microstructure can be enhanced through microfluidization to finely reduce the size of the droplets [22].

For their volatility, essential oils tend to diffuse into headspace or can even directly interact with the surface or get solubilized in the food, depending on the matrix composition. Essential oils derived from spices and herbs such as peppermint, clove, cinnamon, thyme, and oregano are most commonly used as antioxidants and antimicrobials in the active packaging of fruits and vegetables. Since the fruits and vegetables industry generates numerous by-products or wastes like seeds, skin, trimmings, pomace, leaves, and stem, a greater segment of research has explored the extraction of essential oils from these green resources and later use in active packaging to extend the shelf life of fresh fruits and vegetable commodities [90]. The crystallinity level of the film matrix has been observed to impact the release of essential oils. Increased crystallinity results in a more complex diffusion pathway for the essential oils, slowing their release rate. Polar film matrices exhibit strong interactions with polar components of essential oils, including thymol, carvacrol, cinnamaldehyde, and eugenol. This interaction significantly slows down the release of these essential oils. Encapsulation of essential oil is another strategy to achieve a sustained release of active constituents and to improve their solubility in food [67].

13.3.5.7 Pigments

Plant pigments such as carotenoids, betalaine, anthocyanin, and curcumin have been explored as antioxidants and antimicrobials in active packaging for fruits and vegetables. Curcumin extracted from turmeric, when loaded in active packaging, exhibits a potent antimicrobial effect against gram-negative *E. coli* and gram-positive *bacillus cereus*, *Bacillus subtilis*, *Streptococcus aureus*, *micrococcus*, *Listeria monocytogens*, and yeast such as *Candida albicans*, *Candida parapsilosis* and other fungi like *Rhizoctonia solani* and *Penicillium expansum* [8]. While the carrageenan-based film showed a faster release of curcumin, the chitosan-based film gave the slowest release. The alginate-based film had moderate curcumin release characteristics. As a result of strong electrostatic interaction, curcumin loaded with chitosan and Montmorillonite-based nanocomposite gave a sustained release characteristic. The release kinetics of curcumin from gelatin were greater than those of zein fiber due to the greater swelling ability of gelatin. For its high solubility at a low pH, a better release of curcumin was observed at pH 4 than at 7. Overall, the release characteristics of curcumin were reported to depend on the type of matrix, the pore structure of the film, swelling ability, and solubility profiles [83, 8].

Unlike curcumin, anthocyanins are water-soluble pigments and, therefore, have a greater affinity for fruits and vegetables having high water activity. Anthocyanins have been shown to possess significant antioxidant properties mediated through their reducing power, hydrogen donation, singlet oxygen quenching, and metal chelation ability [40]. Additionally, the antimicrobial nature of anthocyanins is linked to their ability to destabilize the cytoplasmic membrane, increase cell membrane permeability, inhibit extracellular enzyme activity, and deplete essential substrates necessary for microbial growth. Anthocyanin extracted from amaranth, bayberry, red cabbage, blackberry seed coat, black carrot, blueberry, cranberry, eggplant, jamun, onion skin, and purple sweet potato possesses strong antioxidant functions and has been used in both active and intelligent packaging applications [117]. For its sensitivity to light, heat, and oxygen, it is important to offer barrier properties in packaging to preserve the functionality of anthocyanin. Gellan gum offers stability to anthocyanin and influences the release rate. Similarly, soy proteins affect the release rate by covalent and non-covalent interactions [112]. However, such pigment-based active packaging systems can cause color staining on fruits and may affect the quality or quality perception of fruits and vegetables.

13.4 Edible Coating Technologies

Fruits and vegetables, being highly perishable, are susceptible to spoilage caused by biochemical processes, enzymatic breakdown, and microbial proliferation on their surfaces. Edible coatings are employed as a method for protecting fruits and vegetables from deleterious effects. These coatings, thin layers of edible substances applied to the surface of produce, create a barrier that regulates the exchange of gases, water vapor, and volatile compounds between the product and its environment. They play a vital role in enhancing the shelf life of fruits and vegetables by decreasing the rate of respiration, enzymatic browning, and moisture loss [33, 49].

In general, edible coatings are prepared from aqueous solutions of various coating agents like polysaccharides (chitosan, alginate, cellulose, carrageenan, pectin, starch, etc.), proteins (whey, casein, soy protein, etc.) and lipids (carnauba, beeswax, and fatty acids). Along with these agents, certain active ingredients like antioxidants, antimicrobials, oxygen scavengers, enzymes, etc., are added to enhance the safety, nutritional, and sensory attributes as it is consumed along with foods [89]. These active agents can be added directly into the edible coating solution or coated onto the product surface for better functionality [17]

Antioxidants in edible coatings help reduce the rate of oxidation reactions in fruits and vegetables, such as color changes, flavor changes, softening, and nutritional losses. Similarly, antimicrobials reduce the risk of texture loss, microbial proliferation, and enzymatic browning reactions. Several studies have been carried out in this field, and an ongoing interest remains in developing edible coatings incorporated with active ingredients.

Edible-coated fresh-cut mango with a combination of alginate and ascorbic acid shows better preservation of color than the uncoated samples [82]. The ripening process of tomato was delayed by coating it with 10% Gum Arabic [7]. The coating acts as a barrier and thus reduces the rate of ethylene production and respiration. Chitosan-coated strawberries show a significant reduction in decay during storage [108]. Composite coating consisting of carboxymethyl cellulose and sodium alginate incorporated with active ingredients such as citric acid and oxalic acid was coated on ready-to-eat pear. This significantly reduced oxidative damage, physiological decay, and browning [59]. The efficacy of edible coatings made from wild sage seeds and pomegranate seed oil (PSO) in prolonging the shelf life of Mexican lime fruit stored under conditions of 20 ± 2 °C and 50–60% relative humidity was examined. These treatments represent a promising approach for enhancing the quality and extending the shelf life of Mexican lime fruit when stored at room temperature [65]. To prevent quick dissolution, researchers have extensively investigated water-insoluble polysaccharides like chitosan. A combination of chitosan and Tween 80 in a composite edible coating effectively preserves the quality characteristics of guava fruits post-harvest by retaining firmness and ascorbic acid content and minimizing weight loss during storage [92].

Combining chitosan coating with antimicrobial agents (grape seed extract, pomegranate, natamycin, nisin) in active packaging enhanced the shelf life of fresh strawberries compared to those without coating [28]. A polysaccharide-protein coating was made by combining chitosan and gelatin with hyaluronic acid to enhance the antioxidant properties. The impact of newly created composite edible coatings on the quality of strawberries was studied during a 15-day storage period. Findings indicated that the amount of hyaluronic acid used influenced the enhancement of inherent characteristics in the coated strawberries, such as reduced weight loss, adjusted pH, titratable acidity, and total solids content [6]. Coating tomatoes with alginate film containing Aloe vera and Garlic oil extended their shelf life because of the film's improved mechanical, thermal, and antimicrobial properties [13]. Edible films and coatings represent an eco-friendly technology with significant benefits for extending shelf life, offering substantial advantages in preservation. Comprehending the technologies and obstacles related to different coating methods for creating active packaging materials will facilitate the successful transition of these technologies into commercial use.

13.5 Effect of Active Packaging on Quality Indicators of Various Fruits and Vegetables

13.5.1 Physical Attributes

Fruits and vegetables, after the harvest, undergo many drastic changes to the quality, which are reflected in the changes to the physical attributes like color, gloss, visual appearance, weight loss, shrinkage/shriveling, surface roughness, etc. Active packaging of fruits and vegetables is known to preserve the physical quality of fresh produce. Tomatoes packed with chitosan and Titanium dioxide exhibited a higher firmness and minimal weight loss compared to chitosan-based film and control. The degradation of ethylene by the film delayed the ripening, as evident from the slow progression in the transformation of the yellow-colored tomato (not fully ripened) to red during storage [48]. A weight loss of around 10% is a commonly accepted threshold limit indicating the spoilage of mangoes. The packing of mangoes in poly-lactic acid and carboxy methyl cellulose-based film kept weight loss less than 5% for 42 days, while the control mangoes had started deteriorating from 21 days. The hydrophilic nature of carboxy methyl cellulose, which absorbs water vapor and controls the relative humidity or fogging in the package, suppressed the respiration rate. This delays senescence and weight loss by transpiration of mangoes. The lower weight loss for active-packaged mangoes can be correlated with their higher firmness and lower shriveling and gloss loss. The development of yellow color is an indication of the ripening of mangoes. Active-packaged mangoes also exhibit a delayed yellow color development in both flesh and peel [51].

In a study by [56] with eugenol-loaded poly-lactic acid-based nanocomposite active packaging, the storage stability of cucumber was evaluated. A cucumber with a water content and turgor pressure exhibits high hardness and minimal stickiness, giving high brittleness to the fruit. Although hardness drops and stickiness increase were noted for active packaged cucumber, the changes were minimal compared to the control. This was evident from the minimal drop in cucumber weight loss during storage. The coating of strawberries with chitosan combined with many different antimicrobials had no significant influence on the lightness and redness of fruits throughout the storage. However, a significant difference was noted with the texture of strawberries. The firmness of uncoated strawberries was highly reduced, but only minimally in cases coated with chitosan. The preserved pectin-rich-middle lamella could be the reason for a higher springiness value for chitosan added with pomegranate extract. Since firmness reflects the strength of the fruit cell walls, the maintenance of structural integrity by minimizing the enzymatic hydrolysis of cell wall components can be interpreted. The adhesiveness of the coated strawberry was lower than that of the uncoated sample. There was a change in the cohesiveness, chewiness, and gumminess attributes of fruit. When beads made from chitosan added with red thyme and lavender oil were used as a sachet in the strawberry box, the weight loss during storage was minimal. This was attributed to the hydrophobic nature of essential oils which vaporise from the sachet to adsorb onto the waxy surface of strawberries and reduce the water vapour trasmission from fruit. This effect has also been reported for cherries and grapes. However, the effect of the usage of a chitosan-essential oil sachet on the color and firmness of strawberries was negligible [86]. Aloui et al. [9], who tested the efficacy of sodium alginate and locust bean gum-based films added with a killer yeast (*Wickerhamomyces anomalus*) to protect orange from green mold rot, found that the film also significantly reduced weight loss. While the control sample observed a 50% reduction in its initial firmness level due to structural deterioration during storage for 15 days, the film-protected fruit had a 21% firmness loss. Both weight loss reduction and firmness preservation are attributed to the water barrier property of the film, which is higher for locust bean gum than sodium alginate due to the former's relatively high surface hydrophobicity and flexibility for greater adhesion to the fruit surface.

López–Gómez et al. [57] evaluated the effectiveness of β-cyclodextrin complexed essential oil-loaded cardboard box/tray type active packaging for preserving berries (grape), stone fruits (nectarine), and green leafy vegetables (lettuce). In grapes, yellowing and browning are the prime color disorders that affect consumers' acceptance of fruit. The active packaged fruit showed a lower color difference value. Essential oils could competitively inhibit browning enzymes like polyphenol oxidase and peroxidase. Essential oils could also create abiotic stress for the grape, generating its own antioxidant enzymes and protecting the fruit from oxidative deterioration. Although essential oils did not avoid color change in nectarines with an anthocyanin-rich peel, the chlorophyll-rich lettuce was protected from the browning effect. Lettuce is expected to be crisp green, with minimal wetness and browning by consumers. Since the color difference of lettuce was within the industrial standards established to correlate color difference with consumer acceptability, active packaging with essential oils was found suitable for preserving the color of the lettuce. Other than the color change, a reduction in weight loss (by 50%), rachis dehydration, and berry shatter was also noted for grapes. Like grapes, nectarines also had 1.5–2 folds and lower weight loss and firmness when actively packed in trays loaded with essential oils. Poly-lactic acid-based active packaging film loaded with thymol gave a reduced weight loss in hot peppers due to the lower water vapor permeability of the film. Since weight loss was low, shriveling of pepper was also minimal. The pepper also retained a higher firmness, crucial for consumer acceptability. Active packaging minimized the changes to the redness of hot pepper. This could mean the active packaging effectively preserves red pigment capsanthin in pepper. However, the yellow-pigmented violaxanthin, which increased upon storage due to maturation, was unaffected by active packaging. Sometimes, consumers mistake the yellow color of pepper as an indication of spoilage [76].

While the yellowness color value of the control papaya sample increased during storage due to maturation, hydroxypropyl methylcellulose film loaded with silver nanoparticles reduced the color change in papaya. Active packaging also reduced >4% weight loss observed in papaya to 2.6–3.25%. This is attributed to the high surface hydrophobicity of film added with nanoparticles [104]. Starch-based biodegradable non-perforated active packaging film loaded with an inorganic ethylene scavenger (Zeolite) reduced the decline in the color (hue) of broccoli during storage. An active film from starch and gluten, along with turmeric and cinnamon essential oils, reduced the weight loss of carrots by 44% during storage [81]. This was attributed to the film's increased hydrophobicity to keep water from getting lost. Warsiki (2017) developed an active wrapping film from chitosan and potassium permanganate for use as an ethylene absorber to extend the shelf life of tomatoes. The active packaged tomato fruit had a significantly higher hardness than the control after storage for four days at room temperature. Potassium permanganate was also used to maintain the color (hue) of tomatoes. In a study by [2], carbon dioxide (sodium carbonate and sodium chloride-based) and oxygen absorbers (iron fillings) were used to preserve the quality of strawberries during storage. The study found that after four weeks of storage, the chewiness, cohesiveness, firmness, and springiness values of active packaged strawberries were higher than the control. Moreover, an increase in the adhesiveness of strawberries was minimal for active packaged strawberries. The oxygen absorber was more effective than the CO_2 absorber in controlling the darkening of strawberries by oxidative browning. It also slightly preserved the redness color value of strawberries, indicating the prevention of anthocyanin oxidation. The electrical conductivity of strawberries packed with a carbon dioxide absorber was lower than that of the control and the one with an oxygen absorber. This is because of the delay in incrementing total soluble solids capable of conducting current.

Mahajan et al. [60] evaluated the effect of using a moisture absorber in modified atmosphere-packed mushrooms. To have an optimal moisture absorption rate from mushrooms, fast-absorbing string desiccants like calcium chloride and sorbitol were mixed with slow-absorbing bentonite. An increase in the amount of desiccant used, from 5 to 15 g (for 250 g of mushroom), increased weight loss in mushrooms from 3.6 to 6.2%. The desiccant absorbed approximately 40% and 70% of its weight in each pack after 3 and 5 days, respectively. Control packages showed water condensation on the film, resulting in poor visibility due to droplets, whereas packages with the moisture absorber exhibited minimal condensation. The package containing 5 g of moisture absorber exhibited the lowest browning index, whereas 10 g and 15 g of desiccants use had a bit higher browning index due to severe moisture loss from the mushroom. The use of desiccant did not significantly affect the composition of oxygen in gas, although a slight change with carbon dioxide was noted.

13.5.2 Chemical Changes

The ascorbate–glutathione cycle in fresh fruits and vegetable cells is crucial for scavenging the reactive oxygen species produced in large amounts during ripening. The inhibition of ROS prevents early senescence by reducing the oxidative deterioration.

Chitosan and alginate-based bilayer film incorporated with active ingredients such as ascorbic acid, citric acid, and phytic acid was tested for the effect on ascorbate–glutathione cycle and the post-harvest shelf life of cherry tomatoes [119]. The added film components exhibited superior antioxidant activity against free radicals, oxygen, and hydroxyl radicals, with a more pronounced effect from citric acid as ascorbic acid than phytic acid. The film inhibited the accumulation of hydrogen peroxide in cherry tomatoes by enhancing the activity of antioxidant enzymes involved in the ascorbate–glutathione cycle. The ascorbate and glutathione levels in the cherry tomatoes packed in active packaging were higher than in control, indicating that the oxidative damage levels are minimal. In a similar study by Kaewklin et al. [48], a delay in the ripening of tomatoes packed in chitosan-TiO_2 film was observed. This was attributed to the TiO_2 producing strong oxidized agents when exposed to UV, which in turn degraded ethylene to CO_2. Without TiO_2 presence, chitosan standalone minimally inhibited ethylene. Chitosan's capacity to scavenge ethylene is likely linked to the oxidizing potential of the hydroxyl and amino groups present in its molecular structure. A lower total soluble sugar found in active-packed tomatoes is an indication of minimal carbohydrate hydrolysis activity by enzymes like polygalacturonase and pectin esterase. There was also a delay in the generation of lycopene, as lycopene synthase activity could have been reduced. Moreover, the active packaging also improved the ascorbic acid levels in tomatoes by preventing them from getting utilized in rapid metabolism during ripening.

Kamthai et al. [51] observed that controlling the internal relative humidity or fogging by carboxy methyl cellulose added poly lactic acid film reduced the ethylene production and respiration rate of mangoes. This ultimately minimises the total soluble sugars in mango. As a result of respiration, the CO2 level increased, and the O_2 level decreased during storage. However, this was more severe for the control sample than for active-packed mangoes. The lowering of respiration rate and oxidation reduced the delayed increase in the concentration of total soluble. This also came with a delayed drop in the levels of organic acids. Cucumber packed in eugenol-loaded polylactic acid-based nanocomposite film has superior chemical stability. Malondialdehyde, an indicator of oxidation in plant cells, was reduced by Eugenol. Moreover, eugenol also increased the concentration of fruit antioxidant enzymes like peroxidase and superoxide dismutase to tackle the reactive oxygen species. These enzymes could have been triggered by a stress response. Moreover, the protein content reduction in cucumber was also minimal than the control sample [56].

The packing of strawberries in antioxidants rich thyme oil-loaded zein nanocomposite film preserved the total phenolic content, titratable acidity, and antioxidant activity of fruits in 15 days of storage [11]. Coating of strawberries with chitosan carrying different antimicrobials, nisin, natamycin, pomegranate, and rapeseed extract on the physicochemical parameters of the fruit was evaluated by Duran et al. [28]. Adding natamycin to the chitosan coating noticeably decreased the oxygen consumption of the fruits compared to the other coating methods. Total soluble solids content reduced due to sucrose hydrolysis. At the end of storage, strawberries treated with chitosan containing both natamycin and nisin displayed reduced CO_2 production in contrast to fruits coated solely with chitosan. This difference may be ascribed to the heightened solubility of CO_2 in chitosan films incorporating natamycin. Since both chitosan and nisin are positively charged, the increased electrostatic repulsion could have modulated the permeability of the film and thereby reduced the fruit respiration rate. Natamycin and pomegranate extract were also found to reduce the water vapor permeability of the film, and therefore, a general decline in water activity of strawberries was minimal for the coated fruit. López–Gómez et al. [57] observed the effect of β-cyclodextrin complexed essential oil-loaded cardboard box/tray of different dimensions on the physicochemical changes with grapes, nectarine, and lettuce. While the grapes had a general increase in the total soluble solids by pectin hydrolysis, nectarine, and lettuce had a lowered total soluble solids and dry matter content due to the use of sugars as an energy pools for metabolism by plant cells. In all fruit and vegetable types studied, the active packaging in a high surface area box minimizes the changes in total soluble solids and dry matter content. A high surface area for a unit weight of fruit/vegetable allows the better release of essential oil to the fruit. The released essential oil could inhibit ethylene synthesis thereby reducing respiration rate and physicochemical changes.

Fresh peppers are abundant in Vitamin C (100 mg/100 g) and are prone to oxidative loss. Active packaging in a poly-lactic-based film loaded with thymol significantly reduced the loss of vitamin C as the internal atmosphere had a minimal oxygen concentration. Thymol is also known to inhibit vitamin C oxidase in fruit. The total phenolic content in the hot pepper had a general increment by maturation during storage. Since active packaging reduces the maturation rate, the total phenolic content in pepper was lower [76]. Papaya wrapped in an active paper packaging loaded with lemon grass essential oil reduced the deterioration of Vitamin C observed in the first 6 days of storage in control. However, on the 12th day, Vitamin C levels were found to be low due to the delay in maturation. It was the same reason for a lower total soluble solids content in active packaged papaya than control [95]. The process of ethylene production in plant tissues is controlled by both l-aminocyclopropane-1-carboxylic (ACC) acid synthase (ACS) and ACC oxidase. ACS plays a pivotal role in converting S-adenosyl-l-methionine into ACC. Certain vegetables, such as broccoli, exhibit high sensitivity to ethylene. This sensitivity is often evident through visual effects like floret yellowing, which is primarily caused by chlorophyll breakdown due to ethylene exposure. Navarro-Martínez et al. [70] evaluated the efficacy of citral vapor in inhibiting the biosynthesis of ethylene in sensitive broccoli and high-ethylene-generating apples. With minimal concentration of citral, ethylene biosynthesis by broccoli and apple declined by 30–40% during storage (22°C). When active-paper sheets loaded with essential oils were tested, active-packaged broccoli had 30–40% and 40–70% ethylene inhibition at 22°C and 2°C, respectively. The mode of action is essential oils vaporizing and diffusing into the broccoli florets and inhibiting the ACC oxidase enzyme. This prevents the formation of ACC, a substrate for ethylene generation. Similarly, Starch-based biodegradable non-perforated active packaging film loaded with an inorganic ethylene scavenger (Zeolite) resulted in a high retention of vitamin C in broccoli during seven days storage at 12°C [63]. Banana carbon fiber, for its high surface area and microporous nature, was more effective than activated charcoal in preserving the chlorophyll level. When banana carbon fiber obtained from pyrolysis of banana fiber is used, it reduces chlorophyll loss from 25 to 1% in broccoli and from 13 to 5% in guava [50]. Deep eutectic solvent (choline chloride, betaine, and glycerol) films based on gelatin were prepared as an active packaging to regulate moisture for preserving cherry tomato fruit. The use of deep eutectic solvents improves the hydrophilicity, moisture sorption capacity, and water vapor permeability of the film. Compared to polyethylene pack, which completely inhibits moisture loss, the moisture regulation by deep eutectic solvent gel reduced the decay rate of tomatoes while also diminishing the drastic changes to the freshness, TSS, and titratable acidity of fruit during 9 days of storage [106].

Iron fillings and sodium polyacrylate filled in a sachet were placed along with Lulo, a tropical fruit, in a poly lactic acid-based packaging. The sachet was able to absorb only 7% of oxygen due to limitations in the oxygen permeability of the sachet from the headspace. This still meets the specifications for a commercial oxygen scavenger. A lowering of oxygen levels in headspace significantly reduced the fruit respiration rate of fruit [29].

13.5.3 Microbial Quality

The Ti-doped CuO nanoparticles reinforced gelatin film inhibited the growth of the bacteria *Staphylococcus aureus* and *E. coli*, as well as the fungi *Botrytis cinerea* and *Penicillium expansum*. Tomatoes wrapped in this film remained fresh and mold-free for 18 days, whereas the control tomatoes rotted in just four days. The tomatoes showed no putrefactive smell or slimy surface. Their structural integrity was preserved, as they did not burst and were free of leakage (sticky juice) [94]. Similarly, Kaewklin et al. [48] found that tomatoes packed in a film made of chitosan and TiO_2 showed no signs of fungal infection after 14 days. Chitosan–gelatin-based films, with or without additives, inhibited surface mold growth in avocados. While mold was visible in uncoated avocados within seven days, coating with chitosan, with or without carbon dots, delayed the appearance of mold to 14 and 21 days, respectively. The collapse and distortion of microbial cell membranes and the leakage of intracellular substances were observed. This can be attributed to the generation of reactive oxygen species (ROS), which cause the breakdown of microbial cell walls and ultimately result in the degradation of cellular components [30].

The deterioration of mangoes after harvest results from anthracnose and stem end-rot diseases caused by *Colletotrichum gloeosporioides* and *Lasiodiplodia theobromae*, respectively. Thyme oil displayed antimicrobial potency against *L. theobromae* and *C. gloeosporioides*, with 66% and 22% inhibition rates, respectively [68]. Cucumbers packaged in eugenol-loaded polylactic acid-based nanocomposite active packaging had minimal total microbial counts, and low yeast and mold count after 15 days of storage. The shelf life of the cucumbers increased from 15 to 21 days at 4 °C [56]. Similarly, the active packaging of strawberries in thyme essential oil-incorporated zein nanocomposite film inhibited microbial growth. Unlike the control samples, which exhibited signs of decay, the strawberries in the thyme-loaded film remained fresh for 15 days [11].

Antimicrobials, such as nisin, natamycin, pomegranate, and grape seed extract as coatings on strawberries reduced the growth of aerobic mesophilic bacteria during 40 days of storage. While nisin and grape extract effectively inhibited aerobic bacteria, natamycin significantly reduced yeast and mold counts. The reduction in coliform counts was also substantial for the coated samples due to changes in internal gas composition [28]. Grey mold rot of strawberries, caused by the growth of *Botrytis cinerea* on the surface, is the main cause of spoilage. In a study by [86], chitosan beads incorporated with essential oils from lavender and red thyme were tested for their efficacy in protecting strawberries. Unika [28] used essential oil-loaded chitosan as a coating, in their study, beads were made from chitosan and placed in sachets to be in close contact with strawberries packed in a clamshell box. The count of infected fruits increased with storage time. While the control sample and chitosan-only treatments exhibited mold incidence starting from day 4, chitosan treated with low levels of essential oil showed mold contamination evident from day 8. An increase in the number of beads improved mold inhibition. On day 12, while the control sample had 40% of fruits infected, the chitosan beads with high levels of lavender and red thyme oil showed infection rates of 5% and 0%, respectively. Therefore, red thyme oil at higher concentrations was more effective in inhibiting mold growth than lavender. The antimicrobial assay also confirmed the potency of red thyme oil at high concentrations to inhibit *Botrytis cinerea* [86].

Green mold rot caused by *Penicillium digitatum* causes a significant post-harvest decay of citrus fruits like oranges. A bioactive sodium alginate and locust bean gum-based film added with killer yeast (*Wickerhamomyces anomalus*) was used to preserve oranges from green mold rot [9]. Numerous factors, such as the characteristics of the entrapped strain, the chemical composition of the material forming the film, and the conditions during drying and storage, can greatly impact the viability of microbial cells confined within biopolymeric matrices. The study found that both sodium alginate and locust bean gum did not affect the viability of added yeast. The film inhibited more than 75% of *Penicillium digitatum* in orange stored for 21 days at 25°C. The antifungal activity of *Wickerhamomyces anomalus* is attributed to their ability to secrete glycoprotein β-glucanase enzyme, which diffuses through the polysaccharide film to hydrolyze glucan, an abundant polymer in the cell wall of fungi like *Penicillium digitatum.* The cell wall lysis and a leak of cytoplasmic contents caused death to the mold and thereby improved the shelf life of the oranges. Since polysaccharide-based film is also a good gas barrier, this can contribute to mold inhibition, which requires aerobic conditions. The type of microbes growing on the substrate may vary depending on the fruit/vegetable composition, packaging, and storage conditions.

In a study by López–Gómez et al. [57], grapes, nectarines, and lettuce were packed in essential oil-loaded cardboard boxes. Active packaging primarily affected the psychrophile count for grapes, reducing it by 0.4 to 0.8 log after 30 days, more than other microbial types. Controlling psychrophiles is crucial for grapes, as these bacteria thrive under refrigerated conditions. The incidence of decay dropped from

42 to 12% after 30 days of storage. A higher surface area of packaging allows for a better release of essential oils, leading to a greater antimicrobial effect. Since essential oils can also inhibit ethylene generation, the lowered respiration rate results in a decline in the enzymatic conversion of complex carbohydrates into simple sugars that are utilizable by microbes. Therefore, essential oils could have both direct and indirect effects on microbes. After 25 days, fungal decay in nectarines was only 2–6%, 1.5-fold lower than in the control sample. Like grapes, psychrophile counts in active-packaged nectarines and lettuce were 0.9-fold lower than in the control. Additionally, a general decline in the growth of enterobacteria was noted in grapes. After harvest, papayas become highly susceptible to anthracnose disease caused by *C. gloeosporioides*. Hydroxypropyl methylcellulose film incorporated with 0.25% silver nanoparticles was found to effectively inhibit *C. gloeosporioides* and extend the shelf life of papaya [104]. Similar to its effect on lettuce, other green leafy vegetables, such as spinach, when packed in thyme essential oil-loaded starch nanocomposite film, exhibited significant reductions in *E. coli* and *Salmonella typhimurium*, whereas the control sample had 4.5 log cfu/g [45]. In another study, active paper packaging loaded with lemongrass essential oil was used for papaya, with the active constituents α-citral and β-citral credited for the effective inhibition of yeast and mold counts in papaya [95]. Pathogen contamination in fresh produce is one of the prime reasons for foodborne outbreaks and gastrointestinal disorders. With *E. coli* O157: H7 iceberg lettuce, Lu et al. [58] studied the effectiveness of various active packaging structures capable of emitting carbon dioxide, allyl isothiocyanate, and chlorine dioxide while scavenging oxygen. All these systems effectively inhibited both *E. coli* O157: H7 and aerobic bacteria under all studied storage conditions (4–22 °C for three weeks). Takala et al. [99] used methylcellulose and alginate matrices to prepare antimicrobial packaging films by loading organic acids, rosemary extract, and spice essential oil. When this film was used to pack broccoli florets, it effectively inhibited *E. coli*, *Listeria monocytogenes*, and *Salmonella typhimurium*. Antimicrobial packaging made from chitosan, with and without casein, inhibited the growth of mesophilic and psychrophilic bacteria in carrots, with bacterial reductions of 2 to 4.5 log. Although ionic interactions between chitosan and casein were noted, they did not influence the antimicrobial effect, as a significant portion of free amino acids remained available to interact with negatively charged bacterial surfaces, affecting cell wall permeability [66]. Haider et al. [38] developed an organic ethylene scavenger pad from rice husk or as a composite with eggshell powder. Using the pad in packed tomatoes resulted in a 40% delay in ripening and an extended shelf life of up to 12 days, attributed to the rice husk rather than the eggshell powder. The microfibrous cellulose with multiple active points adsorbed ethylene during storage, although ethylene adsorption steadily declined. While the levels of β-carotene and lycopene increased in all tomato samples over time, chlorophyll levels decreased. However, these changes were delayed for tomatoes in the presence of the rice husk-based ethylene adsorber. Moreover, total soluble solids in the actively packaged fruit were also minimal, demonstrating the effectiveness of organic rice husk as an ethylene adsorber. Aday and Caner [1] used oxygen absorbers to preserve the quality of strawberries. While the control sample exhibited 66–72% decay due to mold

incidence, the oxygen absorbers reduced the decay rate to 11–15%. Thus, oxygen absorbers were efficient in inhibiting mold growth in strawberries.

13.5.4 Organoleptic Quality

In a study by Kamthai et al. [51], the overall acceptability scores for active-packed mangoes received lower scores than control. However, the low scores were associated with the hardness and sour taste of mangoes. Since active packaging delays ripening, the sensory scores were lower than those of the fully ripened control, which gave a sweet and soft texture. From a shelf-life extension perspective, this is still desirable. López–Gómez et al. [57] stated that the nectarine packed in an essential-oil-loaded cardboard box exhibited a higher freshness score, with no difference in flavor changes during storage. Although color changes could still be perceived with active packaging, it was within the acceptability limit. In the case of lettuce, it was mainly texture and dehydration that influenced the sensory rating. Essential oil did not give any off-flavors to the product. Active-packaged grape, nectarine, and lettuce were found organoleptically acceptable for 30, 25, and 14 days at 2°C, whereas the control group gave bad sensory scores in less than a week. Hot peppers active-packaged in a polylactic acid-based film loaded with thymol received a higher overall quality index than polyethylene-packaged hot peppers. Without thymol, the polylactic acid film gave hot pepper a lower sensory score and was indifferent to polyethylene. Although the differences between packages were not seen in the first week of storage, after 21 days, a significantly higher rating was observed with active packaged hot peppers due to their edibility and no observed off-flavor emissions signs of microbial decay [76]. Wrona et al. [111] reported that consumers preferred the mushrooms preserved with sodium metabisulfate and citric acid combined antioxidant packaging in a ranking test used to determine the effectiveness of different antioxidants loaded into pouches for preserving mushrooms. This is because these antioxidants preserved the white color of mushrooms, although extensive storage caused a yellowing effect by excessive release of sulfur dioxide. Moreover, consumers also preferred green tea extract-preserved mushrooms more than the control mushroom sample. Since purple carrot extract and cinnamon oil left stains on the mushroom's surface and the anthocyanins were ineffective in preserving the storage quality, they received lower scores. The differences between the samples were also confirmed with a paired comparison test. It was revealed in a sensory analysis that the use of CO_2 (sodium carbonate and sodium chloride-based) and O_2 absorbers (iron filling) improved the scores for flavor, texture, global appearance, color, and general acceptability of strawberries [2]. Using activated carbon as an ethylene absorber in modified atmospheric packaging of tomatoes reduced the formation of volatile compounds during storage. While the headspace of the control sample had 22 volatiles, mainly aldehydes (55%), alcohols (17%), hydrocarbons (15%), esters and ethers (11%), and acids (1%), ethylene absorber resulted in only an eight detectable volatiles mainly constituting alcohols. The presence of less diversity of flavor compounds and minimal concentration reduced the odor intensity of both fresh-cut and whole tomatoes. The sensory ratings for flavor, sweetness, and juiciness were higher for tomatoes packed in modified atmospheric packaging along with an ethylene absorber. However, no difference was noted between the control and the ethylene absorber groups regarding sourness and turgidity. The high rating on sweetness perception for the ethylene absorber group had another contributing factor, which is the presence of a relatively high concentration of 2-methyl butanol, giving a typical flavor of sweet, fresh, and ripe tomato [14].

13.6 Integration with Minimal Processing of Fruits and Vegetables

Today's consumers expect convenience in the fruits and vegetables they purchase from the market without any nutritional impairment. Minimal processing techniques encompass operations like washing, disinfection, trimming, peeling, blanching, cutting, shredding, grating, and chilling to produce fruits and vegetables in convenient ready-to-eat or ready-to-use formats. The minimal processing of fruits and vegetables should aim to keep the produce fresh without significant loss of nutritional quality and to have a feasible shelf life to cover extensive distribution and storage [116]. The fruit skin is a natural protective layer that limits quality deterioration by oxidation by inhibiting the access of oxygen and microbes to the edible flesh. However, peeling and cutting removes this protective layer, causing physical injury and adding physiological stress. Naturally, enzymes are compartmentalized to have a controlled series of biochemical reactions, but tissue breakage promotes contact between enzymes and substrates. Although blanching with water or steam can inactivate enzymes and hinder respiratory activity, the loss or inactivation of natural antioxidants makes the produce vulnerable to autooxidation, which can still occur without enzyme participation. The exudation of fluids from bruises and leached nutrients supports microbial growth, thereby reducing shelf life.

Moreover, increased respiration rate, ethylene generation, intracellular oxidative stress damage, and other biochemical reactions occur, the kinetics of which significantly increase as a consequence of these operations. Beyond kinetics, minimally processed fruits or vegetables can have completely different dynamics of reactions, depending on the availability of substrates, enzyme contact, storage conditions, and a wide array of variables that make this process more complicated than with whole fruits or vegetables [12, 116]. This increased complexity adds further weight to the importance of active packaging for preserving freshness and extending shelf life. In a study by Leneveu-Jenvrin et al. [55], active packaging film incorporated with an oxygen scavenger was tested for its efficacy in preserving the quality of cut mango pieces (1 cm × 1 cm) subjected to steam blanching (60 °C, 3 min). Although polyphenol oxidase was inhibited by 70% during blanching, color change could progressively occur during storage at 4 °C due to the residual activity of enzymes, which have better access to polyphenols after heat treatment because of partial cell wall digestion. Installing a hurdle by incorporating an oxygen scavenger in the film reduced the browning-led color change by 30–36%. Moreover, the drop in the hardness of mango pieces was also alleviated. However, the oxygen scavenger had no influence on microbial growth, which was primarily hindered by the steam treatment only.

Mastromatteo et al. [64] studied the effect of active modified atmospheric packaging and active coating with sodium alginate, with or without grapefruit extract, on the storage quality of washed, surface-disinfected, and peeled kiwi fruit. The active coating delayed the dehydration and respiration of fruits. A 2-log reduction in the viable cell count of mesophilic and psychrophilic bacteria was also noted for the active coating. The combination of active coating and modified atmosphere packaging extended the shelf life of minimally processed kiwi fruits up to 13 days. Using an oxygen scavenger (iron fillings) in the packaging of minimally processed oranges reduced the oxygen concentration levels in the headspace of the package. On sensory evaluation, the oranges stored with the oxygen scavenger received a higher rating than the control. Changes to physicochemical quality, such as a drop in pectin levels and an increase in total reducing sugars, were minimized. The oxygen scavenger extended the shelf life of oranges, which started deteriorating due to mold incidence from day 15 in the control sample. Active packaging of fresh-cut carrots in a film loaded with silver nanoparticles reduced the weight loss during storage for ten days at 10 °C. The film exhibited a more pronounced antimicrobial effect against mesophilic aerobic bacteria and coliforms. The bacterial count was within the acceptability limit for consuming uncooked foods. With no influence over the pH and firmness of carrots, the film maintained ascorbic acid levels. Since the migration of nanoparticles to carrot pieces was found below detectable limits, it is considered safe for use [18]. Singh et al. [93] developed a pyrogallic acid (10–20%) based oxygen scavenger to extend peeled garlic's shelf life. A reduction in fungal decay, aerobic bacterial count, and weight loss was noted for both room temperature and refrigerated storage. Unlike the control sample, the garlic packed with the oxygen scavenger remained firmer and did not develop yellow discoloration when stored at 5 °C. However, a slight yellow color appeared at room temperature after 20 days. The shelf life of garlic was extended to 30 days at 5 °C and 20 days at 25 °C.

13.7 SWOT Analysis

The strengths, weaknesses, opportunities, and threats of active packaging for fruits and vegetables are reviewed here. Fruits and vegetables generate more waste than other food categories, and such post-harvest losses can be minimized by active packaging. Active packaging delays ripening and supports international transport. The inhibition or delay of ripening and oxidative browning, along with preserved freshness, improves consumer acceptance of fruits [35]. Active packaging can also serve as a tool to supplement or fortify probiotics and nutrients. Moreover, it prevents the oxidation of nutrients like vitamin C and preserves fruit phytochemicals, enhancing their health value. Fruits and vegetables are a leading cause of food-borne infections, hospitalizations, and deaths. Active packaging has been shown to inhibit numerous pathogens, such as *Listeria monocytogenes* and *Clostridium botulinum* [78]. Therefore, active packaging can ensure the safety of fresh produce. Since active packaging reduces the weight loss of fruits and vegetables by transpiration, the preserved weight generates higher revenue (Fig. 13.4).

Sales and distribution of fresh fruits and vegetables are largely undertaken by farmers or small and medium-scale industries. The higher cost of active packaging makes it difficult for these players to utilize it. Essential oils used in active packaging can oxidize, generating off-flavors. Pigments like anthocyanin can also stain the fruit, affecting consumer acceptability of fresh produce or fresh-cut fruits and vegetables. Moreover, sachets or pads inside a package, which are not intended for consumption, can also impact consumer acceptance. Not all active constituents are compatible with every biopolymer base used in packaging. Some studies report that active constituents can interact complexly with biopolymers, resulting in poor release. Although sustained release is desirable, strong and permanent interactions may cause a loss of functionality in active packaging [16].

Since the quality of fruits and vegetables and the properties of packaging are influenced by multiple variables like temperature and relative humidity, the design stage becomes complex as all variables need to be considered. Recycling active packaging from the supply chain is challenging, as it requires separating the active material layer from the laminate and reactivating the active constituents for reuse. Some active

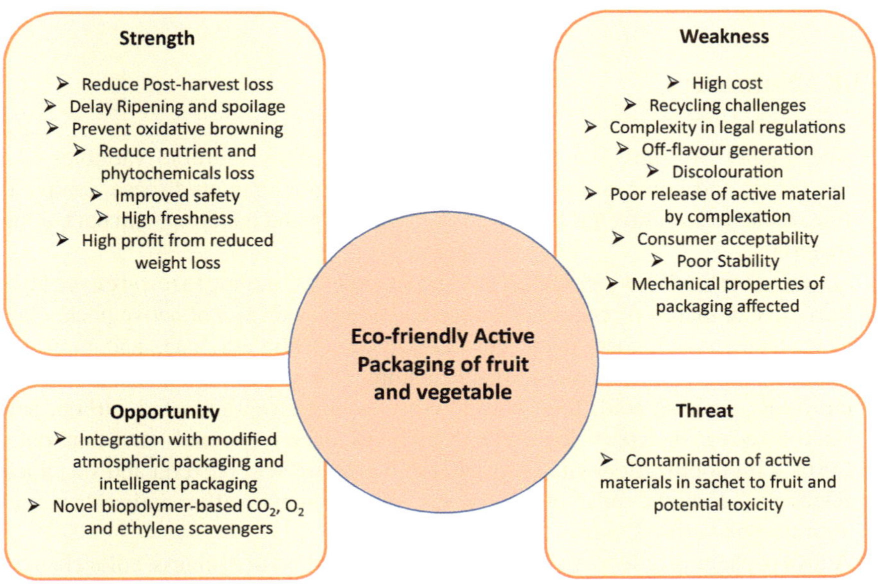

Fig. 13.4 SWOT analysis of eco-friendly active packaging of fruit and vegetable

packaging materials, such as anthocyanins and ascorbic acid, have superior antioxidant and antimicrobial properties but poor thermal or light stability. This necessitates multilayer packaging with suitable barrier materials to protect the functionality of the active constituent, which increases the cost, processability, and versatility of active packaging. Additionally, complex legal and regulatory aspects of active packaging, particularly regarding migration limits and safety, delay the commercialization of this technology [102].

There is a greater scope to integrate active packaging with intelligent and modified atmospheric packaging. Combining edible coatings with active film materials is possible to enhance fresh produce's shelf life. Since these hybridization possibilities within active packaging are nascent, they offer great potential for future exploration [71]. Although many green materials have been explored as antimicrobials and antioxidants, there are few nature-based solutions for ethylene absorbers, CO_2 absorbers/emitters, and O_2 absorbers. Moreover, organic materials have been shown to have lower efficacy than inorganic materials in active packaging. Thus, this segment requires accelerated research to identify or produce sustainable and renewable materials for active packaging. Since sachets containing active constituents (e.g., iron fillings) are in close contact with fruits and vegetables, contamination can occur due to mechanical damage to the sachet or increased permeability upon contact with moisture from the produce [46]. Additionally, nanomaterials increasingly studied for active packaging require extensive toxicity studies before they can be cleared for use. The potential threats of active packaging must be addressed to improve the safety of fruits and vegetables for consumption.

13.8 Summary

- The quality of fruits and vegetables is influenced by multiple variables like the nature and composition of fresh produce, contamination/disease, temperature, relative humidity, gas composition, processing and handling, and packaging materials.
- Unlike other food commodities, fruits and vegetables continue to respire even after harvest and, therefore, require a unique and complex design of active packaging
- The absorbers/adsorbers/scavengers of oxygen, CO_2, ethylene, and moisture could be of an organic or inorganic nature. Novel packaging designs using enzymes, ascorbic acid, food waste/by-products-derived activated carbon, and micronized eggshell particles can be alternatives for conventional active materials.
- Surface immobilization with chitosan and live lactic acid bacteria and sustained release of essential oils, organic acids, and pigments offer antioxidant and antimicrobial effects in fruits and vegetables.
- Active packaging reduces respiration rate, ripening, loss of firmness, color change, and weight loss and preserves nutrients and sensory quality of fresh produce. It has been shown to effectively inhibit both spoilage and pathogenic microbes, thereby improving fresh produce's shelf life and safety.

Active packaging has great scope for integration with intelligent and modified-atmospheric packaging to improve the safety of fruits and vegetables.
Although many green materials have been researched in the context of antioxidant or antimicrobial packaging in the past two decades, there is a greater need to focus on green material-based ethylene, O_2, and CO_2 scavengers and adsorbers.

References

1. Aday, M.S., Caner, C.: The shelf life extension of fresh strawberries using an oxygen absorber in the biobased package. LWT-Food Sci. Technol. **52**(2), 102–109 (2013)
2. Aday, M.S., Caner, C., Rahvalı, F.: Effect of oxygen and carbon dioxide absorbers on strawberry quality. Postharvest Biol. Technol. **62**(2), 179–187 (2011)
3. Ahn, B. J., Gaikwad, K. K., Lee, Y. S.: Characterization and properties of LDPE film with gallic-acid-based oxygen scavenging system useful as a functional packaging material. J. Appl. Polymer Sci. 133(43), (2016)
4. Akram, N.A., Shafiq, F., Ashraf, M.: Ascorbic acid is a potential oxidant scavenger and has a role in plant development and abiotic stress tolerance. Front. Plant Sci. **8**, 238088 (2017)
5. Alemu, K.: Dynamics and management of major postharvest fungal diseases of mango fruits. J. Bio. Agri. healthcare. **4**, (2014)
6. Al-Hilifi, S.A., Al-Ali, R.M., Dinh, L.N., Yao, Y., Agarwal, V.: Development of hyaluronic acid-based polysaccharide-protein composite edible coatings for preservation of strawberry fruit. Int. J. Biol. Macromol. **259**, 128932 (2024)
7. Ali, A., Maqbool, M., Alderson, P.G., Zahid, N.: Effect of gum arabic as an edible coating on antioxidant capacity of tomato (*Solanum lycopersicum* L.) fruit during storage. Postharvest Biol. Technol. **76**, 119–124 (2013)
8. Aliabbasi, N., Fathi, M., Emam-Djomeh, Z.: Curcumin: A promising bioactive agent for application in food packaging systems. J. Environ. Chem. Eng. **9**, 105520 (2021)
9. Aloui, H., Licciardello, F., Khwaldia, K., Hamdi, M., Restuccia, C.: Physical properties and antifungal activity of bioactive films containing Wickerhamomyces anomalous killer yeast and their application for preservation of oranges and control of postharvest green mold caused by *Penicillium digitatum*. Int. J. Food Microbiol. **200**, 22–30 (2015)
10. Alver, B.E.: A comparative adsorption study of C2H4 and SO2 on clinoptilolite-rich tuff: Effect of acid treatment. J. Hazard. Mater. **262**, 627–633 (2013)
11. Ansarifar, E., Moradinezhad, F.: Preservation of strawberry fruit quality via the use of active packaging with encapsulated thyme essential oil in zein nanofiber film. Int. J. Food Sci. Technol. **56**(9), 4239–4247 (2021)
12. Artes, F., Allende, A.: Minimal fresh processing of vegetables, fruits and juices. In: Emerging technologies for food processing, pp 677–716. Academic Press (2005)
13. Aziz, M.S.A., Salama, H.E.: Developing multifunctional edible coatings based on alginate for active food packaging. Int. J. Biol. Macromol. **190**, 837–844 (2021)
14. Bailén, G., Guillén, F., Castillo, S., Serrano, M., Valero, D., Martínez-Romero, D.: Use of activated carbon inside modified atmosphere packages to maintain tomato fruit quality during cold storage. J. Agric. Food Chem. **54**(6), 2229–2235 (2006)
15. Barbosa, A. A. T., de Araújo, H. G. S., Matos, P. N., Carnelossi, M. A. G., de Castro, A. A.: Effects of nisin-incorporated films on the microbiological and physicochemical quality of minimally processed mangoes. Int. J. Food Microbiol. **164**, 135–140 (2013)
16. Barrett, D.M., Beaulieu, J.C., Shewfelt, R.: Color, flavor, texture, and nutritional quality of fresh-cut fruits and vegetables: desirable levels, instrumental and sensory measurement, and the effects of processing. Crit. Rev. Food Sci. Nutr. **50**(5), 369–389 (2010)

17. Bastarrachea, L.J., Wong, D.E., Roman, M.J., Lin, Z., Goddard, J.M.: Active packaging coatings. Coatings **5**(4), 771–791 (2015)
18. Becaro, A.A., Puti, F.C., Panosso, A.R., Gern, J.C., Brandão, H.M., Correa, D.S., Ferreira, M.D.: Postharvest quality of fresh-cut carrots packaged in plastic films containing silver nanoparticles. Food Bioprocess Technol. **9**, 637–649 (2016)
19. Bodbodak, S., Rafiee, Z.: Recent trends in active packaging in fruits and vegetables. In: Eco-friendly technology for postharvest produce quality, pp. 77–125. Academic Press (2016)
20. Bodbodak, S., Rafiee, Z.: Recent trends in active packaging in fruits and vegetables. In: Eco-friendly technology for postharvest produce quality, pp. 77–125. Academic Press (2016)
21. Bovi, G.G., Caleb, O.J., Klaus, E., Tintchev, F., Rauh, C., Mahajan, P.V.: Moisture absorption kinetics of FruitPad for packaging of fresh strawberry. J. Food Eng. **223**, 248–254 (2018)
22. Carpena, M., Nuñez-Estevez, B., Soria-Lopez, A., Garcia-Oliveira, P., Prieto, M.A.: Essential oils and their application on active packaging systems: a review. Resources **10**(1), 7 (2021)
23. Coloma, A., Rodríguez, F.J., Bruna, J.E., Guarda, A., Galotto, M.J.: Development of an active film with natural zeolite as ethylene scavenger. J. Chil. Chem. Soc. **59**(2), 2409–2414 (2014)
24. Cruz, R.S., Camilloto, G.P., dos Santos Pires, A.C.: Oxygen scavengers: An approach on food preservation. Struct. Funct. Food Eng. **2**, 21–42 (2012)
25. de Oliveira, K.Á.R., Fernandes, K.F.D., de Souza, E.L.: Current advances on the development and application of probiotic-loaded edible films and coatings for the bioprotection of fresh and minimally processed fruit and vegetables. Foods **10**(9), 2207 (2021)
26. de Oliveira-Filho, J. G., Lemes, A. C., Braga, A. R. C., Egea, M. B.: Biodegradable eco-friendly packaging and coatings incorporated of natural active compounds. In: Food packaging, pp. 171–206. CRC Press (2020)
27. Duan, C., Meng, X., Meng, J., Khan, M. I. H., Dai, L., Khan, A., ... Ni, Y.: Chitosan as a preservative for fruits and vegetables: a review on chemistry and antimicrobial properties. J. Bioresour. Bioprod. **4**, 11–21 (2019)
28. Duran, M., Aday, M.S., Zorba, N.N.D., Temizkan, R., Büyükcan, M.B., Caner, C.: Potential of antimicrobial active packaging 'containing natamycin, nisin, pomegranate and grape seed extract in chitosan coating'to extend shelf life of fresh strawberry. Food Bioprod. Process. **98**, 354–363 (2016)
29. Escobar, H.J., Garavito, J., Castellanos, D.A.: Development of an active packaging with an oxygen scavenger and moisture adsorbent for fresh lulo (*Solanum quitoense*). J. Food Eng. **349**, 111484 (2023)
30. Ezati, P., Rhim, J.W., Molaei, R., Rezaei, Z.: Carbon quantum dots-based antifungal coating film for active packaging application of avocado. Food Packag. Shelf Life **33**, 100878 (2022)
31. Fasihnia, S. H., Peighambardoust, S. H., Peighambardoust, S. J., Oromiehie, A.: Development of novel active polypropylene based packaging films containing different concentrations of sorbic acid. Food Packag. Shelf Life. **18**, 87–94 (2018)
32. Gaikwad, K.K., Singh, S., Ajji, A.: Moisture absorbers for food packaging applications. Environ. Chem. Lett. **17**(2), 609–628 (2019)
33. Ghidelli, C., Pérez-Gago, M.B.: Recent advances in modified atmosphere packaging and edible coatings to maintain quality of fresh-cut fruits and vegetables. Crit. Rev. Food Sci. Nutr. **58**(4), 662–679 (2018)
34. Giménez, B., Gómez-Estaca, J., Alemán, A., Gómez-Guillén, M.C., Montero, M.P.: Improvement of the antioxidant properties of squid skin gelatin films by the addition of hydrolysates from squid gelatin. Food Hydrocolloids **23**(5), 1322–1327 (2009)
35. Gul, K., Mir, N. A., Singh, P., Wani, A. A.: Postharvest biology and technology of apple. Postharvest Biol. Technol. Temperate Fruits 223–243 (2018)
36. Guo, Y., Chen, X., Gong, P., Long, H., Wang, J., Deng, Z., Chen, F.: Characterization of an active film prepared with Lentinus edodes (shiitake) polysaccharide and its effect on post-harvest quality and storage of shiitake. Int. J. Biol. Macromol. **238**, 123973 (2023)
37. De Wild, H.P.J., Otma, E.C., Peppelenbos, H.W.: Carbon dioxide action on ethylene biosynthesis of preclimacteric and climacteric pear fruit. J. Exp. Bot. **54**(387), 1537–1544 (2003)

38. Haider, Z., Anjum Nawab, F.A., Lutfi, Z., Hasnain, A.: The effect of rice husk based ethylene scavengers on the postharvest quality of tomato fruits. Int. J. Biol. Biotech. **17**(4), 765–775 (2020)
39. Hauser, C., Thielmann, J., Muranyi, P.: Organic acids: usage and potential in antimicrobial packaging. In: Antimicrobial food packaging, pp. 563–580. Academic Press (2016)
40. He, J., Giusti, M. M.: Anthocyanins: natural colorants with health-promoting properties. Annu. Rev. Food Sci. Technol. **1**, 163–187 (2010)
41. Heard, G.M.: Microbial safety of ready-to-eat salads and minimally processed vegetables and fruits. Food Australia **51**, 414–420 (1999)
42. Hodges, R.J., Buzby, J.C., Bennett, B.: Postharvest losses and waste in developed and less developed countries: opportunities to improve resource use. J. Agric. Sci. **149**, 37–45 (2011)
43. Illeperuma, C. K., Nikapitiya, C.: Modified atmosphere packaging of minimally processed avocado cv.'Booth 7'. J. Horticult. Sci. Biotechnol. **81**(4), 607–612 (2006)
44. Imran, M., Klouj, A., Revol-Junelles, A.M., Desobry, S.: Controlled release of nisin from HPMC, sodium caseinate, poly-lactic acid and chitosan for active packaging applications. J. Food Eng. **143**, 178–185 (2014)
45. Issa, A., Ibrahim, S.A., Tahergorabi, R.: Impact of sweet potato starch-based nanocomposite films activated with thyme essential oil on the shelf-life of baby spinach leaves. Foods **6**(6), 43 (2017)
46. Janjarasskul, T., Suppakul, P.: Active and intelligent packaging: the indication of quality and safety. Crit. Rev. Food Sci. Nutr. **58**(5), 808–831 (2018)
47. Jipa, I.M., Stoica-Guzun, A., Stroescu, M.: Controlled release of sorbic acid from bacterial cellulose based mono and multilayer antimicrobial films. LWT **47**(2), 400–406 (2012)
48. Kaewklin, P., Siripatrawan, U., Suwanagul, A., Lee, Y.S.: Active packaging from chitosan-titanium dioxide nanocomposite film for prolonging storage life of tomato fruit. Int. J. Biol. Macromol. **112**, 523–529 (2018)
49. Kaewprachu, P., Rawdkuen, S.: Application of active edible film as food packaging for food preservation and extending shelf life. Microbes Food Health 185–205 (2016)
50. Kalyani, P., Das, P. P., Khandelwal, M.: Utilization of natural fiber–derived active agents for shelf life extension of broccoli (*Brassica oleracea* L.) and guava (*Psidium guajava*). Biomass Convers. Biorefinery 1–12 (2023)
51. Kamthai, S., Magaraphan, R.: Development of an active polylactic acid (PLA) packaging film by adding bleached bagasse carboxymethyl cellulose (CMCB) for mango storage life extension. Packag. Technol. Sci. **32**(2), 103–116 (2019)
52. Kartal, S., Aday, M.S., Caner, C.: Use of microperforated films and oxygen scavengers to maintain storage stability of fresh strawberries. Postharvest Biol. Technol. **71**, 32–40 (2012)
53. Lee, D.S.: Carbon dioxide absorbers for food packaging applications. Trends Food Sci. Technol. **57**, 146–155 (2016)
54. Lee, D.S., Wang, H.J., Jaisan, C., An, D.S.: Active food packaging to control carbon dioxide. Packag. Technol. Sci. **35**(3), 213–227 (2022)
55. Leneveu-Jenvrin, C., Apicella, A., Bradley, K., Meile, J. C., Chillet, M., Scarfato, P., Remize, F.: Effects of maturity level, steam treatment, or active packaging to maintain the quality of minimally processed mango (Mangifera indica cv. José). J. Food Process. Preserv. **45**(7), e15600 (2021)
56. Li, M., Yu, H., Xie, Y., Guo, Y., Cheng, Y., Qian, H., Yao, W.: Effects of double layer membrane loading eugenol on postharvest quality of cucumber. Lwt **145**, 111310 (2021)
57. López-Gómez, A., Ros-Chumillas, M., Buendía-Moreno, L., Martínez-Hernández, G.B.: Active cardboard packaging with encapsulated essential oils for enhancing the shelf life of fruit and vegetables. Front. Nutr. **7**, 559978 (2020)
58. Lu, H., Zhu, J., Li, J., Chen, J.: Effectiveness of active packaging on control of *Escherichia coli* O157: H7 and total aerobic bacteria on iceberg lettuce. J. Food Sci. **80**(6), M1325–M1329 (2015)
59. Magri, A., Landi, N., Capriolo, G., Di Maro, A., Petriccione, M.: Effect of active layer-by-layer edible coating on quality, biochemicals, and the antioxidant system in ready-to-eat 'Williams' pear fruit during cold storage. Postharvest Biol. Technol. **212**, 112873 (2024)

60. Mahajan, P.V., Rodrigues, F.A., Motel, A., Leonhard, A.: Development of a moisture absorber for packaging of fresh mushrooms (*Agaricus bisporous*). Postharvest Biol. Technol. **48**(3), 408–414 (2008)
61. Maneerat, C., Hayata, Y.: Efficiency of TiO_2 photocatalytic reaction on delay of fruit ripening and removal of off-flavors from the fruit storage atmosphere. Trans. ASABE **49**(3), 833–837 (2006)
62. Maneerat, C., Hayata, Y.: Gas-phase photocatalytic oxidation of ethylene with TiO_2-coated packaging film for horticultural products. Trans. ASABE **51**(1), 163–168 (2008)
63. Marzano-Barreda, L.A., Yamashita, F., Bilck, A.P.: Effect of biodegradable active packaging with zeolites on fresh broccoli florets. J. Food Sci. Technol. **58**, 197–204 (2021)
64. Mastromatteo, M., Mastromatteo, M., Conte, A., Del Nobile, M. A.: Combined effect of active coating and MAP to prolong the shelf life of minimally processed kiwifruit (*Actinidia deliciosa* cv. Hayward). Food Res. Int. **44**(5), 1224–1230 (2011)
65. Mohammadi, M., Rastegar, S., Rohani, A.: Enhancing shelf-life and quality of Mexican lime (*Citrus aurantifolia* cv.) fruit: Utilizing edible coating from wild sage seeds enriched with pomegranate seed oils. J. Food Measure. Characteriz. **18**(1), 331–344 (2024)
66. Moreira, M.D.R., Pereda, M., Marcovich, N.E., Roura, S.I.: Antimicrobial effectiveness of bioactive packaging materials from edible chitosan and casein polymers: assessment on carrot, cheese, and salami. J. Food Sci. **76**(1), M54–M63 (2011)
67. Mukurumbira, A.R., Shellie, R.A., Keast, R., Palombo, E.A., Jadhav, S.R.: Encapsulation of essential oils and their application in antimicrobial active packaging. Food Control **136**, 108883 (2022)
68. Nandhavathy, G., Dharini, V., Babu, P.A., Nambiar, R.B., Selvam, S.P., Sadiku, E.R., Kumar, M.M.: Determination of antifungal activities of essential oils incorporated-pomegranate peel fibers reinforced-polyvinyl alcohol biocomposite film against mango postharvest pathogens. Mater. Today: Proc. **38**, 923–927 (2021)
69. Narsaiah, K., Wilson, R.A., Gokul, K., Mandge, H.M., Jha, S.N., Bhadwal, S., Vij, S.: Effect of bacteriocin-incorporated alginate coating on shelf-life of minimally processed papaya (*Carica papaya* L.). Postharvest Biol. Technol. **100**, 212–218 (2015)
70. Navarro-Martínez, A., López-Gómez, A., Martínez-Hernández, G.B.: Potential of essential oils from active packaging to highly reduce ethylene biosynthesis in broccoli and apples. ACS Food Sci. Technol. **1**(6), 1050–1058 (2021)
71. Navina, B., Huthaash, K.K., Velmurugan, N.K., Korumilli, T.: Insights into recent innovations in anti browning strategies for fruit and vegetable preservation. Trends Food Sci. Technol. **139**, 104128 (2023)
72. Nguyenthe, C., Carlin, F.: The microbiology of minimally processed fresh fruit and vegetables. Crit. Rev. Food Sci. Nutr. **34**, 371–401 (1994)
73. Ortiz, C.M., Mauri, A.N., Vicente, A.R.: Use of soy protein based 1-methylcyclopropene-releasing pads to extend the shelf life of tomato (*Solanum lycopersicum* L.) fruit. Innov. Food Sci. Emerg. Technol. **20**, 281–287 (2013)
74. Petkoska, A.T., Daniloski, D., D'Cunha, N.M., Naumovski, N., Broach, A.T.: Edible packaging: sustainable solutions and novel trends in food packaging. Food Res. Int. **140**, 109981 (2021)
75. Qadri, O.S., Yousuf, B., Srivastava, A.K.: Fresh-cut fruits and vegetables: critical factors influencing microbiology and novel approaches to prevent microbial risks—a review. Cogent Food Agric. **1**(1), 1121606 (2015)
76. Qin, Y., Zhuang, Y., Wu, Y., Li, L.: Quality evaluation of hot peppers stored in biodegradable poly (lactic acid)-based active packaging. Sci. Hortic. **202**, 1–8 (2016)
77. Ramaswamy, H. S.: Post-harvest technologies of fruits & vegetables. DEStech Publications, Inc. (2014)
78. Ramos, B., Miller, F.A., Brandão, T.R.S., Teixeira, P., Silva, C.L.M.: Fresh fruits and vegetables—an overview on applied methodologies to improve its quality and safety. Innov. Food Sci. Emerg. Technol. **20**, 1–15 (2013)

79. Rico-Peña, D.C., Torres, J.A.: Sorbic acid and potassium sorbate permeability of an edible methylcellulose-palmitic acid film: water activity and pH effects. J. Food Sci. **56**(2), 497–499 (1991)
80. Ritu Yadav, R. Y., Aradhita Ray, A. R.: Incorporation of oxygen scavenger sachet for active packaging of minimally processed orange (*Citrus sinensis*) (2010)
81. Rivera Leiva, A.F., Hernández-Fernández, J., Ortega Toro, R.: Active films based on starch and wheat gluten (*Triticum vulgare*) for shelf-life extension of carrots. Polymers **14**(23), 5077 (2022)
82. Robles-Sánchez, R.M., Rojas-Graü, M.A., Odriozola-Serrano, I., González-Aguilar, G., Martin-Belloso, O.: Influence of alginate-based edible coating as carrier of antibrowning agents on bioactive compounds and antioxidant activity in fresh-cut Kent mangoes. LWT-Food Sci. Technol. **50**(1), 240–246 (2013)
83. Roy, S., Rhim, J.W.: Preparation of carbohydrate-based functional composite films incorporated with curcumin. Food Hydrocoll. **98**, 105302 (2020)
84. Sadeghi, K., Lee, Y., Seo, J.: Ethylene scavenging systems in packaging of fresh produce: a review. Food Rev. Intl. **37**(2), 155–176 (2021)
85. Sandhya (2010) Modified atmosphere packaging of fresh produce: current status and future needs. LWT—Food Sci. Technol. **43**(3), 381–392 (2010)
86. Sangsuwan, J., Pongsapakworawat, T., Bangmo, P., Sutthasupa, S.: Effect of chitosan beads incorporated with lavender or red thyme essential oils in inhibiting *Botrytis cinerea* and their application in strawberry packaging system. LWT **74**, 14–20 (2016)
87. Scannell, A.G., Hill, C., Ross, R.P., Marx, S., Hartmeier, W., Arendt, E.K.: Development of bioactive food packaging materials using immobilised bacteriocins Lacticin 3147 and Nisaplin®. Int. J. Food Microbiol. **60**(2–3), 241–249 (2000)
88. Schröder, M. J., Schröder, M. J.: Origins and nature of sensory and other performance attributes in foods. Food Quality and Consumer Value: Delivering Food that Satisfies, 137–165 (2003)
89. Sharma, S., Rao, T.R.: Xanthan gum based edible coating enriched with cinnamic acid prevents browning and extends the shelf-life of fresh-cut pears. LWT-Food Sci. Technol. **62**(1), 791–800 (2015)
90. Sharma, S., Barkauskaite, S., Jaiswal, A.K., Jaiswal, S.: Essential oils as additives in active food packaging. Food Chem. **343**, 128403 (2021)
91. Shirazi, A., Cameron, A.C.: Controlling relative humidity in modified atmosphere packages of tomato fruit. HortScience **27**(4), 336–339 (1992)
92. Singh, M., Saroj, R., Kaur, D.: Optimized chitosan edible coating for guava and its characterization. Measurement Food 14, 100145 (2024)
93. Singh, S., Gaikwad, K.K., Lee, Y.S.: Development and application of a pyrogallic acid-based oxygen scavenging packaging system for shelf life extension of peeled garlic. Sci. Hortic. **256**, 108548 (2019)
94. Sooch, B. S., Mann, M. K.: Nanoreinforced biodegradable gelatin based active food packaging film for the enhancement of shelf life of tomatoes (*Solanum lycopersicum* L.). Food Control 130, 108322 (2021)
95. Suwanti, D., Utami, R., Praseptiangga, D., Khasanah, L. U.: The effect of packaging methods (paper, active paper, and edible coating) on the characteristic of papaya MJ9 in ambient temperature storage. In IOP Conference Series: Earth and Environmental Science (Vol. 102, No. 1, p. 012087). IOP Publishing (2018)
96. Svensson, A.: Active food packaging: materials and interactions: a literature review (2004)
97. Suslow, T. V., Cantwell, M.: Tomato-recommendations for maintaining postharvest quality. In: Produce Facts, A. A. Kader, Ed., Postharvest Technology Research & Information Center, Davis, Calif, USA (2009)
98. Taechutrakul, S., Piroonpan, T., Pasanphan, W.: Active film strips to extend the shelf life of fruits: multibranched PLA-gallic acid as an antioxidant/oxygen scavenger in a case study of bananas (Musa AAA group). J. Food Eng. **364**, 111794 (2024)
99. Takala, P.N., Vu, K.D., Salmieri, S., Khan, R.A., Lacroix, M.: Antibacterial effect of biodegradable active packaging on the growth of *Escherichia coli*, *Salmonella typhimurium*

and *Listeria monocytogenes* in fresh broccoli stored at 4 C. LWT-Food Sci. Technol. **53**(2), 499–506 (2013)
100. Tas, C.E., Hendessi, S., Baysal, M., Unal, S., Cebeci, F.C., Menceloglu, Y.Z., Unal, H.: Halloysite nanotubes/polyethylene nanocomposites for active food packaging materials with ethylene scavenging and gas barrier properties. Food Bioprocess Technol. **10**, 789–798 (2017)
101. Trinetta, V., Cutter, C. N.: Pullulan: a suitable biopolymer for antimicrobial food packaging applications. In: Antimicrobial food packaging (pp. 385–397). Academic Press (2016)
102. Tyagi, P., Salem, K.S., Hubbe, M.A., Pal, L.: Advances in barrier coatings and film technologies for achieving sustainable packaging of food products–a review. Trends Food Sci. Technol. **115**, 461–485 (2021)
103. Ummartyotin, S., Pisitsak, P., & Pechyen, C.: Eggshell and bacterial cellulose composite membrane as absorbent material in active packaging. Int. J. Polym. Sci. **2016** (2016)
104. Vieira, A.C.F., de Matos Fonseca, J., Menezes, N.M.C., Monteiro, A.R., Valencia, G.A.: Active coatings based on hydroxypropyl methylcellulose and silver nanoparticles to extend the papaya (*Carica papaya* L.) shelf life. Int. J. Biol. Macromol. **164**, 489–498 (2020)
105. Villalobos, M. C., Ansah, F., Amodio, M. L., Serradilla, M. J., Colelli, G.: Application of modified atmosphere packaging with moisture absorber to extend the shelf life of'Domenico Tauro'breba fruit. In V International Symposium on Fig 1173 (pp. 365–370) (2015)
106. Wan, H., Zhu, Z., Sun, D.W.: Deep eutectic solvents (DESs) films based on gelatin as active packaging for moisture regulation in fruit preservation. Food Chem. **439**, 138114 (2024)
107. Wang, H., Qian, J., Ding, F.: Emerging chitosan-based films for food packaging applications. J. Agric. Food Chem. **66**(2), 395–413 (2018)
108. Wang, S. Y., Gao, H.: Effect of chitosan-based edible coating on antioxidants, antioxidant enzyme system, and postharvest fruit quality of strawberries (Fragaria x aranassa Duch.). LWT-Food Sci. Technol. **52**(2), 71–79 (2013)
109. Warsiki, E.: Application of chitosan as biomaterial for active packaging of ethylene absorber. In: IOP Conference series: earth and environmental science (Vol. 141, No. 1, p. 012036). IOP Publishing (2018)
110. Wei, H., Seidi, F., Zhang, T., Jin, Y., Xiao, H.: Ethylene scavengers for the preservation of fruits and vegetables: a review. Food Chem. **337**, 127750 (2021)
111. Wrona, M., Bentayeb, K., Nerín, C.: A novel active packaging for extending the shelf-life of fresh mushrooms (*Agaricus bisporus*). Food Control **54**, 200–207 (2015)
112. Wu, L. T., Tsai, I. L., Ho, Y. C., Hang, Y. H., Lin, C., Tsai, M. L., Mi, F. L.: Active and intelligent gellan gum-based packaging films for controlling anthocyanins release and monitoring food freshness. Carbohydr. Polym. **254**, 117410 (2021)
113. Ya, Z., Qi-long, S., Xiang-you, W.: Optimization of compound absorbers and its effects on MA storage characteristics of *Agaricus bisporus*. Food Ind. **12**, 47 (2015)
114. Yang, Z., Li, C., Wang, T., Li, Z., Zou, X., Huang, X., Povey, M.: Novel gellan gum-based probiotic film with enhanced biological activity and probiotic viability: application for fresh-cut apples and potatoes. Int. J. Biol. Macromol. **239**, 124128 (2023)
115. Yildirim, S., Röcker, B., Pettersen, M.K., Nilsen-Nygaard, J., Ayhan, Z., Rutkaite, R., Coma, V.: Active packaging applications for food. Comprehens. Rev. Food Sci. Food Safety **17**(1), 165–199 (2018)
116. Yildiz, F., Wiley, R.C. (eds.): Minimally processed refrigerated fruits and vegetables. Springer, US, New York (2017)
117. Yong, H., Liu, J.: Recent advances in the preparation, physical and functional properties, and applications of anthocyanins-based active and intelligent packaging films. Food Packag. Shelf Life **26**, 100550 (2020)
118. Zhao, Y., An, J., Su, H., Li, B., Liang, D., Huang, C.: Antimicrobial food packaging integrating polysaccharide-based substrates with green antimicrobial agents: a sustainable path. Food Res. Int. **155**, 111096 (2022)
119. Zhu, J., Wu, H., Sun, Q.: Preparation of crosslinked active bilayer film based on chitosan and alginate for regulating ascorbate-glutathione cycle of postharvest cherry tomato (*Lycopersicon esculentum*). Int. J. Biol. Macromol. **130**, 584–594 (2019)

Chapter 14
Active Packaging for Convenience Foods: Enhancing Quality and Sustainability

Jyoti Rani, Vibhuti Gulia, Sahil Singh Dhull, Deepak Beniwal, and Deepali

Abstract This chapter explores the ever-evolving state of active packaging for convenience foods, where innovation meets customer desire for quality and sustainability. The introduction establishes the context, emphasizing the need for packaging solutions beyond mere confinement to improve food quality actively. The principles of active packaging are examined, exposing how these dynamic systems interact with food products to preserve freshness and prevent spoilage. A key distinction is made between typical passive packaging and the proactive nature of active packaging techniques, as described in their cutting-edge applications. These include controlled-release technologies that carefully dispense preservatives, flavors, and other agents to protect food integrity. The chapter discusses the most recent breakthroughs in active packaging, including Intelligent Packaging, which uses sensors and indications for real-time monitoring and communication, and how these developments correspond with overarching sustainability aims. It evaluates current market patterns and future estimates, focusing on the growing consumer desire for sustainable packaging solutions. Furthermore, the chapter navigates through the complicated areas of legislative requirements governing active packaging, emphasizing the necessity of compliance and safety measures. Finally, it examines market obstacles and provides a forward-looking perspective on active packaging's potential to change the convenience food sector.

Keywords Active packaging · Convenience food · Controlled-release · Intelligent packaging · Sustainable · Legal regulations · Packaging challenges

14.1 Introduction

Rising affluence, globalization, urbanization, and changes in lifestyle and profession have all contributed to an increase in the demand for packaged meals. Food packaging, essential in a food production system, involves designing, creating, assessing,

J. Rani (✉) · V. Gulia · S. S. Dhull · D. Beniwal · Deepali
Department of Botany, Chaudhary Devi Lal University, Sirsa 125055, India
e-mail: drjyotisangwan@cdlu.ac.in

S. ul Islam and M. Shahid (eds.), *Green Materials for Active Food Packaging*, Engineering Materials, https://doi.org/10.1007/978-981-96-0369-5_14

and producing packages for food distribution, storage, and sale. Its main functions include containing, protecting, communicating, and providing convenience to the end customer. Product packaging also offers details about the contents of the product, labeling, producer, expiration dates, and storage advice. Additionally, food status could be inferred and clarified by packing conditions, including physiological, physical, microbiological, and chemical factors [40].

These days, fresh, wholesome, and nutrient-dense food products are in great demand due to the growing global population, yet their brief-shelf life leads to a substantial amount of wasteful food. While synthetic polymers have historically been favored for their flexibility and durability, with a huge amount of 320 million tons used yearly, their widespread use harms the environment. This includes the creation of microplastics that, when broken down, contaminate food, water, and soil. In response, modern and secure packaging technology is being developed to improve the longevity of perishable items whilst preserving their overall standards and freshness [18]. Sustainable packaging solutions are also being explored, and biodegradable alternatives such as proteins and polysaccharides have been adopted to minimize the detrimental effects of environmental pollution and food waste [63]. However, challenges remain in achieving mechanical, optical, and barrier qualities comparable to synthetic polymers, limiting their widespread commercial use. The packaging sector is a dynamic one. Challenges in the food packaging industry drive the development of innovative packaging technologies [68]. Since the eighteenth century, this industry has evolved significantly, with the most notable advancements occurring in the past century. Food packaging is now more active in marketing and product protection [51]. Emerging ideas in active and intelligent packaging are anticipated to become more significant, particularly by integrating naturally occurring active ingredients, such as antioxidants, antimicrobials into polymeric matrices, since they provide a plethora of creative ways to prolong the "best-before date" or preserve and enhance and keep a focus on the safety and nutritional value of food.

Furthermore, new sustainable alternatives in food packaging systems have been made possible by recent advancements [19]. Systems enabling edible packaging offer the potential to reduce plastic waste, deterioration, and antimicrobial agents while also meeting sustainability goals. A balanced solution to the issues at hand—food safety, shelf-life extension, and the reduction of plastic waste—can be achieved with bio-nanocomposites made from biocompatible and antibacterial materials [7]. This strategic approach strives to improve the functioning of food packaging by using bioactive ingredients. A targeted strategy for preserving food, aligned with the principles of active packaging outlined in European regulation (EC) No. 450/2009, involves the modulated dissipation or absorption of compounds into the preserved food or its surroundings through the polymeric matrix [65]. It emerges as a revolutionary technique that increases bioactive substances' solubility, stability, and bioavailability by providing regulated release and stability of biomolecules.

Moreover, it assists in maintaining the nutritional value of food and keeps the coating systems from deteriorating [57]. Consumer perceptions strongly shape the packaging designs. The rising demand for eco-friendly solutions reflects an increasing consumer awareness and preference for sustainable practices. Intelligent, sustainable, and smart packaging is a proactive response to consumer preferences and environmental responsibilities, driven by advanced technologies impacting the food industry. This marks a revolutionary shift towards a more consumer-focused and sustainable future in food packaging [60].

14.2 Fundamentals of Active Packaging

The term "active packaging" was initially introduced by Labuza in 1987, and it is now increasingly applied in the field of packaging research in developing new forms of food packing [40]. Taking into account EC No. 450/2009 (EU, 2009), some products and articles are made explicitly to retain or enhance the quality of food that has been packaged. Active materials and articles are designed to release or absorb substances into or from the packed foods or their environment, thereby extending the shelf life of the products. This helps ensure that food stays fresh longer and in good condition during storage and transport.

Apprehension of the physicochemical attributes of the packing materials and their surface is key to active systems in food packaging. They could be in the form of multilayer structures or additives (chiefly, sachets, labels, pads, and bottle caps) that are specifically designed to absorb CO_2, oxygen, ethylene, flavors, moisture, odors, and even release carbon dioxide, antioxidants, and microbicidal agents [40]. By utilizing these active systems, packaging can maintain its products' grade, freshness, and overall safeguarding (Fig. 14.1).

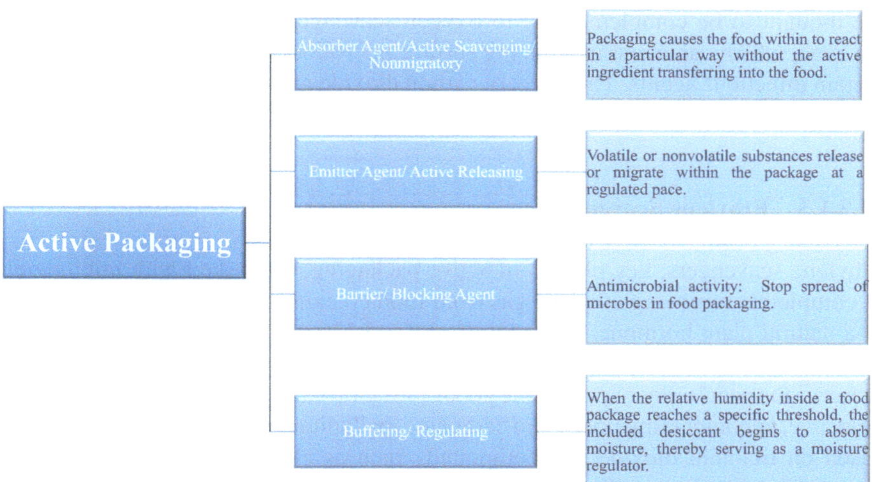

Fig. 14.1 Major classification of active packaging systems with their respective function

14.2.1 Various Active Packaging Systems

14.2.1.1 Oxygen Scavengers

Oxygen scavengers, substances employed in active packaging, diminish or eliminate the oxygen content in food packaging. They are used to prevent or delay deterioration caused by oxidation or the growth of microbes or insects. Oxygen scavengers may be employed independently or alongside modified atmosphere packaging (MAP) to enhance the efficiency of oxygen removal. Iron, ascorbic acid, enzymes, salts, fatty acids, and photosensitive dyes are frequently utilized as oxygen-scavenging agents. These agents can be applied singly or combined to enhance their effectiveness. They are commonly utilized in foodstuff packaging to reduce excess oxygen and enhance the durability period of products. They were beneficial in minimizing discoloration, mold spoilage, rancidness, and flavor change in many food products [56].

14.2.1.2 CO_2 Scavengers and Emitter

CO_2 scavengers and emitters are both key components of food packaging systems, contributing to product quality and safety. CO_2 scavengers find application in food packaging setups to eliminate trapped carbon dioxide from packages, thereby averting package collapse and product deterioration. They are commonly used to remove CO_2 created within the package owing to food deterioration and respiration during storage. However, using CO_2 scavengers to deoxygenate the package may promote the growth of anaerobic microorganisms. In contrast, CO_2 emitters are thought to be complementary to oxygen scavengers. They emit CO_2 into the package, which can assist in reducing surface microflora and prevent the growth of certain infectious agents. The effect of carbon dioxide gas varies with concentration, with greater amounts limiting the growth of aerobic bacteria and certain diseases [4].

14.2.1.3 Ethylene Scavengers

Ethylene scavengers, key players in active packaging, are tasked with minimizing or completely removing ethylene gas from food containers. This gas, ethylene, acts as a natural plant hormone, hastening fresh foods' ripening and decay process. By eliminating ethylene, these scavengers effectively prolong the "use-by" period and preserve the peculiarity of perishable goods [48]. Several methods are utilized for this objective, encompassing potassium carbonate, palladium, natural clays, and systems based on titanium dioxide and compounds such as electron-deficient dienes and trienes. These systems utilize chemical reactions and physical adsorption techniques to trap ethylene, ensuring that food stays fresh and is safe to eat for longer periods [27].

14.2.1.4 Ethanol Emitter

Ethanol emitters introduce ethanol vapor into the packaging environment to inhibit mold growth, thereby extending the shelf life of food products, particularly bakery items. When these emitters are placed within food packaging, they absorb moisture and emit ethanol, which has antibacterial characteristics that eradicate mold and other spoilage organisms. This approach is particularly popular in Japan, where it has greatly extended the conservation period of bakery products by keeping them free of mold. Ethanol emitters can be directly integrated into packing materials or sachet forms, enabling a regulated release of ethanol to protect food quality [69].

14.2.1.5 SO_2 Emitter

SO_2 emitters effectively prevent mold growth and maintain the quality of grapes during storage. It operates through the moisture-induced calcium sulfite reaction and the metabisulfite hydrolysis process. Sulfur dioxide emitter sheets in active packaging are used to study the impact on decay and grape quality. A polymer is used to limit SO_2 release and eliminate fungus without affecting food [40].

14.2.1.6 Antimicrobial Agents

Antimicrobial substances are agents that halt the growth or destroy microorganisms. They are crucial components of active packaging systems, aiding in preventing food degradation by lowering the levels of harmful germs, thus improving the "best-before" of packaged goods. Such agents are derived from both natural and synthetic origins. Natural antimicrobials, for instance, include plant-derived essential oils, animal-sourced enzymes, microbial bacteriocins, organic acids, and polymers found in nature. These agents are integrated into packaging materials using various techniques, such as thermal polymer processing methods like extrusion, co-extrusion, injection molding, or solvent compounding. Biopolymers, known for their solubility in water and compatibility with antimicrobial solvents, are frequently utilized in this context. Antimicrobial films incorporated into active packaging provide durable defense against microbial contamination. It is essential, however, to select antimicrobial agents suitable for food contact and to incorporate them to ensure they remain efficacious at doses higher than the minimum inhibitory concentration (MIC) to guarantee food safety [24].

14.2.1.7 Moisture Absorbance/Humidity Control

Maintaining humidity levels is a crucial element of active packaging, essential for preserving the prime during transportation and storage. It's all about keeping the proper moisture balance to avoid issues like rust, decay, or damage to wood or paper.

In the food industry, it's essential to keep humidity in check to stop mold or fungi from ruining the goods. Active packaging tackles this by bringing in materials that soak up excess moisture or help maintain the proper humidity level. This is particularly crucial for foods susceptible to moisture, such as fruits or meats, as excessive moisture can degrade quality and adversely affect the shelf life of the food. You'll often find things like desiccants—those little packets that drink up moisture—used in various forms like sachets, sheets, trays, or pads to keep the inside of the package just right. They pull moisture from the air, which is a big help in keeping foods from getting spoiled. There are also special trays made from plastics like polypropylene (PP) or polyethylene terephthalate (PET), and these moisture-grabbing materials are built right in, offering a fresh way to handle humidity in food packaging. Choosing the best material or system for controlling moisture depends on what kind of food it is and how much moisture control it needs [26].

14.2.1.8 Antioxidant Release

The purpose of antioxidant releasers is to dispense antioxidants into the surrounding environment of the food, thereby prolonging the product past its use-by date and maintaining its quality. Antioxidants are substances that inhibit food deterioration caused by oxidation. When released from the packaging, antioxidants assist in minimizing the production of free radicals and the oxidation of lipids, oils, and other dietary constituents. This method is particularly effective for products prone to oxidative degradation, such as oils, fats, and certain processed meals. An antioxidant releaser is a type of active packaging that regulates the migration of non-volatile agents, benefiting the food chain while keeping the active ingredient separate from the food [31].

14.3 Distinction from Traditional Packaging

In the past, goods like rice, sugar, and salt were immediately wrapped in bamboo, lotus, and other leaves to preserve fresh food. In Asia, packaging materials, decorations, and styles change over time by productivity, scientific discoveries, and popular trends. Comprehending the foundations of conventional packaging is crucial since it facilitates the differentiation of various package kinds, designs, and approaches. Paper, cardboard, wood, fabric, and even natural fibers like jute or straw are common natural materials used in traditional packing. These materials are commonly selected because they are easy to find, affordable, and biodegradable. For example, cardboard boxes and cartons for shipping and storage, glass bottles and jars for beverages, condiments, and cosmetics, and cans and containers made of metal used for aerosol items and canned goods [44]. Traditionally, laminates and coatings with various properties were manufactured with materials such as glass, metals, papers, and plastics. Standard packaging utilized these materials as well. However, many traditional materials fail to biodegrade and contaminate the environment. The absence of environmental compatibility, as well as short lifespans, is a problem. Traditional packaging protects the product from atmosphere, moisture, and microorganisms by acting as passive barriers. However, there was a shortage of engagement, and current data from traditional materials show resourcefulness, cultural legacy, and a strong bond with the natural world [55].

Active packaging differs from traditional packaging methods in that it is dynamic and interactive. It extends the time a product can be stored, helps preserve or improve its quality, and both contain and protect it. Active packaging materials engage with the environment or the contents actively. These materials release or absorb substances into the package that impact oxygen reactions, moisture content, and microbial development. Various forms of active packaging can be modified to meet specific requirements (such as maintaining freshness, decreasing unpleasant smells, and improving safety). Examples of active packaging include ethylene absorbers, moisture-regulating polymers, oxygen-scavenging films, and antimicrobial packaging. Advanced technologies, including nanotechnology, microencapsulation, controlled release systems, and intelligent packaging sensors, are frequently used in active packaging. Using standard materials and designs, traditional packaging focuses on structural integrity and little protection [70]. Traditional packaging protects a product from physical, chemical, and biological elements by acting as a barrier between it and the outside environment.

Active packaging is a more advanced and specialized approach than standard packaging, offering better functionality and the ability to influence the product's surroundings and attributes. Active packaging represents a dynamic approach that surpasses essential protection by actively engaging with its contents to improve quality and prolong shelf life. It enhances traditional packaging by incorporating innovative elements that improve the conditions within the packaging of a food product. Product protection is one of many purposes of active packaging because specialized materials and processes are needed to accomplish these ends; it is more complex and could be more costly than traditional packaging. Both active and traditional packaging can have an effect on the environment, even though active packaging may make it difficult to dispose of packaging materials containing active substances [9].

Conversely, active packaging significantly benefits preserving product quality and freshness [3]. There are many advantages to active packaging for producers and consumers. Active packaging materials are designed to interact with the items inside the package. For instance, oxygen scavengers remove oxygen from packaging to prevent oxidation and extend the shelf life of products susceptible to oxygen, such as meats and snacks. Moisture absorbers help avoid spoiling due to moisture by assisting in maintaining the proper humidity level. At the same time, traditional packing materials don't have the same direct effect on the contents [13].

14.4 Cutting-Edge Active Packaging Systems and Their Applications in Food Sector

Active packaging systems have transformed worldwide food packaging by providing cutting-edge solutions beyond conventional confinement and protection. This innovative approach has widespread applications across the food industry, each harnessing its potential uniquely. The food sector currently uses several commercially available applications of active packaging technologies [29]. For a comprehensive understanding of these technologies, Table 14.1 offers insights into various active packaging systems, including their working principles and applications in the food sector.

One such application is using active packaging systems for meat products, as illustrated in Fig. 14.2. The figure depicts various commercially available active packaging systems tailored explicitly for meat products. Implementing active packaging systems can protect food items' shelf life, uphold quality standards, and deter spoilage. These systems can also help regulate the release or absorbance of flavors and odors, controlling humidity and providing antimicrobial activity against pathogenic microflora.

In summary, adopting active systems in the food sector represents a promising strategy for enhancing food safety and prolonging the shelf life of food products.

14.5 Active Packaging Associated with Controlled Release

Active packaging with controlled release is a groundbreaking technology that redefines packaging solutions by actively interacting with contents to enhance product quality, safety, and shelf life. By gradually delivering active ingredients like moisture regulators, taste enhancers, antioxidants, and antimicrobials, this innovative packaging prolongs the protection or improves the quality of packaged goods. Its primary advantage lies in reducing bacterial growth and delaying oxidation, thus prolonging the shelf life of perishable goods like fresh produce, meats, and dairy products [24]. Traditionally, active ingredients such as absorbers, sachets, and labels have been integrated into packaging materials through direct integration, coating, or immobilization. Additionally, naturally occurring polymers like chitosan, with antioxidant or antibacterial properties, streamline production processes [39]. Controlled-release packaging excels over conventional techniques by precisely regulating antioxidant activity to preserve food freshness for longer durations. Public acceptance leans towards naturally derived components over synthetic ones due to concerns about sachet splitting and ingestion risk [67].

In the realm of pharmaceuticals and medical equipment, active packaging with controlled release plays a critical role in preserving medications and ensuring patient safety. By incorporating desiccants or moisture absorbers, active packaging safeguards delicate pharmaceutical formulations from moisture-induced deterioration,

Table 14.1 Primary active systems' classification, mechanism of action, and use in the food industry [4, 10, 40]

Active system		Material utilized/ mechanism of action	Applications
Absorber agent/active scavenging/ nonmigratory	Oxygen scavengers	Enzymatic systems: 1. Alcohol oxidase-ethanol vapor 2. Glucose oxidase-glucose	• Extends the shelf life of cheese spreads, cakes, and dark chocolates • Preventing ascorbic acid decomposition in drinks • Quality improvement of dark chocolates
		Chemical Systems: 1. Ascorbate/metallic salts 2. Iron 3. Metal/acid 4. Metal catalyst (e.g., platinum)	
	CO_2 scavengers	Iron powder-calcium hydroxide	• Packaging baked goods and nuts
	Ethylene scavengers	1. Activated carbon, clays, zeolites 2. KMnO4	• Reduce pace of softening in tomatoes
Emitter agent/active releasing	Antioxidant release	1. BHA/BHT 2. Organic acids 3. Tocopherols	• Extending the shelf life of pre-baked buns • Enhancing the quality of apple slices and milk powder
	Ethanol emitter	1. Encapsulated ethanol 2. Alcohol spray	
	CO_2 emitter	Ferrous carbonate—metal halide	
	SO_2 Emitter	Sulfur dioxide	
Barrier/blocking Agent	Antimicrobial activity	1. Antibiotics 2. Ethanol 3. Ozone 4. Peroxide 5. Quaternary ammonium salts 6. Silver-zeolite 7. Sorbates, benzoates, propionates	• Extending the shelf life of fresh beef, bread, and grilled pork • Providing antimicrobial protection against their pathogenic microflora in sliced bread, butter, cooked ham, raw fish products, and salmon

(continued)

Table 14.1 (continued)

Buffering/regulating	Diffusion or absorption of flavors and odors	1. Acetylated paper Activated carbon, clays 2. Cellulose triacetate 3. Citric acid 4. Sodium bicarbonate 5. Ferrous salt, ascorbate	• Quality improvement in roast beef • Prolong shelf-life of cuttle and codfishes
	Moisture absorbance/humidity monitoring	1. Silica gel 2. PVA blanket 3. Activated clays, minerals	

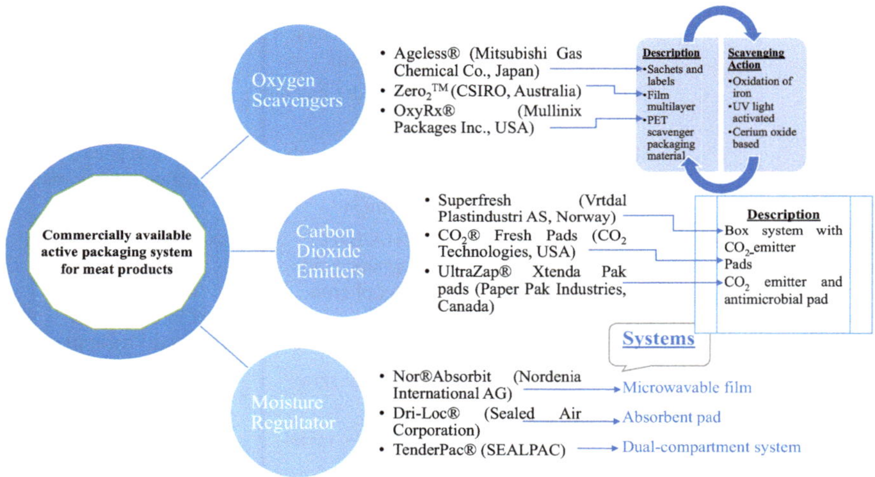

Fig. 14.2 Commercially available active packaging system for meat products [4]

maintaining their effectiveness and stability [32]. Similarly, the controlled release of antimicrobial drugs helps mitigate the risk of infection by preventing microbial contamination of medical equipment. They triggered drug administration and gradual release, presenting promising possibilities for next-generation plastics in food packaging [67]. Controlled release of natural active ingredients into biopolymeric materials mirrors drug delivery methods, requiring precise control systems to ensure regulated release over time [35]. Controlled release in active packaging begins with an initial release triggered by external stimuli such as temperature or pH variations, transitioning to continuous, controlled release [25]. Active substances migrate through the polymeric matrix via diffusion, particularly when encapsulated. Innovative packaging enhances freshness evaluation and spoilage detection using sensors to monitor food conditions. For instance, sensors measuring pH fluctuations serve as indicators of food deterioration. pH variations can trigger release, offering potential for controlled release in food packaging. Environmentally responsive packaging activated at specific pH levels possesses intelligent and active functionalities, enhancing food quality and safety. Leveraging such pathways can enhance antibacterial effectiveness against foodborne microorganisms [49].

Active packaging with controlled release contributes to product safety by real-time monitoring of critical factors such as temperature, humidity, and gas composition while prolonging shelf life and maintaining product quality. Intelligent package labels, for instance, change color in response to temperature abuse or microbial contamination, offering visual cues to consumers and assisting them in making informed choices [65]. To achieve intended functionality while ensuring safety and regulatory compliance, active packaging with controlled release requires careful selection of active ingredients and optimization of release mechanisms. Various methodologies, including encapsulation, immobilization, and coating, integrate active ingredients into packaging materials and regulate their release rates [5]. Encapsulation involves enclosing active components within microcapsules or nanoparticles composed of biodegradable polymers, releasing them in response to specific environmental stimuli such as moisture, pH, or temperature [12]. Immobilization techniques entail attaching active ingredients to the packaging material's surface or embedding them within a matrix to provide sustained release over time. Coating methods involve applying a thin layer of active substances onto the packaging material's surface, gradually diffusing into the packaged environment [9].

The choice of active ingredients and release mechanisms is dictated by the unique requirements of the packaged product, including the type of food or pharmaceutical composition, intended shelf life, and target microorganisms or degradation pathways. Antimicrobial compounds like essential oils or silver nanoparticles may be added to fresh produce packaging to inhibit mold and bacterial growth, while desiccants like silica gel or molecular sieves can be used in pharmaceutical packaging to absorb moisture and prevent degradation of sensitive medication formulations [15, 32]. Adequate active packaging balances gradual release to avoid degradation and minimal interference with food, ensuring that the release rate of active compounds matches the rates at which food breaks down. Environmental characteristics such as UV light and ionic concentration, along with factors influencing migration rates like particle size, dispersion, and molecular weight of the active ingredient, affect release. Control of release heavily depends on compatibility between the active ingredient and biopolymer, influenced by polarity effects. Temperature and humidity can trigger release, offering opportunities for environmentally controlled release. Food attributes like water activity, pH, and fat content also impact release rates [5]. Active packaging with controlled release holds immense potential to elevate packaged goods' quality, safety, and longevity across various industries. By integrating active compounds into packaging materials and managing their release kinetics, active packaging solutions provide prolonged protection or enhancement to the packaged contents, ensuring their freshness, efficacy, and safety throughout the distribution chain. As the demand for safer, more environmentally friendly, and longer-lasting packaging solutions continues to grow, active packaging with controlled release is poised to play an increasingly significant role in packaging technology.

14.6 Advancements in Active Packaging

Food packaging continually evolves to meet changing consumer demands [70]. Most advancements in food technology focus on improving the practicality, efficiency, affordability, and safety of food processing. Traditional food packaging primarily serves as a passive barrier, protecting products during storage, distribution, and transportation [41]. Beyond this, active packaging offers several benefits, actively contributing to food safety and preservation by preventing oxidation, contamination, and other issues [29, 34]. Consumers' concerns over the sustainability and quality of their food products have grown recently. This increased awareness has led to a demand for packaging solutions offering greater convenience and monitoring and preserving food in real-time. Consequently, cutting-edge innovations like intelligent and active packaging have emerged, paving the way for smart packaging technology to meet these evolving consumer needs (Fig. 14.3).

14.6.1 Smart Packaging Solutions for Modern Consumer Needs

In a more comprehensive sense, active and intelligent packaging works in synergy to deliver "smart" packaging [20]. Active packaging systems are designed to extend the shelf life of products by actively interacting with the environment within the package. On the other hand, intelligent packaging systems focus on interacting with the customer or user, providing information, feedback, or other functionalities beyond traditional packaging. [41]. Figure 14.4 underscores each packaging type's distinct role in preserving food quality and giving information.

14.6.2 Intelligent Packaging

Intelligent packaging (IP) incorporates advanced functionalities like communication, detection, recording, sensing, tracing, and the application of scientific reasoning to facilitate informed decision-making, as summarized in Fig. 14.5. It is crucial in prolonging product shelf life, enhancing safety and quality, conveying essential information, and issuing warnings regarding potential complications. Continuous surveillance has a dual purpose: it reduces food waste while protecting consumers from foodborne illnesses. Furthermore, businesses should prioritize enhancing the food

Fig. 14.3 A visual representation of an innovative packaging system for convenience food

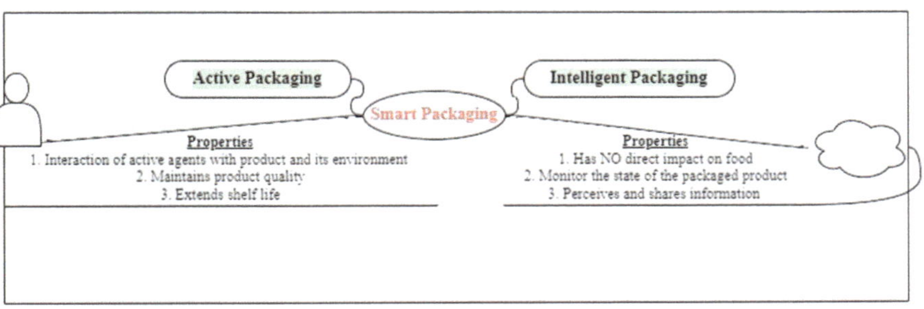

Fig. 14.4 Smart packaging classification with their respective properties

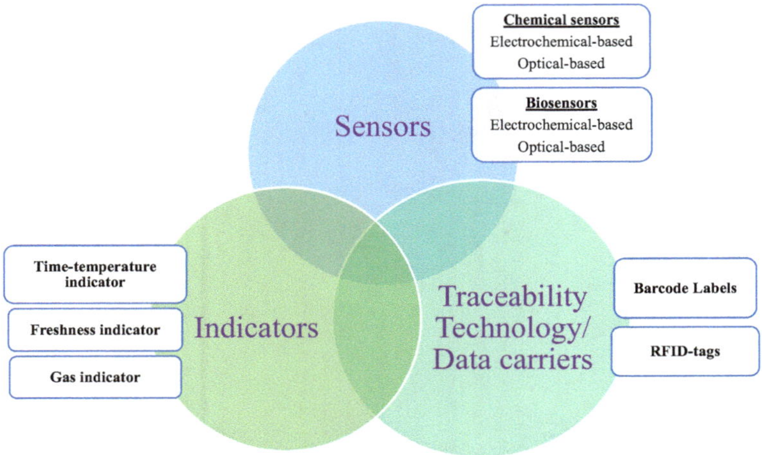

Fig. 14.5 Key target areas of intelligent packaging

industry's efficiency while improving product trackability [46]. Through continuous monitoring of food conditions from production to consumption, intelligent packaging ensures effective communication of quality metrics, thereby upholding the integrity of the food supply chain [30].

14.6.2.1 Sensors

Sensors, sophisticated analytical devices, are adept at detecting and quantifying the physical or chemical alterations a product undergoes. These instruments meticulously convert such variations into electrical signals, visually representing the changes [37]. They can gather and supply quantitative data on concentration, pH, temperature, and other variables. Sensors fall into two major categories: chemical and biosensors, as summarized in Fig. 14.5.

Chemical sensors detect and measure the activity, concentration, composition, or presence of specific chemical analytes. They work by utilizing various technologies, such as non-dispersive infrared (NDIR) sensors, as well as chemical reactions with polymers or solid electrolytes [46].

- NDIR sensors function by measuring the CO_2 content through gas sorbing at a particular wavelength.
- The polymer electrolytes that chemical CO_2 sensors use to function.
- Gas sensors track the concentration of particular gases, such as CO_2 or H_2S, and adjust the sensor's physical characteristics in response to their presence.

Biosensors feature receptors crafted from biological materials like antigens, enzymes, hormones, or nucleic acids. They are resorted to detect the presence, activity, composition, or concentration of specific chemical or physical analytes.

Table 14.2 Indicators for food packaging applications [43, 58]

Indicator	Function/mechanism of action	Application
Gas indicators (GIs)	Measure gas concentration within the package	Detect package integrity
Freshness indicators (FIs)	Detect changes in its chemical or physical properties	Measure food ripening or perishability to assess freshness of the product
Leak indicators (LIs)	Detect leaks or breaches in the packaging	Quality and safety of the product
pH indicators (PHIs)	Monitor the pH level of the product	Indicate microbial growth or spoilage
Time–temperature indicators (TTIs)	Monitor the temperature history of the product	Information about the cold chain break, consequently shelf-life
Thermochromic inks detector	Color change on temperature exposure	Indication of a "ready to serve" message

Biosensors can be used in various applications, including glucose sensors that use glucose oxidases to catalyze the oxidation of glucose [45].

14.6.2.2 Indicators

Devices or materials used in the packaging system as indicators can give consumers a visual representation of the condition and superiority of food products. They are capable of detecting and monitoring changes such as color change, color intensity, or diffusion of color, which can indicate the freshness or deterioration of the food [58]. Depending on their applications and mechanism of action, indicators can be grouped into various categories, as observed in Table 14.2.

14.6.2.3 Data Carriers

The rising frequency of food safety issues has created a demand for traceability. Utilizing data carriers in intelligent packaging is one method to achieve traceability. These data carriers serve as efficient instruments for ensuring food safety. They provide a means of tracking and monitoring various aspects of the food supply chain, including product origin, handling conditions, expiration dates, and more [17]. By utilizing these data carriers, companies can quickly and easily access vital information about their products, thereby enabling them to detect and tackle any potential issues or risks effectively.

2D Barcode A graphical picture that holds product information both vertically and horizontally is called a 2D barcode. Data can be seen using a camera-based reader or appropriate optical scanning equipment.

The RFID tag is a data-carrying device consisting of an antenna connected to a microchip. The system communicates antennas using electromagnetic (EM) waves and displays a reader, which comprises a read/write device consisting of a transmitter and/or a receiver.

A sensor-enabled RFID tag is an automated identification system that utilizes wireless sensors to facilitate recording, tracking, and continuous product information throughout the product's lifecycle [41].

14.7 Sustainable Approach in Active Packaging Development

Food packaging sustainability encompasses ecological, economic, social, and cultural dimensions, considering its impact on food production, distribution, consumption, and waste management. The evaluation of packaging sustainability goes beyond manufacturing and raw material chains to encompass its contribution to preventing food loss and waste, as well as safeguarding products across the entire supply chain. An ideal packaging solution should not only be economical and resource-efficient but also enhance convenience for customers and food chain operators while seamlessly integrating with waste management and recycling systems [28].

In the realm of active packaging, adopting sustainable methods is essential to address environmental concerns and promote responsible production and consumption habits. These methods focus on mitigating the negative environmental impacts of packaging materials, reducing waste generation, and enhancing the overall sustainability of packaging solutions [33]. Active-releasing packaging and non-migratory active packaging are the two primary types. Active-releasing packaging utilizes emitters to control the migration of desired compounds into food products, whereas non-migratory active packaging uses scavengers to remove undesirable components from the package. Achieving sustainability objectives in active packaging entails careful material selection and prioritizing recyclable, biodegradable, and renewable resources to minimize environmental impact. Biopolymers derived from plants, such as PLA and PHA, offer eco-friendly alternatives to traditional petroleum-based plastics. Moreover, maximizing resource efficiency throughout the packaging lifecycle is crucial, necessitating meticulous control of material usage and processing methods. Initiatives like reducing material thickness, optimizing packaging design, and implementing efficient manufacturing techniques contribute to this goal [52].

Comprehensive life cycle assessments (LCAs) are critical for assessing the environmental impact of active packaging solutions across their entire lifespan. LCAs analyze the effects of packaging solutions from raw material extraction to end-of-life disposal, identifying areas for improvement and guiding decisions to enhance sustainability [42]. Prioritizing the biodegradability and compostability of active packaging materials demonstrates a commitment to ecologically responsible disposal methods. Biodegradable polymers like PLA and PHA break down naturally into non-toxic components, reducing plastic waste accumulation in landfills and oceans [11]. Advancing sustainability goals entails embracing circular economy principles and promoting recycling initiatives. Encouraging the recycling of active packaging materials and integrating them into circular economy systems are key components of this strategy. This involves developing infrastructure for efficient packaging waste collection, sorting, and recycling, as well as designing packaging materials with recyclability in mind. Additionally, opting for natural sources for eco-friendly active agents aligns with sustainable practices. Bio-based antioxidants, antimicrobials, and other active substances offer functional benefits without causing harm to the environment [8].

Cutting-edge smart packaging solutions are essential for enhancing supply chain efficiency, maximizing product freshness, and minimizing food waste. Integrating sensors and indicators into packaging enables real-time monitoring of product quality and condition, facilitating timely interventions to prevent waste and spoilage. The advancement of sustainable active packaging solutions relies on collaboration and innovation across the packaging value chain [59]. Promoting collaboration among stakeholders, including suppliers, manufacturers, retailers, and consumers, fosters communication, information exchange, and cooperation to overcome challenges and advance sustainable packaging objectives. Incorporating sustainable practices into the development of new packaging solutions is essential for creating innovative solutions that efficiently safeguard and preserve goods while minimizing environmental impact. Thoughtful material selection, resource optimization, life cycle assessments, biodegradability initiatives, recycling efforts, eco-friendly active agents, smart packaging technologies, and collaborative innovation are integral to achieving a more sustainable future for packaging [63].

14.8 The Legal Regulations on Active Packaging

Legal regulations regarding active food packaging are established to ensure food safety and compliance with specific standards. Different countries and areas possess distinct legal guidelines regarding active packaging for food, which is a reflection of the need to protect consumers, maintain environmental sustainability, and ensure food safety. Active packaging enhances food preservation by interacting with the food or environment to improve quality, preserve nutritional value, and extend shelf life. It does this by releasing antibacterial and antioxidant chemicals or absorbing moisture, oxygen, and unwanted flavors [23].

14.8.1 General Principles of Active Packaging

Understanding the general principles of active packaging is essential for stakeholders involved in food production, packaging, and consumption. It provides a holistic solution to tackle the challenges of food preservation and sustainability.

14.8.1.1 Non-Toxicity and Safety

Active packaging materials must be able to provide proof of their safety and non-toxicity. This includes the materials' initial state as well as any chance that they could leak chemicals into the meal. A good example is Regulation (EC) No. 1935/2004 of the European Union, which states that materials cannot release toxins into food in amounts that could be harmful to human health [69].

14.8.1.2 Labelling Requirements

Per EU's Framework Regulation (EC) No. 1935/2004, general regulations require that the active ingredients in packaging, together with their nature, be labelled properly to warn customers regarding their intended application and potential allergens [22].

14.8.1.3 Limits on Migration

According to the European Commission, strict restrictions have been imposed on the migration of materials from packaging into food to prevent contamination of food products. Under Regulation No. 10/2011, for instance, the EU sets migration limits for materials that come into contact with food. This rule primarily focuses on plastics, but it also applies to the more general idea of integrity and safety of material in food packaging.

14.8.2 Environmental Aspects to Take into Account

The environmental impact of packaging is being addressed by regulations more and more, which call for recyclable and biodegradable choices. The emphasis on reducing the environmental impact of packaging materials is a growing trend, although specific laws may vary. With directives aiming at boosting recycling and minimizing packaging waste, the EU has demonstrated a particularly proactive approach [47].

14.8.3 Overview of the Laws and Regulations of Different Countries About Active Packaging

Food safety, quality, and adherence to certain standards are all dependent on national laws and regulations governing active packaging. These regulations govern the use of chemicals that interact with food or the environment, aiming to preserve nutritional value, enhance quality, and extend shelf life.

14.8.3.1 FDA Regulations

According to the Food and Drug Administration (FDA), the safety of compounds added to foodstuffs and items that come into touch with it, such as additives added to packaging, is regulated. Food producers are in charge of promoting safe products, and additives that are added to food must adhere to safety regulations to guarantee that they won't cause harm when used as intended.

14.8.3.2 Law in Europe

Active and intelligent packaging substances in Europe must comply with Regulation (EC) 450/2009 and the General European Framework (EC) 1935/2004. These regulations establish limitations on the composition of active and intelligent components intended for contact with food, as well as specifications for labelling, declarations of compliance, and accompanying documentation [34].

14.8.3.3 Legal Regulations in India

The Food Safety and Standards Authority of India (FSSAI) monitors' food safety in India, including materials added to food packaging and products directly touching food [53]. In summary, regulations for active food packaging prioritize food safety, adhere to guidelines, and minimize harmful substances in packaging to protect the environment and public health [34].

14.9 Current Market and Future Projections of Active Packaging

The industry is seeing significant expansion because of an increased demand from urban dwellers for convenient and portable food items. The usage of active packaging is greatly increased while processing and chopping vegetables and fruits since they emit gasses within the containers. This leads to a higher demand for active packaging.

In 2023, the worldwide active packaging market was estimated to be worth about 26 billion US$. By 2033, the active packaging market is projected to surpass a value of US$ 64.71 billion [1]. Because more natural and biodegradable polymers are being used, the market share is increasing. The utilization of pullulan in bio-based active packaging for vegetables and fruits is becoming a crucial trend in driving market expansion. Pullulan is a kind of glucan gum that is soluble in water and is extracted from microorganisms. A biodegradable and non-toxic fungus is used in the production process under aerobic circumstances. Its popularity has lately expanded because of its widespread application in the food, chemical, and pharmaceutical industries [2]. Several factors contribute to the market growth of intelligent packaging. Some of them are listed below:

14.9.1 Consumer Awareness

With a rising awareness of the ecological consequences of packaging materials, customers are increasingly seeking sustainable and advantageous environmental products. Intelligent packaging, due to its capacity to prolong the freshness of products, is very compatible with these issues and effectively reduces food waste [38].

14.9.2 Technological Upgrades

The continuous progress in active packaging technology has created new opportunities. Technological advancements such as smart labels that offer real-time updates on product freshness and the development of novel barrier materials are revolutionizing the business and driving its expansion [69].

14.9.3 Legal Initiatives

The introduction of intelligent packaging has been driven by government mandates and actions aimed at reducing food wastage and ensuring the security of drugs. The manufacturers are being motivated to invest in these innovative packaging methods due to the strict quality and safety guidelines [62].

14.9.4 Digital Marketplace Surge

The expansion of e-commerce has increased the demand for packaging that is both economical and dependable. Active packaging solutions enhance the quality of items during transportation, ensuring that they reach customers in the best possible condition. This has led to increased popularity in the field of online shopping [71].

Even though the market is expanding effectively, it has been determined that several industry-wide factors are anticipated to impede growth in the years to come. The stringent policies implemented by emerging country governments prohibiting the use of plastic in convenience stores are probably going to prevent the industry from growing past 2033. Furthermore, a comprehensive investigation has uncovered that active packaging incurs a substantial expense, posing a significant obstacle for the present industry during the estimated duration [3].

14.10 Challenges and Future Outlook of Active Packaging

"Active packaging" specifies methods of enclosing products that interact with the product or its insides to enhance safety, preserve quality, and prolong the storage period of the packed goods. To do this, active chemicals or technologies are typically included in the packaging material or structure. While active packaging has many benefits, it also has certain disadvantages and potential future directions.

14.10.1 Challenges Faced by Active Packaging

14.10.1.1 High Cost

The cost of obtaining materials that are efficient, affordable, and biodegradable remains high despite the increasing availability of active packaging. Even in cases when advance charges offer the potential for long-term benefits, some organizations find it difficult to handle them [59].

14.10.1.2 Compatibility and Stability

Making sure active agents are stable and compatible with packaging materials is a big task. Active ingredients may react with packaging materials, changing their ability to act as a barrier or changing the product in ways that are not needed [54].

14.10.1.3 Market Acceptance

Consumer acceptance and views are crucial for the adoption of active packaging solutions. To overcome obstacles to its use, it is essential to educate customers about the advantages and safety of active packaging.

14.10.1.4 Regulatory Compliance

New chemicals are frequently used in active packaging materials. It's crucial to make sure food safety laws are followed. It is essential to find a balance between safety and creativity [3].

14.10.1.5 Scalability and Manufacturing

It is difficult to increase the production of active packaging materials while preserving quality and performance. The active agents must be distributed uniformly throughout the packing material; manufacturing methods must be improved [3].

14.10.1.6 Recycling Smart Packaging Waste

Smart packaging generates waste that is often not recyclable, posing a significant industry challenge. While recycling policies exist, practical difficulties persist for certain types of packaging waste. A key challenge in designing smart packaging is researching treatment and recycling methods or identifying bio-degradable materials suitable for sensor and communication functionalities [50].

14.10.1.7 Sustainability

The use of artificial chemicals or materials in many active packaging solutions raises questions regarding their sustainability in the environment. It can be difficult to find environmentally friendly substitutes without reducing functionality [6].

14.10.2 Future Outlook to Active Packaging

The future of active packaging holds promise as researchers and industry professionals continue to address these challenges. Innovations in materials, improved manufacturing processes, and consumer education will drive the adoption of active packaging solutions.

14.10.2.1 Nanotechnology

Nanotechnology is poised to play a crucial role in addressing safety concerns related to packaging. Indeed, the performance of nanotechnology can be increased overall when combined with integrated sensors and offer regulated release of active chemicals and improved barrier qualities [16].

14.10.2.2 Biodegradable and Renewable Material

In addressing sustainability challenges, the integration of biodegradable and renewable materials is poised to play a pivotal role in future active packaging solutions. Concurrently, the increasing use of active compounds sourced from natural resources corresponds with the growing adoption of biodegradable packaging materials as carrier polymers. Notably, biopolymers and bio-based compounds have emerged as environmentally fitting alternatives to conventional packaging materials [50].

14.10.2.3 Real-Time Capabilities

The convergence of technology, data, and AI holds promise for transforming packaging systems and optimizing supply chains. Real-time monitoring and control of goods will be enabled through a reliable Industrial Internet of Things (IIoT) infrastructure. Additionally, Artificial Intelligence will play a pivotal role in optimizing processes based on big data insights [64]

14.10.2.4 Antimicrobial Packaging

Antimicrobial packaging has garnered significant interest from researchers and industry professionals, given its potential to improve quality and safety amidst escalating concerns about spoilage bacteria and foodborne diseases. Future research in microbiological active packaging ought to center on naturally derived antimicrobial drugs, preservatives, and biodegradable solutions. Furthermore, researching novel delivery techniques will help to improve the security and reliability of food packing [61].

14.10.2.5 New Business Ventures

As the smart packaging industry hurtles toward a digital horizon, visionary entrepreneurs are crafting novel business models to seize untapped opportunities. These models transcend conventional product-centric approaches, transforming into holistic product-service ecosystems with personalized experiences. The impact on traditional value chains and corporate models will only increase. By combining state-of-the-art technology with creative models, organizations can prosper by utilizing big data and digital innovations [54].

The realm of smart packaging is captivating and offers a complete packaging solution that includes both active and intelligent features. But, integrating these capabilities into a reliable and fully functional system is not an easy feat, considering the numerous challenges that come with it. Nevertheless, this task is expected to pave the way for exciting research opportunities in both foundational and applied disciplines for years to come.

14.11 Conclusion

Active packaging has not just entered the convenience food industry; it has boldly redefined it. Active packaging has allowed perishable goods to have longer shelf lives while maintaining quality and freshness by utilizing controlled-release technology and intelligent packaging. This has not only reduced food waste but also improved food availability around the world. The use of sustainable packaging solutions, such as biodegradable alternatives, has also helped to lessen the negative environmental impact of food packaging. However, the quest to align the mechanical strength, optical clarity, and barrier attributes of synthetic polymers remains a formidable challenge, limiting their full-scale commercial use. As a result, the packaging industry must continue to develop to fulfill consumers' ever-changing wants while also addressing the environmental impact of packaging materials. Compliance with legislative frameworks governing active packaging is paramount to ensuring consumer safety. The outlook for active packaging in the convenience food industry is promising, with significant potential to revolutionize the sector. As the sector navigates the intricate interaction of innovation, consumer expectations, and responsibility towards the environment, the future appears to be both exciting and overwhelming.

14.12 Declaration

The authors declare that they have no known competing financial interests or personal relationships that could have appeared to influence the work reported in this chapter. Generative AI and AI-assisted technologies have been used in the writing process to improve the readability and language of the work only. After using this tool, the authors reviewed and edited the content as needed and took full responsibility for the content of the publication. Microsoft copilot designer was used to create Fig. 14.3 to boost the visual appeal and clarity and project the practicality of the subject in the manuscript for the targeted readership.

References

1. Active Packaging Market: (2022) Future Market Insights. Available at: futuremarketinsights.com/reports/active-packaging-market. Accessed 4 Apr 2024
2. Active Packaging Market Size, Share | Industry Analysis (2024) Fortune Business Insights. Available at: fortunebusinessinsights.com/active-packaging-market-105236. Accessed 4 Apr 2024
3. Ahmed, Md.W., Haque, Md.A., Mohibbullah, Md., et al.: A review on active packaging for quality and safety of foods: current trends, applications, prospects and challenges. Food Packag. Shelf Life **33**, 100913 (2022). https://doi.org/10.1016/j.fpsl.2022.100913
4. Ahmed, I., Lin, H., Zou, L., et al.: A comprehensive review on the application of active packaging technologies to muscle foods. Food Control **82**, 163–178 (2017). https://doi.org/10.1016/j.foodcont.2017.06.009
5. Almasi, H., Jahanbakhsh Oskouie, M., Saleh, A.: A review on techniques utilized for the design of controlled release food active packaging. Crit. Rev. Food Sci. Nutr. **61**, 2601–2621 (2021). https://doi.org/10.1080/10408398.2020.1783199
6. Alves, J., Gaspar, P.D., Lima, T.M., Silva, P.D.: What is the role of active packaging in the future of food sustainability? A systematic review. J. Sci. Food Agric. **103**, 1004–1020 (2023). https://doi.org/10.1002/jsfa.11880
7. Bangar, S.P., Chaudhary, V., Thakur, N., et al.: Natural antimicrobials as additives for edible food packaging applications: a review. Foods 10 (2021)
8. Barone, A.S., Matheus, J.R.V., de Souza, T.S.P., et al.: Green-based active packaging: opportunities beyond COVID-19, food applications, and perspectives in circular economy—a brief review. Comput. Rev. Food Sci. Food Saf. **20**, 4881–4905 (2021). https://doi.org/10.1111/1541-4337.12812
9. Bastarrachea, L., Wong, D., Roman, M., et al.: Active packaging coatings. Coatings **5**, 771–791 (2015). https://doi.org/10.3390/coatings5040771
10. Bhardwaj, A., Alam, T., Talwar, N.: Recent advances in active packaging of agri-food products: a review
11. Boey, J.Y., Mohamad, L., Sen, K.Y., et al.: A review of the applications and biodegradation of polyhydroxyalkanoates and poly(lactic acid) and its composites. Polymers (Basel) **13**, 1544 (2021). https://doi.org/10.3390/polym13101544
12. Boostani, S., Jafari, S.M.: A comprehensive review on the controlled release of encapsulated food ingredients; fundamental concepts to design and applications. Trends Food Sci. Technol. **109**, 303–321 (2021). https://doi.org/10.1016/j.tifs.2021.01.040
13. Brody, A.L., Strupinsky, E.P., Kline, L.R.: Active packaging for food applications. CRC Press (2001)
14. Cai, J., Zhao, X., Pan, C., et al.: Driving e-commerce brand attachment through green packaging: an empirical investigation (2021)
15. Carbone, M., Donia, D.T., Sabbatella, G., Antiochia, R.: Silver nanoparticles in polymeric matrices for fresh food packaging. J. King Saud. Univ. Sci. **28**, 273–279 (2016). https://doi.org/10.1016/j.jksus.2016.05.004
16. Chausali, N., Saxena, J., Prasad, R.: Recent trends in nanotechnology applications of bio-based packaging. J. Agric. Food Res. **7**, 100257 (2022). https://doi.org/10.1016/j.jafr.2021.100257
17. Chuang, M.-L., Shaw, W.H.: RFID: integration stages in supply chain management. IEEE Eng. Manage. Rev. **35**, 80–87 (2007). https://doi.org/10.1109/EMR.2007.899757
18. De Corato, U.: Improving the shelf-life and quality of fresh and minimally-processed fruits and vegetables for a modern food industry: a comprehensive critical review from the traditional technologies into the most promising advancements. Crit. Rev. Food Sci. Nutr. **60**, 940–975 (2020)
19. Deshmukh, R.K., Gaikwad, K.K.: Natural antimicrobial and antioxidant compounds for active food packaging applications. Biomass Convers. Biorefin. **14**, 4419–4440 (2024)
20. Drago, E., Campardelli, R., Pettinato, M., Perego, P.: Innovations in smart packaging concepts for food: an extensive review. Foods 9 (2020)

21. European Commission (2011) Commission Regulation (EU) No 10/2011 of 14 January 2011 on plastic materials and articles intended to come into contact with food. EUR-Lex
22. European Parliament and Council of the European Union (2004) Regulation (EC) No 1935/ 2004 on materials and articles intended to come into contact with food and repealing Directives 80/590/EEC and 89/109/EEC. EUR-Lex
23. Fabech, B.: Active and intelligent food packaging: a Nordic report on the legislative aspects. Nordic Council of Ministers (2000)
24. Fadiji, T., Rashvand, M., Daramola, M.O., Iwarere, S.A.: A review on antimicrobial packaging for extending the shelf life of food. Processes **11**, 590 (2023). https://doi.org/10.3390/pr1102 0590
25. Fuciños, C., Guerra, N.P., Teijón, J.M., et al.: Use of poly(n-isopropylacrylamide) nanohydrogels for the controlled release of pimaricin in active packaging. J. Food Sci. **77**,(2012). https:// doi.org/10.1111/j.1750-3841.2012.02781.x
26. Gaikwad, K.K., Singh, S., Ajji, A.: Moisture absorbers for food packaging applications. Environ. Chem. Lett. **17**, 609–628 (2019). https://doi.org/10.1007/s10311-018-0810-z
27. Gaikwad, K.K., Singh, S., Negi, Y.S.: Ethylene scavengers for active packaging of fresh food produce. Environ. Chem. Lett. **18**, 269–284 (2020). https://doi.org/10.1007/s10311-019-009 38-1
28. Guillard, V., Gaucel, S., Fornaciari, C., et al.: The next generation of sustainable food packaging to preserve our environment in a circular economy context. Front. Nutr. **5**,(2018). https://doi. org/10.3389/fnut.2018.00121
29. Han, J.H.: A review of food packaging technologies and innovations. In: Innovations in food packaging: Second Edition. Elsevier Ltd., pp 3–12 (2013)
30. Heising, J.K., Dekker, M., Bartels, P.V., (Tiny) Van Boekel MAJS,: Monitoring the quality of perishable foods: opportunities for intelligent packaging. Crit. Rev. Food Sci. Nutr. **54**, 645–654 (2014)
31. Hosseinnejad, M.: A r c h i v e o f S I D active packaging for food applications-a review (2014)
32. Jeong, Y., Ansari, J.R., Sadeghi, K., Seo, J.: Applicability of polypropylene/polyethylene glycol/molecular sieve composites as desiccant pharmaceutical packaging materials. Food Packag. Shelf Life **42**, 101266 (2024). https://doi.org/10.1016/j.fpsl.2024.101266
33. Jia, F., Shahzadi, G., Bourlakis, M., John, A.: Promoting resilient and sustainable food systems: a systematic literature review on short food supply chains. J. Clean. Prod. **435**, 140364 (2024). https://doi.org/10.1016/j.jclepro.2023.140364
34. Jiang, Y., Zhang, Y., Deng, Y.: Latest advances in active materials for food packaging and their application. Foods 12 (2023)
35. Kamaly, N., Yameen, B., Wu, J., Farokhzad, O.C.: Degradable controlled-release polymers and polymeric nanoparticles: mechanisms of controlling drug release. Chem. Rev. **116**, 2602–2663 (2016). https://doi.org/10.1021/acs.chemrev.5b00346
36. Kerry, J., Butler, P.: (eds) Smart packaging technologies for fast moving consumer goods. Wiley (2008)
70. Khan, S., Monteiro, J.K., Prasad, A., Filipe, C.D., Li, Y., Didar, T.F.: Material Breakthroughs in Smart Food Monitoring: Intelligent Packaging and On-Site Testing Technologies for Spoilage and Contamination Detection. Advanced Materials, 36(1), 2300875 (2024). https://doi.org/10. 1002/adma.v36.1. https://doi.org/10.1002/adma.202300875
38. Kocetkovs, V., Muizniece-Brasava, S.: Consumer awareness and attitudes towards active and intelligent packaging systems in the Latvian market, pp 222–226 (2019)
39. Kumari, S.V.G., Pakshirajan, K., Pugazhenthi, G.: Recent advances and future prospects of cellulose, starch, chitosan, polylactic acid and polyhydroxyalkanoates for sustainable food packaging applications. Int. J. Biol. Macromol. **221**, 163–182 (2022). https://doi.org/10.1016/ j.ijbiomac.2022.08.203
40. Kuswandi, B., Jumina: Active and intelligent packaging, safety, and quality controls. In: Fresh-cut fruits and vegetables. Elsevier, pp 243–294 (2020)
41. Lloyd, K., Mirosa, M., Birch, J.: Active and intelligent packaging (2018)

42. Lorite, G.S., Rocha, J.M., Miilumäki, N., et al.: Evaluation of physicochemical/microbial properties and life cycle assessment (LCA) of PLA-based nanocomposite active packaging. LWT **75**, 305–315 (2017). https://doi.org/10.1016/j.lwt.2016.09.004
43. Ma, Y., Yang, W., Xia, Y., et al.: Properties and applications of intelligent packaging indicators for food spoilage. Membranes (Basel) **12**, 477 (2022). https://doi.org/10.3390/membranes12050477
44. Mustafa, M., Nagalingam, S., Tye, J., et al.: Looking back to the past: revival of traditional food packaging
45. Müller, P., Schmid, M.: Intelligent packaging in the food sector: a brief overview. Foods **8**, 16 (2019). https://doi.org/10.3390/foods8010016
46. Osmólska, E., Stoma, M., Starek-Wójcicka, A.: Application of biosensors, sensors, and tags in intelligent packaging used for food products—a review. Sensors **22**, 9956 (2022). https://doi.org/10.3390/s22249956
47. Pereira DC de, S., Teixeira, N.S., Zin, I.M., et al.: Active and intelligent packaging: security, legal aspects and global market. Braz. J. Develop. 6:61766–61794 (2020). https://doi.org/10.34117/bjdv6n8-164
48. Prasad, P., Kochhar, A.: Active packaging in food industry: a review (2014)
49. Puligundla, P., Jung, J., Ko, S.: Carbon dioxide sensors for intelligent food packaging applications. Food Control. **25**, 328–333 (2012). https://doi.org/10.1016/j.foodcont.2011.10.043
50. Qian, M., Liu, D., Zhang, X., et al.: A review of active packaging in bakery products: applications and future trends. Trends Food Sci. Technol. **114**, 459–471 (2021)
51. Rundh, B.: The role of packaging within marketing and value creation. British Food J. **118**, 2491–2511 (2016). https://doi.org/10.1108/BFJ-10-2015-0390
52. Sadeghi, K., Seo, J.: Photografting coating: an innovative approach to "non-migratory" active packaging. Adv. Funct. Mater. **31**,(2021). https://doi.org/10.1002/adfm.202010759
53. Sasikumar Nair, S., Mazurek-Kusiak, A.K., Trafialek, J., Kolanowski, W.: Assessing food safety compliance in a small-scale indian food manufacturer: before and after certification of the food safety management system and foreign supplier verification program. Appl. Sci. **13**, 12190 (2023). https://doi.org/10.3390/app132212190
54. Schaefer, D., Cheung, W.M.: Smart packaging: opportunities and challenges. Proc CIRP **72**, 1022–1027 (2018). https://doi.org/10.1016/j.procir.2018.03.240
55. Shameem, R.M., Arivoli, S.: A review on traditional and modern food packaging technologies (2022)
56. Singh, A.K., Ramakanth, D., Kumar, A., et al.: Active packaging technologies for clean label food products: a review. J. Food Measure. Character. **15**, 4314–4324 (2021)
57. Singh, A.K., Kim, J.Y., Lee, Y.S.: Phenolic compounds in active packaging and edible films/coatings: natural bioactive molecules and novel packaging ingredients. Molecules 27 (2022)
58. Siracusa, V., Lotti, N.: Intelligent packaging to improve shelf life. In: Food quality and shelf life. Elsevier, pp. 261–279 (2019)
59. Soltani Firouz, M., Mohi-Alden, K., Omid, M.: A critical review on intelligent and active packaging in the food industry: Research and development. Food Res. Int. 141 (2021)
60. Steenis, N.D., van Herpen, E., van der Lans, I.A., et al.: Consumer response to packaging design: the role of packaging materials and graphics in sustainability perceptions and product evaluations. J. Clean. Prod. **162**, 286–298 (2017). https://doi.org/10.1016/j.jclepro.2017.06.036
61. Suppakul, P., Miltz, J., Sonneveld, K., Bigger, S.W.: Active Packaging Technologies with an Emphasis on Antimicrobial Packaging and its Applications. J. Food Sci. **68**, 408–420 (2003). https://doi.org/10.1111/j.1365-2621.2003.tb05687.x
62. Tiekstra, S., Dopico-Parada, A., Koivula, H., et al.: Holistic approach to a successful market implementation of active and intelligent food packaging. Foods **10**, 1–22 (2021). https://doi.org/10.3390/foods10020465
63. Trajkovska Petkoska, A., Daniloski, D., D'Cunha, N.M., et al.: Edible packaging: Sustainable solutions and novel trends in food packaging. Food Res. Int. **140**, 109981 (2021). https://doi.org/10.1016/j.foodres.2020.109981

64. Vanderroost, M., Ragaert, P., Devlieghere, F., De Meulenaer, B.: Intelligent food packaging: the next generation. Trends Food Sci. Technol. **39**, 47–62 (2014). https://doi.org/10.1016/j.tifs.2014.06.009
65. Vasile, C., Baican, M.: Progresses in food packaging, food quality, and safetyzcontrolled-release antioxidant and/or antimicrobial packaging. Molecules **26**,(2021). https://doi.org/10.3390/molecules26051263
66. Vågsholm, I., Arzoomand, N.S., Boqvist, S.: Food security, safety, and sustainability—getting the trade-offs right. Front Sustain. Food Syst. 4 (2020)
67. Westlake, J.R., Tran, M.W., Jiang, Y., Zhang, X., Burrows, A.D., Xie, M.: Biodegradable active packaging with controlled release: principles, progress, and prospects. ACS Food Sci. Technol. **2**, 1166–1183 (2022). https://doi.org/10.1021/acsfoodscitech.2c00070
68. Wu, F., Misra, M., Mohanty, A.K.: Challenges and new opportunities on barrier performance of biodegradable polymers for sustainable packaging. Prog. Polym. Sci. 117 (2021)
69. Wyrwa, J., Barska, A.: Innovations in the food packaging market: active packaging. Eur. Food Res. Technol. **243**, 1681–1692 (2017). https://doi.org/10.1007/s00217-017-2878-2
70. Yildirim, S., Röcker, B., Pettersen, M.K., et al.: Active packaging applications for food. Compr. Rev. Food Sci. Food Saf. **17**, 165–199 (2018). https://doi.org/10.1111/1541-4337.12322
71. Zhao, X., Pan, C., Cai, J., et al.: Driving e-commerce brand attachment through green packaging: an empirical investigation

Chapter 15
Natural Dyes for the Coloration of Food Packaging

Gazi Farhan Ishraque Toki, Rony Mia, and Anik Das

Abstract In recent years, natural coloring agents from plant resources have drawn scientists into using these hues in food goods. The public's understanding of the negative properties of synthetic dyestuff also raised customer claims for natural coloring agents. Several plant sources can yield different color substances like Anthocyanins, Betalains, Carotenoids, Carminic Acid, and Chlorophylls. Remarkably, these pigments benefit the visual appeal of processed food items and the plant, food, and human well-being. As such, the current state of the art in plant pigment utilization in the food industry is highlighted in this chapter. A synopsis of the uses of plant colorants in food is given first, then explanations of the five main sources of dyestuff and the extraction methods of the natural coloring agent follow. As the value of natural coloring is universal, this chapter also provides a whole understanding of what the food sector may encounter when using natural pigments are also covered.

Keywords Natural resources · Dyestuff · Extraction method · Utilization · Food industry

G. F. I. Toki
Nanotechnology Center, School of Fashion and Textiles, The Hong Kong Polytechnic University, Kowloon 999077, Hong Kong

R. Mia (✉)
Wilson College of Textiles, North Carolina State University, 1020 Main Campus Dr, Raleigh, NC 27606, USA
e-mail: mroni_mia@yahoo.com

Department of Textile Engineering, National Institute of Textile Engineering and Research, University of Dhaka, Dhaka 1000, Bangladesh

A. Das
School of Textile Science and Engineering, Tiangong University, Tianjin 300387, People's Republic of China

S. ul Islam and M. Shahid (eds.), *Green Materials for Active Food Packaging*, Engineering Materials, https://doi.org/10.1007/978-981-96-0369-5_15

15.1 Introduction

The foodstuff manufacturing sector is a vital sector in the global economy, characterized by significant demand in international markets. The majority of food items contain dyes, which can come from either synthetic or natural pigments. A pigment is a chemical that occurs naturally in animal tissue, or plants, or is synthetically generated and imparts a distinct color [1, 2]. Synthetic pigments are widely utilized in the cosmetic, culinary, and pharmaceutical industries due to their favorable characteristics such as a wide spectrum of colors, cost-effectiveness, resistance to oxygen degradation, and solubility [3]. Currently, there is a strong emphasis on creating food-grade pigments from sources that are naturally occurring. This is because synthetic dyes have been found to have negative effects such as toxicity, mutagenicity, and propensity to cause cancer [4, 5]. The public's understanding of ecologically friendly and non-toxic ecological sustainability in bio-based resource colorants has sparked a uprising in natural substance development and study [6, 7]. In the past years, there has been a growing shift from using synthetic dyes to natural pigments. This change is driven by customer desire for more environmentally friendly products, as well as the safety and health benefits associated with natural pigments [8]. Natural colors typically have inherent antibacterial and antioxidant characteristics, which may result in significant therapeutic efficacy [9]. In addition, these colors are composed of a variety of vitamins with varying amounts, which are determined by the specific plant they are derived from. The utilization of natural pigments is regarded as an environmentally suitable and non-toxic procedure [10]. Natural coloring agents have found application in several industries, including food, apparel, medicinal products, and cosmetic sectors, due to their plentiful advantages [11, 12].

The current lack of conclusive evidence regarding the safety and negative consequences of inorganic and synthetic colorants is not very beneficial for the general public. Consumers have another reason to distrust non-natural colorants due to the fact that a number of additives to food that have previously been authorized and considered safe are now banned [13]. This is simply because these additives have been proven to have adverse consequences such as toxic effects over a long period of time and a high occurrence of health disturbances [14]. The questioning of synthetic colorants has spurred the exploration and creation of naturally produced more secure replacements. These colorants are derived from plants, animals, microorganisms, and minerals. Anthocyanins, carotenoids, carminic acid, betalains, chlorophylls, minerals, phycocyanins, curcuminoids, and some vitamins are regularly used and widely available natural colorants and additives [15]. One drawback of utilizing natural coloring agents is that their qualities can impose limitations on their usage [16]. Natural colorants frequently exhibit a lower intensity of color compared to synthetic alternatives, making it challenging to achieve desirable shades, particularly in the case of green and blue [17]. The majority of natural coloring agents are susceptible to the effects of heat, light, and oxygen, which can lead to a loss or alteration of color. Some individuals may have a heightened sensitivity to various factors in the composition of food, such as pH levels, proteins, metal ions, or certain chemical molecules [18].

Dyes are frequently employed in the food business to impart or intensify the vibrant hues of food items, thus increasing their appeal to consumers. Technologically, the use of these additives is justified because many foods lose color during manufacturing and storage, necessitating the inclusion of colors to counteract these changes [19, 20]. Synthetic colorants, in broad terms, offer superior durability and cheaper cost compared to natural dyes, making them more favorable in terms of product uniformity and cost-effectiveness. Nevertheless, there is substantial data indicating that the intake of synthetic colors has health hazards, including the ability to cause cancer and trigger allergies [21]. Food dyes undergo evaluation by both national and international commissions, similar to other food additives. However, there are several limits in this evaluation process, particularly when it comes to assessing their potential to cause allergies. As a result, many nations have attempted to control the use of synthetic coloring agents in order to enhance consumer safety. This is done by imposing restrictions or bans on their use, or by requiring mandatory labeling to notify consumers about their existence [19]. The stricter regulation of synthetic additives is also a response to mounting pressure from different sectors that advocate for the usage of natural compounds to enhance the value of products. These advocates argue that natural foods are entirely safe, despite the lack of sufficient toxicological information [22]. Natural dyes are being promoted and utilized more frequently as a marketing tactic because of their widespread range of organic purposes, including antimicrobial and antioxidant assets. They also offer potential health advantages, such as the ability to prevent chronic non-communicable diseases (NCDs). Additionally, these dyes are commonly found in various foods that are regularly consumed by the population, without posing any potential health hazards [20, 23].

The chapter highlights based on Natural Dyes substances like Anthocyanins, Betalains, Carotenoids, Carminic Acid, and Chlorophylls for the coloration of food packaging. Here, the key features and chemical structures of these dye substances as well as their influences on utilization are described. Next, several extraction methods of such natural colorants such as traditional and prospective methods were mentioned. Following that the utilization of natural pigments in food packaging over the years is described. Finally, the limitations in acquiring and utilizing pigments resultant from natural resources in the food industry is highlighted. Thus, this chapter offers to the scientific community that natural resources for instance Anthocyanins, Betalains, Carotenoids, Carminic Acid, and Chlorophylls can be employed to upsurge stability and progress the possibilities of via natural dyestuff in food industry,

15.2 Colorants Derived from Natural Resources

Pigments are responsible for the vibrant natural coloration of plants. The predominant pigments that exist in vegetables and fruits include green chlorophylls, red-yellow betalains, yellow-orange carotenoids, red–purple anthocyanins, and so on. In addition to their capability to provide hue, these plant pigments also demonstrate probable health-enhancing benefits [24, 25]. The residues of agriculture are an important and environmentally friendly source of naturally occurring pigments. Utilizing these residues for the purpose of recovery of natural pigments can assistance in alleviating the ecological challenges associated with their managing. Seeds and peels of certain vegetables and fruits, which are extracted during the manufacturing process, might still comprehend substantial quantities of pigments, every so often in higher quantities [26, 27]. The natural colorant resources are described below.

15.2.1 Anthocyanins

15.2.1.1 Features and Chemical Structures of Anthocyanins

Anthocyanins are extensively researched natural pigments that belong to the flavonoid family (Fig. 15.1). There are over 700 distinct derivatives of anthocyanins. The substance is capable of dissolving in water, possesses minimal or negligible toxicity, and exhibits a variety of colors, spanning from red to blue [28, 29]. These substances can be found in many plant parts such as leaves, vegetables (eggplant, cabbage), tubers like potatoes (purple), flowers (hibiscus), and fruits (like grapes, cherries, blueberries, pomegranates, and raspberries) that have red, purple, or blue color [30, 31]. Anthocyanins possess a glycosidic composition consisting of sugars and active acidic substances connected to an aglycone called anthocyanidin. Anthocyanidins are flavyl ion constructions, namely 2-phenylbenzopyrilium, with the keto oxygen at the C1 position. They can generate several forms of anthocyanins by undergoing glycosylation or acylation reactions involving diverse combinations of sugar, phenolic, or aliphatic components [30]. The primary anthocyanidins, which account for approximately 90% of all identified anthocyanins, include cyanidin, peonidin, petunidin, pelargonidin, malvidin delphinidin, and malvidin [28, 32].

Anthocyanins in aqueous environments, such as foods, experience reversible changes in structure with pH, resulting in a simultaneous alteration in color [1, 12, 13]. The red flavylium cation is the most abundant species when the pH is below 2. Between pH 3 and 6, there is a quick process of adding water to the flavylium cation at C-2, resulting in the formation of a colorless carbinol pseudobase. The carbinol undergoes ring-opening tautomerization, resulting in the formation of the (Z)-chalcone. This (Z)-chalcone has the ability to isomerize into the (E)-chalcone. Under conditions that are to some extent acidic to neutral, while the flavylium (cation) undergoes deprotonation, resulting in the formation of the bluish quinoidal foundation [34].

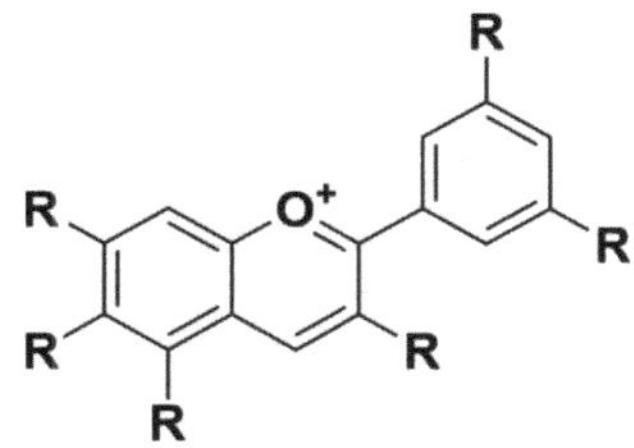

Fig. 15.1 The fundamental composition of an anthocyanidin, whereas the chromophore component of anthocyanins (many anthocyanidins are formed when OH, H, or OCH_3 groups take the place of R radicals) [33]

Various factors, including cultivar/variety, maturation, production procedures, growing region or season/climate, and conditions of storage, influence the anthocyanin content of foods [35–40]. Amarowicz et al. [41] studied the impact of after-harvesting processing and preservation on the levels of flavonoids, which include anthocyanins. They determined that the impact of processing and storage on the polyphenol concentration is insignificant for the majority of the subclasses being studied, as compared to the variations across different plant kinds [41].

15.2.1.2 Factors Affecting Anthocyanins Usage

The hue of anthocyanin is impacted by numerous aspects including its molecular structure, pH level, exposure to Ultraviolet (UV), concentration of the chemical, co-pigmentation, and its existence or shortage of oxygen. To each anthocyanidin has a distinct hue as a result of its chemical composition. Peonidin and cyanidin exhibit a reddish-purple color, while delphinidin along with malvidin display a reddish as well as blue or purplish tint. Petunidin and pelargonidin impart an orange tint to flowers and a reddish color to certain fruits [30].

The ionic structure of the molecules of anthocyanins undergoes modifications at various pH levels, resulting in alterations in both the stability and tone of this pigment. Anthocyanins are present in red hue in acidic conditions, for example. Anthocyanins appear purple in neutral pH conditions, but they undergo a color alteration to green, yellow, and blue under basic pH environments [42, 43]. The amount of anthocyanins is a crucial component that influences the intensity of the color of the extract.

The diversity in color among anthocyanins is caused by a bathochromic shift, which results in a change in hue. This shift occurs when the absorption of light in the observable range shifts from a shorter to a lengthy wavelength [44]. As a result, the color changes from red to purple under acidic pH conditions. The amounts of methoxy groups in the anthocyanin molecule influence this variation. A higher concentration of methoxy groups leads to an enhancement of purple/blue hues, for instance, malvidin. Conversely, a lesser concentration of methoxy groups results in an upsurge in reddish hue, as seen in pelargonidin [32]. Studies demonstrate that anthocyanins possess a significant probability for utilization as a naturally occurring tint, particularly in the realms of food goods and packaging. Additionally, these techniques have been utilized in various other goods, including textiles, humanoid hair, as well as sensitized solar cells [45]. Furthermore, the chemicals may provide advantageous health benefits, including antioxidant, anticancer, and anti-inflammatory activities [23]. Anthocyanins, which are natural food colorings and efficient constituents, have had inadequate application as a consequence of their lower stability and tendency to interact with other molecules in the food matrix. As a result, researchers have investigated other methods, including encapsulation and adsorption, to address these constraints [46].

15.2.2 Betalains

15.2.2.1 Features and Chemical Structures of Betalains

The nitrogen-containing pigments known as betalains include betacyanins, which range in hue since reddish to reddish-violet, and betaxanthins, which are orange-yellow Both molecules are amino acid and cyclo-dopa[cyclo-3-(3,4-dihydroxyphenylalanine)] combines of betalamic acid, correspondingly. Because of the aromatic ring of cyclo-dopa, betacyanins have a bathochromic shift of 50–70 nm when contrasted with betaxanthins. Betacyanin undergoes a hypsochromic shift when betanidin is glycosylated; glucose connected to C-6 is less efficient than glucose associated with C-5 [47, 48]. Adding aliphatic acyl moieties during esterification did not significantly change the maximal absorption of betacyanins [49].

In their assessment of betalain biosynthesis, Khan and Giridhar included all plants known to accumulate betalain, highlighted the presence and pattern of pigment accumulation, and provided a list of these plants [50]. Concerning technology along with impact on human health, Stintzing and Carle projected their work for identifying its impact [51]. Betanin, also known as betanidin 5-O-β-glucoside, was the most abundant pigment for an extended period, and beetroot was thought to be the only place to get betacyanin for food coloring (Fig. 15.2). There has been a rise in the hunt for alternative sources in recent years. Some examples include *Ullucus tuberosus*, a root crop that is both economically and extensively grown in the Andean region of South America [52]; *Basella rubra*, or Malabar spinach, a leafy vegetable that stores pigments in its fruits [53]; red–purple pitaya *Hylocereus polyrhizus* [44]; cactus pear (*O. stricta* and *Opuntia ficus-indica*) [54]; and Amaranthus species [8]. Additionally, the other resources also for instance vegetables (chard), seeds, grains, fruits (red pitaya), roots, leaves, stems (like cacti), and some fungi (*Hygrocybe*, *Amanita muscaria*, along with *Hygrophorus*) [55, 56].

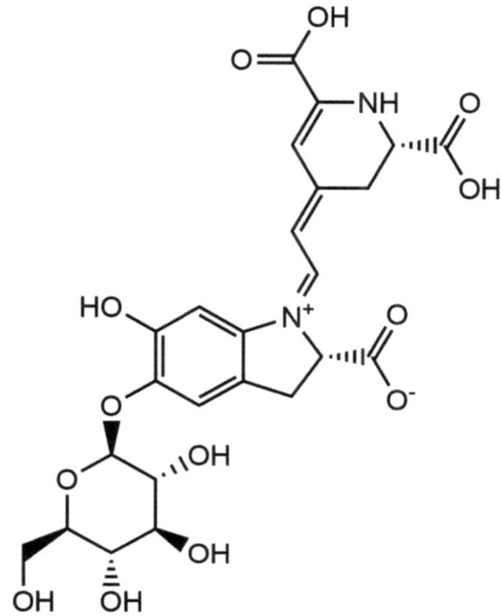

Fig. 15.2 Red beet (*Beta vulgaris L.*) betanine structure

15.2.2.2 Factors Affecting Betalains Usage

When it comes to the extracting procedure and processing of food, betalains are susceptible to changes in stability caused by both internal and external variables. Its pigments, which are soluble in water, are suitable for neutral or low-acid foods and remain stable from pH 3 to 7. The color stability of betalains is higher than that of anthocyanins in the same pH range, and they are also pH-independent [57]. An observable transition from red to blue-violet occurs at pH standards <3, as the anionic betalain structure, which is reddish in tint, is transformed to the cationic version, which is violet. The process of hydrolysis of the aldimine link produces cyclo-dopa-5-O-glycoside and betalamic acid, which causes the color to transition from yellow to brown at pH levels higher than 7 [58]. Light, oxygen, metallic ions, high water activity, along side ambient temperature all degrade the bioactive compound's stability, which limits its application in foods that have a short shelf life or are frozen. Storage of betalain solutions requires low temperatures, little oxygen, and a dark environment since the compounds are biosynthesized [59]. Various approaches are being considered to stabilize betalain and other bioactive compounds, including encapsulation by drying, ionic gelling, and emulsions with a variety of wall materials [60, 61].

15.2.3 Carotenoids

15.2.3.1 Features and Chemical Structures of Carotenoids

The naturally occurring yellow, orange, and red pigments known as carotenoids are a type of lipophilic coloring agent (Fig. 15.3). Carotenoids are tetraterpenoid pigments that have a fundamental structure with 8 isoprene groups. On top of that, they have a C_{40} polyene chain with double bonds of conjugation that are formed by linking two C_{20} geranylgeranyl pyrophosphate molecules, each of which has 4 isoprene units. Carotenoids get their vivid yellow to red hues from their ability to capture light that is visible, which falls between the 400 to 500 nm range, due to the varied number of double bonds with conjugation [62, 63]. In terms of their chemical makeup and assembly, the carotenoids can be categorized into two assemblages: the first is carotene, which is composed solely of carbon and hydrogen atoms (α-carotene, β-carotene, as well as lycopene); the second is xanthophyll, which includes oxygen additionally to carbon and hydrogen (astaxanthin, lutein, zeaxanthin, and β-cryptoxanthin). In addition, apocarotenoids, which have a shorter carbon chain, include bixin (the pigment in annatto) and crocetin (a component in saffron) [63, 64]. Everything from algae and plants to cyanobacteria and even certain non-photosynthetic animals and fungus make carotenoids. Photon absorption is a key aspect of photosynthetic systems, and carotenoids play a role in that process. Carotenoids are involved as pigments in the yellow to red range in animals and tissues that do not engage in photosynthetic processes. Since carotenoids are not biosynthesized by any animal, including humans, they must be obtained from food sources. In food, these substances can be found in leafy greens like broccoli and spinach (which turn yellow when the chlorophylls in them break down), in orange-colored carrots and pumpkins, in red peppers, and in red fruits like tomatoes and watermelon, as well as orange fruits like mango and papaya. Furthermore, the buildup of carotenoids in animal tissues is responsible for the red hue in certain fish (like salmon) and crustaceans (such as boiled lobster, shrimp, and crab), as well as the yellow hue in egg yolks. The four most common carotenoids originate in food items are β-carotene, lutein, lycopene, α-carotene, β-cryptoxanthin, and zeaxanthin [65].

Annatto

The raw red extract produced achiote tree seeds (known as *Bixa Orellana L.* seeds) is called annatto because of the waxy coating that covers them. Tropical countries like Mexico, Brazil, and Peru are home to the little achiote tree or shrub. But South and Central America are also good places to find it growing [66]. The primary components of annatto extract include bixin, which is hydrophobic, and norbixin, which is liposoluble. Other substances discovered in annatto seed extract include isobixin, beta-carotene, and lutein [67].

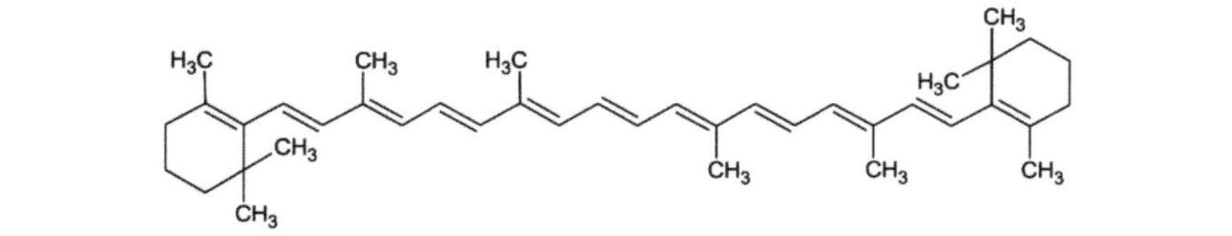

β-carotene

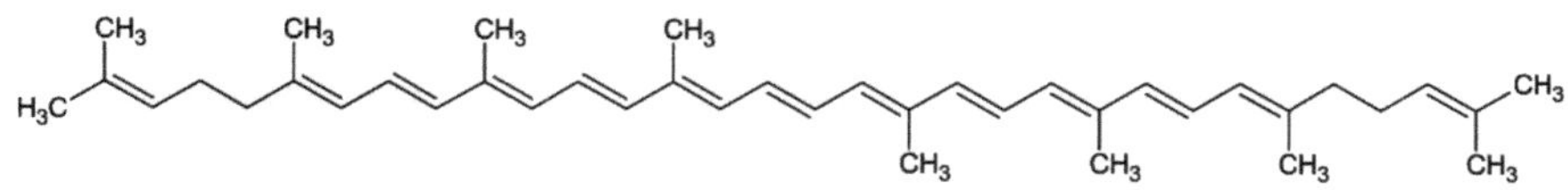

Lycopene

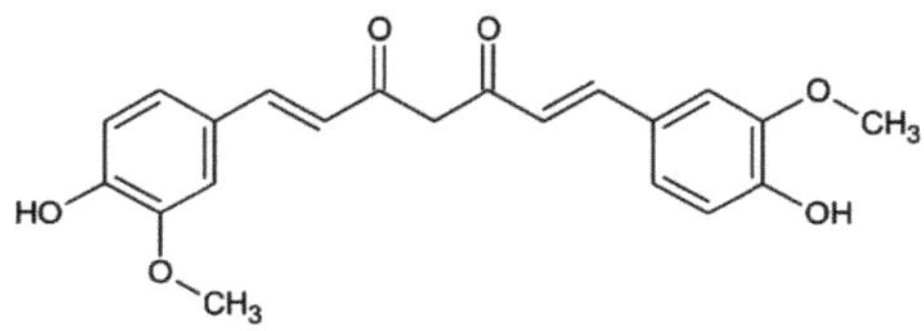

Curcumin

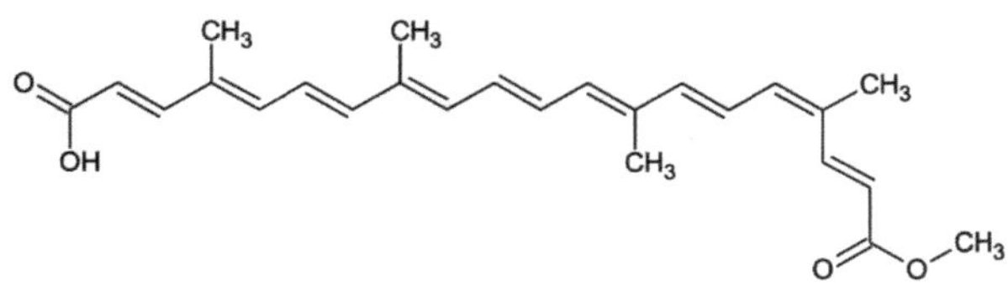

α- bixin

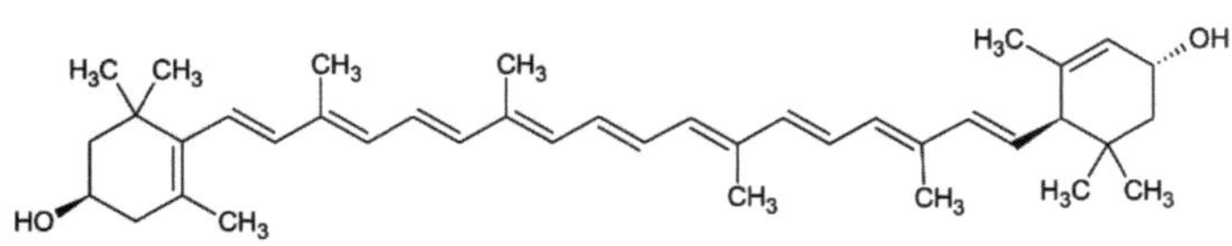

Lutein

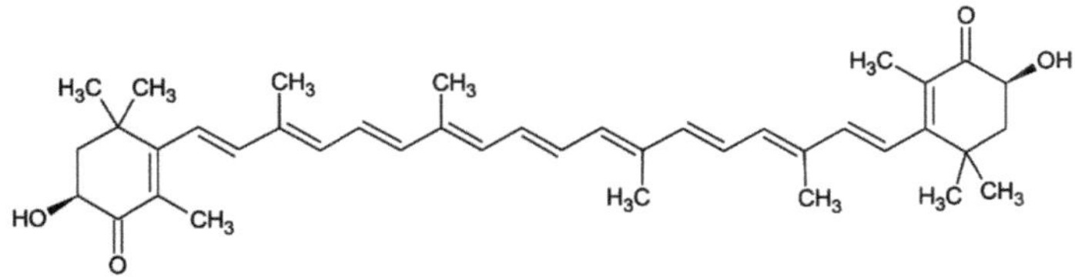

Astaxanthin

Fig. 15.3 Composition of the several carotenoids

Eighty percent of the annatto tints are bixin, the primary colorant. The carotenoid bicinchonin ($C_{25}H_{30}O_4$) has two carboxyl groups and a 9-cis configuration, which is common in oxygenated carotenoids like lutein (xanthophyll group) [68].

β-Carotene

Most people are familiar with β-carotene, a yellowish-orange carotenoid that is mostly found in carrots and fungi, as stated in E 160a (ii) [69]. β-carotene is characterized by its lipophilicity due to its core structure featuring 40 carbon atoms, 9 double bonds of conjugation in the β-ionone rings, and a polyene chain at either end of the molecule ($C_{40}H_{56}$) [70]. Numerous fruits and vegetables, ranging in hue from reddish to yellow, contain this chemical. Some examples include dehydrated red peppers, raw carrots, and raw grape leaves [63].

Fresh vegetables and fruits naturally contain β-carotene in its chemical state as trans-isomers [71]. Yet, marine microalgae species were revealed to have cis- β-carotene, specifically 9 cis-, 13 cis-, and 15 cis-carotene [72].

More specifically, research has shown that natural 9 cis- β-carotene is effective against diseases like atherosclerosis, psoriasis, retinitis pigmentosa, and atherogenesis [73]. Additionally, β-carotene has a significant impact on the well-being of people and is the primary building block of vitamin A, which is formed by combining two β-ionone rings [74]. Being an antioxidant, it prevents lipid peroxidation, which in turn protects against cancer and coronary heart disease and reduces down the aging process [72].

Lycopene

Lycopene, a carotenoid with the chemical formula $C_{40}H_{56}$ and the red coloration found in tomatoes (*Solanum lycopersicum*) and other fruits, is mainly accountable for the red hues of watermelon, papaya, pink grapefruit, and red guava [75]. Mostly discovered as an all-trans isomer, although cis isomers of (5, 9, 13, and 15) were also detected. An individual's natural body is better able to absorb the lycopene cis isomer. Heat, light, oxygen, and acidity all play a role in this cis–trans isomerization [76].

Lutein

Functioning xanthophyll hydroxy, or lutein ($C_{40}H_{56}O_2$), is a type of fat-soluble carotenoid that occurs spontaneously. Fruit and vegetable products rich in this chemical include dark-green vegetables with leaves, flowers, fruits, as well as egg yolks, among many others. Lutein has significant biological activity, with anti-inflammatory, antioxidant, and anticancer properties, and it is categorized as an organic food colorant (INS 161b or E-161b) because of its coloring agent potency.

According to several studies [71–75], the architecture is defined by a lengthy carbon chain that alternates between single as well as double carbon–carbon bonds, with connected methyl side groups [77–79].

15.2.3.2 Factors Affecting Carotenoids Usage

The majority of naturally occurring carotenoids found in food are in their durable trans conformation, whereas a lesser percentage are in the less unstable cis form [63, 80]. On the other hand, isomerization and oxidation can occur in the attached double-bond system while it is being processed or stored. Enzymes, metal ions, light, and oxygen all have a role in accelerating oxidation. Low molecular weight, colorless molecules (volatiles) with little biological action are formed as a byproduct of degradation [81].

Carotenoids are naturally occurring pigments in food that can be extracted from a variety of sources; these pigments can be yellow, orange, or red in color. Various regulatory bodies, including the Food and Drug Administration (FDA) and the European Food Safety Authority (EFSA), have acknowledged carotenoids as GRAS, or Generally Recognized as Safe. Carotenoids find use in a variety of foods, including beef (sausage), vegetable oils, and butter [82]. Water insolubility, instability, and poor bioavailability make their usage as a tint and efficient ingredient hard. Their solubility is improved and they are protected from processing and digestive stressors through encapsulation, which is an effective technique [62].

In addition to its use as a natural dye, carotenoids have numerous positive belongings on human health. The primary advantage of carotenoids for health is their capability to be enzymatically transformed to retinol, or vitamin A. This micronutrient is vital for many bodily functions, including development, reproduction, growth, immunity, and the integrity of the epithelial barrier. Further functions include protecting the cardiovascular system and the nervous system from damage, preventing obesity, lowering blood sugar, and warding off cancer [62, 83].

15.2.4 Carminic Acid

15.2.4.1 Features and Chemical Structures of Carminic Acid

Carminic acid, a water-soluble dye derived from the coconilla species of bugs, is seven-α-D-glycopyranosyl-9,10-dihydro-3,5,6,8-tetrahydroxy-1-methyl-9,10-dioxo-2 anthracenecarboxylic acid (shown in Fig. 15.4). Locations in Mexico, Peru, and the islands of the Canary Islands are where you might spot these bugs [84]. Anthraquinone chromophore, glucose, and a carboxyl group make up carminic acid's chemical composition, which makes it light stable, and non-toxic [85].

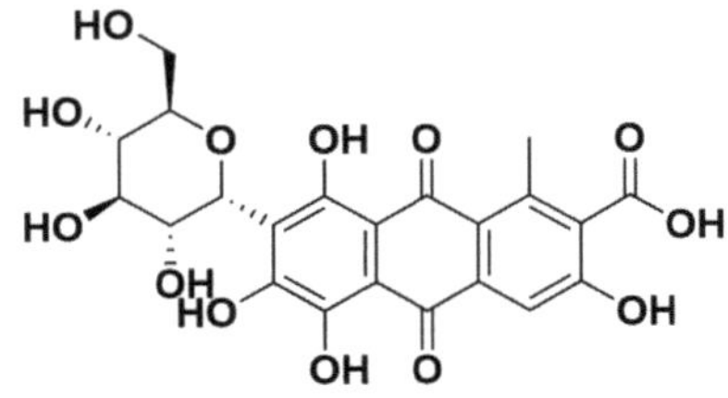

Fig. 15.4 Chemical structure of carminic acid

15.2.4.2 Factors Affecting Carminic Acid Usage

At acidic pH, it appears orange, at moderately acidic conditions to neutral pH levels it turns reddish, and at basic pH phase, it turns violet [86]. Its hue changes depending on the pH of the medium. Even while it remains stable when exposed to light, using it in its pure form leaves it susceptible to photodegradation [87], temperature fluctuations, and acidic pH [88]. There are a lot of moving parts in the purification and extraction of carminic acid, which makes the process time-consuming, difficult, costly, and variable [89].

Cosmetics, medicines, and food all make extensive use of carminic acid. The existence of the statement "cochineal extract" on the foodstuff labeling is required in Italy, by the EFSA, the FDA, and the USA if the content is higher than 1.8% or if the carmine coloring has more than 50% carminic acid concentration. E-120 is the code for it as an additive in food [90, 91].

15.2.5 Chlorophylls

15.2.5.1 Features and Chemical Structures of Chlorophylls

The pigments that give plants their green hue are called chlorophylls as displayed in Fig. 15.5. They are cyclic tetrapyrroles and they have a special ring with five members that helps them undergo photosynthesis [92]. The heart of its architecture contains the porphyrin macrocycle, which is made up of four pyrrole compound rings, and it is coupled to a magnesium (Mg^{2+}) molecule. Connected to the porphyrin ring, the side chain of the phytol chain is very hydrophobic [26]. Because of its solubility in fat, this ingredient is best extracted using an organic or non-polar solvent. Several organic solvents, including chloroform, methanol, acetone, and ethanol, are employed to extract chlorophyll-a [93].

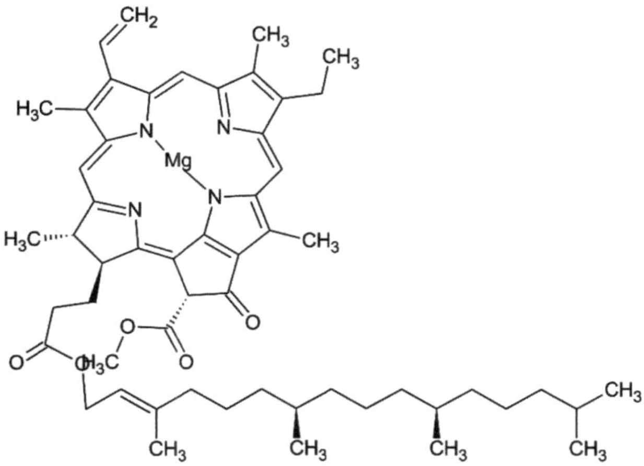

Chlorophyll a

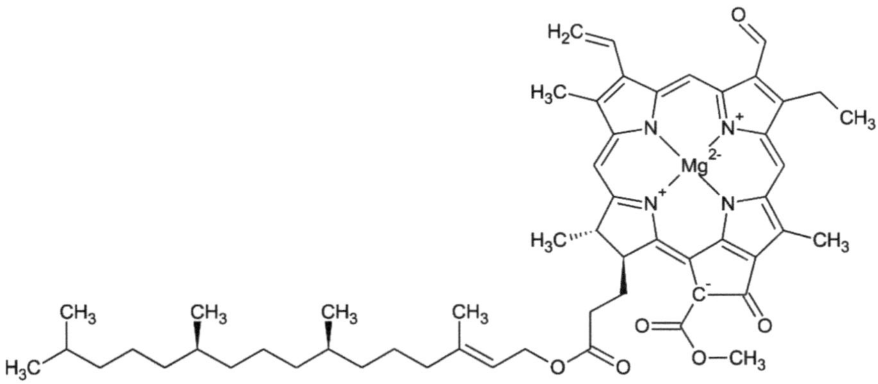

Chlorophyll b

Fig. 15.5 Composition of chlorophylls a and b [33]. Reprinted with permission from Elsevier. Copyright Elsevier, 2021

15.2.5.2 Factors Affecting Chlorophylls Usage

These days, chlorophyll is extracted using solvents and environmentally friendly methods. The implementation of green solvent-based extraction using 2,3-butanediol successfully led to a reduction in specific energy usage while maintaining good yield and antioxidant properties as outcome of the extraction process. The chlorophyll extraction yields were highest (over 70%) with 2,3-butanediol and isopropyl alcohol. Chlorophylla extraction yields were 49% for ethanol, 48% for ethyl lactate, and 36% for methanol. The amounts of chlorophyll that were extracted were 32% by acetone, 37% by 1,3-butanediol, and 16% by 1,3-propanediol [94]. In addition to preserving the stability and biological activity of chlorophyll, other methods, for instance, extraction (like ultrasound-assisted, supercritical, and ionic liquids), also demonstrated extraction yields above 70% [95, 96]. Chlorophyll is a colorant or dye that can be added to food. At present, this pigment is categorized using either the INS or the standard number of the European Commission (E). Chlorophyllins, and chlorophylls copper complexes (INS-141 or E-141) use the same numbering scheme as chlorophylls and copper complexes [97].

A more stable and chelated form of chlorophyll is formed by adding copper salts, which keep the green hue. The manufacturing procedure involves substituting the magnesium ligand with Cu^{2+} to form copper chlorophyllin, which is a further even product compared to the parent chlorophyll [98]. In addition to making the component more chemically stable, this method makes it more thermally stable, which is useful for using it as a food colorant. When it comes to green food colorants, E-141ii—also called copper complexes of chlorophyllins—is by far the most popular choice because of its hydrophilic nature and excellent stability in green color [99]. Several diagrams of structures like Chlorophyll, chlorophyllins, and copper complexes are often shown in Fig. 15.5.

15.3 Extraction Methods of Plant-Based Colorants

It is not possible to employ pigments straight from their natural origins. According to the study, plants mostly consist of chlorophylls, water-insoluble fibers, proteins, carbohydrates, along with tannins, with a negligible amount of colors [100]. The process of extracting natural pigments involves using them in many businesses. To produce higher-quality natural colorants, it is significant to indicate an extraction method that increases extraction yield and prevents the degradation of extracted pigments [58]. Following solid-phase extraction along with accompanying chromatographic separation procedures, the extracted pigments are purified [101].

15.3.1 Traditional Methods

It is common practice to extract crude pigment through natural sources using traditional processes including hydrodistillation, maceration, and Soxhlet [58]. According to the study, these approaches are straightforward, relatively reasonable, and easy to manage [102]. The most common solvents employed to extract both water-soluble and polar chemicals in traditional methods include water, ethanol, methanol, and ethanol [103]. Table 15.1 shows that lipophilic pigments can be extracted using polar solvents for instance acetone, hexane, or trifluoracetic acid [26].

Table 15.1 Usages of natural colorants extracted through several traditional solvent extraction methods

Extraction methods	Colorants	Sources	Usages	Reference
Ethanolic extraction	Zeaxanthin	Goji berries (*Lycium barbarum* L.)	to grow a model for their quantification	[104]
Extraction with a mixture	Carotenoids	Apricots (*Prunus armeniaca* L.)	apricot nutritional value	[105]
Overnight maceration	Lycopene	Tomato (*Solanum lycopersicum* L.) peels	antioxidant action	[106]
Ethanolic extraction and centrifugation	Betalains	Red pitaya (*Hylocereus polyrhizus*)	yogurt	[107]
Maceration	Betalains	Beetroot	Application to soy-based yogurt	[108]
Maceration	Anthocyanins	Red cabbage (*Brassica oleracea* var. *capitata rubra*)	Gummi candy	[109]
Patent extraction process (Grupo Matarromera, ES 2 309 032)	Anthocyanins (Eminol®)	Red pomace of *Vitis vinifera* variety Tempranillo	Milkshake, custard dessert, omelet, and pancakes	[110]
Ethanolic extraction	Anthocyanins	Grape skin and mangosteen	Smart film	[111]
Solvent extraction (aqueous and organic)	Apigenin	Parsley seeds	Dietary supplement	[112]
Ethanolic extraction	Flavonoids	Corn silk	Dietary supplement	[113]

15.3.1.1 Maceration

Pulverized plant samples are combined with solvents to extract crude colorants; the combination is then placed in an extraction vessel with stirring or agitation done every so often. After the procedure is complete, the liquid is filtered and any remaining residue is either mechanically compressed or spun in a centrifuge in order to extract the solvent again until it loses its color [114]. The maceration process has several benefits, such as being inexpensive, quick to use, and not requiring heating for extraction, which could harm bioactive compounds. To get a good extract, you have to use the right solvents for pigment extraction, the right amount of time, the right temperature, and the right mass/solvent ratio [115].

15.3.1.2 Soxhlet Extraction

The original purpose of Soxhlet extraction was to remove lipids, however, it is now also employed to evaluate the efficiency of other extraction methods [114]. Even though Soxhlet extraction is straightforward and easy to work with, the pigment could degrade over the long extraction time because of the heat. This technique involves coarsely grinding the sample and then placing it in a cellulose or filter paper "thimble" that is permeable, and finally placing the whole thing in the chamber of the Soxhlet apparatus. In the lower flask, the extraction solvents are heated until they evaporate. Then, they are condensed in the condenser and dripped back into the sample thimble. According to Patel et al., as the liquid reaches the siphon arm, it drains back into the lower flask and the procedure stays [116]. The amount of solvent needed for Soxhlet extraction is lower than that of the maceration method. The extraction system's usage of very pure Soxhlet extraction solvents, though, might drive up the price. Supercritical fluid extraction and other emerging extraction procedures are less harmful to the environment, although this process is still not ideal [117].

15.3.1.3 Carbon Dioxide Extraction

Carbon dioxide, also known as CO_2, stands out among organic solvent alternatives considering its solvent power may be modulated by modifying its physical state from liquid to supercritical with simple adjustments to extraction pressure and temperature [118]. Supercritical CO_2 has been employed to extract colors from plants. By utilizing CO_2 supercritical extraction, Idham et al. [119] were able to increase the anthocyanin extraction through rose flower calyces to 1,197 mg per 100 g of dried coriacea. The applications of high-pressure carbon dioxide (HPCD) technology in functional compounds have been somewhat restricted so far. The traditional methods of extracting anthocyanins via purple sweet potatoes using water and ethanol were contrasted with HPCD [120]. They utilized 10 MPa HPCD for 20 min, and the anthocyanins relative results went up from 45.58 to 83.20%.

15.3.2 Prospective Extraction Methods

Traditional extraction techniques can lead to heat deterioration of pigments, need excessive amounts of organic solvents, and have poor extraction efficiency. Jurijć et al. noted that there has been a surge in research into more sophisticated extraction techniques that aim to improve intracellular material mass transfer, thereby decreasing extraction time and organic solvent consumption and increasing extract yield and quality [64]. In comparison to conventional extraction methods, new extraction technologies, sometimes called "green extraction techniques," minimize or eliminate the use of organic solvents, shorten the extraction time, reduce negative effects on both humans and the environment, and increase pigment yields [121]. Table 15.2 displays the various extraction methods that have been employed in laboratories and on a wide scale. These methods include enzyme-assisted, high-pressure, microwave-assisted, ultrasound-assisted, pressurized liquid, and supercritical fluid extractions [27]. All of these technologies are more efficient than the old ones, so they're a good option for increasing the availability of colorants without negatively impacting the environment. So, these methods can supplant the more traditional approaches to extraction [122].

15.3.2.1 Ultrasound-Assisted Extraction

The mechanical method of extraction assisted by ultrasound relies on the cavitation phenomena. A process known as microbubble formation and violent collapse occurs when ultrasonic frequencies travel through a liquid. Nirmal et al. found that this process causes the temperature to rise, which in turn triggers the rupture of the cell wall and the subsequent release of the cell contents [137]. Whereas diffusion through the cell wall and rinsing of the contents of the cell following cell wall permeabilization are the physical mechanisms at work during ultrasonic extraction [138]. Ultrasound shortens the solvent extraction time, works well with volatile and heat-sensitive compounds, boosts the extraction yield, and decreases the amount of energy and solvent needed due to its influence on the mass of transport, temperature, and solvent/sample ratio [139]. It is necessary to tune process parameters including time, temperature, amplitude, and frequency in order to maximize the extraction yield while utilizing the ultrasound technology [140].

For the purpose of extracting anthocyanins from red grape pomace, it has been contrasted with microwave-assisted and Soxhlet extraction utilizing various solvent solutions (water and ethanol mixture ratio 1:1). Procyanidins (43,469 ppm) and anthocyanin levels (34,188 ppm) were found to be highest in extracts found by ultrasound in conjunction with water/ethanol when equated to all actions [141]. Ultrasound analysis using ethanol yielded an estimated 3.7 mg/g of betalain from the pulp of *Opuntia dillenii* [142]. When combined with other techniques, like microwave-assisted extraction, it can greatly enhance the extraction process. Melgar et al. found that the betalain concentration in *Opuntia engelmannii* peels was 201.6 mg/g when methanol was present and 132.9 mg/g when it was not [143, 144].

15.3.2.2 High-Pressure Technology

By applying pressure instantly and consistently over a sample without any pressure gradients, high-pressure extraction (HPE) achieves volume-independent results. This idea is known as the isostatic principle. Low or room temperature pressures ranging from 100 to 1000 MPa are utilized in this technique [82]. Because of its shorter extraction durations, less solvent requirements, and higher extraction yields, HPE is seen as a green technique. According to a recent study, excessive pressure severely

Table 15.2 Usages of natural colorants extracted through several prospective solvent extraction methods

Extraction Methods	Colorants	Sources	Disadvantages	References
Ultrasound	Betalains	Red dragon (*Hylocereus undatus*) fruit peel	High-cost, specialized setup required	[123]
	Carotenoids	Carrot (*Daucus carota*) pomace		[124]
	Betacyanin and betaxanthin	*Bougainvillea glabra* flowers		[125]
	Anthocyanins	Jaboticaba (*Plinia cauliflora*) peels		[126]
	Anthocyanins	Blackberry (*Rubus ulmifolius* L.)		[127]
	Betalains	Beetroot (*Beta vulgaris*) peel and pomace		[128]
	Betalains	Dragon fruit (*H. undatus*) peel		[129]
High hydrostatic pressure	Anthocyanins	Jaboticaba (*P. cauliflora*) peels	High cost, possible degradation of thermolabile compounds at high pressures	[126]
	Chlorophyll *a* Chlorophyll *b* Carotenoids	Winter savory (*Satureja montana* L.) leaves		[130]
Supercritical fluid extraction	Lycopene β-Carotene	Tomato (*Solanum lycopersicum* L.) peels	the properties of the fluid used for the extraction	[106]
	β-Carotene	Pumpkin (*Cucurbita* sp.) seeds		[131]
	Carotenoids	Carrot (*D. carota*) peels		[132]
	Carotenoids	Rosemary (*Salvia rosmarinus*)		[133]
	Carotenoids	Sweet potato peels		[134]
		Tomato		
		Apricot		
		Pumpkin		
		Peach peels		
		Pepper wastes		

(continued)

Table 15.2 (continued)

Extraction Methods	Colorants	Sources	Disadvantages	References
Pulsed electric field	Betanin	Beetroot (*Beta vulgaris*)	High energy input	[135]
	Anthocyanin	Blueberries by-products		[136]

damages plant cells, causing solvents to diffuse and enhancing mass transfer and extract release [145]. According to Briones-Labarca et al. [146], HPE is a green technology that has surpassed conventional extraction methods in terms of speed and effectiveness while simultaneously preserving natural fruit and vegetable bioactive ingredients.

The thermal treatment of açaí (*Euterpe oleracea* Martius) pulp resulted in a 16.3% decrease in anthocyanins content when compared to using the HPE procedure. In order to enhance the extraction approach of several colors from agro-waste and foods, it is a realistic and feasible technology. Not to mention that the product that is generated by HPE is completely harmless to microbes [147]. Enzymatic extraction is an alternate method for combining with HPE that increases recovery. Both methods, as compared to samples that were not treated with enzymes, increased the extraction of carotenoids, notably lycopene, through tomato waste by 2–64%, depending on the solvents that were evaluated. Strati et al. found that ethyl lactate had the highest carotenoid extraction yields of 127 mg/kg dry weight and 89.4 mg/kg dry weight, correspondingly [148].

15.3.2.3 Supercritical Fluid Extraction (SFE)

According to the studies, supercritical fluid extraction (SFE) involves separating the extractant from the matrix utilizing supercritical fluids as the solvent for extraction [149]. According to Gallego et al. and Kraujalis and Venskutonis, the final extracts' composition can be significantly changed by adjusting various variables such as time, pressure, CO_2 flow rate, feed, and the utilization of a co-solvent. Both the matrix and the target elements will determine the optimal extraction parameters. It is common practice to use experimental designs for optimizing extraction parameters [150].

With SFE, you can maximize separation by using other parameters like temperature and pressure. The interaction amid the mobile phase and the analytes is boosted as the density of the supercritical fluid improves due to a rise in pressure [133]. In a study conducted by Derrien et al., it was found that supercritical fluid extraction with CO_2 and 93% ethanol recovered more pigments from spinach wastes than the standard method employing acetone [151].

15.3.2.4 Electric Field–Assisted Extraction

When compared to extraction methods, the use of electric fields improves selectivity as well as mass transfer rate while decreasing organic solvent consumption. To improve the dynamics and kinetics of ion transference between incompatible phases, electric fields are used to enhance the electrophoretic movement of charged species [152]. Using a adjustable voltage applied diagonally the solid–liquid or liquid–liquid segment boundaries, this extraction approach makes it possible to regulate several parameters, including selectivity, cleanup rate, and efficiency [153]. In pulsed electric field extraction (PEF), a food product is placed between two electrodes and exposed to high-voltage, short-pulsed electric fields. This process aims to modify the membrane permeability and boost the extraction yield [135]. By testing whether or not the transmembrane potential is surpassed, the electric field travels via the separate cell molecule and cell membrane. This leads to an increase in cell permeability due to the increased repulsion between molecules that carry charges. So, PEF shortens the extraction time and improves pigment extraction effectiveness by mass-transferring into cells and thereby altering their membrane structure [137]. PEF is more cost-effective because it requires less energy and, more importantly, it prevents the sample or extraction medium from being heated above 1 °C, which is necessary for obtaining high-value extracts [138]. In order to improve the extraction amount for a particular solid–liquid system, it is necessary to fine-tune the following PEF parameters: electric field intensity; duty cycle, pulse shape, and frequency; and application time [154].

A rise in food production yield is also attributable to new technology, such as PEF, which facilitates the extraction of pigments. The juice yield of blueberries (*Vaccinium myrtillus L.*) was enhanced with the use of PEF. According to BobinaitĎ et al. [155], the juice production was 28% higher and the total anthocyanin content was 60% higher when PEF was added as a pretreatment before pressing, in comparison to the untreated sample.

15.4 Utilization of Natural Pigments in Food Packaging

Natural colorants are being employed in the formation of smart packaging due to their advantageous characteristics, including ease of preparation, renewable nature low environmental impact, and environmentally friendly characteristics [156, 157]. Anthocyanins, a type of water-soluble flavonoids, have gained significant interest as natural colorants due to their ability to exhibit a color shift response (from red to purple) at varying pH levels [158]. *Solanum lycopersicum* (Purple tomatoes) as a new resource of anthocyanin to create colorimetric indicator films with chitosan matrix using solution casting techniques. The outcomes indicated that the films underwent a hue transition, shifting from a pink hue to a lighter shade of yellowish when the pH value improved from 3 to 11 [159]. Additionally, the hue of the film in milk or fish varied in accordance with the level of freshness or decay of these commodities [160].

The utilization of anthocyanins-chitosan film derived from purple tomatoes shows potential for its application as a smart packaging indication. Liu et al. examined the use of a protein-carbohydrate nanocomplex film loaded with blueberry anthocyanins to assess the cleanness of mushrooms. The film was fabricated using the solution casting technique with blueberry anthocyanins. The hue transitioned from a rosy shade to a deep violet while the mushrooms were being stored. The outcomes indicated that employing such an indicator could be advantageous for instantaneous intensive care of cleanness [161]. Anthocyanin-based freshness indicators extracted from different sources which include red cabbage, purple potato, as well as berries, are utilized to check fluctuations in different kinds of food including meat [162, 163], fish [164, 165], seafood [166], and dairy products [160]. The colorimetric indicator based on anthocyanin exhibits excellent stability. However, its usage in some food products may lead to a decline in product quality [167].

Alizarin, often known as Mordant Red 11 or Turkish Red, is a naturally occurring colored textile dyestuff that has been utilized subsequently the dawn of civilization [168]. The alizarin molecule has a red color under normal pH conditions, but it undergoes a color change to yellow under acidic pH conditions. This change is attributed to the formation of oxygen anions inside the molecular structure. At high pH levels, alizarin molecules undergo intramolecular hydrogen bonding amongst hydroxyl groups and carbonyl oxygen atoms. This interaction causes the color of alizarin to transition from red to purple [168]. The visual color reaction of cellulose nanofibril-polyvinyl alcohol (PVA) films with alizarin when exposed to acetic acid and ammonia was examined by Chen et al. At a temperature of 25 °C, the colorimetric film exhibited greater sensitivity to acid vapor exposure than to ammonia exposure. The film underwent a color transformation from purple to yellow under acidic conditions, whereas it remained purple under basic conditions. Alizarin was included in the development of chitosan-based film to assess the freshness of the seafood. Alizarin exhibited distinct chromatic variations, transitioning from a yellow hue to a purple shade in response to alterations in pH levels alongwith ammonia content. The film's hue has transitioned via a khaki shade to light brown due to the fish's spoilage. Such findings indicated that alizarin showed great potential as a natural colorant for assessing the freshness of seafood [169]. In addition, the alizarin and chitosan-based film offers other advantages, including antioxidant capabilities and antibacterial activities [170]. Alizarin was employed to monitor the deterioration of minced beef. The shade transitioned from yellowish to brown on the second day and to purple on the fourth day, within a pH range of 2 to 11. During a 2-month storage period, it was observed that the hue of the alizarin indicator remained further constant at a temperature of 4 °C compared to a temperature of 25 °C. Color changes in the alizarin-based cellulose-chitosan index were shown to be well correlated with the concentration of total volatile basic nitrogen. As such, this metric is regarded as a useful instrument for determining how fresh minced beef is [171]. In another work, carboxymethyl cellulose (CMC)/agar-based functional halochromic films were made by combining alizarin with grapefruit seed extract. Together with being antioxidant and antibacterial, grapefruit seed extract improved the mechanical strength of the film. The results showed that the CMC/agar film with alizarin and grapefruit seed extract had active and clever packaging properties. One good choice for real-time, ongoing food quality monitoring could be the halochromic film [172].

The naphthoquinone colorant, sometimes referred to as shikonin and alkannin, is a natural colorant that is sensitive to changes in pH. The primary chemical component of Arnebiaeu-chroma root extracts is the enantiomeric naphthoquinone [157]. Dong et al. created a film made of cellulose and naphthoquinone that can detect changes in color caused by rotting in prawns and pork. The color of shikonin underwent a transformation from a rose reddish hue (with pH ranges 5–9) to a blue-violet shade at alkaline pH levels (10–12). The abrupt alteration in color was attributed to the deterioration of the molecular structure of the chromophore under alkaline conditions. In addition to its favorable mechanical and hydrophobic characteristics, this film exhibits a discerning color reaction that can be used to assess the freshness of prawns and pork. In their study, Huang et al. [173] employed a colorimetric film made of naphthoquinone (Arnebiaeuchroma) to monitor fish rotting in real-time. It was observed that colorimetric labels containing lower concentrations of Arnebiaeuchromaextract (AEE) exhibited more pronounced color shifts compared to colorimetric labels with larger AEE quantities. They suggested that using a small amount of naphthoquinone-based colorimetric film would be sufficient for tracking the quality of fish. Despite its extensive usage in traditional Chinese medicine, there has been limited research undertaken on the lipophilic red pigment-based halochromic films derived from Arnebiaeu-chroma [173].

Additionally, Kanatt created a halochromic film out of polyvinyl gelatin and betalain derived from Amaranthus leaves to track when chicken and fish turn rotten. The film's color shifted from red to yellow as the freshness was monitored in real-time. Also, the film exhibited notable mechanical, antioxidant compounds, and antibacterial characteristics; it postponed the sample's spoiling from days 3 to 12 [174]. Betacyanin was isolated from red pitaya flesh and used to create a halochromic film by combining various polysaccharides with a PVA matrix [175]. In contrast to films based on chitosan and carrageenan, the beta cyanin halochromic film made of locust bean gum and PVA shows superior antioxidant stability and color retention for 21 days of room and fridge temperature prawn freshness monitoring. The structural, functional, and physical characteristics of the smart film were supposedly greatly affected by the types of polysaccharides used. A number of variables influence the films' color and antioxidant stability; these include the polysaccharide type as well as exposure to light, temperature, and duration of storage [175].

15.5 Limitations in Acquiring and Utilizing Pigments Derived from Plants in the Food Industry

In recent times, customers have shown a greater inclination towards seeking and utilizing food items that are composed of natural components, mostly because they are perceived as being non-detrimental to one's health [176]. The Natural Products

Association (NPA), based in the United States, has published a guideline for the labeling of natural goods. Hence, the natural goods are required to adhere to the specified rules [177]. First and foremost, the product must include a minimum of 95% natural ingredients, which can either be actually natural or obtained from natural sources. Additionally, it should not contain any compounds that are associated with any suspected dangers to human health. Furthermore, it is imperative that the product is not subjected to any processing methods that substantially or negatively modify the integrity of its natural components, which may include elements obtained from a deliberate and sustainable source in nature, with little processing. The objective is to refrain from using toxic chemicals in order to maintain the material's cleanliness and only include non-natural constituents when there are no viable natural alternatives and they do not offer any possible concerns to human health [178].

The profusion of natural food items available in the marketplace is truly remarkable. Though customers ingest the product, some of it has been supplied without going through necessary physicochemical examinations. Food products containing chemically generated components, such as benzoic acid, butylated hydroxyanisole (BHA), and butylated hydroxytoluene (BHT), may be harmful to health. Apples that produce ethylene can ripen too soon; methylcyclopropene gas is used to stop this. Benzoic acid is a preservative used to stop chemical changes from spoiling food. Antioxidants called BHA and BHT are often used in meals that have additional fats; they prevent rancidity. These substances, however, have been connected to diseases like asthma, childhood hyperactivity, and cancer. The reason the food maker uses these chemicals is that they are easily accessible and less expensive than natural ingredients. Furthermore, while using chemically produced components in food supplements will give users instant benefits, long-term use of these supplements may have unfavorable effects. The maker focuses on these functional effects to enhance the product's marketability and popularity [178].

Although natural pigments have benefits, they are scarce due to their restricted sources, low stability, and expensive price [8]. Natural pigments are inherently unstable due to their vulnerability to oxygen, heat, and light, which leads to hue loss and changes in hue. The requirement of a large quantity of plants in natural pigment synthesis could lead to higher costs of raw materials. Furthermore, the process of extracting flower pigments from seasonal flowers is time-consuming and requires a lengthy cycle to get new sources. In addition, numerous natural colors are prohibited from being used in food products. The FDA and EU have released a list of authorized colorants [179]. Any dye that is not included in this list is deemed illegal due to doubts about its source, particularly if it is derived from plants.

Furthermore, the replication of natural pigments is a significant challenge, as it requires the expertise of professional artisans to manage these substances. Prior to its incorporation into the food product formulation, it is necessary to design a suitable and standardized extraction procedure that does not compromise the desired quality of the pigment. In addition, the process of extracting the colorings is slow, and the colorings themselves are not even, resulting in potential color changes. The occurrence of these reactions is triggered by exposure to sunlight, perspiration, and atmospheric conditions [180]. To obtain stable pigment with suitable colorfastness behavior, it is necessary to develop the proper extraction, identification, and finishing processes from current scientific investigations [9]. There is a need to create a sustainable and cost-effective method for producing plant pigment that can replace synthetic alternatives. This is necessary because the current technologies for industrial-scale pigment manufacturing are expensive. Another important aspect to consider when using natural ingredients is finding a reputable supplier that offers high-quality sources, follows proper procedures, and provides excellent services. It is crucial that the supplier maintains hygienic processing methods and takes measures to prevent cross-contamination. The possibility of microbial contamination during processing should not be disregarded. The numerous uses for natural components generated from plants made an official safety evaluation standard for the natural food product industry quite restricted. One needs to have a suitable metric for various situations in order to evaluate safety more precisely [181]. Any novel food product in Malaysia has to get a JAKIM Halal certificate and approval from the Good Manufacturing Practice (GMP) [182]. There are several uses for certificates in the United States. The FDA regulates cosmetics; the Environmental Protection Agency (EPA) enforces environmental protection laws; the Federal Trade Commission (FTC) addresses unfair trade practices and verifies product performance claims; and the Consumer Product Safety Commission (CPSC) ensures product safety [178].

15.6 Conclusion

When determining the freshness and quality of food, customers mostly consider color. The creation of food goods is using natural colorants more and more because of health concerns about synthetic colorants. Apart from replacing synthetic pigments, these additions offer numerous other advantages including antioxidant and antibacterial properties. Because natural colorants offer so many health benefits, including them in food systems gives a means of fortifying foods. Natural colorants are being investigated as a way to improve the color of many food categories (including dairy, meat-related products, and bakery items).

Furthermore, as naturally derived tints have shown to be a good pH pointer that can be utilized to track the cleanliness of food, their use is enabling the advance of smart packaging machinery.

The color permanence and bulk manufacturing volume are two of the many restrictions on the use of natural colorants, though. Large amounts of acylated petunidin glycoside and acylated delphinidin glycoside can be produced thanks to recent biological engineering methods. Furthermore, gaining popularity for extremely successful colorant extraction from sources that are natural are environmentally friendly solvents and technologies including microwave-, ultrasound-, and enzyme-assisted processes. Additionally, a major source of natural, ecologically beneficial colorants is food sector waste. Advancements in various scientific disciplines are currently leading to an increase in creative possibilities in the field of food packaging. Recent study has demonstrated the substantial impact of several technologies from the fourth industrial revolution, including the IoT, cyber-physical systems, big data, smart sensing, and autonomous robotics, on different aspects of food packaging science. In conclusion, this chapter has showcased the capacity of natural coloring agents in the fields of food coloring and innovative packaging that utilizes naturally derived coloring compounds. Furthermore, these chemicals have the ability to significantly alter several properties of both food and their packaging components. In order to preserve the quality of packaged food for a longer period of time, it is crucial to utilize natural colorants to the greatest extent feasible during the process of food processing and packaging.

Declaration The authors declare that they have no known competing financial interests or personal relationships that could have appeared to influence the work reported in this paper. Generative AI and AI-assisted technologies have been used in the writing process to improve the readability and language of the work only. After using this tool, the author(s) reviewed and edited the content as needed and take(s) full responsibility for the content of the publication.

References

1. Sk, S., Mia, R., Haque, M. A., Shamim, A. M.: Review on extraction and application of natural dyes. Textile Leather Rev **4**(4):218–233 (2021). https://doi.org/10.31881/TLR.2021.09
2. Ahmed, T., et al.: New trends in printing applications of natural dyes and pigments. Renewable Dyes Pigments 139–163 (2024). https://doi.org/10.1016/B978-0-443-15213-9.00003-X
3. Cardoso, L.A., Karp, S.G., Vendruscolo, F., Kanno, K.Y., Zoz, L.I., Carvalho, J.C.: Biotechnological production of carotenoids and their applications in food and pharmaceutical products. Carotenoids **8**, 125–141 (2017)
4. Beata, O., Urbańska, K., Bryś, M.: Selected food colourants with antiplatelet activity as promising compounds for the prophylaxis and treatment of thrombosis. Food Chem. Toxicol. **141**, 111437 (2020). https://doi.org/10.1016/j.fct.2020.111437
5. Sabater-Vilar, M., Maas, R. F. M., Fink-Gremmels, J.: Mutagenicity of commercial Monascus fermentation products and the role of citrinin contamination. Mutat. Res./Genet. Toxicol. Environ. Mutagen. **444**(1), 7–16 (1999). https://doi.org/10.1016/S1383-5718(99)00095-9
6. Yusuf, M., Shabbir, M., Mohammad, F.: Natural colorants: historical, processing and sustainable prospects. Nat. Prod. Bioprospecting **7**(1), 123–145 (2017). https://doi.org/10.1007/s13659-017-0119-9
7. Shahid, M. et al.: Dyeing, fastness and antimicrobial properties of woolen yarns dyed with gallnut (Quercus infectoria Oliv.) extract. Dyes Pigments **95**(1), 53–61 (2012). https://doi.org/10.1016/j.dyepig.2012.03.029
8. Cai, Y.-Z., Sun, M., Corke. H.: Characterization and application of betalain pigments from plants of the Amaranthaceae. Trends Food Sci. Technol. 16(9), 370–376 (2005). https://doi.org/10.1016/j.tifs.2005.03.020
9. Kasiri, M.B., Safapour, S.: Natural dyes and antimicrobials for green treatment of textiles. Environ. Chem. Lett. **12**(1), 1–13 (2014)
10. Roy, K., Gullapalli, S., Roy Chaudhuri, U., Chakraborty, R.: The use of a natural colorant based on betalain in the manufacture of sweet products in India. Int J Food Sci Technol **39**(10), 1087–1091 (2004)
11. Mohammad Azmin, S. N. H., Sulaiman, N. S., Mat Nor, M. S., Abdullah, P. S., Abdul Kari, Z., Pati, S.: A review on recent advances on natural plant pigments in foods: functions, extraction,

importance and challenges. Appl. Biochem. Biotechnol. **194**(10), 4655–4672 (2022). https://doi.org/10.1007/s12010-022-04050-z
12. Banna, B.U., Mia, R., Hasan, M.M., Ahmed, B., Shibly, M.A.H.: Ultrasonic-assisted sustainable extraction and dyeing of organic cotton fabric using natural dyes from *Dillenia indica* leaf. Heliyon **9**(8), e18702 (2023). https://doi.org/10.1016/j.heliyon.2023.e18702
13. Habib, N., Adeel, S., Ali, A., Mia, R., Khan, S.R., Qayyum, M.A.: Ultrasonic-assisted sustainable pollution free advanced method for isolation of colouring material from Amba Haldi (*Curcuma aromatica*) for wool dyeing. Sci. Prog. **106**(3), 00368504231188610 (2023). https://doi.org/10.1177/00368504231188610
14. Martins, N., Roriz, C. L., Morales, P., Barros, L., Ferreira, I. C. F. R.: Food colorants: Challenges, opportunities and current desires of agro-industries to ensure consumer expectations and regulatory practices. Trends Food Sci. & Technol. **52**, 1–15 (2016). https://doi.org/10.1016/j.tifs.2016.03.009
15. Coultate, T., Blackburn, R.S.: Food colorants: their past, present and future. Color. Technol. **134**(3), 165–186 (2018). https://doi.org/10.1111/cote.12334
16. Mia, R., et al.: Natural dye extracted from Triadica sebifera in aqueous medium for sustainable dyeing and functionalizing of viscose fabric. Clean. Eng. Technol. **8**, 100471 (2022). https://doi.org/10.1016/j.clet.2022.100471
17. Mia, R., Bakar, M.A., Islam, M.R., Ahmed, T.: Eco-friendly coloration from mahogany wood waste for sustainable dyeing of organic nonwoven cotton fabric. Results Eng. **17**, 101032 (2023). https://doi.org/10.1016/j.rineng.2023.101032
18. de Boer, F.Y., Imhof, A., Velikov, K.P.: Encapsulation of colorants by natural polymers for food applications. Color. Technol. **135**(3), 183–194 (2019)
19. Commission, C. A.: Discussion paper on the use of colours in foods. Codex committee on food additives and contaminants. Beijing, China: FAO/WHO., vol. Thirty-second Session. (2000)
20. Carle, R., Schweiggert, R.: Handbook on natural pigments in food and beverages: industrial applications for improving food color. Woodhead Publishing (2016)
21. Prado, M.A., Godoy, H.T.: Corantes artificiais em alimentos. Alimentos e Nutrição **14**(2), 237–250 (2003)
22. Araújo, J.: Química de alimentos: teoria e prática," in Química de alimentos: teoria e prática, pp 478–478 (2004)
23. Rodriguez-Amaya, D.B.: Natural food pigments and colorants. Curr. Opin. Food Sci. **7**, 20–26 (2016)
24. Ghosh, S., Sarkar, T., Das, A., Chakraborty, R.: Micro and nanoencapsulation of natural colors: a holistic view. Appl. Biochem. Biotechnol. **193**, 3787–3811 (2021)
25. Zannat, A., Uddin, M. N., Mahmud, S. T., Mia, R., Ahmed, T.: Natural dyes and pigments in functional finishing. In: Renewable Dyes and Pigments. Elsevier, pp. 271–287 (2024)
26. Sharma, M., Usmani, Z., Gupta, V.K., Bhat, R.: Valorization of fruits and vegetable wastes and by-products to produce natural pigments. Crit. Rev. Biotechnol. **41**(4), 535–563 (2021)
27. Ghosh, S., Sarkar, T., Das, A., Chakraborty, R.: Natural colorants from plant pigments and their encapsulation: an emerging window for the food industry. Lwt **153**, 112527 (2022)
28. Wallace, T. C., Giusti, M. M.: Anthocyanins—nature's bold, beautiful, and health-promoting colors. vol. 8, ed: MDPI, p. 550 (2019)
29. Mia, R., Sk, M.S., Oli, Z.B.S., Ahmed, T., Kabir, S., Waqar, M.A.: Functionalizing cotton fabrics through herbally synthesized nanosilver. Clean. Eng. Technol. **4**, 100227 (2021). https://doi.org/10.1016/j.clet.2021.100227
30. Khoo, H. E., Azlan, A., Tang, S. T., Lim, S. M.: Anthocyanidins and anthocyanins: colored pigments as food, pharmaceutical ingredients, and the potential health benefits. Food Nutrit. Res. (2017)
31. Li, H., et al.: Highly pigmented vegetables: anthocyanin compositions and their role in antioxidant activities. Food Res. Int. **46**(1), 250–259 (2012)
32. Koop, B. L., Maciel, A. G., Soares, L. S., Monteiro, A. R., Valencia, G. A.: Natural colorants. Nat. Addit. Foods 87–122 (2022)

33. Ribeiro, J.S., Veloso, C.M.: Microencapsulation of natural dyes with biopolymers for application in food: a review. Food Hydrocolloids **112**, 106374 (2021)
34. Rodriguez-Amaya, D.B.: Natural food pigments and colorants. In: Mérillon, J.-M., Ramawat, K.G. (eds.) Bioactive molecules in food, pp. 1–35. Springer International Publishing, Cham (2017)
35. Ahmadiani, N., Robbins, R. J., Collins, T. M., Giusti, M. M.: Anthocyanins contents, profiles, and color characteristics of red cabbage extracts from different cultivars and maturity stages. J. Agric. Food Chem. **62**(30), 7524–7531 (2014). https://doi.org/10.1021/jf501991q
36. Jin, A. L., Ozga, J. A., Kennedy, J. A., Koerner-Smith, J. L., Botar, G., Reinecke, D. M.: Developmental profile of anthocyanin, flavonol, and proanthocyanidin type, content, and localization in saskatoon fruits (*Amelanchier alnifolia* Nutt.). J. Agric. Food Chem. **63**(5), 1601–1614 (2015). https://doi.org/10.1021/jf504722x
37. Kovačević, D.B., Putnik, P., Dragović-Uzelac, V., Vahčić, N., Babojelić, M.S., Levaj, B.: Influences of organically and conventionally grown strawberry cultivars on anthocyanins content and color in purees and low-sugar jams. Food Chem. **181**, 94–100 (2015)
38. Xu, J., et al.: Characterisation and stability of anthocyanins in purple-fleshed sweet potato P40. Food Chem. **186**, 90–96 (2015)
39. Olsen, H., Aaby, K., Borge, G.I.A.: Characterization, quantification, and yearly variation of the naturally occurring polyphenols in a common red variety of curly kale (*Brassica oleracea* L. convar. acephala var. sabellica cv. 'Redbor'). J. Agric. Food Chem. **58**(21), 11346–11354 (2010)
40. Pervaiz, T., Songtao, J., Faghihi, F., Haider, M.S., Fang, J.: Naturally occurring anthocyanin, structure, functions and biosynthetic pathway in fruit plants. J. Plant Biochem. Physiol **5**(2), 1–9 (2017)
41. Amarowicz, R., et al.: Influence of postharvest processing and storage on the content of phenolic acids and flavonoids in foods. Mol. Nutr. Food Res. **53**(S2), S151–S183 (2009)
42. Roy, S., Rhim, J.-W.: Anthocyanin food colorant and its application in pH-responsive color change indicator films. Crit. Rev. Food Sci. Nutr. **61**(14), 2297–2325 (2021)
43. Castañeda-Ovando, A., de Lourdes Pacheco-Hernández, M., Páez-Hernández, M. E., Rodríguez, J. A., Galán-Vidal, C. A.: Chemical studies of anthocyanins: a review. Food Chem. **113**(4), 859–871 (2009)
44. Islam, S. R., et al.: Natural dyes and pigments as a source of medicine. In: Renewable dyes and pigments. Elsevier, pp 177–232 (2024)
45. Weston, M., Phan, M.A.T., Arcot, J., Chandrawati, R.: Anthocyanin-based sensors derived from food waste as an active use-by date indicator for milk. Food Chem. **326**, 127017 (2020)
46. Koop, B.L., et al.: Flavonoids, anthocyanins, betalains, curcumin, and carotenoids: sources, classification and enhanced stabilization by encapsulation and adsorption. Food Res. Int. **153**, 110929 (2022)
47. Stintzing, F.C., Kammerer, D., Schieber, A., Adama, H., Nacoulma, O.G., Carle, R.: Betacyanins and phenolic compounds from *Amaranthus spinosus* L. and *Boerhavia erecta* L. Zeitschrift für Naturforschung C **59**(1–2), 1–8 (2004)
48. Jiang, H., et al.: Eco-friendly dyeing and finishing of organic cotton fabric using natural dye (gardenia yellow) reduced-stabilized nanosilver: full factorial design. Cellulose **29**(4), 2663–2679 (2022). https://doi.org/10.1007/s10570-021-04401-9
49. Wybraniec, S., et al.: Betacyanins from vine cactus *Hylocereus polyrhizus*. Phytochemistry **58**(8), 1209–1212 (2001)
50. Khan, M.I., Giridhar, P.: Plant betalains: chemistry and biochemistry. Phytochemistry **117**, 267–295 (2015)
51. Stintzing, F.C., Carle, R.: Functional properties of anthocyanins and betalains in plants, food, and in human nutrition. Trends Food Sci. Technol. **15**(1), 19–38 (2004)
52. Cejudo-Bastante, M.J., Hurtado, N., Mosquera, N., Heredia, F.J.: Potential use of new Colombian sources of betalains. Color stability of ulluco (*Ullucus tuberosus*) extracts under different pH and thermal conditions. Food Res. Int. **64**, 465–471 (2014)

53. Kumar, S.S., Manoj, P., Shetty, N., Prakash, M., Giridhar, P.: Characterization of major betalain pigments-gomphrenin, betanin and isobetanin from *Basella rubra* L. fruit and evaluation of efficacy as a natural colourant in product (ice cream) development. J. Food Sci. Technol. **52**, 4994–5002 (2015)
54. Moßhammer, M.R., Stintzing, F.C., Carle, R.: Development of a process for the production of a betalain-based colouring foodstuff from cactus pear. Innov. Food Sci. Emerg. Technol. **6**(2), 221–231 (2005)
55. Belhadj Slimen, I., Najar, T., Abderrabba, M.: Chemical and antioxidant properties of betalains. J. Agric. Food Chem. **65**(4), 675–689 (2017)
56. Herbach, K.M., Maier, C., Stintzing, F.C., Carle, R.: Effects of processing and storage on juice colour and betacyanin stability of purple pitaya (*Hylocereus polyrhizus*) juice. Eur. Food Res. Technol. **224**, 649–658 (2007)
57. Celli, G.B., Brooks, M.S.-L.: Impact of extraction and processing conditions on betalains and comparison of properties with anthocyanins—A current review. Food Res. Int. **100**, 501–509 (2017)
58. Ngamwonglumlert, L., Devahastin, S., Chiewchan, N.: Natural colorants: Pigment stability and extraction yield enhancement via utilization of appropriate pretreatment and extraction methods. Crit. Rev. Food Sci. Nutr. **57**(15), 3243–3259 (2017)
59. Khan, M.I.: Plant betalains: Safety, antioxidant activity, clinical efficacy, and bioavailability. Compr. Rev. Food Sci. Food Saf. **15**(2), 316–330 (2016)
60. Kaimainen, M., Laaksonen, O., Järvenpää, E., Sandell, M., Huopalahti, R.: Consumer acceptance and stability of spray dried betanin in model juices. Food Chem. **187**, 398–406 (2015)
61. Otálora, M.C., Carriazo, J.G., Iturriaga, L., Osorio, C., Nazareno, M.A.: Encapsulating betalains from Opuntia ficus-indica fruits by ionic gelation: pigment chemical stability during storage of beads. Food Chem. **202**, 373–382 (2016)
62. Rehman, A., et al.: Carotenoid-loaded nanocarriers: a comprehensive review. Adv. Coll. Interface. Sci. **275**, 102048 (2020)
63. de Freitas Santos, P. D., Rubio, F. T. V., da Silva, M. P., Pinho, L. S., Favaro-Trindade, C. S.: Microencapsulation of carotenoid-rich materials: a review. Food Res. Int. **147**, 110571 (2021)
64. Jurić, S., et al.: Sources, stability, encapsulation and application of natural pigments in foods. Food Rev. Intl. **38**(8), 1735–1790 (2022)
65. Rodriguez-Concepcion, M., et al.: A global perspective on carotenoids: metabolism, biotechnology, and benefits for nutrition and health. Prog. Lipid Res. **70**, 62–93 (2018)
66. Raddatz-Mota, D., et al.: Achiote (*Bixa orellana* L.): a natural source of pigment and vitamin E. J. Food Sci. Technol. **54**, 1729–1741 (2017). https://doi.org/10.1007/s13197-017-2579-7
67. Vilar, D. d. A., et al.: Traditional uses, chemical constituents, and biological activities of *Bixa orellana* L.: a review. Scientific World J **2014** (2014)
68. Shahid-ul-Islam, R.L., Mohammad, F.: Phytochemistry, biological activities and potential of annatto in natural colorant production for industrial applications–a review. J. Adv. Res. **7**(3), 499–514 (2016)
69. E. P. o. F. Additives and N. S. a. t. Food: Scientific opinion on the re-evaluation of mixed carotenes (E 160a (i)) and beta-carotene (E 160a (ii)) as a food additive. EFSA J. **10**(3), 2593 (2012)
70. Grune, T., et al.: β-Carotene is an important vitamin A source for humans. J. Nutr. **140**(12), 2268S-2285S (2010)
71. Cassani, L., Marcovich, N.E., Gomez-Zavaglia, A.: Valorization of fruit and vegetables agro-wastes for the sustainable production of carotenoid-based colorants with enhanced bioavailability. Food Res. Int. **152**, 110924 (2022)
72. Rammuni, M., Ariyadasa, T.U., Nimarshana, P., Attalage, R.: Comparative assessment on the extraction of carotenoids from microalgal sources: astaxanthin from H. pluvialis and β-carotene from D. salina. Food Chem. **277**, 128–134 (2019)
73. Harvey, P.J., Ben-Amotz, A.: Towards a sustainable Dunaliella salina microalgal biorefinery for 9-cis β-carotene production. Algal Res. **50**, 102002 (2020)

74. Mercadante, A.Z., Rodrigues, D.B., Petry, F.C., Mariutti, L.R.B.: Carotenoid esters in foods-A review and practical directions on analysis and occurrence. Food Res. Int. **99**, 830–850 (2017)
75. Carvalho, G.C., de Camargo, B.A.F., de Araújo, J.T.C., Chorilli, M.: Lycopene: From tomato to its nutraceutical use and its association with nanotechnology. Trends Food Sci. Technol. **118**, 447–458 (2021)
76. Shi, J., Maguer, M.L.: Lycopene in tomatoes: chemical and physical properties affected by food processing. Crit. Rev. Food Sci. Nutr. **40**(1), 1–42 (2000)
77. Chang, Y., Jiao, Y., Li, D.-J., Liu, X.-L., Han, H.: Glycosylated zein as a novel nanodelivery vehicle for lutein. Food Chem. **376**, 131927 (2022)
78. Zeni, A.L.B., Camargo, A., Dalmagro, A.P.: Lutein prevents corticosterone-induced depressive-like behavior in mice with the involvement of antioxidant and neuroprotective activities. Pharmacol. Biochem. Behav. **179**, 63–72 (2019)
79. Gansukh, E., Mya, K.K., Jung, M., Keum, Y.-S., Kim, D.H., Saini, R.K.: Lutein derived from marigold (Tagetes erecta) petals triggers ROS generation and activates Bax and caspase-3 mediated apoptosis of human cervical carcinoma (HeLa) cells. Food Chem. Toxicol. **127**, 11–18 (2019)
80. Manzoor, M., Singh, J., Gani, A., Noor, N.: Valorization of natural colors as health-promoting bioactive compounds: Phytochemical profile, extraction techniques, and pharmacological perspectives. Food Chem. **362**, 130141 (2021)
81. Rodriguez-Amaya, D.B.: Update on natural food pigments-A mini-review on carotenoids, anthocyanins, and betalains. Food Res. Int. **124**, 200–205 (2019)
82. Luzardo-Ocampo, I., Ramírez-Jiménez, A.K., Yañez, J., Mojica, L., Luna-Vital, D.A.: Technological applications of natural colorants in food systems: a review. Foods **10**(3), 634 (2021)
83. De Mejia, E.G., Zhang, Q., Penta, K., Eroglu, A., Lila, M.A.: The colors of health: chemistry, bioactivity, and market demand for colorful foods and natural food sources of colorants. Annu. Rev. Food Sci. Technol. **11**, 145–182 (2020)
84. Rowe, A.: Global cochineal production: scale, welfare concerns, and potential interventions (2020)
85. E. P. o. F. Additives and N. S. a. t. Food: Scientific Opinion on the re-evaluation of cochineal, carminic acid, carmines (E 120) as a food additive. EFSA J **13**(11), 4288 (2015)
86. Zakariya, S.M., et al.: An in-vitro elucidation of inhibitory potential of carminic acid: possible therapeutic approach for neurodegenerative diseases. J. Mol. Liq. **303**, 112692 (2020)
87. Bowers, L.M.R., Sobeck, S.J.S.: Impact of medium and ambient environment on the photodegradation of carmine in solution and paints. Dyes Pigm. **127**, 18–24 (2016)
88. Ghidouche, S., Rey, B., Michel, M., Galaffu, N.: A Rapid tool for the stability assessment of natural food colours. Food Chem. **139**(1–4), 978–985 (2013)
89. Campana, M. G., Robles Garcia, N. M., Tuross, N.: America's red gold: multiple lineages of cultivated cochineal in Mexico. Ecol. Evolut. **5**(3), 607–617 (2015)
90. Cooksey, C.J.: The red insect dyes: carminic, kermesic and laccaic acids and their derivatives. Biotech. Histochem. **94**(2), 100–107 (2019)
91. Yilmaz, U.T., Ergun, F., Yilmaz, H.: Determination of the food dye carmine in milk and candy products by differential pulse polarography. J. Food Drug Anal. **22**(3), 329–335 (2014)
92. Pérez-Gálvez, A., Viera, I., Roca, M.: Development of an accurate and direct method for the green food colorants detection. Food Res. Int. **136**, 109484 (2020)
93. Su, S., Zhou, Y., Qin, J.G., Yao, W., Ma, Z.: Optimization of the method for chlorophyll extraction in aquatic plants. J. Freshw. Ecol. **25**(4), 531–538 (2010)
94. Lee, J., Kwak, M., Chang, Y.K., Kim, D.: Green solvent-based extraction of chlorophyll a from Nannochloropsis sp. Using 2, 3-butanediol. Sep. Purif. Technol. **276**, 119248 (2021)
95. Tavakoli, S., et al.: Ultrasonic-assisted food-grade solvent extraction of high-value added compounds from microalgae Spirulina platensis and evaluation of their antioxidant and antibacterial properties. Algal Res. **60**, 102493 (2021)

96. Vaz, B.M., et al.: Using aqueous solutions of ionic liquids as chlorophyll eluents in solid-phase extraction processes. Chem. Eng. J. **428**, 131073 (2022)
97. E. P. o. F. Additives and N. S. a. t. Food: Scientific Opinion on re-evaluation of copper complexes of chlorophylls (E 141 (i)) and chlorophyllins (E 141 (ii)) as food additives. EFSA J. **13**(6), 4151 (2015)
98. Wang, F., Terazono, Y., Liu, J., Fefer, M., Pelton, R.H.: Adsorption of aqueous copper chlorophyllin mixtures on model surfaces. Colloids Surf A **592**, 124578 (2020)
99. Gandul-Rojas, B., Roca, M., Gallardo-Guerrero, L.: Detection of the color adulteration of green table olives with copper chlorophyllin complexes (E-141ii colorant). LWT-Food Sci Technol **46**(1), 311–318 (2012)
100. Aggarwal, S.: Indian dye yielding plants: Efforts and opportunities. In: Natural Resources Forum, 2021, vol. 45, no. 1: Wiley Online Library, pp 63–86
101. Fernández-López, J.A., Fernández-Lledó, V., Angosto, J.M.: New insights into red plant pigments: more than just natural colorants. RSC Adv. **10**(41), 24669–24682 (2020)
102. Sivarajasekar, S.: Optimization of extraction methods for natural pigment from *Lawsonia inermis*. Int J Green Pharmacy (IJGP) **12**(03) (2018)
103. Liu, Y., Li, J., Fu, R., Zhang, L., Wang, D., Wang, S.: Enhanced extraction of natural pigments from *Curcuma longa* L. using natural deep eutectic solvents. Ind. Crops Prod. **140**, 111620 (2019)
104. Patsilinakos, A., Ragno, R., Carradori, S., Petralito, S., Cesa, S.: Carotenoid content of *Goji berries*: CIELAB, HPLC-DAD analyses and quantitative correlation. Food Chem. **268**, 49–56 (2018)
105. Zhou, W., et al.: Analysis of carotenoid content and diversity in apricots (*Prunus armeniaca* L.) grown in China. Food Chem. **330**, 127223 (2020)
106. Kehili, M., et al.: Supercritical CO2 extraction and antioxidant activity of lycopene and β-carotene-enriched oleoresin from tomato (*Lycopersicum esculentum* L.) peels by-product of a Tunisian industry. Food Bioprod. Process. **102**, 340–349 (2017)
107. Utpott, M., et al.: Evaluation of the use of industrial wastes on the encapsulation of betalains extracted from red pitaya pulp (*Hylocereus polyrhizus*) by spray drying: Powder stability and application. Food Bioprocess Technol. **13**, 1940–1953 (2020)
108. Dias, S., Castanheira, E.M., Fortes, A.G., Pereira, D.M., Gonçalves, M.S.T.: Natural pigments of anthocyanin and betalain for coloring soy-based yogurt alternative. Foods **9**(6), 771 (2020)
109. Esmailian, A., Hosseini, F.: Optimization of anthocyanin extraction from red cabbage and its application in low calorie Gummy candy. J Food Sci. Technol. (Iran) **18**(110), 129–140 (2021)
110. Pineda-Vadillo, C., et al.: The food matrix affects the anthocyanin profile of fortified egg and dairy matrices during processing and in vitro digestion. Food Chem. **214**, 486–496 (2017)
111. Zhu, B., Lu, W., Qin, Y., Cheng, G., Yuan, M., Li, L.: An intelligent pH indicator film based on cassava starch/polyvinyl alcohol incorporating anthocyanin extracts for monitoring pork freshness. J. Food Process. Preserv. **45**(10), e15822 (2021)
112. Al-Rawi, M. I., Almzaien, A. K., Almzaien, K. A.: Hypolipidemic and antioxidant efficacy of apigenin in hydrogen peroxide induced oxidative stress in adult male rats. Medico-Legal Update 21(1) (2021)
113. Zhang, Y., Wu, L., Ma, Z., Cheng, J., Liu, J.: Anti-diabetic, anti-oxidant and anti-hyperlipidemic activities of flavonoids from corn silk on STZ-induced diabetic mice. Molecules **21**(1), 7 (2015)
114. Azmir, J., et al.: Techniques for extraction of bioactive compounds from plant materials: a review. J. Food Eng. **117**(4), 426–436 (2013)
115. Setford, P.C., Jeffery, D.W., Grbin, P.R., Muhlack, R.A.: Factors affecting extraction and evolution of phenolic compounds during red wine maceration and the role of process modelling. Trends Food Sci. Technol. **69**, 106–117 (2017)
116. Patel, M., Patel, P., Kothari, F.: Advance ecological approach to new assessment in red plant pigments. Asian J. Biol. Life Sci. **10**(2), 274–279 (2021)

117. Soumya, S., Sawant, A., Khandetod, Y., Mohod, A., Dhekale, J.: Extraction methods used for extraction of anthocyanin: A review. Pharma Innov J **8**, 280–285 (2019)
118. Sakai, H., Ono, K., Tokunaga, S., Sharmin, T., Aida, T. M., Mishima, K.: Extraction of natural pigments from *Gardenia jasminoides* J. Ellis fruit pulp using CO_2-expanded liquids and direct sonication. Separations **8**(1), 1 (2020)
119. Idham, Z., et al.: Improvement of extraction and stability of anthocyanins, the natural red pigment from roselle calyces using supercritical carbon dioxide extraction. J. CO_2 Utiliz. 56, 101839 (2022)
120. Lao, F., Cheng, H., Wang, Q., Wang, X., Liao, X., Xu, Z.: Enhanced water extraction with high-pressure carbon dioxide on purple sweet potato pigments: Comparison to traditional aqueous and ethanolic extraction. J CO_2 Utiliz **40**, 101188 (2020)
121. Kannan, M.M., Quine, S.D.: Ellagic acid protects mitochondria from β-adrenergic agonist induced myocardial damage in rats; evidence from in vivo, in vitro and ultra structural study. Food Res. Int. **45**(1), 1–8 (2012)
122. Neves, M.I.L., Silva, E.K., Meireles, M.A.A.: Trends and challenges in the industrialization of natural colorants. Food Public Health **9**(2), 33–44 (2019)
123. Li, X., et al.: Improvement of betalains stability extracted from red dragon fruit peel by ultrasound-assisted microencapsulation with maltodextrin. Ultrason. Sonochem. **82**, 105897 (2022)
124. Tiwari, S., Upadhyay, N., Singh, A.K., Meena, G.S., Arora, S.: Organic solvent-free extraction of carotenoids from carrot bio-waste and its physico-chemical properties. J. Food Sci. Technol. **56**, 4678–4687 (2019)
125. Maran, J.P., Priya, B., Nivetha, C.V.: Optimization of ultrasound-assisted extraction of natural pigments from *Bougainvillea glabra* flowers. Ind. Crops Prod. **63**, 182–189 (2015)
126. Nunes Mattos, G., Pessanha de Araújo Santiago, M. C., Sampaio Doria Chaves, A. C., Rosenthal, A., Valeriano Tonon, R., Correa Cabral, L. M.: Anthocyanin extraction from jaboticaba skin (*Myrciaria cauliflora* Berg.) using conventional and non-conventional methods. Foods **11**(6), 885 (2022)
127. Espada-Bellido, E., et al.: Extraction of antioxidants from blackberry (*Rubus ulmifolius* L.): comparison between ultrasound-and microwave-assisted extraction techniques. Agronomy **9**(11), 745 (2019)
128. Vulić, J.J., et al.: In vivo and in vitro antioxidant effects of beetroot pomace extracts. J. Funct. Foods **6**, 168–175 (2014)
129. Mello, F. R. d., et al.: Antioxidant properties, quantification and stability of betalains from pitaya (Hylocereus undatus) peel. Ciência Rural **45**, 323–328 (2014)
130. Moreira, S.A., Pintado, M.E., Saraiva, J.A.: Optimization of antioxidant activity and bioactive compounds extraction of winter savory leaves by high hydrostatic pressure. High Pressure Res. **40**(4), 543–560 (2020)
131. Wang, X., Wang, C., Zha, X., Mei, Y., Xia, J., Jiao, Z.: Supercritical carbon dioxide extraction of β-carotene and α-tocopherol from pumpkin: a Box-Behnken design for extraction variables. Anal. Methods **9**(2), 294–303 (2017)
132. de Andrade Lima, M., Charalampopoulos, D., Chatzifragkou, A.: Optimisation and modelling of supercritical CO_2 extraction process of carotenoids from carrot peels. J Supercritical Fluids **133**, 94–102 (2018)
133. Lefebvre, T., Destandau, E., Lesellier, E.: Sequential extraction of carnosic acid, rosmarinic acid and pigments (carotenoids and chlorophylls) from Rosemary by online supercritical fluid extraction-supercritical fluid chromatography. J. Chromatogr. A **1639**, 461709 (2021)
134. de Andrade Lima, M., Kestekoglou, I., Charalampopoulos, D., Chatzifragkou, A.: Supercritical fluid extraction of carotenoids from vegetable waste matrices. Molecules **24**(3), 466 (2019)
135. Nowacka, M., et al.: The impact of pulsed electric field on the extraction of bioactive compounds from beetroot. Foods **8**(7), 244 (2019)
136. Zhou, Y., Zhao, X., Huang, H.: Effects of pulsed electric fields on anthocyanin extraction yield of blueberry processing by-products. J. Food Process. Preserv. **39**(6), 1898–1904 (2015)

137. Nirmal, N.P., Mereddy, R., Maqsood, S.: Recent developments in emerging technologies for beetroot pigment extraction and its food applications. Food Chem. **356**, 129611 (2021)
138. Bozinou, E., Karageorgou, I., Batra, G., Dourtoglou, V.G., Lalas, S.I.: Pulsed electric field extraction and antioxidant activity determination of *Moringa oleifera* dry leaves: A comparative study with other extraction techniques. Beverages **5**(1), 8 (2019)
139. Chemat, F., Rombaut, N., Sicaire, A.-G., Meullemiestre, A., Fabiano-Tixier, A.-S., Abert-Vian, M.: Ultrasound assisted extraction of food and natural products. Mechanisms, techniques, combinations, protocols and applications. A review. Ultrason. Sonochem. **34**, 540–560 (2017)
140. Oancea, S., Perju, M., Coman, D., Olosutean, H.: Optimization of conventional and ultrasound-assisted extraction of *Paeonia officinalis* anthocyanins, as natural alternative for a green technology of cotton dyeing. Roman. Biotechnol. Lett. **26**(2), 2527–2534 (2021)
141. Drosou, C., Kyriakopoulou, K., Bimpilas, A., Tsimogiannis, D., Krokida, M.: A comparative study on different extraction techniques to recover red grape pomace polyphenols from vinification byproducts. Ind. Crops Prod. **75**, 141–149 (2015)
142. Vázquez-Espinosa, M., González-de-Peredo, A.V., Carrera, C., Palma, M., Barbero, G.F., Aliaño-González, M.J.: Ultrasound-assisted extraction of betalains from Opuntia fruit pulp of different color varieties. Agronomy **12**(11), 2604 (2022)
143. López-Cruz, R., Sandoval-Contreras, T., Iñiguez-Moreno, M.: Plant pigments: classification, extraction, and challenge of their application in the food industry. Food Bioprocess Technol. **16**(12), 2725–2741 (2023)
144. Melgar, B., Dias, M.I., Barros, L., Ferreira, I.C., Rodriguez-Lopez, A.D., Garcia-Castello, E.M.: Ultrasound and microwave assisted extraction of Opuntia fruit peels biocompounds: optimization and comparison using RSM-CCD. Molecules **24**(19), 3618 (2019)
145. Cano-Lamadrid, M., Artes-Hernandez, F.: By-products revalorization with non-thermal treatments to enhance phytochemical compounds of fruit and vegetables derived products: a review. Foods **11**(1), 59 (2021)
146. Briones-Labarca, V., Giovagnoli-Vicuña, C., Cañas-Sarazúa, R.: Optimization of extraction yield, flavonoids and lycopene from tomato pulp by high hydrostatic pressure-assisted extraction. Food Chem. **278**, 751–759 (2019)
147. de Jesus, A. L. T., Cristianini, M., dos Santos, N. M., Maróstica Júnior, M. R.: Effects of high hydrostatic pressure on the microbial inactivation and extraction of bioactive compounds from açaí (Euterpe oleracea Martius) pulp. Food Res. Int. **130**, 108856 (2020). https://doi.org/10.1016/j.foodres.2019.108856
148. Strati, I.F., Gogou, E., Oreopoulou, V.: Enzyme and high pressure assisted extraction of carotenoids from tomato waste. Food Bioprod. Process. **94**, 668–674 (2015)
149. Panja, P.: Green extraction methods of food polyphenols from vegetable materials. Curr. Opin. Food Sci. **23**, 173–182 (2018)
150. Gallego, R., Bueno, M., Herrero, M.: Sub-and supercritical fluid extraction of bioactive compounds from plants, food-by-products, seaweeds and microalgae–An update. TrAC, Trends Anal. Chem. **116**, 198–213 (2019)
151. Derrien, M., Badr, A., Gosselin, A., Desjardins, Y., Angers, P.: Optimization of a green process for the extraction of lutein and chlorophyll from spinach by-products using response surface methodology (RSM). LWT-Food Sci. Technol. **79**, 170–177 (2017)
152. Sousa, D.V., Pereira, F.V., Nascentes, C.C., Moreira, J.S., Boratto, V.H., Orlando, R.M.: Cellulose cone tip as a sorbent material for multiphase electrical field-assisted extraction of cocaine from saliva and determination by LC-MS/MS. Talanta **208**, 120353 (2020)
153. Yamini, Y., Seidi, S., Rezazadeh, M.: Electrical field-induced extraction and separation techniques: promising trends in analytical chemistry–a review. Anal. Chim. Acta **814**, 1–22 (2014)
154. Athanasiadis, V., et al.: Pulsed electric field and Salvia officinalis L. leaves: a successful combination for the extraction of high value added compounds. Foods **10**(9), 2014 (2021)
155. Bobinaitė, R., Pataro, G., Lamanauskas, N., Šatkauskas, S., Viškelis, P., Ferrari, G.: Application of pulsed electric field in the production of juice and extraction of bioactive compounds from blueberry fruits and their by-products. J. Food Sci. Technol. **52**, 5898–5905 (2015)

156. Huang, J., Hu, Z., Li, G., Hu, L., Chen, L., Hu, Y.: Make your packaging colorful and multi-functional: The molecular interaction and properties characterization of natural colorant-based films and their applications in food industry. Trends Food Sci. Technol. **124**, 259–277 (2022). https://doi.org/10.1016/j.tifs.2022.04.028
157. Priyadarshi, R., Ezati, P., Rhim, J.-W.: Recent advances in intelligent food packaging applications using natural food colorants. ACS Food Sci. Technol. **1**(2), 124–138 (2021)
158. Ghorbani, M., et al.: A halochromic indicator based on polylactic acid and anthocyanins for visual freshness monitoring of minced meat, chicken fillet, shrimp, and fish roe. Innov. Food Sci. Emerg. Technol. **74**, 102864 (2021). https://doi.org/10.1016/j.ifset.2021.102864
159. Li, Y., Wu, K., Wang, B., Li, X.: Colorimetric indicator based on purple tomato anthocyanins and chitosan for application in intelligent packaging. Int. J. Biol. Macromol. **174**, 370–376 (2021)
160. Gao, R., et al.: Incorporation of gelatin and Fe2+ increases the pH-sensitivity of zein-anthocyanin complex films used for milk spoilage detection. Curr. Res. Food Sci. **5**, 677–686 (2022)
161. Liu, L., Wu, W., Zheng, L., Yu, J., Sun, P., Shao, P.: Intelligent packaging films incorporated with anthocyanins-loaded ovalbumin-carboxymethyl cellulose nanocomplexes for food freshness monitoring. Food Chem. **387**, 132908 (2022)
162. Guo, Z., Zuo, H., Ling, H., Yu, Q., Gou, Q., Yang, L.: A novel colorimetric indicator film based on watermelon peel pectin and anthocyanins from purple cabbage for monitoring mutton freshness. Food Chem. **383**, 131915 (2022). https://doi.org/10.1016/j.foodchem.2021.131915
163. Hu, D., Liu, X., Qin, Y., Yan, J., Yang, Q.: A novel intelligent film with high stability based on chitosan/sodium alginate and coffee peel anthocyanin for monitoring minced beef freshness. Int. J. Food Sci. Technol. **57**(7), 4673–4686 (2022)
164. Wu, D., Zhang, M., Chen, H., Bhandari, B.: Freshness monitoring technology of fish products in intelligent packaging. Crit. Rev. Food Sci. Nutr. **61**(8), 1279–1292 (2021)
165. Sani, M.A., Tavassoli, M., Hamishehkar, H., McClements, D.J.: Carbohydrate-based films containing pH-sensitive red barberry anthocyanins: application as biodegradable smart food packaging materials. Carbohyd. Polym. **255**, 117488 (2021)
166. Li, B., et al.: A sub-freshness monitoring chitosan/starch-based colorimetric film for improving color recognition accuracy via controlling the pH value of the film-forming solution. Food Chem. **388**, 132975 (2022)
167. Chen, H.-Z., Zhang, M., Bhandari, B., Yang, C.-H.: Novel pH-sensitive films containing curcumin and anthocyanins to monitor fish freshness. Food Hydrocolloids **100**, 105438 (2020)
168. Almasi, H., Forghani, S., Moradi, M.: Recent advances on intelligent food freshness indicators; an update on natural colorants and methods of preparation. Food Packag. Shelf Life **32**, 100839 (2022)
169. Chen, H. z., Zhang, M., Rao, Z.: Development of cellulose nanofibrils reinforced polyvinyl alcohol films incorporated with alizarin for intelligent food packaging. Int. J. Food Sci. Technol. **56**(9), 4248–4257 (2021)
170. Ezati, P., Rhim, J.-W.: PH-responsive chitosan-based film incorporated with alizarin for intelligent packaging applications. Food Hydrocolloids **102**, 105629 (2020)
171. Ezati, P., Tajik, H., Moradi, M.: Fabrication and characterization of alizarin colorimetric indicator based on cellulose-chitosan to monitor the freshness of minced beef. Sens. Actuators, B Chem. **285**, 519–528 (2019)
172. Roy, S., Rhim, J.-W.: Fabrication of carboxymethyl cellulose/agar-based functional films hybridized with alizarin and grapefruit seed extract. ACS Appl. Bio Mater. **4**(5), 4470–4478 (2021)
173. Huang, S., et al.: A novel colorimetric indicator based on agar incorporated with Arnebia euchroma root extracts for monitoring fish freshness. Food Hydrocolloids **90**, 198–205 (2019)
174. Kanatt, S.R.: Development of active/intelligent food packaging film containing Amaranthus leaf extract for shelf life extension of chicken/fish during chilled storage. Food Packag. Shelf Life **24**, 100506 (2020)

175. Yao, X., Liu, J., Hu, H., Yun, D., Liu, J.: Development and comparison of different polysaccharide/PVA-based active/intelligent packaging films containing red pitaya betacyanins. Food Hydrocolloids **124**, 107305 (2022)
176. Kusumawati, I., Indrayanto, G.: Natural antioxidants in cosmetics. Stud. Nat. Prod. Chem. **40**, 485–505 (2013)
177. Joshi, L.S., Pawar, H.A.: Herbal cosmetics and cosmeceuticals: an overview. Nat Prod Chem Res **3**(2), 170 (2015)
178. Azmin, S. M., Sulaiman, N. S., Nor, M., Abdullah, P., Kari, Z. A., Pati, S.: A review on recent advances on natural plant pigments in foods: functions, extraction, importance and challenges. Appl. Biochem. Biotechnol. (2022)
179. Marmion, D., Staff, U. b.: Colorants for foods, drugs, and cosmetics. Kirk-Othmer Encyclopedia of Chemical Technology, pp. 1–46 (2000)
180. Hill, D.: Is there a future for natural dyes? Rev. Prog. Color. Relat. Top. **27**(1), 18–25 (1997)
181. Mohd-Setapar, H.: Natural ingredients in cosmetics from Malaysian plants: a review. Sains Malaysiana **47**(5), 951–959 (2018)
182. M. NATIONAL PHARMACEUTICAL REGULATORY AGENCY MINISTRY OF HEALTH, <Guidelines_for_Cosmetic_Control_.pdf>, (2017)

MIX
Papier aus verantwortungsvollen Quellen
Paper from responsible sources
FSC® C105338

If you have any concerns about our products,
you can contact us on
ProductSafety@springernature.com

In case Publisher is established outside the EU,
the EU authorized representative is:
Springer Nature Customer Service Center GmbH
Europaplatz 3, 69115 Heidelberg, Germany

Printed by Libri Plureos GmbH
in Hamburg, Germany